ESSENTIALS OF
SOLID STATE PHYSICS

ESSENTIALS OF
SOLID STATE PHYSICS

For the undergraduate and postgraduate students of all leading universities; and for NET, SET, GATE, JAM, JEST, IAS, IPS and civil service examinees

Prof. (Dr) S P Kuila MSc, PhD [IIT, Kharagpur]

Senior Professor and Head, Dept of Applied Physical Sciences, and
Chairman, Research and Development Cell
Bengal Institute of Technology and Management, Santiniketan [WB]
Former Professor of Physics, Abhedananda Mahavidyalaya
Sainthia, Birbhum [WB]
Member of Indian Society for Technical Education

AUTHOR OF

Perspective of Quantum Mechanics
Vector Analysis and its Applications
Principles of Engineering Physics [Vols I, II and III]
Fundamentals of Quantum Mechanics
Statistical Mechanics and Solid State Physics, etc.

New Central Book Agency (P) Ltd
LONDON
HYDERABAD ERNAKULAM BHUBANESWAR
NEW DELHI KOLKATA PUNE GUWAHATI

NCBA

REGD OFFICE

8/1 Chintamoni Das Lane, Kolkata 700 009, India
email: ncbapvtltd@ncbapvtltd.com

OVERSEAS

NCBA (UK) Ltd, 149 Park Avenue North, Northampton, NN3 2HY, UK
email: ncbauk@yahoo.com

EXPORT

NCBA Exports Pvt. Ltd, 212 Shahpur Jat, New Delhi 110 049
email: ncbaexports@ncbapvtltd.com

BRANCHES

2/27 Ansari Road, 1st Floor
Daryaganj, **New Delhi** 110 002
email: ncbadel@ncbapvtltd.com

House No. 3-1-315, 1st Floor
Nimboliadda, Kachiguda, **Hyderabad** 500 027
email: ncbahydb@ncbapvtltd.com

Shop Nos. 3 & 4, Vinayak Towers
681/B Budhwar Peth, Appa Balwant Chowk
Pune 411 002
email: ncbapune@ncbapvtltd.com

GSS Shopping Complex, 1st Floor
Opposite College Post Office, Convent Road
Ernakulam 682 035
0484-4021054 / 53

Shop No. 15, Bharati Towers, Ground Floor
Forest Park
Bhubaneswar 751 001
email: ncbabub@ncbapvtltd.com

Radhamayee Complex [Second Floor]
Opp. Axis Bank, Panbazar, Jaswanta Road
Guwahati 781 001
email: ncbaguwahati@ncbapvtltd.com

ESSENTIALS OF SOLID STATE PHYSICS • Prof. (Dr) S P Kuila

First Published: 2013
Reprinted: 2015, 2016, 2018
Second Edition: **2020**

PUBLISHER AND TYPESETTER
New Central Book Agency (P) Ltd
8/1 Chintamoni Das Lane, Kolkata 700 009

PRINTER
New Central Book Agency (P) Ltd
Web-Offset Division, Dhulagarh, Sankrail, Howrah

PROOF CHECKING
Dipan Roy

PROJECT TEAM
Abhijit Sen, Prodip Baidya and Pradip Biswas

ISBN: 978 81 7381 803 5

Price ₹ 615.00

Dedicated to Those eminent
and revered teachers whose
teachings led me to the
world of Absolute Truth.

Preface to the Second Edition

The second edition of the book ESSENTIALS OF SOLID STATE PHYSICS has been throughly revised and enlarged in accordance with the new CBCS syllabus recommend by UGC. At the request of several readers, a new chapter *Elastic Properties of Matter*, has been included in the text along with large number of questions and problems from different Indian universities.

I will consider my labour amply rewarded if this edition of the book receives the same overwhelming popularity as was received by its first edition.

Kolkata
15 August, 2019

S P Kuila

Preface to the First Edition

In the twenty-first century, the research on Solid State Physics has become inevitable for the diversified development and economic viability of every nation of the world. For this purpose is essential an in-depth knowledge regarding the property manifestation of solids originating from the regular periodic arrangement of atoms, ions or molecules in three-dimensional space, the degree of imperfection, the behaviour of electrons and the force acting between the lattice points, the electrons and the nuclei in the solid.

In order to impart a proper knowledge of the property manifestation of solids under different physical conditions, a course on condensed matter of physics has been purposefully incorporated in every university of India and abroad. Though plenty of books have been written so far by eminent scientists and scholars in this field, unfortunately a comprehensive textbook for the students pursuing Solid State Physics at the undergraduate and postgraduate levels is rarely found.

The present work aims to bridge the gap through the exploitation of my long experience of teaching and research over more than four decades. As found in my other books, here also a self-developed technique is in use to present the subject matter with an emphasis on the clarification of the physical principles. The text is written in a simple and lucid language with easy-to-read felicity.

The subject matter is presented here in thirteen chapters. The first chapter is devoted to the review of atomic structure. As it is one of the basic and essential topics, thorough knowledge of this is necessary to follow the topics discussed in the subsequent chapters of the book. At the end of each chapter a separate section titled *Answers of Some Important Questions*, based on basic concepts, has been added. The exercise part is divided into four sections — Section A, deals with Objective Type Questions, Section B: Short Answer Type Questions, Section C: Numerical Problems, and Section D: Broad Answer Type Questions. These are meant to judge the depth of self-understanding of the subject matter.

I shall consider my labour amply rewarded if the book serves the purpose of those for whom it is intended.

I have consulted many a book written by authors and scholars of high repute—both in India and abroad —while preparing the manuscript; for example, while working on Chapter 10 pertaining to *Superconductivity*, the original research papers of several scientists working abroad were collected and extensively utilised for flawless presentation of the topic. I express my unqualified gratitude to all of them. At the same time, I thank my colleagues associated with various institutions and universities in and outside West Bengal. My students, my wife, and my son have all along been sources of constant inspiration behind successful completion of the project.

Every attempt has been made in the book to eliminate all kinds of probable errors that might creep into the text; however, if any mistake is noticed by the careful readers, it may kindly be brought to the knowledge of the author for prompt revision.

Finally, I would like to extend my sincere thanks to Mr Amitabha Sen, Director, Mr Soumen Paul, Mr R M Nath, Mr Dipan Roy, and Mr Prabhat Jash of New Central Book Agency (P) Ltd, Kolkata, for bringing out the book in time.

Any constructive suggestion for further improvement and development of the treatise will be gratefully acknowledged.

The 27th July 2012 S P Kuila

Bengal Institute of Technology and Management
Santiniketan (WB)

General Introduction

In the present century solid state research has become inevitable for the exploration of knowledge to be applied in industrial technology and the related economic development of any nation. This is simply due to the fact that solid state physics provides the basic principles to build up the technology to be applied in the diversified fields of science and engineering. In early centuries, a few metals and alloys (like iron, copper, tin, and brass), a few ceramics (like pottery), and the natural polymers (like wool, cotton, asbestos, and cellulose) were sufficient to fulfil most human needs. But to meet up the present-day requirements, new materials are being developed at a very fast rate. The hitherto unmatched increase in the number of materials, which would be applied in service requirements, have changed the attitudes and views of human beings throughout the world. The advancements in technology—based on the advances of pure science—have prompted the area of Unified Science and Material Technology.

The engineers today also use the traditional age-old materials like glass, rubber, timber, copper and germanium, but with completely different character and properties they possessed in the past. Every engineer is concerned with the material having property suitable to his job at hand. The atomic and molecular structures determine the nature and properties of the materials and, as such, it is up to the engineers to select the suitable material having desired property essential for their jobs.

The recent developments in material science paved the way for fresh understanding of the nature of the materials—and their meaningful applications in technology. Techniques have already been invented to describe the properties of the materials already existing, as well as to predict the same for the new materials in the framework of atomic and molecular structures. This is why a thorough and fundamental understanding of the basic principles of solid state physics has become a prerequisite for the further development of material technology—with a practical view to their successful engineering applications and scientific research.

In the solid state of matter, the atoms and molecules are closely packed together in orderly three-dimensional network in space. This is responsible for property manifestations like magnetism, dielectric conductivity, and superconductivity. The abovementioned properties originate from the atomic and molecular structures, the degree of imperfection in these structures, as well as the multifarious forces and energies associated with them.

In spite of the tremendous progress in solid state science and technology, many problems still remain unsolved and many questions — practically in superconductivity—are yet to be answered. *V L Ginzberg* rightly remarked in his paper *Superconductivity today and tomorrow* in Russian Academy of Science, April 2000, that: "It should be mentioned here that the current situation in solid state theory couldn't be thought of as satisfactory. The progress made over the past century is of course great if we look at the distance passed from the idea of electron motion in conductors, suggested by Drude in 1900, up to today's conditions of physics of metals. But, on the other hand, metallic hydrogen cannot be predicted from *first principle.*"

List of Physical Constants and their Units

No.	Quantity with Symbols	Magnitude	Unit
1.	Charge of electron (e)	1.6021×10^{-19}	coulomb (C)
2.	Rest mass of electron (m_e)	9.109×10^{-31}	kilogram (kg)
3.	Rest mass of proton (m_p)	1.6725×10^{-27}	kilogram (kg)
4.	Rest mass of neutron (m_n)	1.6748×10^{-27}	kilogram (kg)
5.	Atomic mass unit (m_u)	1.6604×10^{-27}	kilogram (kg)
6.	Radius of electron (r_e)	2.818×10^{-15}	metre (m)
7.	Speed of light in vacuum (c)	2.997925×10^{8}	metre/sec (m/s)
8.	Electron-volt (eV)	1.602×10^{-19}	joule (J)
9.	Energy equivalent to electron mass ($m_e c^2$)	0.511	million electron volt (MeV)
10.	Energy equivalent to proton mass ($m_p c^2$)	938.55	million electron volt (MeV)
11.	1 angstrom unit (Å)	10^{-10}	metre (m)
12.	Proton to electron mass (m_p/m_e)	1836.1	—
13.	Specific charge of electron (e/m_e)	1.75879×10^{11}	coulomb/kilogram (C/kg)
14.	Specific charge of proton (e/m_p)	9.583×10^{7}	coulomb/kilogram (C/kg)
15.	Avogadro number (N_A)	6.02252×10^{26}	(kilomole)$^{-1}$ (kmol)$^{-1}$
16.	Boltzmann constant (k)	1.38054×10^{-23}	joule/kelvin (J/K)
17.	Planck's constant (h)	6.626×10^{-34}	joule/second (J/s)
18.	Unit atomic angular momentum (reduced Planck's constant) $\left(\hbar = \frac{h}{2\pi}\right)$	1.05459×10^{-34}	joule-second (J-s)
19.	Permeability of free space (μ_0)	$4\pi \times 10^{-7}$	henry/metre (Hm^{-1})
20.	Permittivity of free space (ϵ_0)	$\frac{10^{-9}}{36\pi} = 8.854 \times 10^{-12}$	farad/metre (Fm^{-1})
21.	Mass of neutral hydrogen atom	1.0081422	atomic mass unit (amu)

No.	Quantity with Symbols	Magnitude	Unit
22.	Radius of Bohr orbit (r_B)	0.529×10^{-10}	metre (m)
23.	Energy of ground state of hydrogen atom (Rydberg energy) (E_0)	2.175×10^{-18}	joule (J)
24.	Rydberg constant for hydrogen (R_H)	1.0967758×10^{7}	$(\text{metre})^{-1}$ (m^{-1})
25.	Bohr magneton (μ_B)	9.2732×10^{-24}	ampere-metre^2 (A-m^2)
26.	Proton magnetic moment (μ_p)	1.410×10^{-26}	ampere-metre^2 (A-m^2)
27.	Faraday constant (F_c)	9.649×10^{-7}	coulomb/kilomole (C/k mole)
28.	Fine structure constant ($\alpha = e^2/\hbar c$)	7.2973×10^{-3}	
29.	Universal gas constant (R)	8.314×10^{7}	joule/kilomole (J.kmol^{-1})
30.	Nuclear magneton $\mu_N = 3\hbar/2m_p c$	5.0508×10^{-27}	J.T^{-1}
31.	1 debye unit (D)	3.33×10^{-30}	coulomb metre (C-m)
32.	Acceleration due to gravity at sea-level on equator (g)	9.8118	metre/sec^2 (m/s^2)

Some Useful Integrals

E.1 Indefinite Integrals

1. $\int (ax+b)dx = \frac{(ax+b)^{n+1}}{a(n+1)}, \; n \neq 1$

2. $\int \frac{dx}{ax+b} = \frac{1}{a}\log(ax+b)$

3. $\int \frac{dx}{x(ax+b)} = \frac{1}{b}\log\frac{x}{(ax+b)}$

4. $\int \sqrt{(ax+b)}dx = \frac{2}{3a}(ax+b)^{3/2}$

5. $\int \frac{dx}{\sqrt{ax+b}} = \frac{2\sqrt{ax+b}}{a}$

E.2 Integrals involving Exponential and Logarithmic Expressions

1. $\int e^{ax}dx = \frac{1}{a}e^{ax}$

2. $\int b^{ax}dx = \frac{b^{ax}}{a\log b}$

3. $\int x^n e^{ax}dx = \frac{1}{a}x^n e^{ax} - \frac{n}{a}\int x^{n-1}e^{ax}dx \; (n>0)$

4. $\int e^{ax}\sin bx dx = \frac{e^{ax}}{a^2+b^2}(a\sin bx - b\cos bx)$

5. $\int \log ax dx = x\log ax - x$

6. $\int \frac{dx}{x\log ax} = \log(\log ax)$

E.3 Definite Integrals

1. $\int_0^{\infty} \frac{\sin^2 x}{x^2}dx = \frac{\pi}{2}$

2. $\int_0^{\infty} \frac{\sin ax}{x}dx = \frac{\pi}{2} \; (a>0)$

3. $\int_0^{\pi} \sin^2 ax dx = \int_0^{\pi} \cos^2 ax dx = \frac{\pi}{2}$

4. $\int_0^{\pi} e^{-a^2x^2}dx = \sqrt{\frac{\pi}{2a}}$

5. $\int_0^{\pi} x^n e^{-ax}dx = \frac{n!}{a^{n+1}} \; (n>0, >0)$

6. $\int_0^{\infty} x^{2n}e^{-ax^2} = \frac{1\cdot 3\cdot 5\cdots(2n-1)}{2^{n+1}a^n}$

7. $\int_0^\infty \sqrt{x}e^{-ax}dx = \frac{1}{2a}\sqrt{\frac{\pi}{a}}$

8. $\int_0^\infty \frac{e^{-ax}}{\sqrt{x}}dx = \sqrt{\frac{\pi}{a}}$

9. $\int_0^1 \frac{\log x}{1-x}dx = -\left(\frac{\pi^2}{6}\right)$

10. $\int_0^1 \frac{\log x}{1-x}dx = -\left(\frac{\pi^2}{12}\right)$

11. $\int_0^1 \frac{\log x}{1-x^2}dx = -\frac{\pi^2}{8}$

12. $\int_0^1 \frac{\log x}{\sqrt{1-x^2}}dx = -\frac{\pi}{2}\log 2$

13. $\int_0^1 \frac{dx}{\sqrt{\log(1/x)}}dx = \sqrt{\pi}$

Contents

[1] V.L. Ginzberg and L.D. Landau, Zh. Eksp. Teor. Fiz. **20,** 1064 (1950).

Chapter 1

A Review of Atomic Structure

1.1 Introduction

In late nineteenth century, scientists believed that the chemical elements consist of atoms which are indivisible and electrically neutral. But with the discovery of electrons by *Sir J.J. Thomson* in 1897 it became apparent that atoms are made up of two parts with equal and opposite electricity in them so that the atom as a whole is neutral.

Now a natural question is how the positive and negative charges are arranged in an atom? J.J. Thomson suggested that atoms are just positively charged lumps of matter in which electrons are embedded similar to raisins in a fruit cake [Fig. 1.1], known as *plumb pudding model* of the atom. But the real atom is something different from *Thomson's model.*

1.2 The Rutherford Atom Model

In 1911, *Hans Geiger and Ernest Marsden* at the suggestion of *Ernest Rutherford* bombard a thin gold foil with α-particles. α-particles are positively charged helium nuclei emitted by certain radioactive elements. It was observed that although most of the α-particles passed through the gold foil undeflected, a few were scattered in the backward direction suffering collision with a central mass (the nucleus) of extremely small dimension $(\sim 3 \times 10^{-5}\,\text{nm})$ [Fig. 1.2]. Rutherford, therefore, proposed that, *an atom consists of a tiny, massive positive charge at the centre, and the negative electrons revolve around it just like the planets revolving round the sun.*

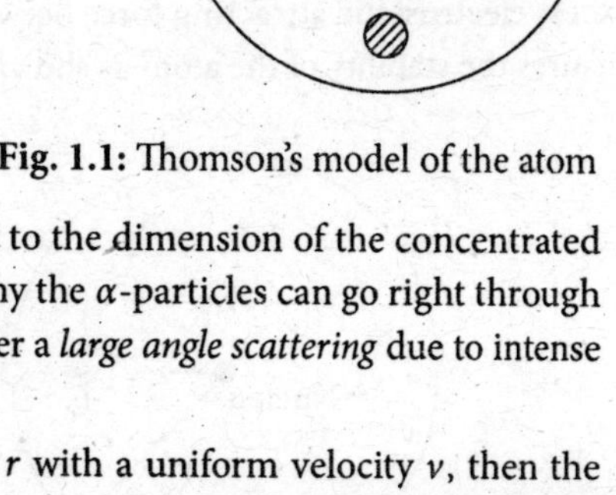

Fig. 1.1: Thomson's model of the atom

The radius of the revolving electron being very large compared to the dimension of the concentrated nuclear mass, the space within an atom is largely empty. That is why the α-particles can go right through the gold foil and when they happen to approach a nucleus, they suffer a *large angle scattering* due to intense Coulombic repulsion.

If an electron of charge e revolves in a circular orbit of radius r with a uniform velocity v, then the Coulombic attractive force is

$$F_e = \frac{(Ze)\,e}{4\pi\epsilon_0 r^2}, \tag{1.1}$$

where ϵ_0 is the *permittivity of the free space* and Z is the *atomic number.*

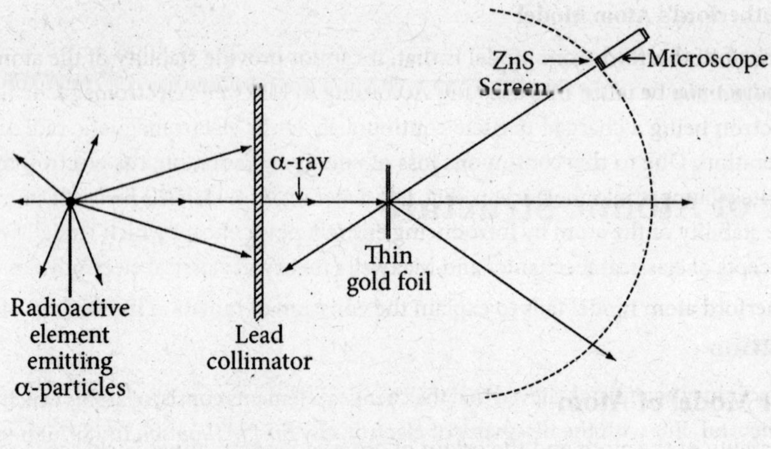

Fig. 1.2: Rutherford α-scattering experiment.

The centrifugal force acting on the electron is given by

$$F_c = \frac{mv^2}{r}, \tag{1.2}$$

where m is the mass of the electron.

The condition for dynamically stable orbit of the electron is

$$F_e = F_c, \text{ i.e., } \frac{Ze^2}{4\pi\epsilon_0 r^2} = \frac{mv^2}{r}. \tag{1.3}$$

Eqn (1.3) represents the force balance in an atom and thus makes the electron orbit dynamically stable.

In conclusion, an atom consists of a positively charged heavy nucleus around which negatively charged electrons are revolving in circular orbits keeping the nucleus at the centre very much like the planets round the sun. The electrons are moving in such a speed that the mechanical centrifugal force just balances the net excess electrostatic attracting force between the electron and the nucleus. This balance of the two forces ensures the stability of the atom as shown in Fig. 1.3.

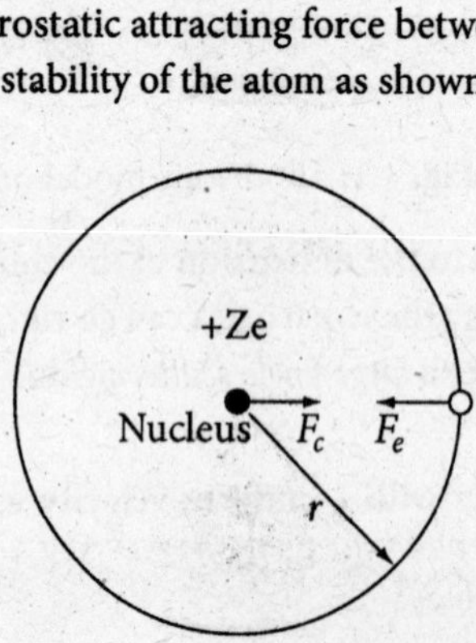

Fig. 1.3: Force balance in an atom

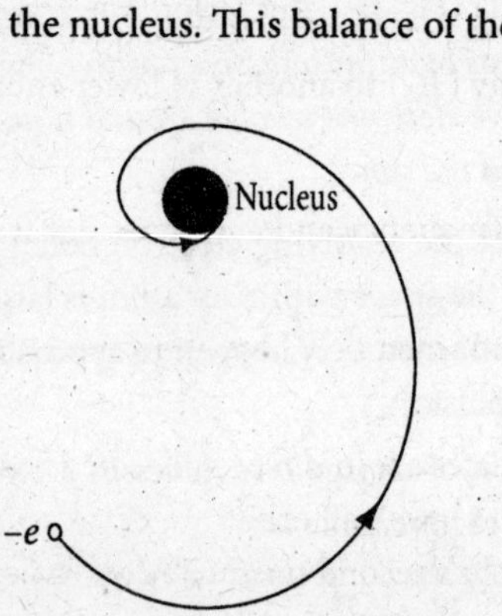

Fig. 1.4: The electron after spiralling plunge into the nucleus

Drawbacks of Rutherford's Atom Model

The first *drawback* of Rutherford atom model is that, it cannot provide stability of the atom if the effect of *classical electrodynamics* be taken into account. According to Maxwell's electromagnetic theory of light, the revolving electron being a charged particle continuously emits electromagnetic radiation due to its centripetal acceleration. Due to this continuous loss of energy by radiation, the electron moves spirally inward and ultimately plunges into the nucleus [Fig. 1.4]. Later on, in 1913, *Niels Bohr* of Denmark, however, could explain the stability of the atom by introducing entirely new concept which straightway rejects the fundamental concepts of classical mechanics and Maxwell's theory of classical electrodynamics.

Secondly, Rutherford atom model fails to explain the emission of radiation in the form of light, X-rays, etc. from the atom.

1.3 The Bohr Model of Atom

To explain the stability of the atom and the origin of observed spectral lines *Niels Bohr* took the case of hydrogen as it is simplest of all elements. He postulated that,

(i) In an atom an electron moves in a circular orbit round the nucleus under the Coulombic force of attraction between the electron and nucleus. The Coulombic force of attraction is balanced by the centrifugal force. Thus,

$$\frac{(Ze)(e)}{4\pi\epsilon_0 r^2} = \frac{mv^2}{r}.$$

(ii) An electron can move round the nucleus in certain definite orbit known as *stationary orbit.* In a stationary orbit the electrons do not radiate energy as predicted by *classical electromagnetic theory of light.*

(iii) The angular momentum of an electron revolving in a stationary orbit is an integral multiple of $\hbar = h/2\pi$, where h is the *Planck's constant.* So

$$L = n\hbar = n\frac{h}{2\pi} \text{ or, } I\omega = n\frac{h}{2\pi} \text{ or, } \left(mr_n^2\right)\frac{v_n}{r_n} = n\frac{h}{2\pi} \text{ or, } mv_n r_n = n\frac{h}{2\pi}, \tag{1.4}$$

where v_n and r_n are the velocity and radius of the electron in the nth orbit respectively and n is a positive number called the *quantum number,* eqn (1.4) is called the *Bohr's quantum condition.*

(iv) Electromagnetic energy is radiated from the atom when an electron jumps from one stationary orbit of higher energy (E_2) to another of lower energy (E_1), then according to *Planck's law,*

$$h\nu = \hbar\omega = E_2 - E_1, \tag{1.5}$$

where ν = the frequency and ω the circular frequency of radiation. Eqn (1.5) is known as *Bohr's frequency condition.*

An electron of energy E_1 can, however, jump to a higher orbit of energy E_2, if it absorbs an energy $h\nu = (E_2 - E_1)$.

Bohr's postulates are in *direct contradiction* to the laws of classical physics. Because according to classical physics (Kepler's laws), all orbits are permissible for the electron just like the planets revolving around the sun. Hence, Bohr's second postulate contradicts the laws of classical physics.

According to classical electromagnetic theory, a charged particle (here the electron) must radiate electromagnetic radiation while it undergoes an acceleration. So, the second postulate of Bohr also contradicts the result of classical electromagnetic theory of light.

For the other hydrogen-like atoms Bohr's postulates are equally applicable.

1.3.1 Bohr's Theory of Hydrogen-like Atoms

In view of Bohr's first postulate, for an electron in the nth orbit of radius r_n,

$$\frac{mv_n^2}{r_n} = \frac{(Ze)(e)}{4\pi\epsilon_0 r_n^2} \text{ or, } v_n^2 = \frac{1}{4\pi\epsilon_0} \cdot \frac{Ze^2}{mr}, \tag{1.6}$$

where ϵ_0 is the *permittivity of vacuum* and has value $(10^{-9}/36\pi)$F/m.

From Bohr's quantum condition

$$L = mr_n^2\omega = mv_n r_n = n\hbar, \tag{1.7}$$

where v_n is the velocity of the electron in the nth orbit.

$$\therefore v_n = \frac{n\hbar}{mr_n}. \tag{1.8}$$

From eqns (1.6) and (1.8)

$$\frac{1}{4\pi\epsilon_0} \cdot \frac{Ze^2}{mr_n} = \frac{m^2\hbar^2}{m^2 r_n^2} \text{ or, } r_n = \frac{4\pi\epsilon_0 n^2\hbar^2}{mZe^2}. \tag{1.9}$$

Eqn (1.9) shows that the radii of the different stationary orbits are proportional to the square of n, called the principal quantum number. For hydrogen atom $(Z = 1)$, the radius for $n = 1$ is called the *Bohr radius* having value

$$a_0 = \frac{4\pi\epsilon_0\hbar^2}{me^2} = 0.529 \times 10^{-10} \text{ m} = 0.529 \text{ Å}, \tag{1.10}$$

where $\epsilon_0 = 8.85419 \times 10^{-12}$ F/m, $m = 9.10953 \times 10^{-31}$ kg, $e = 1.60219 \times 10^{-19}$ C and $\hbar = 1.05459 \times 10^{-34}$ J-s.

Since $r_n \propto n^2$, the radii of successive orbits are in the ratios $1 : 4 : 9 : 16 : 25 : \cdots$.

Now the velocity of the electron in the nth stationary orbit is

$$v_n = \frac{n\hbar}{mr_n} = \frac{Ze^2}{4\pi\epsilon_0 n\hbar}. \tag{1.11}$$

Eqn (1.11) shows that $v_n \propto \frac{1}{n}$, i.e., the electron moves with lower velocity in higher orbits and vice versa. For hydrogen atom, the velocity of the electron in the Bohr orbit is

$$v_1 = \frac{e^2}{4\pi\epsilon_0\hbar} = 2.18 \times 10^6 \text{ m/s}. \tag{1.12}$$

If c be the velocity of light in vacuum, then

$$\frac{v_1}{c} = \frac{2.18 \times 10^6}{3 \times 10^8} = \frac{1}{137}. \tag{1.13}$$

The orbital frequency of the electron in the stationary orbits is given by

$$\omega_n = \frac{2\pi}{T_n} = 2\pi f_n = \frac{v_n}{r_n}.$$

$$\therefore f_n = \frac{1}{2\pi}\left(\frac{v_n}{r_n}\right) = \frac{1}{2\pi}\left(\frac{Ze^2}{4\pi\epsilon_0 n\hbar}\right)\left(\frac{mZe^2}{4\pi\epsilon_0 n^2\hbar^2}\right) \text{ or, } f_n = \frac{mZ^2e^4}{32\pi^3\epsilon_0^2 n^3\hbar^3}. \tag{1.14}$$

Thus, $f_n \propto \frac{1}{n^3}$.

1.3.2 Energy of the Electron in the Stationary Orbits

If E_k and V be respectively the kinetic and potential energies of the electron in a stationary orbit, then the total energy

$$E = E_k + V. \tag{1.15}$$

But

$$E_k = \frac{1}{2}mv^2 = \frac{m}{8\pi\epsilon_0}\frac{Ze^2}{mr} = \frac{1}{8\pi\epsilon_0}\frac{Ze^2}{r}. \tag{1.16}$$

The PE

$$V = \frac{1}{4\pi\epsilon_0}\int_r^\infty \frac{Ze^2}{r^2}\,dr = -\frac{Ze^2}{4\pi\epsilon_0 r}. \tag{1.17}$$

So

$$E = \frac{Ze^2}{8\pi\epsilon_0 r} - \frac{Ze^2}{4\pi\epsilon_0 r} \quad \text{[from eqn (1.15)]}$$

$$= -\frac{Ze^2}{8\pi\epsilon_0 r}. \tag{1.18}$$

Hence, the energy of the nth stationary orbit is

$$E_n = -\frac{mZ^2 e^4}{32\pi^2\epsilon_0^2 n^2 \hbar^2}. \tag{1.19}$$

Replacing $\hbar$ by $h/2\pi$, we get

$$E_n = -\frac{mZ^2 e^4}{8\epsilon_0^2 n^2 h^2}. \tag{1.20}$$

The negative sign in eqn (1.20) represents the bound state of the electron with nucleus and some work must be done to pull it away from it. Since $E_n \propto -\frac{1}{n^2}$, the energy E_n of the electron increases with increase of n (i.e., for the higher orbits) and approaches the limiting value $E_n = 0$ as $n \to \infty$. E_n is the lowest for $n = 1$. The state with $n = 1$ is called the *ground state* which is the most stable state of the electron.

1.3.3 Origin of Spectral Series in Hydrogen

If an electron jumps from an initial orbit n_i to a final orbit n_f, then the energy of radiation emitted is given by

$$\hbar\omega = E_i - E_f = \frac{mZ^2 e^4}{32\pi^2\epsilon_0^2\hbar^2}\left(\frac{1}{n_f^2} - \frac{1}{n_i^2}\right), \tag{1.21}$$

where $n_i > n_f$ so that $E_i > E_f$.

The frequency of the emitted radiation is

$$\nu = \frac{\omega}{2\pi} = \frac{mZ^2 e^4}{64\pi^3\epsilon_0^2\hbar^3}\left(\frac{1}{n_f^2} - \frac{1}{n_i^2}\right). \tag{1.22}$$

If c be the velocity of light, then the *wave number* (i.e., the number of full-waves per unit length) is given by

$$\bar{\nu} = \frac{\nu}{c} = \frac{1}{\lambda} = \frac{mZ^2 e^4}{64\pi^3\epsilon_0^2 c\hbar^3}\left(\frac{1}{n_f^2} - \frac{1}{n_i^2}\right). \tag{1.23}$$

The unit of $\overline{\nu}$ is m^{-1}.

$$\text{or, } \overline{\nu} = R_\infty Z^2 \left(\frac{1}{n_f^2} - \frac{1}{n_i^2}\right), \tag{1.24}$$

where

$$R_\infty = \frac{me^4}{64\pi^3\epsilon_0^2 c\hbar^3} = 1.09737 \times 10^7 \text{ m}^{-1} \tag{1.25}$$

is called the *Rydberg constant.*

$$\therefore E_n = -\frac{2\pi^2 c\hbar R_\infty Z^2}{n^2}. \tag{1.26}$$

When an electron is raised to an orbit of higher energy, then it is called to be *excited.* If the electron is removed from the atomic system, then it is called to be *ionised.* Ionisation of the atom can take place with the supply of sufficient energy called the *ionisation energy* E_i, which can be calculated as follows:

The ionisation energy for hydrogen ($Z = 1$) is

$$E_i = E_\infty - E_1 = \frac{me^4}{32\pi^2\epsilon_0^2 h^2}\left(\frac{1}{1} - \frac{1}{\infty}\right) = \frac{me^4}{32\pi^2\epsilon_0^2 \hbar^2} = 13.6 \text{ eV}. \tag{1.27}$$

An electron, so long revolves in its orbit, no energy is radiated at all. But when it absorbs energy of suitable frequency from outside source it gets excited and jumps to the higher orbit. But the lifetime of the excited state being ~ 10^{-8} s, it reverts back to lower energy state with the emission of electromagnetic radiation of energy $h\nu = \left(E_f - E_i\right)$. In a hydrogen discharge tube, the number of hydrogen atoms is very large which undergo excitation and consequent emission of radiation. In hydrogen atom though there is a single electron in the ground state but there are infinite excited states to which the electron can jump, if excited by the absorption of energy. The excitation of the electron to different lower energy states gives rise to spectral series hydrogen atom.

1.3.4 The Spectral Series of Hydrogen

Balmer in 1883 showed that the experimentally determined *wave number* of the spectral lines of hydrogen can be represented by

$$\overline{\nu} = A\left(\frac{1}{2^2} - \frac{1}{n^2}\right), \tag{1.28}$$

where A is a constant and $n = 3, 4, 5, \ldots$. The value of A found from the experimental data to be $1.09678 \times 10^7 \text{ m}^{-1}$, which is close to the value of R_∞ given by eqn (1.25). This establishes the validity of *Bohr's theory of hydrogen atom.*

Putting $n = 3, 4, 5, \ldots$ in eqn (1.28), we get wave number of different spectral lines of hydrogen in the visible region and are said to belong to the same spectral series called the *Balmer series.* Many other spectral series like *Lyman series, Paschen series,* etc. were discovered later on.

Lyman Series: Lyman series originates when an electron jumps from $n_i = 2, 3, 4, \ldots$ orbits to $n_f = 1$ orbit. The series lies in the UV *region* of the electromagnetic spectrum.

Thus,

$$\overline{\nu} = R_\infty\left(\frac{1}{1^2} - \frac{1}{n_i^2}\right), \quad n_i = 2, 3, 4, \ldots. \tag{1.29}$$

Balmer Series: Balmer series originates when an electron jumps from $n_i = 3, 4, 5, \ldots$ orbits to $n_f = 2$ orbits. This series lies in the *visible region*. The wave number is given by

$$\overline{\nu} = R_\infty\left(\frac{1}{2^2} - \frac{1}{n_i^2}\right), \; n_i = 3, 4, 5, \ldots. \tag{1.30}$$

Paschen Series: This series originates when an electron from $n_i = 4, 5, 6, \ldots$ orbits jumps to $n_f = 3$ orbits. This series lies in the *infrared region* and the wave number is given by

$$\overline{\nu} = R_\infty\left(\frac{1}{3^2} - \frac{1}{n_i^2}\right), \; n_i = 4, 5, 6, \ldots. \tag{1.31}$$

Brackett Series: This series originates when an electron from $n_i = 5, 6, 7, \ldots$ orbits jumps to $n_f = 4$ orbits. This series lies in the far infrared region and

$$\overline{\nu} = R_\infty\left(\frac{1}{4^2} - \frac{1}{n_i^2}\right), \; n_i = 5, 6, 7, \ldots. \tag{1.32}$$

Pfund Series: When an electron jumps from $n_f = 6, 7, 8, \ldots$ orbits to $n_i = 5$ orbits, Paschen series appears in far infrared region. The wave number is given by

$$\overline{\nu} = R_\infty\left(\frac{1}{5^2} - \frac{1}{n_i^2}\right), \; n_i = 6, 7, 8, \ldots. \tag{1.33}$$

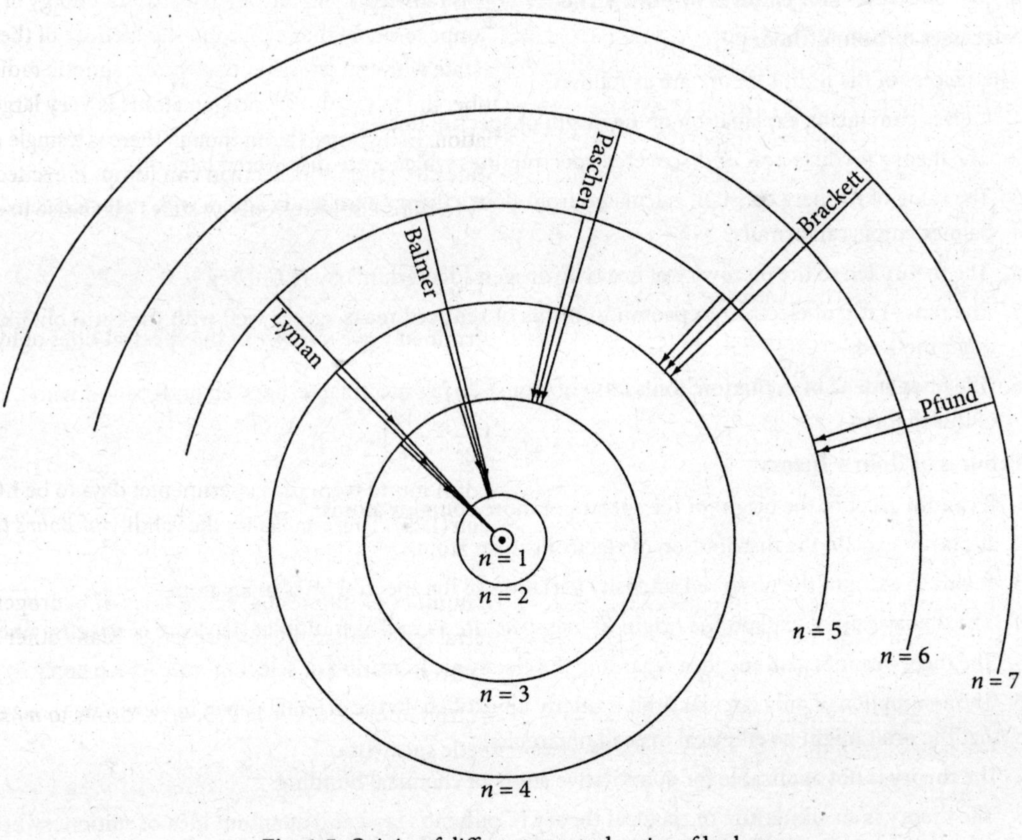

Fig. 1.5: Origin of different spectral series of hydrogen

Thus, the atoms could be taken as consisting of *discrete stationary orbits* corresponding to discrete stationary energy states. The spectral lines appear due to the transition of electrons between them.

The origin of different spectral series associated with emission of radiation is depicted in Fig. 1.5. The arrows indicate the transition from initial to final orbits. The same series in absorption spectra can be represented simply by reversing the arrowheads.

1.3.5 Energy Levels

Energy levels or *energy states* of a quantised system are the allowed energy values of the system. A plot of these energy values as horizontal lines with appropriate spacing is called the *energy level diagram.*

According to Bohr's theory, the energy of an electron in the nth quantum state is given by

$$E_n = -\frac{2\pi c\hbar R_\infty Z^2}{n^2} = -\frac{chR_\infty Z^2}{n^2} = -\frac{R_\infty ch}{n^2} \; [\because Z = 1 \text{ for hydrogen}]. \tag{1.34}$$

The energy values evaluated from eqn (1.34) corresponding to each quantum number are shown in Fig. 1.6. The distance of any level from zero energy level is proportional to the energy of that level. The lowest energy level with $n = 1$ is called the *ground* or *normal state.* With the increase of n the total energy becomes less negative and the atom shifts to more and more excited states. The energy $E_n \to 0$ as $n \to \infty$ and the atom becomes ionised as the electron is completely removed. The region for $E > 0$ represents a continuum.

1.3.6 Successes and Failures of Bohr's Theory

Successes of Bohr's Theory

The success of the Bohr's theory are as follows:

1. It gives convincing explanation of the origin of spectral lines.
2. The theory predicts new undiscovered spectral lines which were discovered later on.
3. The value of Rydberg constant calculated from Bohr's theory excellently agrees with that calculated from Balmer empirical formula.
4. The theory led to the discovery of heavy hydrogen (deuterium) by *H.C. Urey.*
5. The mass ratio of electron to proton in terms of reduced mass agrees well with the ratio obtained by other methods.
6. The phenomena of excitation, ionisation of atoms, X-ray spectra, etc. are well understood with the help of this theory.

Failures of Bohr's Theory

1. It cannot account the origin of the spectra of more complex atoms.
2. It cannot explain the distribution of electrons in an atom.
3. It fails to account the observed intensity variation of the spectral lines in an atom.
4. The theory cannot explain the origin of *fine structure* as well as multiplet structure of spectral lines.
5. The theory cannot give any idea regarding the electronic transition or selection rule which apply to them.
6. The assumption of only circular orbit is utterly unjustified. In the original paper, however, Bohr suggested that the orbit might be elliptical instead of circular.
7. The theory is not applicable for quantitative study of chemical bonding.
8. The theory is an admixture of classical theory (Coulomb's law) and quantum idea of stationary orbits.
9. The theory cannot suggest the state of polarization of the emitted radiation from the atom.

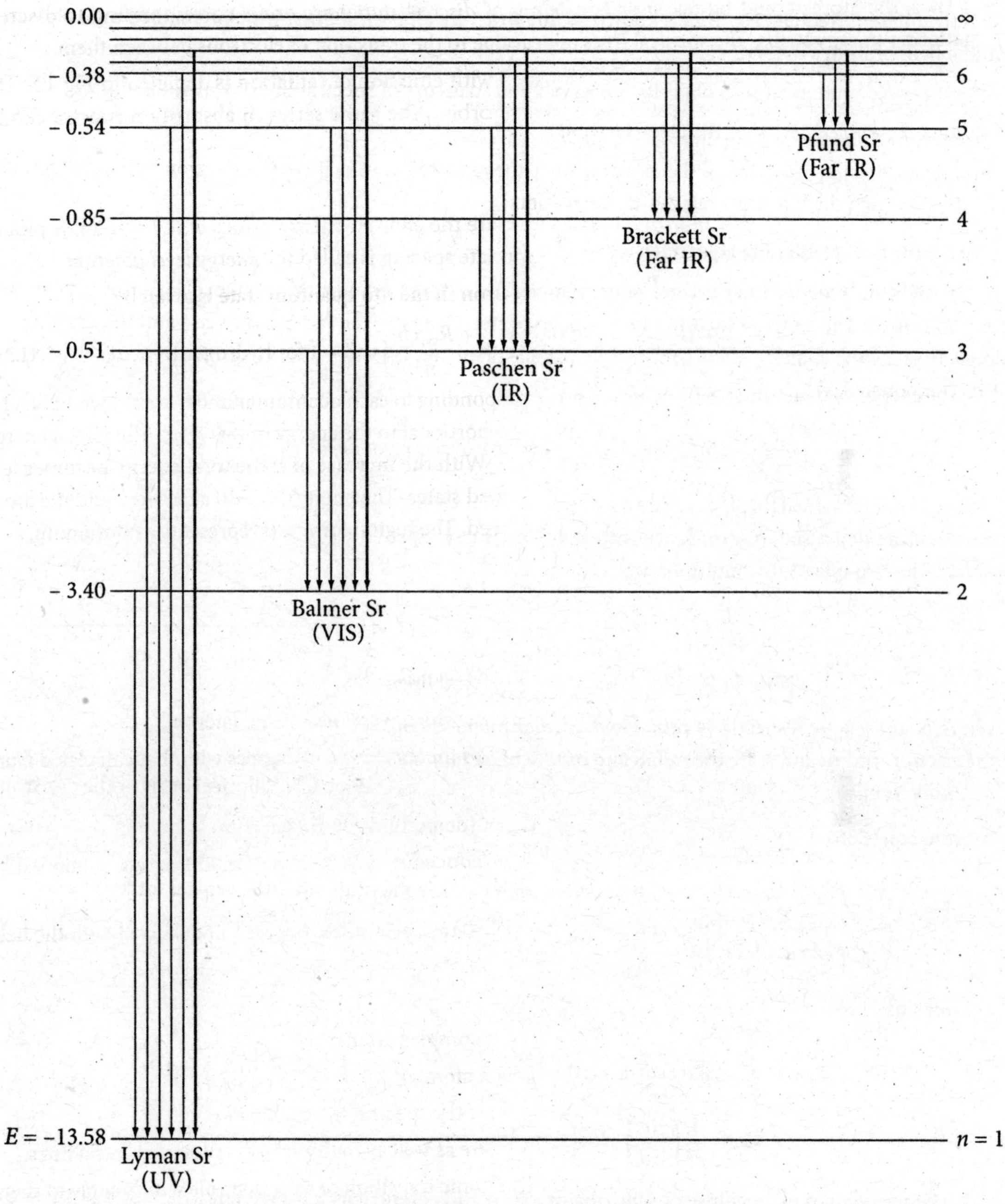

Fig. 1.6: Energy level diagram of hydrogen

1.4 Summerfeld's Extension of Bohr's Theory

In view of Bohr's theory each line of hydrogen-like spectrum should have a well-defined frequency (or wavelength). But Michelson and others observed that H_α, H_β and H_γ lines consist of close components of

slightly different frequencies. This is known as *fine structure of spectral lines,* which remained unexplained in the light of Bohr's theory.

Sommerfeld attempted to explain the observed fine structure of spectral lines by introducing the

(a) idea of *elliptical orbits* for the electrons and

(b) *relativistic mass variation* of the electron since the ratio of the velocity of the electron to that of light is 0.007 which demands the relativistic correction.

1.4.1 Elliptical Orbits for Hydrogen

As the electron moves under inverse square law of force, its path must be an ellipse with the nucleus at one of its focii. The electron then has two degrees of freedom, viz., the radial distance r and the azimuthal angle θ. Since the elliptic path is confined to a single plane the velocity of the electron can be resolved into two components, viz.,

$$\textit{radial component } v_r = \frac{dr}{dt} \text{ and } \textit{transverse component } v_\theta = r\frac{d\theta}{dt}.$$

Sommerfeld assumed that the radial and angular (or azimuthal) momenta are quantised in accordance with Bohr's quantum condition. The two quantum conditions are

$$\oint p_r\, dr = n_r h \tag{1.35}$$

$$\text{and} \oint p_\theta\, d\theta = kh, \tag{1.36}$$

where p_r and p_θ are respectively *radial and angular momenta* of the electron and n_r and k are the *radial* and *orbital* or, *azimuthal quantum numbers.*

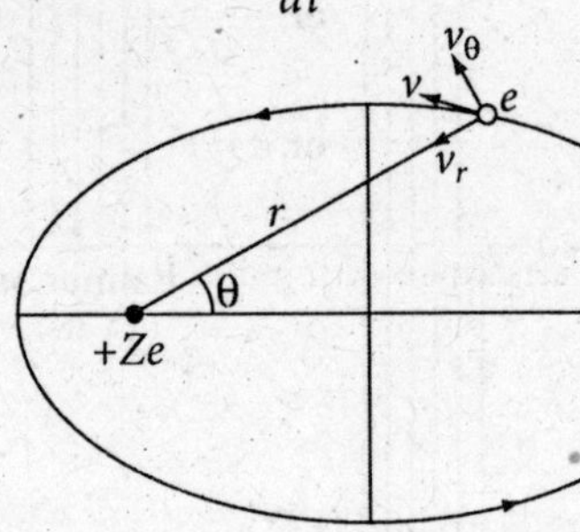

Fig. 1.7: Elliptical orbit of the electron

From eqn (1.36)

$$p_\theta \oint d\theta = kh, \because p_\theta \text{ is constant for a particle moving under the central force}$$

$$\text{or, } p_\theta 2\pi = kh \text{ or, } p_\theta = \frac{kh}{2\pi}. \tag{1.37}$$

From eqn (1.35)

$$n_r h = \oint p_r\, dr = \oint m\dot{r}\, dr = \oint m\frac{dr}{d\theta}\dot{\theta}\frac{dr}{d\theta}\, d\theta = \oint mr^2\dot{\theta}\left(\frac{1}{r}\frac{dr}{d\theta}\right)^2 d\theta$$

$$= p_\theta \oint \frac{1}{r^2}\left(\frac{dr}{d\theta}\right)^2 d\theta \ \left[\because mr^2\dot{\theta} = p_\theta\right]. \tag{1.38}$$

Now the equation of an ellipse of semi-major axis a and eccentricity ϵ is given by

$$r = \frac{a\left(1-\epsilon^2\right)}{1+\epsilon\cos\theta}. \tag{1.39}$$

$$\therefore \frac{dr}{d\theta} = \frac{a\left(1-\epsilon^2\right)\epsilon\sin\theta}{(1+\epsilon\cos\theta)^2} = r\frac{\epsilon\sin\theta}{1+\epsilon\cos\theta} \text{ or, } \frac{1}{r}\frac{dr}{d\theta} = \frac{\epsilon\sin\theta}{1+\epsilon\cos\theta}. \tag{1.40}$$

$\therefore$ to evaluate the integral, let

$$I = \int_0^{2\pi} \frac{1}{r^2}\left(\frac{dr}{d\theta}\right)^2 d\theta = \int_0^{2\pi} \frac{\epsilon^2 \sin^2\theta}{(1+\epsilon\cos\theta)^2}\, d\theta,$$

we put $u = \epsilon \sin\theta$ and $v = \frac{1}{1+\epsilon\cos\theta}$.

$$\therefore I = \int_0^{2\pi} u\, dv = [uv]_0^{2\pi} - \int_0^{2\pi} v\, du = \left[\frac{\epsilon\sin\theta}{1+\epsilon\cos\theta}\right]2\pi_0 - \int_0^{2\pi} \frac{\epsilon\cos\theta}{1+\epsilon\cos\theta}\, d\theta = -\int_0^{2\pi} \frac{\epsilon\cos\theta}{1+\epsilon\cos\theta}\, d\theta$$

$$= \int_0^{2\pi}\left(\frac{1}{1+\epsilon\cos\theta} - 1\right) d\theta = 2\pi\left(\frac{1}{\sqrt{1-\epsilon^2}} - 1\right) \text{ [being a standard integral].}$$

$$\therefore n_r h = 2\pi p_\theta\left(\frac{1}{\sqrt{1-\epsilon^2}} - 1\right) = kh\left(\frac{1}{\sqrt{1-\epsilon^2}} - 1\right) \quad [\because p_\theta = kh/2\pi]$$

$$\text{or, } \frac{1}{\sqrt{1-\epsilon^2}} = 1 + \frac{n_r}{k} = \frac{k+n_r}{k} = \frac{n}{k} \text{ (say)}$$

$$\text{or, } \epsilon = 1 - \frac{k}{n}. \tag{1.41}$$

For an ellipse with a and b as the semi-major and semi-minor axes

$$\frac{b}{a} = \sqrt{1-\epsilon^2} = \frac{k}{n}. \tag{1.42}$$

Eqns (1.40) and (1.41) indicate that the orbits are quantised. *Only those orbits of the electron are possible for which the ratio of the major to minor axis is equal to the ratio of two integers.*

From eqn (1.42), we note that since $b \le a$, $k \le n$.

If $b = a$, $\epsilon = 0$ and $k = n$, i.e., the orbit is *circular.* k cannot be equal to zero, because in that case $b = 0$ and the electron would execute simple harmonic motion through the nucleus which is not permissible. Hence, for a given value of n, k can have n different values which means *there can be n elliptical orbits of different eccentricities.*

Thus, the permissible values of $k = 1, 2, 3, \ldots, n$.

Since, $n = n_r + k$ or, $n_r = n - k$, the corresponding values are

$$n_r = (n-1), (n-2), \ldots, 2, 1, 0.$$

1.4.2 Total Energy of an Electron in an Elliptical Orbit

Total energy of the electron is given by

$$E_n = E_k + E_p = \frac{1}{2}m\left(\dot{r}^2 + r^2\dot{\theta}^2\right) - \frac{Ze^2}{4\pi\epsilon_0 r} = \frac{1}{2m}\left(p_r^2 + p_\theta^2/r^2\right) - \frac{Ze^2}{4\pi\epsilon_0 r} \quad \left[\because p_r = m\dot{r} \text{ and } p_\theta = mr^2\dot{\theta}^2\right]$$

$$\text{or, } \left(\frac{1}{r}\frac{dr}{d\theta}\right)^2 = \frac{2mE_n r^2}{p_\theta^2} + \frac{2mZe^2}{4\pi\epsilon_0 p_\theta^2} r - 1. \tag{1.43}$$

Now from eqn (1.39)

$$1+\epsilon\cos\theta = a\left(1-\epsilon^2\right)/r.$$

$$\therefore \epsilon^2\sin^2\theta = \epsilon^2-\epsilon^2\cos^2\theta = \epsilon^2-\left[\frac{a\left(1-\epsilon^2\right)}{r}-1\right]2 = (\epsilon^2-1)-\frac{a^2\left(1-\epsilon^2\right)}{r^2}+\frac{2a\left(1-\epsilon^2\right)}{r}. \tag{1.44}$$

From eqn (1.40)

$$\left(\frac{1}{r}\frac{dr}{d\theta}\right)^2 = \frac{\epsilon^2\sin^2\theta}{(1-\epsilon\cos\theta)^2}. \tag{1.45}$$

Using eqn (1.44) in eqn (1.45), we get

$$\left(\frac{1}{r}\frac{d\theta}{dr}\right)^2 = \frac{r^2}{a^2(1-\epsilon^2)^2}\left[(\epsilon^2-1)-\frac{a^2(1-\epsilon^2)^2}{r^2}+\frac{2a\left(1-\epsilon^2\right)}{r}\right] = -\frac{r^2}{a^2\left(1-\epsilon^2\right)}+\frac{2r}{a\left(1-\epsilon^2\right)}-1. \tag{1.46}$$

Now comparing the coefficients of like powers of r in eqns (1.43) and (1.46), we get

$$\frac{2mE_n}{p_\theta^2} = -\frac{1}{a^2\left(1-\epsilon^2\right)} \tag{1.47}$$

$$\text{and } \frac{2mZe^2}{4\pi\epsilon_0 p_\theta^2} = \frac{2}{a\left(1-\epsilon^2\right)}. \tag{1.48}$$

Dividing eqn (1.47) by eqn (1.48), we get

$$\frac{4\pi\epsilon_0 E_n}{Ze^2} = -\frac{1}{2a} \text{ or, } E_n = -\frac{Ze^2}{8\pi\epsilon_0 a}. \tag{1.49}$$

From eqn (1.48),

$$a = \frac{4\pi\epsilon_0 p_\theta^2}{mZe^2\left(1-\epsilon^2\right)} = \frac{k^2h^2}{4\pi^2}\cdot\frac{4\pi\epsilon_0}{mZe^2}\cdot\frac{n^2}{k^2}\left[\because p_\theta = \frac{kh}{2\pi} \text{ and } 1-\epsilon^2 = \frac{k^2}{n^2}\right]$$

$$\text{or, } a = \frac{\epsilon_0 n^2h^2}{\pi mZe^2} = \frac{a_0 n^2}{Z}, \tag{1.50}$$

where $a_0 = 4\pi\epsilon_0\hbar^2/me^2$ is the radius of the *first Bohr orbit.* Also the semi-minor axis

$$b = \frac{a_0\, kn}{Z}. \tag{1.51}$$

Substituting the value of a from eqn (1.50) in eqn (1.49), we get

$$E_n = -\frac{1}{n^2}\frac{mZ^2e^4}{8\epsilon_0^2h^2}. \tag{1.52}$$

Thus, the total energy of the electron in the elliptical orbits is the same as that in the Bohr circular orbits. Eqn (1.52) showed that E_n depends on the total quantum number n and not on the azimuthal quantum number k. Hence, *the introduction of the elliptical orbit does not give any new energy levels.* Eqn (1.50) shows that a does not depends on k but depends on the total quantum number n.

1.4.3 Degeneracy of the Orbits

The equality of the energy values for the different elliptic orbits of the electron for a given value of the principal quantum number n but different values of the azimuthal quantum number k is called the *degeneracy of the orbits.* All the orbits with same value of n but different values of k have the same value of the semi-major axis a but different values of the semi-minor axis b. The degeneracy arises due to the equality of the periodic variation of r and θ. So, the value of n and k are dependent on each other.

Since $n = k + n_r$, the different orbits for $n = 4$ are as shown in Table 1.1.

Table 1.1

Values of	4_1	4_2	4_3	4_4
a	$16a_0$	$16a_0$	$16a_0$	$16a_0$
b	$4a_0$	$8a_0$	$12a_0$	$16a_0$
k	1	2	3	4
n_r	3	2	1	0

Thus, for $n = 4$, there are four orbits with different values of eccentricity, but with same value of energy. This means the introduction of elliptic orbits does give no new energy levels, no new transitions. Fig. 1.8 shows the four degenerate orbits with $n = 4$.

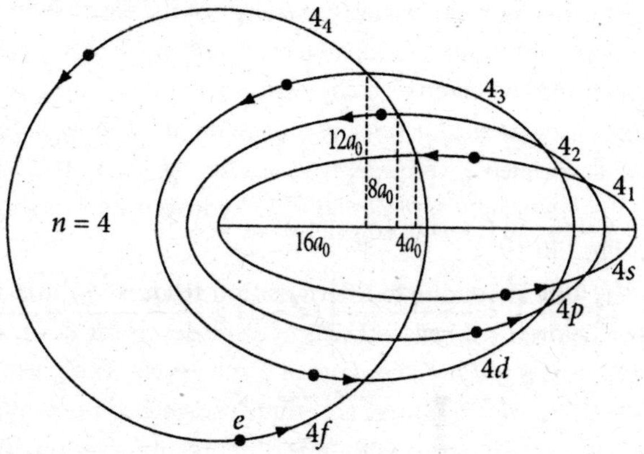

Fig. 1.8: Sommerfeld orbits for $n = 4$

1.4.4 Relativistic correction

Since the velocity of the electron in the elliptic orbit is $v = \frac{c}{137}$, where c is the velocity of light. The relativistic variation of mass m of the electron should be taken into account according to the relation:

$$m = \frac{m_0}{\sqrt{1 - v^2/c^2}},$$

where m_0 = the rest mass of the electron.

It can be shown that the path of the electron orbit as a result of relativistic mass variation of the electron is a *rosette* [Fig. 1.9] given by

$$r = \frac{a\left(1 - \epsilon^2\right)}{1 + \epsilon \cos \gamma\theta}, \tag{1.53}$$

where

$$\gamma^2 = 1 - \frac{Z^2 e^4}{\left(4\pi\epsilon_0\right)^2 p^2 c^2}, \quad p = p_\theta = \text{constant}. \tag{1.54}$$

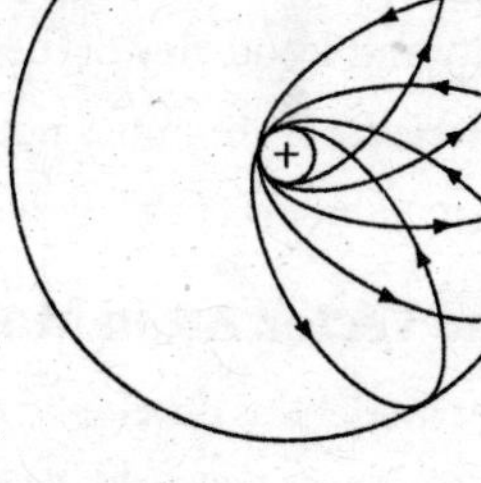

Fig. 1.9: Rosette path of an electron about the nucleus

For $c \to \infty$, $\gamma = 1$ and r repeats to the same value when θ changes by 2π, the periodicity being the same as θ. But for relativistic case, i.e., $\gamma < r$, θ returns to the same value when θ changes by $\frac{2\pi}{\gamma} > 2\pi$ and not 2π.

The angle of precession of the axis of the ellipse is

$$\Delta\theta = \frac{2\pi}{\gamma} - 2\pi. \tag{1.55}$$

It can be shown that the total energy of the electron with principal quantum number n in relativistic theory is

$$E_{n,k} = -\frac{\mu Z^2 e^4}{8\epsilon_0^2 n^2 h^2}\left[1 + \frac{Z^2\alpha^2}{n^2}\left(\frac{n}{k} - \frac{3}{4}\right)\right],$$

where $\alpha = e^2/2\epsilon_0 ch = 1/137$ is a universal constant called the *Sommerfeld's fine structure constant.*

$$\text{or, } E_{n,k} = -\frac{\mu Z^2 e^4}{8\epsilon_0^2 n^2 h^2} - \frac{mZ^2 e^4 \alpha^2}{8\epsilon_0^2 h^2}\left(\frac{n}{k} - \frac{3}{4}\right)\frac{1}{n^4}. \quad (1.56)$$

The first term represents the energy of the electron in the nth Bohr orbit and the second term is very small as $\alpha = 1/137$, which may be taken as a correction term depending on k. Since *for a particular value of n there are n values of k, there will be n closely spaced energy values for a given n*. This leads to the *multiplicity of spectral lines* in hydrogen atom as shown in Fig. 1.10.

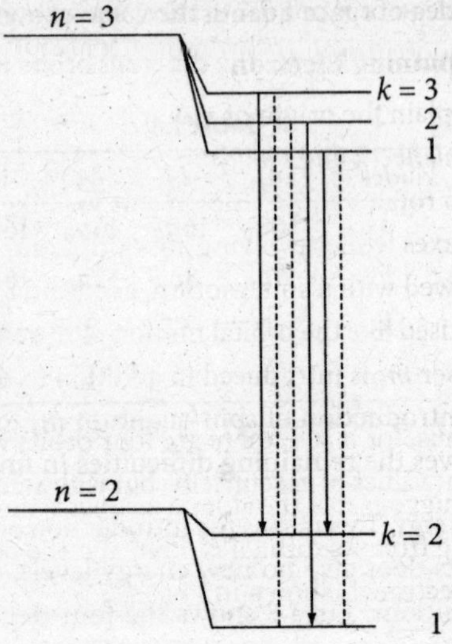

Fig. 1.10: Fine structure levels of H-atom

H_α-line arises due to the transition from $n = 3$ to $n = 2$ level. Now $n = 3$ level consists of 3 sub-levels for $k = 1, 2, 3$. Similarly, $n = 2$ level consists of 2 sub-levels. The selection rule being $\Delta k = \pm 1$, there are (three possible components of H_α-line which agrees with experimental observation. Thus, Sommerfeld theory could explain the fine structure of H_α-line.

1.4.5 Drawbacks of Sommerfeld's Theory

Following are the drawbacks of Sommerfeld's theory:

1. Sommerfeld's theory cannot predict the correct number of observed fine structure lines.
2. The theory cannot explain the anomalous *Zeeman effect* and *Stark effect.*
3. This theory is unable to explain the distribution of electrons in the atom and cannot throw light on the intensity and frequency of the spectral lines.
4. Sommerfeld's theory does not take into account the spin of the electron and Pauli's exclusion principle.

Vector atom model could, however, remove the above drawbacks of Sommerfeld's theory.

1.5 The Vector Atom Model

The vector model was developed by *Bohr, Sommerfeld, Uhlenbeck, Goudsmit, Pauling, Landé, Stern, Gerlach* and others to cover new field of research and remove the limitations of the previous atom model. This model enables us to understand and use the results of quantum mechanics in a convenient way.

The chief characteristics on which vector atom model is based are:

(i) *The idea of space quantisation* and (ii) *The hypothesis of spinning electron.*

(i) Space Quantisation

To explain the observations of *Zeeman effect,* Sommerfeld introduced the concept of *space quantisation.* According to this idea the classically permitted continuous manifold positions of the electron orbits in space are not possible, instead the orbits are quantised in magnitude (i.e., size and form) and spatial orientation. This space quantisation leads the orbits to be treated as vector quantities.

A fixed reference axis is necessary to specify the orientations of the electron orbits. An externally applied magnetic field may serve the purpose of the reference direction. The permitted quantised orbits are those for which the projections of the orbits on the field direction are quantised [Fig. 1.11]. The quantum number associated with space quantisation is called *orbital magnetic quantum number* and is represented by m_l. The idea of space quantisation was proved by *Stern-Gerlach experiment.*

(ii) Spinning Electron

To explain the origin of *multiplet structure of spectral lines,* the *concept of spinning* electron was introduced by *Uhlenbeck* and *Goudsmit in* 1925. According to this theory when an electron revolves round the nucleus it also rotates about its own axis just like the planets spin about their own axes while revolving about the sun. In other words, the electron endowed with a spin motion, according to quantum theory, must be quantised like the orbital motion of the electron. Thus, a spin quantum number m_s is introduced in addition to the angular quantum number m_l. Introduction of spin quantum m_s to the motion of the electron removes the remaining difficulties in fine structure of spectral lines. This suggests that the atomic magnet is the result of two magnets — one arising from the orbital motion and the other from the spin motion of the electron as shown in Fig. 1.11.

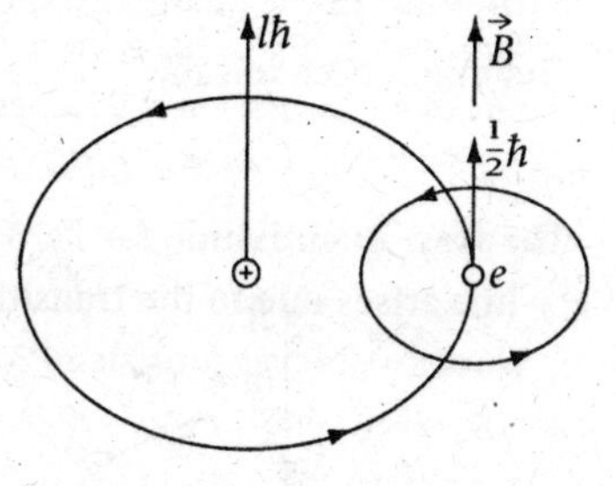

Fig 1.11: The spinning electron

1.6 The Quantum Numbers

In the vector atom model, the quantum numbers associated with each electron are as below:

1. *Principal Quantum Number* (n) : It arises due to the quantisation of energy and determines the size of the quantised electron orbit. It is denoted by n and corresponds to n in *Bohr-Sommerfeld theory* of the atom. The value of $n = 1, 2, 3, \ldots, \infty$, the corresponding electron shells are $K, L, M, \ldots$. The energy of the nth shell is $E_n \propto \frac{1}{n^2}$.

2. *Orbital Quantum Number* (l): It arises from the quantisation of the magnitude of orbital angular momentum (p_l) of the electron and is given by

$$p_l = l\frac{h}{2\pi} = l\hbar, \ (\hbar = h/2\pi).$$

Quantum mechanically

$$p_l = \sqrt{l(l+1)}\,\hbar.$$

The values that l can take are

$$l = 0, 1, 2, 3, \ldots, (n-1)$$

and the corresponding electrons are called $s, p, d, f, \ldots$ electron for $l = 0, 1, 2, 3, \ldots$.

3. *Spin Quantum Number* : It arises from the quantisation of the magnitude of the spin angular momentum (p_s) and is represented by s. Thus,

$$p_s = s\frac{h}{2\pi} = s\hbar = \frac{1}{2}\hbar,$$

since s can take value $\frac{1}{2}$ for electron.

Though classically half integral value of s is contradictory, later on Dirac showed quantum mechanically that $s = \frac{1}{2}$ for the spinning electron.

According to quantum mechanics

$$p_s = \sqrt{s(s+1)}\,\hbar.$$

An electron in an atom can be completely specified by n, l and s, if the external magnetic field (B) is absent.

4. *Magnetic Orbital Quantum Number* (m_l) : It specifies the *space orientation* of the electron orbit and is represented by m_s. For a given value of l, m_l can have $(2l+1)$ values, i.e.,

$$m_l = l, (l-1), (l-2), \ldots, 2, 1, 0, \ldots, -1, -2, \ldots, -(l-1), -l$$

m_l arises due to the quantisation of the direction of p_l. The directions are given by $\cos\theta = m_l/l$.

Quantum mechanically,

$$\cos\theta = m_l/\sqrt{l(l+1)}.$$

The space quantisation for $l = 3$ is shown in Fig. 1.12.

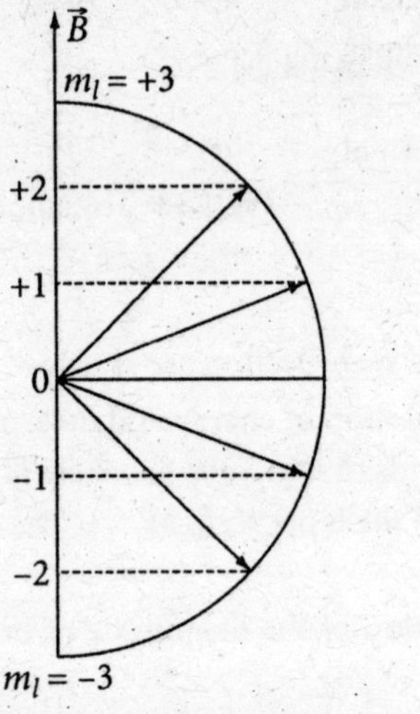

Fig. 1.12: Projection of orbital angular momentum quantum vector l in the direction of $\vec{B}$

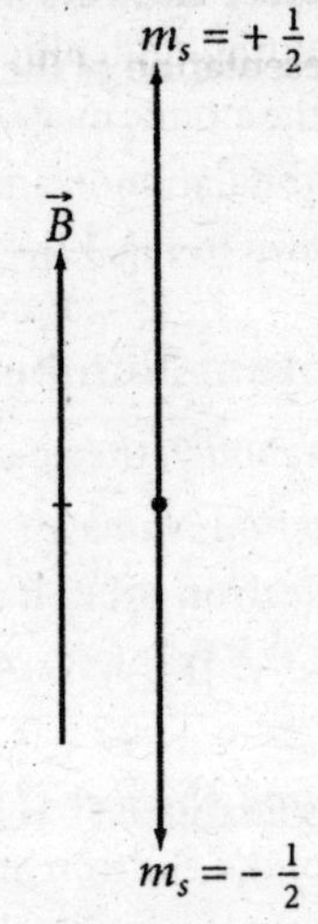

Fig. 1.13: Projection of the spin vector s in the direction of $\vec{B}$

5. *Magnetic Spin Quantum Number* : It arises from the quantisation of the spin angular momentum (p_s) and is represented by m_s, i.e., $p_s = m_s \hbar$.

m_s has values $\pm\frac{1}{2}$ and determines the orientation of the spin either up or down as shown in Fig. 1.13.

6. *Total Angular Quantum Number* (j) : It is the vector sum of the quantum numbers l and s and is represented by j, i.e.,

$$j = l + s = l \pm \tfrac{1}{2} \text{ (as } s = \pm\tfrac{1}{2}).$$

+ and − signs represent respectively s parallel to l and s anti-parallel to l.

The magnitude of the total angular momentum is quantised as $J = j\hbar$.

Quantum mechanically, $J = \sqrt{j(j+1)}\,\hbar$.

Thus, j defines the total angular momentum of the electron and is called the *inner quantum number.*

7. *The total Magnetic Quantum Number* (m_j) : m_j is the vector sum of m_l and m_s, i.e., $m_j = m_l + m_s$. From the knowledge of n, l, m_l and m_s, j and m_j can be found out and an electron in an atom can be completely specified.

Thus, in presence of a magnetic field $\vec{B}$, the four quantum numbers n, l, m_l and m_s can completely specify an electron in an atom.

1.7 The Pauli's Exclusion Principle

The principle states that, *no two electrons in an isolated atom can exist in the same quantum state.*

Wolfgang Pauli in 1925 enunciated the above principle purely on empirical basis which has universal acceptance supported by experimental verification. The implication of the principle is stated as under:

No two electrons in the same atom can have the same values of the four quantum numbers n, l, m_l and m_s. At least one quantum number must be different.

Example: He-atom has two electrons in the $n = 1$ or k-shell. Hence, for one electron: $n = 1, l = 0, m_l = 0, m_s = +\frac{1}{2}$ and for the other electron: $n = 1, l = 0, m_l = 0, m_s = -\frac{1}{2}$.

The quantum representation of the electrons in the k-shell is shown in Table 1.2.

Table 1.2

Quantum state				*Electron*	*Pictorial*
n	l	m_l	m_s	*configuration*	*representation*
1	0	0	$+\frac{1}{2}$	$1s^1$	$\boxed{\uparrow}$
1	0	0	$-\frac{1}{2}$	$1s^1$	$\boxed{\downarrow}$
1	0	0	$\pm\frac{1}{2}$	$1s^2$	$\boxed{\uparrow\downarrow}$

Application of Pauli's Exclusion Principle

(i) *Maximum Number of Electrons in Orbitals* or Sub-shells : We consider an orbit with principal quantum number n and orbital quantum number l. The values of n and l are fixed for an orbit but m_l and m_s can vary.

Now $m_l = -l$ to $+l$ including zero.

So, for a given l, there are $(2l + 1)$ values of m_l. Also for each value of m_l, there are two values of m_s, viz., $+\frac{1}{2}$ and $-\frac{1}{2}$.

Thus, the maximum number of electrons in any sub-shell (or orbital) having orbital quantum number l is $2(2l+1)$, since there can be only $2(2l+1)$ combinations of n, l, m_l and m_s.

Example: For s-sub-shell $l = 0$, and the maximum number of electrons is $2(2 \times 0 + 1) = 2$.

Similarly, for p-sub-shell $l = 1$ and the maximum number of electrons in it is $2(2 \times 1 + 1) = 6$ and so on.

(ii) *Maximum Number of Electrons in Main Shell*: We have already shown that the maximum quantum number n having orbital quantum number l is $2(2l+1), l = 0, 1, 2, 3, \ldots, (n-1)$.

Since a main shell with principal quantum number n contains l number of sub-shells, the maximum number of electrons in it is

$$N = \sum_{l=0}^{(n-1)} 2(2l+1) = 2\sum_{l=0}^{(n-1)} (2l+1) = 2\left[2\sum_{l=0}^{(n-1)} l + \sum_{l=0}^{(n-1)} 1\right] = \left[\frac{2n(n-1)}{2} + n\right] \text{ or, } N = 2n^2. \quad (1.57)$$

(iii) *Distribution of electrons in Various States*: Table 1.3 shows that the electron distribution in various states.

Table 1.3

n	l	m_l	m_s	*No. of electrons in sub-shells*	*Electronic configuration*	*No. of electrons $N = 2n^2$ in main shell*
1	0	0	+1/2	2	$1s^2$	2
1	0	0	−1/2			
2	0	0	+1/2	2	$2s^2$	8
2	0	0	−1/2			
2	1	−1	+1/2	6	$2p^6$	
2	1	−1	−1/2			
2	1	0	+1/2			
2	1	0	−1/2			
2	1	1	+1/2			
2	1	1	−1/2			
3	0	0	+1/2	2	$2s^2$	18
3	0	0	−1/2			
3	1	−1	+1/2	6	$3p^6$	
3	1	−1	−1/2			
3	1	0	+1/2			
3	1	0	−1/2			
3	1	+1	+1/2			
3	1	+1	−1/2			
3	2	−2	+1/2	10	$3d^{10}$	
3	2	−2	−1/2			
3	2	−1	+1/2			
3	2	−1	−1/2			
3	2	0	+1/2			
3	2	0	−1/2			
3	2	+1	+1/2			
3	2	+1	−1/2			
3	2	+2	+1/2			
3	2	+2	−1/2			

Hence, the maximum number of electrons in an orbit of principal quantum number n is $2n^2$.

1.8 Modern Periodic Table

General Features: The empirical observation that, when elements are arranged in increasing order of atomic number (Z), elements with similar chemical and physical properties recur at regular intervals and this phenomenon is known as the *periodic law.* The periodic law was first formulated by the *Russian chemist Dimitri Mendeleev. Periodic table* is the tabular arrangements of the elements showing the recurrence of their physical and chemical properties.

Each vertical column of the periodic table contains elements with similar properties and is numbered from I to VII called *group*. Group I consists of hydrogen and alkali metals having valency +1, which are highly chemically active. Group VII consists of halogens which are volatile, active non-metals with valency 1. Group VIII consists of inert gases which are chemically inactive.

The horizontal rows in the periodic table are called *period*. The elements in the periods are arranged according to the following sequence:

Active metal → less active metal → weakly active non-metal → highly active non-metal → inert gas.

Between groups II and III, in each period after the third, there is a series of elements called *transition elements*. These elements though similar in chemical activity, rarely resemblance with elements in major groups. For example, there are 15 transition elements in period 6 which are called the *lanthanides* or rare earths. There is a similar group of other elements in period 7 and are called *actinides*.

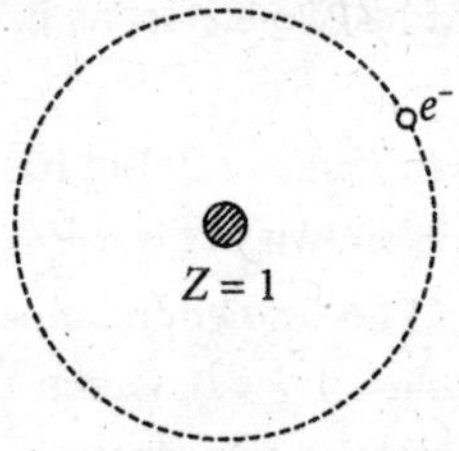

Fig. 1.14(a): H-atom

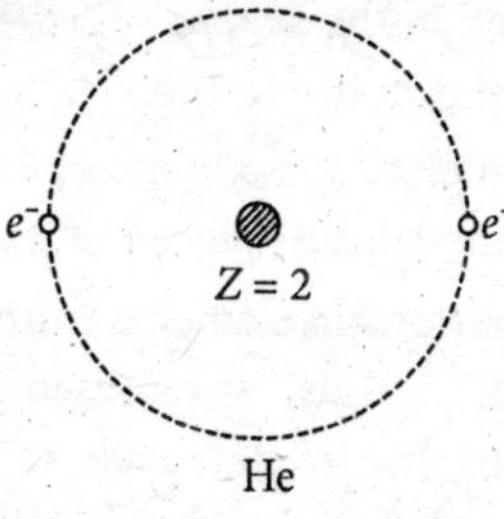

Fig. 1.14(b): He-atom

Explanation of the Pattern in Periodic Table

Table 1.4: Periodic table of elements

Period ↓ / Gr.→	I	II												III	IV	V	VI	VII	VIII
1	1 H																		2 He
2	3 Li	4 Be												5 B	6 C	7 N	8 O	9 F	10 Ne
3	11 Na	12 Mg												13 Al	14 Si	15 P	16 S	17 Cl	18 Ar
4	19 K	20 Ca	21 Sc	22 Ti	23 V	24 Cr	25 Mn	26 Fe	27 Co	28 Ni	29 Cu	30 Zn		31 Al	32 Ge	33 As	34 Se	35 Br	36 Kr
5	37 Rb	38 Sr	39 Y	40 Zr	41 Nb	42 Mo	43 Tc	44 Ru	45 Rh	46 Pd	47 Ag	48 Cd		49 In	50 Sn	51 Sb	52 Te	53 I	54 Xe
6	55 Cs	56 Ba	57 La	72 Hf	73 Ta	74 W	75 Re	76 Os	77 Ir	78 Pt	79 Au	80 Hg		81 Tl	82 Pb	83 Bi	84 Po	85 At	86 Rn
7	87 Fr	88 Ra	89 Ac																
Lanthanides					58 Ce	59 Pr	60 Nd	61 Pm	62 Sm	63 Eu	64 Gd	65 Tb	66 Dy	67 Ho	68 Er	69 Tm	70 Yb	71 Lu	
Actinides					90 Th	91 Pa	92 U	93 Np	94 Pu	95 Am	96 Cm	97 Bk	98 Cf	99 Es	100 Fm	101 Md	102 No	103 Lw	

The pattern in the periodic table can be explained by applying the concept of Pauli's exclusion principle. We know that (i) a sub-shell can contain a maximum of $2(l+1)$ electrons, (ii) maximum number of electrons in a shell is $2n^2$.

To explain the building up of the periodic table we start with the hydrogen atom and add electrons to appropriate energy levels in accordance with the Pauli's exclusion principle. The ground state of hydrogen atom is represented by $n = 1$ and $l = 0$, i.e., the electron of the hydrogen atom is in the energy state $n = 1$, $l = 0$. But this state can contain one more electron according to Pauli's exclusion principle. The electron to be added must have spin quantum number $m_s = -\frac{1}{2}$, if the first electron has $m_s = +\frac{1}{2}$. Thus, there can be two electrons in the state $n = 1, l = 0$. The nuclear charge in that should be made 2. This is nothing but the He-atom [Fig. 1.14(b)].

Lithium atom $(Z = 3)$ contains 3 electrons. As according to Pauli's exclusion principle more than two electrons cannot be placed in the state $n = 1$, there should be two electrons in state $n = 1$ and the third electron in state $n = 2$ as shown in Fig. 1.15.

In the above way, we can place a maximum of 8 electrons in state $n = 2$. But Na-atom has 11 electrons; the eleventh electron, therefore, must be placed in state $n = 3$, as shown in Fig. 1.16.

The electronic configurations of elements from $z = 1$ to 25 are shown in Table 1.4. The letters s, p, d, f, ... refers to $l = 0, 1, 2, 3, \ldots$. Thus, $1s^2$ represents the state $n = 1, l = 0$, which contains two electrons. In this way, $1s^2\,2s^2\,2p^6$ means that $1s^2$ state with $n = 1, l = 0$ contains two electrons, $2s$ state with $n = 2, l = 0$ contains two electrons and $2p^6$ state with $n = 2, l = 1$ contains 6 electrons.

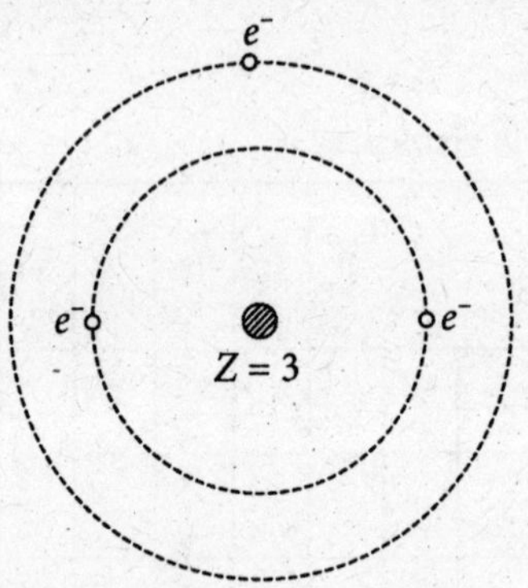

Fig. 1.15: Li-atom

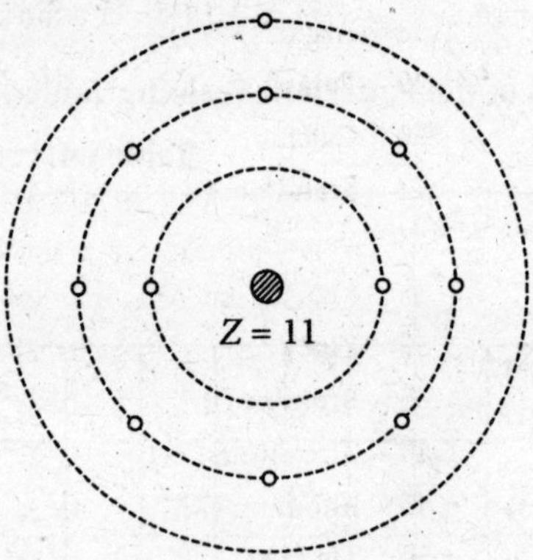

Fig. 1.16: Na-atom

Taking into account the Pauli's exclusion principle the allowed states are filled according to the following order: $1s, 2s, 2p, 3s, 3p, 4s, 3d, 4p, 5s, 4d, 5p, 6s, 5d, 4d, \ldots$

If the states would depend on the principal quantum number n only the ordering of the allowed levels would be $1s, 2s, 2p, 3s, 3p, 3d, \ldots$.

But the energy of an atom in many electron system not only depends on n but slightly on quantum number l. Due to this in the ordering of levels $4s$ occurs before $3d$; $5s$ before $4d$ and so on.

The similarities of *lanthanides* and *actinides* can now be explained. All *lanthanides* have identical $5s^2 5p^2 6s^2$ configurations but they differ only in their incomplete $4f$ sub-shell which does not affect their chemical properties. *Actinides* also have identical configurations $6s^2 6p^6 7s^2$ but differing only in the number of $5f$ and $5d$ electrons.

The *ferromagnetism* in Fe, Co and Ni arises due to the partial occupancy of their 3*d*-sub-shells whose electrons do not pair off to multiply the spin.

Hence, we note that when 3*d* states are filled, we get the *transition elements of group* IV and when the 4*d* states are filled, we get the *transition elements of group* V. By filling the 4*f* states 14 elements of *lanthanide series* in period 6 are obtained. 14 elements of *actinide series* are obtained by filling the 5*s* states of the elements of period 7.

Table 1.5 : Electronic Configurations of 25 elements

z	*Elements*	*Electronic configuration*
1	Hydrogen (H)	$1s^1$
2	Helium (He)	$1s^2$
3	Lithium (Li)	$1s^2 2s^1$
4	Beryllium (Be)	$1s^2 2s^2$
5	Boron (B)	$1s^2 2s^2 2p^1$
6	Carbon (C)	$1s^2 2s^2 2p^2$
10	Neon (Ne)	$1s^2 2s^2 2p^6$
11	Sodium (Na)	$1s^2 2s^2 2p^6 3s^1$
12	Magnesium (Mg)	$1s^2 2s^2 2p^6 3s^2$
13	Aluminium (Al)	$1s^2 2s^2 2p^6 3s^2 3p^1$
17	Chlorine (Cl)	$1s^2 2s^2 2p^6 3s^2 3p^5$
18	Argon (Ar)	$1s^2 2s^2 2p^6 3s^2 3p^6$
19	Potassium (K)	$1s^2 2s^2 2p^6 3s^2 3p^6 4s^1$
20	Calcium (Ca)	$1s^2 2s^2 2p^6 3s^2 3p^6 4s^2$
21	Scandium (Sc)	$1s^2 2s^2 2p^6 3s^2 3p^6 3d^1 4s^2$
23	Vanadium (V)	$1s^2 2s^2 2p^6 3s^2 3p^6 3d^3 4s^2$
25	Manganese (Mn)	$1s^2 2s^2 2p^6 3s^2 3p^6 3d^5 4s^2$
30	Zinc (Zn)	$1s^2 2s^2 2p^6 3s^2 3p^6 3d^{10} 4s^2$
35	Bromine (Br)	$1s^2 2s^2 2p^6 3s^2 3p^6 3d^{10} 4s^2 4p^5$
40	Zirconium (Zr)	$1s^2 2s^2 2p^6 3s^2 3p^6 3d^{10} 4s^2 4p^6 4d^2 5s^2$
45	Rhodium (Rh)	$1s^2 2s^2 2p^6 3s^2 3p^6 3d^{10} 4s^2 4p^6 4d^8 5s^1$
50	Tin (Sn)	$1s^2 2s^2 2p^6 3s^2 3p^6 3d^{10} 4s^2 4p^6 4d^{10} 5s^2 5p^2$
66	Dysprosium (Dy)	$1s^2 2s^2 2p^6 3s^2 3p^6 3d^{10} 4s^2 4p^6 4d^{10} 4f^9 5s^2 5p^6 5d^1 6s^2$

1.9 Spectroscopic Notations

In electronic configuration the orbital angular momentum states of individual electrons are described by small letters as

$$l = 0, 1, 2, 3, 4, 5, \ldots$$
$$\text{letters} = s, p, d, f, g, h, \ldots.$$

The entire electronic state of the atom can also be represented by using capital letters as follows:

$$L = 0, 1, 2, 3, 4, 5$$
$$\text{Letters} = S, P, D, F, G, H.$$

The state of an electron is determined by the three quantum numbers, L, S and J, where

$$L = l_1 + l_2 + l_3 + \cdots$$
$$S = s_1 + s_2 + s_3 + \cdots$$
$$\text{and } J = L + S,$$

where the quantum number J has the possible values, $J = (L+S), (L+S-1), \ldots, (L-S)$.

The state of an atom obeying LS coupling is given by the term $^{n(2s+1)}L_j$.

This is called the *spectroscopic term*, where n is the principal quantum number, $(2s+1)$ represents the multiplicity of states, where $L > S$ as J ranges from $(L+S)$ to $(L-S)$ through O. Thus, for $S = 0$, $(2S+1) = 1$, the multiplicity is 1 and is called the *singlet state* and $J = L$.

For $S = 1/2$, $(2s+1) = 2$, the multiplicity is 2, called the *doublet state* and $J = L \pm \frac{1}{2}$.

For $S = 1$, $(2s+1) = 3$, the multiplicity is 3, called the *triplet state* and $J = L+1, L, L-1$ and so on.

Example: In spectroscopic notation, a $3d$-electron may be represented as $3^2D_{5/2}$ or $3\,^2D_{1/2}$, where $n = 3$, $l = 2$, $2s+1 = 2\left(\frac{1}{2}\right)+1 = 2$ and $j = 2 \pm \frac{1}{2} = \frac{5}{2}, \frac{3}{2}$.

1.10 The Wave Mechanical Model of the Atom

de Broglie proposed that particles like electron, proton, neutron, etc. can have wave properties. If m be the mass of the particle, then the wavelength associated with it is $\lambda = h/mv$, where h is the *Planck's constant* and v is its velocity.

According to *wave mechanical model of the atom*, the electrons revolve round the nucleus as *wave packets*. According to the quantum theory, the electron orbit is an integral multiple of the associated de Broglie wavelength λ. If r be the radius of the electron orbit, then $n\lambda = 2\pi r$.

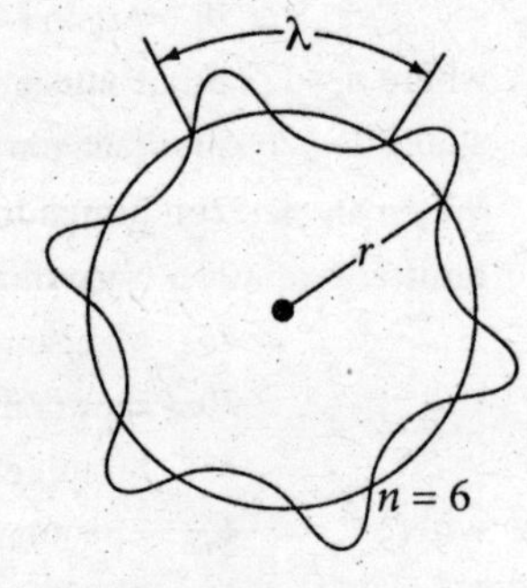

Fig. 1.17: de Brogli's stationary quantised orbit for $n = 6$

But $\lambda = h/p = h/mv$, where p is the momentum; m, the mass and v, the velocity of the electron.

$$\therefore n\frac{h}{mv} = 2\pi r \text{ or, } mvr = n\frac{h}{2\pi},$$

where mvr is the angular momentum of the electron. Hence, *wave mechanical picture leads to Bohr's postulate regarding the quantisation of the orbits.*

1.11 Answers of Some Important Questions

1.1. *What is J.J. Thomson's atom model? Why was it discarded?*

Ans. J.J. Thomson proposed in 1897 that atoms are just positively charged lumps of matter in which electrons are embedded similar to raisins in a fruit cake known as *plumb pudding model* of the atom.

The above atomic model could not explain the large angle scattering of α-particles in *Rutherford's experiment* and hence, discarded.

1.2. *What is Rutherford's atom model? What are its drawbacks?*

Ans. On the basis of large angle scattering of α-particles by thin foils of gold, etc., Rutherford proposed (in 1909) that, *an atom consists of a tiny massive positive charge at the centre and the negative electrons* *revolve around it just like the planets revolving round the sun, the radius of the revolving electron being very large compared to the dimension of the concentrated nuclear mass.*

For drawbacks of Rutherford's atom model, see Art. 1.2.1.

1.3. *What is the peculiarity of Bohr's theory of hydrogen atom?*

Ans. Bohr's theory in one hand rejects the *classical idea* regarding the occurrence of electron orbits of continuously varying radius and the result of *classical electrodynamics.* According to the classical electrodynamics a charged particle undergoing acceleration (or deceleration) emits *electromagnetic radiation.* For explaining the stability of the electron in the orbit again the help of *classical Coulomb's law* and the *centrifugal force* has been taken.

The theory, on the other hand, assumes the *Planck's quantum theory* in proposing the stationary electron orbits and the quantisation of angular momentum. It also assumes Einstein's theory to explain the emission of radiation from the atom.

Thus, the theory is an *admixture of classical and quantum theories.* Here lies the peculiarity of Bohr's theory.

1.4. *Hydrogen atom has only one electron, but many lines are formed in the spectrum of hydrogen. Explain.* **[GNDU 1995; PU 1995]**

Ans. Though hydrogen atom has only one electron there area large number of possible quantised energy states having energy

$$E_n = -\frac{2\pi^2 mZ^2 e^4}{n^2 h^2},$$

where $n = 1, 2, 3, \ldots$ and the other symbols have their usual significances. An electron in these energy states does not radiate energy and as such are called the *stationary states.*

When an electron jumps from a higher energy state n_f to lower energy state n_i, the frequency of the emitted radiation is given by

$$\nu = \frac{2\pi^2 mZ^2 e^4}{h^3}\left[\frac{1}{n_i^2} - \frac{1}{n_f^2}\right] \text{ or, } \frac{1}{\lambda} = \frac{2\pi^2 mZ^2 e^4}{ch^3}\left[\frac{1}{n_i^2} - \frac{1}{n_f^2}\right] = R\left[\frac{1}{n_i^2} - \frac{1}{n_f^2}\right],$$

where $R = \frac{2\pi^2 mZ^2 e^4}{ch^3}$ is known as the *Rydberg constant.*

If $n_i = 1$ and $n_f = 2, 3, 4, \ldots$ we get a series of spectral lines of different wavelength λ.

Practically, in a discharge tube containing hydrogen gas, a large number of hydrogen atoms are present at very low pressure. When a suitable voltage is applied across the discharge tube, different hydrogen atoms absorb different amount of energy and are excited to different excited states (n_f). The lifetime of the excited state being very small ($\sim 10^{-8}$ s) they transit to different lower energy states (n_i) and give rise to radiation of different wavelengths (λ) and hence, different spectral lines.

1.5. *State and prove Bohr's correspondence principle.* **[BU(H) 1999]**

Ans. Statement of Bohr's Correspondence Principle: The principle states that for a large quantum number, the quantum physics gives the same results as those of classical physics.

An electron while revolving in a circular orbit undergoes acceleration and radiates electromagnetic radiation in accordance with classical electrodynamics. The frequency of the emitted radiation is equal to the orbital frequency of the electron.

Now the velocity of an electron revolving round the nucleus is given by

$$v = \frac{e}{\sqrt{4\pi\epsilon_0 mr}}.$$

But the radius of the electron in the nth orbit is

$$r = \frac{n^2h^2\epsilon_0}{\pi me^2} \text{ (for } Z = 1).$$

The frequency of revolution of the electron is

$$\nu = \frac{v}{2\pi r} = \frac{1}{2\pi}\frac{e}{\sqrt{4\pi\epsilon_0 m}r^{3/2}} = \frac{1}{2\pi}\frac{e}{\sqrt{4\pi\epsilon_0 m}\left(\frac{n^2h^2\epsilon_0}{\pi me^2}\right)^{3/2}} \text{ or, } \nu_{Cl} = \nu = \frac{me^4}{4\epsilon_0^2h^3}\cdot\frac{1}{n^3}. \quad \text{(i)}$$

Now according to Bohr's theory, the frequency of the emitted radiation is

$$\nu = \frac{c}{\lambda} = \frac{me^4}{8\epsilon_0^2h^3}\left(\frac{1}{n_i^2} - \frac{1}{n_f^2}\right)\left[\because \frac{1}{\lambda} = \frac{me^4}{8\epsilon_0^2ch^3}\left(\frac{1}{n_i^2} - \frac{1}{n_f^2}\right)\right].$$

For $n \gg 1$, if $n_i = n$ and $n_f = n + 1$, then

$$\nu = \frac{me^4}{8\epsilon_0^2h^3}\left[\frac{1}{n^2} - \frac{1}{(n+1)^2}\right] = \frac{me^4}{8\epsilon_0^2h^3}\left[\frac{2n+1}{n^2(n+1)^2}\right] = \frac{me^4}{8\epsilon_0^2h^3}\cdot\frac{2}{n^3} \; [\because (2n+1) \simeq 2n, \text{ if } n \gg 1].$$

$$\therefore \nu_Q = \nu = \frac{me^4}{4\epsilon_0^2h^3}\cdot\frac{1}{n^3}. \quad \text{(ii)}$$

From eqns (i) and (ii), we note that $\nu_{Cl} = \nu_Q$. *Hence, both classical and quantum physics lead to the same result for a large quantum number* (n). This is Bohr's correspondence principle.

1.6. *Show that Bohr's quantum condition agrees with de Broglie wavelength hypothesis.* **[GNDU 1997]**

Ans. See Art. 1.10.

1.7. *Discuss the merits and demerits of Bohr's theory of hydrogen atom.*

Ans. See Art. 1.3.6.

1.8. *Rydberg constant is not a constant for all nuclei. — Explain.*

Ans. If R_∞ and R_Z be the Rydberg constants for an atom having a nucleus of infinite mass ($M_Z = \infty$) and an atom of finite nuclear mass (M_Z = finite), then

$$R_Z = \frac{R_\infty}{1 + \frac{m}{M_Z}},$$

where m is the electronic mass. From the above relation, it is evident that R_Z is a function of M_Z, the mass of a nucleus of atomic number Z. Since M_Z is different for different nuclei, R_Z cannot be a constant for all nuclei.

1.9. *Discuss how the radii of the electron orbits vary with order number?*

Ans. We have the radius of the electron orbit for the principal quantum number n,

$$r_n = \frac{n^2h^2\epsilon_0}{\pi mZe^2}. \quad \therefore r_n \propto n^2.$$

This shows that, with higher quantum number (n) the radii (r_n) will be greater.

Again,

$$\frac{dr_n}{dn} = \frac{2nh^2\epsilon_0}{\pi nZe^2}, \text{i.e.}, \frac{dr_n}{dn} \propto n,$$

i.e., the spacing of the orbits will be more for higher value of n. In other words, the higher ordered orbits will be more widely spaced.

1.10. *Discuss the nature of energy level diagram of hydrogen atom.*

Ans. According to Bohr's theory the energy of an electron in the nth orbit is

$$E_n = -\left(\frac{1}{n^2}\right)\frac{mZ^2e^4}{8\epsilon_0^2h^2}.$$

Using standard values of m, Z, e, ϵ_0 and h, we get $|E_n| \propto \frac{1}{n^2}$.

This means that the magnitude of energy of higher levels is less.

$$\frac{d|E_n|}{dn} \to -\frac{2dn}{n}.$$

The above relation indicates that the energy levels are more and more closely spaced with the increase of n. for $n \to \infty$, $\frac{d|E_n|}{dn} \to 0$, i.e., the energy levels merge together forming a *continuum.*

1.11. *What is fine structure of spectral lines? On what factor did Sommerfeld concentrate his idea to explain the fine structure of spectral lines?*

Ans. According to Bohr's theory each line of hydrogen-like spectra should have a well-defined frequency (or wavelength). But Michelson and others observed that H_α, H_β and H_γ lines consist of close components of slightly different frequencies. That is each spectral line (like H_α, etc.) consists of a close group of lines of slightly different wavelength. This is known as the *fine structure of spectral lines.*

Sommerfeld concentrated his idea on the following two factors to explain the origin of the fine structure of spectral lines:

(i) the electrons revolve in elliptical orbits keeping the nucleus at one of its focii and

(ii) the mass of the electron varies with velocity according to the Einstein's mass variation principle. Since the ratio $v/c = 0.007$, where v and c are the velocities of the electron and the velocity of light in vacuum, the introduction of relativistic concept of mass variation of the electron is quite justified.

1.12. *Point out the limitations of Bohr-Sommerfeld atom model. What are the basic concepts of vector atom model?*

Ans. For the answer of the first part, see Art. 1.4.5.

Answer to the second part: The basic concepts depending on which *vector atom mode* is proposed are as follows:

(i) **The idea of space quantisation.** According to this idea the plane of the electron orbits can take certain discrete possible orientations in space and not all orientations according to classical idea. This means, the whole space is quantised with reference to the orientation of the plane of the electron orbit.

(ii) **The hypothesis of spinning electron.** According to this the electron has *spin motion* about its own axis in addition to the *orbital motion* around the nucleus.

1.13. *Show that in a Bohr atom if the electron is considered as a wave travelling along the circular path, then the nth orbit will contain n complete de Broglie waves.* **[NBU(H) 2008]**

Ans. If m be the mass and v, the velocity of an electron in the nth orbit in Bohr atom, then from the condition of quantisation of angular momentum $mvr = n\hbar$, where r is the radius of the nth orbit and $\hbar = h/2\pi$. If the electron is associated with *de Broglie wavelength,* then

$$\lambda = \frac{h}{mv} \text{ or, } mv = \frac{h}{\lambda}.$$

$$\therefore \frac{h}{\lambda} r = n\frac{h}{2\pi} \text{ or, } 2\pi r = n\lambda \; [\because mvr = n\hbar].$$

Thus, the nth circular orbit of the electron in Bohr atom will contain *n complete de Broglie waves.*

1.14. *What are the four quantum numbers assigned to an electron in an atom? State the significance of the quantum numbers.* **[VU(H) 2005]**

Ans. The four quantum numbers assigned to an electron in an atom are:

(i) the principal quantum number (n),

(ii) the orbital quantum number (l),

(iii) the magnetic quantum number (m_l) and

(iv) the magnetic spin quantum number (m_s).

These four quantum numbers n, l, m_l and m_s specify an electron in an atom.

For the answer of the second part, see Art. 1.6.

1.15. (a) *State Pauli's exclusion principle.*

(b) *Prove that the maximum number of electrons in a shell is* $2n^2$, *where n is the number of shell.*

Ans. (a) See Art. 1.7; (b) See Art. 1.7.1 (i). **[VU(H) 2005]**

1.16. *What is the electron configuration of the ground state of sodium atom?* **[CU(H) 2005]**

Ans. The electron configuration of the ground state of sodium atom is $1s^2 2s^2 2p^6 3s^1$.

1.17. *Show how one can arrive at Bohr quantisation condition on the basis of de Broglie hypothesis of matter waves.* **[VU(H) 2001]**

Ans. See Art. 1.10.

1.18. *What is the electron configuration of a normal lithium atom?* **[BU(H) 2003]**

Ans. The electron configuration of a normal lithium atom is $1s^2 2s^1$.

1.19. *Why chlorine is chemically active?*

Ans. The electronic configuration of a normal chlorine atom is $1s^2 2s^2 2p^6 3s^2 3p^5$.

Thus, the outer shell $(3p)$ can accommodate one more electron and its valency is –1. An atom is chemically active if it can easily donate or accept an electron.That is why chlorine is chemically active as it can participate in a chemical reaction by accepting an electron from other atom.

1.20. *Why Argon is chemically inactive?*

Ans. The electronic configuration of argon (Ar) is $1s^2 2s^2 2p^6 3s^2 3p^6$.

Thus, the outermost orbit of Ar is completely filled with electrons and the electrons are very strongly bound. It is normally very difficult for Ar-atom to donate one electron from it or accept any electron.

In chemical reaction, electrons rearrange their positions in neighbouring atoms. Since this is not possible for Ar-atom, it does not enter into chemical reaction and behaves as chemically inert element.

1.21. *What are the permitted values of the orbital quantum number l?* **[BU(H) 2004]**

Ans. The permitted values of the orbital quantum number l are: $l = 0, 1, \ldots, (n-1)$, where n is the principal quantum number. Thus, for a particular l, there are $(n-1)$ values of l.

1.22. *What is vector atom model? Briefly explain two elements which characterise the vector atom model.* **[VU(H) 2007]**

Ans. The old quantum theory formulated by Bohr-Sommerfeld could not explain the anomalous *Zeeman effect, Paschen-Back effect, Stark effect,* etc. which were discovered in subsequent years. To provide satisfactory explanation of the above effects a new model, called *Vector Atom Model,* was proposed. The chief contributors to the Vector Atom Model are *Bohr, Sommerfeld, Uhlenbeck, Gaudsmit, Pauli, Stren* and *Gerlach.*

The two elements which characterise the vector atom model are as follows:

(i) **The concept of electron spin.** The idea of electron spin was introduced by *Uhlenbeck* and *Goudsmit* in 1925. To explain the multiple structure of spectral lines it was essential to assume the existence of spin angular momentum of the electron equal to $\hbar/2$. This additional angular momentum is controlled by a third quantum number $s = \frac{1}{2}$.

(ii) **The concept of space quantisation.** This concept was introduced by *Sommerfeld* to explain the magnetic splitting of spectral lines (the *Zeeman effect*). According to this the orbits are to be quantised not only in magnitude but also in spatial orientation. This space quantisation makes the orbits to behave like vector quantities. A feeble magnetic field might offer the reference direction and the projections of the quantised orbits on the field must themselves be quantised. Since the orbital and spin motion of the electron are quantised both in magnitude and direction, the orbital and spin angular momenta are quantised. As all the quantities which determine the state of the atom are quantised vectors, the atom model so designed is called the *Vector Atom Model.*

1.12 Solved Problems

1.1. *Using the physical constants given below, calculate the following for hydrogen atom:* (a) *Velocity of an electron in the ground state,* (b) *Radius of Bohr orbit in the ground state,* (c) *Time taken by the electron to transverse first Bohr orbit,* (d) *Rydberg constant.*

[*Electronic charge,* $e = 1.6 \times 10^{-19}$ C, *Electronic mass,* $m = 9.1 \times 10^{-31}$ kg,
Planck's constant, $h = 6.6 \times 10^{-34}$ J-s, *Speed of light,* $c = 3 \times 10^{8}$ m/s,
and *permittivity of free space,* $\epsilon_0 = 8.86 \times 10^{-12}$ C^2/Nm^2]

Solution: (a) The velocity of the electron in the ground state of hydrogen atom is

$$v = \frac{e^2}{2h\epsilon_0} = \frac{(1.60 \times 10^{-19})^2}{2 \times (6.6 \times 10^{-34}) \times (8.86 \times 10^{-12})} = \mathbf{2.19 \times 10^6\ m/s}.$$

(b) The radius of the Bohr orbit is

$$r_1 = \frac{h^2\epsilon_0}{\pi m e^2} = \frac{(6.6 \times 10^{-34})^2 \times (8.86 \times 10^{-12})}{3.14 \times (9.1 \times 10^{-31}) \times (1.6 \times 10^{-19})^2} = 0.528 \times 10^{-10}\ \text{m} = \mathbf{0.528\ \text{Å}}.$$

(c) Time taken to transverse the Bohr orbit is

$$T = \frac{2\pi r_1}{v} = \frac{2 \times 3.14 \times (0.528 \times 10^{-10})}{2.19 \times 10^6} = \mathbf{1.51 \times 10^{-16}\ s.}$$

(d) Rydberg constant for the hydrogen atom is

$$R_H = \frac{me^4}{8\epsilon_0^2 ch^3} = \frac{(9.1 \times 10^{-31}) \times (1.6 \times 10^{-19})^4}{8 \times (8.86 \times 10^{-12})^2 \times (3 \times 10^8) \times (6.6 \times 10^{-34})^3} = \mathbf{1.101 \times 10^7\ m^{-1}.}$$

1.2. *Calculate the frequency of revolution of the electron in the first Bohr orbit of hydrogen atom.*

Solution: The velocity of the electron in the first Bohr orbit is

$$v = \frac{e^2}{2h\epsilon_0} \quad (\because n = 1).$$

The orbital frequency ν is

$$= \frac{\text{electron velocity}}{\text{orbital circumference}} = \frac{v}{2\pi r} = \frac{e^2}{2h\epsilon_0} \times \frac{1}{2\pi}\frac{\pi me^2}{\epsilon_0 h^2} = \frac{me^4}{4\epsilon_0^2 h^3}$$

$$= \frac{(9.1 \times 10^{-31}) \times (1.6 \times 10^{-19})^4}{4 \times (8.854 \times 10^{-12})^2 \times (6.625 \times 10^{-34})^3} = \mathbf{6530 \times 10^{12}\ Hz.}$$

1.3. *The wavelength of the first member of Balmer series in hydrogen spectrum is* 6563 Å. *Calculate the wavelength of the first member of Lyman series in the same spectrum.*

Solution: The wave number of the first member of the Balmer series is given by

$$\bar{v}_1 = \frac{1}{\lambda_1} = R_H\left(\frac{1}{2^2} - \frac{1}{3^2}\right) = \frac{5}{36}R_H. \quad \text{(i)}$$

The wave number of the first member of *Lyman series* is given by

$$\bar{v}_2 = \frac{1}{\lambda_2} = R_H\left(\frac{1}{2^2} - \frac{1}{2^2}\right) = \frac{3R_H}{4}. \quad \text{(ii)}$$

From eqns (i) and (ii)

$$\frac{\lambda_2}{\lambda_1} = \frac{5}{27} \text{ or, } \lambda_2 = \frac{5}{27}\lambda_1 = \frac{5 \times 6563}{27} = \mathbf{1215.35\ Å.}$$

1.4. *The first line in the Balmer series of hydrogen atom has a wavelength* 6563 Å. *Calculate the wavelength of the second line in Balmer series.* **[NBU(H) 2009]**

Solution: The wavelength of the first line in *Balmer series* in given by

$$\frac{1}{\lambda_1} = R_H\left(\frac{1}{2^2} - \frac{1}{3^2}\right) = \frac{5}{36}R_H. \quad \text{(i)}$$

The wavelength of the second line in *Balmer series* is given by

$$\frac{1}{\lambda_2} = R_H\left(\frac{1}{2^2} - \frac{1}{3^2}\right) = \frac{3}{16}R_H. \quad \text{(ii)}$$

$$\therefore \frac{\lambda_2}{\lambda_1} = \frac{5}{36} \times \frac{16}{3} = \frac{20}{27} \text{ or, } \lambda_2 = \frac{20}{27} \times 6563 = \mathbf{4861.48\ Å.}$$

1.5. *Find the wavelength of a photon that is emitted when a hydrogen atom undergoes a transition from* $n_i = 5$ *to* $n_f = 2$.

Solution: The energy of an electron in the nth orbit is $E_n = -\frac{13.6}{n^2}$ eV.

$$\therefore E_2 = -\frac{13.6}{2^2} = -3.40\,\text{eV and } E_5 = -\frac{13.6}{5^2} = -0.544\,\text{eV}.$$

According to Bohr's postulate, the energy of the emitted photon is

$$E_\gamma = -0.544\,\text{eV} - (-3.40)\,\text{eV} = 2.86\,\text{eV}.$$

The wavelength of the same photon is

$$\lambda = \frac{hc}{E_\gamma} = \frac{12.4 \times 10^3\,\text{eV Å}}{2.86\,\text{eV}} = \mathbf{4340\,Å}.$$

1.6. *According to Bohr's theory how many revolutions will an electron make in the first excited state of hydrogen if the lifetime of that state is 10^{-8} s?*

Solution: If r_2 and v_2 be the radius and velocity of the electron in the first excited state ($n = 2$), then

$$r_2 = 4r_1 = 4 \times 0.529\,Å = 2.12\,Å = 2.12 \times 10^{-10}\,\text{m} \left[\because r_1, \text{ the radius of the Bohr orbit} = 0.529\,Å\right]$$

and

$$v_2 = \frac{v_1}{2} = \frac{c}{2\,(137)} = \frac{3 \times 10^8}{2\,(137)} = \mathbf{1.10 \times 10^6\,m/s}, \left[\because v_1 = c/137\right].$$

The angular velocity is

$$\omega = \frac{v_2}{r_2} = \frac{1.10 \times 10^6\,\text{m/s}}{2.12 \times 10^{-10}\,\text{m}} = 0.52 \times 10^{16}\,\text{rad/s}.$$

Now the total number of revolution is

$$N = \frac{\omega t}{2\pi} = \frac{\left(0.52 \times 10^{16}\right) \times 10^{-8}}{6.28} = \mathbf{8.3 \times 10^6} \text{ revolutions.}$$

1.7. *Show that in a Bohr atom, the nth orbit contains n complete de Broglie (electron) waves.*

[BU(H) 1994, 1997; NBU(H) 2008]

Solution: The radius of the nth Bohr orbit is given by

$$r_n = \frac{n^2 h^2 \epsilon_0}{\pi m e^2}.$$

The circumference of the nth Bohr orbit is

$$2\pi r_n = \frac{2n^2 h^2 \epsilon_0}{m e^2}.$$

The wavelength of the *de Broglie wave* associated with an electron is

$$\lambda = \frac{h}{mv} = \frac{h}{m} \cdot \left(\frac{2nh\epsilon_0}{e^2}\right).$$

[$\because v_n$, the velocity of the electron in the nth orbit $= \frac{e^2}{2nh\epsilon_0}$]

$\therefore$ the number of complete de Broglie waves in the nth orbit is

$$\frac{\text{Circumference}}{\text{Wavelength}} = \frac{\left(\frac{2n^2h^2\epsilon_0}{me^2}\right)}{\frac{h}{m}\left(\frac{2nh\epsilon_0}{e^2}\right)} = n \textbf{ (Proved).}$$

1.8. *What is the velocity of the electron in the ground state of Bohr's hydrogen atom in terms of speed of light? What is this called?*

Solution: We have the velocity of the electron in the nth Bohr orbit

$$v_n = \frac{Ze^2}{2\epsilon_0 nh} = \frac{e^2}{2\epsilon_0 nh} \quad (\because Z = 1).$$

$\therefore$ the velocity of the electron in the first Bohr orbit is

$$v_1 = \frac{e^2}{2\epsilon_0 ch} = \frac{(1.60 \times 10^{-19})^2}{2 \times (8.85 \times 10^{-12}) \times (3 \times 10^8) \times (6.62 \times 10^{-34})} \simeq \frac{1}{137} = \alpha \text{ (say)}.$$

$\alpha = \frac{1}{137}$ is called the *fine structure constant.*

1.9. *Electrons of energies* 10.20 eV *and* 12.09 eV *can cause radiation to be emitted from hydrogen atoms. Calculate in each case, the principal quantum number of the orbit to which electron in the hydrogen atoms is raised and the wavelength of the radiation emitted if it drops back to the ground state.*

Solution: The energy of an electron in the nth Bohr orbit is given by

$$E_n = -\frac{13.6}{n^2} \text{ eV},$$

n is the principal quantum number.

For $n = 1$, $E_1 = -13.6$ eV

$n = 2$, $E_2 = -3.4$ eV

$n = 3$, $E_3 = -1.51$ eV.

Now energy required by an electron to transit from $n = 1$ to $n = 2$ level is $(13.6 - 3.4)$ eV = 10.2 eV.

Energy required for the transition from $n = 1$ to $n = 3$ level is $(13.6 - 1.51)$ eV = 12.09 eV.

This shows that the electron can be raised to $n = 2$ and $n = 3$ from $n = 1$ corresponding to energies 10.2 eV and 12.09 eV respectively.

Again, when the electron jumps from $n = 2$ to $n = 1$ level, we get the first member of *Lyman series.*

$$\therefore \frac{1}{\lambda_1} = R_H \left[\frac{1}{1^2} - \frac{1}{2^2}\right] = \frac{3}{4} R_H \text{ or, } \lambda_1 = \frac{4}{3R_H} = \frac{4}{3 \times (10.97 \times 10^6)} = \mathbf{1216\ \text{Å}}.$$

When the electron jumps from $n = 3$ to $n = 1$ level

$$\frac{1}{\lambda_2} = R_H = \left[\frac{1}{1^2} - \frac{1}{3^2}\right] = \frac{8R_H}{3} \text{ or, } \lambda_2 = \frac{3}{8R_H} = \frac{3}{8 \times (10.97 \times 10^6)} = \mathbf{1026\ \text{Å}},$$

where $R_H = 10.97 \times 10^6\ \text{m}^{-1}$.

1.10. *The amount of energy required to ionise a hydrogen atom in its ground state is* 13.58 eV. *How much energy is needed to excite it from the ground state to the* $n = 4$ *level?* **[BU(H) 2003]**

Solution: The ionisation energy of the hydrogen atom is $(E_H)_1 = 13.58$ eV.

Energy of the atom in the $n = 4$ level is

$$(E_H)_4 = \frac{13.58}{4^2} = \frac{13.58}{16} = 0.849 \text{ eV}.$$

$\therefore$ energy needed to excite the atom from ground state to $n = 4$ level is

$$\Delta E = (E_H)_1 - (E_H)_4 = 13.58 - 0.849 = \mathbf{12.73\ eV}.$$

1.11. *Given the ground state energy of H-atom to be −13.58 eV, estimate the ground state energy effect of nuclear mass.* **[BU(H) 2002]**

Solution: The ground state energy of the H-atom is $(E_H)_1 = -13.58$ eV.

∴ the ground state energy of Li^{++} is

$$(E_{Li^{++}})_1 = 2^2 (E_H)_1 = 4 \times (-13.58) = -54.32\,\text{eV} \left[\because Z = 2 \text{ for } Li^{++}\right].$$

1.12. *Given the value of Rydberg's constant for hydrogen R_H to be $R_H = 109677.6\,\text{cm}^{-1}$, calculate the ground state energy of the hydrogen atom in* eV. **[BU(H) 2004]**

Solution: The energy of an electron in the *n*th state is given by

$$E_n = -\frac{R_\infty chZ^2}{n^2}.$$

For the hydrogen atom, the energy of the electron in the ground state ($n = 1$) is

$$(E_H)_1 = -\frac{R_H ch(1)^2}{1^2} = -R_H ch\ [\because Z = 1 \text{ for hydrogen atom}].$$

Now Rydberg constant for hydrogen .

$$R_H = 109677.6\,\text{cm}^{-1} \simeq 10.97 \times 10^6/\text{m}.$$

Velocity of light $c = 3 \times 10^8$ m/s and Plack's constant $h = 6.6 \times 10^{-34}$ Js.

$$\therefore (E_H)_1 = -\left(10.97 \times 10^6\right) \times \left(3 \times 10^8\right) \times \left(6.6 \times 10^{-34}\right)\text{ J}$$

$$= -\frac{10.97 \times 19.8 \times 10^{-20}}{1.6 \times 10^{-19}}\text{ eV} \left[\because 1\,\text{eV} = 1.6 \times 10^{-19}\,\text{J}\right]$$

$$= -\frac{10.97 \times 19.8}{16} = \mathbf{13.58\ eV}.$$

1.13. *The ground state energy of an electron in the hydrogen atom is −13.6 eV. Compute the energy of (a) $n = 3$ state of He^+, (b) $n = 2$ state of Li^{++}.*

Solution: (a) The energy of an element in the *n*th state of the hydrogen-like atom is

$$E_n = -\frac{RchZ^2}{n^2}.$$

But $Rch = 13.6$ eV

$$\therefore (E_n)_H = -\frac{13.6}{n^2}\ [\because Z = 1 \text{ for hydrogen atom}].$$

For the ground state $n = 1$,

$$(E_1)_H = -13.6\,\text{eV}.$$

Similarly,

$$(E_2)_H = -\frac{13.6}{4} = -3.4\,\text{eV}$$

$$\text{and } (E_3)_H = -\frac{13.6}{9} = -1.5\,\text{eV}.$$

(a) For He^+, $Z = 2$, so for $n = 3$,

$$(E_3)_{He} = 4 \times (E_3)_H = \mathbf{-6.0\ eV}.$$

(b) For Li^{++}, $Z = 3$, so for $n = 2$,

$$(E_2)_{Li^{++}} = 9 \times (E_2)_H = -30.6\text{ eV}.$$

1.14. *Electrons of energy 12.2 eV are fired at hydrogen atoms in a gas discharge tube. Determine the wavelengths of the lines that can be emitted by the hydrogen.*

Solution: The maximum amount of energy that a hydrogen atom can absorb is equal to the electron energy 12.2 eV. This amount of energy can raise the electron from $n = 1$ state to nth state of energy E_n (say).

$\therefore E_n = E_1 + 12.2\text{ eV} = -13.6\text{ eV} + 12.2\text{ eV} = -1.4\text{ eV}.$

But

$$E_n = -\frac{E_1}{n^2} \text{ or, } -1.4\text{ eV} = -\frac{13.6\text{ eV}}{n^2} \text{ or, } n = 3.14.$$

Since n is an integer, the highest state that the electron can reach corresponds to $n = 3$. Hence, the possible transitions that the electron can make are as follows: $3 \to 2$, $2 \to 1$ and $3 \to 1$.

The corresponding wavelengths are:

For $3 \to 2$;

$$\frac{1}{\lambda} = (1.097 \times 10^{-3}/\text{Å})\left(\frac{1}{2^2} - \frac{1}{3^2}\right) \text{ or, } \lambda = \mathbf{6563\ Å}.$$

For $2 \to 1$;

$$\frac{1}{\lambda} = (1.097 \times 10^{-3}/\text{Å})\left(\frac{1}{1^2} - \frac{1}{2^2}\right) \text{ or, } \lambda = \mathbf{1215\ Å}.$$

For $3 \to 1$;

$$\frac{1}{\lambda} = (1.097 \times 10^{-3}/\text{Å})\left(\frac{1}{1^2} - \frac{1}{3^2}\right) \text{ or, } \lambda = \mathbf{1026\ Å}.$$

1.15. *For hydrogen, show that when $n \gg 1$, the frequency of radiation emitted photon in a transition from n to $(n-1)$ equals the orbital frequency of the electron.* **[BU(H) 1994]**

Solution: The orbital frequency of the electron in the nth quantum state is

$$\nu_{Cl} = \frac{\omega}{2\pi} = \frac{v_n}{2\pi r_n} = \frac{2\pi\epsilon_0 e^2/nh}{2\pi n^2 h^2/4\pi\epsilon_0 m e^2} = \frac{4\pi^2\epsilon_0^2 m e^4}{n^3 h^3}. \quad \text{(i)}$$

Again, the frequency of the emitted photon for the transition $n \to (n-1)$ is

$$\nu_Q = \frac{c}{\lambda} = cR_H\left[\frac{1}{(n-1)^2} - \frac{1}{n^2}\right] = cR_H\frac{2n-1}{n^2(n-1)^2}.$$

For $n \gg 1$, $(2n-1) \simeq 2n$ and $(n-1)^2 = n^2$.

$$\therefore \nu_Q = cR_H\frac{2n}{n^4} = c \times \frac{2\pi^2\epsilon_0^2 e^4 m}{h^3 c} \times \left(\frac{2}{n^3}\right)$$

$$\left[\because \text{Rydberg constant for hydrogen is } R_H = \frac{2\pi^2\epsilon_0^2 e^4 m}{h^3 c}\right]$$

$$\text{or, } \nu_Q = \frac{4\pi^2\epsilon_0^2 m e^4}{n^3 h^3}. \quad \text{(ii)}$$

Eqns (i) and (ii) show that for $n \gg 1$, $\nu_{Cl} = \nu_Q$. This forms the basis of *Bohr's correspondence principle.*

1.16. *Use vector atom model to determine the possible values of total angular momentum of an f-electron.*

Solution: The orbital quantum number for an f-electron is $l = 3$. Its spin quantum numbers, $s = \pm 1/2$.

$\therefore$ total angular momentum quantum number of the f-electron is $j = l \pm s = 3 \pm \frac{1}{2} = \frac{7}{2}, \frac{5}{2}$.

The total angular momentum vector is

$$J = \sqrt{j(j+1)}\,\hbar = \frac{3\sqrt{7}}{2}\,\hbar,\ \frac{\sqrt{35}}{2}\,\hbar.$$

1.17. *Find the angular momenta of the electron in the ground state and first excited state of the hydrogen atom.* **[NBU(H) 2009]**

Solution: For the ground state of hydrogen atom, $n = 1$, $l = 0$ and $s = \pm\frac{1}{2}$.

So, for the ground state the allowed values of total angular momentum quantum number are

$$j = |l + \tfrac{1}{2}| \text{ to } j = |l - \tfrac{1}{2}| \text{ or, } j = \tfrac{1}{2}.$$

Now angular momentum vector is

$$J = \sqrt{j(j+1)}\,\hbar = \sqrt{\tfrac{1}{2}\left(\tfrac{1}{2}+1\right)}\,\hbar = \tfrac{\sqrt{3}}{2}\,\hbar.$$

For the first excited state, $n = 2$ and $l = 1$.

$$\therefore j = |l + \tfrac{1}{2}| = |1 + \tfrac{1}{2}| = \tfrac{3}{2}, \ [\because s = \tfrac{1}{2}]$$
$$\text{and } j = |l - \tfrac{1}{2}| = |1 - \tfrac{1}{2}| = \tfrac{1}{2}.$$

$\therefore$ angular momenta for the first excited state are

$$J = \sqrt{j(j+1)}\,\hbar = \sqrt{\tfrac{3}{2}\left(\tfrac{3}{2}+1\right)}\,\hbar = \tfrac{1}{2}\sqrt{15}\,\hbar$$
$$\text{and } J = \sqrt{\tfrac{1}{2}\left(\tfrac{1}{2}+1\right)}\,\hbar = \tfrac{1}{2}\sqrt{3}\,\hbar.$$

1.18. *Ionisation energy of Hydrogen atom is 13.6 eV. Calculate the wavelength of the first line of Balmer series.* **[NBU(H) 2008]**

Solution: The ionisation energy of Hydrogen atom is $E = -13.6$ eV, i.e., the binding energy of the electron in the ground state ($n = 1$).

The magnitude of energy of the electron in $n = 2$ and $n = 3$ states are

$$E_2 = \frac{13.6}{2^2} = \frac{13.6}{4} = 3.4\,\text{eV}$$
$$\text{and } E_3 = \frac{13.6}{3^2} = \frac{13.6}{9} = 1.50\,\text{eV}.$$

$\therefore E_2 - E_3 = 3.40 - 1.51 = 1.89$ eV.

Now the first line of Balmer series corresponds to the transition $n \to 3$ to $n \to 2$ states. If λ be the wavelength of the 1st line of Balmer series, then

$$1.9\,\text{eV} = \frac{hc}{\lambda} \text{ or, } 1.9 \times \left(1.6 \times 10^{-19}\right)\,\text{J} = \frac{hc}{\lambda}.$$

$$\therefore \lambda = \frac{\left(6.62 \times 10^{-34}\right) \times \left(3 \times 10^{8}\right)}{1.89 \times \left(1.6 \times 10^{-19}\right)} = \frac{6.62 \times 3 \times 10^{7}}{1.89 \times 1.6} = 6.568 \times 10^{-7} = 6568 \times 10^{-10}\,\text{m} = \mathbf{6568\,\text{Å}}.$$

1.19. *An electron revolves in a circular orbit around the nucleus of charge +Ze. If r_1 and r_2 be the radii of the two orbits $n_1 = 1$ and $n_2 = 2$ respectively, then its total energy is greater in orbit n = 1. Prove that* (i) $r_1 > r_2$ *and* (ii) *velocity and acceleration in orbit 2 are greater than those in orbit 1.*

Solution: (i) If E be the total energy of the electron, then

$$E = -\frac{Ze^2}{8\pi\epsilon_0 r} = -\frac{1}{2}mv^2.$$

Since, $E_1 > E_2$,

$$\therefore E_1 - E_2 > 0 \text{ or, } -\frac{Ze^2}{8\pi\epsilon_0 r_1} + \frac{Ze^2}{8\pi\epsilon_0 r_2} > 0 \text{ or, } \frac{Ze^2}{8\pi\epsilon_0}\left(\frac{1}{r_2} - \frac{1}{r_1}\right) > 0.$$

$\therefore r_1 > r_2$.

Again,

$$\frac{1}{2}m\left(v_2^2 - v_1^2\right) = \frac{Ze^2}{8\epsilon_0\pi}\left(\frac{1}{r_2} - \frac{1}{r_1}\right) \text{ or, } \left(v_2^2 - v_1^2\right) > 0.$$

$\therefore v_2 > v_1$.

The acceleration of the electron in circular orbit in $n = 1$ state is $f = v^2/r$.

$$\therefore \frac{mv^2}{r} = \frac{Ze^2}{4\pi\epsilon_0 r^2}.$$

$$\therefore f_n = \frac{Ze^2}{4\pi\epsilon_0 m r_n^2}.$$

So

$$f_2 - f_1 = \frac{Ze^2}{4\pi\epsilon_0 m}\left(\frac{1}{r_2^2} - \frac{1}{r_1^2}\right) \text{ or, } f_2 - f_1 > 0 \ (\because r_1 > r_2).$$

$\therefore f_2 > f_1$.

1.20. *Find the magnitude of angular momenta L, S, J of an atom having electronic configuration $1s^2 2s^2 2p^1$.*

[HPU 1995]

Solution: The atom has 2 electrons in 1*s*- and 2*s*-sub-shells, so they are closed sub-shells having angular momentum zero. The angular momentum of the atom is only due to 2*p* electron. For this *p*-electron, $n = 2; l = 1; s = 1/2$.

$\therefore$ the allowed values of j are: $j = l + s = 1 + \frac{1}{2} = \frac{3}{2}$ and $j = l - s = 1 - \frac{1}{2} = \frac{1}{2}$.

$\therefore$ the magnitude of orbital angular momentum

$$L = \sqrt{l(l+1)}\,\hbar = \sqrt{a\,(1+1)}\,\hbar = \sqrt{2}\,\hbar.$$

Magnitude of spin angular momentum

$$S = \sqrt{s(s+1)}\,\hbar = \sqrt{\frac{1}{2}\left(\frac{1}{2}+1\right)}\,\hbar = \frac{\sqrt{3}}{2}\hbar.$$

Magnitude of total angular momentum is $J = \sqrt{j(j+1)}\,\hbar$.

For $j = \frac{3}{2}, J = \sqrt{\frac{3}{2}\left(\frac{3}{2}+1\right)}\,\hbar = \frac{\sqrt{15}}{2}\hbar$

$j = \frac{1}{2}, J = \sqrt{\frac{1}{2}\left(\frac{1}{2}+1\right)}\,\hbar = \frac{\sqrt{3}}{2}\hbar.$

1.21. *Find the quantum number of the Bohr orbit in a hydrogen atom whose radius is* 0.01 mm. *What is the energy of a hydrogen atom in this state?*

Solution: If r_n and a_0 be the radii of the nth orbit and first Bohr orbit of an electron in hydrogen atom respectively, then $r_n^2 = na_0^2$.

$$\therefore n = \sqrt{\frac{r_n}{a_0}}.$$

But $r_n = 1.00 \times 10^{-5}$ m and $a_0 = 5.29 \times 10^{-11}$ m.

$$\therefore n = \sqrt{\frac{1.00 \times 10^{-5}}{5.29 \times 10^{-11}}} = 435.$$

Again, energy of the electron in the orbit of quantum number n is

$$E_n = \frac{E_1}{n^2} = \frac{-13.6\,\text{eV}}{(435)^2} = -7.19 \times 10^{-6}\,\text{eV},$$

where E_1 = the binding energy of the electron in the first Bohr orbit = −13.6 eV.

EXERCISE

1(A): Objective (MCQ) Type Questions

Select the correct answer.

1.1. The dimension of atomic nucleus is of the order of

(a) 10×10^{-8} nm (b) 3×10^{-5} nm (c) 10×10^{-10} nm (d) 10^{-6} nm.

1.2. In Bohr's theory of hydrogen atom if f_n be the frequency of revolution of the hydrogen atom in the nth orbit, then

(a) $f_n \propto n$ (b) $f_n \propto n^2$ (c) $f_n \propto \frac{1}{n^3}$ (d) $f_n \propto \frac{1}{n^2}$.

1.3. Energy of an electron in the nth Bohr orbit is

(a) $E_n \propto n^2$ (b) $E_n \propto -\frac{1}{n^2}$ (c) $E_n \propto \frac{1}{n^3}$ (d) $E_n \propto -\frac{1}{n}$.

1.4. If f_1 and f_2 be the frequencies of the electron in the first and second Bohr orbits respectively, then f_1/f_2 is

(a) 8 (b) 2 (c) 4 (d) 1.

1.5. Lyman series of hydrogen spectrum falls in the

(a) VIS region (b) UV region (c) IR region (d) none of these.

1.6. Pfund series of hydrogen spectrum lies in the

(a) VIS region (b) UV region (c) far IR region (d) IR region.

1.7. The total energy of an electron in the nth orbit of hydrogen atom is

(a) $-\frac{e^2}{8\pi\epsilon_0 r_n}$ (b) $-\frac{e^2}{8\pi\epsilon_0 r_n^2}$ (c) $\frac{e^2}{8\pi\epsilon_0 r_n}$ (d) $\frac{e^2}{8\pi\epsilon_0 r_n}$.

1.8. The ionisation energy of hydrogen atom is

(a) 0.52 eV (b) 13.6 eV (c) −13.6 eV (d) −0.52 eV.

1.9. The value of Rydberg constant of hydrogen atom is

(a) $1.0961 \times 10^{-7}\,\text{m}^{-1}$ (b) 1.0961×10^{6} m (c) $1.0961 \times 10^{-6}\,\text{m}^{-1}$ (d) $1.0961 \times 10^{7}\,\text{m}^{-1}$.

1.10. The radius of the first Bohr orbit in hydrogen is

(a) 0.529 nm (b) 6.29 Å (c) 0.0529 Å (d) 0.0529 nm.

1.11. The value of Bohr magneton μ_B is

(a) $\mu_B = 9.2732 \times 10^{-24}$ Am^2 (b) $\mu_B = 9.2732 \times 10^{-19}$ Am^2

(c) $\mu_B = 9.2732 \times 10^{-34}$ Am^2 (d) $\mu_B = 9.2732 \times 10^{-27}$ Am^2.

1.12. The constant $\alpha = 1/137$ is called

(a) Planck's constant (b) Rydberg constant

(c) Fine structure constant (d) Boltzmann constant.

1.13. In Rutherford atom model, classical theory shows that the electron would reach the nucleus in time $T =$

(a) 4×10^{-8} s (b) 4×10^{-12} s (c) 3×10^{10} s (d) 3×10^{27} s.

1.14. The wavelength of the emitted radiation if the electron jumps from $n = 3$ to $n = 2$ orbit is

(a) 5893 Å (b) 4293 Å (c) 6563 Å (d) 7500 Å.

1.15. The number of de Broglie waves associated with the electron in the nth Bohr orbit is

(a) $n(n+1)$ (b) n (c) n^2 (d) $n(n-1)$.

1.16. Sommerfeld could explain the origin of fine structure of spectral lines by introducing the concept of

(a) elliptical orbits for the electron

(b) relativistic mass variation of the electron moving in elliptic orbit around the nucleus

(c) relativistic mass variation of the electron moving in circular orbit around the nucleus

(d) radial and angular moments for the electron.

1.17. The idea of spinning electron was introduced to explain the origin of

(a) spectral series (b) fine structure of spectral lines

(c) multiplet-structure of spectral lines (d) hyperfine structure of spectral lines.

1.18. The maximum number of electrons in a sub-shell with orbital angular momentum quantum number l is

(a) $2(2l-1)$ (b) $(2l-1)$ (c) $2(2l+1)$ (d) $2(2l+1)$.

1.19. The magnetic moment associated with the first Bohr orbit is

(a) $\frac{ehm}{4\pi}$ (b) $\frac{4\pi m}{eh}$ (c) $\frac{h}{4\pi me}$ (d) $\frac{eh}{4\pi m}$.

1.20. If the azimuthal quantum number $k = 0$, the electron orbit in Sommerfeld model becomes a

(a) circle (b) ellipse (c) straight line (d) none of these.

1.21. For a particular value of the principal quantum number (n) the number of values of l are

(a) n (b) $(l-1)$ (c) $(l+1)$ (d) $l(l+1)$.

1.22. In modern periodic table, the elements have been arranged in order of

(a) atomic weight (b) atomic number (c) chemical properties (d) none of these.

1.23. Ferromagnetism in Fe, Ni and Co arises due to partial vacancy in

(a) $3d$-sub-shell (b) $4p$-sub-shell (c) $2d$-sub-shell (d) $2p$-sub-shell.

1.24. The state of an atom obeying L-S coupling is represented by

(a) $n^{(2s+1)}L_j$ (b) $n^{(s+1)}L_j$ (c) $n^{(2s+1)}L_s$ (d) $n^{2s+1}L_l$.

1.25. If an electron transits from n to $(n-1)$ orbit, where $n \to \infty$, the orbital frequency (ν_e) and frequency of radiation (ν_Q) emitted are related by

(a) $\nu_e > \nu_Q$ (b) $\nu_e < \nu_Q$ (c) $\nu_e \leq \nu_Q$ (d) $\nu_e = \nu_Q$.

1.26. The orbital angular momentum (p) of electron in a closed sub-shell with $Z \to$ large value is

(a) $p > 0$ (b) $p < 0$ (c) $p = 0$ (d) $p = \infty$.

1.27. The number of values of m_l for a particular value of l is

(a) $(2l+1)$ (b) $(l+1)$ (c) $(l-1)$ (d) $(2l-1)$.

1.28. According to vector atom model, the number of quantum numbers necessary to specify an electron is

(a) 3 (b) 4 (c) 6 (d) 5.

1.29. The longest wavelength of the Lyman series is obtained using the relation:

(a) $\frac{1}{\lambda} = R_H\left(\frac{1}{1^2} - \frac{1}{2^2}\right)$ (b) $\frac{1}{\lambda} = R_H\left(\frac{1}{1^2} - \frac{1}{\infty}\right)$ (c) $\frac{1}{\lambda} = R_H\left(\frac{1}{2^2} - \frac{1}{\infty}\right)$ (d) $\frac{1}{\lambda} = R_H\left(\frac{1}{1^2} - \frac{1}{3^2}\right)$.

1.30. The radius of the Bohr orbit for the ground state of hydrogen atom is

(a) 9.50×10^{-8} m (b) 9.63×10^{-10} m (c) 0.53×10^{-8} m (d) 0.53×10^{-13} m.

1.31. The number of allowed quantum number j in hydrogen atom for a given non-zero l is

(a) 0 (b) 1 (c) 2 (d) 3.

ANSWERS

1.1. (b) **1.2.** (c) **1.3.** (b) **1.4.** (a) **1.5.** (b) **1.6.** (c) **1.7.** (a) **1.8.** (b) **1.9.** (d) **1.10.** (d) **1.11.** (a) **1.12.** (c) **1.13.** (b) **1.14.** (c) **1.15.** (b) **1.16.** (b) **1.17.** (c) **1.18.** (d) **1.19.** (d) **1.20.** (c) **1.21.** (b) **1.22.** (b) **1.23.** (a) **1.24.** (a) **1.25.** (d) **1.26.** (c) **1.27.** (a) **1.28.** (b) **1.29.** (a) **1.30.** (b) **1.31.** (c).

1(B): Short Answer Type Questions

1.1. Discuss the drawbacks of Rutherford's atom model.

1.2. Show that in a Bohr atom if the electron is considered as a wave travelling along the circular path, then the nth orbit will contain n complete de Broglie waves. **[NBU(H) 2008; VU(H) 2007]**

1.3. Show that Planck's constant has the dimension of angular momentum. **[BU(H) 2000]**

[*Hint.* $E = h\nu$. $\therefore [h] = \frac{[E]}{[\nu]}$.

But $[E] = [F][L] = [MLT^{-2}][L] = [ML^2T^{-2}]$ and $[\nu] = [T^{-1}]$.

$\therefore [h] = \dfrac{[ML^2T^{-2}]}{[T^{-1}]} = [ML^2T^{-1}] = [MLT^{-1}][L] = [p][L] = [\Omega]$, where $pl = \Omega$ is the angular momentum.]

1.4. What is Bohr magneton? Calculate its value. Give the physical significance of Bohr magneton. **[VU(H) 2007]**

1.5. State Pauli's exclusion principle. Show that the maximum number of electrons in a shell specified by the principal quantum number n is $2n^2$. **[VU(H) 2007, 2005]**

1.6. State the assumptions of Bohr's theory of hydrogen atom.

1.7. Discuss the merits and demerits of Bohr's theory of hydrogen atom.

1.8. An electron is moving in a circular orbit around a nucleus with a radius R and frequency ν. What is the magnetic dipole moment associated with the circulating electron. **[VU(H) 2006, 2007]**

1.9. What are the quantum numbers associated to an electron in an atom? State the significance of the quantum numbers. **[VU(H) 2003, 2005]**

1.10. Show how one can arrive at Bohr's quantisation condition on the basis of de Broglie hypothesis of matter waves. **[VU(H) 2001, 2003]**

1.11. What are the fundamental differences between Bohr's theory and Sommerfeld's theory of atomic structure.

1.12. State and explain Bohr's correspondence principle. **[BU(H) 1999]**

1.13. What is the magnetic moment due to spin of an electron in the unit of Bohr magneton? **[BU(H) 2003]**

1.14. Show that the circumference of a circular Bohr orbit accommodates an integral number of de Broglie wavelength of the orbiting electron. **[BU(H) 1993; 1997; 2001]**

1.15. What are the two basic conceptions on which the idea of vector atom model is based? Explain.

1.16. Give the electronic configurations of sodium and copper.

1.17. What do you understand by ionisation energy? On what factors does it depend?

1.18. Outline the nature of relativistic hydrogen orbit.

1.19. Explain the structure of H_α line of Balmer series.

1.20. Explain why Ar and Ne are chemically inert.

1.21. Point out the salient features of Bohr-Sommerfeld atom model.

1.22. Discuss briefly vector atom model.

1.23. (a) Did Bohr postulate the quantisation of energy in his atom model? If not, what really was quantised?

(b) For Bohr hydrogen atom orbits, the PE is negative and greater in magnitude than the KE. What does it imply? **[BU(H) 1993]**

1(C): Numerical Problems

1.1. Using the Bohr atom model, calculate the energy in eV required to remove the electron from singly ionised helium. **[BU(H) 1993]**

[*Ans.* 54.4 eV]

1.2. Calculate the radius of the first Bohr orbit of the hydrogen atom. What will be the radius of the orbit if the electron is replaced by a negative muon of mass $212m_e$ having electronic charge.

[*Ans.* 0.529×10^{-8} cm; 2.5×10^{-11} cm]

1.3. How many revolutions does an electron in the $n = 2$ state of hydrogen atom make before dropping to the $n = 1$ state? [The average lifetime of an excited state is about 10^{-8} s] [*Ans.* 8.2×10^5]

1.4. The first member of the Balmer series of hydrogen has a wavelength 655.3 nm. Calculate the wavelength of its second member. [*Ans.* 485.5 nm]

1.5. The lowest two excited states of hydrogen atom are 10.2 eV and 12.0 eV above the ground state. Calculate the wavelength of radiation that could be produced by transition between these states and the ground state. [*Ans.* 121.5 nm; 103.5 nm and 690.2 nm]

1.6. A hydrogen atom initially in the $n = 4$ state, finally attains the ground state. In this process, what will be the possible wavelengths of the electromagnetic waves emitted?

[*Ans.* $\lambda = 18.7 \times 10^{-7}$ m, 6.56×10^{-7} m, 1.03×10^{-7} m, 1.21×10^{-7} m, 4.86×10^{-7} m, 0.97×10^{-7} m]

1.7. Show that the force acting on the electron at a point distant r from the centre of the sphere is $F = -\rho e r/3\pi\epsilon_0$ and the motion of the electron is simple harmonic (ρ = the charge density in the sphere).

Apply classical theory to calculate the frequency of light emitted. [Assume the radius of the sphere $= 10^{-10}$ m] [*Ans.* $\nu = 2.54 \times 10^{15}$ Hz]

1.8. Calculate the difference between the wavelengths of a deuterium atom and a hydrogen atom for the transition from the $n = 2$ to the $n = 1$ state. [*Ans.* $\Delta x = 0.327$ Å]

1.9. In a hydrogen atom how much energy must be applied (i) to put the electron in the $n = 6$ state from its ground state? (ii) to ionise the atom from the $n = 3$ state? [*Ans.* (i) $E = 13.2$ eV (ii) $E = 1.51$ eV]

1.10. Show that the frequency of a photon emitted by a hydrogen atom in going from the level $(n+1)$ to the level n is always intermediate between the frequencies of revolution of the electron in the respective orbits. [*Ans.* $f_n/\nu = \frac{2n^2+4n+2}{(2n^2+n)}$, which is greater than 1; $f_n/\nu = \frac{2n^2}{2n^2+3n+1}$, which is less than 1]

1.11. Find the wavelength of the spectral line that corresponds to a transition in hydrogen atom from the $n = 10$ state to the ground state. In what part of the spectrum is this? [*Ans.* $\lambda = 92.1$ nm, UV region]

1.12. The longest wavelength in the Lyman series is 121.5 nm. Use this wavelength together with the values of c and h to find the ionisation energy of hydrogen. [*Ans.* 13.6 eV]

1.13. The ionisation potential of an atom is 13.5 V. With what minimum velocity should an electron strike so that it may be ionised? [*Ans.* 2.19×10^6 m/s]

1.14. A photon of energy 12.1 eV is absorbed by a hydrogen atom originally in the ground state and is raised thereby to an excited state. Find the quantum number of the state. [*Ans.* $n = 3$]

1.15. Mercury has an excitation potential at 4.9 V. Calculate the wavelength of the radiation emitted when a mercury atom is excited to this level and then returns to the ground state in a single quantum jump. [*Ans.* $\lambda = 2533$ Å]

1.16. State and explain Pauli's exclusion principle. Prove that if n be the principal quantum number of a shell, it can at best accommodate $2n^2$ electrons. **[VU(H) 2014; Manipur U 2016]**

1.17. Using vector diagram, determine the possible values of the total angular momentum of an electron system with $L = 2$, $S = \frac{1}{2}$. **[VU(H) 2013]**

1(D): Broad Answer Type Questions

1.1. (a) Mention the salient features of the Bohr model of hydrogen atom and obtain an expression for inverse of wavelength of the Balmer series. **[BU(H) 1997]**

(b) Show that the circumference of a circular Bohr orbit accommodates an integral number of de Broglie wavelength of the orbiting electron. **[BU(H) 2001]**

1.2. (a) Discuss Bohr's theory of hydrogen atom and obtain formula for the Balmer series. Sketch an energy level diagram of the hydrogen atom.

(b) What is Bohr's correspondence principle? Discuss.

1.3. (a) Give an account of Bohr's theory of the spectrum of hydrogen atom explaining the basic assumptions and calculate the quantised energy states of the same on this basis.

(b) What are the shortcomings of Bohr's theory. Discuss the success of the theory.

1.4. Using Bohr's postulates deduce expression for the energy levels and orbital radii of positronium which is a bound system comprising of an electron and a positron.

What is the ionisation energy of the positronium? [BU(H) 1998]

1.5. (a) Explain how Bohr's theory was extended by Sommerfeld to include elliptical orbits and also show that the introduction of elliptical orbits adds no new energy levels.

(b) Show that in a Bohr atom if the electron is considered as a wave travelling along the circular path, then the *n*th orbit will contain *n* complete de Broglie waves. [VU(H) 2008]

1.6. (a) What are the factors which led to the discovery of electron spin? When and by whom the idea of intrinsic spin angular momentum of the electron was proposed? What is its value? What is meant by space quantisation? [GNDU 1992]

(b) Calculate the angular frequency at which an electron's orbital magnetic moment $\vec{\mu}$ precesses in a magnetic field.

1.7. Discuss Bohr's theory of atomic spectra and Sommerfeld's modifications. How far was the modification proved successful?

1.8. (a) State the fundamental postulates on which Bohr based his theory of atomic hydrogen. What is the significance of the negative sign of the energy value?

(b) Explain the origin of the spectral series on the basis of this theory.

1.9. (a) State and explain Bohr's correspondence principle. Illustrate the principle by taking the following transitions of hydrogen atom: $n_i = 11$, $n_f = 10$; $n_i = 51$, $n_f = 50$ and $n_i = 1001$, $n_f = 1000$. What is your conclusion?

(b) Discuss the merits and demerits of Bohr's theory of hydrogen atom.

1.10. (a) Explain the different quantum numbers associated with vector atom model of the atom.

(b) State and explain Pauli's exclusion principle. Give with examples the arrangements of electrons in an atom.

1.11. (a) Justify the introduction of relativistic mass variation of atomic electron.

(b) State the conditions of quantisation of angular momentum of an electron in an atom according to Sommerfeld. Discuss the degeneracy of electron orbits.

(c) Discuss how the origin of five structures of spectral lines was explained.

1.12. (a) Discuss the periodic arrangement of elements.

(b) Give reasons why alkali metals are highly chemically active and Ar, Kr are chemically inert.

1.13. (a) Outline Bohr's theory of hydrogen atom clearly explaining the basic postulates.

(b) Find out expression for radii of Bohr's orbits and energy of the electron in those orbits.

1.14. Write down Sommerfeld's expression for the energy of the electron applying relativistic correction Discuss the significance of the expression with that of Bohr.

1.15. (a) State and explain Pauli's exclusion principle.

(b) Give the electronic configuration of Na(11), Cl(17), Cu(29) and Ge(32).

1.16. (a) Clearly state the postulates of Bohr's atomic model.

(b) Describe the formulae for the radii of Bohr orbits and the energies associated with those orbits.

(c) Show that the circumference of the Bohr orbits are integral multiples of the de Broglie wavelength of the electron. **[BU(H) 2009]**

1.17. (a) Show that in a Bohr atom if the electron is considered as a wave travelling along the circular path, then the *n*th orbit contains *n* complete de Broglie waves.

(b) Explain how Bohr's theory was extended by Sommerfeld to include elliptical orbits and also show that the introduction of elliptic orbits adds to new energy levels. **[NBU(H) 2008]**

Chapter 2

Crystallography

2.1 Introduction

In nature, matter exists in the *solid, liquid* and *gaseous states.* The atoms and molecules in a solid are bound to one another with strong forces of attraction and as such solids have definite *shape* and *size.* A solid deforms elastically under small shear stresses. The branch of physics which deals with the properties of solid is known as *Solid State Physics* or *Physics of the Condensed Matter.*

Solids are classified broadly into two classes, viz., *crystalline* and *non-crystalline* (or *amorphous*).

In *crystalline solids,* the atoms are arranged in periodic manner in all three directions. The crystalline state being the low energy state, most of the solids prefer to be in crystalline state. This is because energy released during the formation of ordered structure is more than that in disordered structure. The crystalline solids are again subdivided into *single crystal* and *microcrystalline* (or *polycrystalline*). The periodicity of atoms extends throughout the volume of the matter in single crystal-like diamond, quartz, etc. In polycrystalline material, there is an aggregate of so many small *crystallites* oriented randomly and separated by *grain boundaries.* The crystallines are called *grains.*

In *amorphous solids,* the atoms or molecules are randomly arranged and the solid exhibits short range order. Amorphous solids are formed when the atoms do not get sufficient time to be arranged in periodic manner. *Glass* is an example of amorphous solid.

The sketch in Fig. 2.1 shows the classification of solids.

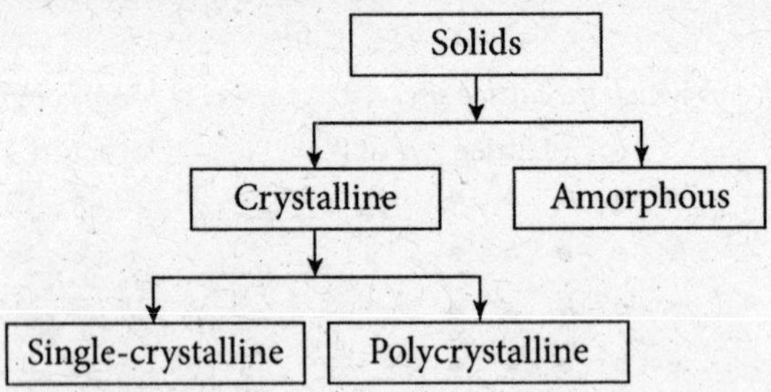

Fig. 2.1: Classification of solids

The physical properties of solids depend on the surroundings of atoms, ions or molecules forming the solid. As the amorphous solids have no directional properties they are known as *isotropic* substances. Whereas a crystal has different periodic arrangements in all the three directions, its physical properties vary with direction and, therefore, the crystal is called *anisotropic* substance. The branch of science dealing with the study of forms and physical properties of crystalline solid is called *crystallography.*

2.1.1 Properties of Crystalline Solids

Crystalline solids like calcium carbonate, ammonium chloride, quartz, etc. have the following specific properties:

(a) a crystalline solid has definite melting point

(b) a crystalline material has a definite geometrical shape and size

(c) crystalline solids are anisotropic

(d) the strength of all the bonds between the ions, atoms or molecules being the same, their chemical composition is uniform

(e) the crystalline solids have definite colour (some are transparent) in visible light.

2.1.2 Properties of Amorphous Solids

Amorphous solids like glass, etc. have the following properties:

(a) amorphous solids have no definite shape

(b) they are isotropic in nature

(c) their chemical composition is not uniform as the bonds between the atoms are not of equal strength

(d) their melting point is not fixed due to the variation of bond strength.

2.2 Space Lattice (or Point Lattice)

An infinite regular and periodic array of points in three dimensions in which every point has surroundings identical to that of every other point in the array is called a space lattice.

The term *identical surroundings* means that, the lattice appears to have the same view when looked from a point $P(\vec{r})$ as from another point $Q(\vec{r}\,')$ in the lattice with reference to an arbitrary origin O [Fig. 2.2].

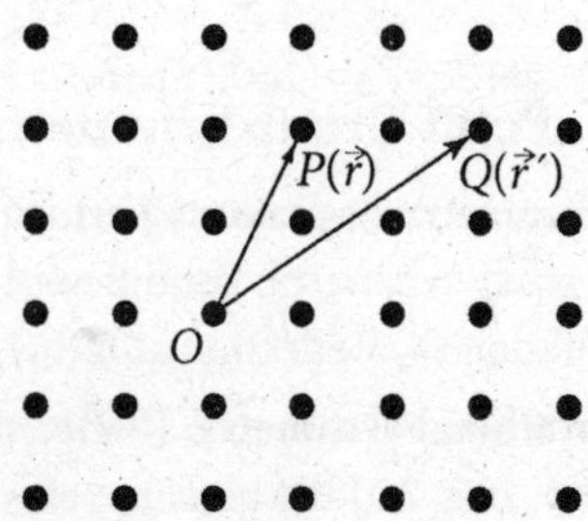

Fig. 2.2: Two-dimensional lattice

2.3 Crystal Lattice

It is defined as a space lattice in which the lattice sites are occupied by atoms or group of atoms known as basis.

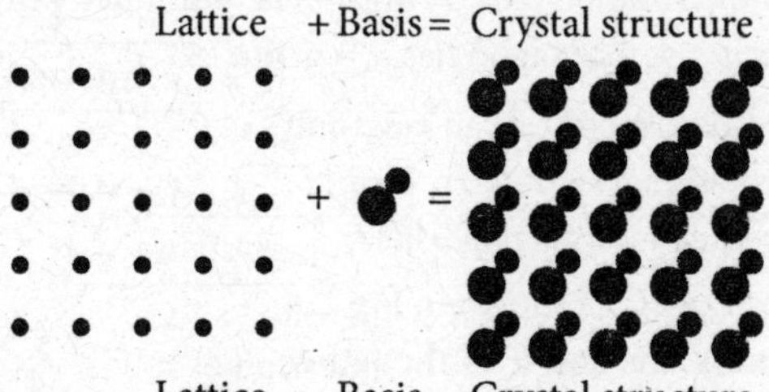

Fig. 2.3: Lattice, Basis and Crystal structure

2.4 Basis

A basis is defined as a cluster of atoms, ions or molecules identical in composition, arrangement and orientation. The periodic repetition of the basis in three dimensions forms the structure of the crystal.

The simple relation between lattice, basis and crystal structure can be represented as in Fig. 2.3.

2.5 Symmetry Considerations

The operation which leaves the crystal and its environment invariant is called *symmetry operation*. External symmetry of most of the crystal shows that, most of them possess *geometrical symmetry*.

The following sketch shows the classification of symmetry operations.

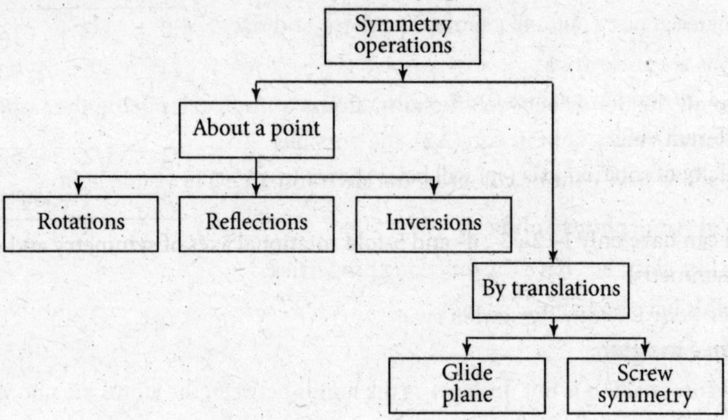

Fig. 2.4: Symmetry operations

2.6 Point Group Symmetry

The symmetry operations performed about a point or a line are called *point group symmetry*. The different point group symmetries are as follows: (i) Rotation (ii) Reflection and (iii) Inversion.

(i) **Rotational symmetry.** If after rotation of a body about an axis by some angle ϕ, [Fig. 2.5] the body appears as it did prior to rotation, i.e., the body is left *invariant* as a result of rotation, then the body is said to possess rotational symmetry. For the body to be left invariant by the rotation, the rotation angle ϕ must be an integral sub-multiple of 2π, i.e., $\phi_n = \frac{2\pi}{n}$, where n is an integer and is called the *multiplicity* of the rotation axis. A point has multiplicity of rotation axis $n = 1$, a square has $n = 4$, a hexagon has $n = 6$, etc.

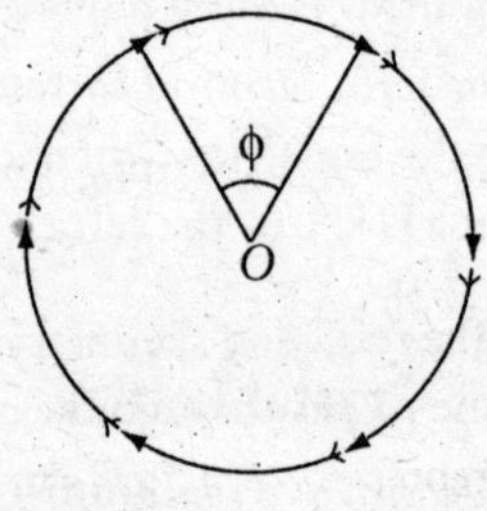

Fig. 2.5: Rotation through an angle ϕ

Absence of five-fold symmetry axis. A crystal can have only 1-, 2-, 3-, 4- and 6-fold axes of rotational symmetry. *It can be proved that 5-fold rotational symmetry axis is impossible.*

Proof. We consider a lattice row *OABC* as shown in Fig. 2.6. Let the lattice has n-fold rotation axis and $\vec{T}$ is the lattice translation vector. The rotation axis passes through the lattice point A and is perpendicular to the plane of the paper. As all the lattice points are equivalent, there must be a similar axis through the tip of B.

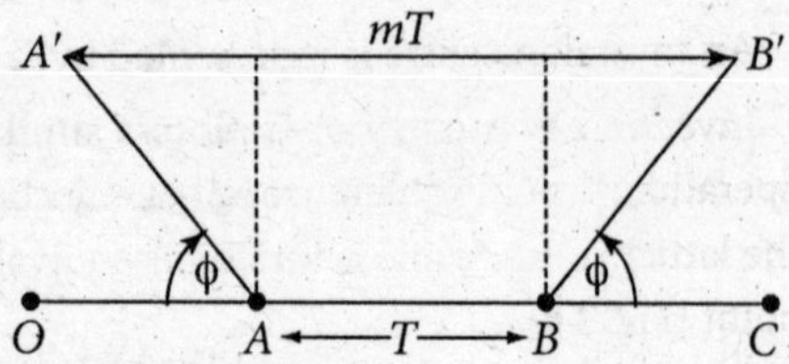

Fig. 2.6: Proof of non-existence of 5-fold symmetry axis of rotation

Let AO be rotated through an angle ϕ, where $\phi = 2\pi/n$ about A so that it produces lattice vector AA'. Similarly, anticlockwise rotation of BC about B through the same angle ϕ produces a new lattice vector BB'. The tips of these vectors A' and B' are the new lattice points. According to the definition of rotation operation it must leave the lattice invariant. Hence, $A'B'$ must be parallel to AB and must be integral multiple of the lattice translation T, i.e.,

$$A'B' = 2T\cos\phi + T = mT, \tag{2.1}$$

where

$$\cos\phi = \frac{m-1}{2}. \tag{2.2}$$

Since $|\cos\phi| \leq 1$, m can have values $0, \pm1, \pm2, \pm3, \pm4, \ldots$

Now using different values of m in eqn (2.2), the possible values of multiplicity of rotation axis (n) will be as shown in Table 2.1.

Table 2.1

m	$\cos\phi$	ϕ	$n = 360/\phi$
0	$-1/2$	120	3
+1	0	90°	4
−1	−1	180°	2
2	1/2	60°	6
3	−1	360° or 0°	1

Thus, a crystal can have only 1-, 2-, 3-, 4- and 5-fold rotational axes of symmetry and cannot have 5-*fold axis* of rotation symmetry.

(ii) **Reflection.** *If a body is left unchanged in every way after being reflection at a plane, then the body is said to have a reflection symmetry.*

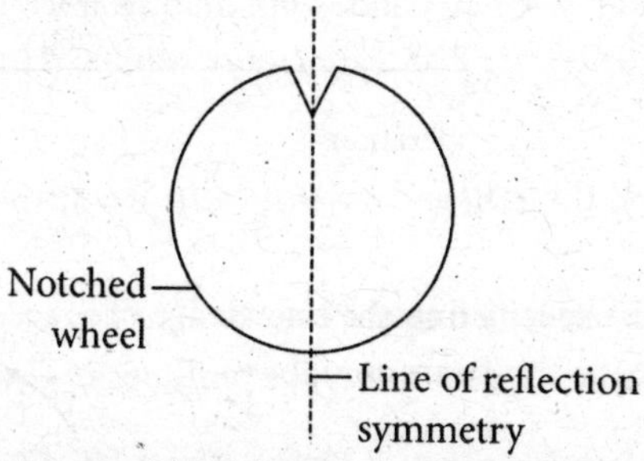

Fig. 2.7: (a) Reflection symmetry using a notched wheel

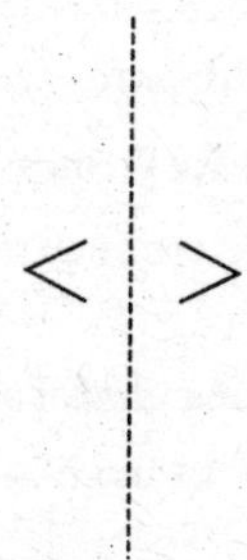
Fig. 2.7: (b) Reflection across a line

The reflection symmetry is illustrated in Fig. 2.7, where the reflection plane passes through the centre of a *notched* wheel. The symbol used to represent reflection symmetry is m.

(iii) **Inversion.** *If for every lattice point represented by the position vector $\vec{r}$ there are corresponding lattice points at the position $-\vec{r}$, then the crystal structure is said to possess inversion symmetry.*

An inversion centre is represented by I.

Inversion symmetry operation is similar to Reflection symmetry operation, the only difference is that reflection occurs at a plane through the lattice point, while inversion is equivalent to reflection through a point [Fig. 2.8].

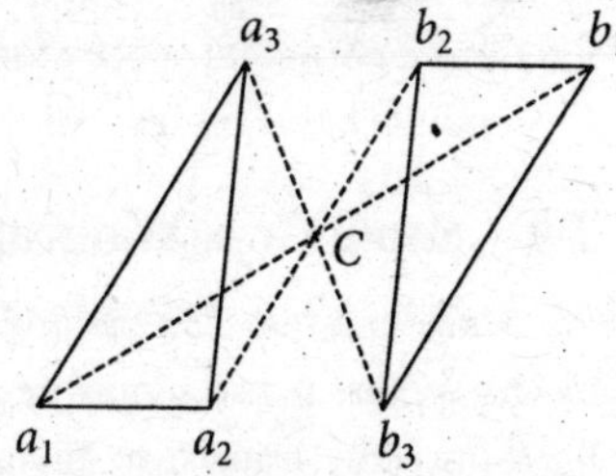

Fig. 2.8: Points a_1, a_2, a_3 of a triangle is reflected through the point C and producing the new points b_1, b_2, b_3

(iv) **Rotation-inversion axis.** *If a crystal is brought into self-coincidence by rotation followed by an inversion about a lattice point through which rotation axis passes, then the crystal structure is said to possess a rotation-inversion axis of symmetry.*

A crystal structure can possess 1-, 2-, 3-, 4- and 6-fold rotation inversion axes denoted by $\bar{1}$, $\bar{2}(\equiv m)$, $\bar{3}$, $\bar{4}$, $\bar{6}(\equiv 3)$ plus normal reflection planes. A *rotation inversion operation* is known as *improper rotation* and the corresponding axis as *improper axis*.

(v) **Combination of symmetry elements**: *Point group.* Symmetry elements like rotation, reflection and inversion can be combined provided they are *compatible.* Different crystal elements possess different symmetries. *The various combinations of allowed rotation and reflection operations is called point group.* There are 42 point groups.

NB : For a detailed discussion on *Point group symmetry,* the reader is referred to *Introduction to Solids,* Azaroff, Tata McGraw Hill (1997).

2.7 Translation Symmetry of Crystals

(i) **Glide plane.** *The combination of a mirror plane with simultaneous translation operation in a crystal is known as Glide plane.* The glide plane in a crystal is always *parallel* to the mirror plane.

(ii) **Screw axis.** *The rotation axis coupled with the translation parallel to the rotation axis gives rise to screw axis.*

The symbol used for screw axis is n_m which is performed by rotation of $\theta = 2\pi/n_m$ and a translation of m/n times the translation vector parallel to the rotation axis in a crystal.

(iii) **Space group.** The combination of *point group symmetry* and *translation symmetry elements,* if they are compatible, give rise to *space group symmetry.* There are 230 *space groups* exhibited by crystals.

Characteristics of Space Group

(i) Space group is determined by the lattice and the basis and it specifies the repetitive pattern of symmetry elements.

(ii) The symmetry of a crystal structure can be specified by the knowledge of space group.

(iii) The space group, through its symmetry elements, determines the positions of equivalent points within the unit cell.

(iv) The space group is characterized by the *Bravais lattice* and the location of the point group and other symmetry elements.

(v) For a particular space group, if only one point in a cell is occupied, then it is necessary that all equivalent points be occupied by identical atoms or molecules.

2.8 Some Crystallographic Terms

We consider a two-dimensional crystal lattice as shown in Fig. 2.9, where the parallelogram $ABCD$ defined by the vectors $\vec{a}$ and $\vec{b}$ may be taken as a *unit cell* of the crystal. It is evident that the whole crystal may be reproduced by translating the parallelogram $ABCD$ along $\vec{a}$ and $\vec{b}$ direction merely by suitably multiplying the vectors $\vec{a}$ and $\vec{b}$. Then any lattice point in the crystal may be represented by a vector $\vec{R}$ such that, $\vec{R} = h\vec{a} + k\vec{b}$, where h and k are integers.

For a three-dimensional lattice, $\vec{R} = h\vec{a} + k\vec{b} + l\vec{c}$.

The three-dimensional crystal lattice may be reproduced by repeated translation of the non-coplanar vectors $\vec{a}$, $\vec{b}$ and $\vec{c}$, known as the *basis vectors.*

(i) *Unit cell.* The unit cell of a crystal lattice is defined as *the region of the crystal represented by the basis vectors $\vec{a}$, $\vec{b}$ and $\vec{c}$ such that a translation of this region by any integral multiple of these vectors will lead to a similar region of the crystal.*

It is also defined as *the fundamental elementary pattern of minimum of atoms, molecules or group of molecules which represent fully all the characteristics of the crystal.*

The choice of a unit cell is not unique.

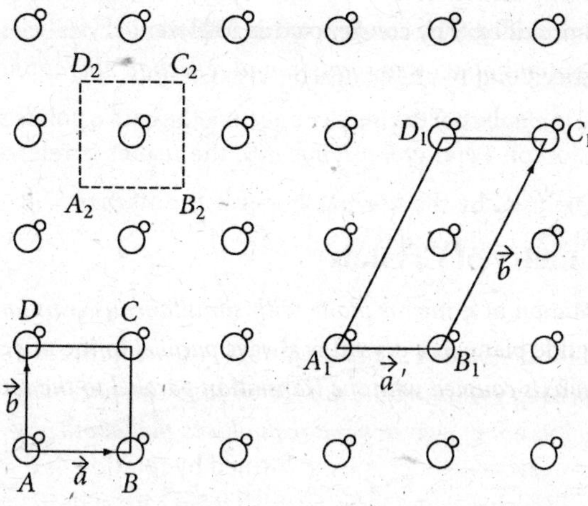

Fig. 2.9: Unit cells in two-dimensional crystal lattice

(ii) *Basis vectors.* The three non-coplanar vectors $\vec{a}$, $\vec{b}$ and $\vec{c}$ used to define a unit cell are called the *basis vectors.*

(iii) *Primitive unit cell. The unit cell occupying the smallest volume in the crystal lattice is termed as a primitive unit cell.* The unit cells $ABCD$, $A_1B_1C_1D_1$, $A_2B_2C_2D_2$ are all primitive unit cells [Fig. 2.10].

(iv) *Primitive basis vectors. The three non-coplanar vectors $\vec{a}$, $\vec{b}$ and $\vec{c}$ which define a primitive unit cell are called primitive basis vectors.*

$\vec{a}'$, $\vec{b}'$, $\vec{c}'$ are also primitive basis vectors for the primitive unit cell, $A_1B_1C_1D_1$.

The primitive basis vectors $\vec{a}$, $\vec{b}$, $\vec{c}$ and interfacial angles α, β, γ are called the *basic lattice parameters* as they determine the form and actual size of the *unit cell.*

Difference between Unit Cell and Primitive Cell

A unit cell, in addition to an atom at its each corner, may or may not contain additional atom or atoms [Fig. 2.10 (a)].

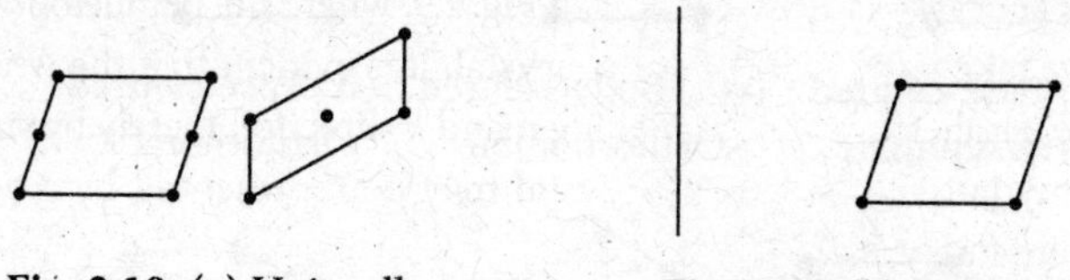

Fig. 2.10: (a) Unit cells **Fig. 2.10:** (b) Primitive unit cells

But a primitive unit cell contains only one atom at its each corner [Fig. 2.10 (b)].

From above we may conclude that *all primitive cells are unit cells but all unit cells may or may not be primitive cells.*

Lattice Constants

The length a, b and c specifying the size of a unit cell are called *lattice constants.* For a cubic crystal, $a = b = c$ and a is the *lattice constant.*

Wigner-Seitz Primitive Cell

Wigner-Seitz primitive unit cell may be constructed as follows:

(a) Connect a given lattice point to all the nearby lattice points.

(b) Draw normals at the *midpoints* of the lines connecting the lattice points as shown in Fig. 2.11.

The smallest volume enclosed by the normal planes is the *Wigner-Seitz primitive unit cell*. The volume of this cell is $v_0 = |\vec{a} \cdot \vec{b} \times \vec{c}|$.

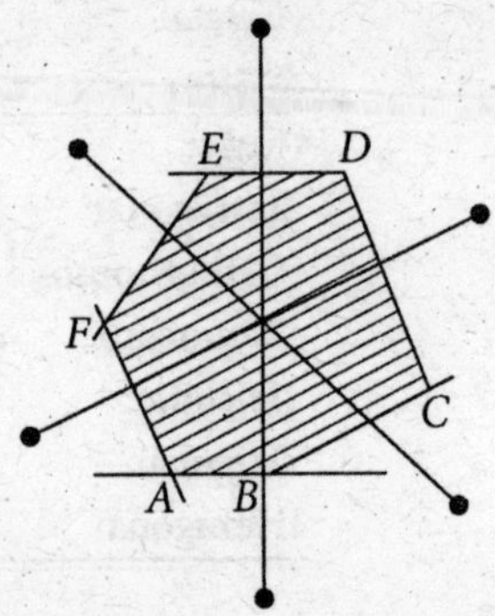

Fig. 2.11: Wigner-Seitz unit cell (*ABCDEF*)

2.9 Bravais Lattice

Bravais in 1884 proposed the concept of the *space lattice. A space lattice is a mathematical concept which represents infinite number of points in space such that the arrangement of points about a given point is similar to that about any other point.*

There are 14 distinguishable ways in which points can be arranged in three-dimensional space. Therefore, Bravais lattices are 14 *in number* as shown in Figs 2.12 (i) to 2.12 (xiv) and belong to 7 *crystal system* [Table 2.1]. The 14 Bravais lattice and their names are listed in Fig. 2.12.

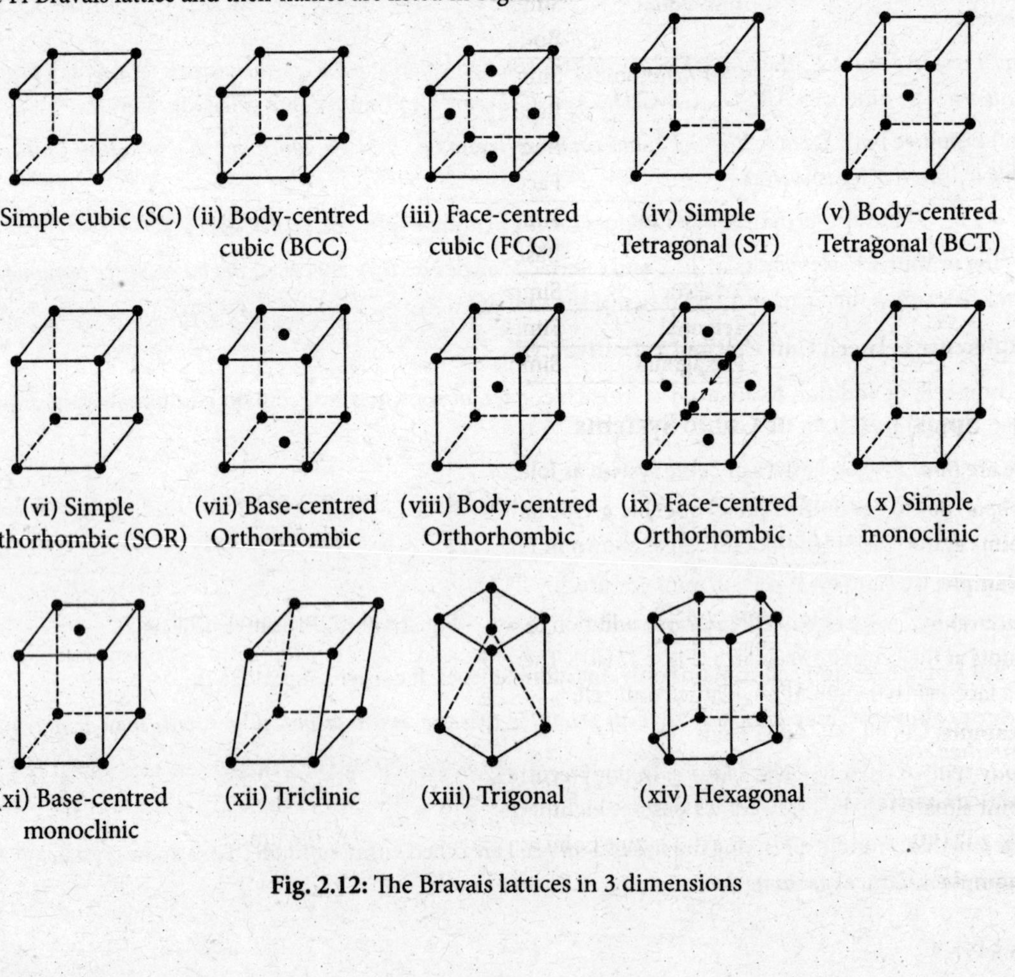

(i) Simple cubic (SC) (ii) Body-centred cubic (BCC) (iii) Face-centred cubic (FCC) (iv) Simple Tetragonal (ST) (v) Body-centred Tetragonal (BCT)

(vi) Simple Orthorhombic (SOR) (vii) Base-centred Orthorhombic (viii) Body-centred Orthorhombic (ix) Face-centred Orthorhombic (x) Simple monoclinic

(xi) Base-centred monoclinic (xii) Triclinic (xiii) Trigonal (xiv) Hexagonal

Fig. 2.12: The Bravais lattices in 3 dimensions

Table 2.2: Crystal Systems

Crystal System	*Lattice Constants*	α, β, γ	*Example*
Cubic	$a = b = c$	$\alpha = \beta = \gamma = 90°$	$NaCl, CaF_2, NaClO_2$
Tetragonal	$a = b \neq c$	$\alpha = \beta = \gamma = 90°$	$NiSO_4, SnO_2$
Orthorhombic	$a \neq b \neq c$	$\alpha = \beta = \gamma = 90°$	$KNO_3, BaSO_4, MgSO_4$
Monoclinic	$a \neq b \neq c$	$\alpha = \beta = 90°, \gamma \neq 90°$	Na_2SO_4, $FeSO_4$
Triclinic	$a \neq b \neq c$	$\alpha \neq \beta \neq \gamma \neq 90°$	$CuSO_4, K_2Cr_2O_7$
Trigonal	$a = b = c$	$\alpha = \beta = \gamma \neq 90°$	$CaSO_4$
Hexagonal	$a = b \neq c$	$\alpha = \beta = 90°, \gamma = 120°$	SiO_2, AgI

Table 2.3: Bravais lattices and crystal systems

Crystal systems	*Bravais lattices*	*Symbols*
Cubic	Simple	*P*
	Body-centred	*I*
	Face-centred	*F*
Tetragonal	Simple	*P*
	Body-centred	*I*
Orthorhombic	Simple	*P*
	Base-centred	*C*
	Body-centred	*I*
	Face-centred	*F*
Monoclinic	Simple	*P*
	Base-centred	*C*
Triclinic	Simple	*P*
Trigonal	Simple	*P*
Hexagonal	Simple	*P*

2.10 Space Lattices of Cubic Systems

There are *three* Bravais lattices of cubic system as follows:

1. *Simple cubic* (or primitive cubic) or cubic *p*-type lattice: In simple cubic (SC) lattice, there are eight lattice points at the corners of the unit cell as shown in Fig. 2.12 (i). There is no other lattice point.

   Example: KCl and Po at certain temperature.

2. *Face-centred cubic or cubic F lattice*: In addition to the lattice points at the eight corners it contains lattice points at the centre of each face [Fig. 2.12 (ii)]. There is an extra lattice point at the centre of each face of the face-centred cubic (FCC) lattice unit cell.

   Example: Cu, Ni, Au, Ag crystals.

3. *Body-centred cubic or cubic I lattice*: In body-centred cubic (BCC) lattice, there is one additional lattice point situated at the body centre besides eight lattice points situated at the eight corners as shown in Fig. 2.12 (iii). Thus, there is an extra lattice point at the body centre of the unit cell.

   Example: Li, Na, K, Cr crystals.

2.10.1 Calculation of Number of Atoms per Unit Cell

1. *Simple cubic* (SC) lattice: Every lattice point in SC lattice situated at the corner of the unit cell is shared by eight unit cells and hence, its contribution to each cell is $\frac{1}{8}$.
   Total contribution of eight corners of the SC lattice is $\frac{1}{8} \times 8 = 1$. Thus, for an SC lattice there is one lattice point per unit cell.
2. *Face-centred cubic* (FCC) lattice: There are eight lattice points at the eight corners of the FCC lattice and contribution of these eight lattice points to the unit cells is $\frac{1}{8} \times 8 = 1$.
   Moreover, there are six-faces of the unit cell each atom of which belongs to two unit cells. So, the contribution of these six lattice points to the unit cell is $\frac{1}{2} \times 6 = 3$.
   $\therefore$ total number of lattice points per FCC cell is $1 + 3 = 4$.
3. *Body-centred cubic* (BCC) lattice: Here there are eight lattice points at the eight corners of the unit cell and each is a member of eight unit cells. In addition, there is one lattice point at the body centre. So, the total number of lattice points per BCC unit cell is $\frac{1}{8} \times 8 + 1 = 2$.

2.10.2 Coordination Number

For a particular lattice, *coordination number* (CN) is defined as the number of equivalent nearest neighbours to a particular atom in the crystal lattice.

2.10.3 Nearest Neighbour Distance

It is defined as the distance between the centres of two nearest neighbouring atoms in the crystal.

2.10.4 Atomic Radius (r)

Atomic radius is defined as half of the distance between nearest neighbour atoms in the crystal.

Here the atoms are supposed to be spherical in size and any two nearest neighbouring atoms touch each other.

1. **SC lattice:** In Fig. 2.13, AB represents the radius r of an atom and $r = a/2$.

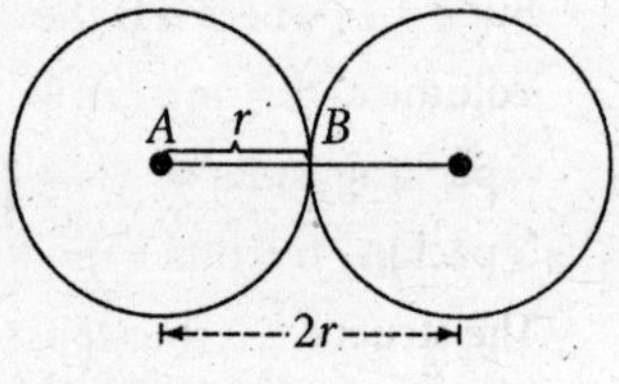

Fig. 2.13: Nearest neighbour distance in SC lattice

The nearest neighbour distance $d = 2r = 2\frac{a}{2} = a$.

2 **BCC lattice:** In Fig. 2.15, a BCC unit cell is shown, where $AB^2 = a^2 + a^2$.

$\therefore AB = \sqrt{2}a$ and $AC^2 = \left(\sqrt{2}a\right)^2 + a^2 = 3a^2$. $\therefore AC = \sqrt{3}a$.

$\therefore$ atomic radius $r = \frac{\sqrt{3}a}{4}$ $\left(\because 4r = \sqrt{3}a\right)$.

So, the atomic radius of a BCC crystal is $r = \frac{\sqrt{3}a}{4}$.

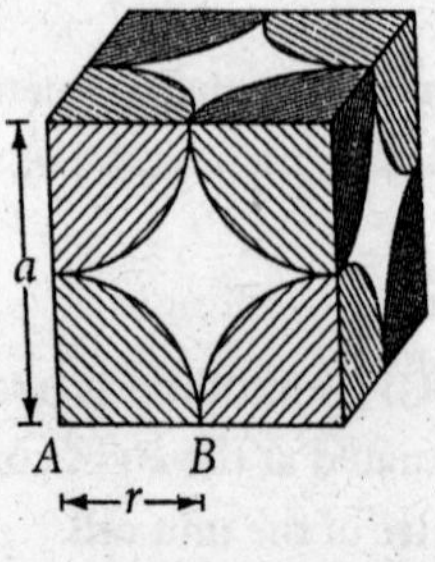

Fig. 2.14: Geometry of a SC unit cell

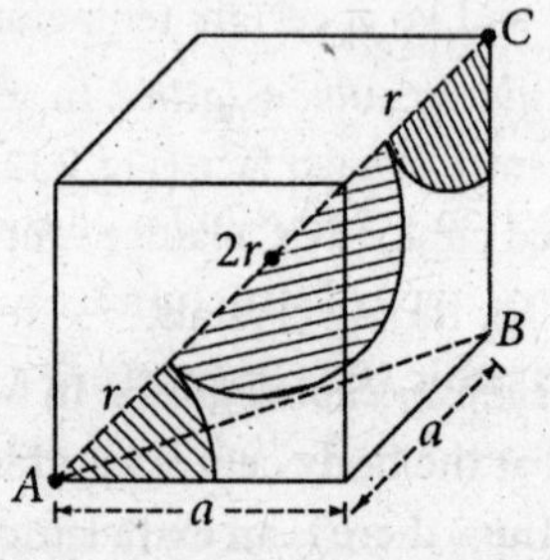

Fig. 2.15: Geometry of a BCC unit cell

The nearest neighbour distance,

$$d = 2r = \frac{\sqrt{3}}{2}a.$$

3. **FCC lattice**: Fig. 2.16 shows an FCC unit cell whose face diagonal is AC.

Face diagonal, $AC = \sqrt{2}a = 4r$.

$$\therefore \text{atomic radius } r = \frac{\sqrt{2}a}{4} \text{ or, } r = \frac{a}{2\sqrt{2}}.$$

The nearest neighbour distance,

$$d = 2r = \frac{a}{\sqrt{2}}.$$

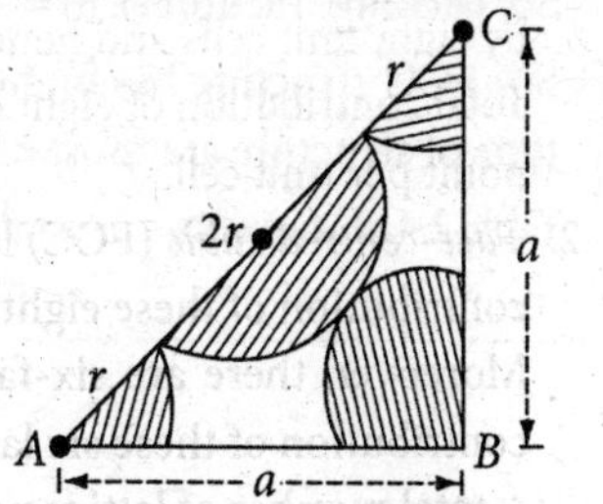

Fig. 2.16: Face diagonal of an FCC unit cell

2.10.5 Atomic Packing Factor

The atomic packing factor of a unit cell is defined as the ratio of the volume occupied by the atoms in the unit cell to the total volume of the unit cell.

If there are n atoms each of volume v, and V is the volume of the unit cell, then Packing Factor (PF) $= \frac{nv}{V}$. Sometimes packing factor (PF) is expressed in %, then % PF = PF × 100.

Atomic packing factor is sometimes called *packing fraction*.

Calculation of Packing Factor

1. **Simple cubic (SC) structure**: In simple cubic structure there is one atom in the unit cell.
   Volume of the atom of radius r is $v = \frac{4}{3}\pi r^3$.
   But $r = \frac{a}{2}$, where a is the lattice constant of SC structure. $\therefore v = \frac{4}{3}\pi\left(\frac{a}{2}\right)^3 = \frac{\pi a^3}{6}$.
   Volume of the unit cell $V = a^3$.
   $\therefore$ packing factor $= \frac{1\times v}{V} = \frac{\pi a^3}{6a^3} = \frac{\pi}{6} = 0.523$.
   $\therefore$ packing fraction $= \frac{\pi}{6} \times 100 = 0.523 \times 100 = 52.3\,\%$.
   The structure is thus loosely packed. Example of this structure is polonium.
2. **Body-centred cubic (BCC) structure**: Here the number of atoms per unit cell is 2.
   $\therefore$ volume of these two atoms $= 2 \times \frac{4}{3}\pi r^3$.
   But atomic radius $r = \frac{\sqrt{3}}{4}a$.
   $\therefore$ total volume occupied by the two atoms is $2 \times \frac{4}{3}\pi\left(\frac{\sqrt{3}}{4}a\right)^3 = \frac{\sqrt{3}\pi}{8}a^3$.
   Volume of the unit cell $V = a^3$.
   $\therefore$ packing factor (PF) $= \frac{\sqrt{3}\pi a^3}{8\times a^3} = \frac{\sqrt{3}\pi}{8} = 0.68$.
   $\therefore$ packing fraction $= \frac{\sqrt{3}\pi}{8} \times 100 = 0.68 \times 100 = 68\,\%$.
   Examples of this type of structure are tungsten, sodium, iron and chromium.
3. **Face-centred cubic (FCC) structure**: Here the number of atoms per unit cell = 4.
   $\therefore$ volume of 4 atoms $= 4 \times v = 4 \times \frac{4}{3}\pi r^3$.
   But atomic radius $r = \frac{\sqrt{2}}{4}a$.
   $\therefore$ total volume occupied by the 4 atoms is $4 \times \frac{4}{3}\pi\left(\frac{\sqrt{2}}{4}a\right)^3 = \frac{\sqrt{2}\pi}{6}a^3$ and volume of the unit cell $V = a^3$.

$\therefore$ packing factor = $\frac{\sqrt{2}\pi a^3}{6 \times a^3} = \frac{\sqrt{2}\pi}{6} = 0.74.$

So, packing fraction (PF) = $\frac{\sqrt{2}\pi}{6} \times 100 = 0.74 \times 100 = 74\,\%$.

So, FCC structure has highest packing fraction compared to SC and BCC structure. Examples of this type of structure are most of the metals like copper, aluminium, lead and silver.

Table 2.4 shows some crystal structure with examples.

Table 2.4: Some crystal structure with examples

Parameters	*Simple Cubic* (SC)	*Body-Centred Cubic* (BCC)	*Face-Centred Cubic* (FCC)	*Hexagonal Close Packed Structure* (HCP)	Diamond Cube (DC)
Atomic radius	$a/2$	$\frac{\sqrt{3}a}{4}$	$\frac{a}{2\sqrt{2}}$	$\frac{a}{2}$	$\frac{\sqrt{3}a}{8}$
Nearest neighbour distance	a	$\frac{\sqrt{3}a}{2}$	$\frac{a}{\sqrt{2}}$	a	$\frac{\sqrt{3}a}{4}$
Coordination number	6	8	12	12	4
Volume of the unit cell	a^3	a^3	a^3	$\frac{3\sqrt{3}}{2}a^2c$	a^3
No. of atoms per unit cell	1	2	4	6	8
Packing factor	0.52	0.68	0.74	0.74	0.34
Example	Polonium	Sodium, Lithium and Chromium	Aluminium, Copper, Silver and Lead	Magnesium, Zinc and Cadmium	Germanium, Silicon and Diamond

2.11 Calculation of Lattice Constant

We consider a cubic crystal of lattice constant a. Then volume of the unit cell = a^3.

If ρ be the density of the crystal, then mass in each unit cell of the crystal = ρa^3.

2.12 Some Other Crystal Structures

2.12.1 Hexagonal Closed Packed (HCP) Structure

The HCP structure is a closest packing structure having packing fraction close to that of FCC space lattice as shown in Fig. 2.17. The unit cell of HCP structure contains one atom at each corner, one atom at the centre of the hexagonal face, in addition to three atoms within the body of the unit cell. An atom in the unit cell touches three atoms in the layers above and below and six atoms in its plane situated at the corners of the hexagon.

If r be the radius of each atom and a the interatomic distance, then $2r = a$. Also the coordination number is 12.

Three atoms are within the body of the cell having full contribution to the cell. The top layer contains seven atoms, each corner atom is shared by surrounding hexagon cells. The centre atom is shared by two adjacent cells.

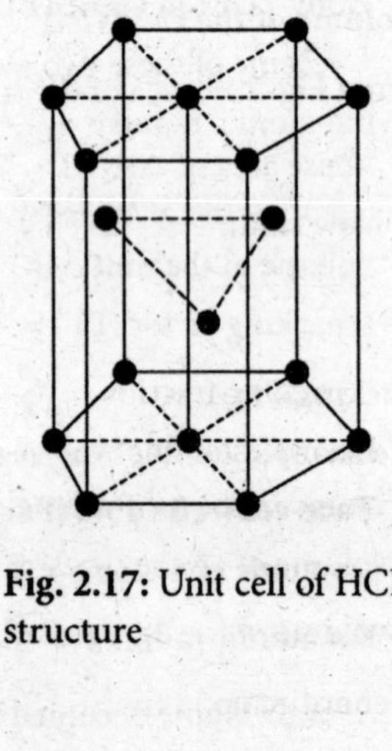

Fig. 2.17: Unit cell of HCP structure

Therefore, in a unit cell the total number of atoms = $3/2 + 3/2 + 3 = 6$.

c/a Ratio for an Ideal HCP Structure

There are three atoms at the bottom and the top face in a horizontal plane at $c/2$ (where c is the height of the cell) from the ortho-centres of alternate equivalent triangles at the top or base of the cell. These three atoms rest on the three corner atoms of the triangle as shown in Fig. 2.18. From the triangle ABY of Fig. 2.19, we have

$$\left.\begin{aligned} \cos 30^\circ = \frac{AY}{AB}, \therefore AY = \frac{a\sqrt{3}}{2} \\ \text{and } AZ^2 = AX^2 + ZX^2 \end{aligned}\right\}. \tag{2.3}$$

From the triangle AXZ,

$$AX = \frac{2}{3}AY = \frac{2a\sqrt{3}}{6} = \frac{a}{\sqrt{3}}.$$

$\therefore$ from the second eqn of eqns (2.3)

$$a^2 = \frac{a^2}{3} + \frac{c^2}{4} \text{ and } \frac{c^2}{4} = a^2 - \frac{a^2}{3} = \frac{2}{3}a^2 \text{ or, } \frac{c^2}{a^2} = \frac{8}{3} \text{ or, } \frac{c}{a} = \sqrt{\frac{8}{3}}. \tag{2.4}$$

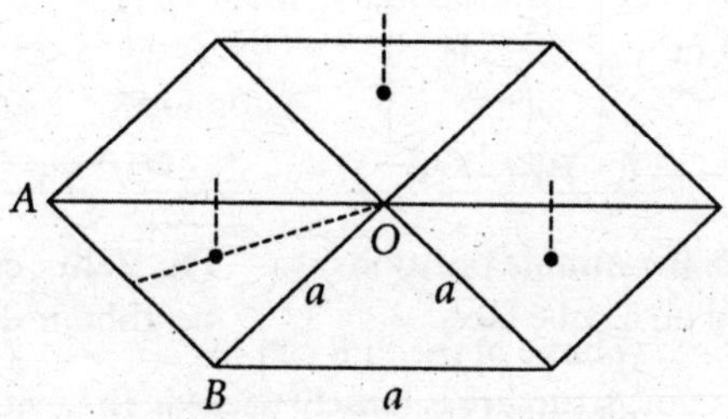

Fig. 2.18: Ideal HCP structure

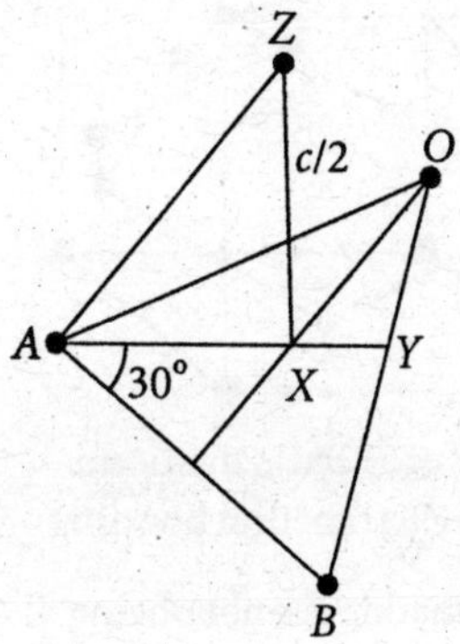

Fig. 2.19

Volume of the Unit Cell

From Fig. 2.19, area of the triangle $AOB = \frac{1}{2}BO \times AY = \frac{1}{2}a\left[a\frac{\sqrt{3}}{2}\right]$.

$\therefore$ base area of $\triangle AOB = \frac{6\times a^2\sqrt{3}}{4} = \frac{3}{2}\sqrt{3}\left(a^2\right)$.

Now volume of the HCP unit cell

$$V = \frac{3}{2}\sqrt{3}(a^2 \times c) = \frac{3\sqrt{3}a^2 \times c}{2}. \tag{2.5}$$

Packing Fraction

We have for the HCP structure, the nearest neighbour distance, $2r = a$, lattice parameter, $a = 2r$, number of atoms in the unit cell = 6,

volume of all the atoms per unit cell, $v = 6 \times \frac{4}{3}\pi r^3 = \frac{24\pi r^3}{3} = \pi a^3$,

volume of the unit cell, $V = \frac{3\sqrt{3}a^2c}{2}$,

coordination number, $CN = 12$, and $\frac{c}{a} = \sqrt{\frac{8}{3}}$.

∴ density of packing fraction

$$PF = \frac{v}{V} = \frac{\pi a^3 \times 2}{3\sqrt{3}a^2 c} = \left(\frac{2\pi}{3\sqrt{3}}\right)\left(\frac{a}{c}\right) = \left(\frac{2\pi}{3\sqrt{3}}\right)\left(\frac{3}{8}\right)^{1/2} = \frac{2\pi}{3\sqrt{3}} \times \frac{\sqrt{3}}{2\sqrt{2}}$$

$$\text{or, } PF = \frac{\pi}{3\sqrt{2}} = 0.74 \text{ or } 74\,\% . \tag{2.6}$$

2.12.2 Diamond Cubic Structure

The unit cell of diamond is similar to that of germanium and silicon as shown in Fig. 2.20 (a). The space lattice is FCC with a basis of two carbon atoms at the lattice points. The projection of atoms in the cubic cell of diamond cubic structure is shown in Fig. 2.20 (b). If a be the cube edge, then the height of the atoms above the base is as represented in the figure. The points at O and $a/2$ are on the face lattice and those at $a/2, 3a/4$ are on a similar lattice displaced along the body diagonal by 1/4 of its length. Thus, the diamond lattice is a combination of two inter-penetrating FCC sub-lattices along the body diagonal displaced relative to each other by $a/4$.

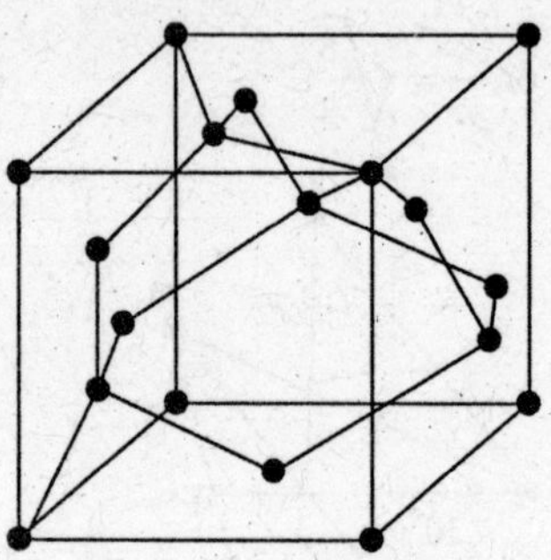

Fig. 2.20: (a) Basic diamond structure-tetrahedral bonding

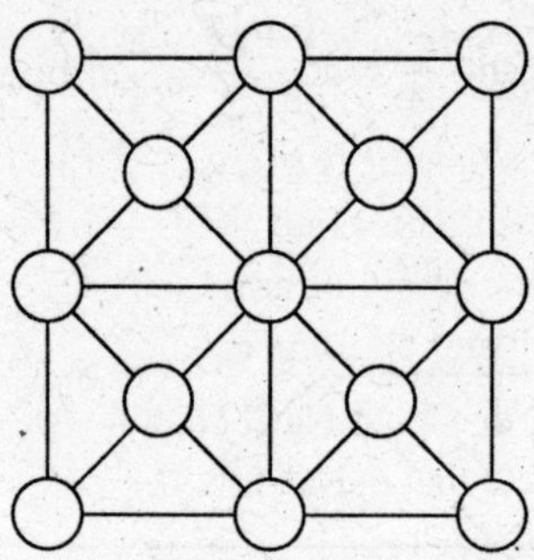

Fig. 2.20: (b) Atomic positions projected on a cube face

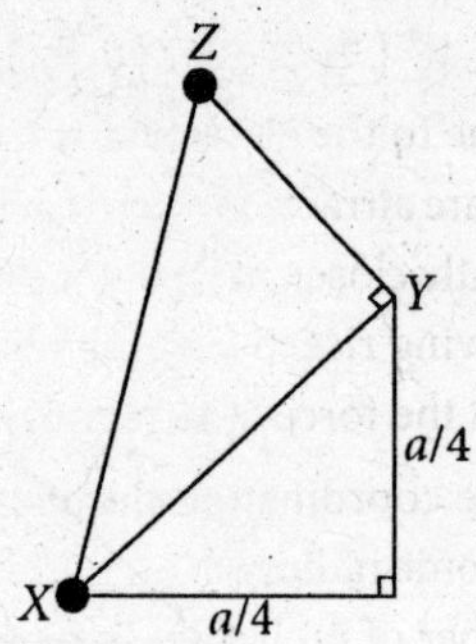

Fig. 2.20: (c) Nearest neighbour distance XZ

The coordination number being 4, the diamond cubic structure is loosely packed. There are eight corner atoms along with six face-centred atoms and four more atoms are presented in the unit cell. Again, each face-centred atom is shared by two unit cells and the corner atoms are shared by eight adjacent unit cells. Hence, the number of atoms per unit cell = $\left(\frac{1}{2} \times 6\right) + \left(\frac{1}{8} \times 8\right) + 4 = 8$.

Packing Fraction

From Fig. 2.20 (c),

$$XY^2 = \left(\frac{a}{4}\right)^2 + \left(\frac{a}{4}\right)^2 = \frac{a^2}{8}.$$

Also

$$XZ^2 = XY^2 + YZ^2 = \frac{a^2}{8} + \frac{a^2}{16} = \frac{3a^2}{16}.$$

But $XZ = 2r$.

∴ nearest neighbour distance

$$2r = \left(\frac{3a^2}{16}\right)^{1/2} \text{ or, } (2r)^2 = \frac{3a^2}{16}.$$

Hence, lattice constant, $a = \frac{8r}{\sqrt{3}}$.

Volume of all atoms in the unit cell, $v = 8 \times \frac{4}{3}\pi r^3$.

Volume of the unit cell, $V = a^3$.

∴ packing fraction

$$\text{PF} = \frac{v}{V} = \frac{\frac{32\pi r^3}{3}}{\left(\frac{8r}{\sqrt{3}}\right)^3} = \frac{32 \times \pi r^3 \times 3\sqrt{3}}{8 \times 8 \times 8 \times r^3} = \frac{\pi\sqrt{3}}{2 \times 8} = 0.34.$$

∴ PF = 34 %.

2.12.3 Sodium Chloride Structure

In sodium chloride crystal the Na and Cl atoms are bound together by an *ionic bond*. The *Bravais lattice* is an FCC with Na^+Cl^- as basis. The Na^+ and Cl^- ions in NaCl crystal are separated by one half of the body diagonal of the unit cell.

The positions of Na^+ and Cl^- in the unit cell as shown in Fig. 2.21 are as follows:

Cl^- : $(0,0,0)$ $\left(\frac{a}{2},\frac{a}{2},0\right)$ $\left(\frac{a}{2},0,\frac{a}{2}\right)$ $\left(0,\frac{a}{2},\frac{a}{2}\right)$

Na^+ : $\left(\frac{a}{2},\frac{a}{2},\frac{a}{2}\right)$ $\left(0,0,\frac{a}{2}\right)$ $\left(0,\frac{a}{2},0\right)$ $\left(\frac{a}{2},0,0\right)$

Due to the electrostatic force between Na^+ and Cl^-, both the ions are attracted toward each other. When the two ions come sufficiently closer, their outer electron shells come into close proximity giving rise to a strong repulsive force. Equilibrium is attained when the force of attraction and repulsion become equal.

The coordination number of Na^+ or Cl^- in NaCl crystal is 6. The ionic radius of Cl^- is 1.18 Å whereas that of Na^+ is 0.92 Å. The number of molecules in a unit cell of NaCl is 4.

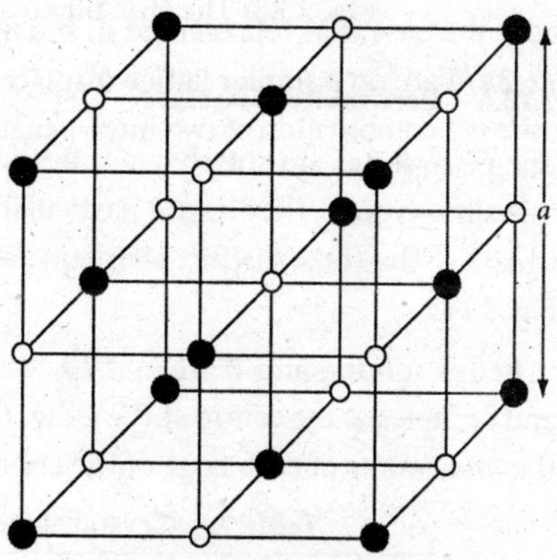

Fig. 2.21: NaCl structure

ONa^+ ●Cl^-

The crystals having the same structure as NaCl are KCl, KBr, MgO, AgBr, etc.

2.12.4 Cesium Chloride Structure

The space lattice of cesium chloride is simple cubic as shown in Fig. 2.22. The ionic radii of both Cs^+ and Cl^- are approximately equal. The basis comprises of Cs^+-ion at $(0,0,0)$ and Cl^--ion at $(a/2, a/2, a/2)$. Each of Cs^+ and Cl^- is situated at the centre of a cube of ions of electrically opposite in nature.

The lattice points of CsCl consist of two interpenetrating simple cubic lattice, one occupied by Cs^+-ion and the other by Cl^--ion. So, the coordination number is 8.

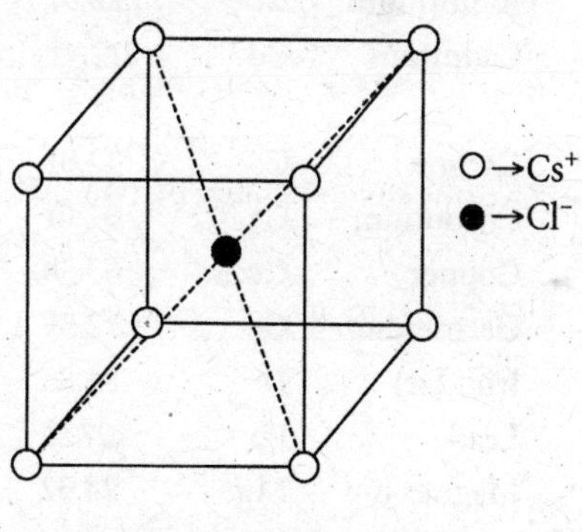

Fig. 2.22: CsCl structure

2.12.5 Zinc Blend (ZnS) Structure

The structure of ZnS is identical to diamond structure except that the two interpenetrating sub-lattices are occupied by two different elements. In this structure the Zn-atoms are placed on one FCC lattice and S-atoms on the other [Fig. 2.23]. The ZnS structure does not possess centre of inversion whereas diamond does.

Other similar structures are CdS, AgI, ZnSe and SiC, $InSb_2$, GaAs and CuCl.

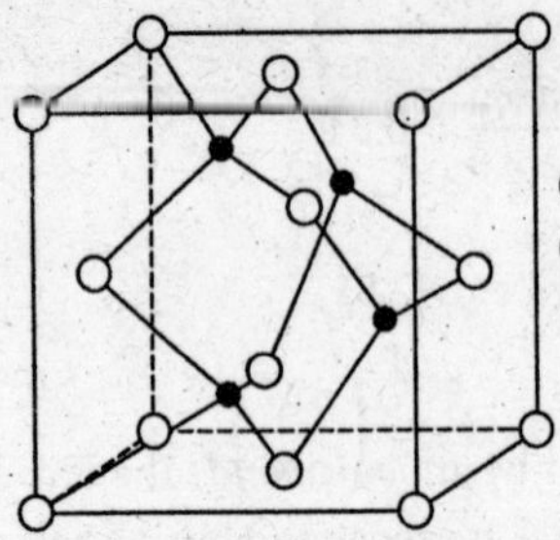

Fig. 2.23: The zinc blend structure

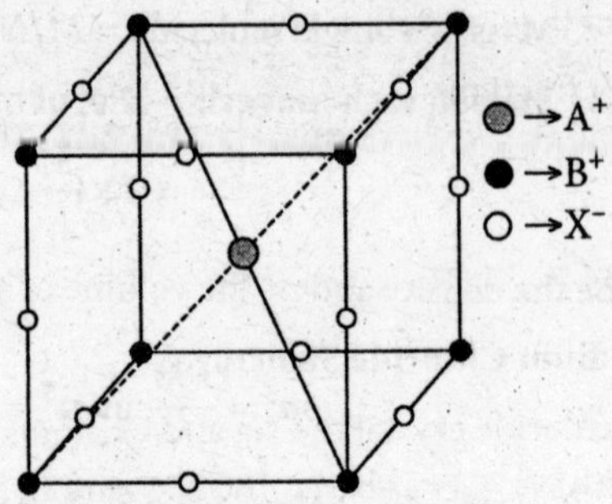

Fig. 2.24: Perovskite structure

2.12.6 Perovskite Structure

The perovskites are simply ceramics which are solids comprising of metallic elements and non-metals (usually oxygen). They have a particular arrangement of atoms in them. The general formula for perovskite is ABX_3. The three distinct elements form a cube and present in the ratio $A : B : X_3 = 1 : 1 : 3$ as shown in Fig. 2.24.

In this solid, A and B are metallic cations and X is non-metallic axioms. The A-cation is largest in size and occupies at the centre of the cube; B-cations occupy the eight corners. The non-metal X-axions occupy the mid-points of each edge of the cube.

Table 2.5: Crystal structure and some properties of representative elements

Elements	*Symbol*	*Atomic weight*	*Density* (kg/m³)	*Crystal structure*	*Lattice constant at* 20 °C (nm)	*Approximate atomic radius*	*Melting* Point (°C)
Aluminium	Al	26.97	27	FCC	0.4049	0.1428	660.2
Cadmium	Cd	112.41	8.65	HCP	$a = 0.2979$ $c = 0.5617$	0.1489	320.9
Carbon	C	12.01	2.22	DC	0.3564	0.0770	3700
Chromium	Cr	52.01	7.19	BCC	0.2885	0.1249	1890
Copper	Cu	63.54	8.96	FCC	0.3615	0.1278	1083
Germanium	Ge	72.59	5.36	DC	0.5658	0.1224	937
Iron (α)	Fe	55.85	7.87	BCC	0.2866	0.1238	1539
Lead	Pb	207.21	11.34	FCC	0.4949	0.1750	327
Magnesium	Mg	24.32	1.74	HCP	$a = 0.3209$ $c = 0.5210$	0.1594	650
Potassium	K	39.10	0.86	BCC	0.5339	0.2312	63.7
Silicon	Si	28.09	2.34	DC	0.5428	0.1176	1430
Silver	Ag	107.88	10.49	FCC	0.4086	0.1444	960.5
Sodium	Na	22.99	0.97	BCC	0.4281	0.1857	97.8
Zinc	Zn	65.38	7.13	HCP	$a = 0.2664$ $c = 0.4945$	0.1332	419.5

Again, if there are n molecules per unit cell, M, the molecular weight and N, the Avogadro number, then

Mass of a single molecule = M/N (2.7)

Mass in each unit cell = (No. of molecules in a unit cell) × (mass of a single molecule)

$$= n \times \frac{M}{N}. \quad (2.8)$$

If ρ be the density and a^3 the volume of a unit cell, then from eqn (2.8)

$$\rho a^3 = \frac{nM}{N} \text{ or, } a^3 = \frac{nM}{\rho N} \text{ or, lattice constant } a = \left(\frac{nM}{\rho N}\right)^{1/3}. \quad (2.9)$$

Thus, if n, M, ρ and N are known, then the value of the lattice constant a of a crystal can be determined from eqn (2.9).

2.13 Lattice Planes and Miller Indices

A crystal lattice may be considered as an aggregate of a set of parallel equidistant planes passing through the lattice points. These set of parallel planes are called *lattice planes* which can be chosen in a number of ways like (1), (2), (3) and (4) as shown in Fig. 2.25. *Miller* discovered a method to designate a plane in a crystal using three numbers h, k and l, known as *Miller indices*.

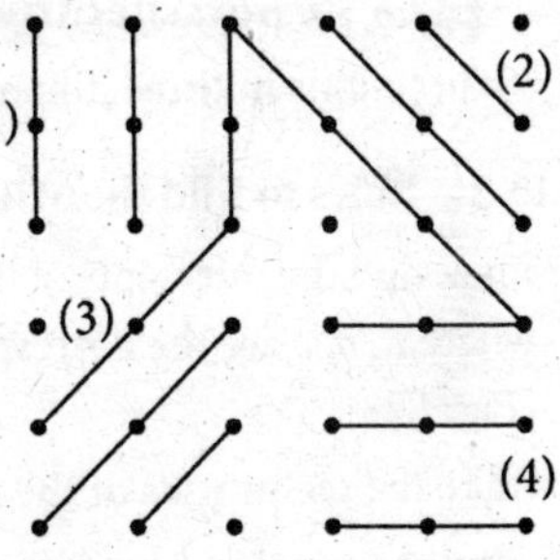

Fig. 2.25: Set of lattice planes

Miller indices are the three smallest possible integers which have the same ratios as the reciprocals of the intercepts of the plane concerned on the three axes.

2.13.1 Crystallographic Directions

In a crystal, the concentration of atoms in different direction is different which leads to the anisotropy of it. For this reason it is necessary to locate different directions and planes for crystallographic analysis.

Definition. The crystallographic direction in a crystal is defined as line joining any two points in the lattice.

In *two dimensions, the crystallographic direction is expressed as a line passing through a point $A(x, y)$ and any other lattice point taken as the origin $O(0,0)$ as shown in* Fig. 2.26, the coordinates x and y being whole numbers.

In Fig. 2.26, the directions of the points A and B with O as the origin are $(6,5)$ and $(7,3)$ respectively.

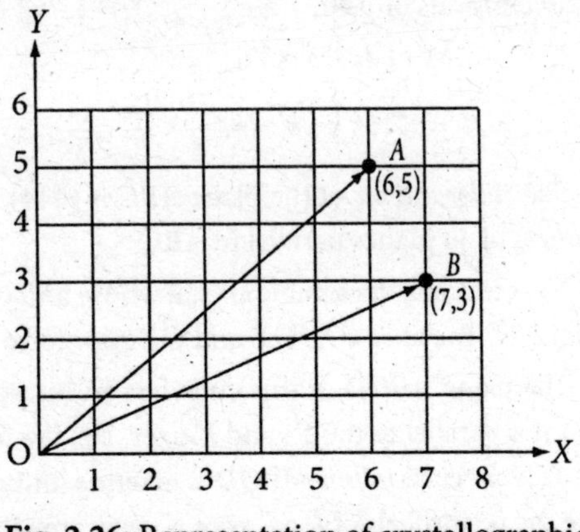

Fig. 2.26: Representation of crystallographic directions in two dimensions

In *three dimensions,* the crystallographic direction of any point A having coordinates (x, y, z), where x, y, z are whole numbers, is conventionally denoted by putting x, y, z in brackets, i.e., $[x, y, z]$.

The crystallographic directions of the corner points of a unit cube are shown in the table below:

Points	$A(100)$	$B(110)$	$C(010)$	$P(001)$	$Q(101)$	$R(111)$	$S(011)$
Direction	[100]	[110]	[010]	[001]	[101]	[110]	[011]

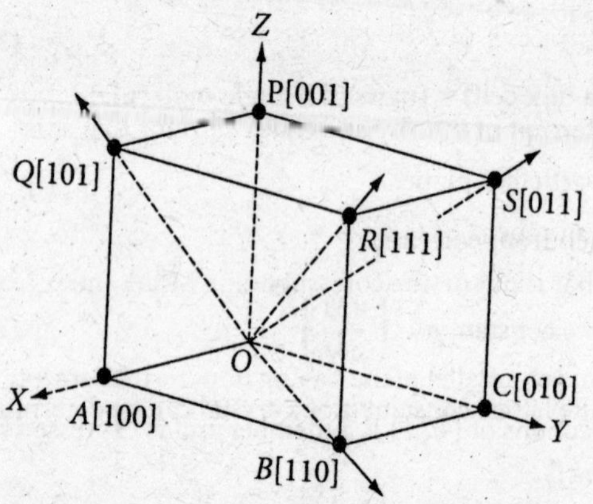

Fig. 2.27: Representation of crystallographic directions in three dimensions

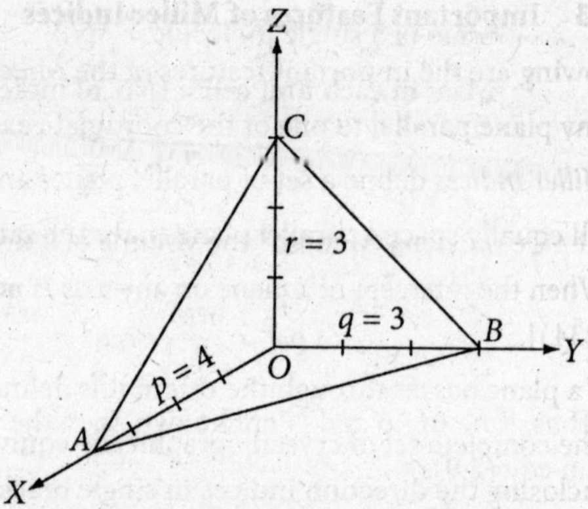

Fig. 2.28: A lattice plane specified by Miller indices (344)

2.13.2 Rules to find the Miller Indices of a Lattice Plane

1. First, find the intercepts of the plane on the three crystal axes (OX, OY, OZ). Let these be pa, qb and rc, where a, b, c are the corresponding *primitives* and p, q, r are the *intercepts* along $\vec{a}$, $\vec{b}$ and $\vec{c}$ respectively [Fig. 2.28].
2. Take the reciprocals of the numbers p, q, r in the form $\frac{1}{p}, \frac{1}{q}, \frac{1}{r}$.
3. Find the smallest possible integers h, k, l by multiplying the reciprocals by the LCM of p, q, r, such that $h : k : l = \frac{1}{p} : \frac{1}{q} : \frac{1}{r}$.

The numbers h, k, l are the *Miller indices*, the corresponding lattice plane is represented by (hkl). It is to be kept in mind that (hkl) *does not refer to a single plane instead a set of parallel planes.*

» ***Example.*** In Fig. 2.28, $p = 4$, $q = 3$ and $r = 3$.

Reciprocals of p, q, r are $\frac{1}{p}, \frac{1}{q}, \frac{1}{r} = \frac{1}{4}, \frac{1}{3}, \frac{1}{3}$.

Now LCM of 4, 3, 3 is 12.

$$\therefore h : k : l = \frac{12}{4} : \frac{12}{3} : \frac{12}{3} = 3 : 4 : 4. \tag{2.10}$$

So, *Miller indices* of the plane ABC is (344). (344) is also the *Miller indices* of all planes parallel to ABC.

Next we consider a cubic crystal whose unit cell is a cube as shown in Fig. 2.29. The axes OX, OY and OZ now form a right-handed system.

The plane $ABCD$ of this cube has an intercept OD on the X-axis, but it is parallel to the Y- and Z-axes, i.e., the intercepts on the Y- and Z-axes are each *infinity.* If OD is taken as unity, then the *Miller indices* are the reciprocals of 1, ∞, ∞, i.e., they are (100).

Similarly, the *Miller indices* of the plane $GADE$ are (110) and those for the plane FAG are (111) [Fig. 2.29].

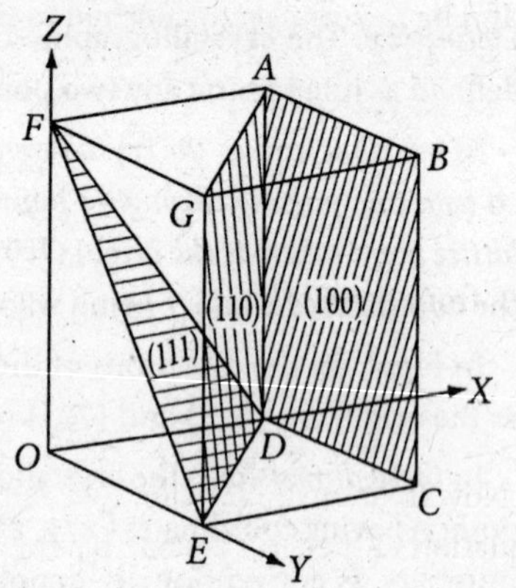

Fig. 2.29: Miller indices of lattice planes in a cubic lattice

2.13.3 Important Features of Miller Indices

Following are the important features of the *Miller indices*:

1. Any plane parallel to one of the coordinate axes has intercept of infinity and index zero for that axis.
2. *Miller indices* define a set of parallel planes and not a particular plane.
3. All equally spaced parallel planes have the same index number (hkl).
4. When the intercept of a plane on any axis is negative a bar is put on the corresponding Miller index [e.g., $(2\bar{3}4)$].
5. If a plane passes through the origin it is defined in terms of parallel plane having non-zero intercept.
6. The complete set of crystallographically equivalent directions of (hkl) is a member and is expressed by enclosing the direction indices in single bracket like $\langle hkl \rangle$.

2.14 Inter-planar Spacing

The spacing d_{hkl} between a family of parallel adjacent planes (hkl) can be calculated by taking any lattice point as an origin and erecting coordinate axes in the $\vec{a}$, $\vec{b}$ and $\vec{c}$ directions and then determining the perpendicular distance between the origin and the plane nearest to this origin.

The intercepts of the (hkl)-plane on $\vec{a}$, $\vec{b}$ and $\vec{c}$ axes are respectively $\frac{a}{h}$, $\frac{b}{k}$ and $\frac{c}{l}$ as shown in Fig. 2.30.

From Fig. 2.30,

$$d_{hkl} = OP = \frac{a}{h}\cos\alpha = \frac{b}{k}\cos\beta = \frac{c}{l}\cos\gamma, \tag{2.11}$$

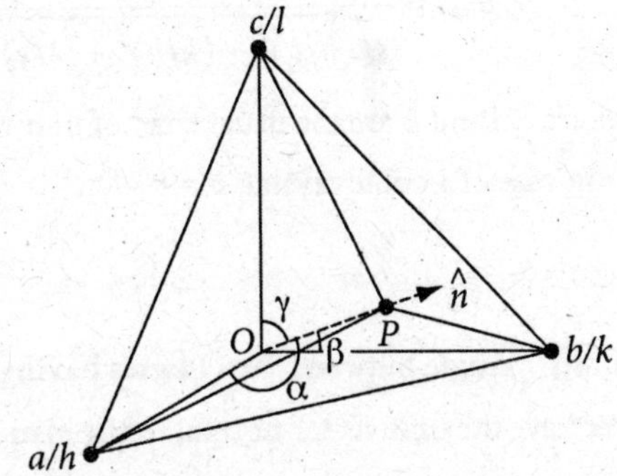

Fig. 2.30: Direction angles of the normal to the (hkl)-plane

where α, β and γ are the angles between the normal to the plane and the $\vec{a}$, $\vec{b}$ and $\vec{c}$ axes respectively.

If $\hat{n}$ be the unit vector normal to the plane, then

$$\hat{n}\cdot\vec{a} = a\cos\alpha$$
$$\hat{n}\cdot\vec{b} = b\cos\beta$$
$$\hat{n}\cdot\vec{c} = c\cos\gamma.$$

$\therefore$ from eqn (2.11)

$$d_{hkl} = \frac{\hat{n}\cdot\vec{a}}{h} = \frac{\hat{n}\cdot\vec{b}}{k} = \frac{\hat{n}\cdot\vec{c}}{l}. \tag{2.12}$$

Now in an orthogonal lattice, taking X-, Y- and Z-axes along $\vec{a}$-, $\vec{b}$- and $\vec{c}$-directions respectively, the equation to the (hkl)-plane having intercepts $\frac{a}{h}$, $\frac{b}{k}$ and $\frac{c}{l}$ is

$$\frac{x}{a/h} + \frac{y}{b/k} + \frac{z}{c/l} = 1 \text{ or, } \frac{hx}{a} + \frac{ky}{b} + \frac{lz}{c} = 1 \text{ or, } \phi(x, y, z) = 1 \text{ (constant)}. \tag{2.13}$$

$$\therefore \nabla\phi = \hat{i}\frac{\partial\phi}{\partial x} + \hat{j}\frac{\partial\phi}{\partial y} + \hat{k}\frac{\partial\phi}{\partial z} = \hat{i}\frac{h}{a} + \hat{j}\frac{k}{b} + \hat{k}\frac{l}{c} = |\nabla\phi|\hat{n}, \tag{2.14}$$

where $\hat{n}$ is the unit vector along the normal to the (hkl) plane and

$$|\nabla\phi| = \sqrt{(h/a)^2 + (k/b)^2 + (l/c)^2}.$$

$$\therefore \hat{n} = \frac{\nabla\phi}{|\nabla\phi|} = \frac{\hat{i}\left(\frac{h}{a}\right) + \hat{j}\left(\frac{k}{b}\right) + \hat{k}\left(\frac{l}{c}\right)}{\sqrt{(h/a)^2 + (k/b)^2 + (l/c)^2}}.$$

Now from eqn (2.12),

$$d_{hkl} = \frac{\hat{n}\cdot\vec{a}}{h} = \frac{1}{h}\left[\frac{\hat{i}\left(\frac{h}{a}\right) + \hat{j}\left(\frac{k}{b}\right) + \hat{k}\left(\frac{l}{c}\right)}{\sqrt{(h/a)^2 + (k/b)^2 + (l/c)^2}}\cdot\hat{i}a\right] = \frac{1}{h}\left[\frac{\frac{h}{a}a}{\sqrt{(h/a)^2 + (k/b)^2 + (l/c)^2}}\right],$$

$$\therefore d_{hkl} = \frac{1}{\sqrt{(h/a)^2 + (k/b)^2 + (l/c)^2}}, \tag{2.15}$$

where $\hat{i}$, $\hat{j}$ and $\hat{k}$ are the usual triad of unit vectors along X-, Y- and Z-directions respectively.

In case of a cubic crystal, $a = b = c$,

$$\therefore d_{hkl} = \frac{a}{\sqrt{h^2 + k^2 + l^2}}. \tag{2.16}$$

2.14.1 Angle between Two Planes having Miller Indices $(h_1k_1l_1)$ and $(h_2k_2l_2)$

We have the unit vector normal to the plane $(h_1k_1l_1)$. Then

$$\hat{n}_1 = \frac{\hat{i}\frac{h_1}{a} + \hat{j}\frac{k_1}{b} + \hat{k}\frac{l_1}{c}}{\sqrt{\left(\frac{h_1}{a}\right)^2 + \left(\frac{k_1}{b}\right)^2 + \left(\frac{l_1}{c}\right)^2}}.$$

For a cubic system, $a = b = c$.

$$\therefore \hat{n}_1 = \frac{(\hat{i}h_1 + \hat{j}k_1 + \hat{k}l_1)}{\sqrt{h_1^2 + k_1^2 + l_1^2}}.$$

Similarly, the unit vector normal to the plane (h_2, k_2, l_2) of a cubic system is

$$\hat{n}_2 = \frac{(\hat{i}h_2 + \hat{j}k_2 + \hat{k}l_2)}{\sqrt{h_2^2 + k_2^2 + l_2^2}}.$$

Now

$$\hat{n}_1\cdot\hat{n}_2 = |\hat{n}_1||\hat{n}_2|\cos\theta \text{ [where } \theta \text{ is the angle between the planes]}$$

$$= \frac{\hat{i}h_1 + \hat{j}k_1 + \hat{k}l_1}{\sqrt{h_1^2 + k_1^2 + l_1^2}}\cdot\frac{\hat{i}h_2 + \hat{j}k_2 + \hat{k}l_2}{\sqrt{h_2^2 + k_2^2 + l_2^2}} = \frac{h_1h_2 + k_1k_2 + l_1l_2}{\sqrt{(h_1^2 + k_1^2 + l_1^2)(h_2^2 + k_2^2 + l_2^2)}}.$$

$$\therefore \theta = \cos^{-1}\left[\frac{h_1h_2 + k_1k_2 + l_1l_2}{\sqrt{(h_1^2 + k_1^2 + l_1^2)(h_2^2 + k_2^2 + l_2^2)}}\right]. \tag{2.17}$$

2.14.2 Inter-Planar Spacing between Lattice Planes of Cubic Lattice

1. **SC lattice:** Here the unit cell is a cube of side a and there is only one lattice point at each corner of the cube.

   The inter-planar spacing of (hkl)-planes for such lattice is

$$d_{hkl} = \frac{a}{\sqrt{h^2 + k^2 + l^2}}.$$

   The miller indices of the three family of planes are

$$(100),\ (110) \text{ and } (111).$$

   For (100)-planes,

$$d_{100} = \frac{a}{\sqrt{1^2 + 0^2 + 0^2}} = a.$$

   For (110)-planes,

$$d_{110} = \frac{a}{\sqrt{1^2 + 1^2 + 0^2}} = \frac{a}{\sqrt{2}}.$$

   For (111)-planes,

$$d_{111} = \frac{a}{\sqrt{1^2 + 1^2 + 1^2}} = \frac{a}{\sqrt{3}}.$$

2. **BCC lattice:** In this cubic lattice apart from one atom at each corner, there is an additional atom at the body centre.

   The family of the lattice planes are represented by (100), (110) and (111). For (100)-planes

$$d_{100} = \frac{a}{2},$$

   since there is an additional plane (100) just half-way between (100)-planes.

   For (110)-planes

$$d_{110} = \frac{a}{\sqrt{2}},$$

   as the (110)-plane of a BCC lattice is similar to (110) plane of an SC lattice.

   For (111)-planes:

   In a BCC lattice, the plane (111) is just midway between (111)-planes of an SC lattice.

$$\therefore d_{111} = \frac{1}{2}\frac{a}{\sqrt{3}}.$$

3. **FCC lattice:** For (100)-plane

$$d_{100} = \frac{a}{2},$$

   since there is an additional plane (100) half-way between (100)-planes of an SC lattice.

   For (110)-planes

$$d_{110} = \frac{1}{2}\frac{a}{\sqrt{2}},$$

   since there is an additional plane (110) midway and parallel to the first set between (110)-planes of an SC lattice.

For (111)-planes

$$d_{111} = \frac{a}{\sqrt{3}}$$

as the (111)-planes of an SC lattice pass through the centres of the faces of an FCC lattice.

2.14.3 Comparison of SC, BCC, FCC Structures

Table 2.6

Properties	*SC*	*BCC*	*FCC*
Position of atoms	Corners of the cube	Corners and body centre of the cube	Corners and face centres of the cube
Volume of the unit cell	a^3	a^3	a^3
Number of atoms per unit cell	1	2	4
Coordination number (CN)	6	8	12
Nearest neighbour distance ($2r$)	a	$\frac{a\sqrt{3}}{2}$	$\frac{a}{\sqrt{2}}$
Atomic radius	$\frac{a}{2}$	$\frac{a\sqrt{3}}{4}$	$\frac{a}{2\sqrt{2}}$
Atomic packing fraction (APF)	0.52	0.68	0.74
Example	KI	NaCl, KCl	$AlCl_2$, $CuSO_4$

2.15 Answers of Some Important Questions

2.1. *What do you mean by single crystal and microcrystallite (or polycrystallites)?*

Ans. If in a solid the periodicity extends throughout the volume of the matter, then the solid is a single crystal, e.g., diamond, quartz, etc.

If there is an aggregate of so many small crystallites oriented at random and separated by *grain boundaries,* then the corresponding solid is called a *microcrystallite.* The crystallites are called *grains.*

2.2. *Amorphous solids are isotropic but crystals are anisotropic. Why?*

Ans. Amorphous solids have no directional properties, hence they are isotropic. But crystals have different periodic arrangements in all three directions which lead to the variation of its physical properties with direction. That is why a crystal is *anisotropic.*

2.3. *All primitive cells are unit cells but all unit cells may or may not be primitive cell. — Justify.*

Ans. In a primitive cell there is one atom at each corner of the unit cell, i.e., the number of atoms per unit cell is one. But in a unit cell there may or may not be additional atom or atoms in addition to an atom at each corner.

Hence, all primitive cells are unit cells but all unit cells may or may not be primitive cell.

2.4. *What are the basic lattice parameters?*

Ans. In a unit cell, the intercepts a, b and c made by three adjacent faces with the crystallographic axes are called *primitives* or *lattice constants.* The angles α, β and γ between the crystallographic axes are known as *interfacial angles.* The quantities a, b, c and α, β, γ are the *basic lattice parameters.* These determine the form and actual size of the unit cell.

2.5. *What do you mean by space lattice?*

Ans. A space lattice is a mathematical concept which represents infinite number of points in space such that the arrangement of points in space about a given point is similar to that about any other point.

2.6. *What are Bravais lattice?*

Ans. The points may be arranged in 14 distinguishable ways in three-dimensional space maintaining perfect periodicity. These 14 space lattices are called *Bravais lattice* and belong to 7-crystal system.

2.7. *Define space lattice and crystal lattice.*

Ans. A *space lattice* is an infinite array of points in three dimensions in which every point has surroundings identical to that of every other point in that array.

A *crystal lattice* is a space lattice in which the lattice sites are occupied by *atoms* or *cluster of atoms.*

2.8. *What do you mean by lattice constant?*

Ans. In a lattice, a unit cell has the length of the edge as well as the angles between the edges are of definite values. Each edge of the unit cell is the distance between the atoms of the same kind and is known as the *lattice constant.*

2.9. *Define coordination number.*

Ans. Coordination number (CN) is defined as the number of nearest neighbours to a particular atom in the crystal lattice.

2.10. *What is atomic packing factor?*

Ans. The atomic packing factor of a unit cell is defined as the ratio of volume occupied by the atoms in the unit cell to the total volume of the unit cell.

If there are n atoms each of volume v, and V is the volume of the unit cell, then

$$\text{Packing Factor (or Fraction) (PF)} = \frac{nv}{V}.$$

2.11. *Find an expression for the density of crystalline material in terms of lattice constant for a cubic lattice.*

Ans. Let ρ be the density of a cubic crystalline material of lattice constant a. If M_A be the atomic weight, then the volume of one mole of the substance is $v_M = \frac{M_A}{\rho}$. v_M is also the volume of Avogadro number (N_A) of atoms.

$\therefore$ the volume of a single atom, $v = \frac{M_A}{N_A\rho}$ and volume of n number of atoms, $v_n = nv = \frac{nM_A}{N_A\rho}$.

But the effective volume of n atoms is equal to the volume of the unit cell a^3.

So

$$a^3 = \frac{nM_A}{N_A\rho} \quad \text{or, } \rho = \frac{nM_A}{a^3 N_A}.$$

2.12. *What are Miller indices?*

Ans. Miller indices are the three smallest possible integers (hkl) which have the same ratios as the reciprocals of the intercepts of the plane concerned on the three axes. A particular set of values of (hkl) represents a family of parallel crystal planes.

2.13. *What are the specialities of the Miller indices?*

Ans. The Miller indices have the following specialities:

(a) Miller indices (hkl) represent a set of parallel planes and not a particular plane.

(b) If the intercepts are at infinity, the corresponding Miller indices are zero. If a plane passes through the origin, it is defined in terms of parallel plane having non-zero intercept.

(c) The Miller indices (hkl) have the same value if the planes are equispaced and parallel.

(d) The Miller index $\bar{h}$ represents that the plane cuts the axis on the negative side of the origin.

2.14. *What are crystallographic directions?*

Ans. The crystallographic direction is a line joining any two points of the space lattice relative to specific point taken as the origin.

2.15. *If $(h_1k_1l_1)$ and $(h_2k_2l_2)$ be the Miller indices of two planes, then what is the angle between them?*

Ans. The angle between the two lattice planes with *Miller indices* $(h_1\,k_1l_1)$ and $(h_2\,k_2l_2)$ is given by

$$\theta = \cos^{-1}\left[\frac{h_1h_2 + k_1k_2 + l_1l_2}{\sqrt{(h_1^2 + k_1^2 + l_1^2)(h_2^2 + k_2^2 + l_2^2)}}\right].$$

2.16. *Find the angle between the plane* (111) *and plane direction* [111].

Ans. The (111)-plane in a cubic unit cell is represented by *PAC*. *PA* is on the (111)-plane and the direction of *PA* is represented by $[10\bar{1}]$. The angle between the direction $[10\bar{1}]$ and plane [111] is the angle between the plane (111) and the direction [111].

If θ be the angle between (111) and [111], then

$$\theta = \cos^{-1}\frac{h_1h_2 + k_1k_2 + l_1l_2}{\sqrt{(h_1^2 + k_1^2 + l_1^2)(h_2^2 + k_2^2 + l_2^2)}}.$$

Here $h_1 = 1 \qquad h_2 = 1$

$k_1 = 0$ and $k_2 = 1$

$l_1 = -1 \qquad l_2 = 1.$

$$\therefore \theta = \cos^{-1}\frac{1\times 1 + 1\times 0 + (-1\times 1)}{\sqrt{1^2 + 1^2 + (-1)^2}\sqrt{(1^2) + (0)^2 + (1)^2}} = 0.$$

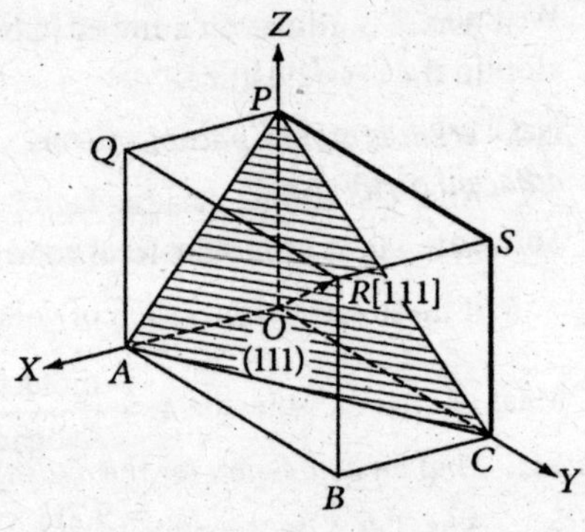

Fig. 2.31

NOTE. If a plane and the direction are represented by same value of h, k, l, then they are always perpendicular to each other.

2.16 Solved Problems

2.1. *Copper has FCC structure and the atomic radius is* 0.1278 nm. *Calculate the density and the interplanar spacing for* (321) *planes. Take the atomic weight of copper as* 63.5. **[WBUT 2005]**

Solution: For a cubic crystal having lattice constant d and number atoms $n(= 4$ for FCC lattice)

$$a = \left(\frac{nM_A}{\rho N}\right)^{1/3},$$

where M_A, ρ and N are respectively the atomic mass, density and Avogadro number.

Again, for an FCC crystal, $a = \frac{4}{\sqrt{2}}r$, where $r = 0.1278\text{ nm} = 0.1278\times 10^{-9}$ m is the atomic radius of the crystal.

$$\text{Now } \rho = \frac{nM_A}{Na^3} = \frac{4 \times 63.5}{6.023 \times 10^{23} \times \left(\frac{4\times0.1278\times10^{-9}}{\sqrt{2}}\right)^3} \text{ gm/cm}^3 = \mathbf{8.926\,gm/cm^3}.$$

and d_{hkl} = inter-planar spacing for (321)-plane = $\dfrac{a}{\sqrt{(h^2+k^2+l^2)}} = \dfrac{\frac{4\times0.1278\times10^{-9}}{\sqrt{2}}}{\sqrt{3^2+2^2+1^2}}$ m

$= 0.0966 \times 10^{-9}$ m = **0.0966 nm.**

2.2. *A cubic crystal has lattice constant* 4.3 Å *and density* 963 kg/m³. *If its atomic weight is 630, then what type of cubic unit cell does it form?*

Solution: Here edge of the unit cell, $a = 4.3$ Å $= 4.3 \times 10^{-10}$ m.

Density of the material of the crystal, $\rho = 963$ kg-m³

Atomic weight, $M_A = 23$ kg/k-mol.

Then

$$n = \frac{a^3\rho N}{M_A},$$

where n = the number of atoms per unit cell and N = the Avogadro number.

$$\therefore n = \frac{\left(4.3\times10^{-10}\right) \times 963 \times 6.02 \times 10^{26}}{23} = 2.$$

We know that in a body-centred cubic (BCC) crystal the number of atoms per unit cell is 2. So, the atomic crystal is of BCC type.

2.3. NaCl *crystal has FCC structure and is of density* 2.18 gm/cm³. *Calculate the distance between the two adjacent atoms.*

Solution: We have molecular weight of NaCl

$$M = 23 + 35.5 = 58.5.$$

$$\text{Mass of a NaCl molecule } m = \frac{\text{molecular weight } (M)}{\text{Avogadro number } (N)} = \frac{58.5 \text{ gm-mol}}{6.023 \times 10^{23} \,(\text{mol}^{-1})}$$

$$= 9.713 \times 10^{-23} \text{ gm}.$$

$$\text{Number of molecules per unit volume} = \frac{\text{mass of unit volume of NaCl } (\rho)}{\text{mass of a NaCl molecule } (m)}$$

$$= \frac{2.18}{9.713 \times 10^{-23}}$$

$$= 2.244 \times 10^{22} \text{ molecule/cm}^3.$$

Now number of atoms per unit volume of NaCl = $\mathbf{2 \times 2.244 \times 10^{22}}$ ($\because$ NaCl is diatomic).

If a and n represent respectively the distance between adjacent atoms and number of atoms along the edge of 1 cm cube, then the length of the edge of a unit cube = na.

Volume of the unit cube = $n^3a^3 = 1$ cm³.

$$\text{or, } \left(4.49 \times 10^{22}\right)(a)^3 = 1 \left[\because n^3 = 4.49 \times 10^{22}\right]$$

$$\therefore a = \left(\frac{1}{4.49 \times 10^{22}}\right)^{1/3} = 2.81 \times 10^{-8} \text{ cm} = \mathbf{2.81\,Å}.$$

2.4. *Calculate the lattice constant of rock-salt* (NaCl) *from the following data:*
Molecular weight by NaCl = 58.45 kg-mol
Density of NaCl = 2.17×10^3 kg m^3
Avogadro number $N = 6.023 \times 10^{23}$/gm-mol = 6.023×10^{26}/kg-mol. **[BU(H) 1996; CU(H) 2004]**

Solution: The number of lattice point (n) per unit cell of NaCl = 4 (as NaCl is an FCC crystal)

Now lattice constant

$$a = \left(\frac{4M}{\rho N}\right)^{1/3} = \left[\frac{4 \times 58.45}{2.17 \times 10^3 \times 6.023 \times 10^{26}}\right]^{1/3} \text{m} = \mathbf{5.64\ Å}.$$

2.5. KCl *is of FCC crystal structure having density* 1.98×10^3 kg/m^3. *If its molecular weight is* 74.6, *calculate* (i) *the distance from one atom to the next atom of same kind* (ii) *the distance between adjacent atoms.* **[cf. BU(H) 1995]**

Solution: If a be the lattice constant of the crystal, then

$$a = \left(\frac{nM_A}{\rho N}\right)^{1/3},$$

where n = number of ions in a unit cell = 4
M_A = molecular weight = 74.6 kg/k-mol
ρ = density = 963 kg/m^3
and N = Avogadro number = 6.02×10^{26} atoms/kg-mol.

$$\therefore a = \left(\frac{4 \times 74.6}{1.98 \times 10^3 \times 6.023 \times 10^{26}}\right)^{1/3} \simeq \mathbf{6.3\ Å}.$$

Thus, the distance between two atoms of the same kind is a = 6.3 Å.
Hence, the distance between two adjacent atoms is

$$\frac{a}{2} = \mathbf{3.15\ Å}.$$

2.6. *Calculate the lattice constant of potassium bromide. Given, density of potassium bromide* = 2.7×10^3 kg/m^3 *and it belongs to* FCC *lattice.*
Molecular weight of potassium bromide = 199 *and*
Avogadro No. = 6.02×10^{23}/gm-mol = 6.02×10^{26}/kg-mol. **[CU(H) 1992, cf. 91]**

Solution: Mass of a unit cell = $\dfrac{4 \times 119}{6.02 \times 10^{26}}$ kg.
[∵ KBr is an FCC crystal having 4 atoms/cell.
Again, mass of a unit cell of KBr can be written as $a^3 \times 2.7 \times 10^3$ kg [a = lattice constant]
So

$$a^3 \times 2.7 \times 10^3 = \frac{4 \times 119}{6.02 \times 10^{26}}$$

$$\text{or, } a = \left[\frac{4 \times 119}{2.7 \times 10^3 \times 6.02 \times 10^{26}}\right]^{1/3} = 6.57 \times 10^{-10} \text{ m} = \mathbf{6.57\ Å}.$$

2.7. *In a crystal, a lattice plane cuts intercepts of* 2a, 3b *and* 6c *along the three axes, where* a, b *and* c *are primitive vectors of the unit cell. Determine the Miller indices of the given plane.*

Solution: Let pa, qb and rc be respectively the intercepts of the plane on the three crystal axes, then

$$pa : qb : rc = 2a : 3b : 6c$$

$$\text{or, } p : q : r = 2 : 3 : 6$$

$$\text{or, } \frac{1}{p} : \frac{1}{q} : \frac{1}{r} = \frac{1}{2} : \frac{1}{3} : \frac{1}{6} = \mathbf{3 : 2 : 1.}$$

Hence, the Miller indices of the given plane are **(321)**.

2.8. *In an orthorhombic crystal a lattice plane cuts intercepts of lengths* $3a$, $-2b$ *and* $3c/2$ *along three axes. Deduce the Miller indices of the plane, where* $\vec{a}, \vec{b}, \vec{c}$ *are vectors of the unit cell.*

Solution: If the intercepts of the plane in the three crystal axes are pa, qb and rc respectively, then

$$pa : qb : rc = 3a : (-2b) : \frac{3c}{2}$$

$$\text{or, } p : q : r = 3 : (-2) : \frac{3}{2}$$

$$\text{or, } \frac{1}{p} : \frac{1}{q} : \frac{1}{r} = \frac{1}{3} : \left(-\frac{1}{2}\right) : \frac{2}{3} = 2 : (-3) : 4.$$

∴ Miller indices of the plane are $\mathbf{(2\bar{3}4)}$.

2.9. *Find the Miller indices of a plane having intercepts of* $8a$, $4b$ *and* $2c$ *on the a-, b- and c-axes respectively.*

[BU(H) 2001]

Solution: If the intercepts of the plane on a-, b- and c-axes respectively be pa, qb and rc, then

$$pa : qb : rc = 8a : 4b : 2c$$

$$\text{or, } p : q : r = 8 : 4 : 2$$

$$\text{or, } \frac{1}{p} : \frac{1}{q} : \frac{1}{r} = \frac{1}{8} : \frac{1}{4} : \frac{1}{2} = 1 : 2 : 4.$$

∴ the Miller indices of the plane are **(124)**.

2.10. *In an orthorhombic crystal a plane makes intercepts* 2.93 mm, 4.47 mm *and* 2.35 mm *along the three crystallographic axes, the corresponding primitives being* 3.05 Å, 6.99 Å *and* 4.90 Å. *Deduce the Miller indices of the cleavage plane.*

Solution: We have

$$p \times 3.05 : q \times 6.99 : r \times 4.90 = 2.93 : 4.47 : 2.35$$

$$\text{or, } p : q : r = \frac{2.93}{3.05} : \frac{4.47}{6.99} : \frac{2.35}{4.90}$$

$$= 0.96 : 0.64 : 0.48$$

$$= 6 : 4 : 3.$$

$$\therefore \frac{1}{p} : \frac{1}{q} : \frac{1}{r} = \frac{1}{6} : \frac{1}{4} : \frac{1}{3} = \mathbf{2 : 3 : 4}$$

Hence, Miller indices of the plane are **(234)**.

2.11. *Calculate the number of atoms per unit area of* (111)*-plane in an* SC *structure.*

Solution: Area of the (111)-plane is

$$\frac{1}{2} \times \text{base} \times \text{height} = \frac{1}{2} \times a\sqrt{2} \times EO.$$

Now from Fig. 2.32 (b)

$$\cos \angle DEO = \cos 30° = \frac{EO}{a\sqrt{2}}$$

$$\text{or, } EO = a\sqrt{2} \times \cos 30° = a\sqrt{2} \times 0.866.$$

∴ area of the (111)-plane is

$$\frac{1}{2} \times \left(a\sqrt{2}\right) \times \left(a\sqrt{2} \times 0.866\right) = 0.866a^2.$$

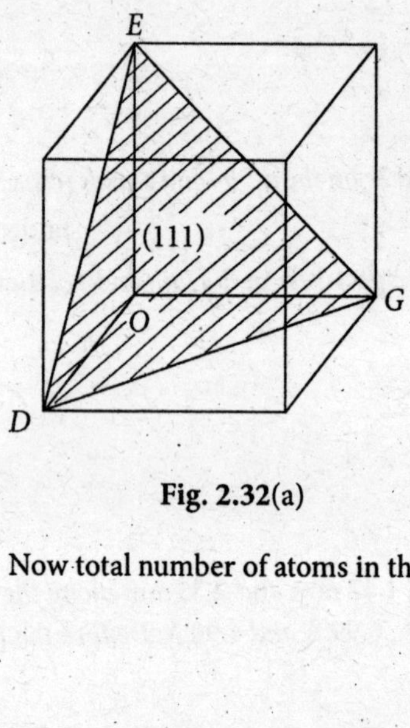

Fig. 2.32(a)

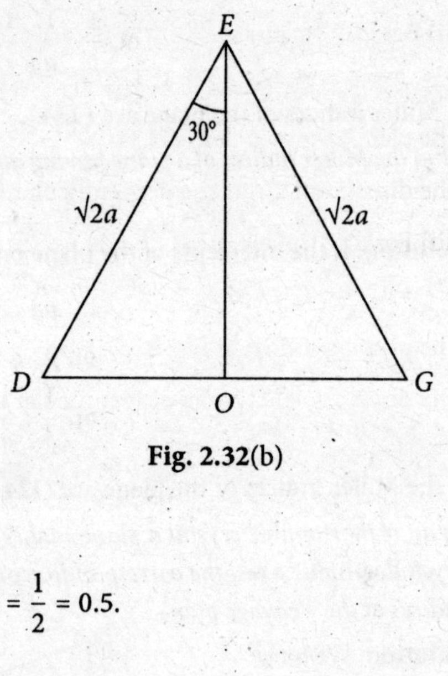

Fig. 2.32(b)

Now total number of atoms in this plane is

$$\left(\frac{1}{360} \times 60\right) \times 3 = \frac{1}{2} = 0.5.$$

Hence, number of atoms per unit area

$$= \frac{0.5}{0.866a^2} = \frac{\mathbf{0.58}}{\boldsymbol{a^2}} \textbf{ atoms/unit area.}$$

2.12. *Draw the following planes and directions in the case of an FCC structure* (i) (100) (ii) (110) (iii) (112).

Solution: (i) For (100)-plane, $h = 1,\ k = 0,\ l = 0$.

The reciprocals are $\frac{1}{1}, \frac{1}{0}, \frac{1}{0}$, i.e., 1, ∞, ∞.

The plane is shown in Fig. 2.33.

The direction [100] is given by the normal to (100)-plane and passing through the origin, i.e., *SR*.

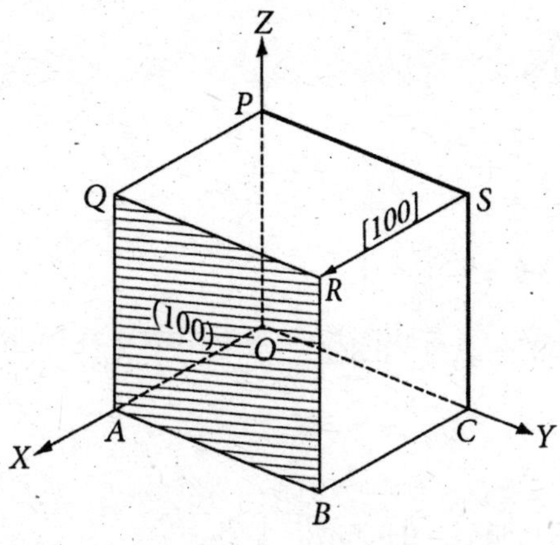

Fig. 2.33: (100)-plane

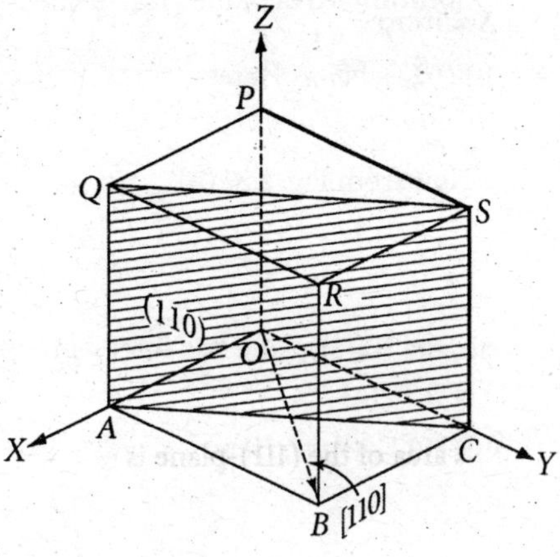

Fig. 2.34: (110)-plane

(ii) For (110)-plane, $h = 1,\ k = 1,\ l = 0$.

$\therefore \frac{1}{h} = 1,\ \frac{1}{k} = 1,\ \frac{1}{l} = \infty.$

The plane is shown in Fig. 2.34.

The direction [110] is the perpendicular to (110)-plane and passing through O, i.e., OB.

(iii) For (112)-plane, $h = 1,\ k = 1,\ l = 2$.

$\therefore \frac{1}{h} = 1,\ \frac{1}{k} = 1,\ \frac{1}{l} = 0.5.$

The plane is shown in Fig. 2.35.

The direction [112] is the perpendicular to (112)-plane and passing through O, i.e., OR.

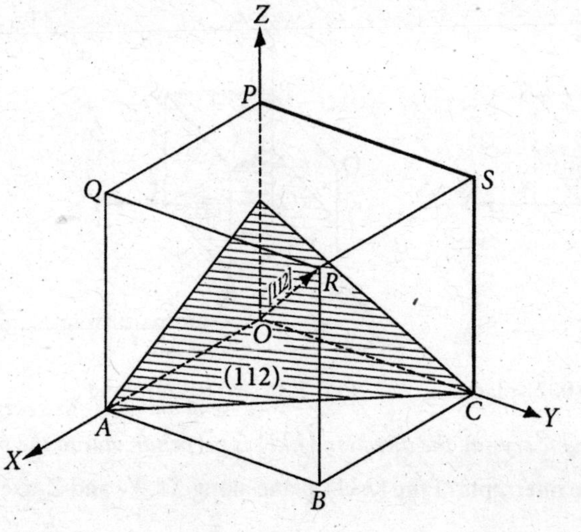

Fig. 2.35: (112)-plane

2.13. *Draw the crystal planes* (002), (012) *and* (022) *in an FCC lattice.*

Solution:

(i) (002) *Plane*: Here

$$h = 0, \therefore \tfrac{1}{h} = \tfrac{1}{0} = \infty$$
$$k = 0, \therefore \tfrac{1}{k} = \tfrac{1}{0} = \infty$$
$$\text{and } l = 2, \therefore \tfrac{1}{l} = \tfrac{1}{2} = 0.5.$$

So, the (002)-plane has intercepts $\infty, \infty, 0.5$ along X-, Y-, Z-directions respectively as shown in Fig. 2.36 (a).

(ii) (012)-*Plane*: Here

$$h = 0, \therefore \tfrac{1}{h} = \tfrac{1}{0} = \infty$$
$$k = 1, \therefore \tfrac{1}{k} = \tfrac{1}{1} = 1$$
$$\text{and } l = 2, \therefore \tfrac{1}{l} = \tfrac{1}{2} = 0.5.$$

So, the (012)-plane has intercepts ∞, 1 and 0.5 along the X-, Y-, Z-directions respectively [Fig. 2.36 (b)].

(iii) (022)-*Plane*: Here

$$h = 0, \therefore \tfrac{1}{h} = \tfrac{1}{0} = \infty$$
$$k = 2, \therefore \tfrac{1}{k} = \tfrac{1}{2} = 0.5$$
$$\text{and } l = 2, \therefore \tfrac{1}{l} = \tfrac{1}{2} = 0.5.$$

So, the (022)-plane has intercepts ∞, 0.5 and 0.5 along the X-, Y- and Z-directions respectively [Fig. 2.36 (c)].

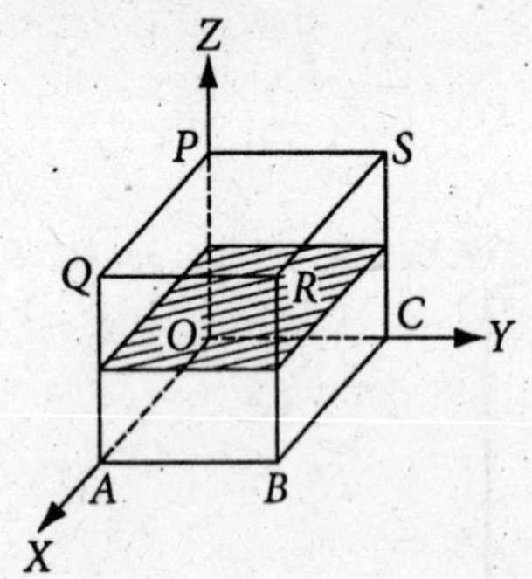

Fig. 2.36: (a) (002)-plane

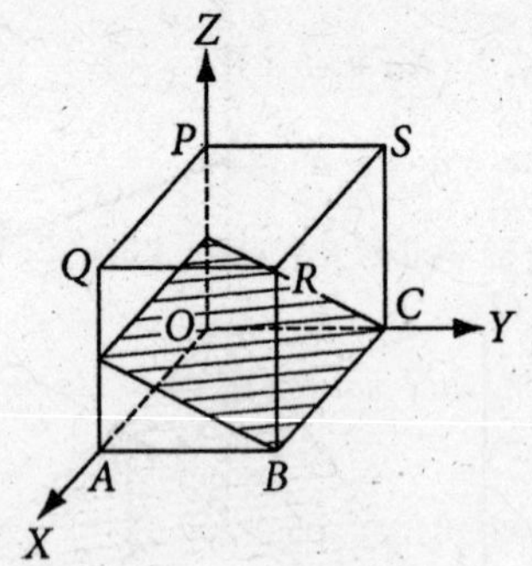

Fig. 2.36: (b) (012)-plane

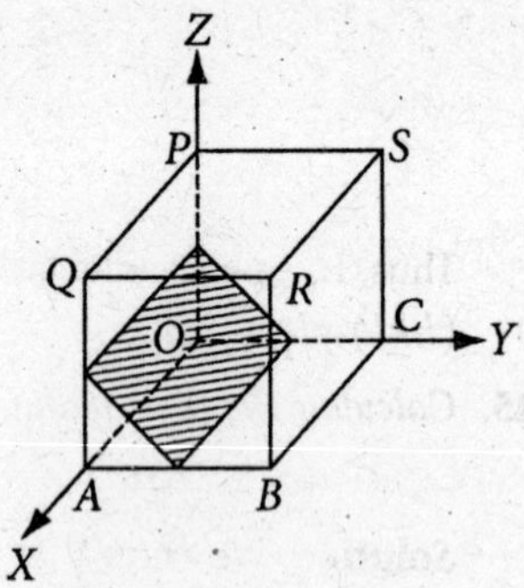

Fig. 2.36: (c) (022)-plane

2.14. *Prove that in a cubic crystal, the direction $[hkl]$ is perpendicular to the plane (hkl).*

Solution: Let the intercepts of the (hkl)-plane along X-, Y- and Z-axes of the cubic unit cell be

$$\vec{R}_1 = \frac{\vec{a}}{h},\ \vec{R}_2 = \frac{\vec{b}}{k} \text{ and } \vec{R}_3 = \frac{\vec{c}}{l},$$

where $\vec{a}$, $\vec{b}$ and $\vec{c}$ are basis vectors along X-, Y- and Z-directions respectively, Fig. 2.37.

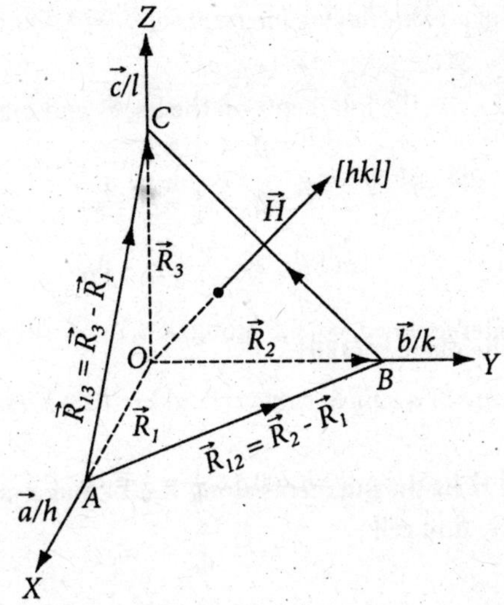

Fig. 2.37: (hkl)-plane in a cubic crystal

The vector $\vec{H} = h\vec{a} + k\vec{b} + l\vec{c}$ represents the direction $[hkl]$ and $\vec{R}_{12}, \vec{R}_{13}$ lie in the plane (hkl).

Now

$$\vec{H} \cdot \vec{R}_{12} = \left(h\vec{a} + k\vec{b} + l\vec{c}\right) \cdot \left(\vec{R}_2 - \vec{R}_1\right) = \left(h\vec{a} + k\vec{b} + l\vec{c}\right) \cdot \left(\frac{\vec{b}}{k} - \frac{\vec{a}}{h}\right)$$

$$= \vec{b} \cdot \vec{b} - \vec{a} \cdot \vec{a} = \vec{a} \cdot \vec{a} - \vec{a} \cdot \vec{a}$$

$$= 0 \; [\because a = b = c \text{ for a cubic crystal}].$$

Also

$$\vec{H} \cdot \vec{R}_{13} = \left(h\vec{a} + k\vec{b} + l\vec{c}\right) \cdot \left(\frac{\vec{c}}{l} - \frac{\vec{a}}{h}\right) = \vec{c} \cdot \vec{c} - \vec{a} \cdot \vec{a}$$

$$= \vec{a} \cdot \vec{a} - \vec{a} \cdot \vec{a} = 0.$$

Thus, $\vec{H}$ is perpendicular to $\vec{R}_{12}$ as well as $\vec{R}_{13}$ which lie on (hkl)-plane. Hence, $\vec{H}$ is perpendicular to (hkl)-plane.

2.15. *Calculate the inter-planar spacing d of planes* (111) *in a simple cubic lattice of side a.*

[WBUT 2003, 2004, 2007]

Solution: We know that the inter-planar spacing d_{hkl} between the consecutive (hkl)-planes is

$$d_{hkl} = \frac{1}{\sqrt{\dfrac{h^2}{a^2} + \dfrac{k^2}{b^2} + \dfrac{l^2}{c^2}}}.$$

In this case, $a = b = c$ and $h = k = l = 1$.

$$\therefore d_{hkl} = \frac{1}{\sqrt{\dfrac{1}{a^2} + \dfrac{1}{a^2} + \dfrac{1}{a^2}}} = \frac{a}{\sqrt{3}}.$$

2.16. *Obtain the Miller indices of a plane having intercepts of a, b/2 and ∞ on the a-, b- and c axes respectively.* **[BU(H) 2004]**

Solution: Let pa, qb and rc be the intercepts on the a-, b- and c-axes respectively, then

$$pa : qb : rc = a : \frac{b}{2} : \infty \text{ or, } p : q : r = 1 : \frac{1}{2} : \infty.$$

$$\therefore \frac{1}{p} : \frac{1}{q} : \frac{1}{r} = \mathbf{1 : 2 : 0}.$$

So, the Miller indices of the plane is **(120)**.

2.17. *The Miller indices of a plane in a simple cubic crystal are* (321). *Find the intercepts of the plane and sketch it.* **[BU(H) 2000]**

Solution: Let pa, qb and rc be the intercepts along X-, Y- and Z-axes respectively, where a, b, c are the primitive vectors of the unit cell.

Then

$$pa : qb : rc = 3a : 2b : c \text{ or, } p : q : r = 3 : 2 : 1.$$

As the crystal is SC, $a = b = c$ and the intercepts on the X-, Y- and Z-axes are 3, 2 and 1 respectively as shown in Fig. 2.38.

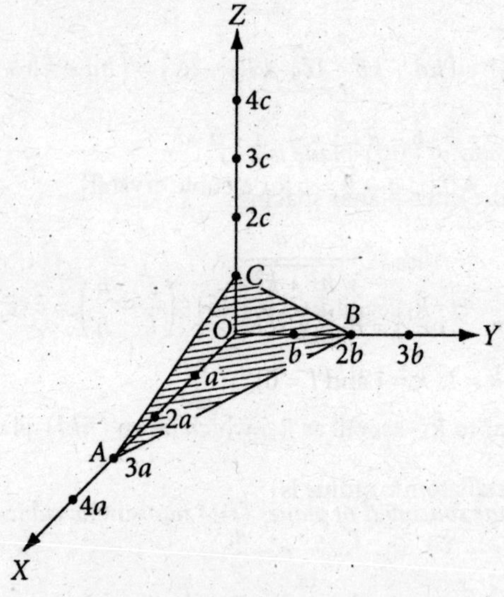

Fig. 2.38: The intercepts of (321)-plane

2.18. *Find the inter-planar spacing for a* (321)*-plane in a, simple cubic lattice whose lattice constant is* 4.2 × 10⁻⁸ cm.

Solution: Here $a = b = c = 4.2 \times 10^{-8}$ cm $= 4.8 \times 10^{-10}$ m and

$$d_{hkl} = \frac{a}{\sqrt{h^2 + k^2 + l^2}} = \frac{4.2 \times 10^{-10}}{\sqrt{3^2 + 2^2 + 1^2}} \quad [\because h = 3,\ k = 2 \text{ and } l = 1]$$

$$= \mathbf{1.1 \times 10^{-10}\ m}.$$

2.19. *In a tetrahedral lattice, $a = b = 2.5$ Å and $c = 1.8$ Å. Find the lattice spacing between (111)-planes.*

Solution: For a given set of parallel planes (hkl) the lattice spacing is given by

$$d_{hkl} = \left[\frac{h^2}{a^2} + \frac{k^2}{b^2} + \frac{l^2}{c^2}\right]^{-1/2}.$$

In this case, $h = 1$, $k = 1$, $l = 1$ and $a = b = 2.5$ Å, $c = 1.8$ Å.

$$\therefore d_{hkl} = \left[\frac{1^2}{(2.5)^2} + \frac{1^2}{(2.5)^2} + \frac{1^2}{(1.8)^2}\right]^{-1/2} = \left[\frac{2}{6.25} + \frac{1}{3.24}\right]^{-1/2} = \textbf{1.26 Å}.$$

2.20. *The distance d_{100} between (100)-planes in BCC structure is 0.232 nm. What is the size of the unit cell? What is the radius of the atom?* **[WBUT 2006]**

Solution: For a BCC crystal the radius of the atom is given by

$$r = \frac{\sqrt{3}a}{4},$$

where a is the lattice constant.

Now for a BCC structure

$$d_{100} = a/2 \text{ or, } 0.232 = \frac{a}{2} \quad [\because d_{100} = 0.232\,\text{nm}]$$

$$\text{or, } a = 0.464\,\text{nm}.$$

$$\therefore r = \frac{\sqrt{3}a}{4} = \frac{\sqrt{3}}{4} \times 0.464 = \mathbf{0.20\,nm}$$

2.21. *The inter-planar spacing of (110)-plane is 2.2 Å for a cubic crystal. Calculate the atomic radius.*

Solution: We have the inter-planar spacing for a cubic crystal between (110)-planes

$$d_{100} = \frac{a}{\sqrt{h^2 + k^2 + l^2}}$$

$$\text{or, } a = d_{100} \times \sqrt{h^2 + k^2 + l^2} = 2.2 \times \sqrt{1^2 + 1^2 + 0^2}.$$

$[\because d_{100} = 2.2$ Å and $h = 1$, $k = 1$ and $l = 0]$

$$\therefore a = 3.11 \text{ Å}.$$

Now for an SC crystal atomic radius is

$$r = \frac{a}{2} = \frac{3.11}{2} = \textbf{1.555 Å}.$$

For a BCC crystal

$$r = \frac{\sqrt{3}}{4}a = 3.11 \times \frac{\sqrt{3}}{4} = \textbf{1.347 Å}.$$

and for an FCC crystal

$$r = \frac{\sqrt{2}}{4}a = 3.11 \times \frac{\sqrt{2}}{4} = \textbf{1.099 Å}.$$

2.22. *In a simple cubic crystal find the ratio of* (a) *spacing of the* (110)- *and* (111)-*planes;* (b) *nearest neighbour distance to the next nearest neighbour distance.*

Solution: (a) For an SC lattice

$$d_{hkl} = \frac{a}{\sqrt{h^2 + k^2 + l^2}}.$$

For (110)-planes

$$d_{110} = \frac{a}{\sqrt{1^1 + 1^2 + 0^2}} = \frac{a}{\sqrt{2}}.$$

For (111)-planes

$$d_{111} = \frac{a}{\sqrt{1^2 + 1^2 + 1^2}} = \frac{a}{\sqrt{3}}.$$

$$\therefore d_{110} : d_{111} = \frac{a}{\sqrt{2}} : \frac{a}{\sqrt{3}} = \frac{1}{\sqrt{2}} : \frac{1}{\sqrt{3}}.$$

(b) In an SC lattice the nearest neighbour distance $d_1 = a$, where a is the side of the unit cube.

The next nearest neighbour is situated at the end of the face diagonal. Hence, the next nearest neighbour distance is

$$d_2 = \sqrt{a^2 + a^2} = \sqrt{2}a.$$

$$\therefore d_1 : d_2 = a : \sqrt{2}a = \mathbf{1 : \sqrt{2}}.$$

2.23. *If the radius of* Ni *atom is* 1.245 Å, *calculate the inter-planar distance of* (100)- *and* (111)-*planes of* Ni *crystal.*

Solution: If a and r be respectively the lattice constant and radius of the Ni atom, then

$$a = 2\sqrt{2}r = \frac{4}{\sqrt{2}}r = \frac{(4 \times 1.245)}{\sqrt{2}} = \mathbf{3.52\ Å}.$$

Now the inter-planar distance between (100)-planes is

$$d_{100} = \frac{a}{\sqrt{1^2 + 0^2 + 0^2}} = a = \mathbf{3.52\ Å}$$

and the inter-planar distance between (111)-planes is

$$d_{111} = \frac{a}{\sqrt{1^2 + 1^2 + 1^2}} = \frac{3.52}{\sqrt{3}} = \mathbf{2.03\ Å}.$$

2.24. *Copper has an* FCC *structure and the atomic radius is* 0.1278 nm. *Calculate the inter-planar spacing for* (111)-*planes.* **[BU(H) 1998]**

Solution: If a be the lattice constant and r the atomic radius, then for FCC cubic crystal,

$$a = \frac{4}{\sqrt{2}}r = \frac{4}{\sqrt{2}} \times 0.1278 \times 10^{-9}\ \text{m}.$$

$$\left[\because r = 0.1278\ \text{nm} = 0.1278 \times 10^{-9}\ \text{m}\right]$$

Now inter-planar spacing for (hkl)-planes is

$$d_{hkl} = \frac{a}{\sqrt{h^2 + k^2 + l^2}}.$$

Here $h = k = l = 1$ and

$$a = \frac{4}{\sqrt{2}} \times 0.1278 \times 10^{-9}\ \text{m}.$$

$$\therefore d_{111} = \frac{\frac{4}{\sqrt{2}} \times 0.1278 \times 10^{-9}}{\sqrt{1^2+1^2+1^2}} = \frac{4 \times 0.1278 \times 10^{-9}}{\sqrt{6}} = \frac{0.5112 \times 10^{-9}}{2.4495}$$

$$= 0.209 \times 10^{-9}\ \text{m} = \mathbf{0.209\ nm.}$$

2.25. *Find the Miller indices for a plane with the intercepts $6\vec{a}, 2\vec{b}, 3\vec{c}$.* **[BU(H) 2003]**

***Solution*:** If pa, qb and rc be the intercepts of the plane along X-, Y- and Z-axes respectively, then

$$pa : qb : rc = 6a : 2b : 3c$$

$$\text{or, } p : q : r = 6 : 2 : 3$$

$$\text{or, } \frac{1}{p} : \frac{1}{q} : \frac{1}{r} = \frac{1}{6} : \frac{1}{2} : \frac{1}{3} = 1 : 3 : 2$$

$$\text{or, } h : k : l = 1 : 3 : 2.$$

$\therefore$ the required Miller indices of the plane are **(132).**

2.26. *The density of BCC iron is 7900 kg-m*^{3} *and its atomic weight is 56. Calculate the side of the unit cell and the interatomic spacing.* **[BU(H) 1993]**

***Solution*:** The side of a unit cell of a BCC lattice is

$$a = \left(\frac{nM}{\rho N}\right)^{1/3},$$

where n = number of atoms per unit cell

M = atomic weight

ρ = the density of the material and

N = Avogadro number.

Here $n = 2$, $M = 56$, $\rho = 7900\ \text{kg-m}^3$ and $N = 6.023 \times 10^{26}/\text{kg-mol}$.

$$\therefore a = \left[\frac{2 \times 56}{7900 \times 6.023 \times 10^{26}}\right]^{1/3} = \left[\frac{224}{47581.7 \times 10^{26}}\right]^{1/3}$$

$$\simeq 2.86 \times 10^{-10}\ \text{m} = \mathbf{2.86\ Å.}$$

Now interatomic spacing of a BCC lattice is

$$a_1 = \frac{\sqrt{3}}{2} a = \frac{\sqrt{3}}{2} \times 2.86 = \frac{1.732 \times 2.86}{2} = \mathbf{2.48\ Å}$$

2.27. *Write down Miller indices of a plane having intercepts $2a$, ∞ and $c/2$ on the a-, b- and c-axes respectively.* **[BU(H) 2002]**

***Solution*:** Let pa, qb and rc be the intercepts of the plane along a-, b- and c-axes respectively.

Then

$$pa : qb : rc = 2a : \infty : \frac{c}{2}$$

$$\text{or, } p : q : r = 2 : \infty : \frac{1}{2}$$

$$\text{or, } \frac{1}{p} : \frac{1}{q} : \frac{1}{r} = \frac{1}{2} : \frac{1}{\infty} : 2$$

$$\text{or, } h : k : l = 1 : 0 : 4.$$

Hence, the Miller indices of the plane is **(104)**.

2.28. *Find the Miller indices of a plane having intercepts of 8a, 4b and 2c on the a-, b- and c-axes respectively.*

[BU(H) 2001]

Solution: If pa, qb and rc be the intercepts of the plane along a-, b- and c-axes respectively, then

$$pa : qb : rc = 8a : 4b : 2c$$

$$\text{or, } p : q : r = 8 : 4 : 2$$

$$\text{or, } \frac{1}{p} : \frac{1}{q} : \frac{1}{r} = \frac{1}{8} : \frac{1}{4} : \frac{1}{2} = 1 : 2 : 4$$

$$\text{or, } h : k : l = 1 : 2 : 4.$$

So, the Miller indices of the plane is **(124)**.

2.29. *Show that the maximum radius of the sphere that can just fit into the void at the body centre of the FCC structure coordinated by the facial atoms is 0.414r, where r is the radius of the atom.*

Solution: Let R be the radius of the sphere that can just fit into the void as shown in Fig. 2.39.

$$\therefore 2r + 2R = a \text{ or, } r + R = \frac{a}{2} \text{ or, } R = \frac{a}{2} - r.$$

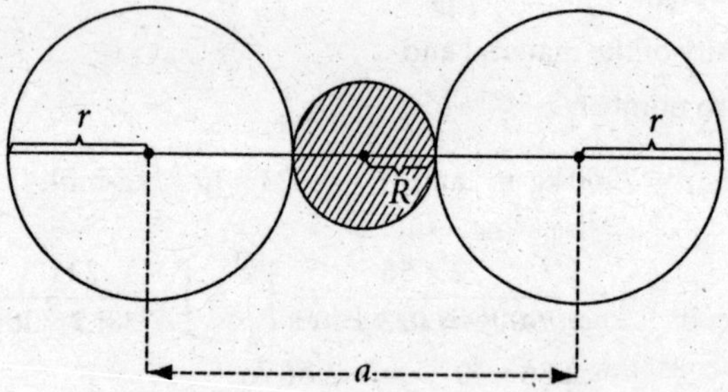

Fig. 2.39

For FCC structure

$$a = \frac{4r}{\sqrt{2}}.$$

$$\therefore R = \frac{4r}{2\sqrt{2}} - r = \frac{2r}{\sqrt{2}} - r = r\left(\sqrt{2} - 1\right) = \mathbf{0.414r}.$$

2.30. *Copper has FCC structure and the atomic radius is 0.1278 nm. Calculate the inter-planar spacing for (110)- and (212)-planes.*

Solution: We know that inter-planar spacing d_{hkl} of a lattice having cubic structure of lattice constant a is

$$d = \frac{a}{(h^2 + k^2 + l^2)^{1/2}}.$$

For FCC structure $a = \frac{4r}{\sqrt{2}}$, where the atomic radius $r = 0.1278 \times 10^{-9}$ m.

$$\therefore a = \frac{4 \times 0.1278 \times 10^{-9}}{\sqrt{2}} = 0.3615\,\text{nm}.$$

Now inter-planar spacing for (110)-plane is

$$d_{110} = \frac{0.3615}{(1^2 + 1^2 + 0^2)^{1/2}} = \frac{0.3615}{\sqrt{2}}\,\text{nm} = \mathbf{0.2556\,nm}.$$

Inter-planar spacing for (212)-plane is

$$d_{212} = \frac{0.3615}{(2^2 + 1^2 + 2^2)^{1/2}} = \frac{0.3615}{9^{1/2}} = \mathbf{0.1205\,nm}.$$

2.31. *Show that in simple cubic lattice,*

$$d_{100} : d_{110} : d_{111} = 1 : 0.71 : 0.58.$$ **[cf. CU(H) 2005]**

Solution: The inter-planar spacing for (hkl)-planes in an SC structure of lattice constant a is

$$d = \frac{a}{(h^2 + k^2 + l^2)^{1/2}}.$$

$$\therefore d_{100} = \frac{a}{\sqrt{1^2 + 0^2 + 0^2}} = \frac{a}{1} = a,$$

$$d_{110} = \frac{a}{(1^2 + 1^2 + 0^2)^{1/2}} = \frac{a}{\sqrt{2}}$$

$$\text{and } d_{111} = \frac{a}{(1^2 + 1^2 + 1^2)} = \frac{a}{\sqrt{3}}.$$

$$\therefore d_{100} : d_{110} : d_{111} = 1 : \frac{1}{\sqrt{2}} : \frac{1}{\sqrt{3}} = \mathbf{1 : 0.71 : 0.58}.$$

2.32. KBr *crystal has a cubic structure. If its density is* 2.75×10^3 kg-m^3 *and molecular weight is* 119.01, *calculate the lattice constant.* **[CU(H) 2000, '06]**

Solution: Given molecular weight of KBr is

$$= 119.01\,\text{gm-wt} = 119.01 \times 10^{-3}\,\text{kg-wt}.$$

Mass of KBr molecule

$$M = \frac{\text{Molecular weight}}{\text{Avogadro number}} = \frac{119.01 \times 10^{-3}}{6.023 \times 10^{23}}\,\text{kg} = 19.76 \times 10^{-26}\,\text{kg}.$$

Number of KBr molecule/volume

$$= \frac{\text{Density}}{M} = \frac{2.75 \times 10^3}{19.76 \times 10^{-26}} = 0.139 \times 10^{29}.$$

The number of atoms/volume of KBr

$$= 2 \times 0.139 \times 10^{29} \; [\because \text{KBr is diatomic}]$$
$$= 27.8 \times 10^{27} \text{ atoms/m}^3.$$

If a be the distance between two adjacent atoms, the volume a^3 will contain 1 atom due to the cubic structure.

So, the total number of atoms/volume = $\frac{1}{a^3}$.

$$\therefore \frac{1}{a^3} = 27.8 \times 10^{27} \text{ or, } a = 3.30 \times 10^{-16} \text{ m.}$$

$\therefore$ lattice constant, $2a = 6.6 \times 10^{-10}$ m = **6.6 Å.**

2.33. *Calculate the fraction of volume unoccupied in the unit cell of body-centred cubic lattice.*

Solution: If a be the lattice constant, then volume of the unit cell is $V = a^3$.

If r be the radius of the atom, then for BCC structure, $4r = \sqrt{3}a$ or, $r = 3/4a$.

Number of atoms in the BCC unit cell = $1 + 8 \times \frac{1}{8} = 2$.

$$\therefore \text{volume occupied by atom} = 2 \times \frac{4}{3}\pi r^3 = \frac{8}{3}\pi\frac{3\sqrt{3}}{64}a^3 = \frac{\sqrt{3}}{8}\pi a^3.$$

$$\therefore \text{fraction of volume unoccupied in the unit cell is } (a^3 - \sqrt{3}\pi a^3/8)a^3 = \frac{8-\sqrt{3}\pi}{8}.$$

2.34. *A solid with* FCC *structure has a lattice crystal* 4.0 Å. *Each lattice point has an atom of mass* 4.0×10^{-26} kg. *Calculate* (i) *the* (*mass*) *density of the solid,* (ii) *the number of conduction electrons in* 1 m^3 *volume of the solid, if each of the solid contributes two electrons to the conduction band.*

Solution: A unit cell of FCC lattice has 4 atoms. So mass of unit cell m = 4× mass of one atom at the lattice point = $4 \times 4 \times 10^{-26} = 1.6 \times 10^{-25}$ kg.

Volume of the unit cell, $V = a^3 = 64 \times 10^{-30}$ m^3.

$$\therefore \text{mass density of solid} = \frac{m}{V} = \frac{1.6 \times 10^{-25}}{64 \times 10^{-30}} \text{ kg/m}^3 = \mathbf{2.5 \times 10^3 \text{ kg/m}^3}.$$

(ii) Number of atoms per unit volume

$$= \frac{\text{No. of atoms in unit cell}}{\text{Volume of unit cell}} = \frac{4}{64 \times 10^{-30}} = 6.25 \times 10^{28} \text{ atoms/m}^3$$

Each atom contributes 2 electrons to the conduction band.

$\therefore$ no. of conduction electrons in 1 m^3 volume, $N = 2 \times 6.25 \times 10^{28} = \mathbf{1.25 \times 10^{29}}$.

2.35. *An* FCC *lattice is formed by atoms having radius r.* (a) *Obtain the expression, in terms of r, for the areal density of atoms (atoms/m*^{2}*) for the* (111) *plane.* (b) *Obtain the relationship between r and the maximum radius R of a sphere that can just occupy the centre of the unit cell* (Fig. 2.40).

Solution: (a) We know that for FCC lattice, $4r = \sqrt{2}a$ or, $a = 2\sqrt{2}r$. No. of corner atoms covered by the plane = $3 \times \frac{1}{6} = \frac{1}{2}$ atom.

No. of face-centred atoms contained by the plane = $\frac{1}{2} \times 3 = \frac{3}{2}$ atom.

Total no. of atoms contained by the plane = $\frac{1}{2} + \frac{3}{2} = 2$.

$$\text{Areal density (atoms/m}^3) = \frac{\text{No. of atoms}}{\text{Area of the plane}} = \frac{2}{\sqrt{3}a^2/2} = \frac{2}{\sqrt{3}(8r^2)/2} = \frac{1}{2\sqrt{3}r^2}.$$

(b) Let R be the maximum radius of the sphere. From Fig 2.40, we have $OD = \frac{1}{3}AD = \frac{1}{3}\frac{\sqrt{3}}{2}\sqrt{2}a = \frac{1}{\sqrt{6}}a$.

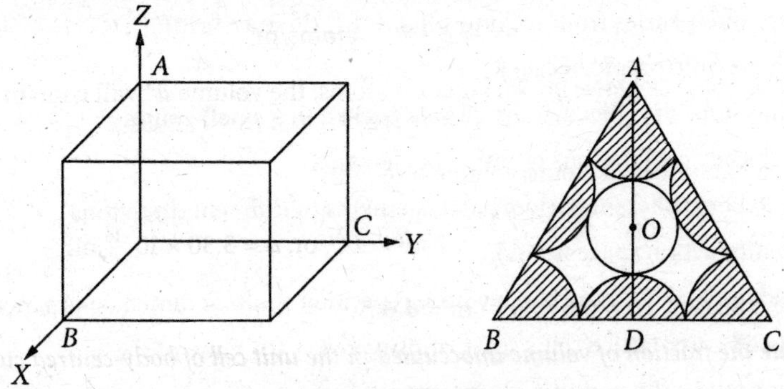

Fig. 2.40: Figure for problem 2.35

Also $OD = R + r$.

$$\therefore \frac{1}{\sqrt{6}}a = R + r \text{ or, } \frac{1}{\sqrt{6}}2\sqrt{2}r = r + r \text{ or, } \left(\frac{2}{\sqrt{3}} - 1\right) r = R \text{ or, } \left(2 - \sqrt{3}\right) r = \sqrt{3}R$$

This is the relationship between r and R.

EXERCISE

2(A): Objective (MCQ) Type Questions

Select the correct answer.

2.1. Solids have definite shape and size because
- (a) density of solids is larger compared to liquid or gas
- (b) the atoms or molecules of solid are bound with strong forces of attraction
- (c) solids are generally of high rigidity
- (d) solids can be transformed into liquids and gases.

2.2. Most of the solids prefer to be in the crystalline state, because
- (a) volume is generally minimum in crystalline state than liquid or gas state
- (b) crystalline state is the low energy state
- (c) crystal has definite shape and size
- (d) none of these.

2.3. In amorphous solids the atoms and molecules are
- (a) regularly arranged in periodic manner
- (b) randomly arranged
- (c) partly regularly and partly randomly arranged
- (d) none of these.

2.4. The physical properties of solids depend on
- (a) shape and size of the solids
- (b) the atoms, ions or molecules forming the solid
- (c) the presence of impurities
- (d) the irregularities in the lattice.

2.5. Amorphous solids are isotropic because
(a) they do not have directional property (b) they exhibit short range order
(c) their colour varies from solid to solid (d) they are brittle.

2.6. Crystals are anisotropic because
(a) atoms, ions or molecules are closely packed in a small volume
(b) often they contain ions of different elements
(c) crystal has different periodic arrangements in different directions
(d) the interaction space is void.

2.7. The fundamental building block whose repetition in three dimensions generates a crystal is called
(a) unit cell (b) lattice cell (c) primitive cell (d) none of these.

2.8. In a plane lattice the periodic arrangement of atoms occur
(a) in three dimensions (b) in two dimensions (c) in a line (d) both in 2 and 3 dimensions.

2.9. The number of basic crystal system is
(a) 2 (b) 3 (c) 4 (d) 7.

2.10. Basis is defined as
(a) an atom about a lattice point (b) a group of atoms about a lattice point
(c) one or a group of atoms about a lattice point (d) none of these.

2.11. Lattice + basis is
(a) unit cell (b) primitive cell (c) crystal structure (d) molecular structure.

2.12. The number of Bravais lattices in three dimensions are
(a) 7 (b) 9 (c) 14 (d) 19.

2.13. The magnitudes of basis vectors $\vec{a}, \vec{b}, \vec{c}$ are called
(a) lattice parameters (b) lattice vectors (c) crystal vectors (d) unit vectors.

2.14. In a primitive unit cell, the number of lattice points is
(a) 2 (b) 5 (c) 4 (d) 1.

2.15. For a cubic crystal systems,
(a) $a = b = c;\ \alpha = 90°, \beta \neq \gamma$ (b) $a \neq b = c;\ \alpha = \beta = \gamma = 90°$
(c) $a = b = c;\ \alpha = \beta = \gamma = 90°$ (d) $a = b \neq c;\ \alpha = \beta \neq \gamma$.

2.16. If $\vec{a}, \vec{b}$ and $\vec{c}$ represent three basis vectors, then the volume of Wigner-Seitz cell is
(a) $|\vec{a} \times \vec{b} \times \vec{c}|$ (b) $|(\vec{a} + \vec{b}) \times \vec{c}|$ (c) $|\vec{a} \cdot (\vec{b} + \vec{c})|$ (d) $|\vec{a} \times \vec{b} \cdot \vec{c}|$.

2.17. The effective number of lattice points in an SC lattice is
(a) 4 (b) 2 (c) 5 (d) 1.

2.18. In a BCC structure, the effective number of lattice points per unit cell is
(a) 2 (b) 4 (c) 1 (d) 3.

2.19. In an FCC lattice, the number lattice point per unit cell is
(a) 2 (b) 4 (c) 5 (d) 1.

2.20. NaCl crystal is

(a) SC (b) BCC (c) FCC (d) none of these.

2.21. If the number of Bravais lattice is three, then the crystal is

(a) monoclinic (b) hexagonal (c) cubic (d) tetragonal.

2.22. The coordination number of an SC structure is

(a) 6 (b) 12 (c) 8 (d) 4.

2.23. The FCC structure is

(a) primitive (b) non-primitive

(c) may be either primitive or non-primitive (d) none of these.

2.24. The unit cell of a BCC structure is

(a) primitive (b) non-primitive

(c) may be either primitive or non-primitive (d) none of these.

2.25. Miller indices (hkl) represent

(a) a set of parallel planes (b) a particular plane

(c) a set of arbitrarily oriented planes (d) none of these.

2.26. The Miller indices of a plane having intercepts 2, ∞, ∞ units on the $\vec{a}$, $\vec{b}$ and $\vec{c}$ axes respectively are

(a) 101 (b) 001 (c) 200 (d) 020.

2.27. Miller indices of a plane having intercepts 2, 3, 4 units along X-, Y- and Z-axes are

(a) (643) (b) (436) (c) (231). **[WBUT 2006]**

2.28. The inter-planar distance d_{100} for (100)-plane of SC lattice is

(a) $\frac{a}{\sqrt{3}}$ (b) $\frac{a}{\sqrt{2}}$ (c) a (d) $\frac{a}{\sqrt{5}}$.

2.29. The inter-planar distance d_{111} for a (111)-plane of an SC cubic lattice is

(a) a (b) $\frac{a}{\sqrt{3}}$ (c) $\frac{a}{\sqrt{2}}$ (d) $\frac{a}{\sqrt{0.5}}$.

2.30. The inter-planar distance (d_{100}) for (100)-plane of a BCC lattice is

(a) $\frac{a}{\sqrt{2}}$ (b) $\frac{a}{\sqrt{3}}$ (c) a (d) $\frac{a}{2}$.

2.31. The inter-planar spacing d_{111} for (111)-plane of an FCC lattice is

(a) $\frac{a}{\sqrt{2}}$ (b) a (c) $\frac{a}{\sqrt{3}}$ (d) $\frac{a}{2}$.

2.32. If a and r be respectively the lattice constant and radius of an atom in an SC structure, then

(a) $a = \frac{4}{\sqrt{3}}r$ (b) $a = 2r$ (c) $a = \frac{2}{\sqrt{2}}r$ (d) $a = \frac{r}{2}$.

2.33. If a and r be respectively the lattice constant and radius of an atom in a BCC structure, then

(a) $r = \frac{\sqrt{3}}{4}a$ (b) $\frac{\sqrt{2}}{4}a$ (c) a (d) none of these. **[WBUT 2007]**

2.34. For an FCC structure the relation between lattice constant (a) and radius (r) of the atom is

(a) $r = \frac{\sqrt{3}}{4}a$ (b) $r = \frac{a}{2}$ (c) $r = a$ (d) $r = \frac{\sqrt{2}}{4}a$.

2.35. The coordination number of SC structure is

(a) 6 (b) 3 (c) 2 (d) 4.

2.36. The coordination number of BC structure is

(a) 1 (b) 8 (c) 6 (d) 2.

2.37. For FCC cubic structure the coordination number is

(a) 6 (b) 8 (c) 12 (d) 10.

2.38. For SC structure atomic packing factor is

(a) 74 % (b) 52 % (c) 68 % (d) 47 %.

2.39. If the atomic packing factor for a crystal is 68 %, then the structure of the crystal is

(a) SC (b) FCC (c) BCC (d) none of these.

2.40. The atomic packing factor for FCC structure is

(a) 74 % (b) 52 % (c) 68 % (d) 49 %.

2.41. For NaCl crystal having lattice constant 5.6 Å the interatomic distance is

(a) 2.8 Å (b) 1.4 Å (c) 5.6 Å (d) 7.2 Å.

2.42. If for a crystal d_{hkl} = 1.414 nm and a = 1 nm, then the Miller indices of the planes are

(a) (111) (b) (101) (c) (110) (d) (100).

2.43. (111)-plane of a cubic crystal is

(a) perpendicular to X-axis (b) perpendicular to Y-axis

(c) perpendicular to Z-axis (d) none of these. **[WBUT 2005]**

2.44. $d_{100} : d_{110} : d_{111}$ for an SC structure is

(a) $1 : \sqrt{3} : \sqrt{6}$ (b) $\sqrt{2} : \sqrt{5} : \sqrt{6}$ (c) $\sqrt{6} : \sqrt{3} : \sqrt{2}$ (d) $\sqrt{2} : \sqrt{6} : \sqrt{5}$.

2.45. The (111)- and (222)-planes in a lattice are

(a) perpendicular to each other (b) parallel to each other

(c) inclined at 60° to each other (d) none of these.

2.46. The element which crystallizes in a BCC structure is

(a) Sodium (b) Zinc (c) Copper (d) Iron.

2.47. If the interatomic distance of NaCl crystal is 0.28 nm, then the lattice parameter is

(a) 0.14 nm (b) 0.28 nm (c) 0.56 nm (d) none of these.

2.48. The crystal structure of aluminium is

(a) SC (b) BCC (c) FCC (d) HCP.

2.49. If the primitives $a \neq b \neq c$ and interfacial angles $\alpha \neq \beta \neq \gamma \neq 90°$, then it belongs to the system

(a) monoclinic (b) orthorhombic (c) tetragonal (d) cubic.

2.50. In a cubic unit cell, the angles between the normals to the planes (111) and (101) is

(a) 50° (b) 64.5° (c) 54.69° (d) 70°.

2.51. Volume of a unit cell in an FCC crystal is

(a) $16\sqrt{2}r^3$ (b) $4\sqrt{2}r^3$ (c) $\frac{16\sqrt{2}}{r^3}$ (d) $2\sqrt{2}r^3$.

2.52. The density of an FCC structure is

(a) $\rho = \frac{M}{Na^3}$ (b) $\rho = \frac{MN}{a^3}$ (c) $\rho = \frac{4M}{Na^3}$ (d) $\rho = \frac{Ma^3}{4N}$,

the different letters have their usual meanings.

2.53. The nearest neighbour distance in case of BCC structure is

(a) $\frac{a\sqrt{3}}{2}$ (b) $\frac{a\sqrt{2}}{\sqrt{2}}$ (c) $\frac{2a}{\sqrt{3}}$ (d) $\frac{2a}{\sqrt{3}}$. **[WBUT 2008]**

2.54. The primitives of a crystal are 1.2 Å, 1.8 Å, 2 Å, 2 Å along the three axes, where a plane with Miller indices (231) cuts intercept 1.2 Å along x-axis. What will be the lengths of intercept along y- and z-axes?

(a) 1.5 Å and 5 Å (b) 1.9 Å and 2.5 Å (c) 1.2 Å and 4 Å (d) 1.5 Å and 4 Å. **[WBUT 2008]**

2.55. A simple cubic crystal with lattice parameter a_c undergoes transition into a tetragonal structure with lattice parameter $a_t = b_t = \sqrt{2}a_c$ and $c_t = 2a_c$ below a certain temperature. The ratio of inter-planar spacings of (101)-planes for the cubic and the tetragonal structure is

(a) $\sqrt{1/6}$ (b) 1/6 (c) $\sqrt{3/8}$ (d) 3/8. **[GATE 2012]**

2.56. A crystal grows by stacking of unit cells of 10 × 20 × 5 nm size. How many unit cells will make a crystal of 1 cm^3 volume?

(a) 10^6 (b) 10^9 (c) 10^{12} (d) 10^{18}. **[cf. NET June 2013]**

[*Hint.* 1 $cm^3 \div (10 \text{ nm} \times 20 \text{ nm} \times 5 \text{ nm}) = 10^{18}$]

2.57. Which of the following statements is correct for NaCl crystal structure?

(a) It is a simple cubic lattice with one atom basis

(b) It is a face-centred lattice with one atom basis

(c) It is a simple cubic lattice with two atoms basis

(d) It is a face-centred cubic lattice with two atoms basis. **[JAM 2005]**

2.58. Metallic copper is known to form cubic crystal and the lattice constant is measured from X-ray diffraction study to be about 0.36 nm. If the specific gravity of copper is 8.96 and its atomic weight is 63.5, one can conclude that

(a) the crystals are simple cubic type

(b) the crystals are BCC type

(c) the crystals are FCC type

(d) the crystals are mixture of FCC and BCC type

(e) there is insufficient data to distinguish between the previous questions.

2.59. Consider the atomic packing fraction (APF) of the following crystals structure:

P. Simple cubic Q. Body-centred cubic R. Foce-centred cubic S. Diamond T. Hexagon closed pocket.

Which of the above structures have equal APFs?

(a) P and Q (b) S and T (c) R and S (d) R and T.

2.60. A cubic crystal has lattice constant 4.3 Å and density 963 kg m^{-3}. If its atomic weight is 12, then what type of cubic unit cell does form?

(a) SC (b) BCC (c) FCC (d) None of these.

ANSWERS

2.1. (b) **2.2.** (b) **2.3.** (b) **2.4.** (b) **2.5.** (a) **2.6.** (c) **2.7.** (a) **2.8.** (b) **2.9.** (d) **2.10.** (c) **2.11.** (c) **2.12.** (c) **2.13.** (b) **2.14.** (d) **2.15.** (c) **2.16.** (d) **2.17.** (d) **2.18.** (a) **2.19.** (b) **2.20.** (c) **2.21.** (c) **2.22.** (d) **2.23.** (b) **2.24.** (b) **2.25.** (a) **2.26.** (c) **2.27.** (a) **2.28.** (c) **2.29.** (b) **2.30.** (d) **2.31.** (c) **2.32.** (b) **2.33.** (a) **2.34.** (d) **2.35.** (a) **2.36.** (b) **2.37.** (c) **2.38.** (b) **2.39.** (c) **2.40.** (a) **2.41.** (a) **2.42.** (c) **2.43.** (d) **2.44.** (c) **2.45.** (b) **2.46.** (a) **2.47.** (b) **2.48.** (c) **2.49.** (a) **2.50.** (c) **2.51.** (a) **2.52.** (c) **2.53.** (a) **2.54.** (c) **2.55.** (c) **2.56.** (d) **2.57.** (d) **2.58.** (c) **2.59.** (d) **2.60.** (b).

2(B): Short Answer Type Questions

2.1. Why most of the solids prefer to be in the crystalline state? Why the solids have definite shape and size?

2.2. Why crystalline solids are isotropic in nature but the amorphous are not?

2.3. Distinguish between crystalline and amorphous solids. **[CU(H) 1996]**

2.4. What do you mean by crystal lattice? What are plane lattice and space lattice?

2.5. Explain the terms *lattice translation operation* and *fundamental translation vectors.* When are the translation vectors are called *primitive*? **[BU(H) 1999]**

2.6. What is unit cell? **[WBUT (Old) 2005]**

2.7. If a unit cell has the following characteristics $a = b = 10.5$ Å, $c = 6$ Å, $\alpha = \beta = \gamma = 90°$, identify to which crystal system does the unit cell belong. **[BU(H) 2003]**

2.8. What is a Wigner-Seitz primitive cell? How can it be constructed?

2.9. What is the difference between a primitive cell and unit cell? Show that all primitive cells are unit cells but all unit cells may or may not be primitive cell.

2.10. What are Bravais lattice? What is the number of Bravais lattice in three dimensions?

2.11. What is the basis of dividing the crystal systems into different types?

2.12. Draw the diagrams of SC, FCC and BCC structures.

2.13. Find the number of atoms per unit cell of SC, FCC and BCC types of crystals.

2.14. What is meant by Miller indices. Explain with examples. **[WBUT 2007, '03]**

2.15. What is the utility of Miller indices? **[CU(H) 1990]**

2.16. Define a crystal lattice. What is a primitive cell? Is it unique? **[BU(H) 1997]**

2.17. What are basis vectors? How many different *Bravais lattices* are shown in crystal structure? **[BU(H) 2006]**

2.18. For an FCC lattice what are

(a) the volume of the primitive cell (b) lattice points per unit volume

(c) nearest neighbour distance and (d) the number of nearest neighbours? **[BU(H) 1999]**

2.19. Deduce the inter-planar spacing of a simple cubic lattice of side a. **[WBUT 2007]**

2.20. Show that in a cubic crystal of side a the inter-planar spacing between consecutive parallel planes of Miller indices (hkl) is $d_{hkl} = \frac{a}{\sqrt{h^2+k^2+l^2}}$. **[WBUT 2005, 2004]**

2.21. Calculate atomic packing fraction and atoms per unit cell in crystals having

(a) simple cubic

(b) body-centred cubic and

(c) face-centred cubic structure considering the atoms as hard spheres. **[BU(H) 2003; WBUT 2003]**

2.22. Calculate atomic radius for SC, FCC and BCC structures.

2.23. What do you mean by coordination number of atoms. Calculate the coordination number of SC, FCC and BCC structure.

2.24. Prove that in a cubic crystal, the direction $[hkl]$ is perpendicular to the plane (hkl).

2.25. Write down short note on Bravais lattice and Miller indices. **[WBUT 2003]**

2.26. What do you mean by lattice? How many types of Bravais lattice are there? Show the possible cubic structures with a schematic drawing. [CU(H) 2012]

2.27. Show that for a cubic lattice, the lattice constant a given by

$$a = \sqrt{\frac{nM}{N\rho}}$$

where the symbols have their usual meaning.

2.28. Obtain an expression for packing fraction of a crystal with (i) SC (ii) BCC. [Kakatia U 2015]

2.29. For a simple cubic lattice, compare the density of points in (111) and (110) planes. [BU(H) 2011]

2.30. Diamond crystallites are in FCC Bravais lattice with $C(000)$ and $C(\frac{1}{4}\frac{1}{4}\frac{1}{4})$ as basis. Build the crystal. [CU(H) 2003]

2.31. Find the Miller indices for planes with intercepts of $(6a, 2b, 3c)$ and $(a, 2b, \infty)$. [CU(H) 2003]

2.32. Why are most solids crystalline in nature?

2(C): Numerical Problems

2.1. Lead is face-centred cubic with an atomic radius $r = 1.746$ Å. Find the inter-planar spacing for (220)-plane. [WBUT 2005]

2.2. Aluminium has FCC structure. If the unit cell dimension and atomic weight of Aluminium is 4.048 Å and 26.98 respectively, then calculate its density. [*Ans.* 2700 kg/m^3]

2.3. In a cubic unit cell, find the angle between the normals to the planes (111) and (101).

2.4. Copper has an FCC structure and the atomic radius is 0.1278 nm. Calculate the inter-planar spacing for (111)-planes. [BU(H) 1998]

[*Ans.* 0.209 nm]

2.5. Silicon crystallizes in the diamond cubic structure. The radius of silicon atom is 0.1176 nm. The atomic weight and density of silicon are respectively 28.09 and 2.3×10^3 kg/m^3. What is the number of atoms present in a unit cell. [*Ans.* 8]

2.6. Compare the densities of lattice points in (111)- and (110)-planes in an SC lattice.

[*Ans.* $\rho_{111} : \rho_{110} = \sqrt{2} : \sqrt{3}$]

2.7. Aluminium is an FCC crystal with lattice constant $a = 0.405$ nm.

(a) How many unit cells are there in an aluminium foil 0.005 cm thick and side 25 cm square?

(b) If its weight is 0.0085 kg, then how many atoms are present?

(c) What is the number of atoms per unit cell? [*Ans.* 4.7×10^{22}; 1.9×10^{23}; 4]

2.8. Copper has FCC structure having atomic radius 1.278 Å. Find its density. [*Ans.* 8.94 gm/cc]

2.9. Calculate the inter-planar spacing for a (321)-plane in an SC lattice whose lattice constant is 4.2×10^{-8} cm. [*Ans.* 1.1×10^{-8} cm]

2.10. In a general lattice, $a = b = 2.5$ Å, $c = 1.8$ Å, what is the lattice spacing between (111)-planes.

[*Ans.* 3.96 Å]

2.11. A lattice plane in an SC structure cuts the $\vec{a}$, $\vec{b}$ and $\vec{c}$ axes at distances $6a$, $4b$ and $3c$. Find the Miller indices of the plane. [*Ans.* (234)]

2.12. Find the minimum distance of the Na and Cl^- ions in NaCl crystal. [Given, atomic weights of Na and Cl are 23 and 35.5, density of NaCl is 2.17 gm/cm^3.] **[CU(II) 2017]**

2.13. KBr crystal has a cubic structure. Assuming it to be an ionic crystal, calculate its lattice constants. [Given, density = 2.75×10^3 kg m^{-3}, Molecular weight = 119.0] **[CU(H) 2016]**

2.14. Show that in a cubic crystal the distance between adjacent planes with Miller indices (h, k, l) is given by

$$d_{hkl} = \frac{a}{\sqrt{h^2 + k^2 + l^2}}$$

where a is the lattice constant. **[CU(H) 2016, 2013, 2011]**

2.15. (a) Find the Miller indices for planes in each of the following sets which intercept $\vec{a}$, $\vec{b}$ and $\vec{c}$ axes at (i) $3a, 3b, 2c$ (ii) $a, 25, \alpha$.

(b) A three-dimensional lattice has the basis vectors $\vec{a} = 2\vec{x}$; $\vec{b} = \vec{x} + 2\vec{y}$; $\vec{c} = \vec{z}$. Find the reciprocal lattice vectors. **[BU(H) 2007]**

2.16. Draw a BCC lattice structure. Show that it has packing fraction of 0.68. Compare it with that of a simple cubic lattice structure. **[CU (H) 2015]**

2.17. Draw an FCC structure. Show that it has a packing fraction 0.74. **[CU(H) 2014]**

2.18. (a) Find out the number of atoms per square millimetre on a plane (100) of lead whose interatomic distance is 3.449 Å. Lead has FCC structure.

(b) In case of certain crystal the maxima of reflected X-rays are obtained at glancing angles $5°, 12°, 7°24''$ and $9°1'$ respectively from three different reflecting planes. Determine the type of cubic crystal.

[Pune U 2013]

2.19. (a) Calculate the Miller indices of crystal planes which cut through the crystal axes at (i) $(6a, 3n, 3c)$ and (ii) $(2a, 3b, 3c)$.

(b) The distance between (111) planes in a face centred cubic crystal is 2 Å. Determine the lattice parameter and atomic diameter. **[Pune U 2013]**

2.20. (a) Explain Miller indices in details and derive formula for inter-planar distance.

(b) Find out Miller indices for intercepts $x = 3a, y = 2b, z = 1c$. **[Bhabnagar U 2017]**

2.21. Show that for an orthorhombic crystal with three perpendicular axes of unequal lengths

$$d = \frac{1}{\sqrt{(h/a)^2 + (k+b)^2 + (l+c)^2}}$$

where the symbols have their usual meanings.

2.22. How may nearest neighbours are there in a simple cubic lattice? Write down their coordinates.

[KU(H) 2013]

2.23. Mention the type of variation of a cubic crystal. Draw the structure of KCl crystal and explain.

[KU(H) 2013]

2(D): Broad Answer Type Questions

2.1. (a) Define unit cell.

(b) Show that the spacing d of planes (hkl) in an SC lattice of side a is $d = \frac{a}{\sqrt{h^2+k^2+l^2}}$. **[CU(H) 2008]**

(c) Lead is face-centred cubic with an atomic radius of $r = 1.746$ Å. Find the inter-planar spacing for (220)-planes. **[WBUT 2005]**

2.2. (a) Show that in a cubic crystal of side a the spacing between consecutive parallel planes of Miller indices (hkl) is $d_{hkl} = \frac{a}{\sqrt{h^2+k^2+l^2}}$.

(b) Copper has FCC structure and atomic radius is 0.1278 nm. Calculate the density and the inter-planar spacing for (321) planes. Take atomic weight of copper as 63.5.

2.3. (a) What are Miller indices? **[WBUT 2005]**

(b) Find the packing factor for SC structure and BCC lattice. **[WBUT 2004]**

(c) Calculate the inter-planar spacing d of planes (111) in an SC lattice of side a.

[WBUT 2003, 2004, 2007]

2.4. (a) What are the differences between crystalline and amorphous solids? **[CU(H) 1996]**

(b) Show that an FCC structure is more closely packed than a BCC structure.

(c) Illustrate the SC, FCC and BCC structures with diagrams.

2.5. (a) Find the atomic radii of SC, FCC and BCC structures of lattice constant a.

(b) What are Miller indices? How can they be determined?

(c) What are the characteristics of Miller indices?

2.6. (a) Explain the terms *lattice translation operation* and *fundamental translation vectors*? When are the translation vectors called *primitive*?

(b) For an FCC lattice, what are (i) the volume of the primitive cell (ii) the lattice points per unit volume (iii) the nearest neighbour distance, (iv) the number of nearest neighbours?

(c) What fraction of total volume is occupied by atoms in an FCC structure? **[BU(H) 1999]**

2.7. (a) Find out expression for the atomic radius and coordination number of SC, FCC and BCC structures of lattice constant a.

(b) Illustrate the SC, FCC and BCC structures with diagrams.

(c) Find out the packing fraction for FCC structure. Find out the expression for the separation between the planes with Miller indices h, k, l and lattice constant a for a simple cubic structure.

[WBUT 2008]

2.8. (a) What are Miller indices? Determine the inter-planar spacing between two parallel planes with Miller indices (h, k, l) in a cubic crystal of side a.

(b) Find the Miller indices of a plane with intercept at $\left(\frac{a}{2}, b, x\right)$. Draw a diagram showing the plane.

[NBU(H) 2008]

2.9. (a) What is a crystal? Define the terms: lattice and basis. Explain the statement, lattice + basis = crystal.

(b) Mention seven types of crystal system structure. How many Bravais lattice are there?

(c) What do you mean by *atomic packing fraction*? Find the atomic packing fraction of BCC structure.

(d) Show that for a simple cubic system $d_{100} : d_{110} : d_{111} = \sqrt{6} : \sqrt{3} : \sqrt{2}$. **[NBU(H) 2008]**

2.10. (a) Distinguish between amorphous and crystalline solids with suitable example.

(b) What do you mean by Miller indices? Show that in a cubic crystal, the distance (h, k, l) is given by

$$d_{hkl} = \frac{a}{\sqrt{h^2 + k^2 + l^2}}$$

where a is the lattice constant. [CU(H) 2001; BU(H) 2012; Pune U 2013]

(c) Find the number of atoms per mm^2 of the plane (110) in a BCC lattice having lattice constant 0.287 nm. [CU(H) 2009]

2.11. (a) What are symmetry operations? Prove that the crystal cannot have five-fold rotational symmetry.

(b) Why are simple cubic crystal structure rarely found in nature compared to other cubic structures? [BU(H) 2012]

2.12. (a) Explain concepts of lattice and basis. Discuss fundamental translation vector, primitive translation vector and non-primitive translation vector with necessary diagram.

(b) Obtain packing factor of BCC, FCC and SC structures. [Bhabnagar U 2017]

Chapter 3

Crystal Structure Analysis by X-Rays

3.1 Introduction

Von Laue in 1912 discovered that X-rays are diffracted by single crystal samples. The crystal acts as a three-dimensional space grating for the diffraction of X-rays. This X-ray diffraction technique can be successfully applied to determine the structure of the single crystals. In this technique the incident parallel monochromatic beam of X-rays is diffracted by the atoms of the single crystal. The intensity and the direction of the diffracted beam is related to the lattice structure and chemical composition of the crystal.

The three principal methods of crystal structure analysis are:

(i) Laue Method,

(ii) Rotating Crystal Method and

(iii) Powder Method or Debye-Scherrer Method.

Professional crystallographers nowadays prefer to use Complicated Precision Camera Method.

3.2 Why X-rays are used for the Analysis of Crystal Structure?

There are several reasons to use X-rays for the analysis of crystal structure as stated below:

(i) For diffraction effect to be observed *the wavelength of the radiation should be of the same order of the diffracting elements. Incidentally, the wavelength of the X-rays is comparable to the interatomic distances in crystals.*

(ii) X-rays are scattered elastically without any change of wavelength by the charged particles.

(iii) The wavelength of X-rays can be determined very accurately.

(iv) X-rays of suitable wavelength for diffraction by a specific crystal can be obtained by changing the voltage of the X-ray tube.

(v) Radiation of very longer wavelength (e.g., visible light) and shorter wavelength (shorter than X-rays) are unsuitable for producing diffraction effect in single crystals.

3.3 Bragg's Law for X-ray Diffraction

W.L. Bragg established a relationship between the wavelength of X-rays diffracted by the atomic planes of a crystal and the interatomic distance of the same.

His work is based on a simple model with the following assumptions:

(i) *The monochromatic X-rays are reflected by successive parallel planes of the atoms in the crystal, so that angle of incidence is equal to the angle of reflection.*

(ii) *Constructively interfering beams which are reflected by a set of parallel atomic planes, give rise to a strong diffracted beam.*

(iii) *The scattering of X-rays by the atoms is perfectly elastic, so that the wavelength of X-rays remains unchanged on reflection.*

Derivation of the law

Let a crystal be made up of equidistant parallel planes of atoms with inter-planar distance d as shown in Fig. 3.1. Let a monochromatic ray of X-rays PA of a wavelength λ be incident at an angle θ with the atomic planes. The corresponding reflected ray AQ must also be inclined at an angle θ to the plane. RB is another ray incident on the next plane at the same angle θ and the corresponding reflected ray be BS inclined also at θ with atomic plane.

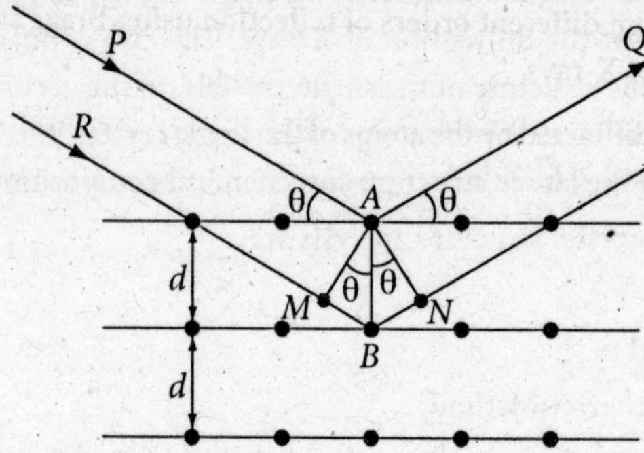

Fig. 3.1: Diffraction of X-rays by crystal planes

The path difference between the two reflected rays AQ and BS is

$$\Delta = MB + BN, \tag{3.1}$$

where AM and AN are perpendiculars from A on RB and BS respectively.

In ΔMAB,

$$\sin\theta = \frac{MB}{AB} = \frac{MB}{d} \quad \text{or,}\ MB = d\sin\theta.$$

Similarly, in ΔNAB

$$BN = d\sin\theta.$$

∴ from eqn (3.1),

$$\Delta = d\sin\theta + d\sin\theta = 2d\sin\theta. \tag{3.2}$$

If the two consecutive crystal planes scatter X-rays in the same phase, then the path difference must be integral multiple of λ, i.e.,

$$\Delta = n\lambda, \tag{3.3}$$

where $n = 0, 1, 2, \ldots$ is the *order of reflection*.

Hence, the condition of constructive interference of the reflected X-rays by the planes in a crystal is

$$2d\sin\theta = n\lambda. \tag{3.4}$$

Since the maximum possible value of $|\sin\theta| = 1$,

$$\frac{n\lambda}{2d} \leq 1. \tag{3.5}$$

Hence, the wavelength λ must not exceed twice the inter-planar spacing d for Bragg diffraction to occur.

Definition. Bragg's law is defined as the X-rays reflected from different parallel planes of a crystal interfere constructively when the path difference is integral multiple of the wavelength of X-rays.

NOTE. Bragg's law of X-ray reflection has the following utilities:

1. If λ of X-rays is known, then by measuring θ experimentally the value of inter-planar spacing d and hence, lattice constant can be determined.

2. If d is supplied, then measuring θ we can find out the wavelength λ of the unknown X-rays.

Correction of Bragg's Law Due to Refraction

Stenstrom (1919) observed that the wavelength of the same X-ray line does not give *identical values* for the wavelength, when computed from different orders of reflection using Bragg's equation. This discrepancy was attributed to the *refraction* of X-rays.

Consider an X-ray of wavelength λ in air enters a crystal through the surface XX' at a glancing angle θ and be incident on the Bragg plane YY' at θ'.

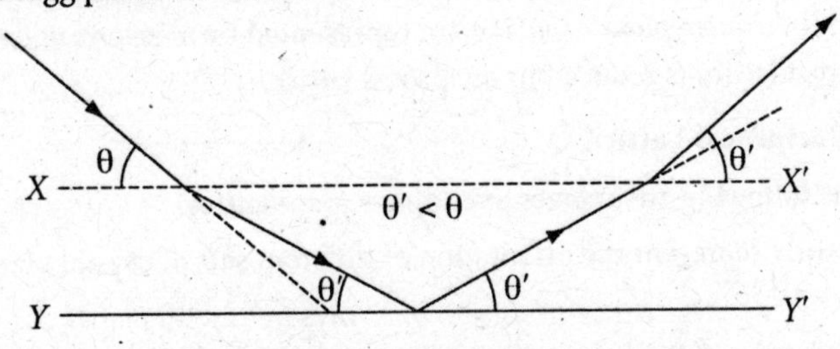

Fig. 3.2: Refraction of X-rays by a crystal

From Bragg's equation

$$n\lambda' = 2d \sin \theta', \tag{3.6}$$

where λ' is the wavelength of X-rays in the crystal.

If μ be the refractive index of the crystal, then

$$\mu = \frac{\lambda}{\lambda'}, \text{ also } \mu = \frac{\cos \theta}{\cos \theta'}. \tag{3.7}$$

Putting $\lambda' = \frac{\lambda}{\mu}$ in eqn (3.6), we get

$$\frac{n\lambda}{\mu} = 2 \sin \theta' = 2d\left(1 - \cos^2 \theta'\right)^{1/2}$$

$$\text{or, } n\lambda = 2\mu d\left(1 - \cos^2 \theta'\right)^{1/2} = 2d\left(\mu^2 - \mu^2 \cos^2 \theta'\right)^{1/2}$$

$$= 2d\left(\mu^2 - \cos^2 \theta\right)^{1/2} \text{ [from eqn (3.7)]}$$

$$= 2d\left[\mu^2 - (1 - \sin^2 \theta)\right]^{1/2} = 2d \sin \theta \left[1 - \frac{1 - \mu^2}{\sin^2 \theta}\right]^{1/2}$$

$$\text{or, } n\lambda \simeq 2d \sin \theta \left[1 - \frac{1 - \mu^2}{2 \sin^2 \theta}\right] \tag{3.8}$$

$$\text{or, } n\lambda = 2d \sin \theta \left[1 - \frac{1 - \mu}{\sin^2 \theta}\right] \left[\because 1 - \mu^2 = (1 + \mu)(1 - \mu) \simeq 2(1 - \mu), \because \mu \simeq 1\right]. \tag{3.9}$$

Again, from $n\lambda = 2d \sin\theta$,

$$\sin^2\theta = \frac{n^2\lambda^2}{4d^2}.$$

$\therefore$ from eqn (3.9), we get

$$n\lambda = 2d\sin\theta\left[1 - \frac{4d^2(1-\mu)}{n^2\lambda^2}\right]. \tag{3.10}$$

Eqn (3.10) is corrected form of *Bragg's equation*. The correction term $\frac{4d^2(1-\mu)}{n^2\lambda^2}$ is usually very small and is progressively smaller for higher values of n. Thus, uncorrected Bragg's law can be used without serious error.

3.4 Reciprocal Lattice

In the analysis of crystal structure by using X-rays, we have to deal with planes of several slopes in the same problem and it is difficult to visualize several slopes of such planes. In reciprocal lattice *the orientation of a set of parallel planes* and their *inter-planar spacing* are represented by a *reciprocal lattice point* and as such the problem faced in direct lattice is reduced in reciprocal lattice.

3.4.1 Advantages of Reciprocal Lattice

Reciprocal lattice has the following advantages over direct space lattice.

1. We can very conveniently represent the orientation of different sets of crystal planes and their spacings in reciprocal lattice.
2. The diffraction photographs of crystals can be understood and analysed easily.
3. It provides a simple method of obtaining the diffraction condition.
4. The wave mechanical behaviour of electrons in periodic lattice can be conveniently understood in terms of reciprocal lattice.

These are the importance of reciprocal lattice in solid state physics.

3.4.2 Geometrical Construction of the Reciprocal Lattice

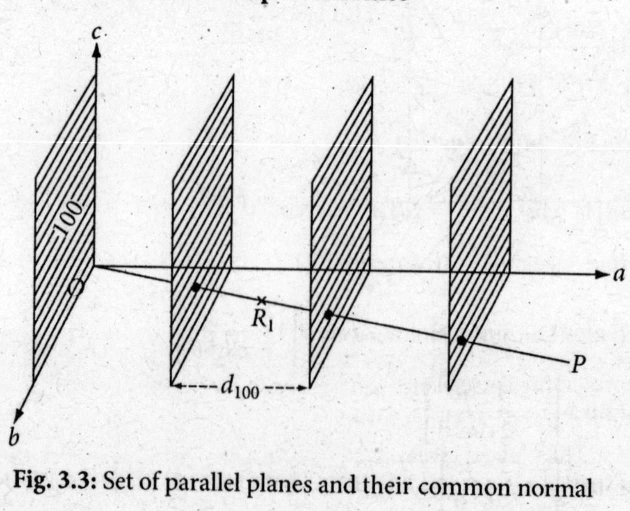

Fig. 3.3: Set of parallel planes and their common normal

Let us consider a set of parallel planes (100) as shown in Fig. 3.3. The *reciprocal point* for these number of planes is formed by drawing a common normal to the planes from an origin taken on any lattice point and locating a point on this line at a distance which is inversely proportional to the inter-planar spacing. In Fig. 3.3 the normal is represented by OP and the reciprocal lattice point by R_1, where $OR_1 = \frac{M}{d_{100}}$, where M is a constant called the *scale (or magnification) factor* for OR_1 and d_{100} is the *inter-planar spacing*. Thus, the entire set of parallel plane is represented by a single point R_1 in Fig. 3.3.

3.4.3 Rules for Constructing the Reciprocal Lattice

(i) Draw normal to every set of planes from the origin point in the direct lattice.

(ii) For a particular set of planes, set the length of each normal equal to the reciprocal of the inter-planar spacing.

(iii) Put a point at the end of each normal.

The collection of points so obtained gives us a new lattice array called the *reciprocal lattice*, because the distances in this lattice are *reciprocal* to those in the direct lattice.

3.4.4 Graphical Demonstration of Reciprocal Lattice in Two-Dimensional Space

Fig. 3.4 shows a monoclinic lattice ($a \neq b \neq c; \alpha = \beta = 90° \neq \gamma$) looking along the unique axis namely c-axis, called the first axis. We suppose all the lattice planes parallel to the c-axis chosen perpendicular to the plane of the figure. OX and OY represent respectively the unit lattice parameters a and b along x- and y-axes of the direct lattice. The normals to all the planes in the zone described above will lie in the plane of the diagram. Fig. 3.4 shows several sets of planes of the said zone. Only that plane of each family of planes nearest to the origin has been shown in the figure for the sake of simplicity. The Fig. 3.4 shows the traces of (100), (010), (120), etc. planes which are perpendicular to the plane of the paper.

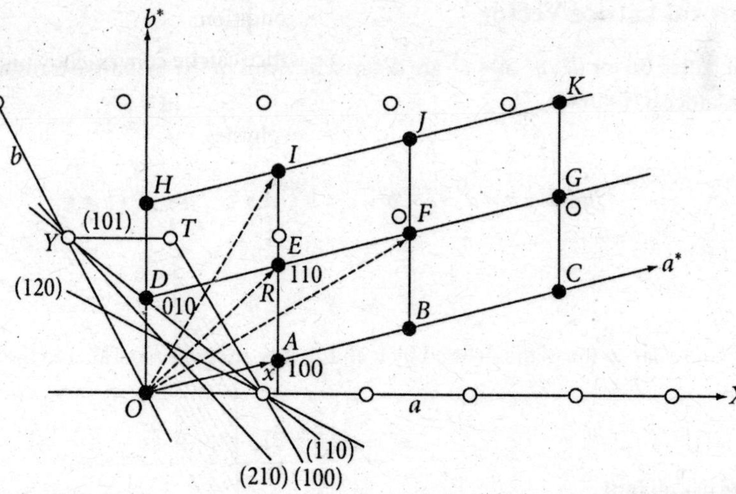

Fig. 3.4: Graphical representation of a reciprocal lattice in two dimensions

The reciprocal points representing these planes can be formed by drawing normal to each family of planes and then locating a point at a distance M times the reciprocal of the inter-planar spacing. The normal OA to the a plane (bc-plane), i.e., (100)-plane gives a^* (or x^*) axis and OD which is normal to the p-plane (or ac-plane), i.e., (010)-plane defines the b^* (or y^*) axis.

Corresponding to the family of equidistant parallel planes with Miller indices (100), the reciprocal lattice points lie along the normal OA to these planes at a distance M/d_{100} from the origin where d_{100} is the interplanar distance for (100)-planes. The reciprocal lattice point represented by the coordinates 100 (without parenthesis) is shown at A. Similarly, the planes (010) are represented by reciprocal point D which lies on the normal to these planes at a distance M/d_{010} and is assigned coordinates 010. Also the reciprocal lattice point E for (110)-planes is located on the normal at a distance M/d_{110} and is represented by the coordinates 110.

Applying the same geometrical contraction throughout, we can locate all the reciprocal lattice points relating to the crystal.

3.4.5 Definition of Reciprocal Lattice

The reciprocal lattice corresponding to a Bravais lattice point $\vec{R}$ is defined as the set of all wave vectors $\vec{k}$ that give plane waves of the form $e^{i\vec{k}\cdot\vec{r}}$ with the periodicity $(\vec{R})$ of the Bravais lattice, i.e.,

$$e^{i\vec{k}\cdot(\vec{r}+\vec{R})} = e^{i\vec{k}\cdot\vec{r}} \tag{3.11}$$

for any $\vec{r}$ and all $\vec{R}$ in the *Bravais lattice*

$$\text{or, } e^{i\vec{k}\cdot\vec{r}} \times e^{i\vec{k}\cdot\vec{R}} = e^{i\vec{k}\cdot\vec{r}}, \text{ i.e., } e^{i\vec{k}\cdot\vec{R}} = 1. \tag{3.12}$$

So, the reciprocal lattice vector is represented by a set of lattice vectors $\vec{k}$ satisfying eqn (3.12), for all $\vec{R}$ in the Bravais lattice.

Since $e^{i2\pi n} = 1$, where n is an integer, we can write,

$$e^{i\vec{k}\cdot\vec{R}} = e^{i2\pi n}, \text{ i.e., } \vec{k}\cdot\vec{R} = 2\pi n. \tag{3.13}$$

3.5 Reciprocal Lattice Vector

The reciprocal lattice vector $\vec{a}^*$, $\vec{b}^*$ and $\vec{c}^*$ are defined in terms of the primitive translation vectors $\vec{a}$, $\vec{b}$ and $\vec{c}$ of the direct lattice as follows:

$$\vec{a}^*\cdot\vec{a} = \vec{b}^*\cdot\vec{b} = \vec{c}^*\cdot\vec{c} = -1 \tag{3.14}$$

$$\text{and } \vec{a}^*\cdot\vec{b} = \vec{a}^*\cdot\vec{c} = \vec{b}^*\cdot\vec{c} = \vec{b}^*\cdot\vec{a} = \vec{c}^*\cdot\vec{a} = \vec{c}^*\cdot\vec{b} = 0. \tag{3.15}$$

From eqn (3.15)

$$\vec{a}^*\cdot\vec{b} = \vec{a}^*\cdot\vec{c} = 0,$$

i.e., $\vec{a}^*$ is perpendicular to the plane defined by $\vec{b}$ and $\vec{c}$. It is, therefore, parallel to the vector $\vec{b}\times\vec{c}$ and can be expressed as

$$\vec{a}^* = A\left(\vec{b}\times\vec{c}\right), \tag{3.16}$$

where A is a scalar quantity.

Now from eqn (3.14), we have

$$\vec{a}^*\cdot\vec{a} = A\left(\vec{b}\times\vec{c}\right)\cdot\vec{a} = 1. \tag{3.17}$$

$$\therefore A = \frac{1}{\vec{a}\cdot\left(\vec{b}\times\vec{c}\right)}. \tag{3.18}$$

∴ from eqns (3.16) and (3.18),

$$\vec{a}^* = \frac{\vec{b} \times \vec{c}}{\vec{a} \cdot (\vec{b} \times \vec{c})}. \tag{3.19}$$

In a similar way,

$$\vec{b}^* = \frac{\vec{c} \times \vec{a}}{\vec{a} \cdot (\vec{b} \times \vec{c})} \tag{3.20}$$

$$\text{and } \vec{c}^* = \frac{\vec{a} \times \vec{b}}{\vec{a} \cdot (\vec{b} \times \vec{c})}. \tag{3.21}$$

Some authors prefer to define *primitive translation vectors* $\vec{a}, \vec{b}, \vec{c}$ of the direct lattice and *primitive translation vectors* of the reciprocal lattice in the following way:

$$\vec{a}^* \cdot \vec{a} = \vec{b}^* \cdot \vec{b} = \vec{c}^* \cdot \vec{c} = 2\pi \tag{3.22}$$

$$\text{and } \vec{a}^* \cdot \vec{b} = \vec{a}^* \cdot \vec{c} = \vec{b}^* \cdot \vec{c} = \vec{b}^* \cdot \vec{a} = \vec{c}^* \cdot \vec{a} = \vec{c}^* \cdot \vec{b} = 0.$$

Then we obtain in the above way

$$\left.\begin{aligned} \vec{a}^* &= 2\pi\frac{\vec{b} \times \vec{c}}{\vec{a} \cdot (\vec{b} \times \vec{c})} \\ \vec{b}^* &= 2\pi\frac{\vec{c} \times \vec{a}}{\vec{a} \cdot (\vec{b} \times \vec{c})} \\ \vec{c}^* &= 2\pi\frac{\vec{a} \times \vec{b}}{\vec{a} \cdot (\vec{b} \times \vec{c})} \end{aligned}\right\}. \tag{3.23}$$

The common denominator in each case is

$$\vec{a} \cdot (\vec{b} \times \vec{c}) = V, \tag{3.24}$$

where V is the volume of the primitive unit cell in the direct lattice.

The vectors $\vec{a}$, $\vec{b}$ and $\vec{c}$ in the direct lattice have the dimension of [length], whereas the reciprocal lattice vectors $\vec{a}^*$, $\vec{b}^*$ and $\vec{c}^*$ have the dimension of $[\text{length}]^{-1}$. The crystal is a lattice in the *real space* but the reciprocal lattice is a lattice in the *Fourier space* [cf. eqn (3.12)].

Choosing $\vec{a}^*$, $\vec{b}^*$ and $\vec{c}^*$ as the three reciprocal axes, any arbitrary set of primitive vectors $\vec{a}, \vec{b}, \vec{c}$ of a given crystal leads to the same set of reciprocal lattice points represented by the *reciprocal lattice vector*

$$\vec{G} = h\vec{a}^* + k\vec{b}^* + l\vec{c}^*, \tag{3.25}$$

where h, k, l are integers and $\vec{a}^*, \vec{b}^*, \vec{c}^*$ are the *primitive translation vectors* of the reciprocal lattice.

The reciprocal lattice vector $\vec{G}$ represents a set of points in the reciprocal lattice planes which are called the *reciprocal points*.

If $\vec{T}$ be the *crystal translation vector*, then

$$\vec{G} \cdot \vec{T} = 2\pi \times \text{integer} = 2\pi N \tag{3.26}$$

$$\text{or, } e^{i\vec{G}\cdot\vec{T}} = 1. \tag{3.27}$$

But

$$e^{i\vec{G}\cdot\vec{T}} = \cos\left(\vec{G}\cdot\vec{T}\right) + i\sin\left(\vec{G}\cdot\vec{T}\right) = \cos(2\pi N) + i\sin(2\pi N) = 1 + 0 = 1.$$

$$\therefore \cos\left(\vec{G}\cdot\vec{T}\right) = 1 = \cos 0°.$$

$$\therefore \vec{G}\cdot\vec{T} = 0. \tag{3.28}$$

Eqn (3.28) shows that $\vec{G}$ *is perpendicular to* $\vec{T}$.

3.5.1 Properties of the Reciprocal Lattice

Reciprocal lattice is an invariant geometrical object and its properties are fundamental in the theory of solids. Some of the geometrical properties are as follows:

(i) *Every reciprocal lattice vector is normal to a lattice plane of the crystal lattice.*

Proof. Let $\vec{G} = h'\vec{a}^* + k'\vec{b}^* + l'\vec{c}^*$ be a reciprocal lattice vector and (hkl) be a set of parallel lattice planes in the direct lattice, then

$$\frac{h'}{h} = \frac{k'}{k} = \frac{l'}{l} = n. \tag{3.29}$$

From Fig. 3.5

$$\overrightarrow{AC} = -\frac{\vec{a}}{h} + \frac{\vec{c}}{l} \text{ and } \overrightarrow{AB} = -\frac{\vec{a}}{h} + \frac{\vec{b}}{k},$$

where the vectors $\overrightarrow{AB}$ and $\overrightarrow{BC}$ lie on the (hkl)-plane.

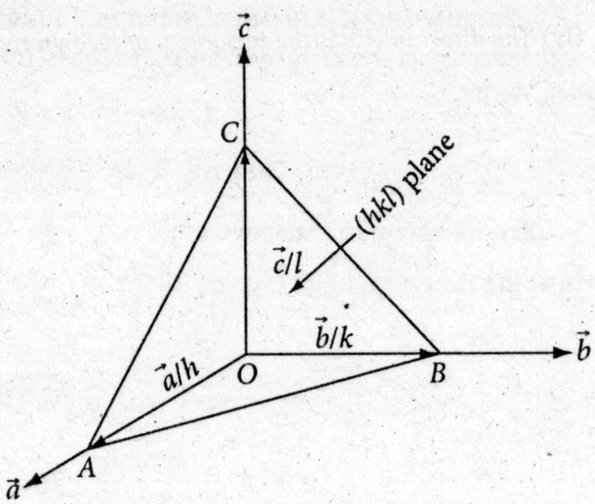

Fig. 3.5: An (*hkl*) plane in the direct lattice

Now

$$\vec{G}\cdot\overrightarrow{AC} = \vec{G}\cdot\left[-\frac{\vec{a}}{h} + \frac{\vec{c}}{l}\right] = \left(h'\vec{a}^* + k'\vec{b}^* + l'\vec{c}^*\right)\cdot\left[-\frac{\vec{a}}{h} + \frac{\vec{c}}{l}\right] = 2\pi\left(-\frac{h'}{h} + \frac{l'}{l}\right) = 0. \tag{3.30}$$

Similarly,

$$\vec{G}\cdot\left[-\frac{\vec{a}}{h} + \frac{\vec{b}}{k}\right] = 0 \text{ [using eqns (3.14) and (3.15)]}. \tag{3.31}$$

(ii) *If the components of* $\vec{G}$ *have no common factor, then* $|\vec{G}|$ *is inversely proportional to the spacing of the lattice planes normal to* $\vec{G}$.

The vectors $\overrightarrow{AC}$ and $\overrightarrow{AB}$ are in the (hkl)-plane. Hence, $\vec{G}$ is perpendicular to the (hkl) plane.

If $\hat{n}$ be a unit vector normal to (hkl)-plane, then $\hat{n} = \frac{\vec{G}}{G}$, where $\vec{G}$ is normal to the (hkl)-plane.

$$\therefore d = \frac{\vec{a}\cdot\hat{n}}{h} = \frac{\vec{a}\cdot\vec{G}}{hG} = \frac{\vec{a}\cdot\left(h'\vec{a}^* + k'\vec{b}^* + l'\vec{c}^*\right)}{hG}$$

$$= \frac{2\pi h'}{hG} = \frac{2\pi n}{G}\left(\because \frac{h'}{h} = n\right), \tag{3.32}$$

where d is the spacing between adjacent (hkl)-planes. Eqn (3.32) shows that $|\vec{G}| \propto 1/d$.

(iii) *The volume of a unit cell in the reciprocal lattice is inversely proportional to the volume of a unit cell of the direct lattice.*

Proof. If $\vec{a}^*, \vec{b}^*$ and $\vec{c}^*$ be the reciprocal lattice vectors, then the volume of its unit cell is

$$V^* = \left(\vec{a}^* \cdot \vec{b}^* \times \vec{c}^*\right) = (2\pi)^3 \frac{\vec{b} \times \vec{c}}{\left(\vec{a} \cdot \vec{b} \times \vec{c}\right)} \cdot \frac{\left[(\vec{c} \times \vec{a}) \times \left(\vec{a} \times \vec{b}\right)\right]}{\left(\vec{a} \cdot \vec{b} \times \vec{c}\right)\left(\vec{a} \cdot \vec{b} \times \vec{c}\right)} = (2\pi)^3 \frac{\left(\vec{b} \times \vec{c}\right) \cdot \left[(\vec{c} \times \vec{a}) \times \left(\vec{a} \times \vec{b}\right)\right]}{\left(\vec{a} \cdot \vec{b} \times \vec{c}\right)^3}$$

$$= (2\pi)^3 \frac{\left(\vec{b} \times \vec{c}\right) \cdot \left[\left\{\vec{c} \cdot \left(\vec{a} \times \vec{b}\right)\right\} \vec{a} - \left\{\vec{a} \cdot \left(\vec{a} \times \vec{b}\right)\right\} \vec{c}\right]}{\left(\vec{a} \cdot \vec{b} \times \vec{b}\right)} = \frac{(2\pi)^3}{\left(\vec{a} \cdot \vec{b} \times \vec{c}\right)^3} = \frac{(2\pi)^3}{V}, \quad (3.33)$$

where V is the volume of the unit cell in the direct lattice. $\therefore V^* \propto 1/V$.

(iv) *The direct lattice is the reciprocal of its own reciprocal lattice:*

Proof. We have

$$\vec{a}^* = 2\pi \frac{\vec{b} \times \vec{c}}{\vec{a} \cdot \vec{b} \times \vec{c}}.$$

Now the reciprocal of $\vec{a}^*$ is given by

$$(\vec{a}^*)^* = 2\pi \frac{\vec{b}^* \times \vec{c}^*}{\left(\vec{a}^* \cdot \vec{b}^* \times \vec{c}^*\right)} = 2\pi \left(\vec{a} \cdot \vec{a}^*\right) \cdot \frac{\vec{b}^* \times \vec{c}^*}{\vec{a}^* \cdot \vec{b}^* \times \vec{c}^*}, \quad [\because \vec{a} \cdot \vec{a}^* = 1]$$

$$= 2\pi \vec{a} \frac{\vec{a}^* \cdot \vec{b}^* \times \vec{c}^*}{\vec{a}^* \cdot \vec{b}^* \times \vec{c}^*} = 2\pi \vec{a}. \quad (3.34)$$

Hence, it is proved that *the direct lattice is the reciprocal of its own reciprocal lattice.*

3.5.2 Reciprocal Lattice to SC Lattice

If $\hat{i}, \hat{j}$ and $\hat{k}$ be taken as the unit vectors along primitive translation vectors $\vec{a}, \vec{b}$ and $\vec{c}$ of a simple cubic lattice, then $\vec{a} = \hat{i}a,\ \vec{b} = \hat{j}b,\ \vec{c} = \hat{k}c$.

Volume of the simple cubic cell is $\vec{a} \cdot \vec{b} \times \vec{c} = a^3 \left(\hat{i} \cdot \hat{j} \times \hat{k}\right) = a^3$.

Now the reciprocal lattice vectors to the SC lattice are

$$\left.\begin{aligned} \vec{a}^* &= 2\pi \frac{\vec{b} \times \vec{c}}{\left(\vec{a} \cdot \vec{b} \times \vec{c}\right)} = 2\pi \frac{a\hat{j} \times a\hat{k}}{a^3} = \frac{2\pi}{a}\hat{i} \\ \text{Similarly, } \vec{b}^* &= 2\pi \frac{\vec{c} \times \vec{a}}{\vec{a} \cdot \vec{b} \times \vec{c}} = \frac{2\pi}{a}\hat{j} \\ \text{and } \vec{c}^* &= 2\pi \frac{\vec{a} \times \vec{b}}{\vec{a} \cdot \vec{b} \times \vec{c}} = \frac{2\pi}{a}\hat{k} \end{aligned}\right\}. \quad (3.35)$$

It is evident from eqn (3.35) that all the three reciprocal lattice vectors are equal in magnitude which means the *reciprocal lattice to SC lattice is also SC but with lattice constant equal to* $\frac{2\pi}{a}$.

3.5.3 Reciprocal Lattice to BCC Lattice

The primitive translation vectors of a body-centred cubic lattice [Fig. 3.6] are

$$\left.\begin{aligned}\vec{a}' &= \frac{a}{2}\left(\hat{i}+\hat{j}-\hat{k}\right)\\ \vec{b}' &= \frac{a}{2}\left(-\hat{i}+\hat{j}+\hat{k}\right)\\ \vec{c}' &= \frac{a}{2}\left(\hat{i}-\hat{j}+\hat{k}\right)\end{aligned}\right\}, \qquad (3.36)$$

where $\hat{i}$, $\hat{j}$ and $\hat{k}$ are the *orthogonal unit vectors* along the edges of the cube and a is the length of the cube edge.

The volume of the primitive unit cell is

$$V = \vec{a}'\cdot\vec{b}'\times\vec{c}' = \frac{a}{2}\left(\hat{i}+\hat{j}-\hat{k}\right)\cdot\left[\frac{a^2}{4}\left(-\hat{i}+\hat{j}+\hat{k}\right)\times\left(\hat{i}-\hat{j}+\hat{k}\right)\right]$$

$$= \frac{a}{2}\left(\hat{i}+\hat{j}-\hat{k}\right)\cdot\frac{a^2}{2}\left(\hat{i}+\hat{j}\right) = \frac{a^3}{2}.$$

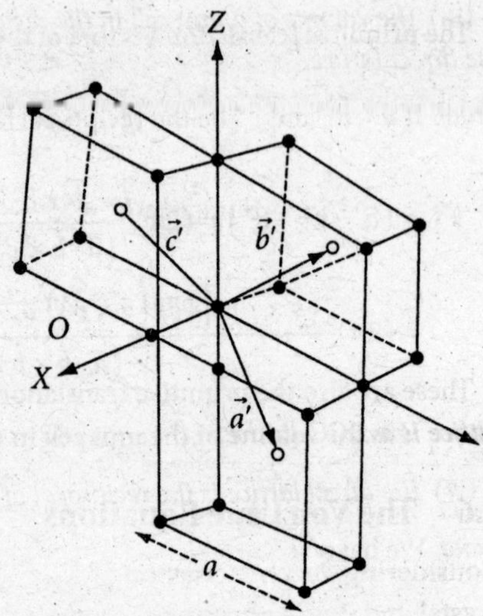

Fig. 3.6: Primitive translation vectors of a BCC lattice

∴ the reciprocal lattice vectors are

$$\left.\begin{aligned}\vec{a}^* &= 2\pi\frac{\vec{b}'\times\vec{c}'}{\vec{a}'\cdot\vec{b}'\times\vec{c}'} = \frac{2\pi\left(\frac{a^2}{2}\right)}{\frac{a^3}{2}}\left(\hat{i}+\hat{j}\right) = \frac{2\pi}{a}\left(\hat{i}+\hat{j}\right)\\ \text{Similarly, } \vec{b}^* &= 2\pi\frac{\vec{c}'\times\vec{a}'}{\vec{a}'\cdot\vec{b}'\times\vec{c}'} = \frac{2\pi}{a}\left(\hat{j}+\hat{k}\right)\\ \text{and } \vec{c}^* &= 2\pi\frac{\vec{a}'\times\vec{b}'}{\vec{a}'\cdot\vec{b}'\times\vec{c}'} = \frac{2\pi}{a}\left(\hat{k}+\hat{i}\right)\end{aligned}\right\}. \qquad (3.37)$$

These are also the *primitive translation vectors of an FCC lattice.* Hence, *the reciprocal lattice to a BCC lattice is FCC lattice.*

3.5.4 Reciprocal Lattice to FCC Lattice

The primitive translation vectors of an FCC lattice are shown in Fig. 3.7 and given by

$$\left.\begin{aligned}\vec{a}' &= \frac{a}{2}\left(\hat{i}+\hat{j}\right)\\ \vec{b}' &= \frac{a}{2}\left(\hat{j}+\hat{k}\right)\\ \vec{c}' &= \frac{a}{2}\left(\hat{k}+\hat{i}\right)\end{aligned}\right\}. \qquad (3.38)$$

Volume of the unit cell is given by

$$V = \vec{a}'\cdot\vec{b}'\times\vec{c}' = \frac{a}{2}\left(\hat{i}+\hat{j}\right)\cdot\frac{a^2}{4}\left[\left(\hat{j}\times\hat{k}\right)\times\left(\hat{k}\times\hat{i}\right)\right]$$

$$= \frac{a}{2}\left(\hat{i}+\hat{j}\right)\cdot\frac{a^2}{4}\left(\hat{i}+\hat{j}-\hat{k}\right) = \frac{a^3}{4}.$$

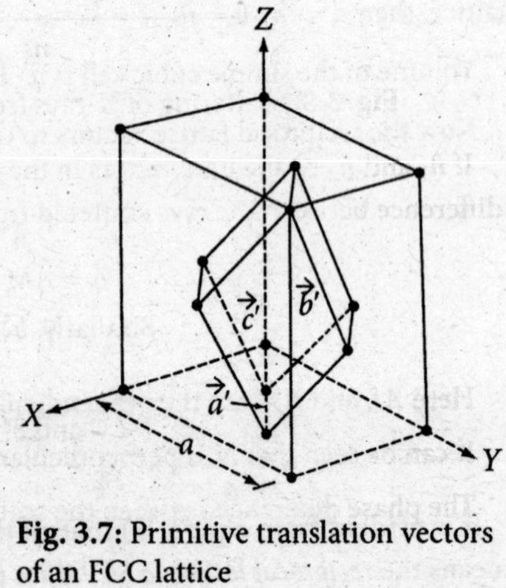

Fig. 3.7: Primitive translation vectors of an FCC lattice

The primitive translation vectors of the reciprocal lattice are

$$\left.\begin{aligned} \vec{a}^* &= 2\pi\frac{\vec{b}'\times\vec{c}'}{\vec{a}'\cdot\vec{b}'\times\vec{c}'} = 2\pi\frac{\frac{a^2}{4}\left(\hat{i}+\hat{j}-\hat{k}\right)}{\frac{a^3}{4}} = \frac{2\pi}{a}\left(\hat{i}+\hat{j}-\hat{k}\right) \\ \text{Similarly, } \vec{b}^* &= 2\pi\frac{\vec{c}'\times\vec{a}'}{\vec{a}'\cdot\left(\vec{b}'\times\vec{c}'\right)} = \frac{2\pi}{a}\left(-\hat{i}+\hat{j}+\hat{k}\right) \\ \text{and } \vec{c}^* &= 2\pi\frac{\vec{a}'\times\vec{b}'}{\vec{a}'\cdot\left(\vec{b}'\times\vec{c}'\right)} = \frac{2\pi}{a}\left(\hat{i}-\hat{j}+\hat{k}\right) \end{aligned}\right\}. \tag{3.39}$$

These are also the primitive translation vectors of an FCC lattice. Thus, *the reciprocal lattice to an FCC lattice is a BCC lattice.*

3.6 The Von Laue Equations

Considering the phenomenon of X-ray diffraction as the scattering of X-rays by individual atoms in the crystal and their subsequent recombination, *Von Laue* obtained the directions of diffraction maxima in a more general way. The diffraction maxima occur in some specific direction obeying certain conditions known as *Laue's equations. Bragg's laws* of X-ray diffraction can be obtained from *Laue equations.*

Let a beam of X-rays be incident on two identical scattering centres A and B separated by a distance $\vec{r}$ in a crystal [Fig. 3.8].

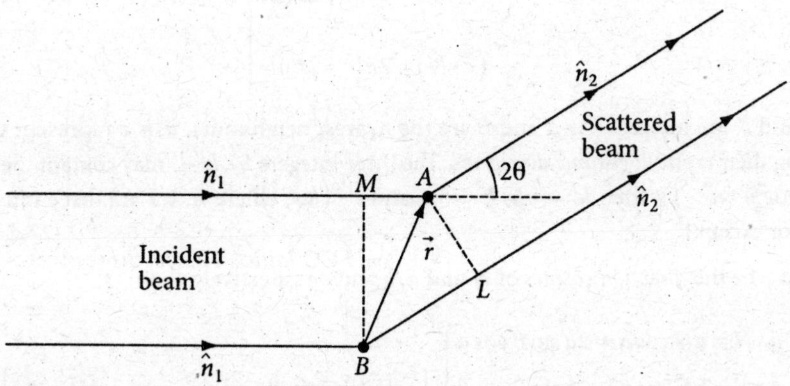

Fig. 3.8: Scattering of X-rays from two identical scattering centres separated by a distance $\vec{r}$

If $\hat{n}_1$ and $\hat{n}_2$ be the unit vectors in the direction of incident and scattered beam respectively, then the path difference between the rays scattered from A and B is

$$\begin{aligned} \Delta &= AM - BL = \vec{r}\cdot\hat{n}_1 - \vec{r}\cdot\hat{n}_2 = \vec{r}\cdot(\hat{n}_1-\hat{n}_2) \\ &= \vec{r}\cdot\vec{N}\ [\text{where } \vec{N} = \hat{n}_1-\hat{n}_2]. \end{aligned} \tag{3.40}$$

Here AL and BM are the perpendiculars to the directions of incident and scattered beam respectively. It can be seen that $\vec{N}$ is perpendicular to the reflecting plane [Fig. 3.9].

The phase difference between the scattered and incident beams from B and A is

$$\phi = \frac{2\pi}{\lambda}\left(\vec{r}\cdot\vec{N}\right). \tag{3.41}$$

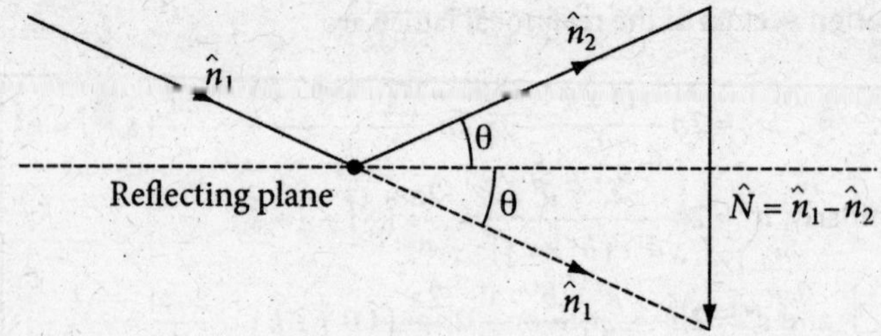

Fig. 3.9: Geometrical relationship between incident beam, scattered beam, reflecting plane and the normal

The X-rays scattered from A and B will interfere constructively, if the phase difference ϕ be integral multiple of 2π. Due to periodicity of the atoms in the crystal radiation scattered from the other atoms may be in the same direction having exactly the same phase as those scattered from A and B. In a three-dimensional lattice $\vec{r}$ may coincide with any of the lattice vectors $\vec{a}$, $\vec{b}$ and $\vec{c}$.

So, for *constructive interference*

$$\left.\begin{aligned} \frac{2\pi}{\lambda}\left(\vec{a}\cdot\vec{N}\right) &= 2\pi h' = 2\pi nh \\ \frac{2\pi}{\lambda}\left(\vec{b}\cdot\vec{N}\right) &= 2\pi k' = 2\pi nk \\ \frac{2\pi}{\lambda}\left(\vec{c}\cdot\vec{N}\right) &= 2\pi l' = 2\pi nl \end{aligned}\right\}, \tag{3.42}$$

where h', k' and l' are integers. As A and B are the nearest neighbours, a, b, c represent the interatomic distances along their *crystallographic directions*. The three integers h', k', l' may contain the largest integral common factor n $(n > 1)$, then $h' = nh$, $k' = nk$ and $l' = nl$; where h, k, l are three integers having no common factor except 1.

Let α, β and γ be the *direction cosines* of $\vec{N}$ and $\vec{a}$, $\vec{b}$ and $\vec{c}$ respectively, then

$$\left.\begin{aligned} \vec{a}\cdot\vec{N} &= aN\cos\alpha = 2a\sin\theta\cos\alpha \\ \vec{b}\cdot\vec{N} &= bN\cos\beta = 2a\sin\theta\cos\beta \\ \text{and } \vec{c}\cdot\vec{N} &= cN\cos\gamma = 2a\sin\theta\cos\gamma \end{aligned}\right\} \left[\because |\vec{N}| = 2\sin\theta \text{ and } \alpha = \cos\left(\vec{a},\vec{N}\right), \text{ etc.}\right]. \tag{3.43}$$

Then from the above set of equations

$$\left.\begin{aligned} \vec{a}\cdot\vec{N} &= 2a\alpha\sin\theta = h'\lambda = nh\lambda \\ \vec{b}\cdot\vec{N} &= 2\beta b\sin\theta = k'\lambda = nk\lambda \\ \text{and } \vec{c}\cdot\vec{N} &= 2\gamma c\sin\theta = k'\lambda = nk\lambda \end{aligned}\right\}. \tag{3.44}$$

The set of eqns (3.44) is known as the *Laue's equations* representing *the conditions of X-ray diffraction.* These equations give the direction of *constructive interference* for a particular wavelength (λ). Eqns (3.44) can be written as

$$\left.\begin{aligned} \frac{2a\alpha}{nh} &= \frac{\lambda}{\sin\theta} \\ \frac{2b\beta}{nk} &= \frac{\lambda}{\sin\theta} \\ \frac{2c\gamma}{nl} &= \frac{\lambda}{\sin\theta} \end{aligned}\right\}. \tag{3.45}$$

$$\left[\because \frac{2\pi}{\lambda}\left(\vec{a}\cdot\vec{N}\right) = 2\pi nh \text{ or, } \frac{1}{\lambda}2a\alpha\sin\theta = nh \text{ or, } \frac{2a\alpha}{n\lambda} = \frac{\lambda}{\sin\theta}\right]$$

$$\text{or, } \frac{a\alpha}{h} = \frac{b\beta}{k} = \frac{c\gamma}{l} = \frac{n\lambda}{2\sin\theta}. \tag{3.46}$$

Eqns (3.45) can be expressed as

$$\left.\begin{aligned} \frac{2\alpha\sin\theta}{n\lambda} &= \frac{h}{a} \\ \frac{2\beta\sin\theta}{n\lambda} &= \frac{k}{b} \\ \frac{2\gamma\sin\theta}{n\lambda} &= \frac{l}{c} \end{aligned}\right\}. \tag{3.47}$$

This shows that in a direction of constructive interference, the direction cosines are proportional to $\frac{h}{a}$, $\frac{k}{b}$ and $\frac{l}{c}$. α, β and γ are also the direction cosines $\vec{N}$, the normal to the reflecting plane.

The set of eqns (3.44) uniquely determine the values of θ and N and thus determine the scattering direction $(\hat{n}_2)$. Again, from eqn (3.44) the direction cosines of $\vec{N}$ are proportional to $\frac{h}{a}$, $\frac{k}{b}$ and $\frac{l}{c}$. So (hkl)-planes are parallel to the *reflecting plane*.

3.6.1 Bragg's Law from Laue Equations

Since the inter-planar spacing (Art. 2.13) d is given by

$$d = \frac{a\alpha}{h} = \frac{b\beta}{k} = \frac{c\gamma}{l},$$

$\therefore$ from eqn (3.44),

$$\begin{aligned} & 2a\alpha\sin\theta = nh\lambda \\ \text{or, } \quad & 2hd\sin\theta = nh\lambda \qquad \left[\because d = \frac{a\alpha}{h}\right]. \\ \text{or, } \quad & 2d\sin\theta = n\lambda. \end{aligned} \tag{3.48}$$

This is *Bragg's law of reflection*, where n is the *order of reflection*. Thus, one may consider different order reflection from (hkl)-plane or equivalently first order reflection from (nh, nk, nl)-planes. Putting $n = 1$, one gets rid of the factor n in structure determination of crystals.

3.6.2 Comparison of Bragg and Laue Approaches

(i) Bragg considered the crystal as a set of equispaced parallel planes of ions or atoms.

Laue did not consider a particular set of parallel lattice planes, rather he considered crystal ions or atoms placed at the lattice points.

(ii) Bragg considered that X-rays are specularly reflected by the atoms or ions in one lattice plane.

According to Laue, incident X-rays are considered to be scattered elastically by the ions or atoms in all directions.

(iii) In Bragg reflection, the reflected ray from successive plane interferes constructively and produces maxima.

In Laue's consideration constructive interference is equivalent to the Bragg assumption of specular reflection.

Thus, both the Bragg and Laue equations are based on the same physical principle.

3.7 Bragg's Law in Reciprocal Lattice: Ewald Construction

Bragg's law of X-ray diffraction can be expressed as a relation between vectors in the reciprocal lattice.

In Fig. 3.10, $\overrightarrow{AO}$ is a vector of length $\frac{1}{\lambda}$ in the direction of the incident X-ray beam and terminates at the origin O of the reciprocal lattice. With A as the centre we draw a sphere of radius $|\overrightarrow{AO}| = \frac{1}{\lambda}$, which intersects some point B in the reciprocal lattice.

The coordinates of B is (h', k', l') which may have a highest common factor n, i.e., $h' = nh$, $k' = nk$ and $l' = nl$, where h, k, l have common factor unity. $\overrightarrow{OB}$ then represents a vector joining the origin O of the reciprocal lattice and the point $B\ (h', k', l')$. $\overrightarrow{OB}$, therefore, must be normal to the plane $(h'k'l')$ and is of length

$$\left|\overrightarrow{OB}\right| = \frac{n}{d_{hkl}}. \tag{3.49}$$

Fig. 3.10 shows that AE is such a plane. Now from the ΔAOB,

$$OB = 2OE = 2OA \sin\theta = \frac{2\sin\theta}{\lambda} \quad \left[\because OA = \frac{1}{\lambda}\right], \tag{3.50}$$

where $\angle EAO = \theta$.

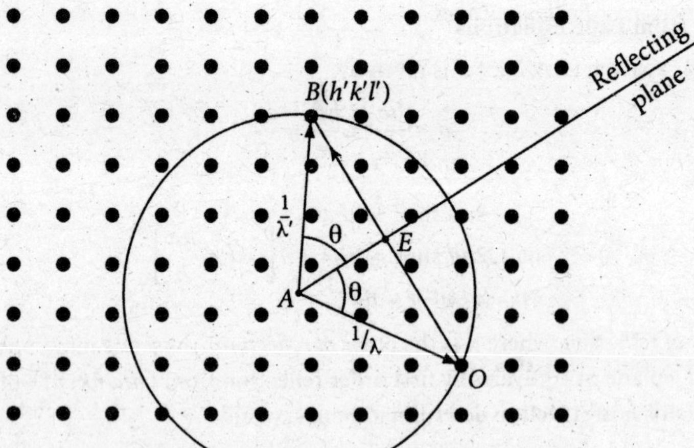

Fig. 3.10: Ewald Construction in the reciprocal lattice

From eqns (3.49) and (3.50)

$$\frac{2\sin\theta}{\lambda} = \frac{n}{d_{hkl}} \quad \text{or,}\ 2d_{hkl}\sin\theta = n\lambda. \tag{3.51}$$

Eqn (3.51) is the *Bragg's law* and *n is the order of reflection.*

NOTE. (i) Since the spacing of the reciprocal lattice is $\frac{1}{a}$, if $\frac{1}{\lambda} < \frac{1}{2a}$ or $\lambda > 2a$, the sphere will not touch any other point and as such no diffraction will take place, if $\lambda > 2a$.

(ii) If $\frac{1}{\lambda} \varepsilon 2a$ or $\lambda \leq 2a$, the diffraction is possible.

The above construction in the reciprocal lattice giving Bragg's condition of X-ray reflection is called *Ewald construction.*

Vector Form of Bragg's Law

Let the vectors $\overrightarrow{A'O'}$, $\overrightarrow{O'B'}$ and $\overrightarrow{A'B'}$ be the magnified forms by a constant factor of 2π such that $A'O' = 2\pi(OA)$, $O'B' = 2\pi(OB)$ and $A'B' = 2\pi(AB)$ [Fig. 3.11].

Since $OA = 1/\lambda$,

$\therefore A'O' = \frac{2\pi}{\lambda}$ can represent the wave vector $\vec{k}$ by the vector $\overrightarrow{A'O'}$. Then $\overrightarrow{O'B'}$ is a reciprocal vector $\vec{G}$.

Now from the *triangle law of vectors*, $\overrightarrow{A'B'} = \overrightarrow{A'O'} + \overrightarrow{O'B'} = \vec{k} + \vec{G}$.

For the diffraction to take place

$$\left(\vec{k} + \vec{G}\right)^2 = k^2 \text{ or, } k^2 + 2\vec{k} \cdot \vec{G} + G^2 = k^2$$

$$\text{or, } 2\vec{k} \cdot \vec{G} + G^2 = 0. \qquad (3.52)$$

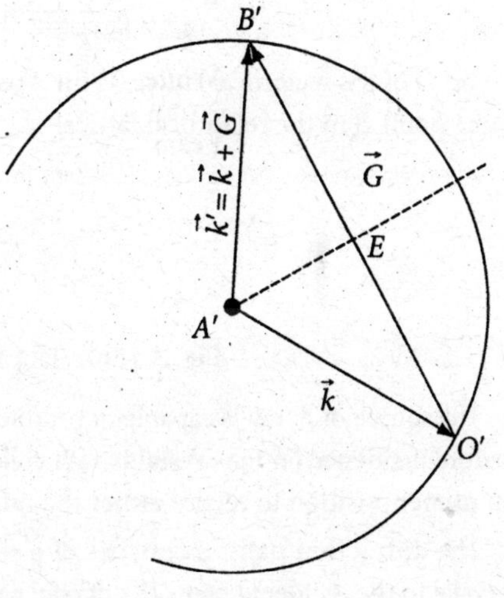

Fig. 3.11: Magnified Ewald Construction

Eqn (3.52) represents the *vector form of Bragg's law* which is used in the construction of *Brillouin Zones.*

Let $\overrightarrow{A'B'} = \vec{k}'$, then

$$k'^2 = k^2 \qquad (3.53)$$

$$\text{and } \vec{k}' - \vec{k} = \Delta\vec{k} = \vec{G}. \qquad (3.54)$$

This shows that

(i) *scattering changes the direction of $\vec{k}$ and not its magnitude.*

(ii) *the scattered wave differs from the incident wave by a reciprocal lattice vector $\vec{G}$.*

3.8 X-ray Diffraction Methods

To determine the structure of solids and for the study of X-ray spectroscopy the phenomenon of X-ray diffraction is employed. The underlying principle in the above cases is the *Bragg's law.*

Here we shall describe *three* standard methods of X-ray diffraction.

3.8.1 The Laue Method

In this method a single crystal is held stationary and a beam of *white radiation* is made incident on it at a fixed glancing angle θ, i.e., θ is fixed and λ varies. The different wavelengths present in the incident beam select the appropriate reflecting planes out of the numerous ones present in the crystal so that the *Bragg's condition of diffraction* is satisfied.

The experiment

Fig. 3.12 shows the experimental arrangement of the Laue method. An X-ray beam of *continuous wavelength* (*white radiation*) is allowed to fall on a stationary single crystal (C). The crystal selects and diffracts the appropriate wavelength λ for which a family of planes of inter-planar spacing d and glancing angle θ, Bragg's condition, $2d \sin\theta = n\lambda$ is satisfied.

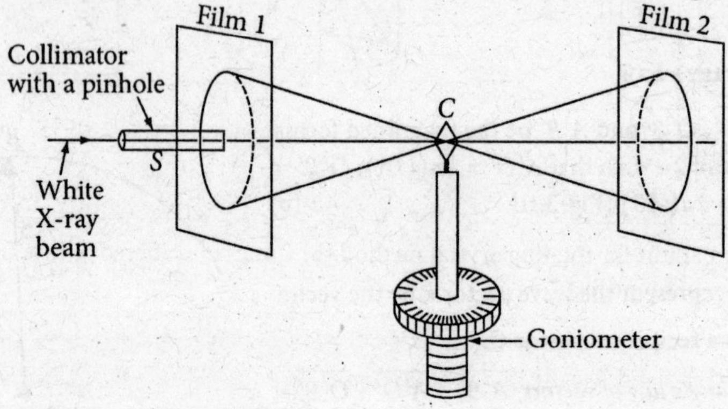

Fig. 3.12: A flat plate camera used in Laue diffraction method

The source of X-ray is capable of producing X-rays of wide range of wavelengths (0.2 Å to 2 Å). The beam before incidence on the crystal is well collimated by a pinhole (S) arrangement. Film 1 and Film 2 are held in proper position to record either the reflected beam or transmitted diffracted beam.

The diffraction pattern consists of a series of spot whose position is related to the orientation of the crystal to the incident beam. The crystal can be rotated with the help of goniometer to change its orientation w.r.t. the incident beam when a new set of atomic planes satisfies the condition of diffraction. In this way different types of pattern is obtained on the film. In each pattern, different types of symmetry of the crystal is reflected.

Laue method is used to determine (i) the *crystal symmetry* and (ii) *imperfection or strains in the crystal.*

3.8.2 Rotating Crystal Method

In this method, a single crystal C is rotated about a fixed vertical axis in a beam of *mono-energetic X-ray* [Fig. 3.13 (a)]. The photographic film is mounted on the inner wall of a cylinder co-axial with the mount. One of the crystallographic axes is kept vertical and normal to the incident beam of X-rays. The X-rays are diffracted from a given crystal plane, when in the course of rotation, the value of θ satisfies the Bragg's condition $2d \sin\theta = n\lambda$.

The planes parallel to the axis of rotation diffract the X-rays in the horizontal plane in which the incident X-rays lie. So, these diffracted patterns cannot be observed. The planes which are inclined to the rotation axis produce diffraction patterns which lie above or below the horizontal plane containing incident X-rays, depending upon the angle of inclination.

If the c-axis of the crystal is made to coincide with the axis of rotation, then all planes of the type (hko) will produce the central *layer lines*. Similarly, the planes (hkl) and $(hk\bar{l})$ will produce the layer line above and below the central line. The vertical distance between the layer lines depends on the distance between the lattice points along c-axis of the crystal. Thus, the distance c can be measured from the film. In a similar way the lattice vectors $\vec{a}$ and $\vec{b}$ can be measured and the size of the unit cell can be determined.

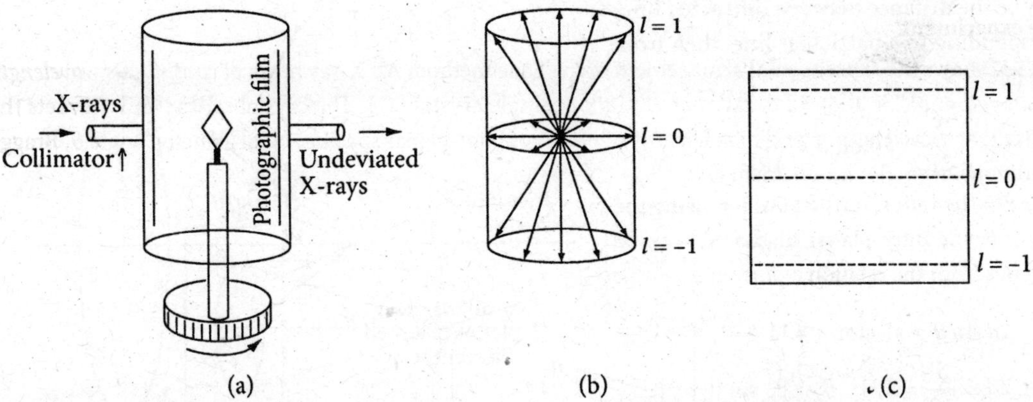

Fig. 3.13: (a) Apparatus for rotating crystal method (b) Cones of scattered X-rays from (*hkl*) planes (c) Layer lines on photoplate

If $\vec{c}$-axis of the crystal coincides with the axis of rotation of the crystal, then $S_n = R\cos\phi_n$, where S_n = separation of layer lines and R = the radius of the camera and $\phi_n = (90^\circ - \theta_n)$, θ_n = the angle of diffraction for the nth order.

$$\therefore S_n = \frac{\cos\phi_n \cdot R}{\sin\phi_n} = \frac{\cos\phi_n \cdot R}{\sqrt{1-\cos^2\phi_n}} = \frac{\frac{n\lambda}{c}\cdot R}{\sqrt{1-\left(\frac{n\lambda}{c}\right)^2}} \text{ or, } c = \frac{n\lambda}{\sin}\sqrt{(R^2+S_n^2)}.$$

Knowing the values of n, λ, R and S_n the value of c can be found out. In a similar way, the values of b and a can be determined. Thus, the parameters a, b and c of the unit cell of the crystal can be determined.

3.8.3 Powder Method

In this method, a large number of small crystallites are exposed to X-rays. This lattice planes of the crystallites which satisfy the Bragg's condition, $2d\sin\theta = n\lambda$, diffract the X-rays. Here d and θ may have different values for a fixed value of λ.

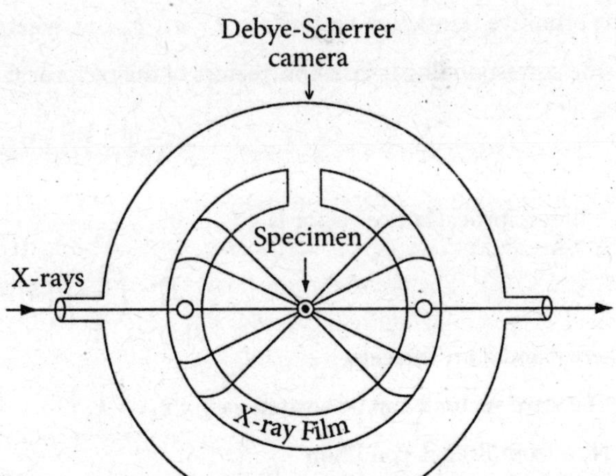

Fig. 3.14: X-ray diffraction camera for powder samples

A finely powdered specimen is packed in a thin-walled capillary tube which is held on a movable mount at the centre of a cylindrical camera [Fig. 3.14] called *Debye-Scherrer camera*. For an arbitrary orientation of the sample w.r.t. the incident *monochromatic X-rays*, some set of the crystal planes satisfy the condition of Bragg reflection, $2d\sin\theta = n\lambda$. For a particular value of λ the reflected rays for a particular set of lattice plane lie on the surface of a cone of semi-vertical angle 2θ. Additional cones of different semi-vertical angles arise due to other set of planes. These cones intersect the photographic film in concentric rings or arcs of circles with a sharp circular spot at the centre due to the direct beam. The crystal is slowly rotated during exposure of X-rays to ensure that all possible sets of planes face the incident X-rays.

If S be the distance between diffracted lines corresponding to a particular line, then from Fig. 3.15

$$\frac{S}{2R} = 2\theta \text{ or, } \theta = \frac{S}{4R},$$

where R = the radius of the film. Knowing the value of θ, the inter-planar distance d can be found out from the relation:

$$2d \sin\theta = \lambda, \text{ (for } n = 1)$$

$$\text{or, } 2d \times \frac{S}{4R} = \lambda, \ [\because \theta \to 0, \ \sin\theta \simeq \theta]$$

$$\text{or, } d = \frac{2R\lambda}{S}. \tag{3.55}$$

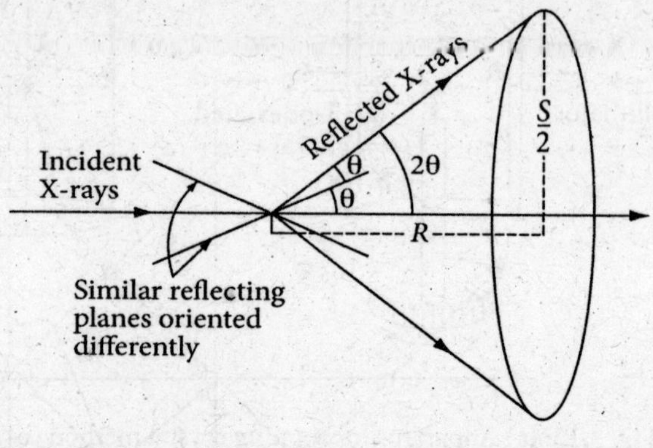

Fig. 3.15: X-ray diffracted cone of semi-vertical angle 2θ

The d values are used to determine the space lattice of the crystal structure.

3.9 Brillouin Zones

We have seen in Art. 2.6 that all the k-values for which the reciprocal lattice points intersect the Ewald sphere are Bragg reflected.

A Brillouin Zone is, therefore, the locus of all those k-values in the reciprocal lattice which are Bragg reflected.

3.9.1 Brillouin Zones for a Simple Square Lattice

The primitive translation vectors are $\vec{a} = a\hat{i}$; $\vec{b} = a\hat{j}$, where a is the side of the square lattice.

The corresponding translation vectors of the reciprocal lattice are

$$\vec{a}^* = \left(\frac{1}{a}\right)\hat{i}; \ \vec{b}^* = \left(\frac{1}{a}\right)\hat{j}.$$

∴ the reciprocal lattice vector is

$$\vec{G} = \left(\frac{2\pi}{a}\right)\left(h\hat{i} + k\hat{j}\right),$$

where h and k are integers.

The wave vector $\vec{k}$ can be written as $\vec{k} = k_x\hat{i} + k_y\hat{j}$.

Now from Bragg's condition

$$2\vec{k}\cdot\vec{G} + G^2 = 0 \text{ or, } \frac{4\pi}{a}\left[\left(k_x\hat{i} + k_y\hat{j}\right)\cdot\left(h\hat{i} + k\hat{j}\right)\right] + \frac{4\pi^2}{a^2}\left(h^2 + k^2\right) = 0$$

$$\text{or, } hk_x + kk_y = -\left(\frac{\pi}{a}\right)\left(h^2 + k^2\right). \tag{3.56}$$

Taking all possible combinations of h and k, we can get the values of k which will be Bragg reflected: If $h = \pm 1$, $k = 0$, then $k_x = \pm\frac{\pi}{a}$, k_y is arbitrary. Also, if $h = 0$ and $k = \pm 1$, then $k_y = \pm\frac{\pi}{a}$ and k_x is arbitrary.

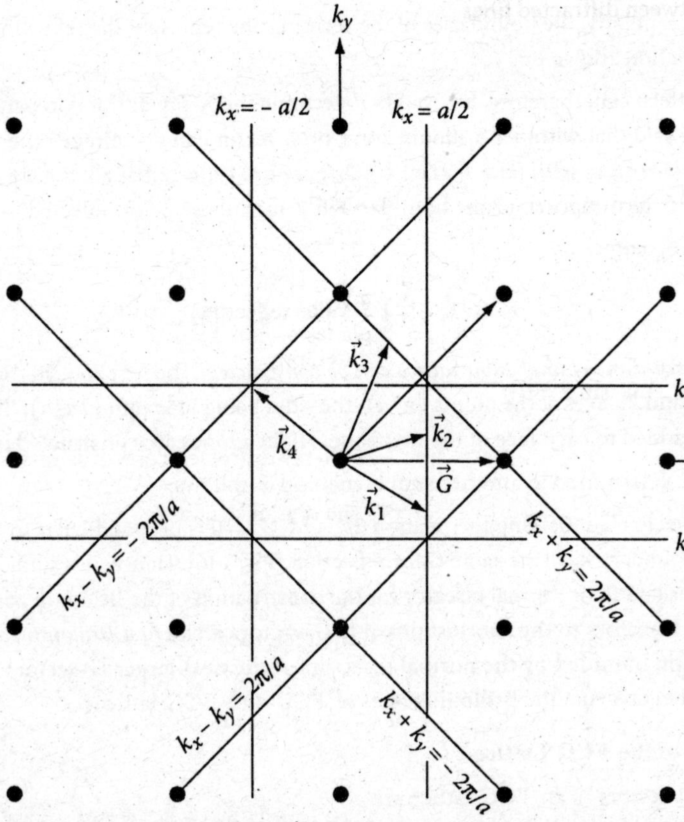

Fig. 3.16: Brillouin zones of a square lattice in its reciprocal lattice. The vectors $\vec{k}_x$, $\vec{k}_y$ and $\vec{k}_z$ are Bragg reflected whereas $\vec{k}_4$ is not. The vectors $\vec{k}_1$ and $\vec{k}_2$ have the same reciprocal lattice vector $\vec{G}_1$ whereas $\vec{G}_2$ is the reciprocal vector of $\vec{k}_3$.

These four lines, viz.,

$$\left.\begin{aligned} k_x &= +\frac{\pi}{a}; \quad & k_x &= -\frac{\pi}{a} \\ k_y &= +\frac{\pi}{a}; \quad & k_y &= -\frac{\pi}{a} \end{aligned}\right\} \tag{3.57}$$

are plotted in Fig. 3.16. All $\vec{k}$-vector originating at the origin and terminating on these lines will produce Bragg reflection.

Some other set of lines, viz.,

$$\pm k_x \pm k_y = \frac{2\pi}{a} \tag{3.58}$$

are possible. The regions included in such lines are called *Brillouin zones.* The square bounded by $k_x = \pm\frac{\pi}{a}$ and $k_y = \pm\left(\frac{\pi}{a}\right)$ is the *first Brillouin zone* which is a square of side $(2\pi/a)$. The additional area included by $\pm k_x \pm k_y = (2\pi/a)$ is the *second Brillouin zone* and so on.

The boundaries of the Brillouin zones represent the loci of $\vec{k}$-values which are Bragg reflected and may be considered as the reflecting planes. Thus, the boundaries of the first zone represent the reflecting planes for the first order reflection. Similarly, the boundaries of the second zone represent the reflecting planes for the second order Bragg reflection and so on.

The Brillouin zone pattern can, therefore, be used to determine the X ray diffraction pattern of a crystal and vice versa. It may be said that within a Brillouin zone, no $\vec{k}$ vector can give Bragg reflection. This does not mean that even the $\vec{k}$-vectors with $|\vec{k}| > \frac{\pi}{a}$ (i.e., $\lambda < 2a$) are not reflected at all, but this simply means that if these vectors do not terminate at a zone boundary only, then these go unreflected.

Using the generalized equation

$$hk_x + kk_y + lk_z = -\left(\frac{\pi}{a}\right)\left(h^2 + k^2 + l^2\right), \tag{3.59}$$

the *Brillouin zones for three-dimensional cubic lattice can be constructed.* The first zone in this case is a cube intersecting the k_x-, k_y- and k_z-axes at the points (π/a), the sides being of length $(2\pi/a)$. The second zone is made up of pyramids added to each face of the first zone. Higher zones are constructed in a similar way.

Brillouin zones can be determined in another simple method as follows:

The reciprocal lattice vector $\vec{G}$ satisfying the relation $2\vec{k}\cdot\vec{G} + G^2 = 0$ is perpendicular to zone boundary and all $\vec{k}$'s lying on this boundary have the same $\vec{G}$ for reflection [Fig. 3.16]. Hence, it is sufficient to consider only the allowed $\vec{G}$-values and their normal bisector for the construction of the Brillouin zones. The region bounded by the normal bisectors to the shortest possible $\vec{G}$-vectors is the *first Brillouin zone.* The *second Brillouin zone* is the region bounded by the normal bisectors of the next larger $\vec{G}$-vectors and so on. We shall apply this method to construct the Brillouin zones of FCC and BCC lattices.

3.9.2 Brillouin Zones of the FCC Lattice

The primitive translation vectors of an FCC lattice are

$$\vec{a} = \frac{a}{2}(\hat{i} + \hat{j})$$
$$\vec{b} = \frac{a}{2}(\hat{j} + \hat{k})$$
$$\vec{c} = \frac{a}{2}(\hat{k} + \hat{i}).$$

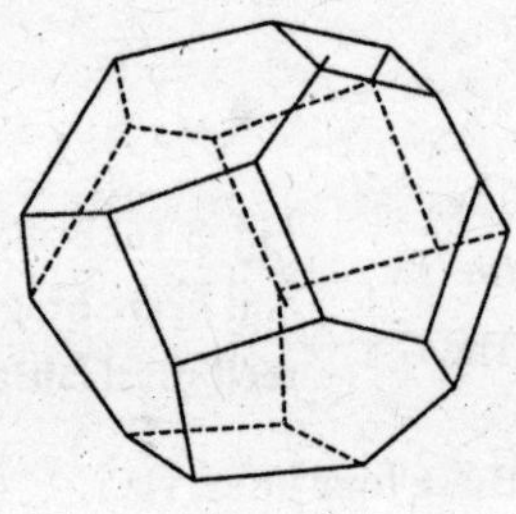

Fig. 3.17: Brillouin zones of an FCC lattice

The primitive translation vectors of its reciprocal lattice are

$$\vec{a}^* = \left(\frac{2\pi}{a}\right)(\hat{i} + \hat{j} - \hat{k})$$
$$\vec{b}^* = \left(\frac{2\pi}{a}\right)(-\hat{i} + \hat{j} + \hat{k})$$
$$\vec{c}^* = \left(\frac{2\pi}{a}\right)(\hat{i} - \hat{j} + \hat{k}).$$

If h, k, l are integers, then the G-type vector is

$$\vec{G} = 2\pi\left(\vec{a}^* h + \vec{b}^* k + \vec{c}^* l\right) = \frac{2\pi}{a}\left[(h - k + l)\hat{i} + (h + k - l)\hat{j} + (-h + k + l)\hat{k}\right]. \tag{3.60}$$

The shortest non-zero $\vec{G}$'s are the following eight vectors:

$$(2\pi/a)\left(\pm\hat{i} \pm \hat{j} \pm \hat{k}\right).$$

The first zone boundaries are determined mostly by the eight planes normal to these vectors at their midpoints. But the corners of the octahedron so formed are truncated by the planes which are perpendicular bisectors of the six reciprocal lattice vectors:

$$\frac{2\pi}{a}\left(\pm 2\hat{i}\right);\ \frac{2\pi}{a}\left(\pm 2\hat{j}\right);\ \frac{2\pi}{a}\left(\pm 2\hat{k}\right).$$

Then the first Brillouin zone is the truncated octahedron for the FCC lattice as shown in Fig. 3.17. This is also one of the primitive unit cells of a BCC lattice.

3.9.3 Brillouin Zones of BCC Lattice

The primitive translation vectors of a BCC lattice are

$$\vec{a} = \left(\frac{a}{2}\right)\left(\hat{i}+\hat{j}-\hat{k}\right)$$
$$\vec{b} = \left(\frac{a}{2}\right)\left(-\hat{i}+\hat{j}+\hat{k}\right)$$
$$\vec{c} = \left(\frac{a}{2}\right)\left(\hat{i}-\hat{j}+\hat{k}\right).$$

The primitive translation vectors of its reciprocal lattice are

$$\vec{a}^* = \left(\frac{2\pi}{a}\right)\left(\hat{i}+\hat{j}\right)$$
$$\vec{b}^* = \left(\frac{2\pi}{a}\right)\left(\hat{j}+\hat{k}\right)$$
$$\vec{c}^* = \left(\frac{2\pi}{a}\right)\left(\hat{k}+\hat{i}\right).$$

These are the primitive translation vectors of an FCC lattice. Thus, *the FCC lattice is the reciprocal lattice of the BCC lattice.*

The $\vec{G}$-type lattice vector is

$$\vec{G} = 2\pi\left(h\vec{a}^* + k\vec{b}^* + l\vec{c}^*\right) = \frac{2\pi}{a}\left[(h+l)\hat{i} + (h+k)\hat{j} + (k+l)\hat{k}\right]. \tag{3.61}$$

The shortest non-zero $\vec{G}$'s are the following twelve vectors:

$$\left(\frac{2\pi}{a}\right)\left(\pm\hat{i}\pm\hat{j}\right);\ \left(\frac{2\pi}{a}\right)\left(\pm\hat{j}\pm\hat{k}\right);\ \left(\frac{2\pi}{a}\right)\left(\pm\hat{k}\pm\hat{i}\right).$$

The first Brillouin zone is the region enclosed by the normal bisector planes to these twelve vectors. This zone has the shape of a regular twelve faced solid and is called *rhombic dodecahedron* [Fig. 3.18].

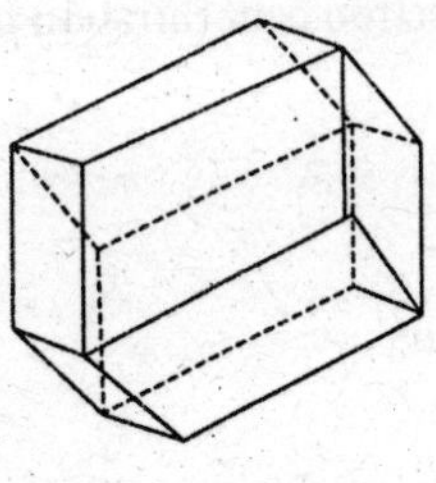

Fig. 3.18: First Brillouin zone of a BCC lattice

3.10 Amplitude of the Scattered Wave

Let a plane wave of X-rays be incident on a small crystal with $\vec{a}$, $\vec{b}$ and $\vec{c}$ as the *primitive vectors* of the space lattice. If the beam is not disturbed due to the refractive index of the crystal or due to loss of energy in the scattering, then the incident wave at $\vec{R}$ can be written as [Fig. 3.19]

$$F\left(\vec{R}, t\right) = F_0 \exp\left(\vec{k}\cdot\vec{R} - \omega t\right), \tag{3.62}$$

where $\vec{k}$ is the *incident wave vector* and ω is the *angular frequency.*

Putting $t = 0$ in eqn (3.62), we get the spatial variation of $F(\vec{R})$ as

$$F(\vec{R}) = F_0 \exp(i\vec{k}\cdot\vec{R}). \tag{3.63}$$

Now the magnitude of the scattered X-rays at P, outside the crystal is proportional to

$$F_0 \exp(i\vec{k}\cdot\vec{R})\left(\frac{e^{iKr}}{r}\right). \tag{3.64}$$

The factor $F_0 \exp(i\vec{k}\cdot\vec{R})$ contains the amplitude and phase of the incident wave and (e^{iKr}/r) represents the spatial variation of the scattered X-rays from $\vec{R}$ in the crystal.

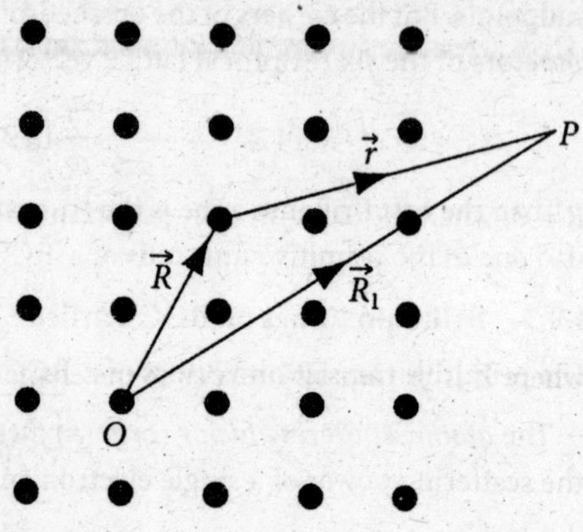

Fig. 3.19: Amplitude of the scattered wave

From Fig. 3.19

$$r^2 = |\vec{R}_1 - \vec{R}|^2 = R_1^2 + R^2 - 2R_1R\cos(\vec{R}_1,\vec{R}) \tag{3.65}$$

$$\simeq R_1\left[1 - \frac{2R}{R_1}\cos(\vec{R},\vec{R}_1)\right]^{1/2}$$

$$\simeq R_1 - R\cos(\vec{R},\vec{R}_1), \text{ (when } R_1 \gg R). \tag{3.66}$$

Eqn (3.64), then reduces to

$$\frac{F_0}{r}\exp\left[i\vec{k}\cdot\vec{R} + ikR_1 - ikR\cos(\vec{R},\vec{R}_1)\right]. \tag{3.67}$$

Neglecting the interaction of the incident radiation with the nucleus, the electrons in an atom are the major scatterer. So, the amplitude of the scattered wave from the element of volume dV around $\vec{R}$ is proportional to the electron concentration $\rho(\vec{R})$ at $\vec{R}$. Thus, the amplitude of the scattered radiation at P is proportional to

$$\int \rho(\vec{R})\exp\left[i\vec{k}\cdot\vec{R} - ikR\cos(\vec{R},\vec{R}_1)\right]dV. \tag{3.68}$$

[The factor $\exp(ikR_1)$ being constant over the volume is omitted]

Again,

$$i\vec{k}\cdot r - ikR\cos(\vec{R},\vec{R}_1) = i\vec{R}\cdot(\vec{k}-\vec{k}') = -i\vec{R}\cdot\vec{K}, \tag{3.69}$$

where $(\vec{k}-\vec{k}') = \vec{K}$ and $\vec{k}'$ is in the direction of scattering $\vec{R}_1$.

Therefore, the scattered amplitude by the electron distribution is proportional to f, where

$$f = \int \rho(\vec{R})\, e^{-i\vec{R}\cdot\vec{K}}\, dV. \tag{3.70}$$

Atomic Scattering Factor

In the foregoing discussion if the angle between $\vec{R}$ and $\vec{K}$ be α, then $\vec{R}\cdot\vec{K} = RK\cos\alpha$. If the electron distribution is assumed to be *spherically symmetrical* about the origin O [Fig. 3.19], then

$$dV = 2\pi R^2 dR \sin\alpha\, d\alpha.$$

Then

$$f = 2\pi \int_{\alpha=\pi}^{0} \int_{R} R^2 dR \sin\alpha \, d\alpha \rho(R)\, e^{-iRK\cos\alpha} = 2\pi \int_{R} \int_{\alpha=0}^{\pi} R^2 dR d(\cos\alpha) \rho(R)\, e^{-iRK\cos\alpha}$$

$$= 2\pi \int_{R} R^2 \rho(R) \frac{e^{iRK} - e^{-iRK}}{iRK} dR = 2\pi \int_{R} R^2 \rho(R) \frac{\sin KR}{RK} dR. \quad (3.71)$$

$R = 0$, when the electron concentration is localized at O, then

$$f_0 = 4\pi \int \rho(R) R^2 dR = Z, \left[\because \frac{\sin RK}{RK} \to 1 \text{ as } R \to 0\right] \quad (3.72)$$

where Z is the *atomic number* of the element.

The *atomic scattering factor* (or *form factor*) f describes the scattering power of a single atom vis-à-vis the scattering power of a single electron and is given by

$$f = \frac{\text{amplitude of radiation scattered by the actual electron distribution}}{\text{amplitude of radiation scattered from an electron}}.$$

In general, $f < Z$.

3.11 Geometrical Structure Factor

The intensity of the X-ray beam diffracted from a crystal depends upon the contents of the unit cell, type and distribution of atoms in the cell in addition to the atomic scattering factors of various atoms involved. The phase of the scattered X-rays from different atoms may also differ.

To find out an expression for the *geometrical structure factor* (S), let the nucleus of the jth atom be at $\vec{r}_j$ given by

$$\vec{r}_j = u_j\vec{a} + v_j\vec{b} + w_j\vec{c}. \quad (3.73)$$

Now the total scattered amplitude is proportional to

$$F(hkl) = \sum_j f_j e^{i\vec{r}_j \cdot \vec{K}}. \quad (3.74)$$

The summation extends over all the atoms in the unit cell which is defined as

$$\frac{\text{[the amplitude of the wave scattered by all the atoms in the unit cell]}}{\text{[the amplitude scattered by a free electron for the same incident beam]}}.$$

If (hkl) be the Miller indices of the reflecting plane, then

$$\vec{r}_j \cdot \vec{K} = \left(u_j\vec{a} + v_j\vec{b} + w_j\vec{c}\right) \cdot \left(h\vec{a}^* + k\vec{b}^* + l\vec{c}^*\right)$$
$$= 2\pi\left(u_j h + v_j k + w_j l\right). \quad (3.75)$$

From eqns (3.74) and (3.75), the total scattered amplitude is

$$F(hkl) = \sum f_j e^{i2\pi(u_j h + v_j k + w_j l)} \quad (3.76)$$
$$= fS, \quad (3.77)$$

where

$$S = \sum_j e^{2\pi i(u_jh+v_jk+w_jl)} \tag{3.78}$$

is called the *geometrical structure factor.*

The intensity of radiation is given by

$$I = |F|^2 = F^* F$$

$$= \left[\sum_j f_j \cos 2\pi\left(u_jh' + v_jk' + w_jl'\right)\right]^2 + \left[\sum_j f_j \sin 2\pi\left(u_jh' + v_jk' + w_jl'\right)\right]^2, \tag{3.79}$$

where F^* *is the complex conjugate of* F.

3.12 Geometrical Structure Factor for Some Simple Crystals

3.12.1 Simple Cubic Crystal

In SC crystal the effective number of atoms in a unit cell is one. If the atom is at the origin, then the *structure factor* from eqn (3.78) is unity. Then the total diffraction amplitude from eqn (3.77) is $F(hkl) = f$.

Hence, all the diffraction lines predicted by *Bragg's law* will appear in the diffraction pattern.

3.12.2 Body-Centred Cubic Crystal

A body-centred cubic crystal contains two atoms per unit cell. The contributions from eight corner-atoms give rise to one atom and the other is situated at the centre of the unit cell. If the coordinates of the corner-atom be $(0, 0, 0)$, then the coordinates of the central atom is $\left(\frac{1}{2}, \frac{1}{2}, \frac{1}{2}\right)$.

The diffraction amplitude from eqn (3.77) is then

$$F(hkl) = f\sum_j e^{2\pi i(u_jh+v_jk+w_jl)} = f\left[1 + e^{i\pi(h+k+l)}\right], \tag{3.80}$$

since both the atoms in the unit cell are identical.

It is evident from eqn (3.80), that the geometrical structure factor becomes zero for *odd* values of $(h+k+l)$, since $e^{i\pi n} = -1$, if n is odd. If $(h+k+l)$ is *even*, then $F(hkl) = 2f$ and the intensity from eqn (3.79) is $4f^2$. Thus, the reflection from (100)-, (111)-, (210)-planes, etc., are missing and those from (110), (200), etc., are present in a BCC lattice. It is to be noted that the presence or absence of a reflection is considered only in terms of *first order* reflection.

3.12.3 Face-Centred Cubic Crystal

A face-centred cubic crystal has *four* atoms per unit cell. If the coordinates of a corner atom be $(0, 0, 0)$, then the remaining *three* atoms of the unit cell have the coordinates $\left(\frac{1}{2}, 0, \frac{1}{2}\right)$, $\left(\frac{1}{2}, \frac{1}{2}, 0\right)$ and $\left(0, \frac{1}{2}, \frac{1}{2}\right)$. Then the diffraction amplitude becomes

$$F(hkl) = f\left[1 + e^{\pi i(h+l)} + e^{\pi i(h+k)} + e^{\pi i(k+l)}\right]. \tag{3.81}$$

Thus, we note from eqn (3.81) that $F(hkl)$ is non-zero, when h, k, l are *all even* or *all odd.* Therefore, the reflections from (111), (200), (220) are present but those from (100), (110), (211) are absent.

3.13 Liquid Crystals

Liquid crystals show *intermediate mesomorphic (liquid crystalline) phase* of crystals and liquids. These are primarily *organic compounds resembling partly an ordered crystal and a disordered liquid.* They exhibit *anisotropic* properties like dielectric, optical, magnetic, etc. which are possessed by crystals. At the same time the property of fluidity, droplet forming capacity of liquid are also exhibited.

The transition from solid phase to meso-phase takes place at a specific temperature, called the melting or (solidification) point (T_s). The liquid crystal converts to the isotropic transparent liquid at T_t, called the *transparency temperature.* The values of T_s and T_t are different for different crystals and the interval $\Delta T = (T_t - T_s)$ is the range of temperature in which the liquid crystalline phase exists. The value of ΔT differs from crystal to crystal. Liquid crystals are organic materials made of rigid and moderately large rod like molecules having length 1 to 3 nm and breadth 0.3 to 1.0 nm. A high degree of polarizability is exhibited by a cluster of atoms situated at the ends of the molecules. Weak *van der Waals' force* makes the ordered molecular arrangement in the liquid crystal.

The electric displacement vector $\vec{D}$ and electric field $\vec{E}$ are not in the same direction in the anisotropic liquid crystal medium.

The following chart shows the classification of the liquid crystal:

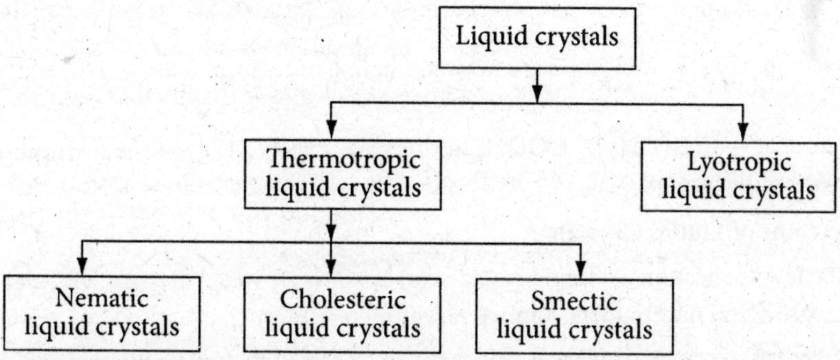

3.13.1 Nematic Liquid Crystals

The molecules of the *nematic liquid crystals* have a natural tendency to orient their long axes along a preferred *director axis* $(\vec{L})$ as shown in Fig. 3.20. The molecules align parallel to a unique axis in different small regions of this crystal. The direction of orientation generally varies in different regions of the liquid crystal. This property of the crystal is responsible for the *high optical inhomogeneity* of the crystal because of which it appears *opaque* both in reflected and transmitted light.

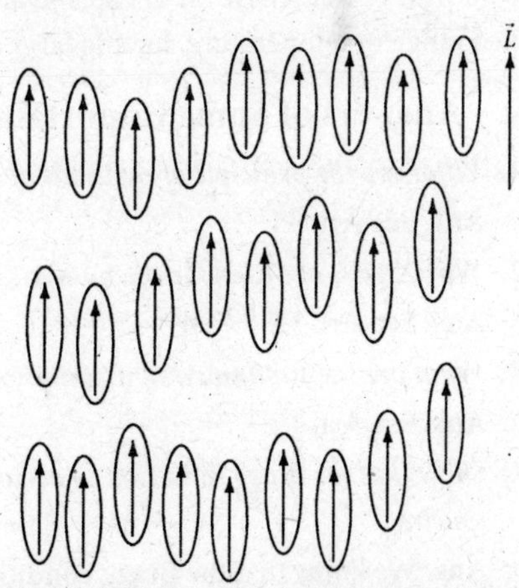

Fig. 3.20: Orientation of molecules along a preferred director axis $(\vec{L})$

Smectic liquid crystals

In *smectic liquid crystal,* the molecules orient themselves in parallel layers gliding one relative to other. This promotes fluidity of the system. The long axis of the molecules are either perpendicular to the planes [Fig. 3.21] or rotated through an angle ϕ with the normal to the planes [Fig. 3.22]. The rotation angle ϕ depends on the temperature which can be altered by applying an external electric field. The *smectic axis* represented by the vector $\vec{L}$ and the shear plane *AA* is shown in the Figs (3.21) and (3.22).

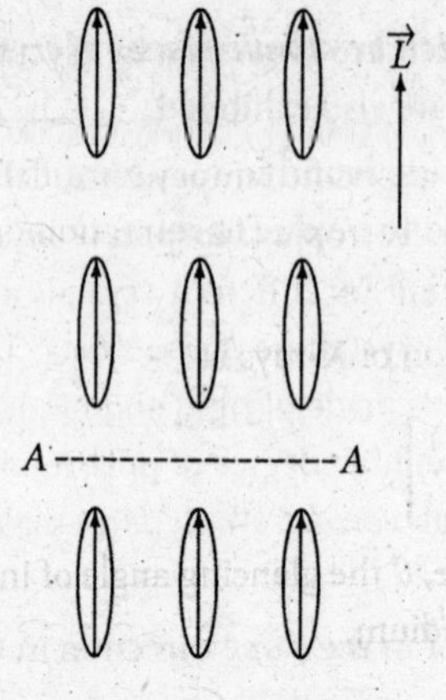

Fig. 3.21: The long axis of the molecule is perpendicular to the plane

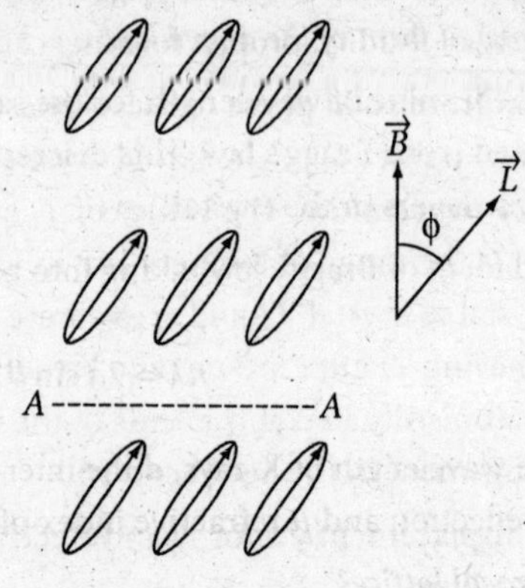

Fig. 3.22: The long axis of the molecule is rotated through an angle

The two parts of *amphiphilic* molecules are (a) water hating or hydrophobic part (b) water loving part or hydrophilic part.

Example. Stearic acid — $CH_3(CH_2)_{16}$ COOH. In this case, $CH_3(CH_2)_{16}$ is the hydrophobic group and COOH is the hydrophilic group.

3.13.2 Applications of Liquid Crystals

Following are the few applications of liquid crystals: (a) Low power visual displays, viz., electronic watches, calculators, etc. (b) Digital multimeters, ammeters, voltmeters, etc.

Advantage of the liquid crystal displays is that they can be manufactured in any size and shape easily and consume less power. These are cheap and are of variety of colours. The *disadvantage* is that they operate in a limited range of temperature, having large response time, etc.

3.14 Answers of Some Important Questions

3.1. *What are the principle of methods of crystal structure analysis?*

Ans. See Art. 3.1.

3.2. *Why X-rays are used for the analysis of crystal structure?*

Ans. See Art. 3.2.

3.3. *What are the fundamental assumptions in deriving Bragg's law of X-ray diffraction?*

Ans. See Art. 3.3.

3.4. *Show that for Bragg reflection to occur the wavelength of X-rays must not exceed twice the inter-planar spacing.*

Ans. We know that the Bragg condition for X-ray reflection by atomic planes of crystals is $2d \sin\theta = n\lambda$, where θ is the glancing angle of incidence, d is the inter-planar spacing, λ is the wavelength of X-rays and n is the order of reflection.

Since $|\sin\theta| \leq 1$, $\frac{n\lambda}{2d} \leq 1$.

For the first order reflection $n = 1$. $\therefore$ $\lambda \leq 2d$.

Thus, λ must not exceed twice the inter-planar spacing (d) for Bragg reflection to occur.

3.5. *What led to introduce correction in Bragg's law of X-ray reflection by atomic planes of crystals. Write down the corrected form of the above law.*

Ans. Experiment value of wavelength for the same X-ray line was found to vary for different orders when computed from Bragg's law. This discrepancy was found to be due to refraction of X-rays at different atomic layers in the crystal.

The corrected form of Bragg's law taking into account, refraction of X-rays is

$$n\lambda = 2d\sin\theta\left[1 - \frac{4d^2(1-\mu)}{n^2\lambda^2}\right],$$

where λ is the wavelength of X-rays, d the inter-planar distance, θ the glancing angle of incidence, n the order of reflection and μ refractive index of the crystal-medium.

3.6. *What is reciprocal lattice?*

Ans. See first para of Art. 3.4. Then add: The reciprocal lattice corresponding to a Bravais lattice point $\vec{R}$ is defined as the set of all wave vectors that give plane waves of the form $e^{i\vec{k}\cdot\vec{r}}$ with the periodicity $(\vec{R})$ of the Bravais lattice.

3.7. *What is the importance of reciprocal lattice in solid state physics?*

Ans. See Art. 3.4.1.

3.8. *Write down Laue equation representing the conditions of X-ray diffraction by crystals.*

Ans. See Art. 3.6 [Eqn (3.44)].

3.9. *Obtain Bragg's Law from Laue equations.*

Ans. See Art. 3.6.1.

3.10. *Compare Bragg and Laue approaches as regards X-ray reflection is concerned.*

Ans. See Art. 3.6.2.

3.11. *Give the vector form of Bragg's Equation.*

Ans. Art. 3.7.1.

3.12. *What are Brillouin Zones?*

Ans. All the values of wave vector $\vec{k}$ for which the reciprocal lattice points intersect the Ewald sphere are Bragg reflected. A Brillouin zone is, therefore, the locus of all those k-values in the reciprocal lattice which are Bragg reflected.

3.13. *What are liquid crystals?*

Ans. Liquid crystals are organic compounds exhibiting mesomorphic phase of crystals and liquids. They exhibit anisotropic properties like dielectric, optical, magnetic, etc.

3.14. *How does anisotropy arise in liquid crystals?*

Ans. In liquid crystal, the direction of the electric field $\vec{E}$ and the electric displacement vector $\vec{D}$ are not in the same direction which give rise to anisotropy.

3.15. *Mention some applications of liquid crystals.*

Ans. Various liquid crystal displays are used in electronic watch, calculators, digital electronic meters, etc.

3.15 Solved Problems

3.1. *The spacing between principal planes of NaCl crystal is 2.82 Å. It is found that first order Bragg reflection occurs at an angle of 10°. What is the wavelength of λ-rays?*

Solution: The Bragg's equation is $2d \sin\theta = n\lambda$.

Here the inter-planar distance $d = 2.82 \times 10^{-10}$ m, the order number $n = 1$ and the glancing angle of incidence is $\theta = 10°$, $\lambda = ?$

From Bragg's equation,

$$\lambda = \frac{2d \sin\theta}{n} = \frac{2 \times (2.82 \times 10^{-10}) \times \sin 10°}{1} = 0.98 \times 10^{-10}\ \text{m} = \mathbf{0.98\ Å}.$$

3.2. *A beam of X-rays of wavelength 0.842 Å is incident on a crystal at a glancing angle of 8°35′ when the first order Bragg reflection occurs. Calculate the glancing angle of the third order reflection.* **[WBUT 2002]**

Solution: The Bragg's equation is $2d \sin\theta_n = n\lambda$.

For the first order spectrum, $n = 1$, $\lambda = 0.842$ Å, $\theta_1 = 8°35' \simeq 8.6°$.

$$\therefore d = \frac{\lambda}{2\sin\theta_1} = \frac{0.842}{2 \times 0.1347} = \mathbf{3.125\ Å}.$$

If θ_3 be the glancing angle for the third order ($n = 3$) reflection, then

$$\sin\theta_3 = \frac{3\lambda}{2d} = \frac{3 \times 0.842}{2 \times 3.125} = \mathbf{0.40416}.$$

$$\therefore \theta_3 = \sin^{-1}(0.40416) = \mathbf{26°29'}.$$

3.3. *X-rays of wavelength 0.71 Å are reflected from the (110)-plane of a rock salt crystal of lattice constant $a = 2.82$ Å. Calculate the corresponding glancing angle for second order reflection.*

Solution: The inter-planar spacing between (111)-planes is

$$d = \frac{a}{\sqrt{h^2 + k^2 + l^2}}.$$

Here $a = 2.82$ Å, $h = 1$, $k = 1$, $l = 0$

$$\therefore d = \frac{2.82}{\sqrt{1^2 + 1^2 + 0^2}} = \frac{2.82}{\sqrt{2}} = \frac{2.82}{1.41} = \mathbf{2\ Å}.$$

Now from Bragg's law, $2d \sin\theta = n\lambda$.

Here $d = 2$ Å, $n = 2$ and $\lambda = 0.71$ Å.

$$\therefore 2d\sin\theta = 2 \times 2\sin\theta = 2 \times 0.71 \text{ or, } \sin\theta = \frac{2 \times 0.71}{2 \times 2} = 0.355 \text{ or, } \theta = \mathbf{20°42'}.$$

3.4. *The spacing between successive (100)-planes of NaCl is 2.82 Å. X-rays incident on the surface of the crystal are found to give rise to first order Bragg reflection at glancing angle 8.8°. Calculate the wavelength of X-rays.* **[BU(H) 2003]**

Solution: The Bragg's equation is $2d \sin\theta = n\lambda$.

Here $d = 2.82$ Å, and $n = 1$, $\lambda = ?$

From Bragg's equation $1 \times \lambda = 2 \times 2.82 \times \sin(8.8°)$ or, $\lambda = 2 \times 2.82 \times 0.1530 = \mathbf{0.8629\ Å}$.

3.5. *Using a crystal of KCl, Bragg found the following glancing angles for intense reflection from the lattice planes (100), (110) and (111); $\theta_1 = 5.22°$, $\theta_2 = 7.30°$ and $\theta_3 = 9.05°$ respectively. Show that KCl has SC structure.* **[cf. CU(H) 1996]**

Solution: Bragg's law is $2d \sin\theta = n\lambda$.

$$\therefore \frac{1}{d} = \frac{2\sin\theta}{\lambda} \quad [\because n = 1 \text{ for the first order reflection}].$$

$$\therefore \frac{1}{d_{100}} : \frac{1}{d_{110}} : \frac{1}{d_{111}} = \sin 5.22° : \sin 7.30° : \sin 9.05° \quad \left[\because \tfrac{2}{\lambda} \text{ is constant}\right]$$

$$= 0.0910 : 0.1272 : 0.1570 = 1 : 1.40 : 1.73 \simeq \mathbf{1 : \sqrt{2} : \sqrt{3}}.$$

$\therefore$ KCl *has SC structure.*

3.6. *The first order Bragg reflection in polonium appears at $\sin\theta = 0.225, 0.317$ and 0.389 for reflection from (100)-, (110)- and (111)-planes respectively. The wavelength of X-ray used is 1.54 Å. Show that the unit cell of polonium is cubic. What is the length of the side of the unit cell?*

Solution: We have $\sin\theta = 0.225, 0.317$ and 0.389.

So, the values of $\sin\theta$ are in the ratio $0.225 : 0.317 : 0.389 = 1 : 1.41 : 1.73 = \mathbf{1 : \sqrt{2} : \sqrt{3}}$.

Thus, *the unit cell is cubic.*

If a be the length of the side of the cubic unit cell, then from Bragg's law for reflection at (100)-plane

$$a = d_{100} = \frac{\lambda}{2\sin\theta} = \frac{1.54}{2 \times 0.225} = \mathbf{3.42\ Å}.$$

3.7. *If V is the volume of a primitive unit cell in the direct lattice, show that the volume of a primitive unit cell in the reciprocal lattice is $8\pi^3/V$.* **[BH(H) 1996]**

Solution: If $\vec{a}^*, \vec{b}^*$ and $\vec{c}^*$ be the primitive vectors defining the unit cell in the reciprocal lattice, then its volume is

$$V^* = \left(\vec{a}^* \cdot \vec{b}^* \times \vec{c}^*\right) = (2\pi)^3 \frac{\vec{b} \times \vec{c}}{\left(\vec{a} \cdot \vec{b} \times \vec{c}\right)} \cdot \frac{\left\{(\vec{c} \times \vec{a}) \times (\vec{a} \times \vec{b})\right\}}{\left(\vec{a} \cdot \vec{b} \times \vec{c}\right)\left(\vec{a} \cdot \vec{b} \times \vec{c}\right)} = (2\pi)^3 \frac{\left(\vec{b} \times \vec{c}\right) \cdot \left\{(\vec{c} \times \vec{a}) \times (\vec{a} \times \vec{b})\right\}}{\left(\vec{a} \cdot \vec{b} \times \vec{c}\right)^3}$$

$$= (2\pi)^3 \frac{\left(\vec{b} \times \vec{c}\right) \cdot \left[\left\{\vec{c} \cdot (\vec{a} \times \vec{b})\right\}\vec{a} - \left\{\vec{a} \cdot (\vec{a} \times \vec{b})\right\}\vec{c}\right]}{\left(\vec{a} \cdot \vec{b} \times \vec{c}\right)^3} = \frac{(2\pi)^3}{\vec{a} \cdot \vec{b} \times \vec{c}} = \frac{(2\pi)^3}{V^3},$$

where $V = \vec{a} \cdot \vec{b} \times \vec{c}$ is the volume of the unit cell of the direct lattice. Thus, $V^* \propto 1/V$.

3.8. *Prove that the direct lattice is the reciprocal of its own reciprocal lattice.*

Solution: We consider the reciprocal of the reciprocal lattice vector

$$\vec{a}^* = 2\pi \frac{\vec{b} \times \vec{c}}{\vec{a} \cdot \vec{b} \times \vec{c}} \text{ as } \left(\vec{a}^*\right)^* = (2\pi) \frac{\vec{b}^* \times \vec{c}^*}{\vec{a}^* \cdot \vec{b}^* \times \vec{c}^*}.$$

Multiplying the RHS of the above equation by $(\vec{a} \cdot \vec{a}^*)$, we get

$$\left(\vec{a}^*\right)^* = (2\pi)(\vec{a} \cdot \vec{a})^* \cdot \frac{\vec{b}^* \times \vec{c}^*}{\vec{a}^* \cdot \vec{b}^* \times \vec{c}^*} = 2\pi\vec{a} \cdot \frac{\vec{a}^* \cdot \vec{b}^* \times \vec{c}^*}{\vec{a}^* \cdot \vec{b}^* \times \vec{c}^*} = \mathbf{2\pi\vec{a}}.$$

Thus, the statement is proved.

3.9. *Prove that the reciprocal lattice of a BCC lattice is an FCC lattice.* **[BU(H) 2000]**

Solution: The primitive basis vectors of a BCC lattice are

$$\vec{a}' = \frac{a}{2}(\hat{i}+\hat{j}-\hat{k}),\ \vec{b}' = \frac{a}{2}(-\hat{i}+\hat{j}+\hat{k}),\ \vec{c}' = \frac{a}{2}(\hat{i}-\hat{j}+\hat{k}),$$

where $\hat{i}$, $\hat{j}$ and $\hat{k}$ are the unit vectors along the edge of the cube whose a is the cube edge. The corresponding reciprocal lattice vectors are

$$\vec{a}^* = 2\pi\frac{\vec{b}'\times\vec{c}'}{\vec{a}'\cdot\vec{b}'\times\vec{c}'} = 2\pi\frac{a^2/2}{a^3/2}(\hat{i}+\hat{j}) = \frac{2\pi}{a}(\hat{i}+\hat{j})$$

$$\vec{b}^* = 2\pi\frac{\vec{c}'\times\vec{a}'}{\vec{a}'\cdot\vec{b}'\times\vec{c}'} = \frac{2\pi}{a}(\hat{j}+\hat{k})$$

$$\text{and } \vec{c}^* = 2\pi\frac{\vec{a}'\times\vec{b}'}{\vec{a}'\cdot\vec{b}'\times\vec{c}'} = \frac{2\pi}{a}(\hat{k}+\hat{i}).$$

For an FCC lattice, the primitive basis vectors are

$$\vec{a}' = \frac{a}{2}(\hat{i}+\hat{j}),\ \vec{b}' = \frac{a}{2}(\hat{j}+\hat{k}),\ \vec{c}' = \frac{a}{2}(\hat{k}+\hat{i}).$$

Comparing $\vec{a}^*, \vec{b}^*, \vec{c}^*$ and $\vec{a}', \vec{b}', \vec{c}'$, we find that $\vec{a}^*, \vec{b}^*, \vec{c}^*$, are the primitive basis vectors of an FCC lattice having cube edge $(4\pi/a)$. *This proves that the reciprocal lattice of a BCC lattice is an FCC lattice.*

3.10. *Starting from $2\vec{K}\cdot\vec{G} + G^2 = 0$, obtain the diffraction condition, $2d \sin\theta = n\lambda$.*

Solution: We know that spacing d_{hkl} between parallel lattice planes normal to the direction $\vec{G} = h\vec{A} + k\vec{B} + l\vec{C}$ is

$$d_{hkl} = \frac{2\pi}{|\vec{G}|}.$$

Now $2\vec{K}\cdot\vec{G} = G^2$ may be written as

$$2\left(\frac{2\pi}{\lambda}\right)\sin\theta = \frac{2\pi}{d_{hkl}} \quad \text{or, } \mathbf{2d_{hkl}\ sin = \lambda}.$$

Here θ is the angle between the incident beam and lattice plane. The integers that define $\vec{G}$ may contain a common factor n. Then we obtain $2d \sin\theta = n\lambda$.

3.11. *The wavelength of the k_α line of* Ag *is* 0.563 Å. *The radiation from a* Ag *target is analyzed with a Bragg spectrometer using a calcite crystal (a simple cube of lattice constant* 3.02945 Å*). Determine the angle of reflection of the first order. What is the highest order for which this may be observed?* **[CU(H) 1990]**

Solution: Bragg's law for the first order ($n = 1$) is $2d \sin\theta = \lambda$.

Here $d = 3.02945$ Å and $\lambda = 0.563$ Å.

$$\therefore \sin\theta = \frac{\lambda}{2d} = \frac{0.563}{6.0589} = 0.0929.$$

Hence, the angle of reflection for the first order is $\theta = \sin^{-1}(0.0929) = \mathbf{5.33°}$.

For the nth order reflection, $2d \sin\theta = n\lambda$.

For the highest volume of θ, i.e., 90°

$$n = \frac{2d}{\lambda} = \frac{6.0589}{0.563} = 10.8.$$

So, the highest order for which this reflection may be observed is **10**.

3.12. *A simple cubic crystal illuminated with X-rays of wavelength* 0.09 nm *is rotated and the first order Bragg reflection occurs at a minimum glancing angle of* 8.8°. *Which set of crystal planes are responsible for this reflection? Find the spacing between these planes and the angle for the first order reflection from the* (110) *crystal planes.*

Solution: Since the glancing angle is the minimum for the planes having largest spacing, the planes in question will be (100)-planes.

$\therefore 2d_{100} \sin\theta = \lambda$, (for $n = 1$).

$$\therefore d_{100} = \frac{\lambda}{2\sin\theta} = 0.294\,\text{nm}.$$

The spacing between the (110)-planes is

$$d_{110} = \frac{d_{100}}{\sqrt{1^2+1^2+0^2}} = \frac{d_{100}}{\sqrt{2}} = \frac{0.294}{\sqrt{2}} = \mathbf{0.208\,nm}.$$

So

$$2d_{110}\sin\theta_{110} = \lambda \text{ or, } \sin\theta_{110} = \frac{0.09}{2\times 0.208} = 0.2163.$$

$\therefore \theta_{110} = \mathbf{12.5°}$.

3.13. *Show that greater the diffraction angle, greater is the accuracy in determining the lattice parameters.*

Solution: We have the Bragg's equation:

$$2d\sin\theta = n\lambda \text{ or, } d\sin\theta = \frac{n\lambda}{2}.$$

Differentiating

$$d\cos\theta\, d\theta + \sin\theta\, d(d) = 0 \text{ or, } \frac{d(d)}{d} = -\cot\theta\, d\theta.$$

The above equation shows that as $\theta \to \frac{\pi}{2}$, $\cot\theta \to 0$, i.e., the error in the measurement of $d \to 0$. *This shows that greater the angle of diffraction, greater is the accuracy in determining the lattice parameters.*

3.14. *Given a square piece of X-ray film* 10 cm × 10 cm. *With Copper radiation of* $\lambda = 1.52$ Å *and powdered* NaCl (cubic) *with lattice parameter* 4.57 Å, *devise a diffraction experiment in such fashion that the rays from the* (111)*-planes will produce a circle of diameter* 10 cm (sin 16.75° = 0.288, tan 33.5° = 0.662).

[CU(H) 1999]

Solution: See Art. 3.8.3: (*Powder Method*).

Bragg's law for the first order diffraction is $2d\sin\theta = \lambda$.

Here

$$d = \frac{4.57}{\sqrt{3}}\,\text{Å} \quad\text{and}\quad \lambda = 1.52\,\text{Å}.$$

$$\therefore \sin\theta = \frac{\lambda}{2d} = \frac{1.52\times\sqrt{3}}{2\times 4.57} = 0.288 = \sin 16.75°.$$

$\therefore \theta = \mathbf{16.75°}$.

The X-ray film is placed normally to the incident X-radiation at a distance D from the powdered sample such that the diffraction cone intersects the film in a circle of 5 cm.

Then

$$\tan 2\theta = \frac{5}{D} \text{ or, } D = \frac{5}{\tan 2\theta} = \frac{5}{\tan 33.5^\circ} = \frac{5}{0.662} = \mathbf{7.55\ cm}.$$

3.15. *Monochromatic X-rays of wavelength 1.4 Å are incident on a crystal having 1.5 Å as interatomic spacing. Find the maximum order in which the diffraction takes place.*

Solution: We have Bragg's equation:

$$2d \sin\theta = n\lambda \text{ or, } n = \frac{2d\sin\theta}{\lambda}.$$

Here $d = 1.5$ Å and $\lambda = 1.4$ Å

$$\therefore n = \frac{3}{1.4}\sin\theta \text{ or, } n = 2.14 \sin\theta.$$

$\because |\sin\theta| \le 1$, the maximum order of the spectrum that can be observed is not more than 2.

$\therefore \boldsymbol{n \le 2}$.

EXERCISE

3(A): Objective (MCQ) Type Questions

Select the correct answer.

3.1. X-rays are

(a) mechanical waves (b) electromagnetic waves of short wavelength

(c) matter waves (d) none of these.

3.2. X-rays are used for diffraction studies in crystals because

(a) the wavelength of X-rays is of the same order as that of interatomic spacing

(b) X-rays penetrate the crystals

(c) Crystals have atoms/molecules capable of scattering X-rays

(d) none of these.

3.3. Monochromatic X-rays are used in

(a) Laue method (b) powder method (c) rotating crystal method (d) none of these.

3.4. Every reciprocal lattice vector is

(a) parallel to a lattice plane (b) normal to a lattice plane

(c) inclined at an angle of 45° to the lattice plane (d) none of these.

3.5. The reciprocal lattice to an FCC lattice is

(a) an SC lattice (b) a BCC lattice (c) an FCC lattice (d) none of these.

3.6. The reciprocal lattice to a BCC lattice is

(a) an SC lattice (b) an FCC lattice (c) a BCC lattice (d) none of these.

3.7. The reciprocal lattice to an SC lattice is

(a) an SC lattice (b) a BCC lattice (c) an FCC lattice (d) none of these.

3.8. Bragg's law in vector form is

(a) $2\vec{k}\cdot\vec{G}=0$ (b) $2\vec{k}\cdot\vec{G}+G^2=0$ (c) $2\vec{k}\cdot\vec{G}-G^2=0$ (d) $2\vec{k}\cdot\vec{G}+G^2=k^2$.

3.9. Bragg's law is modified due to

(a) reflection of X-rays (b) refraction of X-rays

(c) polarization of X-rays (d) total internal reflection of X-rays.

3.10. The modified Bragg's law due to refraction of X-rays is

(a) $n\lambda = 2d\sin\theta\left[\frac{4d^2(1-\mu)}{n^2\lambda^2}\right]$ (b) $n\lambda = 2d\sin\theta\left[1-\frac{4d^2(1-\mu)}{\lambda^2}\right]$

(c) $n\lambda = 2d\sin\theta\left[1-\frac{4d^2(1-\mu)}{n^2\lambda^2}\right]$ (d) $n\lambda = 2d\sin\theta\left[1-\frac{4d^2(1-\mu)^2}{n^2\lambda^2}\right]$.

3.11. The refractive index (μ) of X-rays is given by

(a) $\mu>1$ (b) $\mu=0$ (c) $\mu<1$ (d) $\mu\geq 1$.

3.12. If V and V^* be the volumes of a unit cell in direct and reciprocal lattice respectively, then

(a) $V=V^*$ (b) $V=\frac{V^*}{8\pi^3}$ (c) $V=8\pi^3V^*$ (d) $V=\frac{8\pi^3}{V^*}$.

3.13. Liquid crystals are

(a) inorganic compounds (b) organic compounds (c) mixture of (a) and (b) (d) none of these.

3.14. In anisotropic liquid crystals the directions of $\vec{D}$ and $\vec{E}$ are

(a) same (b) different (c) both (a) and (b) (d) none of these.

3.15. An externally applied electric field converts the cholesteric liquid crystal into

(a) nematic type (b) lyotropic type (c) smectic type liquid crystal (d) none of these.

3.16. A thermal neutron of rest mass m_0 has kinetic energy given by $\frac{3kT}{2}$, where k = Boltzmann constant and T = the room temperature. What is the de Broglie wavelength associated with such a neutron?

(a) $\frac{h}{\sqrt{3m_0kT}}$ (b) $\frac{3hkT}{2\sqrt{m_0}}$ (c) $\frac{2h}{3kT\sqrt{m_0}}$ (d) $\frac{2h}{\sqrt{3m_0kT}}$.

3.17. In Bragg scattering of a beam of electrons each of mass m and velocity v by a nickel crystal, the maximum is observed at $\theta = 30°$, (θ being the angle the beam makes with the crystal plane). What is the inter-planar distance for the crystal?

(a) $\frac{h}{mv}$ (b) $\frac{2h}{mv}$ (c) $\frac{h}{2mv}$ (d) $\frac{mv}{h}$. **[IAS Preliminary 2005]**

3.18. A lattice has the following primitive vectors (in Å):

$$\vec{a}=\frac{a}{2}(\hat{j}+\hat{k}),\ \vec{b}=\frac{a}{2}(\hat{k}+\hat{i}),\ \vec{c}=\frac{a}{2}(\hat{i}+\hat{j})$$

The reciprocal lattice corresponding to the above lattice is

(a) BCC lattice with cube edge of $\left(\frac{4\pi}{a}\right)$ Å^{-1} (b) BCC lattice with cube edge of $\left(\frac{2\pi}{a}\right)$ Å^{-1}

(c) FCC lattice with cube edge of $\left(\frac{4\pi}{a}\right)$ Å^{-1} (d) FCC lattice with cube edge of $\left(\frac{2\pi}{a}\right)$ Å^{-1}.

[cf. GATE 2013]

3.19. In Bragg experiment of diffraction by crystals using X-rays of wavelength x Å, the glancing angle of the first order is 8°. If the lattice spacing is 2.82 Å, then using $\sin 8° = 0.1392$ we can have the value of x as

(a) 78.51 nm (b) 7.851 nm (c) 0.7851 nm (d) 0.07851 nm. **[IAS Preliminary 2003]**

3.20. A metal with body-centred cubic (BCC) structure shows the first (i.e., the smallest angle) diffraction peak at a Bragg angle of $\theta = 30°$. The wavelength of X-ray used is 2.1 Å. The volume of the primitive unit cell of the metal is

(a) 26.2(Å)^3 (b) 13.1(Å)^3 (c) 9.3(Å)^3 (d) 4.6(Å)^3. **[GATE 2009]**

3.21. The volume of the primitive cell of the FCC reciprocal lattice is

(a) $4\left(\frac{2\pi}{a}\right)^3$ (b) $4\left(\frac{\pi}{a}\right)^3$ (c) $4\left(\frac{\pi}{2a}\right)^3$ (d) $4\left(\frac{3\pi}{a}\right)^3$. **[GATE 2009]**

3.22. X-ray beam of wavelength 1.54 Å is diffracted from the (110) planes of a solid with a cubic lattice of lattice constant 3.08 Å. The first order Bragg diffraction occurs at

(a) $\sin^{-1}\frac{1}{4}$ (b) $\sin^{-1}\frac{1}{2\sqrt{2}}$ (c) $\sin^{-1}\frac{1}{2}$ (d) $\sin^{-1}\frac{1}{\sqrt{2}}$. **[JAM 2012]**

3.23. In a Davisson-Germer experiment, a collimated beam of electrons of energy 54 eV at normal incident on a given crystal shows a peak at a reflection angle of 50°. If the electron beam is replaced by a neutron beam and the peak appears at the same value of reflection angle, then the energy of the neutrons must be

(a) 330 eV (b) 33 eV (c) 0.3 eV 0.033 eV. **[cf. TIFR 2013]**

3.24. The primitive translation vectors $\vec{A}$, $\vec{B}$ and $\vec{C}$ of the reciprocal lattice are

(a) $\vec{A} = \frac{2\pi}{a}(\hat{x} - \hat{y})$, $\vec{B} = \frac{2\pi}{a}(\hat{y} + \hat{z})$, $\vec{C} = \frac{2\pi}{a}(\hat{x} - \hat{z})$

(b) $\vec{A} = \frac{2\pi}{a}(\hat{x} + \hat{y})$, $\vec{B} = \frac{2\pi}{a}(\hat{y} - \hat{z})$, $\vec{C} = \frac{2\pi}{a}(\hat{x} + \hat{z})$

(c) $\vec{A} = \frac{2\pi}{a}(\hat{x} + \hat{y})$, $\vec{B} = \frac{2\pi}{a}(\hat{y} + \hat{z})$, $\vec{C} = \frac{2\pi}{a}(\hat{x} - \hat{z})$

(d) $\vec{A} = \frac{2\pi}{a}(\hat{x} + \hat{y})$, $\vec{B} = \frac{2\pi}{a}(\hat{y} + \hat{z})$, $\vec{C} = \frac{2\pi}{a}(\hat{x} + \hat{z})$. **[GATE 2008]**

3.25. The inter-planar separation in a crystal is d. If the incoming electrons have a constant and the momentum makes an angle θ with the reflecting planes, the constructive interference observed at θ equal to

(a) $\pi/6$ (b) $\pi/4$ (c) $\pi/8$ (d) $\pi/3$.

3.26. Of the following $h^2 + k^2 + l^2$ values, which reflection is not observed?

(a) 29 (b) 30 (c) 31 (d) 32.

3.27. The reciprocal lattice vector $\sigma_{hkl} = h\vec{a} + k\vec{b} + l\vec{c}$ is related to the normal $\hat{n}$ of (hkl) plane in which of the following way?

(a) $\vec{\sigma}_{hkl}$ and $\hat{n}$ are parallel (b) $\vec{\sigma}_{hkl}$ and $\hat{n}$ are perpendicular

(c) $\vec{\sigma}_{hkl}$ and $\hat{n}$ are arbitrarily oriented (d) none is correct.

ANSWERS

3.1. (b) **3.2.** (a) **3.3.** (b) and (c) **3.4.** (b) **3.5.** (b) **3.6.** (b) **3.7.** (a) **3.8.** (b) **3.9.** (b) **3.10.** (c) **3.11.** (c) **3.12.** (d) **3.13.** (b) **3.14.** (b) **3.15.** (a) **3.16.** (a) **3.17.** (a) **3.18.** (a) **3.19.** (d) **3.20.** (b) **3.21.** (a) **3.22.** (a) **3.23.** (d) **3.24.** (d) **3.25.** (a) **3.26.** (c) **3.27.** (a).

3(B): Short Answer Type Questions

3.1. What are the principal methods for crystal structure analysis?

3.2. Why X-rays are used for the analysis of crystal structure?

3.3. Point out the underlying assumptions in the derivation of Bragg's law.

3.4. What are the utilities of Bragg's law for X-rays diffraction?

3.5. Give the modification of Bragg's law due to refraction of X-rays. [CU(H) 2001]

3.6. What is a reciprocal lattice? How does it differ from direct lattice?

3.7. Why is the concept of reciprocal lattice in solid state physics so important?

3.8. Show that the direct lattice is the reciprocal of its own reciprocal lattice.

3.9. State Bragg's law of diffraction and discuss its importance in crystal structure analysis.

3.10. How does the Laue approach differ from the Bragg approach? [CU(H) 2004]

3.11. On the basis of Ewald's construction, find the geometrical significance of the Bragg's equation in the reciprocal lattice space.

3.12. Define (a) atomic scattering factor (b) geometrical structure factor.

3.13. Define Brillouin zone for a lattice. How would you construct the first Brillouin zone for a BCC lattice?

3.14. Prove that a reciprocal lattice vector is normal to a lattice plane of the crystal lattice.

3.15. What is liquid crystal? Discuss the principle of action of liquid crystal display.

3.16. Distinguish between reciprocal lattice and direct lattice. Show that the reciprocal lattice for an FCC lattice is a BCC structure. [BU(H) 2014]

3.17. Show that greater the diffraction angle, the greater is the accuracy in determining the lattice parameter. [BU(H) 2012]

3.18. Write down Bragg's law of crystal diffraction in reciprocal lattice. [BU(H) 2007]

3.19. Show that every reciprocal lattice vector is perpendicular to direct lattice plane. [Pune U 2013]

3.20. Explain the significance of Laue experiment. [KU(H) 2013]

3.21. Write down Bragg's law. How is the law modified due to refraction at crystal surface? [KU(H) 1999]

3.22. How can it be established by X-ray diffraction that NaCl is faced-centred cubic? [CU(H) 1998; JU(H) 2002]

3.23. State and explain Bragg's law. [CU(H) Electronics 2009]

3.24. Why are X-rays most useful for analysis of crystal structure? [CU(H) Electronics 2010]

3.25. Explain the condition for Bragg reflection from a crystal. [CU(H) Electronics 2010]

3.26. What is meant by reciprocal lattice? [BU(H) 2010; CU(H) Electronics 2005; KU(H) 2010, 2009]

3.27. How does Laue approach differ from Bragg approach? [CU(H) 2010, 2008]

3.28. Crystals act as natural three-dimensional grating for X-rays. Justify the statement. [KU(H) 2004]

3(C): Numerical Problems

3.1. Calculate the longest wavelength that can be analyzed by a rock salt crystal of spacing $d = 2.81$ Å in the first order. [*Ans.* 5.64 Å]

3.2. Using a Bragg's X-ray spectrometer, the glancing angle for first order spectrum was observed to be equal to 6°. Find the wavelength of X-rays, if $d = 2.82 \times 10^{-8}$ cm. [*Ans.* 0.59×10^{-8} cm]

3.3. Using a crystal of KCl, Bragg found the following glancing angles for intense reflection from the lattice planes (100), (110) and (111) respectively: $\theta_1 = 5.22$ Å, $\theta_2 = 7.30$ Å and $\theta_3 = 9.05$ Å. Show that KCl has a simple cubic crystal structure. [*Ans.* $\frac{1}{d_{100}} : \frac{1}{d_{110}} : \frac{1}{d_{111}} \simeq 1 : \sqrt{2} : \sqrt{3}$]

3.4. X-rays of wavelength 1 Å are incident on the cube face (111) of a NaCl crystal ($a = 1.702$ Å). Deduce the value of glancing angle corresponding to the first order diffraction maximum. [*Ans.* 30°]

3.5. The line A of an X-ray beam gives a first order reflection maximum when the glancing angle is 30° to the smooth face of a crystal. Line B with $\lambda = 0.97$ Å gives a third order reflection maximum at an angle of 60° from the same face of the same crystal. Determine the wavelength of the line A. **[KU 2004]**

[*Ans.* 1.68 Å]

3.6. The primitive basis vectors of a lattice are $\vec{a} = \hat{i} + 2\hat{j}$, $\vec{b} = 4\hat{j}$ and $\vec{c} = \hat{k}$. What are the primitive translation vectors of its reciprocal lattice? [*Ans.* $\vec{a}^* = 2\pi\hat{i}$, $\vec{b}^* = \frac{\pi}{2}\left(-2\hat{i} + \hat{j}\right)$, $\vec{c}^* = 2\pi\hat{k}$]

3.7. A beam of X-rays is incident on a NaCl crystal with lattice spacing 0.282 nm. Calculate the wavelength of X-rays, if the first order Bragg reflection takes place at a glancing angle of 8°35′. Also calculate the maximum order of diffraction possible. [*Ans.* $\lambda = 0.0842$ nm, $n = 6$]

3.8. A set of lattice planes reflects X-rays of 1.32 Å at glancing angle of 9.30°. Deduce the possible spottings of this set of planes. [*Ans.* 4 Å, 9 Å, 12 Å]

3.9. Calculate the wavelength of X-ray beam incident at 12° for the first order reflection from calcite crystal if grating constant for the crystal is 3.035 Å. [*Ans.* 126 Å]

3.10. (a) Show that the reciprocal lattice corresponding to simple cubic lattice is another simple cubic lattice. Find the relation between volumes of the unit cells of the two lattices.

(b) Write down Bragg's equation and hence, argue greater is the accuracy in determining the lattice parameter.

(c) A diffractometer data of a crystal of an element show peaks at 2θ angles of 44.46 Å, 64.7 Å, 82 Å and 93.22 Å. If the wavelength of X-rays used is 1.543, assign Miller indices to the peaks and determined the lattice constant. Can you identify the crystal structure? **[CU(H) 2017]**

3.11. (a) Show that the reciprocal lattice corresponding to a BCC lattice is an FCC lattice in the reciprocal space.

(b) Write down Bragg's law. How is this law modified due to refraction at crystal surface? **[CU(H) 2016]**

3.12. (a) Draw a BCC lattice structure. Show that it has packing fraction 0.68. Compare it with that of a simple cubic lattice structure.

(b) A crystal could not possibly diffract ultraviolet radiation. Explain.

3.13. The spacing between successive (110) planes in NaCl is 2.82 Å. A given X-ray when shone upon the surface is found to give rise to first order Bragg reflection at grazing angle of 8°55′. Find the wavelength of the X-ray and the angle at which the second order Bragg reflection would occur.

[CU(H) 2014]

3.14. On an X-ray powder photograph of a cubic structure with copper k_α radiation ($\lambda = 1.542$ Å) lines are observed at the following Bragg angles (θ in degrees): 12.3, 14.1, 20.2, 24.0, 25.1, 29.3 and 32.2. Assign Miller indices to these lines and calculate the unit cell dimension. **[CU(H) 2013]**

3.15. Find Bragg angles and the indices of diffraction for the three lowest angle lines on the powder photograph of a cubic crystal having lattice constant 6 Å. The wavelength of the X-rays (Cu-k_α) used is 1.54 Å.

[CU(H) 2011]

[*Ans.* $\theta = 7.37°, 14.87°, 22.64n°$ for $n = 1, 2, 3$]

3.16. (a) A plane intercepts at $(a, 2a, a/3)$ in a simple cubic unit cell. What are the Miller indices of the plane?

(b) A diffraction pattern of a cubic crystal of lattice parameter 3.16 Å is obtained with a monochromatic X-ray beam of wavelength 1.54 Å. The first four lines on this pattern were observed to have the following values:

Line	θ *(in degrees)*
1	20.3
2	29.2
3	36.7
4	43.6

Determine the inter-planar spacing and the Miller indices of the reflecting planes. **[CU(H) 2019]**

3.17. NaCl is a cubic crystal having density 3.13×10^3 kg/m^3. Find the length of a unit cell (At. wt. of of Na = 23, Cl = 35.5). **[CU(H) 2010]**

3.18. The spacing between successive (100) planes in NaCl is 2.820 Å, X-ray, incident upon the surface of this crystal, is found to give rise to first order Bragg reflection at grazing angle of 8°35′. Calculate the wavelength of the X-ray and angle at which the second order Bragg reflection would occur. **[BU(H) 2011]**

3.19. The k_α line of a palladium target gives a first order maximum at an angle of 6° using the (100)-planes of rock-salt. Calculate the wavelength of this radiation. [Given, d_{100} = 2.82 Å and sin 6° = 0.104] **[BU(H) 2010]**

3.20. Calculate the glancing angle of the cube (100) of a rock salt crystal (a = 2.84 Å) corresponding to second order diffraction maximum of X-rays for wavelength 0.710 Å. **[DU 2014]**

3.21. In Bragg diffraction condition if d = 1.5 Å, then what is the upper limit of λ for obtaining the first order reflection? **[Pune U 2014]**

3.22. Calculate the distance between two lattice planes which give first order diffraction at an angle of 26.42° with X-rays of wavelength 0.75 Å. **[Pune U 2013]**

3.23. Calculate the glancing angle at which the electrons of energy 100 eV must be incident on the lattice plane (having inter-planar spacing 2.15 Å) of a metal crystal in order to give a strong Bragg reflection in the first order. **[VU(H) 2010]**

[*Ans.* 16.6°]

3.24. The first order reflection angle by (100)-planes in a cubic crystal using X-rays of wavelength 1.54 Å is 18°. Calculate the inter-planar spacing for the (110)-, (111)-planes. **[VU(H) 2003; cf. CU (H) 2005]**

[*Ans.* 2.49 Å, 1.44 Å]

3.25. The lattice constant of a cubic crystal is 2.25 Å. Find the inter-planar spacing for the set of crystallographic planes having Miller indices (100). **[CU(H) 2001; CU(H) Electrons 2010]**

[*Ans.* 2.25 Å]

3.26. What are Brillouin zones? Illustrate your answer by constructing two Brillouin zones of a square lattice. **[CU(H) Electrons 2000]**

3.27. The spacing between principal planes of a NaCl crystal is 2.84 Å. If the third order Bragg reflection occurs at an angle of 15°, find the wavelength of X-rays. **[KU(H) 2013]**

[*Ans.* 0.49 Å]

3.28. On an X-ray powder photograph of a cubic substance with copper k_α radiation ($a = 1.542$ Å) lines are observed at the following Bragg angle θ (in degrees): 12.3, 14.1, 20.2, 24.0, 25.1, 29.3, 32.2, 33.1. Assign Miller indices to these lines and calculate the unit cell dimension. **[CU(H) 2013; KU(H) 1999]**

[*Ans.* FCC: (111), (200), (220), (311), (222), (400); 6.3 Å]

3.29. A beam of X-rays of wavelength 0.842 Å is incident on a crystal at glancing angle of 8°35′, when the first order Bragg reflection occurs. Calculate the glancing angle of the third order reflection.

[WBUT 2002]

[*Ans.* $\theta = 26°36'$]

3.30. A simple cubic crystal illuminates with X-rays of wavelength 0.09 nm is rotated and the first order Bragg reflection occurs at minimum glancing angle of 8.8°. Which set of crystal planes are responsible for such reflection? Find the spacing between these planes. [*Ans.* (100)-planes, $d_{100} = 0.294$ nm]

3(D): Broad Answer Type Questions

3.1. (a) State Bragg's law of diffraction and discuss its importance in crystal structure analysis. **[cf. CU 1998]**
(b) Why X-rays are used for the analysis of crystal structure analysis?

3.2. (a) State the condition under which Bragg reflection can occur.
(b) Derive Bragg's law of X-ray diffraction clearly stating the underlying assumptions.
(c) Give the corrected form of Bragg's law taking into account the correction due to refraction.

3.3. (a) What is reciprocal lattice? Discuss its importance in solid state physics.
(b) Prove that reciprocal lattice vector is normal to a lattice plane of the crystal lattice.
(c) Prove that a direct lattice is the reciprocal of its own reciprocal lattice.

3.4. (a) Describe Ewald's construction and explain its importance.
(b) On the basis of Ewald's construction, find the geometrical significance of the Bragg's equation in the reciprocal lattice space. **[BU(H) 1996]**

3.5. (a) Derive Laue equations and hence, deduce Bragg's law of X-ray diffraction.
(b) How does the Laue approach differ from the Bragg approach? **[CU(H) 2004]**

3.6. Discuss (a) Laue method and (b) the rotating crystal method of X-ray diffraction. Mention specific applications of each of them.

3.7. (a) What are Brillouin zones? Explain their significance.
(b) Discuss how Brillouin zones of FCC and BCC lattices can be constructed.

3.8. Explain what do you mean by Atomic Scattering Factor and Geometrical Scattering Factor. Find out their expressions.

3.9. (a) What are liquid crystals? Mention the name of different types of liquid crystals.
(b) Describe a cholesterol liquid crystal and mention some applications of LCD.

3.10. (a) Draw an FCC lattice structure. Show that it has a packing fraction of 0.74.
(b) Write down Laue's condition for constructive interference in crystal. Derive Bragg's law from it for a simple cubic lattice. **[CU(H) 2014]**

3.11. (a) Show that the reciprocal lattice corresponding to a BCC lattice is an FCC lattice in the reciprocal space.
(b) Write down Bragg's law. How is this law modified due to refraction at crystal surface? **[CU(H) 2013]**

3.12. (a) Set up Laue's equation for X-ray diffraction. Derive Bragg's law from it.

(b) Write down Bragg's law in reciprocal lattice. **[CU(H) 2012]**

3.13. Derive Laue's equations of diffraction of X-rays and obtain Bragg's law from them. **[BU(H) 2013, 2011]**

3.14. (a) What is reciprocal lattice? Prove that the volume of a unit cell in reciprocal lattice is inverse of the volume of a unit cell in direct lattice.

(b) Explain the Ewald construction. Discuss, reciprocal lattice concept is useful in X-ray diffraction study. **[Mahatma Gandhi U 2012]**

3.15. (a) Deduce the Bragg's sine law from the diffraction condition $2\vec{k}\cdot\vec{G}+\vec{G}^2=0$ and show that its geometrical interpretation leads to the concept of Brillouin zones in crystals.

(b) Prove that 5-fold rotation axis cannot exist in a crystal lattice.

(c) Given the basis vector of hcp structure:

$$a=\left(\sqrt{\frac{3}{2}}a\right)\hat{i}+\left(\frac{a}{\sqrt{2}}\right)\hat{j},\ b=\left(-\frac{\sqrt{3}}{2}a\right)\hat{i}+\left(\frac{a}{2}\right)\hat{j},\ c=c\vec{k}$$

Find the reciprocal lattice of d_{210} for a cubic crystal with side a?

3.16. What are the properties of reciprocal lattice? Construct the reciprocal lattice of a cubic crystal of lattice constant a. **[CU(H) Electronics 2009, 2004]**

3.17. What are Brillouin zones? Illustrate your answer by constructing two Brillouin zones for a square lattice. **[CU(H) Electronics 2000]**

3.18. (a) Describe Bragg X-ray spectrometer and explain how it can be used to determine the wavelength of X-rays. **[KU(H) 2008, 2007]**

(b) Discuss the structure of NaCl crystal. **[KU(H) 2011, 2008]**

Chapter 4

Imperfections in Crystals

4.1 Introduction

An *ideal* (or *perfect*) crystal is that in which the lattice arrays of atoms persists without any irregularity in all directions of the crystal. Strictly speaking such conception is valid only at absolute zero temperature where *entropy* of the system tends to zero. As the temperature increases the entropy also increases introducing certain structural *defects* and the crystal becomes *structurally imperfect. Thus, any irregularity in the crystallographic structure of a crystal is called imperfection.*

4.2 Lattice Imperfections and Properties of Solids

The imperfections in crystals give rise to the following important properties:

1. The defects may scatter conduction electrons and phonons in solids and greatly reduce the thermal and electrical conductivities.

2. Impurities present in the lattice of semiconducting material, considerably increases the conductivity by creating localised energy levels in the forbidden gap between valence and conduction bands.

3. Optical properties, e.g., colour of some salts and sensitivity of photographic emulsion are due to the presence of imperfections in the lattice.

4. Hysteresis loss of ferromagnetism is due to the lattice imperfections in the material.

5. Some imperfections largely decrease the mechanical strength of metals and alloys.

Thus, imperfections in crystal lattice lead to *structure sensitive properties* as discussed above.

4.3 Classification of Imperfections (or Defects)

Any deviation from regular atomic arrangement in the crystal lattice is called imperfection or defects. Table 4.1 shows different possible imperfections with their description.

All the above defects arise in the geometrical structure of the lattice and are called *static imperfections.* Apart from the above defects there are other different types of defects as below:

(a) *Electronic Imperfections*: This arises due to errors in charge distribution or energy in the crystal.

(b) *Excitation States of Crystal*: These are imperfections as they cause deviation from perfect crystal symmetry.

Examples: (i) Photons and magnons (ii) Conduction electrons and holes (iii) Excitons — the quantised electron-hole pair (iv) Quantised plasma waves.

(c) *Transient Imperfections*: These imperfections are introduced from the outside source, e.g., (i) Photons (ii) Electrons, protons, mesons and ions (iii) Neutrons and neutral atoms.

Table 4.1: Imperfections in crystals

Types of imperfection	*Examples*	*Description of imperfection*
(1) Point defects	(i) Interstitial	Extra atom (either a foreign atom or an atom of the same type) in an interstitial site.
	(ii) Schottky defect	Atom missing from normal site.
	(iii) Frenkel defect	Atom displaced to interstitial site creating nearby vacancy.
	(iv) Substitutional impurity atom	A foreign atom either in interstitial or regular lattice site.
(2) Line defects	(i) Edge dislocation	The periodicity of the atomic lattice array is interrupted along certain directions in a crystal.
	(ii) Screw dislocation	Row of atoms about which a normal crystallographic plane appears to be spiral.
(3) Plane defects	(i) Lineage boundary	Boundary between two adjacent perfect regions in the same crystal that are slightly tilted w.r.t. each other.
	(ii) Grain boundary	The boundary between two crystals in a polycrystalline solid.
	(iii) Stacking fault	The plane separating two incorrectly juxtaposed layers.

4.4 Point Defects

The vacancy defects in a solid are of two types:

1. *Schottky defect*: This type of vacancy defect is created due to the migration of the displaced positive ion in successive steps which eventually settles at the surface of the crystal [Fig. 4.1].

2. *Frenkel defect*: In this type of defect the displaced positive ion occupies an interstitial position and both the vacancy and positive ion together form the Frenkel defect [Fig. 4.1].

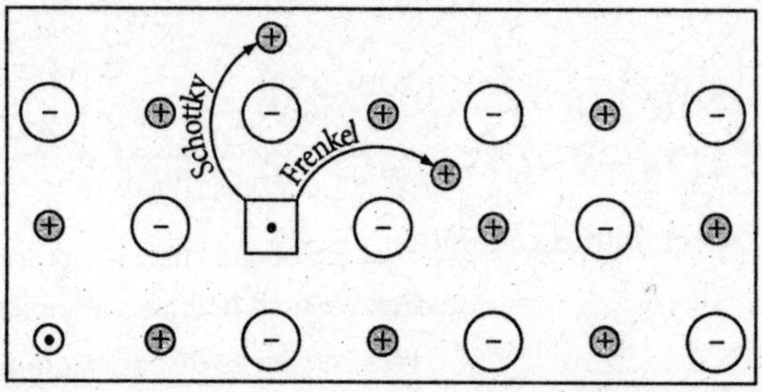

Fig. 4.1: Schottky and Frenkel defects in an ionic solid

4.4.1 Formation of Point Defects

The lattice atoms in a crystal execute *elastic vibration* due to the thermal energy of the solid. Large displacement of vibration may cause the atoms to jump from normal to the interstitial positions. This creates both interstitial atom and vacant lattice sites (Frenkel defect). At any suitable temperature some of the lattice atoms jump to the surface through successive migration and eventually settle down creating Schottky defect.

4.4.2 Energy of Formation of Vacancy

At any temperature T, the number of vacancies n is given by

$$n = f(N, E_v, T), \tag{4.1}$$

where N is the total number of atoms/ions and E_v, the average energy required to create a vacancy at temperature.

Above $T = 0$ K, there is always disorder or entropy S in a crystal. The relation between defect concentration and the change in entropy ΔS is given by Boltzmann law:

$$\Delta S = k \ln p, \tag{4.2}$$

where p is the number of ways in which defects may be arranged among the available number of sites and $k = 1.38 \times 10^{-23}$ J/K, the Boltzmann constant.

Out of N lattice sites n atoms can be chosen in $\frac{N!}{(N-n)!\, n!}$, i.e.,

$$p = \frac{N!}{(N-n)!\, n!}. \tag{4.3}$$

The increase in entropy ΔS, due to the pressure of n vacancy defects is given by

$$\Delta S = k \ln \left[\frac{N!}{(N-n)!\, n!} \right] = k\,[\ln N! - \ln(N-n)! - \ln n!]. \tag{4.4}$$

Now since E_v is the average energy to create a vacancy, the energy required to create n such vacancies is

$$\Delta E = nE_v. \tag{4.5}$$

Therefore, the change in *Helmholtz free energy* is

$$\Delta F = \Delta E - T\Delta S = nE_v - kT\,[\ln N! - \ln(N-n)! - \ln n!]. \tag{4.6}$$

Using *Stirling's approximation*,

$$\ln x! = x \ln x - x \text{ (for } x \gg 1),$$

we get

$$\begin{aligned} \Delta F &= nE_v - kT\,[N \ln N - N - (N-n)\ln(N-n) + (N-n) - n\ln n + n] \\ &= nE_v - kT\,[n \ln N - (N-n)\ln(N-n) - n \ln n]. \end{aligned}$$

Free energy is minimum at equilibrium condition, i.e.,

$$\frac{\partial \Delta E}{\partial n} = 0 \tag{4.7}$$

$$\begin{aligned} \text{or, } \frac{\partial \Delta E}{\partial n} &= E_v - \frac{\partial}{\partial n} kT\,[N \ln N - (N-n)\ln(N-n) - n \ln n] \\ &= E_v - kT \frac{\partial}{\partial n}\,[-(N-n)\ln(N-n) - n\ln n] \quad \left[\because \frac{\partial}{\partial N}(N \ln N) = 0, \text{ as } N \text{ is constant} \right] \\ &= E_v - kT \left[\left(\frac{N-n}{N-n} \right)(-1) + \ln(N-n)\,(-1) + \frac{n}{n} + \ln n \right] \\ &= E_v - kT \ln \left(\frac{N-n}{n} \right). \end{aligned}$$

But $\dfrac{\partial \Delta F}{\partial n} = 0$.

$$\therefore E_v = kT \ln\left(\frac{N-n}{n}\right). \tag{4.8}$$

Eqn (4.8) represents the enthalpy of formation of the point defect. *It is equal to the energy required to remove an atom from an inside lattice site to the lattice site on the surface of the crystal.*

4.4.3 Number of vacancies at any Temperature

Since, the number of vacancies is much smaller than the total number of atoms, $n \ll N$. Under this condition

$$E_v = kT \ln\left(\frac{N-n}{n}\right)$$

reduces to

$$E_v = kT \ln\frac{N}{n} \text{ or, } \frac{E_v}{kT} = \ln\frac{N}{n} \text{ or, } \frac{n}{N} = e^{-Ev/kT} \text{ or, } n = Ne^{-E_v/kT}. \tag{4.9}$$

Eqn (4.9) can be used to determine the number of vacancies n where $n \ll N$ at T much below the melting point.

4.4.4 Equilibrium concentration of Schottky Defects in Ionic Crystals

Schottky defect in ionic crystals is due to the missing of cation-anion pairs (represented by A^+ and B^-) from the respective lattice site. The energy absorbed during the formation of this defect is more than compensated by the resultant order in the structure.

If E_s be the average energy required to create a Schottky defect and n be the number of Schottky defect produced by removing nA^+ cations and nB^- anions, then the total number of different ways in which n Schottky defects can be produced is given by

$$p = \left[\frac{N!}{(N-n)!\,n!}\right]^2, \tag{4.10}$$

where N is the total number of cation-anion pairs. The power 2 in the right-hand of eqn (4.10) is due to the fact that as the number of ways in which a cation or an anion can be removed is given by

$$\left[\frac{N!}{(N-n)!\,n!}\right],$$

the different number of ways in which n cations as well as n anions can be removed is given by the compound probability which is just the product

$$\left[\frac{N!}{(N-n)!\,n!}\right] \times \left[\frac{N!}{(N-n)!\,n!}\right].$$

The change in entropy of the system in the above process is

$$\Delta S = 2k \ln\left[\frac{N!}{(N-n)!\,n!}\right]. \tag{4.11}$$

But

$$\ln\left[\frac{N!}{(N-n)!\,n!}\right] = n \ln N - (N-n)\ln(N-n) - n \ln n.$$

[By Stirling's approximation].

∴ the change in free energy,

$$\Delta F = \Delta E - T\Delta S = nE_s - 2kT\,[N\ln N - (N-n)\ln(N-n) - n\ln n].$$

At equilibrium condition ΔF is minimum.

So, $\dfrac{\partial \Delta F}{\partial n} = 0.$

But

$$\frac{\partial \Delta E}{\partial n} = \frac{\partial}{\partial n}[nE_S] - \frac{\partial}{\partial n}2kT[N\ln N - (N-n)\ln(N-n) - n\ln n] = E_S - 2kT\ln\left(\frac{N-n}{n}\right)$$

$$\text{or, } 0 = E_S - 2kT\ln\left(\frac{N-n}{n}\right) \text{ or, } E_S = 2kT\ln\left(\frac{N-n}{n}\right). \tag{4.12}$$

Eqn (4.12) gives the average energy required to create a Schottky defect.

If $n \ll N$,

$$E_s = 2kT\ln\left(\frac{N}{n}\right) \text{ or, } n = Ne^{-E_S/2kT}. \tag{4.13}$$

Eqn (4.13) shows that, the number of Schottky defect depends on: (i) the total number of ion pairs (N) (ii) the average energy (E_S) required to create a Schottky defect (iii) the temperature (T) of the crystal. Eqn (4.13) indicates the *number of Schottky defect increasing exponentially with increase of temperature.*

Using eqn (4.13), we can calculate the number of Schottky defect in binary ionic crystal (e.g., NaCl and MgO) by knowing the values of N, E_S and T.

4.4.5 Equilibrium concentration of Frenkel defects in Ionic Solids

If E_f be the average energy required to displace a cation from its normal lattice position to an interstitial site and N_i be the number of interstitial sites, then the number of different ways in which an ion can be removed is given by

$$\frac{N(N-1)(N-2)\cdots(N-n+1)}{n!} = \frac{N!}{(N-n)!\,n!}, \tag{4.14}$$

where N is the total number of cations (in an ionic crystal A^+, B^-) and n is the number of Frenkel defects.

If N_i be the total number of interstitial sites out of which n are occupied, then the number of different ways in which n interstitial defects can be formed are given by

$$p = \frac{N!}{(N-n)!\,n!} \times \frac{N_i!}{(N_i-n)!\,n!}. \tag{4.15}$$

Hence, the *change in entropy* of the system is

$$\Delta S = k\ln p = k\ln\left[\frac{N!}{(N-n)!\,n!}\right]\times\left[\frac{N_i!}{(N_i-n)!\,n!}\right]$$

$$= N\ln N + N_i\ln N_i - (N-n)\ln(N-n) - (N_i-n)\ln(N_i-n) - 2n\ln n.$$

[By Stirling's approximation]

Now *change in free energy,*

$$\Delta F = \Delta E - T\Delta S = nE_f - kT\,[N\ln N + N_i\ln N_i - (N-n)\ln(N-n) - (N_i-n)\ln(N_i-n) - 2n\ln n].$$

But at equilibrium

$$\left[\frac{\partial \Delta F}{\partial n}\right] = 0 = E_f - kT\ln\left[\frac{(N-n)\,(N_i-n)}{n^2}\right] \;[\because N \text{ and } N_i \text{ are constants}]. \quad (4.16)$$

If $N \gg n$ and $N_i \gg n$, then

$$\frac{E_f}{kT} = \ln\left(\frac{NN_i}{n^2}\right) \text{ or, } \frac{E_f}{2kT} = \frac{1}{2}\ln\left(\frac{NN_i}{n^2}\right) = \ln\frac{(NN_i)^{1/2}}{n} \text{ or, } e^{\frac{E_f}{2kT}} = \frac{(NN_i)^2}{n}$$

$$\text{or, } n = (NN_i)^{1/2}\, e^{-\frac{E_f}{2kT}}. \quad (4.17)$$

Eqn (4.17) shows that *at low temperature, the number of Frenkel defects is small.*

4.4.6 Interstitial (or Substitutional Impurity)

If an impurity atom/ion be introduced in an otherwise pure lattice, interstitially or substitutionally, the physical and electrical properties are highly effected.

Vacancies produced as a result of thermal equilibrium process are called *intrinsic vacancies,* where those produced in ionic crystals by the introduction of proper types of impurities are called *extrinsic vacancies.* Extrinsic vacancies in diatomic alkali halides are produced by adding a molecule with a missing ion. For example, KCl grown in presence of controlled amounts of triatomic molecule like $CaCl_2$. It has a missing positive ion as there are two Cl^- ions for every Ca^{2+}. When the molecule goes into solution in the crystal, a positive ion vacancy is created so that overall charge neutrality and thermal equilibrium be maintained.

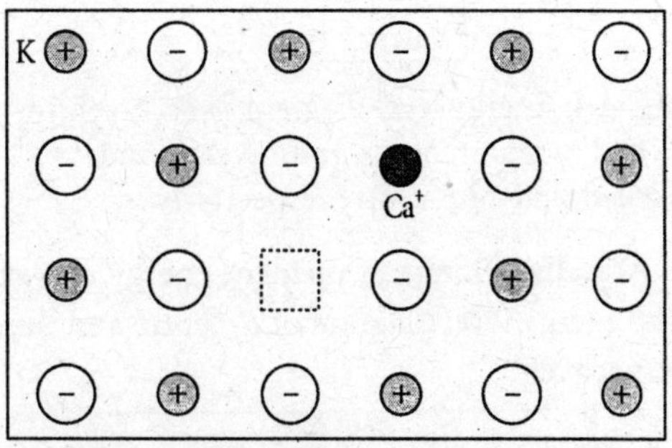

Fig. 4.2: Production of extrinsic vacancy by adding $CaCl_2$ in KCl

If a $CaCl_2$ molecule be placed on the surface of a KCl crystal, it will diffuse inward. Two Cl^- ions will occupy two anion vacancies and the Ca^{2+} ion will occupy a cation vacancy and an excess cation vacancy will be left in the crystal. The anion vacancy concentration, however, has been reduced by two below the value at equilibrium by introducing $CaCl_2$. Hence, two additional pairs are to be formed. The net result is that an extra cation vacancy is left behind.

4.4.7 Colour Centre (*F*-centre)

F-centre is a negative ion vacancy with one excess electron bound at the vacancy and is the simplest of all the colour centres. The name *F*-centre is derived from the German word *Farbzentrum* meaning colour. By absorption and electron spin resonance studies it has been established that an *F-centre is an electron trapped in negative ion vacancy* as shown in Fig. 4.3. This type of defect imparts colour to an otherwise colourless crystal-like NaCl, KCl, KBr, etc.

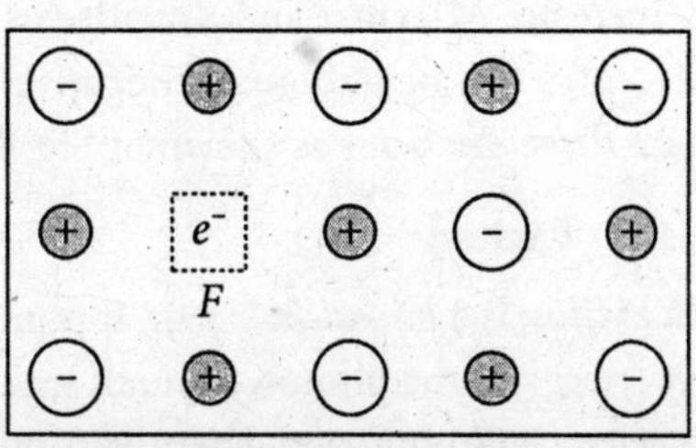

Fig. 4.3: An F-centre

F-centre can be produced by heating alkali halide crystals in an atmosphere containing excess of the alkali metal vapour. The excess metal atoms deposit on crystal surface. The halogen ions from interior, then diffuse to the surface and combine with metal atoms. By losing the valence electron, the metal atoms become ionised. This electron, then diffuse into the crystal and get trapped in a negative ion vacancy forming an *F*-centre. Similar result is obtained when an excess of alkali metal is added to the melt during the crystal growth.

The quantised energy states of the *F*-centre lie between the valence band and conduction band, i.e., in the forbidden region. When the crystal is irradiated with white light, some component with appropriate energy is absorbed and the *F*-centre is excited to higher energy state.

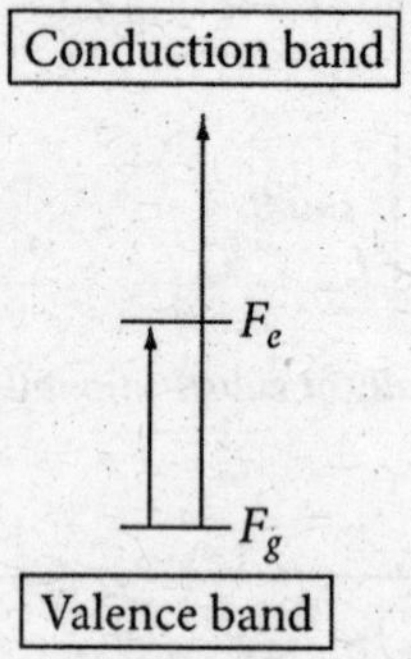

Fig. 4.4: Energy level diagram of an F-centre; F_g and F_e represent the ground state and excited state of F-centre respectively.

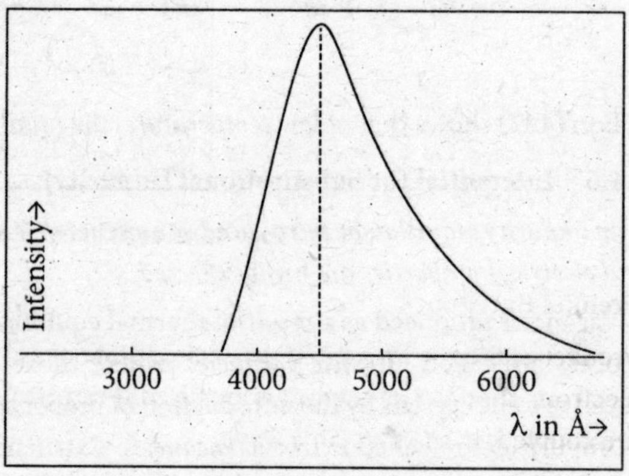

Fig. 4.5: Absorption band of KCl F-centre

Actually, a range of incident energy can excite the electron so that an absorption band is observed. The energy level diagram of *F*-centre and the *F*-band (absorption) for KCl are shown in Figs 1.4 and 1.5 respectively.

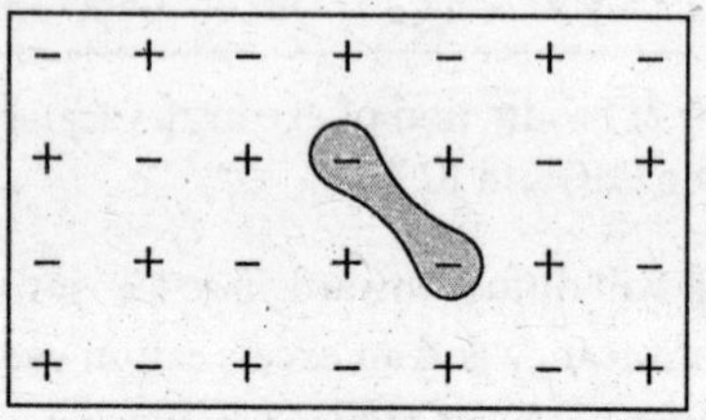

Fig. 4.6: (a) M-centre in (100)-plane

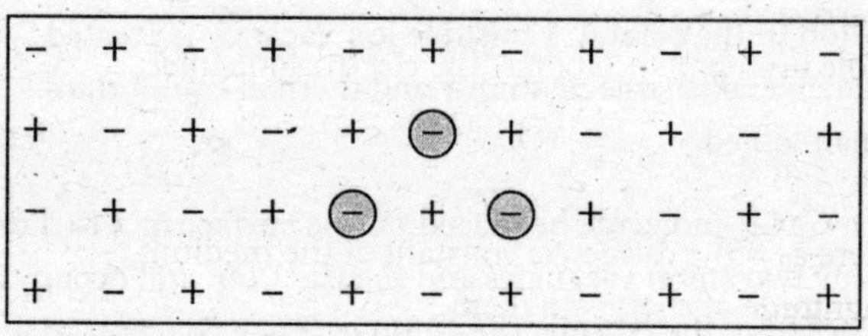

Fig. 4.6: (b) R-centre in (111)-plane

There are *M*-centre and *R*-centre which add to the colouration of the crystal in addition to *F*-centre. In *M*-centre two neighbouring negative ions in a (100)-plane trap two electrons. In *R*-centre, a (111)-plane traps three electrons as shown in Fig. 4.6.

4.4.8 Exciton

An exciton is a *bound-hole* pair. It is produced due to the photon-electron interaction in the lattice. Though the energy of the photon is much smaller than electron's energy, the photon-electron collision may raise the electron to sufficient energy producing an empty energy state, called a hole. Now the excited electron and the hole lose their identity and produce an *exciton*. Excitons are actually electronic point defects which move through the crystal transporting excitation energy.

Excitons are of the following two types: (i) Frenkel exciton — these are small and tightly bound and (ii) Wannier-Mott exciton — these are weakly bound.

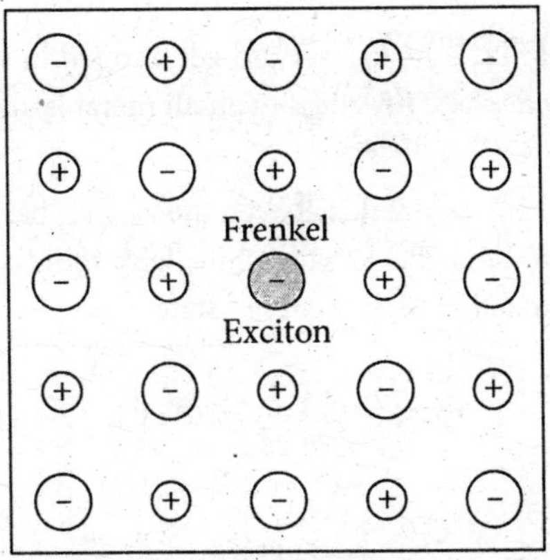

Fig. 4.7: Frenkel exciton in an alkali halide crystal

Frenkel Excision

Frenkel exciton is essentially an excited state of an atom, the hole being usually on the same atom as the electron. Though the exciton is localised it can travel by hopping from one atom to the next, as the atoms are coupled, in the form of a wave.

As in an ionic crystal the coupling between the ion's outer electron shell is strong, the exciton energy can be transferred from ion to ion without causing shift in them. Because of this reason Frenkel exciton [Fig. 4.7] is more mobile than vacancies, interstitial or substitutional impurities.

Wannier-Mott Exciton

To arrive at an expression for the energy of the exciton, we consider an electron in the conduction band and a hole in the valence band. The electrostatic interaction energy between them is given by

$$V(r) = \frac{e^2}{4\pi\epsilon_0 r}, \tag{4.18}$$

where ϵ_0 = the dielectric constant of the medium, r = the distance between the electron and hole and e = magnitude of charge on them.

The quasi-atom comprising an electron and hole (similar to a *positronium* as both the positive and negative charges are sufficiently light) is called a *Wannier-Mott exciton.*

The energy of the exciton E_x will be smaller than the energy $(E_c - E_h)$ as it is in the bound state.

Now

$$E_x = \frac{\mu e^4}{2\hbar^2 \epsilon^2 n^2}, \tag{4.19}$$

where n is the principal quantum number, μ is the reduced mass of the exciton given by

$$\mu = \frac{m_e m_h}{m_e + m_h}. \tag{4.20}$$

m_e and m_h being the masses of the electron and hole respectively.

$\therefore$ energy of the exciton in the nth state is

$$E_n = E_g - E_x. \tag{4.21}$$

The exciton ground state energy is given by

$$E_x = \frac{\mu e^4}{2\hbar^2 \epsilon^2} \text{ (for } n = 1)$$

$$= \left(\frac{\mu}{m}\right)\frac{1}{2}\frac{me^2}{\hbar^2}\frac{e^2}{\epsilon^2} = \frac{\mu}{m}\frac{e^2}{2a_0\epsilon^2}, \tag{4.22}$$

where $\frac{\hbar^2}{me^2} = a_0$ is the *Bohr radius*

$$\text{or, } E_x = \frac{\mu}{m}\frac{1}{\epsilon^2}\,(13.6\,\text{eV}), \tag{4.23}$$

where

$$\frac{e^2}{2a_0} = 13.6\,\text{eV}. \tag{4.24}$$

Thus, the lowest bound state extends over a radius

$$a_x = \frac{\hbar^2}{\mu(e^2/\epsilon)} = \frac{\epsilon m}{\mu}\frac{\hbar^2}{m} = \epsilon\left(\frac{m}{\mu}\right)a_0 \tag{4.25}$$

[replacing m by μ and e by effective charge e^2/ϵ].

Eqn (4.25) shows that $a_x \gg a_0$.

Table 4.2: Binding energy of excitons

Material→	Ge	Si	GaAs	GaP	CdS	CdSe	BaO	InP	InSb	KI	KCl	KBr	RbCl	LiF	AgBr	AgCl	TiCl
E_x(meV)	3.8 to 4.1	14.7	4.2	3.5	29	15	56	4.0	0.4	480	400	400	440	1000	20	30	6

The energy levels of exciton is very much similar to that of hydrogen atom's energy levels. Fig. 4.8 shows the energy level of an exciton created indirect process. The binding energy of the exciton E_x is equal to the difference in energy of a free electron and free hole.

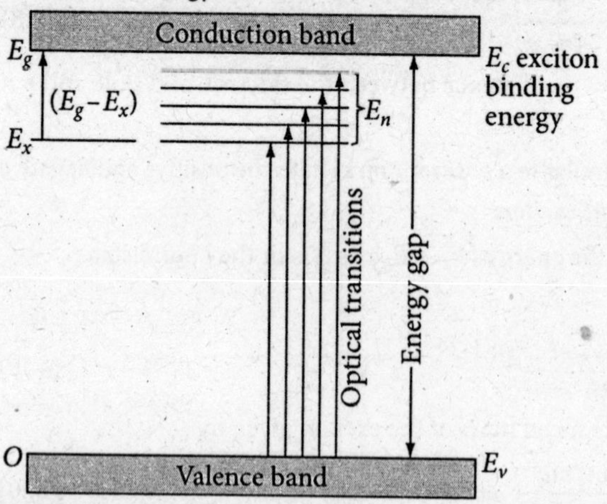

Fig. 4.8: Energy level diagram of an exciton created in direct process

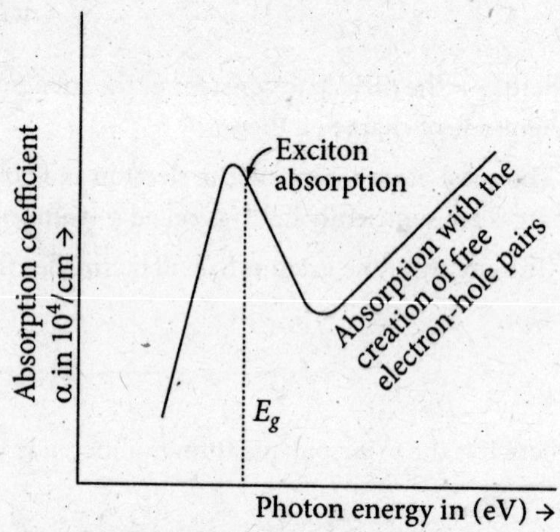

Fig 4.9: Absorption of exciton

Fig. 4.9 shows the absorption of the exciton when the crystal is irradiated with radiation of proper frequency.

4.5 Dislocation

If a crystal is deformed elastically by the application of external stress, it returns to its original configuration after the removal of the applied stress. When the same crystal is subjected to a large stress beyond elastic limit a certain amount of deformation is left after the removal of the applied stress. This plastic deformation generates a plastic flow of crystal creating another type of deformation known as *dislocation*.

If one part of a crystal slides relative to another due to plastic flow, the length of the crystal undergoes a change due to the action of the tension therein. Fig. 4.10 shows the increase in length of a crystal due to the above process.

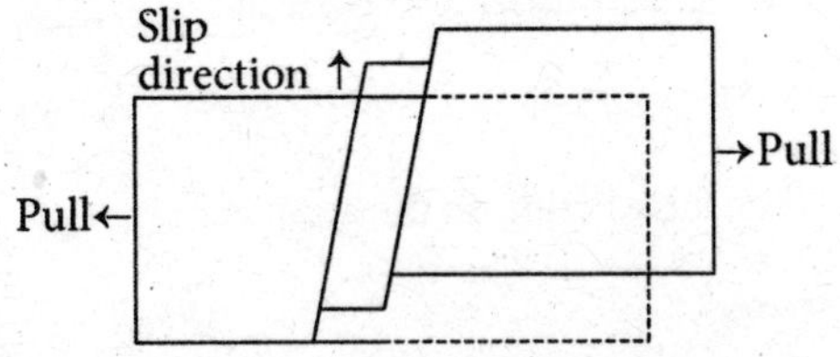

Fig. 4.10: Process of slip due to tensile stress

Slip is a type of lattice imperfection in which an atomic plane glides over another by integral multiples of an interatomic distance.

Here a relatively small number of atoms, namely, the atoms on the surface layer of the crystal on either side of the slip plane are involved. In FCC lattice, slip occurs along (111)-planes in which number of atoms is high. In contrast to slip, all atoms take part in elastic deformation and is inholomogeneous and anisotropic.

4.5.1 Interpretation of Slip: Dislocation

Practically, the strength of single crystals is less than that calculated from the theory of elasticity. In other words, the theoretically critical stress of perfect crystals is much larger than the observed stress for pure crystals.

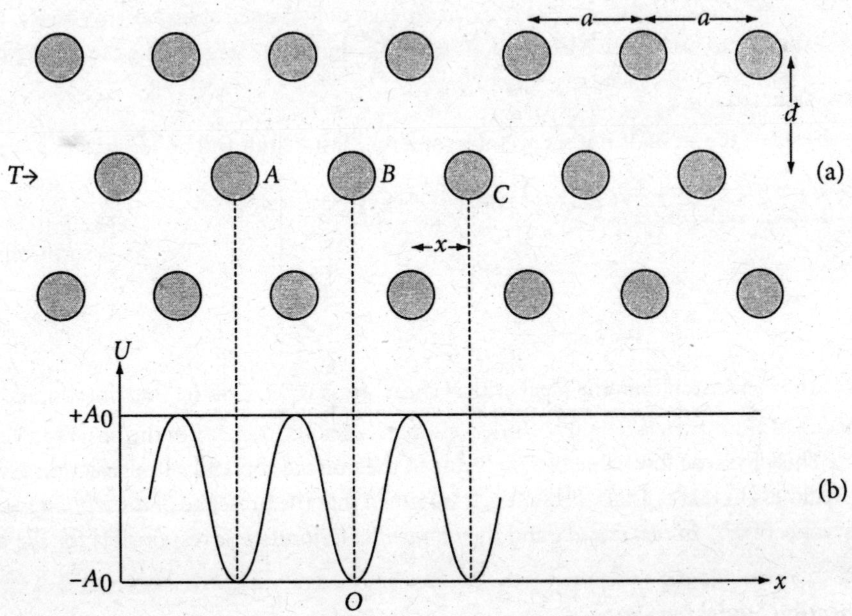

Fig. 4.11: (a) Displacement of middle plane of atoms by x due to a shear stress T (b) Potential energy U as a function of x

The reason can be understood from the simplified method of calculating the theoretical shear strength of perfect crystals as proposed by J. Frenkel in 1926.

Fig. 4.11 shows a closed packed structure and arrangement of atoms in successive layers. The variation of PE (U) for the atoms A, B, C due to displacement from their equilibrium position is shown in Fig. 4.11 (b).

The PE, according to Frenkel, is

$$U(x) = -A_0 \cos\left(\frac{2\pi x}{a}\right). \tag{4.26}$$

∴ the force acting on the atom is

$$F(x) = -\frac{\partial U}{\partial x} = \frac{2\pi A_0}{a} \sin\left(\frac{2\pi x}{a}\right). \tag{4.27}$$

Now the shear stress T acting on the crystal,

$$T = N^{2/3}F(x) = N^{2/3}\frac{2\pi A_0}{a} \sin\left(\frac{2\pi x}{a}\right). \tag{4.28}$$

N is the number of atoms per unit volume.

For small displacement, i.e., $x \to 0$, $\sin\frac{2\pi x}{a} \to \frac{2\pi x}{a}$.

$$\therefore T = N^{2/3}\frac{2\pi A_0}{a} \times \frac{2\pi x}{a} = N^{2/3}\frac{4\pi^2 A_0}{a^2} d\left(\frac{x}{d}\right) \text{ or, } T = G\left(\frac{x}{d}\right), \tag{4.29}$$

where d is the inter-planar distance.

Eqn (4.28) represents *Hooke's law*, where $\left(\frac{x}{d}\right)$ represents the *strain* and G, the *shear modulus.*

When x is not small, then

$$T = \frac{G}{2\pi} \sin\left(\frac{2\pi x}{a}\right), \tag{4.30}$$

where we have assumed $a = d$.

For large value of x, the strain is not recoverable, and the elastic limit will be reached at $x = a/4$.

$$\therefore T_{\text{crit}} = \frac{G}{2\pi} \sin\left(\frac{2\pi}{a} \times \frac{a}{4}\right) \quad (\because x = a/4)$$

$$= \frac{G}{2\pi} \sin\frac{\pi}{2}$$

$$\text{or, } T_{\text{crit}} = \frac{G}{2\pi}. \tag{4.31}$$

This leads to the conclusion that, the ideal critical shear stress is ~ 1/6 of the shear modulus.

A typical value of G for metals is ~ 10^{11} N/m^2, yielding $T_{\text{crit}} \simeq 10^{10}$ N/m^2. But the observed value of T_{crit} is ~ 10^6 N/m^2. This observed low value of T_{crit} is due to the pressure of lattice imperfection giving rise to mechanical weakness in real crystals. Presently, it is known that the presence of *dislocation* is responsible for the lower value of T_{crit} in *real crystals* and the *moment of dislocation* is responsible for *slip* at very low applied stresses.

4.5.2 Dislocation Model for Slip

In Fig. 4.12 (a), ABC is an arbitrary closed curve in the slip plane PQR. In Fig. 4.12 (b), the inside of the curve ABC is hatched. Let the material over the hatched area in the upper half of the crystal is displaced somehow by an amount represented by the vector $\vec{b}$ relative to the lower half of the crystal, the material in the upper half lying outside of area ABC remains undisplaced. Thus, a fraction of the upper half of the real crystal suffers slip relative to the lower half.

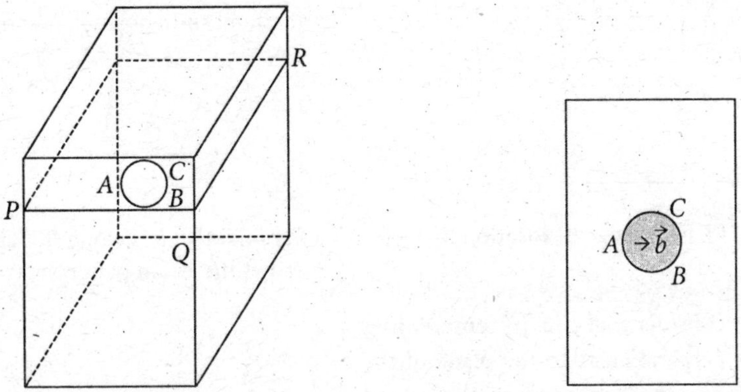

(a) (b)

Fig. 4.12: (a) Representation of a dislocation ABC in a slip-plane PQR
(b) Slip occurred only in the hatched area

Now the fraction of the slip is given by

$$f = \frac{\text{area } ABC}{\text{area } PQR}.$$

If the area ABC is made to increase, then at a time f will be equal to unity (i.e., $f = 1$) when the whole of the upper half of the crystal would suffer a displacement $\vec{b}$ relative to the lower half. The average displacement of the upper half relative to the lower half is $f\vec{b}$ for $f < 1$.

The line ABC representing the boundary between the slipped and unslipped material is called the *dislocation line.* The vector $\vec{b}$ defining the magnitude and direction of the slip is the *Burger's vector.* As the atoms tends to the condition of minimum energy for equilibrium $\vec{b}$ represents the separation of two equilibrium positions. The dislocation is of unit strength when the displacement equals a, the lattice spacing of the crystal.

4.5.3 Edge and Screw Dislocations

Another type of lattice imperfection is dislocation. It involves a large number of atoms along a line and is a most complicated type of lattice defects. Dislocation is of two basic types: (a) *Edge dislocation* and (b) *Screw dislocation.*

(a) *Edge dislocation*: A pure dislocation may be defined as a dislocation for which the Burger's vector $\vec{b}$ is everywhere perpendicular to the line of dislocation. Edge dislocation may be considered as a result of partial interval slip process as shown in Fig. 4.13 (a).

We consider a piece of the material in which a block is cut across $ABEF$ in such a manner that the upper and lower parts are separated. The upper half is then pushed sideways such that the line AB is shifted to $A'B'$ so that $\vec{BB'} = \vec{b}$. If the two halves are glued together, then what is formed is called *dislocation.* In this case, the upper and lower halves will be under *compression* and *extension* respectively.

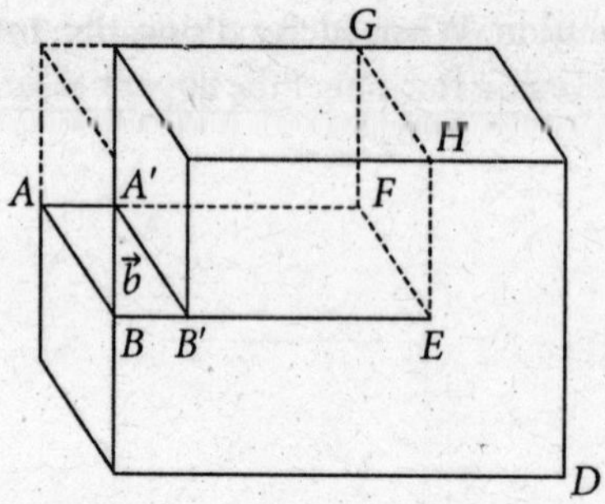

Fig. 4.13(a): *EF* is the edge dislocation line

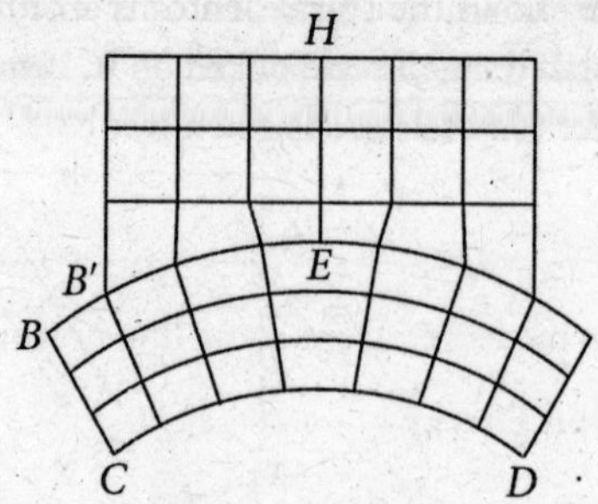

Fig. 4.13(b): $\vec{b}$ is along *BB′* and the figure shows the strain pattern

Alternatively, we consider Fig. 4.13 (b) representing the rows of atoms perpendicular to the plane of the paper. By cutting the piece along the plane *EFGH* and putting the half plane of atoms initially above *AB* inside the cut we can have the edge dislocation. This produces an extra half plane atoms as represented by the line *HE*. Here *EF* represents the dislocation line. The distorted configuration extends all along the edge of the crystal. As the region of maximum distortion is centred around the edge of the incomplete plane, the distortion is named as *edge dislocation*. Edge dislocations are represented by *L* or *T* according as the incomplete plane starts from the bottom or top of the crystal. The configurations *L* and *T* are referred to as *positive* or negative *edge dislocations.*

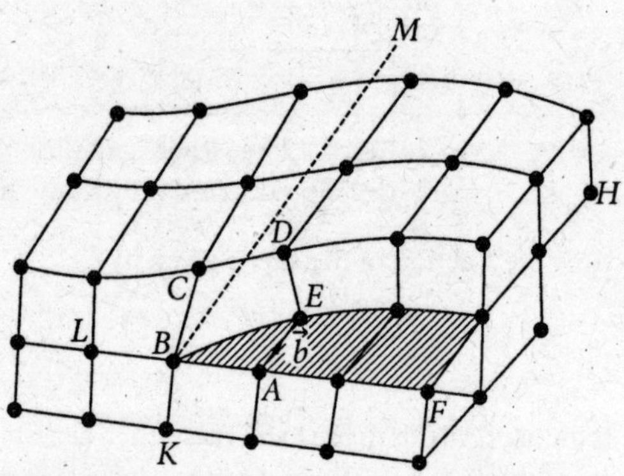

Fig. 4.14: Schematic representation of Screw dislocation when *BM* is the dislocation line

(b) *Screw Dislocation: Burger* in 1939 introduced the concept of screw dislocation (also called *Burger dislocation*). *The screw dislocation results from a displacement of the atoms in one part of a crystal relative to the rest of the crystal, forming a spiral ramp around the dislocation line as shown in* Fig. 4.14. The atomic configuration may be obtained by cutting the block across the area *BFHM* and then pushing the upper part backward along the direction of the Burger's vector $\vec{b}$. If we move around the dislocation line *BM* (which is parallel to $\vec{b}$) along the path *AKLCDE*, we advance a distance $\vec{b}$ along *BM* for each turn. This justifies the name *screw dislocation*.

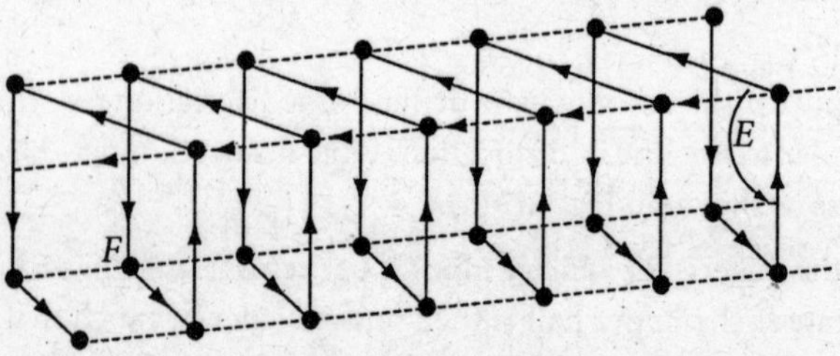

Fig. 4.15: Spiral of atoms around the screw dislocation

Fig. 4.15 shows the spiral of atoms around the screw dislocation. When such a dislocation intersects a crystal surface a step is produced on it. If z-axis represents the dislocation line, then displacement of atoms in the direction of z-axis is given by

$$\Delta Z = \frac{b\theta}{2\pi}, \tag{4.32}$$

where θ is measured from some line normal to z. If $\theta = 2\pi$, then $\Delta z = b$.

Dislocation Energy

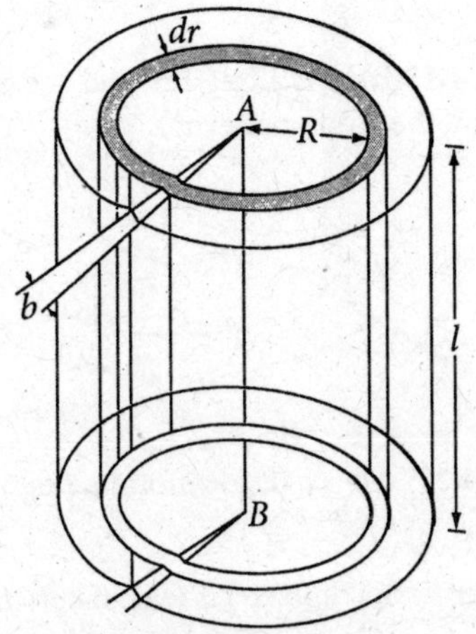

Fig. 4.16: (a) Distorsion in an elastic cylinder

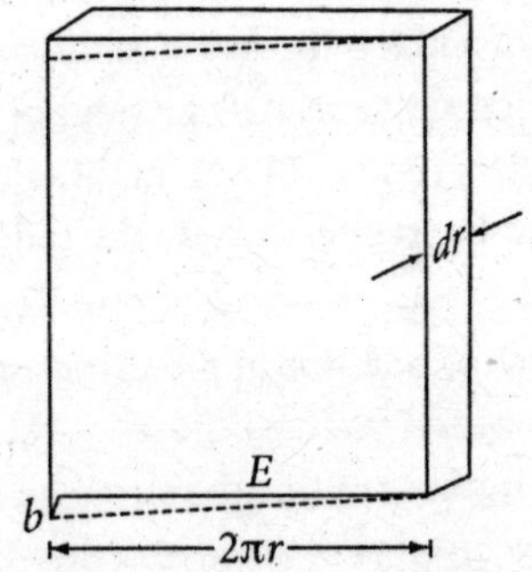

Fig. 4.16: (b) Unrolled cylinder showing the shear strain

We consider an *elastically isotropic medium* in which screw dislocation is present. The screw dislocation and the edge dislocation have different strain pattern as the screw dislocation forms a right-handed or left-handed screw of pitch equal to the Burger's vector $|\vec{b}|$. The elastic distortion can be represented by a radial slit which is parallel to the axis of a hollow cylinder of the elastic material, the hollow cylinder being parallel to the line of dislocation. Then the free surfaces undergo a displacement w.r.t. one another as shown in Fig. 4.16 by an amount b along the z-direction. The shear strain in this case is symmetrical and directed along the z-axis if the dislocation is distributed uniformly over the circumference of the cylinder.

The shear strain of screw dislocation may be defined as

$$\epsilon = \frac{b}{2\pi r}, \tag{4.33}$$

where r is the radius of the cylinder and b, the *Burger's vector.*

The formation of the screw dislocation line is associated with the strain energy stored in the dislocation region. For a screw dislocation this elastic energy is equal to the strain energy per unit volume.

The shear stress of a screw dislocation is given by

$$\sigma = GE = \frac{Gb}{2\pi r}, \tag{4.34}$$

where G is the *shear modus of elasticity.*

The strain energy density dU is given by

$$dU_{sc} = \frac{1}{2} \times \text{stress} \times \text{strain} \times dV = \frac{1}{2}\epsilon\sigma \times 2\pi r\, dr$$

$$[\because dV = 2\pi r\, dr]$$

$$= \frac{1}{2} \times \frac{Gb}{2\pi r} \times \frac{b}{2\pi r} \times 2\pi r l\, dr = \frac{1}{4\pi} G b^2 l \frac{dr}{r}.$$

$$\therefore U_{sc} = \frac{1}{4\pi} G b^2 l \int_{r_0}^{R} \frac{dr}{r} \text{ or, } U_{sc} = \frac{1}{4\pi} G b^2 l \ln\left(\frac{R}{r_0}\right), \quad (4.35)$$

where r_0 is the radius at which the elastic approximation breaks down and R is the radius of the cylinder of length l. For a typical single crystal, the value of $R/r_0 \simeq 10^5$. The factor $\ln(R/r_0)$ in eqn (4.34) implies that the energy is spread over the region outside the radius as shown in Fig. 4.17.

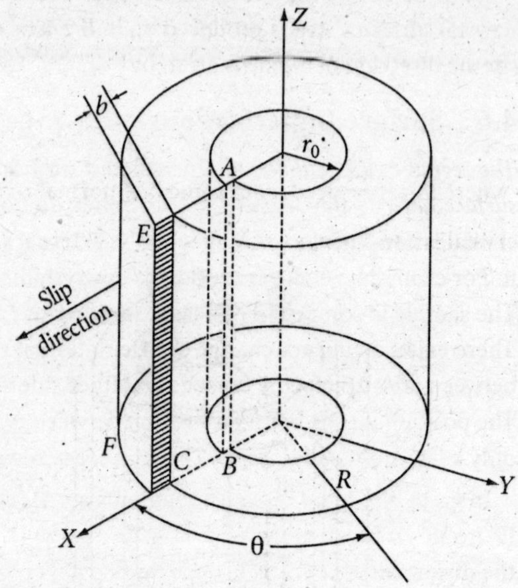

Fig. 4.17: *AB* is a dislocation line perpendicular to $\vec{b}$

The atoms of one side of the slip plane in edge dislocation are *compressed* and those on the other side are *expanded.* Because of this the stress field dislocation is *symmetrical* but that of an edge dislocation is not.

Since both extensional and compressional strains are present *Poisson's ratio* (μ) is to be taken into account in calculating the shear stress. Thus, shear stress is given by

$$\sigma = \frac{Gb}{2\pi(1-\mu)} \frac{\cos\theta}{r}. \quad (4.36)$$

If a slit be considered cut along the slip line (i.e., $\cos\theta = 1$), then it can be shown that the energy of edge dislocation is given by

$$U_{ed} = \frac{Gd^2}{4\pi(1-\mu)} \ln\left(\frac{R}{r_0}\right). \quad (4.37)$$

From eqns (4.36) and (4.37), it is clear that, U_{ed} is greater than U_{sc} by a factor $\left(\frac{1}{1-\mu}\right)$.

The high strain energy associated with the dislocation has the following two consequences: (i) *the dislocation line will tend to attain a lower energy by shortening its length,* (ii) *the dislocation lines will increase the free energy of the crystal equal to strain energy.*

The dislocation lines cease to exist in thermal equilibrium.

Dislocation Density

Let dl be the element of length of a dislocation line such as ABC in Fig. 4.12 (a) and v, the velocity of the element along the normal to dl in the slip plan. If H be the height of the crystal and A, the area of the slip plane, the rate of increase in strain due to the motion of the element dl is

$$\frac{dy}{dt} = \frac{v\, dl\, b}{AH} \text{ (where } y = x/d). \quad (4.38)$$

Now the rate of flow resulting from all directions in planes parallel to the plane PQR [Fig. 4.12 (a)] is

$$\frac{dy}{dt} = \frac{\bar{v} L b}{V} = \rho b \bar{v}, \quad (4.39)$$

where ρ is the dislocation density and $\langle \bar{v} \rangle$ is the average velocity and L, the total length of all the dislocations

$$\text{or, } \rho = \frac{1}{b\bar{v}} = \frac{dy}{dt}. \tag{4.40}$$

4.6 Surface Imperfections

The region of dislocations that lie about a surface having a thickness of a few atomic diameters is known as surface imperfection. As the atomic bonds do not extend beyond the surface, the finite dimension of the crystal is itself a defect. Along with the external surface, the crystal may have surface imperfections within it. For example, when iron solidifies or crystallises, the new crystal is not a single crystal but poly-crystal. The sample is composed of a large number of crystal pieces or grains, joined together to form one solid. There exists an abrupt change of orientation at the boundary leading to *surface defect.* The atoms held in between are attracted by crystals on either side and occupy equilibrium position depending on the forces. The positions at the boundary region between two grains are *distorted.* The thickness of distorted region is only a few atomic diameters. This boundary region is called *grain boundary.*

In grain boundary the adjacent atoms are inclined towards one another. If the angle included is large, i.e., 10° to 15°, it is called *wide-angle grain boundary,* and when the included angle is small, i.e., less than 10°, the distortion is not drastic and is called *low-angle grain boundary.*

There is a change from one orientation to the other through simple or complex dislocation arrays. *It is to be remembered that grain boundary dislocations are different from lattice dislocations and the atoms along the boundary have greater energy than those inside the grains.*

If $\vec{b}_A$ and $\vec{b}_B$ represent respectively the lattice vector of grains A and B at the boundary, then the Burger's vector is given by

$$\vec{b} = \vec{b}_A + \vec{b}_B. \tag{4.41}$$

4.6.1 Effect of Grain Boundaries

The properties of polycrystalline solids are affected in different ways due to the pressure of *grain boundaries*:

(i) *Crystallisation in solid phase usually starts at the grain boundaries.*

(ii) *Grain boundary segregation occurs due to the accumulation of impurities in metal in grain boundaries.*

(iii) *The strength of the metals increases enormously as the grain boundaries provide strong obstacle to the motion of dislocation.*

4.6.2 Tilt Boundaries

If two crystallites are mutually inclined at an angle less than 10°, then it is the case of a low angle boundary (or tilt boundary). It can be described by a regular arrangement of dislocation. Two such crystals placed close together with a small relative inclination are shown in Fig. 4.18 (a). The plane of each of the two figures in Fig. 4.18 (a) is parallel to the cube face and normal to the axis of the crystal. By rotating each crystal about its axis, one to the left and the other to the right by an angle $(\theta/2)$, the two figures are joined together as in Fig. 4.18 (b). Thus, a tilt boundary is formed whose plane contains a crystal axis common to both the crystals. The crystal orientation on both sides of boundary plane are symmetrical with each other.

The misfit in the bi-crystal formed due to the joining of the grains is accommodated in the following ways: (i) *by atomic misfit on the boundary of the crystal,* (ii) *by electric deformation which extends over areas of many spacings.*

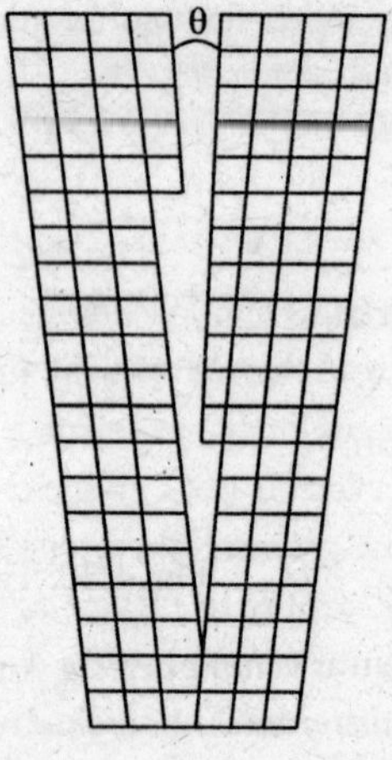

Fig. 4.18: (a) Two grains have a common cube axis and angular difference in orientation θ

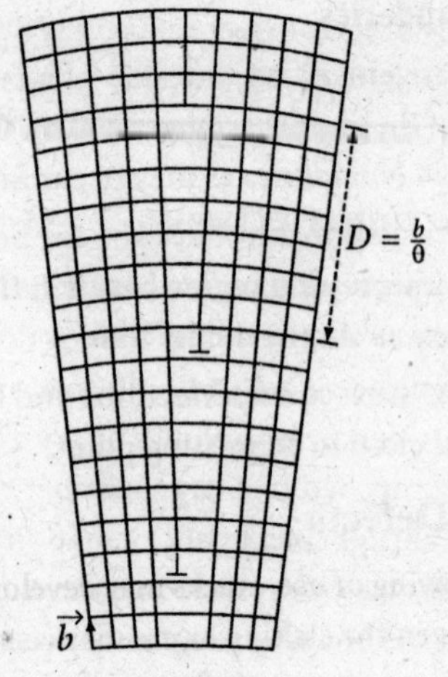

Fig. 4.18: (b) The grains are jointed to form a bi-crystal

In a pure tilt boundary the orientation can be described by a small rotation θ about the common axis of one part of the crystal relative to the other. The tilt boundary is given by an array of edge dislocations having spacing,

$$D = b/\theta, \tag{4.42}$$

where b is the Burger's vector of dislocation.

4.6.3 Stacking Faults

The planar surface imperfection caused by fault in the stacking sequence of atomic planes in crystals is called *stacking fault.* In FCC crystal the stacking sequence may be represented by *ABC* but in HCP stacking we have only two different layers *BC*. Thus, the stacking during the growth of the FCC crystal is

$$\cdots ABC\ \overset{\downarrow}{A}BC\ ABC\ ABC\ldots$$

If the plane *A* (indicated by arrow) is missed, then we get the sequence of stacking as

$$\cdots ABC\ A\boxed{BC\ BC}\ ABC\cdots$$

The stacking in the missing region becomes HCP. This region is thin and represents a surface imperfection called *stacking fault.*

Fig. 4.19 represents some stacking fault in FCC crystal.

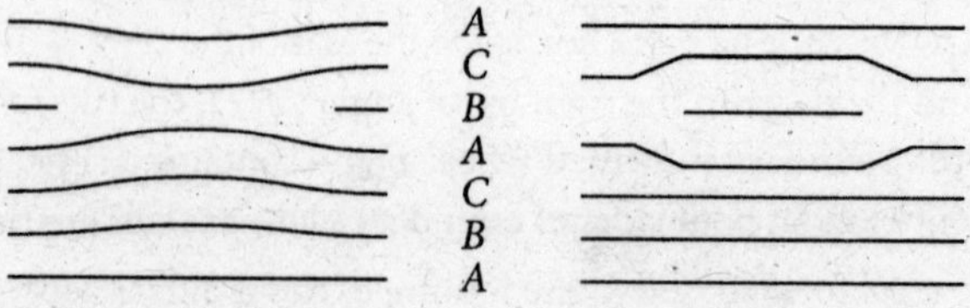

Fig. 4.19: (a) Instrinsic staking fault

Fig. 4.19: (b) Extrinsic staking fault

4.6.4 Twin Boundaries

If the atomic arrangement on one side of a twin boundary is a *mirror reflection* of the atomic arrangement on the other side, then they are called *twin boundaries* as they occur in pair.

The change of orientation introduced by one boundary is restored by the other. The intervening region between the two boundaries called *twined region* as shown in Fig. 4.20.

These are planar surface imperfections and may disappear on heating the crystal close to its melting point.

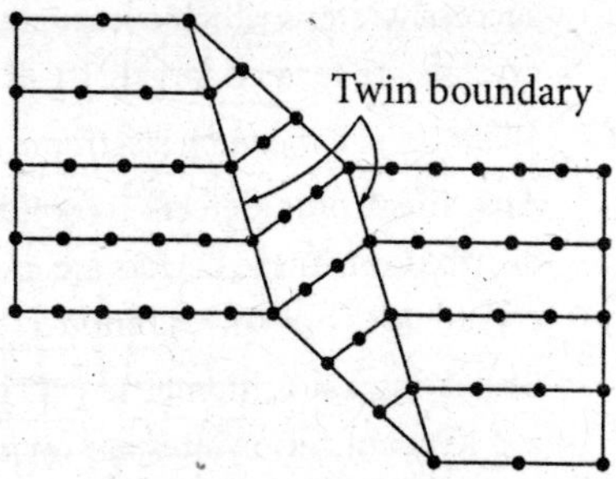

Fig. 4.20: Twin boundary

4.7 Volume Defects

During crystal growing or use, cracks may develop which are known as *volume defects.* Any small electrostatic dissimilarity between the staking layers may cause a crack during its growth. In crystal, when used for some device application, cracks are generated due to sudden thermal waves. Presence of large vacancy or void, viz., cluster of atoms missing falls under the category *volume defect.*

4.8 Answers of Some Important Questions

4.1. *What do you mean by lattice imperfection?*

Ans. Any irregularity in the crystallographic structure of a crystal is called *lattice imperfection* (or *defect*). With increase of entropy, certain structural defects are introduced in the lattice of the crystal which are called *structural defects.* At absolute zero of temperature, entropy is zero and a crystal lattice may be free from such defects at that temperature.

4.2. *What are structure sensitive properties of crystals?*

Ans. Some properties of the crystal are greatly affected by lattice defects or atomic imperfections. Strength of crystal, luminescence, colour of crystals, diffusion, crystal growth, plasticity, thermal and electric properties, magnetic properties, etc. are some of these properties which are called *structure sensitive properties of the crystals.*

4.3. *What are the different types of point defects?*

Ans. Point imperfections or defects are of three types namely, (a) vacancy (b) substitutional impurity and (c) interstitial impurity.

A *vacancy* refers to an atomic site from where the atom is missing.

A *substitutional* impurity is a point imperfection which refers to a foreign atom that substitutes for or replaces a parent atom in the crystal lattice.

In *interstitial impurity* a small sized atom is occupying the void space in the parent crystal without disturbing the parent atoms from their regular sites in an interstitial impurity.

4.4. *What are Frenkel and Schottky defects?*

Ans. Frenkel and Schottky defects in ionic crystals are point imperfections in the crystal lattice.

In ionic crystals, an ion displaced from the regular site to an interstitial site is called *Frenkel imperfection.* Generally, cations are displaced into the void space because of their smaller size. The void space being too small for the anions, they do not get displaced. Frenkel defect does not alter the overall electrical neutrality of the crystal [Fig. 4.1].

Schottky defects are vacancy defects created due to the migration of the displaced positive ion in successive steps which eventually settles at the surface of the crystal as shown in Fig. 4.1. This type of crystals are formed largely in alkali halide crystals.

4.5. *What do you mean by electronic defects in crystals?*

Ans. Electronic defects in solids result from the error in charge distribution. These defects are created if the composition of an ionic crystal does not correspond to the exact *stoichiometric formula.*

By heating a stoichiometric FeO in oxygen atmosphere Fe_xO, where $x < 1$ vacant cation sites are formed as shown in Fig. 4.21. The two electrons required by each excess oxygen atom are donated by two Fe^{2+} ions which then become Fe^{3+} ions as in Fig. 4.21.

Fe^{2+}	O^{2-}	Fe^{2+}	O^{2-}	Fe^{2+}
O^{2-}	Fe^{3+}	O^{2-}	Fe^{2+}	O^{2-}
Fe^{3+}	O^{2-}		O^{2-}	Fe^{2+}
O^{2-}	Fe^{3+}	O^{2-}	Fe^{2+}	O^{2-}
	O^{2-}	Fe^{3+}	O^{2-}	Fe^{2+}

Fig. 4.21: Electronic defects in FeO

4.6. *What is the energy of formation of vacancy in a crystal? Write down the expression for the number of vacancy at any temperature T.*

Ans. The energy of formation of vacancy in a crystal is given by

$$E_v = kT \ln\left(\frac{N-n}{n}\right),$$

where n = the number of vacancy at temperature T, N = the total number of lattice sites in the crystal and k = the Boltzmann constant.

The expression for the number of vacancy at temperature T is given by $n = Ne^{-E_v/kT}$.

4.7. *Write down expressions for the number of Schottky defects and Frenkel defects in a crystal.*

Ans. The number of Schottky defects in a crystal is given by $n = Ne^{-E_s/kT}$ and that Frenkel defect is $n = (NN_i)^2 e^{-E_f/kT}$, where N_i is the total number of interstitial sites and N is the total number of cations. E_s is the average energy required to create a Schottky defect and E_f is the average energy required to displace a cation from its normal lattice position to an interstitial site.

4.8. *What is an F-centre? Describe with suitable diagram.*

Ans. See Art. 4.4.7.

4.9. *What is an exciton? Describe different types of excitons with suitable diagram.*

Ans. See Art. 4.4.8.

4.10. *What do you mean by dislocation and slip?*

Ans. See Art. 4.5.

4.11. *Discuss in brief the edge dislocation and screw dislocation.*

Ans. See Art. 4.5.3.

4.12. *What are planar or surface defects?*

Ans. The regions of dislocations that lie about a surface having a thickness of a few atomic diameters are known as surface imperfections.

The surface defects are basically of four types, viz., (i) grain boundaries (ii) tilt boundaries (iii) twin boundaries and (iv) stacking faults.

[For their detail discussion see Art. 4.6.]

4.9 Solved Problems

4.1. *In a particular crystal the energy required to move an atom from the interior of the crystal to the surface is 1 eV. What is the proportional of vacancies present in the crystal at 1000 K and 300 K?*

Solution: If n and N be the number of vacancies and N the total number of sites in the crystal, then $n = Ne^{-E_v/kT}$.

Here 1 eV = 1.6×10^{-19} J.

$\therefore$ at $T = 1000$ K,

$$\frac{n}{N} = \exp\left[-\frac{1.6 \times 10^{-19}}{1.38 \times 10^{-23} \times 1000}\right] = e^{-11.6} \simeq \mathbf{10^{-5}}.$$

At $T = 300$ K,

$$\frac{n}{N} = \exp\left[-\frac{1.6 \times 10^{-19}}{1.38 \times 10^{-23} \times 300}\right] = e^{-3.05} \simeq \mathbf{1.66 \times 10^{-17}}.$$

4.2. *For a metal, the fraction of vacancy sites at 500 °C is 1×10^{-10}. Find the fraction of vacancy sites of the same metal at 1000 °C.* **[JNTU 2004]**

Solution: We have $\frac{n}{N} = e^{-(E_f/kT)}$.

At $T_1 = 500 + 273 = 773$ K, $\frac{n}{N} = 1 \times 10^{-10}$.

Let $\frac{n}{N} = x$ at $T = 1000 + 273 = 1273$ K.

$$\therefore \ln\frac{n_1}{N} = -\frac{E_f}{kT_1} \text{ and } \ln\frac{n_2}{N} = -\frac{E_f}{kT_2}.$$

Hence,

$$\frac{\ln\frac{n_1}{N}}{\ln\frac{n_2}{N}} = \frac{T_2}{T_1} \text{ or, } \frac{\ln\left(1 \times 10^{-10}\right)}{\ln x} = \frac{1273}{773}$$

$$\text{or, } \ln x = \frac{773}{1273}\ln\left(10^{-10}\right) = 0.6072 \times (-10) \times 2.303 = -10 \times 1.3984 = -13.984.$$

$$\therefore x = e^{-13.894} = \frac{1}{e^{13.984}} = 8.4494 \times 10^{-7}.$$

Hence, the fraction of vacancy sites at 1000 °C = $\mathbf{8.4494 \times 10^{-7}}$.

4.3. *The vacancy formation energy for silver is 1.1 eV. (i) What is the number of vacancies per cubic centimetre at 300 °C? (ii) What is the fraction of lattice sites that are vacant at the melting temperature of silver (i.e., 960 °C)? [At. wt. of silver A = 107.9, density of silver = 10.5 kg/m³]*

Solution: We have $n = Ne^{-E_v/kT}$ and the number of atomic sites/m³ is

$$N = N_A\frac{\rho}{A} = 6.02 \times 10^{23} \text{ atoms/mol} \times \frac{10500}{1079 \times 10^{-3}\text{ kg/m}} = 5.86 \times 10^{28} \text{ atoms/m}^3.$$

Boltzmann constant $k = 1.38 \times 10^{-23}$ J/K = 8.63×10^{-5} eV/K.

Now at $T = 300$ °C,

$$n = \left(5.86 \times 10^{28}\right) \times \exp\left[-\frac{1.1\text{ eV}}{8.63 \times 10^{-5}\frac{\text{eV}}{\text{K}} \times 1233}\right] = \mathbf{1.096 \times 10^{19} \text{ vacancies/m}^3}.$$

At 900 °C,

$$\frac{n}{N} = \frac{Ne^{-E_v/kT}}{N} = \exp\left[\frac{1.1\,\text{eV}}{8.63\times10^{-5}\frac{\text{eV}}{\text{K}}\times1233}\right] = e^{-10.34} = 3\times10^{-5}.$$

4.4. *In an ionic crystal, the average energy required to create a Frenkel defect is 1.4 eV. What is the ratio of the number of Frenkel defects at 300 °C and 200 °C per gram of the crystal.*

Solution: The number of Frenkel defect at temperature T is given by

$$n = (NN_i)^{1/2}\,e^{-E_f/2kT},$$

where N is the total number of cations and N_i, the total number of interstitial sites in the ionic crystal. E_f and k are respectively the average energy required to create a Frenkel defect and the Boltzmann constant.

When $T = 300\,°\text{C} = (300+273) = 573\,\text{K}$,

$$n_{573} = (NN_i)^{1/2}\,e^{-\frac{1.4}{1146k}}.$$

When $T = 20\,°\text{C} = (20+273) = 293\,\text{K}$,

$$n_{293} = (NN_i)^{1/2}\,e^{-\frac{1.4}{586k}}.$$

$$\therefore \frac{n_{293}}{n_{573}} = \exp\left[-\frac{1.4}{k}\right]\left(\frac{1}{586}-\frac{1}{1146}\right) = \exp\left[-\frac{1.4}{8.625\times10^{-5}}\right]\frac{560}{586\times1146}$$

$$\text{or, } \ln\left(\frac{n_{573}}{n_{293}}\right) = \frac{1.4\times560}{(8.625\times10^{-5})\times(586\times1146)}$$

$$\text{or, } \log_{10}\left(\frac{n_{573}}{n_{293}}\right) = \frac{13.5354}{2.303} = 5.8773$$

$$\text{or, } \frac{n_{573}}{n_{293}} = \mathbf{7.54\times10^5}.$$

4.5. *Calculate the average energy required to create a Schottky defect in* NaCl *if the density of Schottky defect in it is* $5\times10^{11}/\text{m}^3$ *at* 300 K. [*The inter-ionic separation* = 2.82 Å.]

Solution: If n be the density of Schottky defect at T and N the ion pairs/m³ of NaCl, then $n/N = e^{-E_s/2kT}$, where E_s is the average energy to create a Schottky defect.

Here $n = 10^{11}/\text{m}^3$, $T = 300$ K and interatomic separation $= 2.82\times10^{-10}$ m.

The volume of the unit cell in NaCl without defect is $V = (2\times2.82\times10^{-10})^3 = 1.794\times10^{-28}\,\text{m}^3$ (since there are four ion pairs/unit cell of NaCl).

$$\therefore N = \frac{4}{1.794\times10^{-28}} = 2.23\times10^{28}\text{ ion pairs.}$$

Now

$$\frac{N}{n} = e^{\frac{2kT}{E}} \text{ or, } E = 2kT\ln\left(\frac{N}{n}\right) = 2\times(8.625\times10^{-5})\times300\times\ln(4.46\times10^{16})$$

$$= 2\times8.625\times10^{-5}\times300\times38.336 = \mathbf{1.984\,eV}.$$

4.6. *A Burger's lineage boundary in a simple cubic crystal creates a line of etch pits* 2×10^{-4} cm *apart when the crystal is etched. Assuming that the lattice constant of the crystal is* 4×10^{-8} cm, *find the angle of misfit between the sections of the crystal on opposite sides of the lineage boundary.*

Solution: A tilt boundary orientation is represented as an array of edge dislocations of spacing, $D = b/\theta$, b = Burger's vector and θ = angle of misfit.

Since, the minimum value of b = lattice constant,

$$\therefore \theta = b/D = \left(4\times10^{-8}\right)/\left(2\times10^{-4}\right) = \mathbf{2\times10^{-4}\ radians}$$

4.7. *If* 1 eV *is the energy required to move an atom from the crystal's interior to the surface, what is the proportion of vacancies present in the crystal at* 1000 K *and* 300 K? **[BU(H) 2012]**

Solution: If n be the number of vacancies per defects, N the total number of atoms/ions or available lattice sites, T the temperature of the lattice, ϵ, the average energy required to create a vacancy at temperature T, then vacancy concentration is given by $n = Ne^{-\epsilon/kT}$.

$$\therefore n_1 = Ne^{-\epsilon/kT_1},\ n_2 = Ne^{-\epsilon/kT_2} \text{ or, } \frac{n_1}{n_2} = \frac{Ne^{-\epsilon/kT_1}}{Ne^{-\epsilon/kT_2}} = \exp\left[\frac{\epsilon}{k}\left(\frac{1}{T_1}-\frac{1}{T_2}\right)\right].$$

Given, $\epsilon = 1\text{ eV}$, $k = 1.38\times10^{-23}\ \text{JK}^{-1}$

$$\text{or, } k = \frac{1.38\times10^{-23}\ \text{eV}}{1.6\times10^{-19}\ \text{K}} = 86.25\times10^{-6}\ \text{eV/K or, } \frac{e}{k} = \frac{1\text{ eV}}{86.25\times10^{-6}\ \text{eV/K}^{-1}} = \frac{1}{86.25\times10^{-6}\ \text{K}^{-1}}.$$

Again, for $T_1 = 1000$ K, $T = 300$ K

$$\frac{n_1}{n_2} = \exp\left[\frac{1}{86.25\times10^{-6}\ \text{K}^{-1}}\left(\frac{1}{300\ \text{K}-\frac{1}{1000\ \text{K}}}\right)\right] = e^{27.053} = \mathbf{5.6\times10^{11}}.$$

4.8. *Find the average energy required to create Frenkel defects in ionic solid.*

Solution: The number of Frenkel defects is given by $n = \sqrt{NN_i}\,e^{-\epsilon_F/2kT}$ [where $N, N_i \to$ constant]

$$\therefore \frac{\sqrt{NN_i}}{n} = e^{\epsilon_F/2kT} \text{ or, } \frac{NN_i}{n^2} = e^{E_F/kT} \text{ or, } \ln\frac{NN_i}{n^2} = \frac{\epsilon_F}{kT}.$$

$$\therefore \boldsymbol{\epsilon_F = kT\ln\frac{NN_i}{n^2}}.$$

4.9. *A metal is rapidly quenched from* 800 K *to* 400 K. *Find the ratio of the number of vacancies in equilibrium.* [Energy for vacancy defect = 1 eV]

Solution: The number of vacancies/defect is $n = Ne^{-\epsilon/kT}$.

$$\therefore \frac{n_1}{n_2} = \frac{e^{-\epsilon/kT_1}}{e^{-\epsilon/kT_2}} = \exp\left[\frac{\epsilon}{k}\left(\frac{1}{T_2}-\frac{1}{T_1}\right)\right] = \exp\left[\frac{1.6\times10^{-19}}{1.38\times10^{-23}}\left(\frac{1}{800}-\frac{1}{40}\right)\right] = \mathbf{5\times10^{-7} < 1},$$

i.e., vacancy decreases.

4.10. *The average energy needed to create a Frenkel defect at ionic crystal is* 1.4 eV. *Find the ratio of Frenkel defects at* 23 °C *and* −23 °C.

Solution: The number of vacancy/defect is $n = \sqrt{NN_i}e^{-\epsilon_F/2kT}$. At temperatures T_1, T_2,

$$\frac{n_1}{n_2} = \frac{\sqrt{NN_i}\,e^{-\epsilon_F/2kT_1}}{\sqrt{NN_i}\,e^{-\epsilon_F/2kT_2}} = \exp\left[\frac{\epsilon_F}{2k}\left(\frac{1}{T_2}-\frac{1}{T_1}\right)\right]$$

$$= \exp\left[\frac{1.4\text{ eV}}{2(1.38\times10^{-23}\ \text{JK}^{-1})}\left(\frac{1}{-23+273}-\frac{1}{23+273}\right)\frac{1}{\text{K}}\right]$$

$$= \exp\left[\frac{1.4\times1.6\times10^{-19}\ \text{J}}{2(1.38\times10^{-23}\ \text{J})}\left(\frac{1}{250}-\frac{1}{296}\right)\right] = \mathbf{155}.$$

EXERCISE

4(A): Objective (MCQ) Type Questions

Select the correct answer.

4.1. Lattice imperfections are generated when the entropy (S) of the crystal is
(a) $S = 0$ (b) $S < 0$ (c) $S > 0$ (d) $S \geq 0$.

4.2. The lattice imperfections affect the
(a) thermal properties of the crystal (b) electrical properties
(c) both thermal and electrical properties (d) none of these.

4.3. Mechanical strength of metals and alloys due to lattice imperfections
(a) increases (b) decreases (c) remains unaltered (d) none of these.

4.4. In Schottky defect the displaced positive ion settles at the
(a) interstitial position (b) surface of the crystal
(c) outside surface of the crystal (d) none of these.

4.5. In Frenkel defect the displaced positive ion settles at the
(a) interstitial position (b) surface of the crystal
(c) anywhere in the crystal (d) none of these.

4.6. If n and N be respectively the number of vacancy and total number of atoms in a crystal, then
(a) $n = Ne^{-E_u/kT}$ (b) $n = Ne^{E_u/kT}$ (c) $n = N/e^{-E_u/kT}$ (d) $n = N/e^{E_u/kT}$.

4.7. The number of Frenkel defects in a crystal at temperature T is given by
(a) $n = (NN_i)^{1/2} \times e^{-E_f/2kT}$ (b) $n = (N/N_i)^{1/2}\, e^{-E_f/2kT}$
(c) $n = (NN_i)\, e^{-E_f/2kT}$ (d) $n = (NN_i)^{1/2}\, e^{-E_f/kT}$.

4.8. The F-centre is an electron trapped at
(a) positive ion vacancy (b) negative ion vacancy
(c) interstitial position (d) none of these.

4.9. An exciton is
(a) an electron in the excited state (b) bound electron-hole pair
(c) an exciting radiation (d) none of these.

4.10. The binding energy of exciton in KCl in eV is
(a) 380 (b) 200 (c) 100 (d) 400.

4.11. The critical stress of a perfect crystal is given by
(a) $T_{crit} = G/2\pi$ (b) $T_{crit} = 2\pi G$ (c) $T_{crit} = 2\pi/G$ (d) $T_{crit} = 1/2\pi G$.

4.12. If z-axis represents the dislocation line, then displacement of atoms in the direction of z-axis is given by
(a) $\Delta z = \frac{b}{2\pi\theta}$ (b) $\Delta z = 2\pi/b\theta$ (c) $\Delta z = b\theta/2\pi$ (d) $\Delta z = 2\pi b\theta$.

4.13. The elastic energy of edge dislocation (U_{ed}) is greater than that of screw dislocation (U_{sc}) by a factor
(a) $\frac{1}{(1+\mu)}$ (b) $(1-\mu)$ (c) $\frac{1}{(1-\mu)}$ (d) μ.

4.14. In tilt boundary two crystals are mutually inclined at an angle θ, where
(a) $\theta > 10°$ (b) $\theta < 10°$ (c) $\theta = 0°$ (d) $\theta = 180°$.

4.15. The tilt boundary is given by an array of edge dislocations having spacing
(a) $D = b\theta$ (b) $D = \theta/b$ (c) $D = b/\theta$ (d) $D = 1/b\theta$.

4.16. The plastic deformation in a crystal is possible due to
(a) point defect (b) Frenkel defect
(c) Schottky defect (d) the motion of dislocation in the crystal.

4.17. The zero-dimensional defect is
(a) point defect (b) volume defect (c) surface defect (d) line defect.

4.18. Interstitial impurity is a
(a) surface defect (b) volume defect (c) line defect (d) point defects.

4.19. Doping of semiconductor results in
(a) Frenkel defect (b) Schottky defect (c) interstitial defect (d) substitutional defect.

4.20. In edge dislocation, Burger's vector is
(a) parallel (b) perpendicular
(c) at any orientation w.r.t. the dislocation line (d) none of these.

4.21. In an ionic crystal with N number of cation-anion pairs at temperature T, if E_s be the average energy required to create a Schottky defect, the number of Schottky defect is given by
(a) $n = Ne^{E_s/kT}$ (b) $n = Ne^{-E_s/kT}$ (c) $n = Ne^{-E_s/2kT}$ (d) $n = Ne^{E_s/2kT}$.

4.22. In ionic crystal, missing of a pair of a cation and an anion results in
(a) Schottky defect (b) electronic defect (c) Frenkel defect (d) none of these.

4.23. In ionic crystal, errors in charge distribution are called
(a) volume defects (b) point defects (c) electronic defects (d) line defects.

ANSWERS

4.1. (c) **4.2.** (c) **4.3.** (b) **4.4.** (b) **4.5.** (a) **4.6.** (a) **4.7.** (a) **4.8.** (b) **4.9.** (b) **4.10.** (d) **4.11.** (a) **4.12.** (c) **4.13.** (c) **4.14.** (b) **4.15.** (c) **4.16.** (d) **4.17.** (a) **4.18.** (d) **4.19.** (d) **4.20.** (b) **4.21.** (c) **4.22.** (a) **4.23.** (c).

4(B): Short Answer Type Questions

4.1. What do you mean by imperfection in crystals? Explain. [JNTU 2003]

4.2. How the mechanical and electrical properties of crystal are affected by the pressure of defects?

4.3. What are the different possible types of defects in crystals?

4.4. What are Schottky defect and Frenkel defect in crystals? Show diagrammatically. [JNTU 2005]

4.5. Discuss the temperature dependence of Schottky and Frenkel defects in a crystal.

4.6. Explain the following: (a) edge dislocation (b) screw dislocation. [JNTU 2004]

4.7. Explain the significance of Burger's vector. [JNTU 2004]

4.8. Distinguish between Frenkel and Schottky defects. [JNTU 2003]

4.9. Distinguish between edge and screw dislocations. What is Burger's vector?

4.10. Discuss the effect of grain boundaries on the properties of polycrystalline solids.

4.11. What are excitons? Discuss various types of excitons.

4.12. Discuss edge and screw dislocation in brief.

4.13. What are grain boundaries? What do you mean by tilt boundaries?

4.14. Discuss stacking faults.

4.15. What is Wannier-Mott exciton? Draw the energy level diagram of an exciton created in direct process.

4.16. What is meant by point defects in crystal lattice? What are the different types of point defects? How are they caused? **[Bhabnagar U 2017]**

4(C): Numerical Problems

4.1. Aluminium metal is quenched rapidly form 800 K to 300 K. Calculate the ratio of number of vacancies in equilibrium at the two temperatures. [*Ans.* 3.99×10^{-8}]

4.2. Find out the ratio of vacancies in a metal at 100 K and 500 K if the average energy required to produce a vacancy in the metal be 1 eV. [*Ans.* 1.082×10^{5}]

4.3. If the average energy to create a Schottky pair in a metal be $E_s = 1$ eV at $T = 1000$ K, then calculate the ratio of Schottky defects to the total number of atoms per cc in the metal. [*Ans.* 10^{18}/cc]

4(D): Broad Answer Type Questions

4.1. (a) What are lattice imperfections? How do they effect the properties of solids?

(b) What are different classes of lattice imperfections? How point defects are formed?

4.2. Calculate the energy required for the formation of vacancy. Find out an expression for the number of vacancies at a given temperature.

4.3. (a) Derive the equation relating the number of vacancies (n) in a non-atomic crystal to the energy E required to remove one atom to the crystal's exterior.

(b) What are the factors on which Schottky defect depends?

4.4. (a) Show that the average energy required to create Frenkel defect in ionic solids is

$$E_f = kT \ln\left(\frac{NN_i}{n^2}\right),$$

where N = total number of cations, N_i = total number of interstitial sites and n = the number of interstitial defects.

4.5. Hence, show that $n = (NN_i)^{1/2}\, e^{-E_f/2kT}$ at any temperature T. Discuss the effect of T on n.

4.6. (a) What are F-centres? Discuss the quantised energy states of F-centre.

(b) Show diagrammatically the configuration of M-centre and R-centre.

4.7. (a) What are Frenkel excitons?

(b) Show that the lowest bound state of Wannier-Mott exciton extends over a radius given by

$$a_x = \epsilon\left(\frac{m}{\mu}\right) a_0,$$

where μ is the reduced mass of the exciton and a_0 is $\frac{\hbar^2}{me^2}$, the Bohr radius.

4.8. (a) What are dislocation and slip?

(b) Show that theoretically critical stress of perfect crystal is much larger than the observed stress for pure crystals.

4.9. (a) What is Burger's vector? Use it to explain dislocation model for slip.

(b) Describe with suitable diagram the edge and screw dislocation.

(c) What combinations of dislocations can you postulate around which the Burger's circuit closes upon itself, i.e., Burger's vector is zero? Use sketches to illustrate your answer.

4.10. (a) Calculate the dislocation of energy of an elastic isotropic crystal.

(b) Show that edge dislocation energy (U_{ed}) is greater than screw dislocation energy by a factor $\left(\frac{1}{1-\mu}\right)$, where μ is the Poisson's ratio.

4.11. (a) Find out an expression for dislocation density of a crystal.

(b) What are surface imperfections? What is meant by grain boundaries.

(c) Discuss the effect of grain boundaries on the properties of polycrystalline solids.

4.12. Write notes on: (a) Tilt boundary (b) Stacking fault (c) Twin boundary and (d) Volume defects.

Chapter 5

Bonding in Solid

5.1 Introduction

The closer collection of atoms resulting in bulk materials is called *solid.* In solids the atoms are bound together by very strong force. These forces are classified on the basis of the nature of electrostatic interaction between the neighbouring atoms. The crystals are generally classified into the following *five types* depending on the *binding forces* holding the atoms of the crystal together in the solid:

(i) Ionic crystal (e.g., alkali halides, alkaline oxides, etc.)

(ii) Covalent crystals (diamond, silicon, etc.)

(iii) Metallic crystals (various metals and alloys)

(iv) Molecular crystals (O_2, H_2, solid He, Kr, Xe)

(v) Hydrogen-bonded crystals (some fluorides and compounds having water of crystalisation).

5.2 Types of Bonding

Bonds in solid are due to *two* types of forces, viz.,

(i) The *attractive force* arising from the interaction of the negative electron cloud of one atom with the positive nuclear charge of the other.

(ii) The *repulsive forces* arising due to the overlapping of electron clouds of the atoms when they are brought closer. The positive nuclei of the two atoms also contribute to the *repulsive force.*

The magnitude of the *attractive* and *repulsive forces* varies as some power of interatomic distance (r).

On the basis of these forces following *five* types of bond arise:

(i) *Ionic bond* (due to transfer of valence electrons as in NaCl)

(ii) *Covalent bond* (due to sharing of valence electrons as in diamond)

(iii) *Metallic bond* (due to free nature of valence electron as in Ag, Cu)

(iv) *Hydrogen bond* (as in ice)

(v) van der Waals' bond (electrons remain associated with original molecules).

The first *three* are called *primary bonds* with bond energy 0.1 to 10 eV/bond and the last *two* are called *secondary bonds* with bond energy 0.01 to 0.5 eV/bond. There are crystals which show *mixed bonding* also.

5.3 Ionic Bonding (or Electrovalent Bonding)

The *ionic bonding* results between two atoms by the *total transfer of valence electrons* from one atom to the other, so that each atom acquires a stable electronic configuration. The resulting positive and negative ions attract each other by *Coulomb's law force* which dominates the *Coulombic repulsive force* between positive or negative ions. After the transfer of electrons, the ions attain electronic configuration as *inert gas* atoms. That is why the crystals bound by ionic bond are called *ionic crystals.*

The *cohesive energy* of the ionic crystals is of the order of 5 to 10 eV which is very large.

Example. If sodium is burnt in an atmosphere of chlorine, the valence electron of sodium is transferred to chlorine. Sodium and chlorine ions then attain stable filled shell of outer electrons[1] and a strong electrostatic attraction is set up which binds Na^+ and Cl^- forming a stable NaCl molecule at the equilibrium spacing (r_0). At r_0 the PE of NaCl is the minimum. The process can be represented by

$$Na + 5.1\,eV \rightarrow Na^+ + e^-$$

as the ionization energy, i.e., the energy required to remove the valence electron of Na is 5.1 eV.

Again, the electron affinity of chlorine being 3.6 eV, this much energy is released when the electron is added to chlorine. Then chlorine becomes negatively charged ion:

$$Cl + e^- \rightarrow Cl^- + 3.6\,eV$$

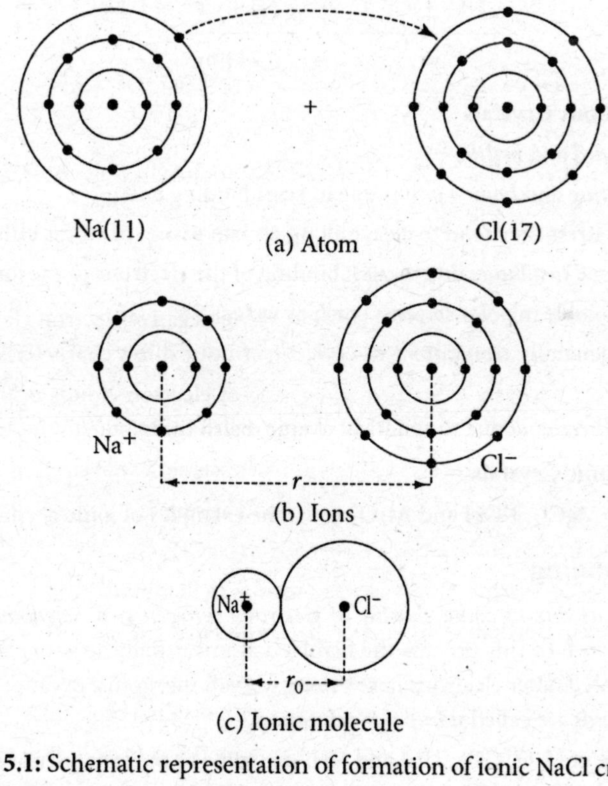

Fig. 5.1: Schematic representation of formation of ionic NaCl crystal

So, the net energy spent in ionizing both Na and Cl atoms is

$$(5.1 - 3.6)\,eV = 1.5\,eV$$

$$Na + Cl + 1.5\,eV \rightarrow Na^+ + Cl^-$$

[1] $Na(11) = 1s^2 2s^2 2p^6 3s^1$; $Na^+ = 1s^2 2s^2 2p^6$ (Ne-structure); $Cl(17) = 1s^2 2s^2 2p^6 3s^2 3p^5$; $Cl^- = 1s^2 2s^2 2p^6 3s^2 3p^6$ (Ar-structure)

Now the electrostatic attraction between Na^+ and Cl^- brings them closer to r_0, r_0 is the *equilibrium distance* ($r_0 = 0.24\,\text{nm}$), where the PE is the minimum. Hence, when the NaCl molecule is formed, energy (called *bond energy*) is released.

5.3.1 Calculation of Bond Energy

The PE of the NaCl molecule at equilibrium position is

$$V = -\frac{e^2}{4\pi\varepsilon_0 r_0} = -\left[\frac{(1.602\times10^{-19})}{4\pi\,(8.85\times10^{-12})\,(2.4\times10^{-10})}\right]\text{J} = -\frac{9.612\times10^{19}}{1.602\times10^{-19}}\,\text{eV} = \mathbf{-6\,eV}.$$

Hence, the net energy released due to the formation of NaCl molecule is

$$(5.1 - 3.6 - 6)\,\text{eV} = \mathbf{-4.5\,eV}.$$

Thus, when NaCl molecule is formed, the energy released is 4.5 eV. This is also the energy required to dissociate NaCl molecule to Na^+ and Cl^- ions. This means to dissociate NaCl molecule in Na^+ and Cl^-, 4.5 eV energy will have to supply. The following represents the scheme

$$Na^+ + Cl^- \;\rightarrow\; Na + Cl \;\rightarrow\; NaCl$$

$$\xleftrightarrow{\quad\infty\quad} \qquad \xleftrightarrow{\quad 0.24\,\text{nm}\quad}$$

5.3.2 Properties of Ionic Crystals

1. These crystals are *hard* and *brittle*.
2. They have *high melting* and *boiling points* due to large binding energy.
3. When subjected to stress they tend to cleave along certain planes of atoms rather than to deform.
4. Ionic crystals are *good insulators* due to tight binding of the electrons to the ions.
5. These crystals are soluble in *polar solvents* (such as water).
6. Ionic crystals are generally *transparent to visible light* but exhibit characteristic absorption peaks in infrared region.
7. Ionic bond is *non-directional* due to spherical charge distribution.

5.3.3 Examples of Ionic Crystals

NaCl, KCl, KBr, MgO, $MgCl_2$, KOH and Al_2O_3 are some examples of ionic crystals.

5.4 Covalent Bonding

Covalent bonding results due to equal sharing of electrons between two neighbouring atoms each having incomplete outermost shell. In this process the involved atoms usually attain *noble gas* configuration and become stable. The shared outer electrons here belong to both the atoms. Because of this reason, the solids formed by *covalent bonds* are called *non-polar substances*.

Example. The bonding of H_2O, CO_2, HCl and NH_3 are some of the examples of *covalent molecules*.

Let us study the formation of covalent bonds in hydrogen molecule. During the formation of a hydrogen molecule, the electron clouds of two hydrogen atoms, when brought together, overlap and interact with each other [Fig. 5.4]. The probability of finding the $1s^1$ electron in the two atoms is the maximum. Following *Pauli's exclusion principle* the two electrons must have opposite spin. This anti-spin interaction of $1s^1$ electrons of the two atoms give rise to a strong force of attraction which is balanced by electrostatic repulsion between the atoms. The molecule becomes stable when the two atoms come to a separation r_0, called the *equilibrium separation.* At this position the bond is *saturated* and the PE of the system becomes the minimum. The covalent bonding is also known as the *homopolar* or *electron-pair* bonding.

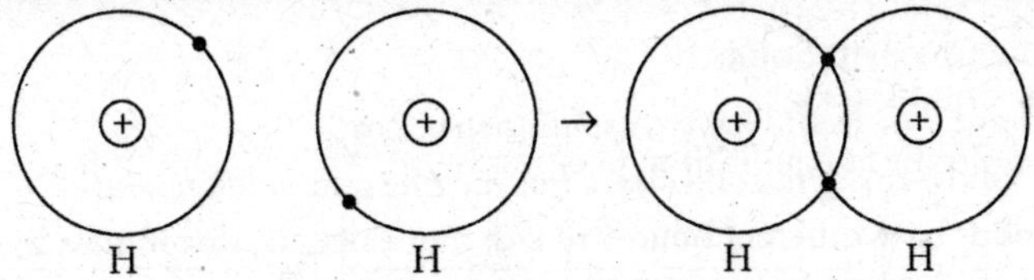

Fig. 5.2: (a) Hydrogen molecule

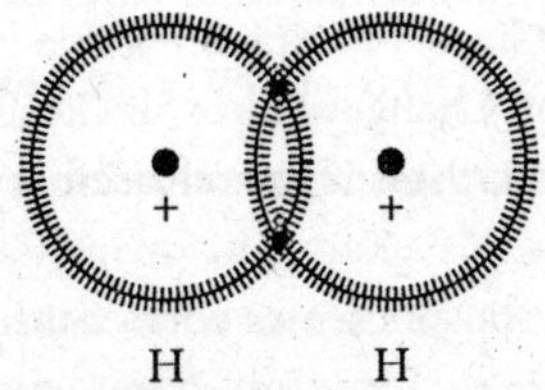

Fig. 5.2: (b) Covalent bonding in hydrogen molecule.

The covalent bonding is always *directional in character.* This is due to the restricted orbital motion of the electrons. Covalent bonding arises not only due to the overlap of pure *s*-orbital or pure *p*-orbital but also due to the overlap of *s*- and *p*-orbitals. The bonding due to overlap of *s*- and *p*-orbitals is called *hybrid bonding.*

Hybrid bonding exists in carbon. A normal carbon atom has the electronic configuration

$$1s^2\, 2s^2\, 2p^2.$$

The electron spin distribution of such a carbon atom is

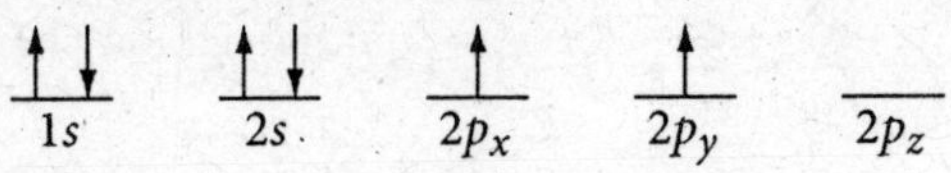

Here 1*s*- and 2*s*-orbitals are spin paired but 2*p*-orbitals are not. As unpaired electrons are expected to form bond, two bonds are possible to be formed. But when two carbon atoms come closer, one electron of 2*s*-orbital is excited to 2*p* orbit and the resulting spin distribution becomes

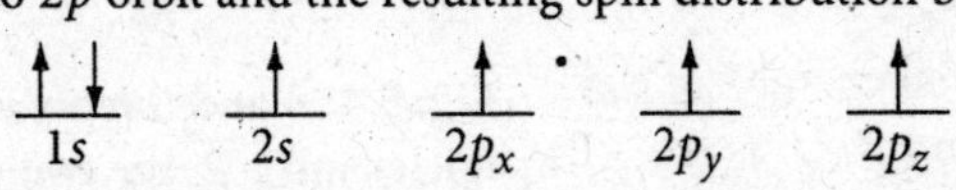

In the above representation the number of unpaired electrons is four—one in 2*s*- and three in 2*p*-orbitals. These four bonding directions are directed along the four corners of a *regular tetrahedron;* the bond angle being 109.5° [Fig. 5.3]. The arrangement of this sp^3 hybrids is called the *hybridization.* Carbon atoms, in diamond, exhibit sp^3 tetrahedral covalent bonding. Si, Ge, α-Sn, etc. are other examples of covalent crystals of this structure.

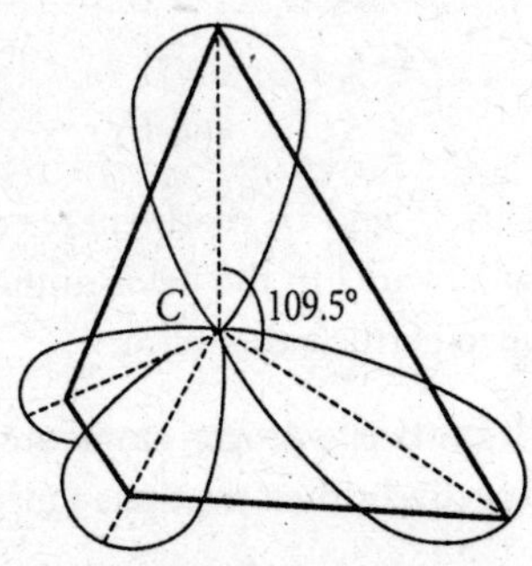

Fig. 5.3: sp^3-orbitals directed symmetrically towards the corners of a tetrahedron

Double covalent bond may be formed (as in O_2) if the participating atom has a deficiency of two electrons. Triple covalent bond is formed in nitrogen molecule. Actually, the number covalent bonds is given by

$(8-N)$ rule, where N is the number of column in the periodic table to which the atom belongs. For example, oxygen belongs to VI group and according to $(8-N)$ rule the number of covalent bond is $8-6=2$.

5.4.1 Properties of the Covalent Solids

1. Covalent solids are *hard, brittle* and have *crystalline structure.*
2. Compared to ionic solids, covalent solids have *low melting* and *boiling points.*
3. Different covalent solids have different bond strength. For example, diamond is the hardest material and has high melting point. Electrically, it is a good *insulator.* But tin is soft and has low melting point. It is a good *electrical conductor.*
   Covalent solids like Ge, Si are *semiconductors* having *moderate binding force.*
4. Covalent bond has *directional character.* Change of direction leads to the formation of different substance.
5. The *electrical conductivity* of covalent crystal is low at ordinary temperature and increases with the increase of temperature.
6. Electrical conductivity of covalent crystals increases when *doped* with certain materials.
7. Covalent crystals are *transparent* for $\lambda \geq \chi$ for infrared but *opaque* for shorter λ. The transition from transparent to opaque occurs abruptly at a characteristic λ in the visible or near infrared region.

5.4.2 Heitler and London Theory of Covalent Bonding

Heitler and London theory can be understood in the light of bonding existing in the hydrogen molecule (H_2).

In Fig. 5.4, let A and B represent two positively charged nuclei and 1 and 2 two electrons of two hydrogen atoms. The distances of the electrons from A and B as well as that between A and B are as shown in the same figure.

Let ψ_{1A} and ψ_{2B} be the eigenfunctions of the *atoms* A and B respectively when they are in the normal states. The PE of the system is

$$U = \frac{e^2}{r_{AB}} - \frac{e^2}{r_{1A}} - \frac{e^2}{r_{1B}} - \frac{e^2}{r_{2A}} + \frac{e^2}{r_{2B}} + \frac{e^2}{r_{12}} = U' + U_1 + U_2, \quad (5.1)$$

where

$$U' = \frac{e^2}{r_{AB}} + \frac{e^2}{r_{12}} - \frac{e^2}{r_{1B}} - \frac{e^2}{r_{2A}} \quad (5.2)$$

$$\text{and } U_1 = -\frac{e^2}{r_{1A}}, \; U_2 = -\frac{e^2}{r_{2B}}, \quad (5.3)$$

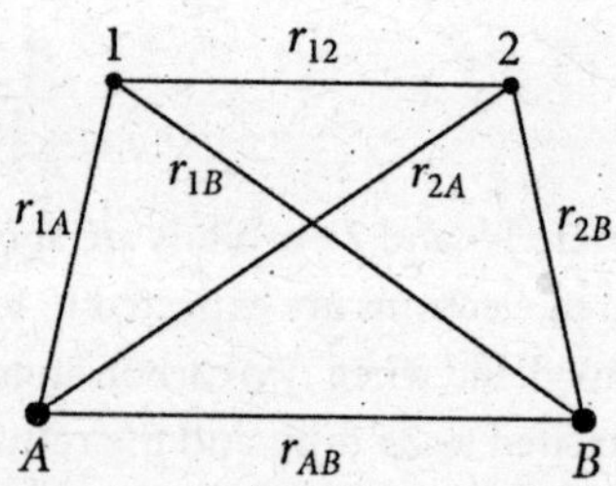

Fig. 5.4: A and B represent two nuclei and 1, 2 the two electrons of two hydrogen atoms

where U_1 and U_2 represent respectively the potential energies of electrons 1 and 2 in the fields of their own nuclei A and B. U' is the PE due to perturbation term.

If the two hydrogen atoms are assumed to be far apart and the electrons 1 and 2 belong to nuclei A and B respectively, then the wave function of the two electrons is given by

$$\psi(x_1, y_1, z_1, s_1; x_2, y_2, z_2, s_2), \quad (5.4)$$

where $\psi_A(1)$ and $\psi_B(2)$ depend on position and spin coordinates. The *Schrödinger equation* for the two electrons is given by

$$H\psi = E\psi \text{ or, } -\frac{\hbar^2}{2m}\left(\nabla_1^2 + \nabla_2^2\right)\psi + V\psi = E\psi, \quad (5.5)$$

where E represents the energy eigenvalue relating to ψ.

If $r_{AB} \rightarrow \infty$, we can treat the system as two hydrogen atom and then eqn (5.5) can be written as

$$\nabla_1^2 \psi_{1A} + \frac{2m}{\hbar^2}(E_1 - V_1)\psi_{1A} = 0 \tag{5.6}$$

$$\text{and } \nabla_2^2 \psi_{2B} + \frac{2m}{\hbar^2}(E_2 - V_2)\psi_{2B} = 0. \tag{5.7}$$

The solutions of eqns (5.6) and (5.7) are

$$\psi_{1A} = \frac{1}{\sqrt{\pi a_0^3}} e^{-r_{1A}/a_0}, \; E_1 = \frac{e^2}{a_0} \tag{5.8}$$

$$\text{and } \psi_{2B} = \frac{1}{\sqrt{\pi a_0^3}} e^{-r_{2B}/a_0}, \; E_2 = \frac{e^2}{a_0}, \tag{5.9}$$

where a_0 is the *radius of the Bohr orbit.*

Since, the probability of finding the electron 1 near nucleus A and electron 2 near nucleus B for $r_{AB} \rightarrow 0$ are very small, we can write

$$E = E_1 + E_2 = 2E_0. \tag{5.10}$$

The situation is described in Fig. 5.5 where $|\psi|^2$ is the *probability of finding an electron.* The wave function of the system is then $\psi = \psi_{1A}\,\psi_{2B}$.

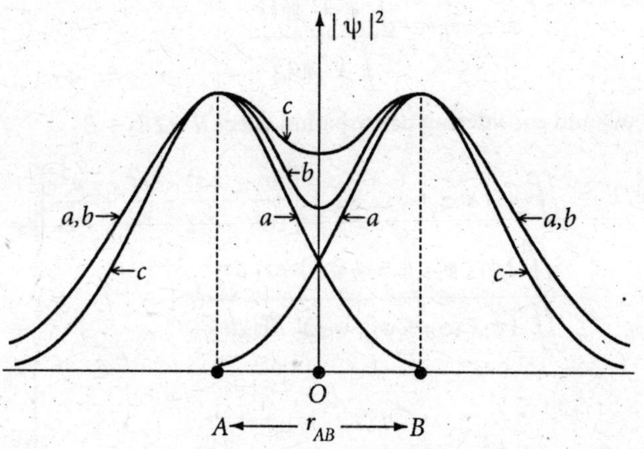

Fig. 5.5: a→represents electron density distribution in independent H-atom, b→electron density distribution when two H-atoms are separated by r_{AB}, c→actual electron density distribution in H_2 molecule

As the electrons are *indistinguishable*, we can write, $\psi' = \psi_{2A}\,\psi_{1B}$ by exchange degeneracy and then the total energy will also be $E = 2E_0$. The superposition of ψ and ψ' will also be a solution of the system. Now the superposition of ψ and ψ' may be either *symmetric* (ψ_S) or *antisymmetric* (ψ_A), i.e.,

$$\psi_S = P_S\left[\psi_{1A}\,\psi_{2B} + \psi_{2A}\,\psi_{1B}\right] \tag{5.11}$$

$$\text{and } \psi_A = P_A\left[\psi_{1A}\,\psi_{2B} - \psi_{2A}\,\psi_{1B}\right], \tag{5.12}$$

where

$$P_3 = \frac{1}{\sqrt{2+2S}} \tag{5.13}$$

$$\text{and } P_A = \frac{1}{\sqrt{2-2S}}. \tag{5.14}$$

Here

$$S = \iint \psi_{1A}\, \psi_{1B}\, \psi_{2A}\, \psi_{2B}\, dr_1\, dr_2$$

represents the *overlap integral.*

Fig. 5.5(b) shows the distribution of electron density when the atoms are brought closer. Since in this case r_{AB} is not negligible, the PE of the system is

$$V = V_0 + V', \tag{5.15}$$

where

$$V_0 = V_1 + V_2 \text{ and } H = H_0 + H'. \tag{5.16}$$

The interaction energy E' can be calculated from

$$E' = \frac{\int \psi^* H' \psi dV}{\int \psi^* \psi dV}, \tag{5.17}$$

where $\psi = \psi_{1A}\, \psi_{2B} \pm \psi_{2A}\, \psi_{1B}$ and considering perturbation effect $E = 2E_0 + B$

$$\text{or, } E' = \iint \psi_{1A}\, \psi_{2B} \pm \psi_{2A}\, \psi_{1B} \times \left[\frac{e^2}{r_{AB}} - \frac{e^2}{r_{12}} - \frac{e^2}{r_{1B}} - \frac{e^2}{r_{2A}}\right]$$

$$= \frac{\iint (\psi_{1A}\, \psi_{2B} \pm \psi_{2A}\, \psi_{1B})\, dr_1\, dr_2}{\iint (\psi_{1A}\, \psi_{2B} \pm \psi_{2A}\, \psi_{1B})^2\, dr_1\, dr_2}. \tag{5.18}$$

Now

$$\iint (\psi_{1A}\, \psi_{2B} \pm \psi_{2A}\, \psi_{1B})^2\, dr_1\, dr_2$$

$$= \iint (\psi_{1A}\, \psi_{2B})^* (\psi_{1A}\, \psi_{2B})\, dr_1\, dr_2 + \iint (\psi_{2A}\, \psi_{1B})^* (\psi_{2A}\, \psi_{1B})\, dr_1\, dr_2$$

$$+ 2 \iint \psi_{1A}\, \psi_{1B}\, \psi_{2A}\, \psi_{2B}\, dr_1\, dr_2$$

$$= 2 \pm 2S, \tag{5.19}$$

where

$$S = \iint \psi_{1A}\, \psi_{1B}\, \psi_{2A}\, \psi_{2B}\, dr_1\, dr_2$$

represents the overlap integral.

The numerator of eqn (5.18) is equal to

$$\iint \psi_{1A}^* \psi_{2B}^* H' \psi_{1A} \psi_{2B}\, dr_1\, dr_2 \pm \iint \psi_{1A}^* \psi_{2B}^* H' \psi_{2A} \psi_{1B}\, dr_1\, dr_2$$
$$\pm \iint (\psi_{2A}^* \psi_{1B}^* H' \psi_{1A} \psi_{2B})\, dr_1\, dr_2$$
$$+ \iint (\psi_{2A}^* \psi_{1B}^* H' \psi_{2A} \psi_{1B})\, dr_1\, dr_2$$
$$= 2\xi + 2\eta, \tag{5.20}$$

where

$$\xi = \iint \psi_{1A}^* \psi_{2B}^* H' \psi_{1A} \psi_{2B}\, dr_1\, dr_2. \tag{5.21}$$

Eqn (5.21) is the *Coulomb integral* representing the energy of electrostatic interaction of electrons with the nuclei, of two electrons with each other and of the two nuclei over all the possible states of the electrons.

$$\eta = \iint \psi_{1A}^* \psi_{2B}^* H' \psi_{2A} \psi_{1B}\, dr_1\, dr_2 \tag{5.22}$$

represents the *exchange interaction* due to the atoms exchanging electrons.

The energy eigenvalues are

$$E_S = 2E_0 + \frac{\xi + n}{1 + S} \tag{5.23}$$

$$\text{and } E_A = 2E_0 + \frac{\xi - n}{1 - S}, \tag{5.24}$$

where $\xi < 0, \eta < 0$ and $S \leq 1$.

E_S represents the *symmetric energy state with even parity* (i.e., spins anti-parallel) and E_A, the antisymmetric energy state with odd parity (i.e., spin parallel).

Fig. 5.6 shows the variation of PE of the system with r_{AB} both symmetric and antisymmetric modes. The figure shows that a hydrogen molecule at equilibrium distance r_0 is possible to be formed by attractive interaction if the eigenfunction is symmetric. An antisymmetric eigenfunction, however, cannot form a hydrogen molecule as the two atoms repel each other in ψ_A state.

Hence, *the two participating electrons in a covalent bond pair up with parallel spins and the bonding arises through an exchange interaction.*

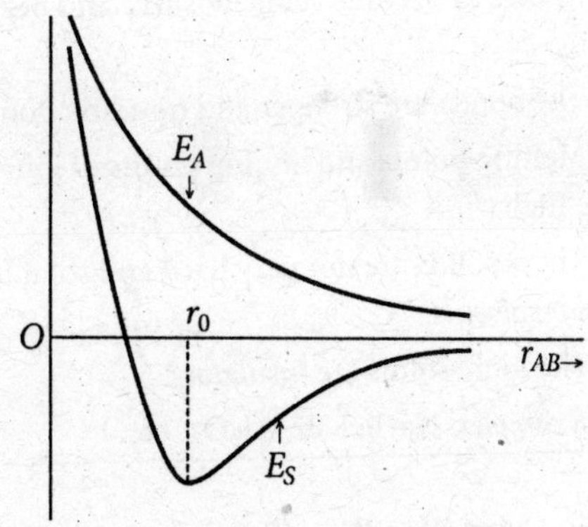

Fig. 5.6: Interaction energy of two H-atoms in symmetric and antisymmetric states

Table 5.1 shows the interatomic separation (r_0) and cohesive energy of some of the common covalent solids.

Table 5.1

Crystal	*Interatomic separation* (r_0) *in* $(10^{-10}$ m)	*Cohesive energy in* eV/atom
C	1.54	3.68
Si	2.35	2.32
Ge	2.44	1.94
GaAs	2.45	1.63
InP	2.54	1.74
ZnS	2.34	1.59
CdTe	2.81	1.03

Table 5.2: Covalent bond length and the binding energies

Bond	*Bond length* (nm)	*Binding energy in* 10^3 kJ/mole	*Bond*	*Bond length* (nm)	*Binding energy in* 10^3 kJ/mole
H – H	0.074	436	O = O	0.121	498
C – H	0.108	414	Cl – Cl	0.199	242
C – C	0.154	348	N – N	0.100	390
C = C	0.134	615	H – Cl	0.128	431
C ≡ C	0.120	840	C – F	0.136	486
C – O	0.143	358	C – Si	0.193	301
C = O	0.122	745	Si – H	0.145	293
C – Cl	0.177	327	Si – Si	0.234	176
N ≡ N	0.109	945	Si – Cl	0.216	360
O – H	0.097	463	Si – O	0.183	368

5.4.3 Comparison between Ionic and Covalent Solids

Table 5.3

Ionic Solids	*Covalent Solids*
1. Ionic solids are formed by the *actual transfer of electrons* from one atom to other in different elements to become stable positive and negative ions.	1. Covalent solids are formed by *equal sharing of electrons* between two neighbouring atoms of the same element each having incomplete outermost shell.
2. The bonds are stronger and nondirectional.	2. The bonds are weaker but directional.
3. Melting points and boiling points of ionic solids are high.	3. Melting points and boiling points of covalent solids are low.
4. These solids are not very hard and soluble in *polar solvents.*	4. These are very hard and soluble in *non-polar solvents.*
5. The ionic solids are *insulators.*	5. Covalent solids are *semiconductors.*
6. *Examples*: NaCl, KBr, Al_2O_3, etc.	6. *Examples*: Diamond, Ge, Si, etc.

5.5 Metallic Bonding

Partial sharing of valence electrons by the neighbouring atoms is the basis of metallic bonding. As the sharing of electrons in this bonding is not localized, it is considered as *delocalized* or *unsaturated covalent bond.* The metallic bond is the result of simultaneous attraction of an electron by two or more positive ions.

According to *Drude*, a metal contains a large number of *free electrons* at ordinary temperatures. These free electrons are actually the *valence electrons* of the atom and move about at random chaotic motion in the framework of positive ion cores. The free electrons form an *electron cloud or gas* and behave like a *mobile glue* which binds the ion cores together through *Coulombic attractive force.* The atoms in this bonding can share the required number of electrons with their neighbours through the common *sea* of electrons to complete their octet. A metal may be pictured as in Fig. 5.7.

As the free electrons in metal can wander from atom to atom due to so many unoccupied states in each atom, *the thermal and electrical conductivities of metals are high.*

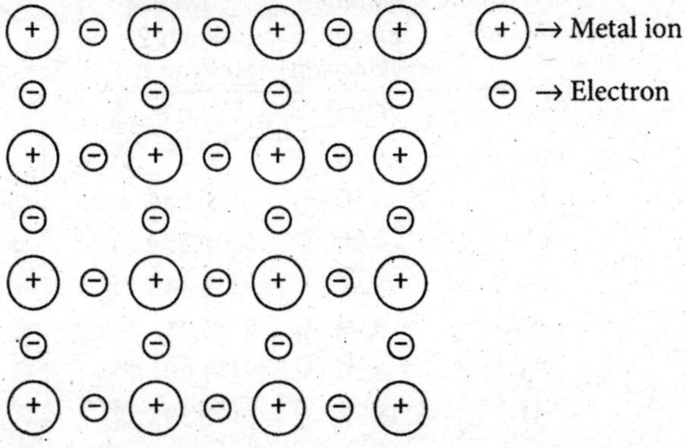

Fig. 5.7: Distribution of free electrons and ions in a metal

The atoms in a metal are bound together by a cloud of free electrons and not by covalent bonds, the metal can easily undergo *shear*. The energy of the electrons changes considerably by a change in volume of the metal so that there is a good *cleavage strength.* As the energy of the electrons is not affected by mere movement of the atoms, the metals are *malleable and ductile.*

The *opacity* and *lustre* of metals are due to the presence of free electrons in metallic bonding. Light and other electromagnetic radiation is first absorbed by the free electrons and then radiate it producing *lustre.* Again, metals are *opaque* to all wavelength because of their very high *reflectivity.*

Properties of Metallic Bond

Following are the important properties of metallic solids:

1. Metallic bond is *weaker* than ionic or covalent bond. The binding energies range from 0.7 eV/atom for *mercury* to 8.8 eV/atom for *tungsten.*
2. Due to symmetrical arrangements of the positive ions in a space lattice, metals are highly *crystalline.*
3. Metallic bond is *weaker* than ionic or covalent bond. The melting point of metallic solids are *lower* than ionic or covalent solids but *stronger* than *van der Waals' bond.*
4. Due to the presence of a large number of free electrons, metallic solids have *high electrical* and *thermal conductivities.*
5. Metallic solids are *ductile* and *malleable.*
6. Metallic solids have *high optical reflectivity* and *lustre.*
7. Metallic solids are *not soluble* in *polar* and *non-polar* solvents.
8. Metallic bond is *nondirectional.*
9. Due to *unsaturated nature* of the bonds metal crystals are of *unlimited size.*
10. Metals dissolve other elements and form alloys (e.g., C in α–Fe to form steel, Zn in Cu to form brass. Here the solute atoms either occupy vacant lattice sites in the solvent crystal (e.g., Zn in Cu) or go into the interstices between the solvent metal atoms. Such alloys are known respectively as *substitutional* or *interstitial alloys* or *solid solutions.*

Example. *Sodium, Copper, Gold, Silver, Aluminium, Iron, Tin,* etc.

Table 5.4: Binding energy and cohesive energy of some metals

Metal	*Binding energy* (eV/electron)	*Cohesive energy* (eV/atom)
Li	6.94	1.63
Na	6.26	1.113
K	5.30	0.934
Cs	4.90	0.804
Ca	9.91	1.84
Sr	9.92	1.73
Ba	8.56	1.90
Al	18.77	3.39
Zn	14.28	1.35
Mg	12.10	1.51

5.6 van der Waals' Bonding

In inert gas atoms (like He and Ar) the outermost orbits are completely filled and there is no valence electron. The electron distribution around the nucleus is spherically symmetric [Fig. 5.8]. If the spherical distribution of electrons be disturbed, the centres of positive and negative charges no longer coincide and a weak fluctuating dipole results [Fig. 5.9]. The electric field originating from this imbalance charge distribution induces a dipole moment in the neighbouring atom so that the atom gets attracted to it. The electric field of the second dipole produces another dipole by inducing dipole moment in the neighbouring atom and so on. This dipole induced dipole interaction is termed as *van der Waals' bond* or *dispersion bond* which is *nondirectional.* van der Waals' force is weak (energy 0.1 eV) which allows inert gas atoms to condense to liquid or crystal at low temperature. The crystals formed due to van der Waals' bond are called *molecular crystals.* In He, Ne, Ar, etc. and other gases like O_2, Cl_2, CO_2, CH_4, etc. are some of molecular crystals.

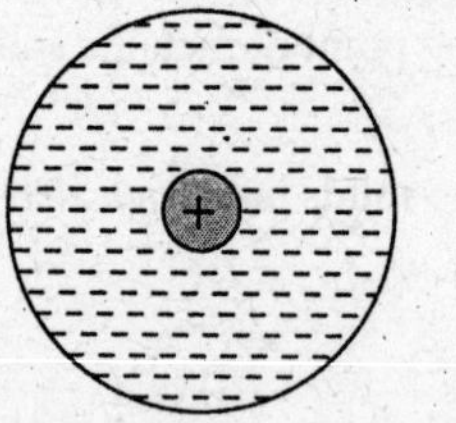

Fig. 5.8: Symmetrical electron cloud of an isolated inert gas atom

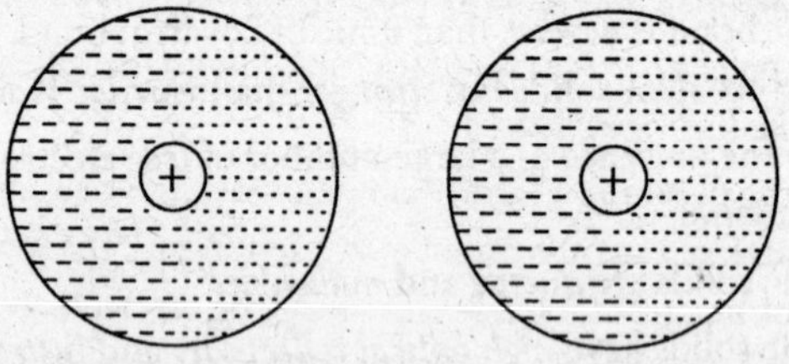

Fig. 5.9: Distorted electronic distribution around two neighbouring inert gas atoms generated fluctuating dipoles

As the electrons are localized the solids formed due to *van der Waals' bonding* are electrical insulators. Such solids are *poor conductors of electricity* and *heat.* They have low melting and boiling points and also low mechanical strength and easy deformability.

van der Waals' forces arise due to *dipole-dipole* and *dipole-induced dipole interaction* and depend on the following factors:

(i) *Number of electrons in the molecule*: Greater is the force, greater the number of of electrons in a molecule.

(ii) *Size of molecules*: The surface of larger size molecule is greater and the van der Waals' force is higher.

(iii) *Effect of pressure*: With the increase of pressure, the molecules are brought closer resulting a stronger attraction.

(iv) *Effect of temperature*: At low temperature the molecular motion is slow and the van der Waals' interaction is strong.

Properties of Inert Gas Crystals

The characteristic properties of inert gas crystals are as below:

1. The inert gas solids exist both in crystalline as well as non-crystalline states.
2. The crystals have very low melting and boiling points due to weak van der Waals' force.
3. These crystals show properties like *friction viscosity, surface tension, cohesion* and *adhesion*, etc.
4. As no valence electron exists, the inert gas crystals are good insulators.
5. The crystals are transparent to light and soluble both in *polar* and *non-polar liquids.*
6. The inert gas crystals are brittle, lack strength and have high ionization energy.

Example: *Solid argon, solid neon.*

5.7 Hydrogen Bonding

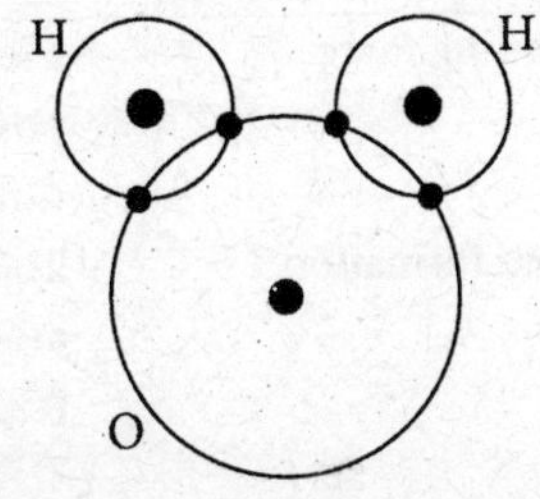

Fig. 5.10: Sharing of electrons between H and O atoms in H_2O

In a *covalent bond* formed between a hydrogen atom and a highly electronegative atom (e.g., O, F, Cl, etc.), the shared electron-pair is more attracted towards the electronegative atom than the hydrogen atom, making the electronegative atom more negative and the hydrogen atom equally positive. The molecule formed is polarized and acts like a permanent dipole. Then the Coulombic attraction binds many more dipoles together.

The interaction between the oppositely charged ends of permanently polarized molecules each containing a hydrogen atom is called the hydrogen bond.

In hydrogen bonding, the hydrogen atom can be regarded as a proton fixed to one end of a covalent bond, the positive charge not being shielded by surrounding electrons like covalent bond. For this reason the positively charged hydrogen atom can attract more strongly the negative ends of other molecules compared to other positively charged atom. *That is why the hydrogen bonds are stronger than dipole bonds.* Hydrogen bonded materials have higher melting and boiling points than molecular solids having van der Waals' bonding.

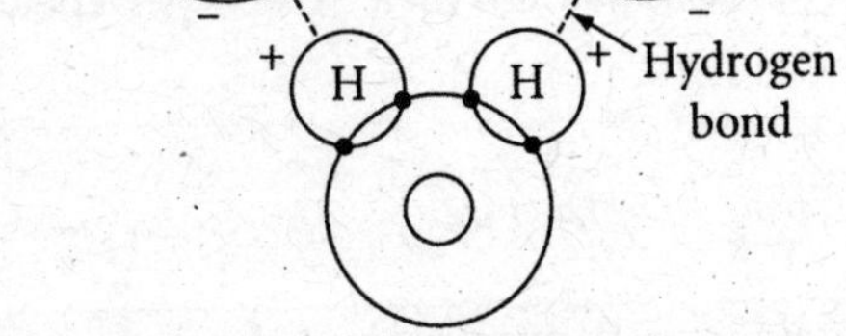

Fig. 5.11: Hydrogen bonding among water molecules

Fig. 5.11 illustrates the existence of hydrogen bond among water molecules. The binding energy of water or ice is ~ 0.5 eV. If hydrogen bond were absent, then the boiling point of water would be −80 °C instead of 100 °C at normal atmospheric pressure. The hydrogen bond is directional as it is formed between two oppositely charged ends of two permanent dipoles. Hydrogen bonding is very important in many biological molecules like DNA, etc.

Properties of Hydrogen Bonded Solids

Following are some important properties of hydrogen bonded solids:

1. Hydrogen bonded solids have *low but higher melting points* than molecular solids.
2. Hydrogen bonds are *directional.*
3. This bond is *stronger* than other dipole-dipole interaction.
4. Hydrogen bonded solids are *soluble* in both *polar* and *non-polar solvents.*
5. These solids are *transparent* to light and have *low density.*
6. The density of ice is less than that of water because in water, the molecules come closer and form a closely packed liquid.

Example. *Water molecules in ice, Ammonia molecules,* etc.

5.8 Comparison between Ionic Bond and Metallic Bond

Table 5.5

Properties	*Ionic Bond*	*Metallic Bond*
Bonding force	Electrostatic attraction between positive and negative ions of different elements.	Electrostatic attraction between the electron cloud of valence electrons and positive ions form this bond.
Bond formation	This is easily formed when one of the atoms has less number of electrons (e.g., alkali metals).	This type of bond is formed with elements having less number of loosely bound electrons (i.e., valence electrons).
Conductivity	Ionic solids are poor conductors of electricity.	Metallic crystals are good conductor of electricity and heat.
Mechanical properties	Ionic solids are very hard in nature.	These solids are good ductile and malleable.
Bond strength	Bond strength is more than metallic bond.	These bonds are generally less stronger than ionic bonds.

Types of Solids and their Structure Dependent Properties

Table 5.6

Types	*Structure*	*Property*	*Example*	*Cohesive energy* kJ/mole
Ionic	Positive and negative ions	Brittle, high melting point and non-conducting	NaCl LiF	184
Covalent	Bonded to one another by mutual sharing of electrons	Hard, non-conducting high melting point	Diamond SiC	170 244
Metallic	Positive ions surround by electron cloud	High electrical and thermal conductivities	Na Fe	26 94

Types	Structure	Property	Example	Cohesive energy kJ/mole
Hydrogen	Hydrogen bonding holds the molecules together	Insulators, low melting point	H_2O HF	12 7
van der Waals'	Atoms and molecules	Soft, low melting point, volatile, insulator	Argon	2

5.9 Variation of Interatomic Force with Interatomic Spacing

The interatomic force existing among the atoms of a crystal holds the atoms together to form solid. The existence of solid leads us to conclude the following:

(a) *There must be an attractive force between the atoms or molecules forcing them to form a solid.*

(b) *There must also be a repulsive force which comes into play when a solid is compressed.*

When two atoms approach each other the negatively charged electron shells come closer than their positively charged nuclei. The two atoms at a certain separation r_0 become stable as the attractive and repulsive forces become equal at r_0. r_0 is called the *equilibrium separation* of the atoms. At r_0, since the two atoms come to stable condition, the PE is the minimum.

If the PE due to attraction U_{att} varies as $\frac{1}{r^m}$ and that due to repulsion U_{rep} as $\frac{1}{r^n}$, then the resultant PE is

$$U = U_{att} + U_{rep} = -\frac{A}{r^m} + \frac{B}{r^n}, \tag{5.25}$$

where A, B, m and n are constants depending on the nature of the atoms. A and B are known as *attraction* and *repulsion constants* respectively.

Cohesive Energy

The force between the two atoms is

$$F = -\frac{dU}{dr} = \frac{mA}{r^{m+1}} - \frac{nB}{r^{n+1}}. \tag{5.26}$$

At $r = r_0$, the attractive and repulsive forces become equal.

So $$F(r_0) = 0 \text{ or, } \frac{mA}{r_0^{m+1}} - \frac{nB}{r_0^{n+1}} = 0 \text{ or, } \frac{m}{n} = \frac{B}{A} r_0^{m-n} \text{ or, } r_0^{n-m} = \left(\frac{B}{A}\right)\left(\frac{n}{m}\right)$$

$$\text{or, } r_0 = \left(\frac{Bn}{Am}\right)^{1/(n-m)}. \tag{5.27}$$

Eqn (5.27) gives the *equilibrium separation* of the atoms.

Hence, the PE at equilibrium position is

$$U(r_0) = -\frac{A}{r_0^m} + \frac{B}{r_0^n} = -\left(\frac{A}{r_0^m}\right)\left(1 - \frac{m}{n}\right). \tag{5.28}$$

Since $m \neq n$, $U(r_0) \neq 0$, i.e., at equilibrium position though attractive force is equal to the repulsive force, the attractive energy is not equal to the repulsive energy.

The energy $U(r_0)$ corresponding to the equilibrium position ($r = r_0$) is called the *binding energy* or *the energy of cohesion*. This much energy is required to *dissociate* the atoms of the molecule and hence called the *energy of dissociation*.

For $n \gg m$,

$$U(r_0) = -\frac{A}{r_0^m}. \tag{5.29}$$

This means, the total energy is essentially the energy of attraction.

Again, for $U(r)$ to be minimum at r_0,

$$\left.\frac{d^2U}{dr^2}\right|_{r=r_0} > 0 \text{ or, } -\frac{m(m+1)A}{r_0^{m+2}} + \frac{n(n+1)B}{r_0^{n+2}} > 0,$$

$$\text{i.e., } -\frac{m(m+1)A}{r_0^{m+2}} + \frac{n(n+1)B}{r_0^{n+2}} > 0 \text{ or, } -m-1+n+1>0 \text{ or, } n-m>0 \text{ or, } n>m.$$

Hence, from eqn (5.26), we find that the force acting between the atoms is *mostly electrostatic.*

The variation of PE (U) and interatomic force (F) with interatomic distance (r) are shown in Figs 5.12 (a) and (b) respectively.

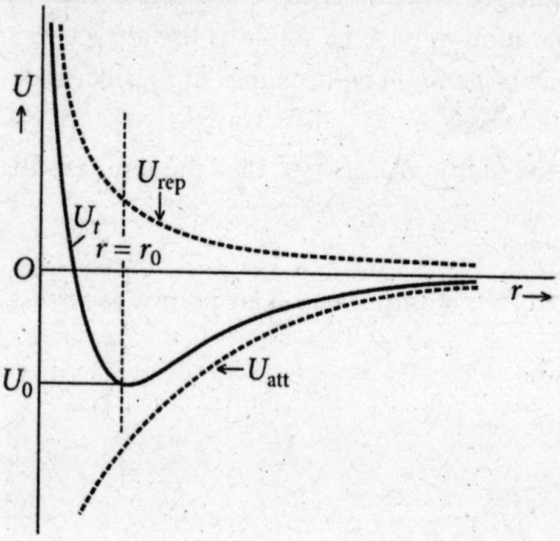

Fig. 5.12: (a) Variation of potential energy with interatomic distance

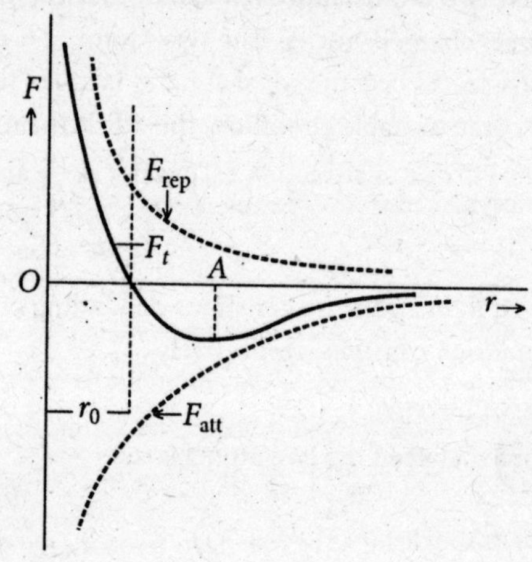

Fig. 5.12: (b) Variation of interatomic force with distance

5.10 Calculation of Cohesive Energy of Ionic Crystals

The cohesive energy of a solid is defined as the energy which will be given out in the process of the formation of the crystal by bringing the neutral atoms from infinity to the position of equilibrium separation. The cohesive energy (or binding energy) of ionic crystals was first calculated in 1910 by *Born and Madelung* which was subsequently modified by *Born and Mayer.*

Born and *Madelung's theory* was based on the following assumptions:

1. Ionic crystals are formed by positive and negative ions each having spherical charge distribution very much like inert gas atoms.
2. The attracting force between the ions is independent of the direction of approach but depends only on the distance between the ions.

3. The contribution to the binding energy arises mainly from electrostatic interaction and is called the *Madelung energy.*
4. Two types of interaction exist in ionic crystals:
   (a) The long range *Coulombic interaction* may be either *attractive* or *repulsive* in nature. The Coulomb's electrostatic *repulsive energy* between the charges $+q$ and $-q$ is $+\frac{q^2}{4\pi\varepsilon_0 r}$ and *attractive energy* is $-\frac{q^2}{4\pi\varepsilon_0 r}$.
   (b) A short range repulsive energy, β/r^n, where β and n are constants, also exists between the positive and negative ions, which was subsequently replaced by $\lambda\, e^{-r/\rho}$, where λ and ρ represent respectively the *strength* and *range* of the interaction. $\lambda\, e^{-r/\rho}$ is called the *central field repulsive potential.*

According to *Born and Madelung's theory* the interaction energy (U_i) on the ith ion due to all other ions is

$$U_i = \sum_j U_{ij}, \tag{5.30}$$

where U_{ij} is the interaction energy between the ions i and j, the summation extends over all values of j except $j = i$, U_i is also called the *cohesive energy* of the ith ion.

The interaction energy between the ith and jth ions can be expressed as

$$U_{ij} = \lambda\, e^{-r_{ij}/\rho} \pm \frac{q^2}{r_{ij}}. \tag{5.31}$$

If the crystal contains N molecules, i.e., $2N$ ions, then the total energy of the crystal is

$$U_{\text{tot}} = NU_i. \tag{5.32}$$

In the above equation we have taken N instead of $2N$ because each pair of ij interaction must be considered only once.

If R be the nearest neighbour separation in the crystal, then let

$$r_{ij} = p_{ij}R, \tag{5.33}$$

where p_{ij} is the distance between any two ions in terms of nearest neighbour distance R and is a dimensionless quantity. If the ions i and j be the nearest neighbours, then

$$r_{ij} = R \text{ or, } p_{ij} = 1.$$

From eqns (5.31) and (5.33), we write,

$$U_{ij} = \lambda\, e^{-(p_{ij}R/\rho)} \pm \frac{q^2}{(p_{ij}R)}.$$

Now eqn (5.30) becomes

$$U_i = \sum_j \left[\lambda\, e^{-(p_{ij}R/\rho)} \pm \frac{q^2}{(p_{ij}R)}\right]. \tag{5.34}$$

If the repulsive interaction $\lambda\, e^{-(p_{ij}R/\rho)}$ is effective for the nearest neighbours only, then

$$U_i = Z\lambda\, e^{-(p_{ij}R/\rho)} \pm \sum_j \frac{q^2}{(p_{ij}R)} = Z\lambda\, e^{-(p_{ij}R/\rho)} - \alpha\frac{q^2}{R}, \tag{5.35}$$

where Z = the number of nearest neighbours of the ith ion. Then the lattice energy is

$$U_{tot} = NU_i = N\left[Z\lambda e^{-R/\rho} - \alpha q^2/R\right]. \tag{5.36}$$

Here

$$\alpha = \sum \pm \frac{1}{p_{ij}} \tag{5.37}$$

and is called the *Madelung constant* or *Coulomb energy constant.*

If the reference ion be taken as the negative charge, then positive sign for positive ions and negative sign for negative ions are to be used.

At the nearest equilibrium distance R_0,

$$\frac{dU_{tot}}{dR} = 0.$$

$\therefore$ from eqn (5.36)

$$N\left.\frac{dU_i}{dR}\right|_{R=R_0} = -\frac{NZ\lambda}{\rho}e^{-R_0/\rho} + \frac{N\alpha q^2}{R_0^2} = 0 \text{ or, } e^{-R_0/\rho} = \frac{\rho\alpha q^2}{Z\lambda R_0^2}. \tag{5.38}$$

So, from eqn (5.36),

$$U_{tot} = N\left[Z\lambda\frac{\rho\alpha q^2}{Z\lambda R_0^2} - \frac{\alpha q^2}{R_0}\right] = -\frac{N\alpha q^2}{R_0}\left[1 - \frac{\rho}{R_0}\right]. \tag{5.39}$$

The term $-\frac{N\alpha q^2}{R_0}$ is the *Madelung energy* and the term $\left[1 - \frac{\rho}{R_0}\right]$ represents the contribution of the *short range repulsive interaction.*

Now the cohesive energy of the ith ion is

$$U_i = \frac{U_{tot}}{N} = -\frac{\alpha q^2}{R_0}\left[1 - \frac{\rho}{R_0}\right]. \tag{5.40}$$

In SI units,

$$U_i = -\frac{\alpha q^2}{4\pi\varepsilon_0 R_0}\left[1 - \frac{\rho}{R_0}\right]. \tag{5.41}$$

When $\rho \sim 0.1R_0$, U_i is dominated by Madelung contribution. U_i decreases rapidly as $\rho/R_0 \to 0$, indicating that the *repulsive interaction is very short range.*

For NaCl, $\frac{\rho}{R_0} \sim \frac{1}{9}$, $q = e = 1.6 \times 10^{-19}$ C and $\alpha = 1.748$ and $R_0 = 2.81$ Å. Using these values, the lattice energy (U_i) per ion pair comes out to be $U_i = -7.97$ eV, which agrees closely with the experimental value -7.96 eV per ion pair.

5.10.1 Evaluation of Madelung Constant

We consider a one-dimensional lattice of oppositely charged ions [Fig. 5.13]. Taking a negatively charged ion as the reference and inter-ionic distance R, we have

$$\alpha = \sum_j \pm \frac{1}{p_{ij}} = 2\left[1 - \frac{1}{2} + \frac{1}{3} - \frac{1}{4} + \cdots\right]. \tag{5.42}$$

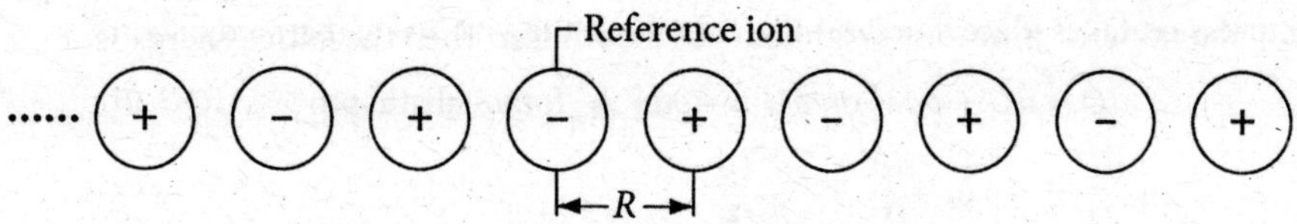

Fig. 5.13: One-dimensional ionic crystal

The factor 2 is due to the presence of similar ions on both sides of the reference ion.

Now the series expansion

$$\ln(1+x) = x - \frac{x^2}{2} + \frac{x^3}{3} - \frac{x^4}{4} + \cdots.$$

Putting $x = 1$,

$$\ln 2 = 1 - \frac{1}{2} + \frac{1}{3} - \frac{1}{4} + \cdots$$

$$\therefore \alpha = 2\left[1 - \frac{1}{2} + \frac{1}{3} - \frac{1}{4} + \cdots\right]$$

$$= 2\ln 2 = 2 \times 0.69 = 1.38. \quad (5.43)$$

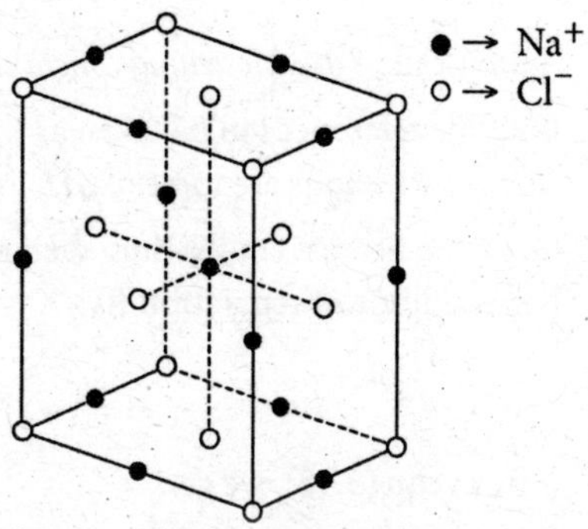

Fig. 5.14: Equilibrium positions of ions in NaCl crystal

5.10.2 Madelung Constant of NaCl Crystal

In three-dimensional NaCl structure [Fig. 5.14] the nearest neighbours to the negative reference ion at the corner of the unit cell are six positive ions at $p_{ij} = p = 1$, giving a positive contribution to α of $\frac{6}{1}$. There are twelve negative ions at $p = \sqrt{2}$, giving a negative contribution of $-\frac{12}{\sqrt{2}}$.

Also there are eight positive ions at $p = \sqrt{3}$, giving positive contribution of $\frac{8}{\sqrt{3}}$; six negative ions at $p = \sqrt{4}$, giving negative contribution of $\frac{6}{\sqrt{4}}$ and so on.

$$\therefore \alpha = \frac{6}{1} - \frac{12}{\sqrt{2}} + \frac{8}{\sqrt{3}} - \frac{6}{\sqrt{4}} + \frac{24}{\sqrt{5}} \cdots$$

$$= 6.000 - 8.485 + 4.619 - 3.000 + 10.733 = 1.748. \quad (5.44)$$

If the value of α is higher, then the *cohesive energy* is also higher and stability of the structure is greater. The *Madelung constant* for CsCl = 1.763 and that for ZnS = 1.638.

5.10.3 Determination of Range: Expression for Bulk Modulus of Elasticity of Ionic Crystals

The *range of repulsive interaction* ρ is given by

$$B = \frac{\alpha q^2}{18R_0^4}\left(\frac{R_0}{\rho} - 2\right),$$

where B is the bulk modulus of elasticity of the ionic solid.

If dp be the change in pressure required to produce a volume change dV, then the volume strain is dV/V and

$$B = -\frac{dp}{dV/V}. \quad (5.45)$$

If the compression takes place *adiabatically*, then from the *first law of thermodynamics.*

$$dQ = dU + pdV \text{ or, } dU = -pdV \; [\because \text{ for adiabatic process } dQ = 0]$$

$$\text{or, } \frac{dU}{dV} = -p \text{ or, } \frac{dp}{dV} = -\frac{d^2U}{dV^2}. \tag{5.46}$$

From eqns (5.45) and (5.46), we get

$$B = V\frac{d^2U}{dV^2}. \tag{5.47}$$

But

$$\frac{dU}{dV} = \frac{dU}{dR}\cdot\frac{dR}{dV} \text{ and } \frac{d^2U}{dV^2} = \frac{d^2U}{dR^2}\left(\frac{dR}{dV}\right)^2 + \frac{dU}{dR}\cdot\frac{d^2R}{dV^2}. \tag{5.48}$$

For equilibrium separation R_0,

$$\left.\frac{dU}{dR}\right|_{R_0} = 0.$$

$\therefore$ from eqn (5.48), we get

$$\frac{d^2U}{dV^2} = \frac{d^2U}{dR^2}\left(\frac{dR}{dV}\right)^2. \tag{5.49}$$

Hence, from eqn (5.47),

$$B = V\left(\frac{d^2U}{dR^2}\right)\left(\frac{dR}{dV}\right)^2, \text{ at } R = R_0. \tag{5.50}$$

For NaCl molecule with lattice constant $a = 2R$, the volume of the unit cell is $8R^3$, as there are 4 molecules. Then the volume occupied by N molecules is

$$V = 2NR^3 \tag{5.51}$$

$$\text{or, } \frac{dV}{dR} = 6NR^2. \tag{5.52}$$

From eqns (5.50) and (5.52),

$$B = 2NR_0^3\left(\frac{1}{6NR^2}\right)^2\left.\frac{d^2U}{dR^2}\right|_{R=R_0} = 2NR_0\left(\frac{1}{36N^2R_0^4}\right)\left.\frac{d^2U}{dR^2}\right|_{R=R_0} \text{ or, } B = \left(\frac{1}{18NR_0}\right)\left.\frac{d^2U}{dR^2}\right|_{R=R_0}. \tag{5.53}$$

Again, total interaction energy of the ionic solid is

$$U_{\text{tot}} = N\left[Z\lambda\, e^{-R/\rho} - \alpha q^2/R\right].$$

$$\therefore \frac{dU_{\text{tot}}}{dR} = -\frac{NZ\lambda}{\rho}e^{-R/\rho} + \frac{N\alpha q^2}{R^2} \text{ and } \frac{d^2U_{\text{tot}}}{dR^2} = \frac{NZ\lambda}{\rho^2}e^{-R/\rho} - \frac{2N\alpha q^2}{R^3}. \tag{5.54}$$

Substituting the value of $\frac{d^2U_{\text{tot}}}{dR^2}$ from eqn (5.54) in eqn (5.53), we get

$$B = \frac{1}{18NR_0}\left[\left(\frac{NZ\lambda}{\rho^2}\right)e^{-R_0/\rho} - \frac{2N\alpha q^2}{R_0^3}\right] = \frac{1}{18NR_0}\left[\frac{NZ\lambda}{\rho^2}\left(\frac{\rho\alpha q^2}{Z\lambda R_0^2}\right) - \frac{2N\alpha q^2}{R_0^3}\right]\left[\because e^{-R_0/\rho} = \frac{\rho\alpha q^2}{Z\lambda R_0^2}\right]$$

$$\text{or, } B = \frac{\alpha q^2}{18R_0^4}\left(R_0/\rho - 2\right). \tag{5.55}$$

The value of ρ can be calculated using the observed values of R_0 and B.

5.11 Binding Energy of Inert Gas Crystals

In inert gas atoms the binding energy originates due to the following *two* types of interactions: (a) *Attractive interaction* known as the *van der Waals-London interaction.* (b) *Repulsive interaction.*

(a) *Attractive interaction.* This type of attraction originates due to dipole-dipole interaction between two neighbouring atoms to produce a weak force.

Let some fluctuation of charge distribution on jth atom induces an instantaneous dipole moment p_i on the ith atom which produces an electrostatic field E at the centre of the jth atom of magnitude

$$E = \frac{2p}{r_{ij}^3} \text{ (in CGS units).} \tag{5.56}$$

The instantaneous dipole moment p_j at the jth atom produced by E is

$$p_j = \alpha E = 2\alpha_e p_i / r_{ij}^3, \tag{5.57}$$

where α_e is the *electronic polarizability.* These two dipoles p_i and p_j interact with each other to produce a net attractive force [Fig. 5.15]. The PE due to the interaction is

$$U_{\text{att}} = -2p_i p_j / r_{ij}^3 = -4\alpha_e p_i^2 / r_{ij}. \tag{5.58}$$

Eqn (5.57) shows that α_e is a *dimensional quantity* of dimension $[\text{length}]^3$. So, the dimensions of $\alpha_e p_i^2$ is $[\text{length}]^5 \times [\text{charge}]^2$. Hence, $\alpha_e p_i^2$ may be replaced by $e^2 r_0^5$, where r_0 is the atomic radius.

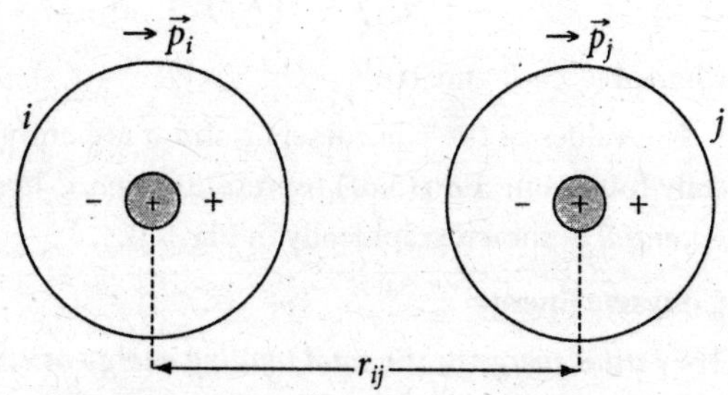

Fig. 5.15: Instantaneous $\vec{p}_i$ induces $\vec{p}_j$

From eqn (5.58),

$$U_{\text{att}} \simeq -4e^2 r_0^5 / r_{ij}^6 \simeq -10^{-58} / r_{ij}^6 \text{ ergs} = -B / r_{ij}^6, \tag{5.59}$$

where $r_0 = 10^{-8}$ cm and $B = 10^{-58}$ erg-cm^6.

Since $U_{\text{att}} \propto \frac{1}{r_{ij}^6}$, the *van der Waals' interaction is of very short range.*

(b) *Repulsive Interaction.* When two atoms of inert gas approach each other their charge distributions gradually overlap [Fig. 5.16 (b)] and the electrostatic energy of the system is changed. When the atoms are sufficiently closer the energy due to overlap becomes repulsive because of *Pauli's exclusion principle.*

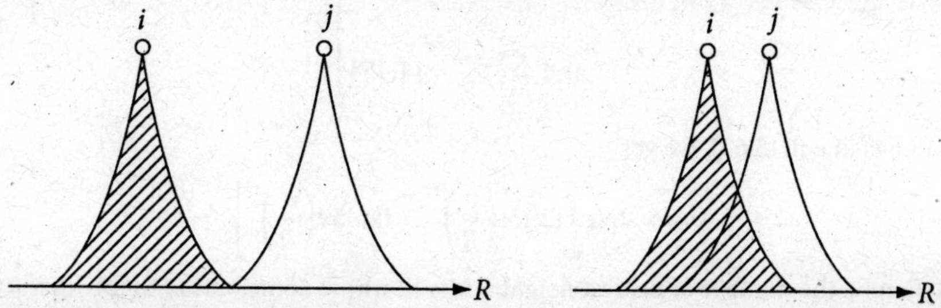

Fig. 5.16: (a) i and j atoms without electron cloud overlap

Fig. 5.16: (b) Overlap of electron clouds of atoms i and j

The repulsive interaction energy due to the overlap of electron clouds of ith and jth atoms may be expressed by the empirical relation

$$U_{rep} = \frac{C}{r_{ij}^{12}}, \tag{5.60}$$

where C is a constant.

Then the total interaction energy between these two atoms is

$$U_{ij} = -\frac{B}{r_{ij}^6} + \frac{C}{r_{ij}^{12}}$$

$$\text{or, } U_{if} = 4\varepsilon\left[\left(\frac{\sigma}{r_{ij}}\right)^{12} - \left(\frac{\sigma}{r_{ij}}\right)^{6}\right], \tag{5.61}$$

where $4\varepsilon\sigma^6 = B$ and $4\varepsilon\sigma^{12} = C$.

The values of the parameters ε and σ are empirically found out. Eqn (5.61) represents *Lenard-Jones Potential* as shown graphically in Fig. 5.17.

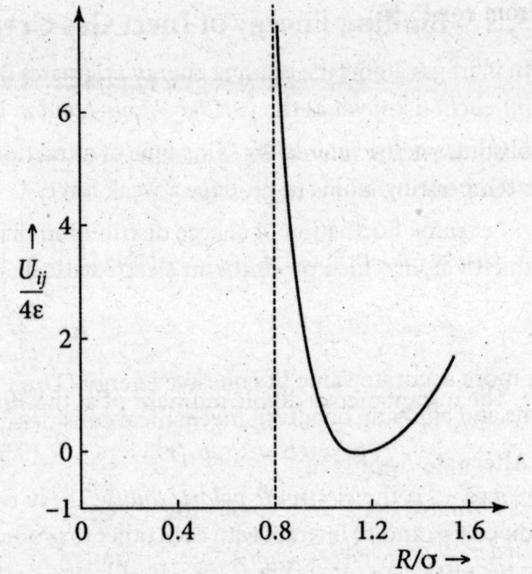

Fig. 5.17: Plot of Lenard-Jones potential versus R/σ. The minimum is at $R/\sigma = 1.122$

Cohesive Energy

The *lattice energy* or the *total binding energy* or *cohesive energy* of the crystal is given by

$$U_{tot} = \frac{1}{2}N(4\varepsilon)\sum_j\left[\left(\frac{\sigma}{r_{ij}}\right)^{12} - \left(\frac{\sigma}{r_{ij}}\right)^{6}\right]. \tag{5.62}$$

Here the KE of the atoms has been neglected.

Putting $r_{ij} = p_{ij}R$ in eqn (5.62), we get

$$U_{tot} = \frac{1}{2}N(4\varepsilon)\sum_j\left[\left(\frac{\sigma}{p_{ij}R}\right)^{12} - \left(\frac{\sigma}{p_{ij}R}\right)^{6}\right]. \tag{5.63}$$

For FCC structure,

$$\left.\begin{aligned}\sum_j p_{ij}^{-12} &= 12.131\\ \text{and } \sum_j p_{ij}^{-6} &= 14.454\end{aligned}\right\} \tag{5.64}$$

Using eqn (5.64) in eqn (5.63), we get

$$U_{tot} = 2N\varepsilon\left[12.131\left(\frac{\sigma}{R}\right)^{12} - 14.454\left(\frac{\sigma}{R}\right)^{6}\right]. \tag{5.65}$$

In FCC structure, the number of nearest neighbour is 12 whose contribution to the interaction energy is the major part in inert gas crystal.

The equilibrium distance R_0 is obtained from the condition,

$$\left.\frac{dU_{tot}}{dR}\right|_{R=R_0} = 0.$$

From eqn (5.65),

$$\frac{dU_{tot}}{dR} = -2N\varepsilon\left[12 \times 12.131\left(\frac{\sigma^{12}}{R_0^{13}}\right) - 6 \times 14.454\left(\frac{\sigma^6}{R_0^7}\right)\right] = 0 \text{ or, } \frac{R_0}{\sigma} = 1.09. \tag{5.66}$$

Substituting this value of R_0/σ in eqn (5.65), we get the *cohesive energy* of the inert gas crystal at absolute zero temperature and zero pressure, i.e.,

$$U_{tot} = \frac{1}{2}N(4\varepsilon)\left[12.13\left(\frac{\sigma}{R_0}\right)^{12} - 14.454\left(\frac{\sigma}{R_0}\right)^6\right] = \frac{1}{2}N(4\varepsilon)\left[12.13 \times \left(\frac{1}{1.09}\right)^{12} - 14.454\left(\frac{1}{1.09}\right)^6\right]$$
$$= -2.15(4N\varepsilon). \tag{5.67}$$

A more accurate value of cohesive energy (U_{tot}) can be obtained *by taking into account the* KE of the atoms and applying quantum mechanical consideration.

An Alternate Approach

van der Waals-London Interaction and Cohesive Energy of Inert gas crystals:

van der Waals-London Interaction: To find out an expression of van der Waals' interaction (or London interaction or the induced dipole-dipole interaction) we consider two linear harmonic oscillators (1) and (2) representing the two identical atoms of *inert gas* as shown in Fig. 5.18. The oscillators are placed along the X-axis separated by a distance R.

Let the charge on each oscillator be $\pm e$ with separations x_1 and x_2 and the atoms oscillate along the X-axis.

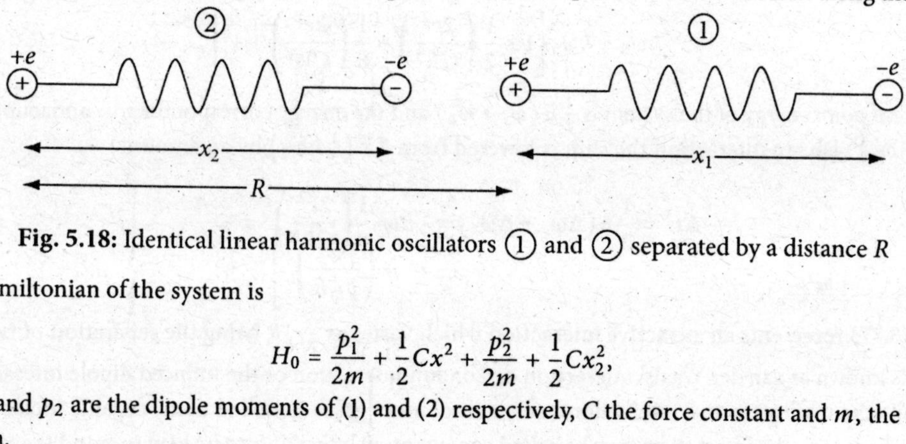

Fig. 5.18: Identical linear harmonic oscillators ① and ② separated by a distance R

The Hamiltonian of the system is

$$H_0 = \frac{p_1^2}{2m} + \frac{1}{2}Cx^2 + \frac{p_2^2}{2m} + \frac{1}{2}Cx_2^2, \tag{5.68}$$

where p_1 and p_2 are the dipole moments of (1) and (2) respectively, C the force constant and m, the mass of each atom.

Each uncoupled oscillator is assumed to have the frequency ω_0, the resonate frequency of the simple harmonic oscillator. Then $C = m\omega_0$.

If H_1 be the Coulomb interaction energy of the two oscillators, then

$$H_1 = \frac{e^2}{R} + \frac{e^2}{R + x_1 - x_2} - \frac{e^2}{R + x_1} - \frac{e^2}{R - x_1} \tag{5.69}$$
$$= \frac{1}{R}\left[e^2 + e^2\left(1 + \frac{x_1 - x_2}{R}\right) - e^2\left(1 + \frac{x_1}{R}\right)^{-1} - e^2\left(1 - \frac{x_1}{R}\right)^{-1}\right]$$
$$\simeq -\frac{2e^2 x_1 x_2}{R} \text{ [where } |x_1|, |x_2| \ll R]. \tag{5.70}$$

Making use of normal mode transformation, we write,

$$x_s = \frac{1}{\sqrt{2}}(x_1 + x_2),\ x_a = \frac{1}{\sqrt{2}}(x_1 - x_2) \tag{5.71}$$

$$\text{and } x_1 = \frac{1}{\sqrt{2}}(x_s + x_a),\ x_2 = \frac{1}{\sqrt{2}}(x_s - x_a), \tag{5.72}$$

where subscripts *s* and *a* represent respectively *symmetric and antisymmetric modes of motion.*

The dipole-moments p_1 and p_2 relating to the two modes are

$$p_1 = \frac{1}{\sqrt{2}}(p_s + p_a),\ p_2 = \frac{1}{\sqrt{2}}(p_s - p_a). \tag{5.73}$$

Now the the total Hamiltonian $H_0 + H_1$ from eqns (5.68) and (5.69) is

$$H = \left[-\frac{1}{2m}p_s^2 + \frac{1}{2}\left(C - \frac{2e^2}{R^3}\right)x_s^2\right] + \left[+\frac{1}{2m}p_s^2 + \frac{1}{2}\left(C + \frac{2e^2}{R^3}\right)x_a^2\right]. \tag{5.74}$$

The two frequencies of the coupled oscillators are found from eqn (5.74) as

$$\omega = \left[\left(C \pm \frac{2e^2}{R^2}\right)\Big/ m\right]^{1/2} \simeq \left[1 \pm \frac{1}{2}\left(\frac{2e^2}{CR^3}\right) - \frac{1}{8}\left(\frac{2e^2}{CR^3}\right)^2 + \cdots\right] \tag{5.75}$$

with $\omega_0 = \sqrt{C/m}$.

$$\therefore \hbar\omega_0 \simeq \hbar\omega_0\left[1 \pm \frac{1}{2}\left(\frac{2e^2}{CR^3}\right) - \frac{1}{8}\left(\frac{2e^2}{CR^3}\right) + \cdots\right]. \tag{5.76}$$

The *zero point energy* of the system is $\frac{1}{2}\hbar(\omega_s + \omega_a)$ and the energy corresponding to unoccupied level is $2 \times \left(\frac{1}{2}\hbar\omega_0\right)$. Due to interaction the sum is lowered from $2 \times \left(\frac{1}{2}\hbar\omega_0\right)$ by an amount

$$\Delta U = \frac{1}{2}\hbar(\Delta\omega_s + \Delta\omega_a) = -\hbar\omega_0\frac{1}{8}\left(\frac{2e^2}{CR^3}\right)^2 = -\frac{A}{R^6}, \tag{5.77}$$

where $A = -\frac{1}{2}\frac{\hbar e^4\omega_0}{C^2}$.

Eqn (5.77) represents an attractive interaction which varies as $\frac{1}{R^6}$, R being the separation of two atoms.

This is known as van der Waals' interaction or London interaction or the induced dipole interaction. This is actually the main attractive interaction in crystals of inner gases as well as many organic molecules. Since $\Delta U \to 0$ as $\hbar \to 0$, the interaction may be called as a quantum effect. It is to be kept in mind that the van der Waals' interaction is independent of any overlap of the charge densities of the relevant atoms.

The repulsive energy arising out of the repulsive forces is given by B/R^n, where B is a constant and n is an integer. The experimental value of n is 12. The interatomic PE in the form of *Lenard-Jones potential* is

$$\phi(R) = \phi_0\left[\left(\frac{\sigma}{R}\right)^{12} - \left(\frac{\sigma}{R}\right)^6\right], \tag{5.78}$$

where $B = \phi_0\,\sigma^{12}$ and $A = D_0\sigma^6$.

Now differentiating eqn (5.78) w.r.t. R, we get

$$\frac{d\phi}{dR} = \phi_0\left[-12\sigma^{12}R^{-13} + 6\sigma^6 R^{-7}\right].$$

But

$$\left(\frac{d\phi}{dR}\right)_{R=R_0} = 0.$$

$$\therefore 2\sigma^{12}R^{-13} = \sigma^6 R^{-7} \text{ or, } R^6 = 2\sigma^6. \tag{5.79}$$

Table 5.7

Gas	σ (Å)	ϕ_0 (eV)	*Cohesive energy* in (eV/atom)
Ne	2.74	0.012	−0.02
Ar	3.40	0.042	−0.08
Kr	3.65	0.056	−0.116
Xe	3.98	0.080	−0.17

Thus, $R_0 = R^{1/6}\sigma = 112\sigma$ and the value of ϕ/R at $R = R_0$ is $-\phi_0/4$.

According to *Bernardes,* the experimental value of *Lenard-Jones parameters* of some inert gases are as shown in Table 5.7.

5.12 Answers of Some Important Questions

5.1. *What do you mean by crystal bonding?*

Ans. Bonding is the ability to hold the atoms or ions of a crystal together and is called *crystal bonding.* The atoms or ions (or molecules) in a solid are strongly held together by *interatomic* or *intermolecular forces* known as *bonding force* which is electrostatic in nature. Under the bonded condition the PE is the minimum.

5.2. *What are the different types of bonding?*

Ans. The bondings are classified into the following *five* classes:

(a) *Ionic Bonding*: Ionic bonding arises due to transfer of valence electrons.

Example: NaCl, LiF, etc.

(b) *Covalent Bonding*: Covalent bonding is due to the sharing of valence electron.

Example: Diamond, SiC, etc.

(c) *Metallic Bonding*: Metallic bonding arises due to the electrostatic interaction between the positively charged metal ions and the negatively charged electron clouds.

Example: Cu, Ag, Fe, etc. are metallic bonded solids.

(d) *van der Waals' Bonding*: This type of bonding arises due to the fluctuating dipole nature of an atom with all occupied electron shell filled.

Example: Solid-argon.

This type of classification is quite *arbitrary* because there are many solids which fall under the intermediate types of bonding.

(e) *Hydrogen Bonding*: In hydrogen bonding, bonds arise due to electrostatic attraction between the atoms with hydrogen atom as the positive end of a dipole.

Example: Ice.

5.3. *Give a comparative study regarding the magnitude of bonding energy and melting point of different types of bonding.*

Ans. The following table gives a comparative study of the magnitude of different types of bonding.

Table 5.8

Types of bonding	*Bonding energy (eV* per atom or molecule)	*Melting point* (°C)
Ionic	3.3 (NaCl)	801
	5.2 (MgO)	2800
Covalent	4.7 (Si)	1410
	7.4 (diamond)	3600
Metallic	0.7 (Hg)	−39
	4.2 (Fe)	1538
van der Waals	0.08 (Ar)	−189
	032 ((Cl_2)	−101
Hydrogen	0.36 (NH_3)	−72
	0.52 (H_2O)	0

5.4. *What is cohesive Fnergy?*

Ans. The cohesive energy (or the energy of dissociation) of a solid is defined as the energy which will be given out in the process of the formation of a crystal by bringing neutral atoms from infinity to the position of equilibrium separation. It is also defined as the energy required to separate the atoms of a diatomic molecule to an infinite distance.

5.5. *Show that the repulsive forces must be of shorter range than the attractive ones for the formation of a chemical bond.*

Ans. For the existence of a solid the following two forces must exist:

(a) *An attractive force* between the atoms or molecules of the solid which holds them together.

(b) A repulsive force between the atoms or molecules since a large external pressure is required to compress a solid to any appreciable extent.

When two atoms are brought from infinity to a close proximity they must attract each other. If they are brought closer than a definite distance (r_0) they repel each other, i.e., a kind of spring action takes place between them. The force of interaction is $F = -\frac{dV}{dr}$, where V is the potential and is a function of the separation r. The variation of $V(r)$ with r is shown in Fig. 5.19.

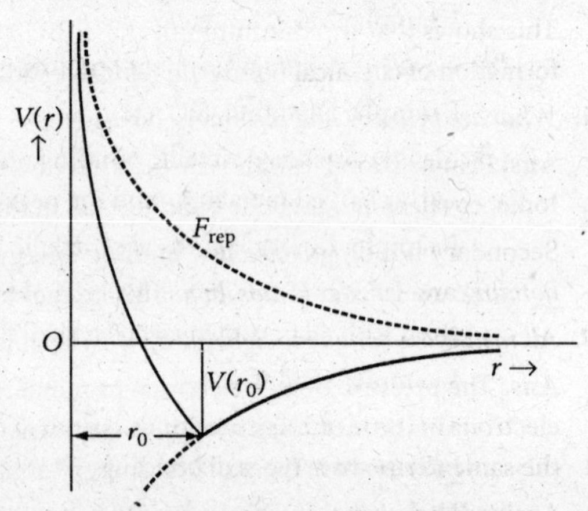

Fig. 5.19: Variation of $V(r)$ with r

In Fig. 5.19 the variation of *attractive* and *repulsive forces* are shown by *dotted lines* and the resultant force by the *solid line*. The PE has a minimum value $V(r_0)$ at r_0, at equilibrium position which represents the *binding energy* or *cohesive energy of the solid*. $V(r_0)$ is negative and $-V(r_0)$ is positive, called the *dissociation energy* of the solid.

If the V_a and V_r represent respectively the *attractive* and *repulsive potentials*, then

$$V_a \propto \frac{1}{r^m} = -\frac{A}{r^m} \text{ and } V_r \propto \frac{1}{r^n} = \frac{B}{r^n}.$$

The resultant potential is

$$V = V_a + V_r = -\frac{A}{r^m} + \frac{B}{r^n},$$

where A and B are constants of proportionality.

The *force of attraction* between the two atoms is

$$F(r) = -\frac{dV}{dr} = \frac{mA}{r^{m+1}} - \frac{nB}{r^{m+1}}.$$

At $r = r_0$, $F(r_0) = 0$,

$$\therefore \frac{mA}{r_0^{m+1}} - \frac{nB}{r_0^{n+1}} = 0 \text{ or, } \frac{m}{n} = \left(\frac{B}{A}\right) r_0^{m-n} \text{ or, } r_0^{m-n} = \left(\frac{A}{B}\right)\frac{m}{n}.$$

Hence,

$$V(r_0) = -\frac{A}{r_0^m} + \frac{B}{r_0^n} = -\frac{A}{r_0^m}\left[1 - \frac{B}{A} r_0^{m-n}\right] = -\frac{A}{r_0^m}\left[1 - \frac{m}{n}\right].$$

Now for $V(r_0)$ to be minimum,

$$\left|\frac{d^2V(r)}{dr^2}\right| > 0.$$

Now

$$\frac{d^2V}{dr^2} = -\frac{m(m+1)A}{r^{m+2}} + \frac{n(n+1)B}{r^{n+2}} > 0.$$

At $r = r_0$,

$$-\frac{m(m+1)A}{r_0^{m+2}} \times \frac{Bn}{Am} + \frac{n(n+1)B}{r_0^{n+2}} > 0 \text{ or, } -m-1+n+1 > 0 \text{ or, } n-m > 0 \text{ or, } n > m.$$

This shows that the minimum of the PE curve can exist, only if $n > m$. In other words, for the formation of chemical bonds the repulsive force must be of shorter range than the attractive force.

5.6. *What are primary and secondary bonds?*

Ans. Primary bonds are those which have bonding energy in the range between 0.1 to 10 eV/bond. Ionic, covalent and metallic bondings fall in this category.

Secondary bondings have low binding energy in the range from 0.01 to 0.05 eV/bond. *Hydrogen bonding* and *van der Waals' bondings* are the examples of secondary bonding.

5.7. *All metals are opaque to light for all wavelengths and have high lustre. Explain.* **[Bang U 1994; HPU 1994]**

Ans. The pressure of free electrons in metallic bonding is the origin of lustre of metals. The free electrons in the metal first absorb the incident energy and other electromagnetic energy and radiate the same afterwards. The radiated energy (of higher wavelength) produces lustre.

Again, due to the *high reflectivity* of the metal surfaces, light is completely reflected from it and gives rise to opacity.

5.8. *Metal crystals shear easily. Explain.* **[Bang U 1994; HPU 1994]**

Ans. Metallic crystals are held together by cloud of free electrons and not by strong covalent bonds. A change in volume causes a remarkable change in energy of these electrons so that there is a good *cleavage strength*. The movement of the atoms does not change the volume and the energy. This property makes the metals *malleable* and *ductile*.

Again, since the positive and negative ions in metal are uniformly distributed, any displacement of ions produces the same environment and hence, are sheared and deformed easily.

5.9. *How many covalent bonds can be formed in oxygen?*

Ans. The number of covalent bonds that an atom can form is determined by the $(8 - N)$ rule, where N is the column number of the periodic table in which the atoms belong.

For oxygen, $N = 6$, i.e., it belongs to VI group of the periodic table. Hence, the number of covalent bond that an oxygen atom can form is $(8 - 6) = 2$.

5.10. *Explain why inert gas crystals exist at low temperature only and not at room temperature.* **[Pbi U 1997]**

Ans. The inert gas crystals are formed due to *van der Waals' bonding which is very weak*, the bond energy being of the order of 0.1 eV/bond. The thermal energy acquired by the inert gas atom or molecules at room temperature dominates the binding force and as such the inert gas solids exist in gaseous state at rom temperature.

But at low temperatures van der Waals' forces dominate the thermal forces and change of state from gas to liquid and solid becomes possible. That is why inert gas crystal can exist only at low temperatures.

5.11. *What makes the metallic bond nondirectional?*

Ans. In case of metallic bond the electrons are spread uniformly throughout the crystal and so the directional nature of metallic bond is not exhibited.

5.12. *What is Madelung Constant?*

Ans. The main contribution to the cohesive energy of an ionic crystal is the electrostatic potential V_a given by

$$V_a = -\frac{Z_1 Z_2 e^2}{4\pi\varepsilon_0 r} = -\frac{\alpha\, e^2}{4\pi\varepsilon_0 r}.$$

$Z_1 e$ and $Z_2 e$ are the charge on the positive and negative ions respectively separated by a distance r. Then $\sigma = Z_1 Z_2$ is called the *Madelung constant* which is a dimensionless quantity having no units.

5.13 Solved Problems

5.1. *Determine whether a gaseous molecule A^+B^- will be stable w.r.t. the isolated A and B gaseous atoms. Given: First ionization energy of A* = 502 kJ/mol, *Electron affinity of B atom* = −335 kJ/mol, *Inter ionic* (A^+B^-) *separation* = 0.3 nm, *Permittivity free space* = 8.85×10^{-12} C/N-m.

Solution: The electrostatic attraction energy of A^+B^- molecule is

$$U_{\text{att}} = -\frac{e^2}{4\pi\varepsilon_0 r} = \frac{(1.602 \times 10^{-19})^2}{4\pi \times (8.85 \times 10^{-12}) \times (3 \times 10^{-10})}$$

$$= -7.689 \times 10^{-19} \text{ J/ion pair}$$

$$= -7.689 \times 10^{-19} \times 6.002 \times 10^{23} \times 10^{-3} \text{ kJ/mol}$$

$$= -463 \text{ kJ/mol}.$$

Net change in energy per mole is 502 – 335 – 463 = **–296 kJ/mol.**

As the change of energy is negative, A^+B^- molecule will be stable.

5.2. *Assuming the equilibrium separation of two atoms to be 2.8 Å and the dissociation energy* 8×10^{-19} J, *calculate the value of A and B in the expression of* PE $U(r) = \frac{A}{r^9} - \frac{B}{r}$. **[CU(H) 1986]**

Solution: We have

$$U(r) = \frac{A}{r^9} - \frac{B}{r}.$$

At $r = r_0 = 2.8$ Å, $U(r) = U(r_0) = -8 \times 10^{-19}$ J = –5 eV.

$$\therefore -5 = \frac{A}{(2.8)^9} - \frac{B}{2.8}. \qquad \text{(i)}$$

But at $r = r_0 = 2.8$ Å,

$$\frac{dU}{dr} = 0 \text{ i.e., } -\frac{9A}{r_0^{10}} + \frac{B}{r_0^2} = 0 \text{ or, } B = \frac{9A}{r_0^8} = \frac{9A}{(2.8)^8}. \qquad \text{(ii)}$$

From eqns (i) and (ii), we have

$$\frac{8A}{(2.8)^9} = 5 \text{ or, } A = \frac{5 \times (2.8)^9}{8} = \mathbf{6611\ eV.Å} \text{ and } B = \frac{9 \times 2.8 \times 5}{8} = \mathbf{15.75\ eV.Å}.$$

5.3. *Calculate the energy required to form* (a) K^+ *and* Cl^- *ions pair from a pair of* K *and* Cl *atoms.* (b) *Calculate the separation between* K^+ *and* Cl^- *ions pair if their total energy be zero.* [*Ionization energy of K* = 4.1 eV *and the electron affinity of* Cl *is* 3.6 eV]

Solution: We know that net energy required to produce an ion-pair is ΔE = I.P. – E.A.

Here I.P. of K = 4.1 eV, E.A. of Cl = 3.6 eV.

$\therefore \Delta E = 4.1 - 3.6 = 0.5$ eV.

For their total energy to be zero, $\Delta E = E_c$, where E_c = the Coulomb energy.

$$\therefore E_c = 0.5 = -\frac{e^2}{4\pi\varepsilon_0 r} \text{ J} = -\frac{e}{4\pi\varepsilon_0 r} \text{ eV}.$$

Hence, the separation (r) of K^+ and Cl^- ions is

$$R = \left|\frac{e}{4\pi\varepsilon_0 \times 0.5}\right| = \frac{(1.6 \times 10^{-19}) \times (9 \times 10)^9}{0.5} = 28.8 \times 10^{-10} \text{ m} = \mathbf{2.88\ nm}.$$

5.4. *If the* PE *function is expressed as* $U(r) = -\frac{A}{r^6} + \frac{B}{r^{12}}$, *then show that* (a) *The intermolecular distance* r_0 *for which the* PE *is minimum is given by* $\left(\frac{2B}{A}\right)^{1/6}$ (b) *the intermolecular distance* r_0' *for which the* PE *is zero, is given by* $r_0 \times (2)^{-1/6}$ (c) *the minimum* PE *is given by* $U_{\min} = -\frac{A^2}{4B}$.

Solution: (a) The condition for the minimum in the PE is

$$\left.\frac{dU}{dr}\right|_{r=r_0} = 0.$$

$$\therefore \left(\frac{dU}{dr}\right)_{r=r_0} = 0 = \left[\frac{6A}{r^7} - \frac{12B}{r^{13}}\right]_{r=r_0} \text{ or, } \frac{6A}{r_0^7} = \frac{12B}{r_0^{13}} \text{ or, } r_0 = \left(\frac{2B}{A}\right)^{1/6}. \qquad \text{(i)}$$

(b) The intermolecular distance r'_0 for which the PE is zero, i.e.,

$$\left[-\frac{A}{r_0'^6}+\frac{B}{r_0'^{12}}\right]=0 \text{ or, } (r'_0)^6=\frac{B}{A}=\frac{(r'_0)^6}{2} \text{ [from (i)]}$$

$$\text{or, } r'_0=r_0\left(\frac{1}{2}\right)^{1/6}.$$

(c) The minimum PE is

$$U_{\min}=\left[-\frac{A}{r_0^6}+\frac{B}{r_0^{12}}\right]=-\frac{A}{r_0^6}\left[1-\frac{B}{A}\frac{1}{r_0^6}\right]=-\frac{A}{\left(\frac{2B}{A}\right)}\left[1-\frac{B}{A}\cdot\frac{A}{2B}\right],\ \left[\because r_0=\left(\frac{2B}{A}\right)^{1/6}\right]$$

$$=-\frac{A^2}{2B}\left[1-\frac{1}{2}\right]=-\frac{\mathbf{A^2}}{\mathbf{4B}}.$$

5.5. *Prove that the value of repulsive exponent n in the following expression for the energy of the* NaCl *crystal,* $U(r)=-\frac{\alpha}{r}+\frac{\beta}{r^n}$ *is given by* $\frac{1}{k_0}=\frac{\alpha(n-1)}{18Nr_0^4}$, *where* k_0 *is the compressibility of* NaCl *solid at equilibrium separation* r_0 *and N is the number of atoms in the* NaCl *crystal.*

Solution: The compressibility k_0 is the change of volume per unit volume produced by unit change of pressure, i.e.,

$$k_0=-\frac{1}{V}\frac{dV}{dp} \text{ or, } \frac{1}{k_0}=B=-V\frac{dp}{dV}, \tag{i}$$

where B is called the *bulk modulus.*

At 0 K, from the first law of thermodynamics, $dQ=dU+pdV$ or, $dU=-pdV$ [$dQ=0$ as the process is adiabatic].

$$\therefore \frac{d^2U}{dV^2}=-\frac{dp}{dV}. \tag{ii}$$

From (i) and (ii), we get

$$B=V\frac{d^2U}{dV^2}. \tag{iii}$$

We are given that

$$U=-\frac{\alpha}{r}+\frac{\beta}{r^n}=-\frac{\alpha}{r}\left[1-\left(\frac{r}{r^n}\right)\frac{\beta}{\alpha}\right].$$

But at $r=r_0$ and

$$\left.\frac{dU}{dr}\right|_{r=r_0}=0,$$

we get

$$U=-\frac{\alpha}{r}\left[1-\frac{1}{n}\left(\frac{r_0}{r}\right)^{n-1}\right] \text{ or, } \frac{dU}{dr}=\frac{\alpha}{r^2}-\frac{\alpha r_0^{n-1}}{r^{n+1}} \text{ and } \frac{d^2U}{dr^2}=-\frac{2\alpha}{r^3}+\frac{\alpha r_0^{n-1}(n+1)}{r^{n+2}}.$$

At $r=r_0$,

$$\left.\frac{d^2U}{dr^2}\right|_{r=r_0}=-\frac{2\alpha}{r_0^3}+\frac{(n+1)\alpha}{r_0^3}. \tag{iv}$$

For NaCl, $V = 2Nr^3$.

$$\therefore dV = 6Nr^2 dr \text{ or, } \frac{dV}{dr} = \frac{1}{6Nr^2} \text{ and } \left(\frac{dV}{dr}\right)^2_{r=r_0} = \frac{1}{36N^2 r_0^4}.$$

But

$$\frac{dU}{dV} = \frac{dU}{dr}\frac{dr}{dV} \text{ and } \frac{d^2U}{dV^2} = \frac{dU}{dr}\cdot\frac{d^2r}{dV^2} + \frac{d^2U}{dr^2}\left(\frac{dr}{dV}\right)^2.$$

At $r = r_0$,

$$B = V\frac{d^2U}{dr^2}\left(\frac{dr}{dV}\right)_{r=0} = 2Nr_0^3\left[-\frac{2\alpha}{r_0^3} + \frac{(n+1)\alpha}{r_0^3}\right] \times \frac{1}{36N^2 r_0^4}$$

$$= 2Nr_0^3\left[\frac{(n-1)\alpha}{r_0^3}\right] \times \frac{1}{36N^2 r_0^4}$$

$$\text{or, } B = \frac{\alpha(n-1)}{18Nr_0^4} \tag{v}$$

$$\text{i.e., } \boldsymbol{\frac{1}{k_0} = \frac{\alpha(n-1)}{18Nr_0^4}}. \tag{vi}$$

5.6. *Calculate the cohesive energy of* NaCl *from the following data: Equilibrium separation between the pair* = 0.281 nm, *ionization energy of* Na = 5.14 eV, *electron affinity of* Cl = 3.61 eV, *Born repulsive exponent* = 9, *Madelung constant* = 1.748.

Solution: Cohesive energy per molecule of NaCl is

$$\frac{Ae^2}{4\pi\varepsilon_0 r_0}\left(1 - \frac{1}{n}\right).$$

Putting $e = 1.602 \times 10^{-19}$ C, $\varepsilon_0 = 8.85 \times 10^{-12}$, $r_0 = 0.281$ nm $= 0.281 \times 10^{-9}$ m, $A = 1.748$, $n = 9$.

$\therefore$ cohesive energy/molecule of NaCl

$$= \frac{1.748 \times (1.602 \times 10^{-19})^2 \times \frac{8}{9}}{4\pi \times (8.85 \times 10^{-12}) \times (0.281 \times 10^{-9})} = \frac{3.98762 \times 10^{-38}}{31.2632 \times 10^{-21}}$$

$$= 0.12755 \times 10^{-17}\text{ J} = \frac{0.12755 \times 10^{-17}}{1.602 \times 10^{-19}}\text{ eV} = \mathbf{7.96\ eV}.$$

5.7. *Calculate the nearest neighbour separation* r_0 *at which* $U(r_0) = -N\left[\frac{\alpha e^2}{r} - \frac{\beta}{r^n}\right]$ *remains unchanged if* $\frac{\beta}{r^n}$ *is replaced by* $Ce^{-r/\rho}$.

Solution: We have $\frac{\beta}{r^n} = Ce^{-r/\rho}$ at $r = r_0$ or, $\beta = r^n Ce^{-r/\rho}$.

Differentiating $r^n = \frac{\beta}{Ce^{-r/\rho}}$, we get

$$nr^{n-1} = \frac{\beta/\rho}{Ce^{-r/\rho}} = \frac{r^n Ce^{-r/\rho}}{Ce^{-r/\rho}\rho} \text{ or, } \frac{n}{r_0} = \frac{1}{\rho} \text{ or, } \boldsymbol{r_0 = n\rho}.$$

5.8. *The* PE *of a diatomic molecule is given in terms of the interatomic distance r by the expression* $U(r) = -\frac{A}{r^2} + \frac{B}{r^{10}}$. *Calculate the equilibrium spacing of the two atoms and the dissociation energy.* $\left[A = 1.440 \times 10^{-39}\text{ Jm}^2 \textit{ and } B = 2.19 \times 10^{-115}\text{ Jm}^{10}\right]$.

Solution: The electrostatic PE of a diatomic molecule is given by

$$U(r) = -\frac{A}{r^2} + \frac{B}{r^{10}}.$$

$$\therefore \left[\frac{dU}{dr}\right]_{r=r_0} = \frac{2A}{r_0^3} - \frac{10B}{r_0^{11}} = 0 \text{ or, } r_0^8 = \frac{5B}{A} = \left[\frac{5 \times 2.19 \times 10^{-115}}{144 \times 10^{-39}}\right] \text{ or, } r_0 = \mathbf{0.408\ nm}.$$

Again,

$$U_{\min} = U_d = -\frac{A}{r_0^m}\left[\frac{(n-m)}{n}\right] = -\frac{144 \times 10^{-39}}{[0.408 \times 10^{-19}]^2} \times \left[\frac{10-2}{10}\right] = \mathbf{-4.33\ eV.}$$

$\therefore$ dissociation energy of the molecule is **4.33 eV.**

5.9. *Calculate the binding energy per molecule of* NaCl *crystal if* $r_0 = 2.81$ Å, $\alpha = 1.75$, $\rho = 0.298$ Å, $e = 4.8 \times 10^{-10}$ esu. **[HPU 1995]**

Solution: In CGS unit the binding energy per ion pair of NaCl crystal is

$$U_{\text{tot}} = -\frac{\alpha e^2}{r_0}\left(1 - \frac{\rho}{r_0}\right) = -\frac{1.75 \times \left(4.8 \times 10^{-10}\right)^2}{2.81 \times 10^{-8}}\left[1 - \frac{0.298}{2.81}\right]$$

$$= -\frac{1.75 \times 4.8 \times 4.8 \times 0.89 \times 10^{-12}}{2.81} = 12.77 \times 10^{-12} \text{ ergs}$$

$$= \frac{12.77 \times 10^{-12}}{1.6 \times 10^{-12}} = 7.98 \text{ eV.}$$

$\therefore$ binding energy per molecule of NaCl crystal = **7.98 eV.**

5.10. *Total energy of an ionic crystal is expressed as*

$$U = -\frac{A|e|^2}{r} + \frac{b}{r^n}.$$

Show that equilibrium energy is proportional to $\left(1 - \frac{1}{n}\right)$. *Also establish that it is indeed a minimum.*

Solution: Given,

$$U = -\frac{a}{r} + \frac{b}{r^n}, \quad a = A|e|^2.$$

At equilibrium, $\frac{dU}{dr} = 0$ at $r = r_0$.

$$\frac{dU}{dr} = \frac{a}{r^2} - \frac{bn}{r^{n+1}} = 0 \text{ or, } r^{n-1} = \frac{nb}{a} \text{ or, } r = \left(\frac{nb}{a}\right)^{1/(1-n)} = r_0.$$

$$U(r_0) = -\frac{a}{r_0} + \frac{b}{r_0^n} = -\frac{a}{r_0} + \frac{b}{r_0 r_0^{n-1}} = -\frac{a}{r_0} + \frac{b}{r_0} \cdot \frac{a}{nb} = -\frac{a}{r_0}\left(1 - \frac{1}{n}\right).$$

$$\therefore U(r_0) = -\frac{A|e|^2}{r_0}\left(1 - \frac{1}{n}\right).$$

So, the equilibrium energy $U(r_0)$ is proportional to $\left(1 - \frac{1}{n}\right)$.

Now

$$\frac{d^2U}{dr^2} = \frac{d}{dr}\left(\frac{dU}{dr}\right) = \frac{dU}{dr} = \frac{d}{dr}\left(\frac{a}{r^2} - \frac{bn}{r^{n+1}}\right) = -\frac{2a}{r^3} + \frac{bn(n+1)}{r^{n+2}}.$$

$$\therefore \left(\frac{d^2u}{dr^2}\right)_{r=r_0} = -\frac{2a}{r_0^3} + \frac{bn(n+1)}{r_0^{n+1}} = \frac{2a}{r_0^3} + \frac{bn(n+1)}{r_0^{n-1}r_0^3} = -\frac{2a}{r_0^3} + \frac{bn(n+1)}{r_0^3} \cdot \frac{a}{nb}$$

$$= -\frac{2a}{r_0^3} + \frac{a(n+1)}{r_0^3} = \frac{a(n-1)}{r_0^3} > 0.$$

$\therefore U(r_0)$ is minimum.

5.11. *Calculate the binding energy per atom of* NaCl *crystal from the following data:*

Equilibrium distance between ions is 2.82. *Madelung constant* = 1.7475, *ionisation energy of* Na = 5.1 eV, *electron affinity* 3.6 eV *and Born repulsive component* 9.

Solution: If $-U_0^L$ and $-U_0^C$ be the lattice energy and atomic cohesive energy of NaCl crystal, then $-U_0^L = -U_0^C - (1-E)$, E = ionisation energy of Na.

Taking $-U_0^L = -8\,\text{eV}$, $-8\,\text{eV} = -U_0^C - (5.1\,\text{eV} - 3.6\,\text{eV})$ or, $\mathbf{-U_0^C = -6.5\,eV.}$

5.12. *The Madelung constant* α_M *for* NaCl *structure is an infinite series* (i) *Show that the first three terms yield* $\alpha_M = 2.13$. (ii) *Calculate the 4th and 5th terms and comment on the convergence of the series.*

Solution: (i) Considering the first three terms Madelung constant is given by

$$\alpha_M = 6 - \frac{12}{\sqrt{2}} + \frac{8}{\sqrt{3}} = 2.13.$$

(ii) The fourth term of the series is $-\frac{6}{\sqrt{4}} = -3$. This gives $\alpha_M = 2.13 - 3 = -0.87$.

Fifth term of the series is $\frac{24}{\sqrt{5}} = 10.73$. This gives $\alpha_M = -0.87 + 10.73 = 9.86$.

$\therefore$ the series does not converge.

5.13. *The potential energy of a diatomic molecule as a function of interatomic distance is given by*

$$U = -\frac{A|e|^2}{r} + \frac{b}{r^n}.$$

If the repulsive interaction be $Ce^{r/\rho}$ (ρ = *constant*), *the nearest neighbour separation* r_0 *is same. Find* r_0.

Solution: We have

$$U = -\frac{A|e|^2}{r} + \frac{b}{r^n} = -\frac{A|e|^2}{r} + Ce^{-r/\rho}.$$

$$\therefore \frac{b}{r^n} = Ce^{-r/\rho} \text{ or, } r^n = \frac{b}{C}e^{r/\rho} \text{ or, } n\ln r = \ln\frac{b}{C} + \frac{r}{\rho}.$$

Differentiating, $\frac{n}{r} = \frac{1}{\rho}$ or, $r = n\rho = r_0$ = equilibrium separation.

5.14. *The potential energy of a diatomic molecule as a function of interatomic distance is*

$$U = -\frac{a}{r^2} + \frac{b}{r^{10}}$$

with $a = 1.44 \times 10^{-39}\,\text{Jm}^2$, $b = 2.19 \times 10^{-115}\,\text{JM}^{10}$. Evaluate equilibrium spacing and dissociation energy.

Solution: We have

$$U(r) = -\frac{a}{r^2} + \frac{b}{r^{10}}.$$

$$\therefore \frac{d}{dr}\left(-\frac{a}{r^2}+\frac{b}{r^{10}}\right)=0 \text{ or, } r=\left(\frac{5b}{a}\right)^{1/8}=r_0=\text{equilibrium separation.}$$

Minimum potential energy,

$$U(r_0)=-\frac{a}{r_0^2}+\frac{b}{r_0^{10}}=-\frac{a}{r_0^2}+\frac{b}{r_0^2}\cdot\frac{a}{5b}\;[\because r_0=(5b/a)^{1/8}]$$

$$=-\frac{a}{r_0^2}+\frac{a}{5r_0^2}=-\frac{4a}{5r_0^2}.$$

EXERCISE

5(A): Objective (MCQ) Type Questions

Select the correct answer.

5.1. In solids the equilibrium condition of the atoms is characterized by
(a) maximum KE (b) minimum PE (c) minimum KE (d) maximum total energy.

5.2. When a solid is formed
(a) some energy is gained (b) some energy is lost
(c) energy remains constant (d) none of these.

5.3. At equilibrium condition of a diatomic molecule the total force is
(a) minimum (b) maximum (c) unity (d) zero.

5.4. The net energy required for the creation of a Na^+ and a Cl^--ion is
(a) 2.2 eV (b) 3.6 eV (c) 1.5 eV (d) 5.1 eV.

5.5. Electrostatic bond in solids is a
(a) primary bond (b) secondary bond (c) hydrogen bond (d) van der Waals' bond.

5.6. In solids the strongest bond is
(a) hydrogen bond (b) metallic bond (c) covalent bond (d) ionic bond.

5.7. Between the ice molecules, the bond is
(a) covalent bond (b) hydrogen bond (c) metallic bond (d) none of these.

5.8. The bond in diamond is
(a) metallic (b) ionic (c) dipole (d) covalent.

5.9. Ionic crystals have melting point
(a) high (b) low (c) intermediate (d) none of these.

5.10. The bond energy of primary bond is in the range
(a) 0.01–0.5 eV (b) 10–100 eV (c) 0.1–10 eV (d) 0–0.01 eV.

5.11. The bond energy of the secondary bond is in the range
(a) 0.1–10 eV (b) 10–100 eV (c) 0–0.01 eV (d) 0.01–0.5 eV.

5.12. The formation of primary bond in solid is due to
(a) interatomic forces (b) intermolecular forces (c) van der Waals' forces (d) none of these.

5.13. Ionic bonds are generally formed in
(a) organic compounds (b) metals (c) inorganic compounds (d) none of these.

5.14. The formation of metallic bond is due to

(a) transfer of electrons

(b) sharing of electrons

(c) sharing of many electrons by larger number of atoms

(d) none of these.

5.15. Covalent bond is formed due to

(a) induced dipole (b) sharing of electrons (c) transfer of electrons (d) none of these.

5.16. The melting points of molecular solids are

(a) fairly low (b) very low (c) very high (d) none of these.

5.17. The bonds of Ge and Si are

(a) ionic crystals (b) covalent crystals (c) molecular crystals (d) none of these.

5.18. The bond of CH_4 is

(a) ionic (b) metallic (c) covalent (d) van der Waals.

5.19. Metallic bonds are

(a) directional (b) nondirectional (c) may be both (a) and (b) (d) none of these.

5.20. Hydrogen bonds are

(a) directional (b) nondirectional (c) both (a) and (b) (d) none of these.

5.21. Ionic solids are soluble in

(a) polar solvents (b) non-polar solvents

(c) both in polar and non-polar solvents (c) none of these.

5.22. Ionic solids are

(a) good insulators (b) conductors of electricity

(c) semiconductors (d) none of these.

5.23. Covalent crystals are soluble in

(a) non-polar solvent (b) polar solvent (c) both (a) and (b) (d) none of these.

5.24. The dissociation energy of NaCl molecule into Na and Cl ions is

(a) 3.6 eV (b) 1.5 eV (c) 4.5 eV (d) 5.1 eV.

5.25. Cohesive energy of NaCl molecule is

(a) 4.5 eV (b) −4.5 eV (c) 5.1 eV (d) 7.95 eV.

5.26. Metallic solids are generally

(a) malleable and ductile (b) hard and brittle (c) soft and deformable (d) none of these.

5.27. Ionic solids are

(a) soft and brittle (b) hard and brittle (c) malleable and ductile (d) none of these.

5.28. van der Waals' interaction varies as

(a) $\frac{1}{r^2}$ (b) $\frac{1}{r^3}$ (c) $\frac{1}{r^6}$ (d) $\frac{1}{r^{12}}$.

5.29. Semiconductors are most likely produced by the crystals of

(a) ionic bond (b) covalent bond (c) metallic bond (d) van der Waals' bond.

5.30. Which of the following has hydrogen bonding?

(a) CH_4 (b) CsCl (c) NaCl (d) HF.

5.31. Which of the bonds are directional?

(a) covalent (b) ionic (c) metallic (d) van der Waals.

ANSWERS

5.1. (b) **5.2.** (c) **5.3.** (d) **5.4.** (d) **5.5.** (a) **5.6.** (d) **5.7.** (b) **5.8.** (d) **5.9.** (a) **5.10.** (c) **5.11.** (d) **5.12.** (a) **5.13.** (c) **5.14.** (c) **5.15.** (b) **5.16.** (a) **5.17.** (b) **5.18.** (d) **5.19.** (b) **5.20.** (a) **5.21.** (a) **5.22.** (a) **5.23.** (a) **5.24.** (c) **5.25.** (d) **5.26.** (a) **5.27.** (b) **5.28.** (c) **5.29.** (b) **5.30.** (d) **5.31.** (a).

5(B): Short Answer Type Questions

5.1. What do you mean by primary and secondary bonds? Explain with examples.

5.2. Explain how interatomic forces are responsible for the existence of solid structure.

5.3. What are the different types of bonds? Give examples.

5.4. What is ionic bonding? Give the properties of ionic crystals.

5.5. What is meant by Madelung constant and Madelung energy? Give the significance of Madelung constant?

5.6. What is cohesive energy of a solid? Find an expression of it for ionic crystals.

5.7. Inert gas crystals exist only at very low temperature. Why?

5.8. Show that for the formation of a chemical bond the repulsive forces must be of shorter range than the attractive forces.

5.9. Explain the formation of van der Waals' bonding in crystals. What are the properties of inert gas crystals?

5.10. Discuss the difference between ionic bond and covalent bond.

5.11. Give a comparative study between ionic bond and metallic bond.

5.12. Metals are good conductor of electricity and heat but ionic crystals are not. Explain why.

5.13. Explain reason of opacity and lustre of metals.

5.14. Discuss the properties of metallic bonding.

5.15. Explain why ionic crystals are hard but brittle.

[*Hints.* Ionic crystals are hard due to strong electrostatic forces of attraction between anions and cations. But when a shearing force is applied to an ionic crystal the ions tend to slip past one another with relatively more ease, because of the nondirectional character of ionic bond. At a certain stage due to the strong repulsive forces originating from the violation of Pauli's exclusion principle, the crystal undergoes fracture. That is why ionic crystals are hard but brittle.]

5.16. Mention the types of bonding in the following crystals: LiF, SiC, Na, H_2O (ice), Ar and HF.

5.17. Evaluate Madelung constant for NaCl.

5.18. The PE of a system of two atoms is given by

$$U(r) = -\frac{\alpha}{r^4} + \frac{\beta}{r^{12}}.$$

Calculate the amount of energy released when the atoms form a stable bond. Determine the bond length. **[VU(H) 2009]**

5.19. Briefly discuss the different types of bonding in solids. **[BU(H) 2003]**

5.20. Explain the origin of van der Waals' interaction between two atoms of the inert gas in the solid state. **[BU(H) 1996]**

5.21. Write down the nature of bonding in diamond form carbon and CO_2 and explain briefly the mechanism of such bonding. **[CU(H) 1998]**

5.22. Mention the different types of bondings in crystalline solids. Describe their properties briefly. **[CU(H) 2017]**

5.23. Argue that the van der Waals attraction between two molecules at a distance r is proportional to r^{-7}. **[CU(H) 2015]**

5.24. What are covalent bonds and hydrogen bonds in crystal binding? How does hydrogen bond differ from dipole bond? **[BU(H) 2010]**

5.25. What is the nature of interatomic in silicon? **[BU(H) 2007]**

5.26. Differentiate between covalent and van der Waals bonding with examples. **[DU 2014]**

5(C): Numerical Problems

5.1. If an atom A has an ionization energy 5 eV and another atom B of an electron affinity 4 eV are at a distance 5 Å, then find the energy required to transfer an electron from A to B.

[*Hints.* Bond energy = $-V - IP + EA$, where

$$V = \frac{-e}{4\pi\varepsilon_0 r_0} = -2.9\,\text{eV}\ \left[\because r_0 = 5\,\text{Å}\right]$$

Ionization energy IP of A = 5 eV and electron affinity of B = 4 eV, $\because$ Bond energy = −1.9 eV]

5.2. The PE of a pair of atoms is

$$U = -\frac{A}{r^4} + \frac{B}{r^{12}},$$

where r is the interatomic distance. Find the value of r where a stable bond is formed. Also calculate the energy released when the atoms form a stable bond. [*Ans.* $\left(\frac{3B}{A}\right)^{1/3}, \left(\frac{4A^3}{27B}\right)^{1/2}$]

5.3. If the interaction energy between atoms is given by an expression of the form:

$$U(r) = -\frac{\alpha}{r^n} + \frac{\beta}{r^m},$$

where $n = 2,\ m = 10$ and the two atoms form a stable nucleus with an internuclear distance of 3 Å and dissociation energy of 4 eV, calculate the values of α and β.

5.4. Show that the value of the repulsive potential exponent n in the total energy expression of $2N$ ions of NaCl

$$U(r) = -N\left(\frac{\alpha e^2}{r} - \frac{\beta}{r^n}\right) \text{ is 9.4,}$$

where the compressibility of NaCl is $k = 3.3 \times 10^{12}$ cm²/dyne, Madelung constant $\alpha = 1.75$ and equilibrium nearest neighbour spacing $r = 2.81$ Å.

5.5. The energy of two particles in the field of each other at a separation r is given by

$$U(r) = \frac{\alpha}{r} + \frac{\beta}{r^8},$$

where α and β are constants. At what separation they will form a stable compound?[*Ans.* $r_0 = \left(\frac{8\beta}{\alpha}\right)^{1/7}$] **[HPU 1994]**

5.6. The cohesive energy and the nearest neighbour distance for a LiF molecule are 1.68×10^{-18} J and 2.014 Å respectively. The structure of LiF is the same as that of NaCl. Calculate the bulk modulus of LiF using the following expression for the PE, $U = -\alpha q^2/R + \beta/R^n$.
The value of $\alpha = 1.75$. [*Ans.* 7.15×10^{10} N/m^2]

5.7. Calculate the binding energy of NaI where the nearest neighbour distance is 0.324 nm. Express the energy in eV and also in kJ/mol.
[Madelung constant for NaI = 1.748 and $n = 9.5$]
[*Hints.* $V_0 = \dfrac{Ae^2}{4\pi\varepsilon_0 r_0}\left(\dfrac{n-1}{n}\right)$ J $= \dfrac{Ae}{4\pi\varepsilon_0 r_0}\left(\dfrac{n-1}{n}\right)$ eV $= 6.9$ eV.
Binding energy per k-mole of NaI is $U_0 N = 667 \times 10^3$ kJ/mol.]

5.8. The interaction energy of two atoms is given by

$$U = -\frac{A}{r^6} + \frac{B}{r^{12}},$$

where $A = 4.19 \times 10^{-76}$ Jm^6 and $B = 1.53 \times 10^{-133}$ Jm^{12}. Calculate the force required to reduce the interatomic distance by 5 % of the value at equilibrium. [*Ans.* 5.97×10^{-9} N]

5.9. Show that the first five terms in the series for the Madelung constant for NaCl (FCC) are

$$\alpha = 6 - 12/\sqrt{2} + 8/\sqrt{3} - 6/2 + 24/\sqrt{5}.$$

How can the convergence of the series be improved? **[BU(H) 2001]**

5.10. The potential energy of a system of two atoms is given by

$$U = -\frac{A}{r^6} + \frac{B}{r^{12}}.$$

The atoms form a cubic bond having bond length 3 Å and bond energy 1.8 eV. Calculate the force required to break the molecule and the critical interatomic distance for which it occurs. **[BU(H) 2017]**

5.11. A two-body interaction is specified by the potential

$$\mu(r) = -\frac{a}{r^m} + \frac{b}{r^n},$$

where a and b are positive constants. Show that the system is stable for $n > m$. Also find the equilibrium separation. Find the corresponding ratio of the attractive to repulsive energy terms at equilibrium. **[CU(H) 2014]**

5.12. The potential energy of a diatomic molecule is given by

$$U(r) = -\frac{a}{r^2} + \frac{b}{r^{10}},$$

where r is the interatomic distance in terms of a and b. **[CU(H) 2010]**

5.13. Calculate the lattice energy per ion on N-crystal of which the nearest neighbour distance is 2.81 Å. Madelung constant for NaCl is 1.748 and the repulsive exponent is 9. **[BU(H) 2014]**

5.14. Calculate the equilibrium spacing of the two atoms and the dissociation energy in problem 5.13. [Given, $a = 1.44 \times 10^{-39}$ Jm^2 and $b = 2.19 \times 10^{-15}$ Jm^{10}] **[BU(H) 2013]**

5.15. (a) What is Madelung constant? What is its unit?
(b) Calculate the value of Madelung constant for a linear ionic crystal and show that for a one-dimensional array of ions of alternate signs, its value is −1.38. **[BU(H) 2009]**

5.16. (a) What is Madelung constant? What are its units?

(b) Calculate the value of Madelung constant for a linear ionic crystal and show that for a one-dimensional array of ions of alternate signs, its value is 1.38.

(c) Explain why inert gas crystals exist only at low temperatures? **[BU(H) 2007]**

5.17. The interatomic force between two atoms is given by the following function of distance r between the centre of the atom:

$$F(r) = \frac{A}{r^m} - \frac{B}{r^n} \text{ with } n > m.$$

(a) Discuss the significance of the two atoms with suitable graph.

(b) Obtain an expression for the cohesive energy. **[Bhabnagar U 2017]**

5.18. Write down the expression of cohesion energy of a crystal lattice and show that it shifts only by 1 to 2 per cent if the repulsive exponent shifts by unity.

[Hint. $U_1 = -k\left(1 - \frac{1}{n}\right)$, $U_2 = -k\left(1 - \frac{1}{n+1}\right)$, k = constant

$$\frac{U_2 - U_1}{U_1} = \frac{1}{n^2 - 1} \rightarrow \frac{100}{n^2 - 1}\ \%.$$

For $n = 7$, $\dfrac{U_2 - U_1}{U_1} \sim 2\%$. For $n = 9$, $\dfrac{U_2 - U_1}{U_1} \sim 1\%$.]

5(D): Broad Answer Type Questions

5.1. (a) Plot and explain the variation of (i) attractive PE (ii) repulsive PE (iii) resultant PE with interatomic distance when two atoms are brought together.

(b) Describe how crystals have been classified according to their bonding.

5.2. (a) How are solids classified on the basis of interatomic forces?

(b) What is the dominant binding mechanism in the following solids:

Ge, Xe, Cu, NaF, Ice?

Which of these solids will have the lowest melting point? **[BU(H) 1993]**

5.3. (a) How do you explain cohesion in inert gas crystals? Using semiquantitative arguments obtain dependence of the van der Waals' interaction on the interatomic separation.

(b) Show that the total lattice energy of an ionic crystal assuming a repulsive interaction of the form $\lambda e^{-R/\rho}$, operating between nearest neighbours only, is given by

$$U = -\frac{1}{4\pi\varepsilon_0}\frac{N\alpha e^2}{R_0}\left(1 - \frac{\rho}{R_0}\right),$$

where $2N$ is the number of ions in the crystal, R_0 is the equilibrium separation between nearest neighbours and α is the Madelung constant. **[BU(H) 1994]**

5.4. (a) Discuss briefly the principal types of binding forces in crystals.

(b) Define cohesive energy. Explain how cohesion is brought about in a simple metal like Na. **[BU(H) 1995]**

5.5. How do you account for cohesion in an ionic crystal? Assuming a repulsive interaction of the form λ/R^n between nearest neighbours, find the total lattice energy of an ionic crystal. Also derive an expression for the bulk modulus of such a crystal. **[BU(H) 1996]**

5.6. (a) Discuss how ionic bond between two atoms is formed by the transfer of electrons from one atom to the other. Is it nondirectional?

(b) Discuss how a covalent bond results from the localised sharing of an equal number of valence electrons of neighbouring atoms. Is it a directional bond?

5.7. (a) Explain qualitatively how a covalent bond is formed. What are the important characteristics of a covalent bonded solid?

(b) In an ionic crystal the interaction energy between ions i and j is given by

$$U_{ij} = \begin{cases} \lambda\, e^{-R/\rho} - \dfrac{q^2}{R} & \text{(nearest neighbour)} \\ \pm \dfrac{q^2}{R_{ij}} & \text{(otherwise)} \end{cases}$$

where R_{ij} is the distance between i and j; R is the nearest neighbour separation; $\pm q$ is the charge on each ion; λ and ρ are constants.

Find the total lattice energy of a crystal of $2N$, such ions at their equilibrium separation, if the number of nearest neighbours is Z and the Madelung constant of the structure is α. Explain how we can experimentally determine the parameter ρ of the repulsive interaction. **[BU(H) 1999]**

5.8. (a) What is the nature of forces binding atoms and molecules in solids?

(b) Determine Madelung constant. Evaluate the Madelung constant for a one-dimensional chain of ions of alternative signs. **[BU(H) 1997]**

5.9. (a) How are the atoms bound together in an inert gas crystal? Find the order of magnitude of the attractive interaction energy between two atoms in such a crystal. What is the origin of repulsive force between two inert gas atoms?

(b) Deduce an expression for the bulk modulus of an ionic crystal assuming a repulsive interaction between nearest neighbours of the form λ/R^n. **[BU(H) 1998]**

5.10. (a) Classify each of the following materials according to crystalline bonding: (i) NaCl, (ii) Diamond, (iii) Ice, (iv) Al.

(b) Why do inert gas atoms are found in condensed form (liquid/solid) at a very low temperatures, but not at room temperature?

(c) Explain clearly the differences between ionic and covalent types of bonding. **[BU(H) 2000]**

(d) Explain why the inert gas crystals exist only at very low temperature and not at room temperature.

(e) All metals are opaque to visible light and have high lustre. Explain. **[BU(H) 2002]**

5.11. (a) What type of interaction is responsible for binding among inert gas atoms in the inert gas crystals? Explain the origin of both attractive and repulsive part of the interaction.

(b) Discuss the Lenard-Jones PE between two inert gas atoms. **[GNDU 1997; PU 1994, '92; HPU 1992]**

5.12. (a) What is the Madelung constant? What is its unit?

(b) Calculate the value of the Madelung constant for a linear ionic crystal and show that for a one-dimensional array of ions of alternative signs its value is 1.38. **[GNDU 1992; HPU 1996; BU(H) 2003]**

5.13. (a) What type of interaction is responsible for the binding among inert gas atoms in inert gas crystals? Explain the origin of both the attractive and repulsive parts of the interaction.

(b) Give a simplified derivation of the form of the attractive interaction.

5.14. (a) What are ionic crystals? Describe their formation.

(b) Derive an expression for lattice energy from Born's theory. Discuss the result by comparing the calculated and experimental values in a typical case. **[KU(H) 1991]**

5.15. (a) Discuss the formation of hydrogen bond. Is it directional or nondirectional? Justify your answer.

(b) Discuss in brief the Born's theory of the lattice energy. What are the refinements made in the theory by Born and Mayer?

5.16. (a) Consider the cohesive energy of two atoms at a separation R

$$U(R) = 2N\epsilon\left[(12.13)\left(\frac{\sigma}{R}\right)^{12} - (12.45)\left(\frac{\sigma}{R}\right)^{6}\right],$$

where σ, ϵ are constants and N is the total number of atoms.

Suppose, there are N atoms in an FCC lattice of lattice constant b and R is the nearest neighbour distance.

(i) What is the volume occupied by N atoms?

(ii) Show that the above cohesive energy can be expressed as

$$U(V) = \frac{b_{12}}{V^4} - \frac{b^4}{V^2},$$

where b_{12} and b_4 are constants in terms of N, ϵ and σ.

(iii) Hence, find the bulk modulus B of the system.

(b) Write down two important differences between ionic and metallic bonds. **[CU(H) 2016]**

5.17. (a) What are ionic crystals?

(b) Derive an expression for binding energy for an ionic crystal and obtain the expression for the Madelung constant.

(c) Evaluate the Madelung constant for an infinitely long one-dimensional ionic crystal consisting of singly charged alternative positive and negative ions. **[BU(H) 2014]**

5.18. If the potential energy of a system of two atoms is expressed as

$$U(r) = -\frac{A}{r^6} + \frac{B}{r^{12}}.$$

(a) Show that the interatomic distance r_0 for which the potential energy is zero is given by $r_0 \times (2)^{-1/6}$.

(b) Show that the stored energy in the polarised atom is equal to $\frac{1}{2}\alpha E^2$, where α is the polarisability. **[BU(H) 2011]**

Chapter 6

Lattice Vibration

6.1 Introduction

A two-dimensional crystal lattice may be regarded as a regular arrangement of atoms connected by *elastic springs* as shown in Fig. 6.1. In such a model the vibration of any single atom is shared by all other atoms and the system executes *coupled vibration*. These lattice vibrations may be looked upon as a series of superimposed lattice waves with a frequency spectrum depending on the elastic properties of the crystal.

The vibration of the lattice may be of the following *two* types:

(i) It may vibrate freely in its *normal modes* due to its own internal energy, yielding thermal properties like *specific heat* and *thermal conductivity*.

(ii) It may vibrate under the action of dynamical external forces. These forces may be either *mechanical* or *electromagnetic* in nature. Because of this type of forced vibration *acoustical* and certain *optical properties* arise.

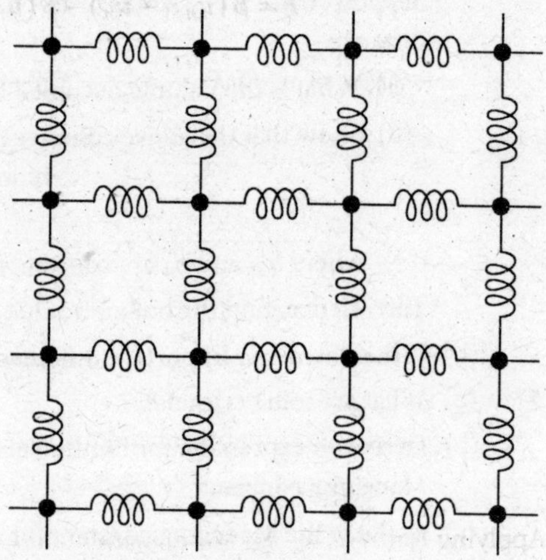

Fig. 6.1: Two-dimensional model of a lattice in which atoms are joined together by elastic springs

6.2 Vibrations of Monatomic Linear Lattice

To investigate the vibrational motion of a *monatomic linear crystal lattice* we assume the following:

(i) Each atom has mass m and is joined to another atom by massless elastic springs. The arrangement of atoms constitutes a *non-homogeneous medium*.

(ii) The displacements of atoms obey *Hooke's law*.

(iii) The force exerted by a spring on a nearest neighbour atom is

$$F = \beta u, \tag{6.1}$$

where β is the *force constant* and u the *displacement of the spring*. This means the force required to produce an atomic displacement is proportional to the displacement itself signifying that, the motion of the atoms is *simple harmonic*.

(iv) The atoms may vibrate either *parallelly* or *perpendicularly* to the direction of the wave vector of the propagating elastic wave but we shall consider only the longitudinal elastic vibrations of the atoms.

Fig. 6.2 shows that in equilibrium state, the atoms are situated at equally spaced sites along the x-axis. The sites of the atoms are represented by $n-1$, n, $n+1$, etc. and the X-coordinates of the corresponding atoms are $(n-1)a$, na, $(n+1)a$, where a is the *interatomic spacing.* As the atoms vibrate simple harmonically, they become the source of elastic waves propagating through the medium. At any instant of time during vibration, let the displacements of the nth, $(n-1)$th, $(n-2)$th, ... atoms be represented by u_n, u_{n-1}, u_{n-2}, ... respectively.

As the nth atom is attached to $(n-1)$th and $(n+1)$th atoms by two springs, it experiences two opposite forces. The net force on the nth atom by eqn (6.1) is

$$F = \beta(u_{n+1} - u_n) - \beta(u_n - u_{n-1}) = \beta(u_{n+1} + u_{n-1} - 2u_n). \tag{6.2}$$

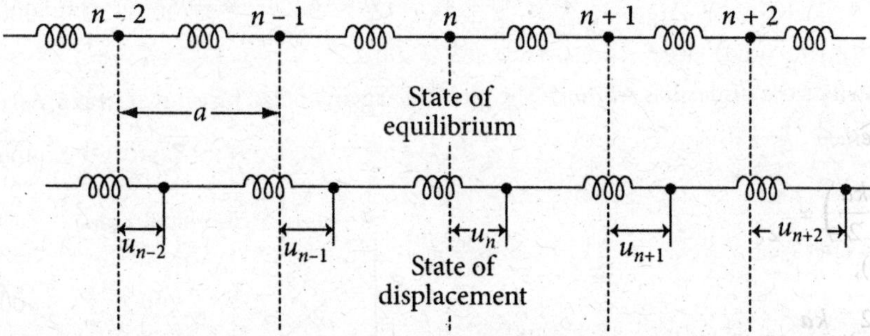

Fig. 6.2: Equilibrium and displaced state of a one-dimensional monatomic lattice $(u_{n+2} > u_{n+1} > u_n > u_{n-1} > u_{n-2})$

Applying *Newton's second law of motion,* the equation of motion of the nth atom is

$$m\frac{d^2u_n}{dt^2} = \beta(u_{n+1} + u_{n-1} - 2u_n), \tag{6.3}$$

where $\frac{d^2u_n}{dt^2}$ represents the *acceleration* of the nth atom.

The solution of eqn (6.3) is

$$u_n = u_0 e^{i(\omega t - kna)} \tag{6.4}$$

where $k = \frac{2\pi}{\lambda}$ is the *wave vector; na*, the X-coordinate of the nth atom and ω, the *angular frequency* of the wave.

Similarly, for $(n+1)$th and $(n-1)$th atoms, we can write,

$$u_{n+1} = u_0\, e^{i[\omega t - k(n+1)a]}$$
$$u_{n-1} = u_0\, e^{i[\omega t - k(n-1)a]}. \tag{6.5}$$

Using eqns (6.4) and (6.5) in eqn (6.3), we get

$$-m\omega^2 = \beta\left(e^{ika} - 2 + e^{-ika}\right) = \beta\left[e^{ika/2} - e^{-ika/2}\right]^2 = -4\beta\sin^2\left(\frac{ka}{2}\right) \tag{6.6}$$

$$\text{or, } \omega = \pm\sqrt{(4\beta/m)}\sin\left(\frac{ka}{2}\right)\left[\because \sin\alpha = \frac{e^{i\alpha} - e^{-i\alpha}}{2i}, \therefore \sin^2\alpha = -\frac{1}{4}\left(e^{i\alpha} - e^{-i\alpha}\right)^2\right] \tag{6.7}$$

If C and ρ be the longitudinal stiffness and the mass per unit length of the line of atoms, then $C = \beta a$ and $\rho = m/a$.

∴ from eqn (6.7),

$$\omega = \pm\frac{2}{a}\sqrt{\frac{C}{\rho}}\sin\left(\frac{ka}{2}\right) = \pm\frac{2}{a}v_s\sin\left(\frac{ka}{2}\right), \tag{6.8}$$

where

$$v_s = \sqrt{C/\rho}. \tag{6.9}$$

For a given lattice, since C and ρ are constants, v_s is also a constant and has dimension of velocity. It is the velocity of the longitudinal sound wave in the solid.

Since the frequency is a positive quantity,

$$\omega = \frac{2}{a}v_s\left|\sin\left(\frac{ka}{2}\right)\right|. \tag{6.10}$$

Eqn (6.10) is known as the *dispersion relation*. The plot of ω against (π/a) is shown in Fig. 6.3.

(i) *At Low Frequencies, $k \to 0$*:

For $k \to 0$, $\sin\left(\frac{ka}{2}\right) \simeq \frac{ka}{2}$.

∴ from eqn (6.10),

$$\omega = \frac{2}{a}v_s\frac{ka}{2} \quad \text{or,}\ \omega = v_s k. \tag{6.11}$$

For this case, *phase velocity*

$$v_p = \frac{\omega}{k} \tag{6.12}$$

and *group velocity*

$$v_g = \frac{d\omega}{dk}. \tag{6.13}$$

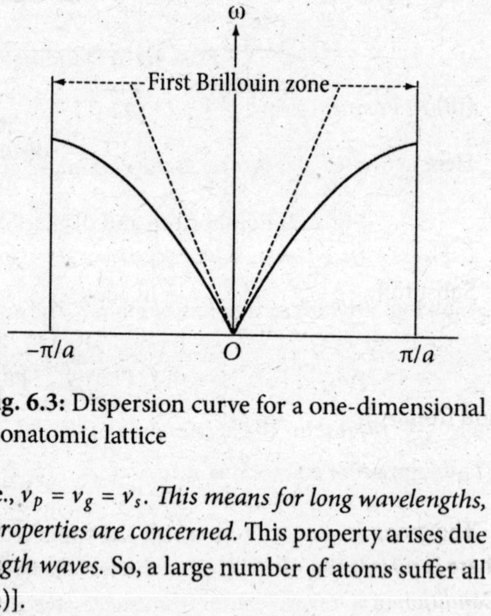

Fig. 6.3: Dispersion curve for a one-dimensional monatomic lattice

Thus, we see that both v_p and v_g for $\lambda \to \infty$ are equal, i.e., $v_p = v_g = v_s$. *This means for long wavelengths, the atomic nature of solids is insignificant when dynamical properties are concerned.* This property arises due to *insensitiveness of the discrete medium to the long wavelength waves.* So, a large number of atoms suffer all types of displacements as in a *homogeneous line* [Fig. 6.4 (a)].

(ii) *At High Frequencies*: From eqn (6.10),

$$v_p = \frac{\omega}{k} = \frac{2v_s}{ka}\sin\left(\frac{ka}{2}\right) \tag{6.14}$$

$$\text{and } v_g = \frac{d\omega}{dk} = v_s\cos\left(\frac{ka}{2}\right). \tag{6.15}$$

Thus, for high frequency (low wavelength) both v_p and v_g are functions of frequency (ω). Similar situation arises when light passes through a medium whose refractive index is a function of frequency and is responsible for *dispersion* of light. By analogy, the medium under consideration is called a *dispersive medium.*

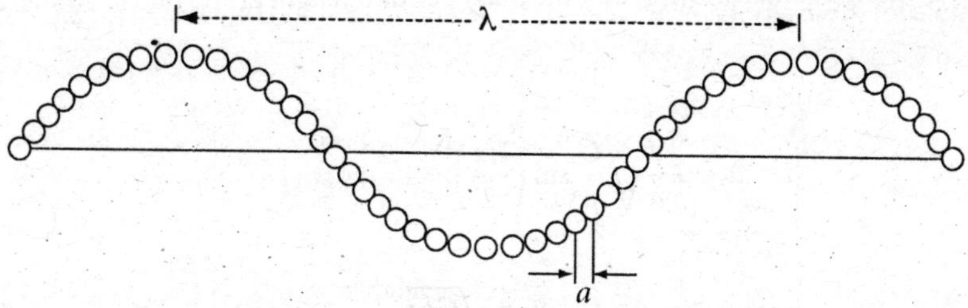

Fig. 6.4: (a) In case of long wavelength the motion approaches in a homogeneous line

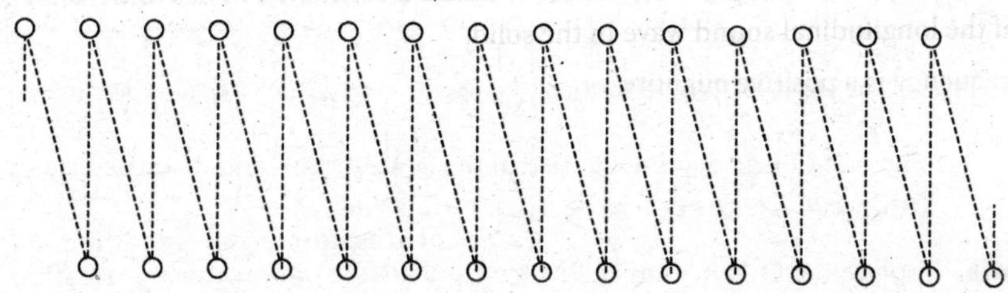

Fig. 6.4: (b) Standing wave formtion at $\omega = \sqrt{4\beta/m}$; $\lambda = 2a$

(iii) *At Frequency* $\omega = \sqrt{\frac{4\beta}{m}}$:

Here $\omega = \sqrt{4\beta/m}$ represents the maximum angular frequency of vibration. From eqn (6.10),

$$\sin\left(\frac{ka}{2}\right) = 1 \text{ or, } \frac{ka}{2} = \frac{\pi}{2} \text{ or, } k = \frac{\pi}{a} \text{ or, } \frac{2\pi}{\lambda} = \frac{\pi}{a} \text{ or, } \lambda = 2a. \tag{6.16}$$

Substituting $k = \frac{\pi}{a}$ and $\frac{ka}{2} = \frac{\pi}{2}$ in eqn (6.14), we get

$$v_p = \frac{2}{\left(\frac{\pi a}{a}\right)} v_s \sin\frac{\pi}{2} = \frac{2v_s}{\pi} \tag{6.17}$$

and from eqn (6.15), we get

$$v_g = v_s \cos\frac{\pi}{a} = 0. \tag{6.18}$$

The *group velocity* v_g represents the transfer of energy or signal. Since $v_g = 0$ in this frequency limit, there is no transfer of energy and the wave behaves as a *standing wave.* This situation is similar to the *Bragg Condition* of X-ray reflection from successive atomic planes in a crystal and is given by

$$2d\sin\theta = n\lambda \text{ or, } \lambda = 2d \text{ [for first order reflection, } n = 1]$$

for *normal incidence.* This shows that at the edges of the Brillouin Zone, $k = -\frac{\pi}{a}$ and $k = +\frac{\pi}{a}$, $v_g = 0$, which is a necessary condition for the standing waves.

Thus, a linear chain can propagate waves of frequency, $\omega \le \frac{2v_s}{a}$ and as such acts as a *low pass filter* which allows the waves of $\omega = 0$ to $\omega = \frac{2v_s}{a}$.

6.3 Vibration of One-dimensional Diatomic Lattice

We consider a diatomic crystal in which there are two atoms per unit primitive cell. The masses of the atoms are m and M $(m < M)$ and placed alternately along the X-axis with an interatomic distance a. In the equilibrium state let these atoms be at $(2n-3)$, $(2n-2)$, $(2n-1)$, $2n$, $(2n+1)$, $(2n+2)$, $(2n+3)$, ... as shown in Fig. 6.5.

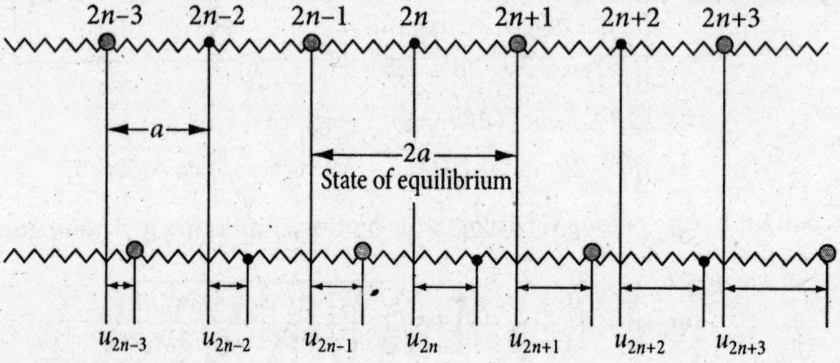

Fig. 6.5: Linear diatomic lattice in the equillibrium and disturbed states $[u_{2n+3} > u_{2n+2} > u_{2n+1} > u_{2n} > u_{2n-1} > u_{2n-2} > u_{2n-3}]$

Let u_{2n} be the displacement of an atom at $2n$ site at any instant of time during vibration. Then using the assumptions similar to the case of monatomic case the equations of motion of atoms of masses m and M become respectively

$$F_{2n} = m\frac{d^2u_{2n}}{dt^2} = \beta\left(u_{2n+1} + u_{2n-1} - 2u_{2n}\right)$$

$$\text{and } F_{2n+1} = M\frac{d^2u_{2n+1}}{dt^2} = \beta\left(u_{2n+2} + u_{2n} - 2u_{2n+1}\right).$$

The solutions of these equations representing travelling waves are

$$u_{2n} = A\,e^{i(\omega t-2nka)} \tag{6.19}$$

$$u_{2n+1} = B\,e^{i[\omega t-(2n+1)ka]} \tag{6.20}$$

$$\text{and } u_{2n+2} = A\,e^{i[\omega t-(2n+2)ka]} \tag{6.21}$$

$$u_{2n-1} = B\,e^{i[\omega t-(2n-1)ka]}, \tag{6.22}$$

where k is the *wave vector* for a particular mode of vibration. The frequencies of vibration of both types of atoms have been assumed to be the same because both types of atoms participate in the same wave motion. Since $m \neq M$, the amplitudes A and B are different.

Now

$$\frac{d^2u_{2n}}{dt^2} = -A\omega^2 e^{i[\omega t-(2n)ka]} \text{ or, } m\frac{d^2u_{2n}}{dt^2} = -Am\omega^2 e^{i[\omega t-(2n)ka]}$$

$$\therefore\ -Am\omega^2 e^{i[\omega t-(2n)ka]} = \beta B\left[e^{i\{\omega t-(2n+1)ka\}} + e^{i\{\omega t-(2n-1)ka\}}\right] - 2\beta A\,e^{i\{\omega t-(2n)ka\}}$$

$$\text{or, } -m\omega^2 A = \beta B\left(e^{ika} + e^{-ika}\right) - 2\beta A \text{ or, } \left(2\beta - m\omega^2\right)A - (2\beta\cos ka)B = 0. \tag{6.23}$$

Likewise

$$M\frac{d^2u_{2n+1}}{dt^2} = -BM\omega^2 e^{i[\omega t-(2n+1)ka]}$$

$$\text{or, } -M\omega^2 B\,e^{i[\omega t-(2n+1)ka]} = \beta A\left[e^{i\{\omega t-(2n+2)ka} + e^{i\{\omega t-(2n)ka\}}\right] - 2\beta B\,e^{i\{\omega t-(2n+1)ka\}}$$

$$\text{or, } (-2\beta\cos ka)A + \left(2\beta - \omega^2 M\right)B = 0\ \left[\because e^{ix} + e^{-ix} = 2\cos x\right]. \tag{6.24}$$

We know that, a set of homogeneous linear equations would give rise to non-zero solutions for A and B if the determinant of the coefficients A and B is zero, i.e.,

$$\begin{vmatrix} 2\beta - m\omega^2 & -2\beta\cos ka \\ -2\cos ka & 2\beta - Ma^2 \end{vmatrix} = 0$$

$$\text{or, } \left(2\beta - M\omega^2\right)\left(2\beta - m\omega^2\right) - 4\beta^2\cos^2 ka = 0$$

$$\text{or, } 4\beta^2 - 2\beta m\omega^2 - 2\beta M\omega^2 + Mm\omega^4 - 4\beta^2\cos^2 ka = 0$$

$$\text{or, } \omega^4 - 2\beta\frac{(m+M)}{mM}\omega^2 + \frac{4\beta^2}{mM}\left(1 - \cos^2 ka\right) = 0$$

$$\text{or, } \omega^4 = \beta\left[\frac{1}{m} + \frac{1}{M}\right] \pm \sqrt{\left(\frac{1}{m} + \frac{1}{M}\right)^2 - \frac{4\sin^2 ka}{mM}}. \tag{6.25}$$

Eqn (6.25) represents the *dispersion relation* for a *linear diatomic lattice.* The corresponding relation for a monatomic lattice has been found to be

$$\omega^2 = \frac{4\beta}{m}\sin^2\frac{ka}{2}.$$

Since frequency ω is always positive, we find that there is only one value of ω for a single k for a monatomic lattice whereas for a diatomic lattice, there are two values of ω, say ω_+ and ω_-, where

$$\left.\begin{aligned} \omega_+^2 &= \beta\left[\frac{1}{m} + \frac{1}{M}\right] + \beta\sqrt{\left(\frac{1}{m} + \frac{1}{M}\right)^2 - \frac{4\sin^2 ka}{mM}} \\ \text{and } \omega_-^2 &= \beta\left[\frac{1}{m} + \frac{1}{M}\right] - \beta\sqrt{\left(\frac{1}{m} + \frac{1}{M}\right)^2 - \frac{4\sin^2 ka}{mM}} \end{aligned}\right\}. \tag{6.26}$$

Difference between the two branches of dispersion relation curve

The graph of ω against k leads to two branches. One branch corresponds to ω_+ and is called the *optical branch*, the other branch corresponds to ω_- and is called the *acoustical branch* [Fig. 6.6].

(i) ***Optical Branch* (*when* $\omega = \omega_+$):** In case of ω_+ for $k \to 0$, $\sin ka \to 0$, we get from first equation of (6.26)

$$\omega_+ = \sqrt{2\beta\left(\frac{1}{m} + \frac{1}{M}\right)}. \tag{6.27}$$

For $k \to \frac{\pi}{2a}$, $\sin ka \to 1$ and then

$$\omega_+^2 = \beta\left[\frac{1}{m} + \frac{1}{M}\right] + \beta\sqrt{\left[\frac{1}{m} + \frac{1}{M}\right]^2 - \frac{4}{mM}} = \beta\left[\frac{1}{m} + \frac{1}{M}\right] + \beta\left[\frac{1}{m} - \frac{1}{M}\right] = \frac{2\beta}{m}.$$

$$\therefore \omega_+ = \sqrt{\frac{2\beta}{m}}. \tag{6.28}$$

(ii) ***Acoustical Branch*** (when $\omega = \omega_-$): We note for $k \to 0$, $\sin ka \simeq ka$ and $\sin ka \neq 0$, for if we take the latter case, then ω_- will be zero which is not acceptable.

Then

$$\omega_-^2 = \beta\left[\frac{1}{m}+\frac{1}{M}\right]-\beta\left[\frac{1}{m}+\frac{1}{M}\right]\sqrt{1-\frac{4k^2a^2}{mM}\left(\frac{mM}{m+M}\right)^2} = \beta\left[\frac{1}{m}+\frac{1}{M}\right]\left[1-\sqrt{1-\frac{mM}{(m+M)^2}4k^2a^2}\right]$$

$$= \beta\left[\frac{1}{m}+\frac{1}{M}\right]\left[1-1+\frac{mM}{(m+M)^2}2k^2a^2+\cdots\right] \text{ (using Binomial theorem)}$$

$$\simeq \frac{2\beta}{m+M}k^2a^2.$$

$$\therefore \omega_- = ka\sqrt{\frac{2\beta}{m+M}}. \tag{6.29}$$

If $k \to \pi/2a$, the solution becomes same as eqn (6.28) for

$$\omega_-^2 = \beta\left[\frac{1}{m}+\frac{1}{M}\right]-\beta\sqrt{\left(\frac{1}{m}+\frac{1}{M}\right)^2-\frac{4}{mM}} = \beta\left[\frac{1}{m}+\frac{1}{M}\right]-\beta\left[\frac{1}{m}-\frac{1}{M}\right] = 2\beta/M$$

$$\text{or, } \omega_- = \sqrt{\frac{2\beta}{M}}. \tag{6.30}$$

The graph of dispersion relation is obtained by plotting ω_+ and ω_- against k as shown in Fig. 6.6.

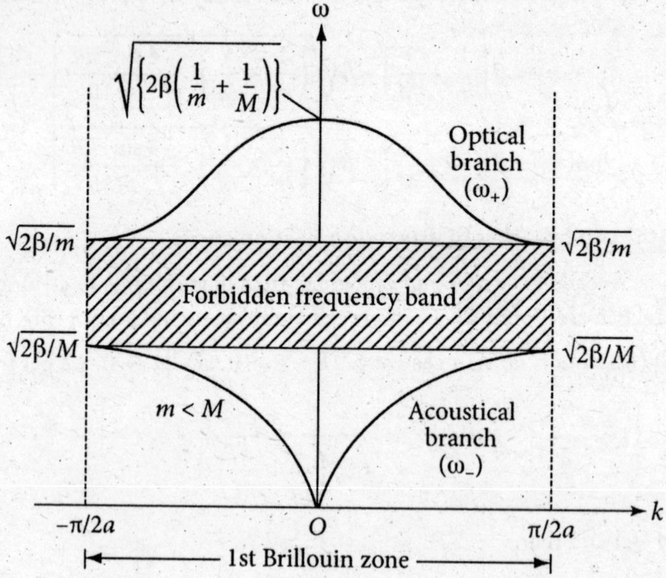

Fig. 6.6: Acoustical and optical branches due to a linear diatomic lattice

Forbidden Band: Fig. 6.6 shows that wave-like solutions do not exist between the frequency range $\sqrt{2\beta/m}$ and $\sqrt{2\beta/M}$. This means that vibrations at frequencies lying between these frequencies are not possible to excite in the lattice. This band of frequency is known as *forbidden frequency band.* The width of the forbidden band depends on the mass ratio M/m. The larger the value M/m, greater is the forbidden bandwidth. For $M/m = 1$, i.e., $M = m$, the *optical* and *acoustical branches* coincide at $k = \pm\pi/2a$, then the forbidden band vanishes.

The *first Brillouin Zone* is defined by the values of k lying in the range $-\frac{\pi}{2a} < k < \frac{\pi}{2a}$.

The smallest possible wavelength of this zone is $4a$ corresponding to $k = \frac{\pi}{2a}$.

Difference between two branches: For the optical branch, as $k \to 0$, i.e., $\cos ka \to 1$, we get from eqns (6.23) and (6.24)

$$-\omega^2 mA = 2\beta B - 2\beta A \text{ or, } \frac{\omega^2}{2\beta} = -\frac{A-B}{MB}$$

$$\text{and } -\omega^2 MB = 2\beta A - 2\beta B \text{ or, } \frac{\omega^2}{2\beta} = -\frac{A-B}{mB}.$$

$$\therefore 1 = -\frac{MB}{mA} \text{ or, } A/B = -M/m. \tag{6.31}$$

Eqn (6.31) shows that the two atoms move in opposite directions, their amplitudes being inversely proportional to their masses so that the centre of mass of the unit cell does not change. Fig. 6.7 (a) shows such a mode of vibration.

Fig. 6.7 (a): Optical mode   **Fig. 6.7 (b):** Acoustical mode

For $m = M$, $A/B = -1$, the atoms thus move in opposite directions even if their masses are equal.

For the acoustical branch, $k \to 0$.

$$\therefore \cos ka \to \left(1 - \frac{k^2a^2}{2}\right) \left[\because \cos\theta = 1 - \frac{\theta^2}{2} + \cdots\right].$$

$\therefore$ from eqn (6.23),

$$(2\beta - m\omega^2)A - 2\beta B\left[1 - \frac{k^2a^2}{2}\right] = 0 \text{ or, } -m\omega^2 A = 2\beta B\left[1 - \frac{k^2a^2}{2}\right] - 2\beta A. \tag{6.32}$$

Similarly, from eqn (6.24),

$$-2\beta A\left[1 - \frac{k^2a^2}{2}\right] + (2\beta - \omega^2 M)B = 0 \text{ or, } -M\omega^2 B = 2\beta A\left[1 - \frac{k^2a^2}{2}\right] - 2\beta B. \tag{6.33}$$

Adding eqns (6.32) and (6.33), we get

$$-\omega^2(mA + MB) = -k^2a^2\beta(A+B).$$

But

$$\omega = ka\sqrt{\frac{2\beta}{m+M}} \text{ or, } \omega^2 = k^2a^2\frac{2\beta}{m+M}.$$

$$\therefore -k^2a^2.\frac{2\beta}{m+M}(mA+MB) = -k^2a^2\beta(A+B) \text{ or, } 2(mA+MB) = (m+M)(A+B)$$

$$\text{or, } mA + MB = MA + mB \text{ or, } (m-M)A = (m-M)B.$$

$$\therefore A/B = +1. \tag{6.34}$$

This shows that, *the two atoms of different masses move in the same direction with same amplitude* [Fig. 6.7 (b)]. Their centre of mass also undergoes movement.

Thus, we conclude that in *acoustical case* the neighbouring atoms move in opposite directions whereas in *optical case,* they move in the same direction. This feature of acoustical and optical branches is true for all other values of k.

In *acoustical branch,* the vibrations can be excited by a force causing the movement of the atoms in the lattice in the same direction. By directing a beam of sound waves on the crystal surface, such effect may be observed. That is why it is called *acoustical branch.* Monatomic crystals show such vibrations.

Again, in *optical branch* the neighbouring atoms can be made to move in opposite directions by the application of a force. The electric intensity of light waves can produce this effect and as such these vibrations are called *optical vibrations.* The hcp as well as diamond lattices contain two atoms per primitive cell and show optical branches.

In summary,

(i) The allowed frequency in diatomic molecules constitutes two frequency bands, viz., the *acoustical mode* and *optical mode of vibration.*

(ii) The frequency range in both the *monatomic* and *diatomic lattice* is the same if $m = M$. There is no *forbidden band.*

(iii) At the Brillouin zone edge ($k = \pi/2a$) the lighter atom of mass m vibrates in the *optical branch* and the heavier atom of mass M vibrates in the *acoustical branch.*

(iv) With the increase of heavier mass M, the optical branch flattens and the acoustical branch yields downwards. For $M \to \infty$, the acoustical branch coincides with the k-axis and the optical branch becomes parallel to k-axis [Fig. 6.8].

(v) With the decrease of the lighter mass m, the optical branch moves upwards and acoustical branch remains unaffected. For $m \to 0$, the optical branch disappears.

(vi) At $k = \pi/2a$, the heavy atom is stationary and the natural frequency of vibration of the lighter atom is $\sqrt{2\beta/m}$.

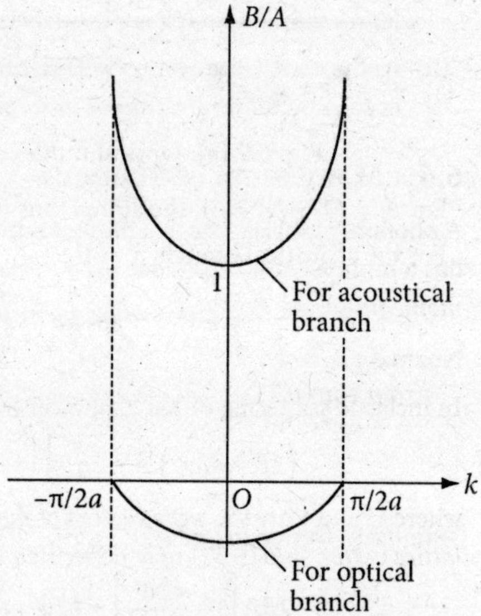

Fig. 6.8: The plot of B/A against k

6.4 Phonons

Phonon is the quantum of energy of *lattice vibration* just as *photon* is the quantum of *electromagnetic energy.* Most of the concepts which apply to photons are also valid for phonons. Depending on the mode of excitation, phonons are of different kinds, e.g., *thermal phonons, acoustical phonons and optical phonons.*

Phonons obey *wave-particle duality concept* of *quantum theory.* The energy of a phonon is given by $\varepsilon = \hbar\omega$, where ω is the angular frequency of a mode of vibration. The total energy of a particular mode of vibration is

$$E = n\varepsilon = n\hbar\omega, \tag{6.35}$$

where n is the number of phonons in that mode of vibration and can have value zero or any positive integer.

The vibrational spectrum of phonon wave ranges from 10^4 to 10^{12} Hz. The low frequency part lies in the *acoustic region* and the high frequency part lies in the *infrared region.*

6.5 Statistics Obeyed by Phonons

Phonons are indistinguishable and have zero spin like photons and as such obey *Bose-Einstein statistics* and *Bose-Einstein distribution law*:

$$n = \frac{1}{e^{\hbar\nu/kT} - 1}, \tag{6.36}$$

where k is the *Boltzmann's constant* and T is the absolute temperature.

Due to particle nature, a phonon interacts with another phonon or electron and the process may be considered as a scattering collision between the two particles.

Evidence of Existence of Phonons

The following experimental facts suggest the existence of phonons:

(i) The lattice heat capacity tends to zero as the temperature approaches zero. This can be explained if the lattice vibrational energy is quantized in terms of phonons.

(ii) In the elastic scattering of X-ray by photon or neutron in crystal, energy and momentum change according to the *creation* or *absorption* of one or more phonons.

6.6 Momentum of Phonons

A phonon, physically, has no momentum. But it interacts with photons, electrons, neutrons and fields as if it has a momentum $\hbar\vec{K}$, whose $|\vec{K}| = \frac{2\pi}{\lambda}$ is the wave vector related to the phonon wave. $\hbar\vec{K}$ is called *crystal momentum.*

Normal process

In inelastic scattering of X-ray photon (i.e., Bragg diffraction) the wave vector conservation law is given by

$$\vec{k}' = \vec{k} + \vec{G}, \tag{6.37}$$

where $\vec{k}'$ and $\vec{k}$ are the wave vectors of the incident and scattered photons respectively and $\vec{G}$ is the *reciprocal lattice vector.* Eqn (6.37) may be written as

$$\hbar\vec{k}' = \hbar\vec{k} + \hbar\vec{G}$$

$$\text{and } \hbar\omega_{ph} = \hbar\omega'_{ph}. \tag{6.38}$$

So, the frequencies of the incident and scattered photons are equal, i.e., $\omega_{ph} = \omega'_{ph}$.

Here the crystal as a whole recoils with momentum $-\hbar G$ and the frequency of the incident photon remains unchanged, i.e., $|\vec{k}'| = |\vec{k}|$, which holds good for X-ray diffraction. *Such a process in which the frequency of the incident photon (ω_{ph}) is the same as the scattered photon (ω'_{ph}) is called normal or N-process.*

Umklapp process (or $\dot{U}$-process)

When the scattering of photon is inelastic with the *creation* of a phonon of wave vector $\vec{K}$ [Fig. 6.9 (a)], the wave vector conservation law takes the form:

$$\vec{k}' + \vec{K} = \vec{k} + \vec{G}, \tag{6.39}$$

where $\vec{k}$ and $\vec{k}'$ are the wave vectors corresponding to the *incident and scattered photon respectively.*

$$\hbar\vec{k}' + \hbar\vec{K} - \hbar\vec{k} + \hbar\vec{G} \tag{6.10}$$

$$\text{and } \hbar\omega'_{ph} + \hbar\omega = \hbar\omega_{ph}, \tag{6.41}$$

ω is the frequency of the generated phonon.

Eqns (6.40) and (6.41) give the *conservation law of momentum* and *energy* respectively.

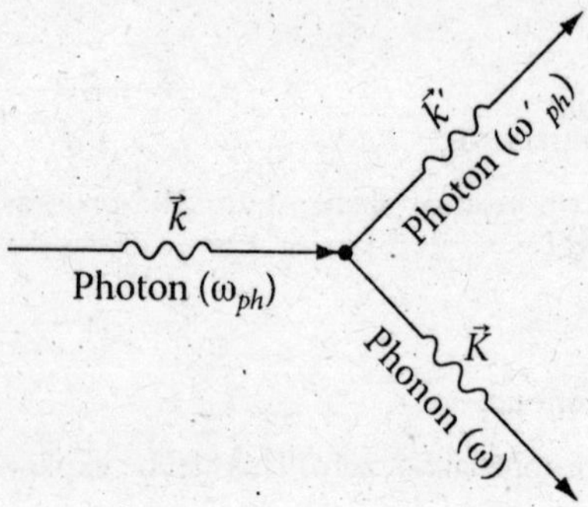

Fig. 6.9: (a) Inelastic scattering of a photon of wave vector $\vec{k}$ to produce scattered photon of wave vector $\vec{k}'$ along with emission of a phonon of wave vector ($\vec{K}$)

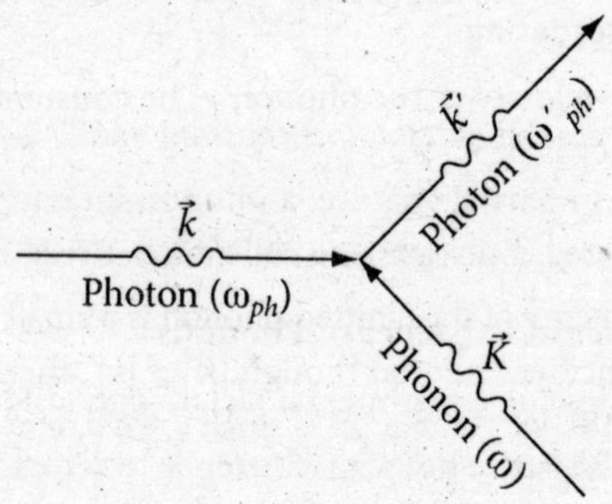

Fig. 6.9: (b) Inelastic scattering of a photon to produce a photon of wave vector ($\vec{k}'$) with the absorption of a phonon of wave vector ($\vec{K}$)

If a phonon is *absorbed* in the scattering process [Fig. 6.9 (b)] the wave vector conservation law gives

$$\vec{k}' = \vec{k} + \vec{K} + \vec{G} \tag{6.42}$$

and the *momentum and energy conservation law* becomes

$$\hbar\vec{k}' = \hbar\vec{k} + \hbar\vec{K} + \hbar\vec{G} \tag{6.43}$$

$$\hbar\omega'_{ph} = \hbar\omega_{ph} + \hbar\omega. \tag{6.44}$$

From eqns (6.42) and (6.44) it is clear that, $\omega'_{ph} \neq \omega_{ph}$.

Such a process in which the frequency of the photon undergoes a change is called an Umklapp process. In this U-process a photon is created or absorbed along with *Bragg reflection.* The momentum in this process is transferred to the crystal.

N.B. *Umklapp* is a German word which means *flipping over.*

6.7 Inelastic Scattering of Photons by Phonons

Let a photon of frequency $\nu = \frac{\omega_{ph}}{2\pi}$ be propagated in a crystal which is supposed to be a continuum of refractive index μ. If c be the velocity of light in vacuum, then the velocity of photon in the medium is c/μ. The wave vector of the photon is

$$k = \frac{2\pi}{\lambda} = \frac{2\pi}{c/\mu} \cdot \frac{\omega}{2\pi} = \frac{\omega\mu}{c}. \tag{6.45}$$

If this photon is scattered *inelastically* by a phonon (sound wave), a phonon may either be created or absorbed. If the wave vector of the photon changes from $\vec{k}$ to $\vec{k}'$, frequency from ω_{ph} and ω'_{ph} and a phonon of wave vector $\vec{K}$ and frequency ω be created [Fig. 6.10], then from the conservation of energy and momentum,

$$\hbar\omega_{ph} = \hbar\omega'_{ph} + \hbar\omega \tag{6.46}$$

$$\hbar\vec{k} = \hbar\vec{k}' + \hbar\vec{K}. \tag{6.47}$$

Eqn (6.47) holds in absence of *Bragg's diffraction* along with that of scattering.

If the velocity of the phonon v_s be constant, then its angular frequency is

$$\omega = v_s k. \tag{6.48}$$

The energy of the emitted phonon is a small fraction of energy of the incident photon though $|\vec{K}| \simeq |\vec{k}|$, then from eqns (4.45) and (4.48), we get $\omega_{ph} \gg \omega$ since $v_s \ll c$ or, $\frac{c}{\mu}$.

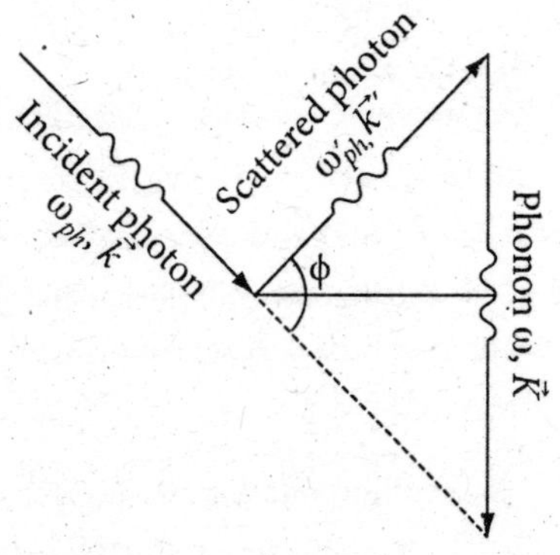

Fig. 6.10: Selection rule diagram for the emission of a phonon when $\vec{k} = \vec{k}'$

Now from eqns (6.46) and (6.47), we find

$$\omega_{ph} \simeq \omega'_{ph} \; [\because \omega_{ph} \gg \omega]$$

$$\text{and } \vec{k} \simeq \vec{k}'.$$

For $\vec{k} = \vec{k}'$, the triangle of Fig. 6.10 becomes isosceles.

$$\therefore K = 2k \sin\frac{\phi}{2}. \tag{6.49}$$

Hence, when a photon is scattered *inelastically* at an angle ϕ from the direction of incidence, a phonon is created whose frequency is

$$\omega = v_s K = 2v_s k \sin\frac{\phi}{2} = 2\frac{v_s \omega_{ph} \mu}{c} \sin\frac{\phi}{2}. \tag{6.50}$$

When photons of *visible light* from a laser source are scattered inelastically in quartz and sapphire, phonons are created in the *microwave frequency range*. The experimentally observed change in frequency agrees well with the shift calculated from eqn (6.50).

6.8 Specific Heat

The specific heat of a substance is defined as the amount of heat required to raise the temperature of one gram-molecule of the substance through one degree centigrade.

If the temperature of a gram-mole of the substance changes from T to $T + dT$ due to the application of dQ amount of heat, then the *molar specific heat* is given by

$$C = \frac{dQ}{dT}. \tag{6.51}$$

The amount of heat required to raise the temperature of one gram-mole of a substance through one degree centigrade depends on the condition of heat supply and as such there are two specific heats, viz., the *specific heat at constant pressure* (C_p) and the *specific heat at constant volume* (C_v).

Now molar specific heat at *constant pressure* is represented as

$$C_p = \left(\frac{dQ}{dT}\right)_p \tag{6.52}$$

and that at *constant volume,*

$$C_v = \left(\frac{dQ}{dT}\right)_v, \tag{6.53}$$

where $C_p > C_v$ for gases. For solids, due to small change in volume, $C_p \simeq C_v$, particularly at *low temperature.*

If dU be the increase in internal energy and dV the increase in volume of a system due to the addition of dQ amount of heat at constant pressure p, then according to the *first law of thermodynamics,*

$$dQ = dU + p\,dV. \tag{6.54}$$

For solids almost all the amount of heat supplied is utilised to increase the internal energy. This increase in internal energy occurs in the following *two* ways:

(i) *The atoms assumed to be free, vibrate vigorously about their equilibrium position. This is manifested by temperature rise.*

(ii) *In metals and semiconductors, the free electrons get thermally excited to the higher energy states.*

If U_{lattice} and U_{electron} denote respectively the internal energy due to *atomic vibrations* and *thermal excitation* of electrons, then

$$U = U_{\text{lattice}} + U_{\text{electron}}. \tag{6.55}$$

Then

$$C_v = \left(\frac{dQ}{dT}\right)_v = \left(\frac{dU}{dT}\right)_v = \frac{d}{dT}(U_{\text{lattice}}) + \frac{d}{dT}(U_{\text{electron}}) = C_{\text{lattice}} + C_{\text{electron}}. \tag{6.56}$$

If there is no free electrons within the lattice, then

$$C_{\text{electron}} = 0 \text{ and } C_v = C_{\text{lattice}}, \tag{6.57}$$

where C_{lattice} is the specific heat of the solid due to atomic vibrations in the solid.

6.9 Classical Theory of Lattice Heat Capacity

The term *heat capacity* is used for specific heat and is defined as *the heat required to raise the temperature of the complete mass of a solid through one degree centigrade.* Thus, the heat capacity (C_H) of a body of mass M is

$$C_H = M\,C_V. \tag{6.58}$$

Classical theory assumes that, each atom of the solid behaves as a three-dimensional oscillator vibrating about their mean position of rest. Corresponding to Cartesian component of vibration there are two degrees of freedom, viz., *translational* and *rotational.* Again, according to the *theorem of equipartition of energy,* the mean KE associated with each degree of freedom is $\frac{1}{2}kT$, where k is the *Boltzmann constant* and T is the absolute temperature. Hence, the energy associated with each component of vibration is

$$\tfrac{1}{2}kT + \tfrac{1}{2}kT = kT. \tag{6.59}$$

So, the mean energy associated with the oscillator is

$$\langle \varepsilon \rangle = 3kT. \tag{6.60}$$

The total energy per gm-mole of the solid is.

$$E = N\langle E\rangle = 3NkT = 3RT, \tag{6.61}$$

where N is the *Avogadro number* and $R = Nk$ is the *universal gas constant.* The *molar specific heat of the solid* is, therefore,

$$C_v = \frac{dE}{dT} = 3R \sim 6 \text{ cal/}^\circ\text{C/mole}. \tag{6.62}$$

This is *Dulong-Petit's law* for specific heat of solid which holds good at higher temperature only.

Experiments on Ag, Ge and Si show that C_v decreases with the decrease of temperature and tends to zero as $T \to 0$, as shown in Fig. 6.11. The relation of C_v with T for various substances in the temperature range 0 – 20 K is as below:

For *non-metals,* $C_v = \alpha T^3$. For *metals,* $C_v = \alpha T^3 + \gamma T$, where γ is the *volume expansion coefficient* and α is a constant.

For *layered compounds,* $C_v = \alpha T^3$ and for linear chained compound $C_v = \alpha T$.

The above discrepancies between *Dulong* and *Petit's law* and the experimental results on specific heat of solids were removed by *Einstein* and then *Debye* using *quantum theory.*

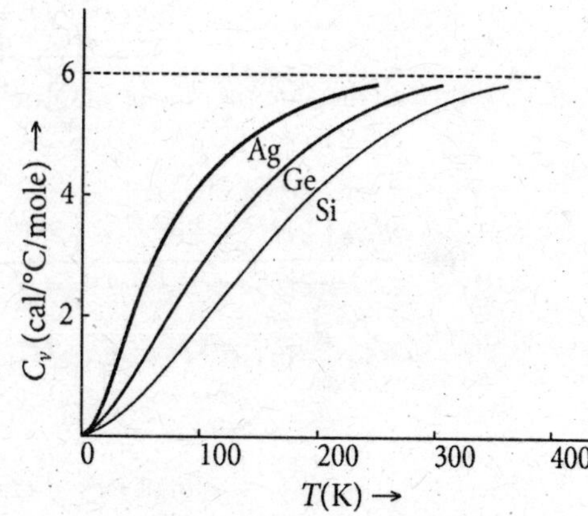

Fig. 6.11: Variation of C_v with T

6.10 Einstein's Theory of Lattice Heat Capacity

Einstein replaced the idea of classical harmonic oscillator by quantum harmonic oscillator which can have only *discrete energy values* and retains all assumptions of classical theory. The main assumptions of *Einstein's theory* were as follows:

(i) *The atoms of a crystal are regarded as identical and independent harmonic oscillators.*

(ii) *A solid having N atoms is equivalent to having 3N one-dimensional harmonic oscillators.*

(iii) *Due to identical environment each oscillator vibrates with the same natural frequency ω.*

(iv) *As the oscillators are quantum mechanical, they can have discrete energy spectrum.*

(v) *Any quantum state can have any number of oscillators.*

(vi) *The oscillators are distinguishable and obey Maxwell-Boltzmann distribution law of energy.*

In view of *Planck's quantum theory,* the discrete energy values of an oscillator having frequency ν are

$$\varepsilon_n = nh\nu = n\hbar\omega, \tag{6.63}$$

where $n = 0, 1, 2, 3, \ldots$ is the quantum number. Using *wave mechanical consideration,* the energy of a harmonic oscillator is

$$\varepsilon_n = \left(n + \tfrac{1}{2}\right)\hbar\nu_0 = \left(n + \tfrac{1}{2}\right)\hbar\omega_0. \tag{6.64}$$

Here $\frac{1}{2}\hbar\omega_0$ is the *zero point energy* contribution to the internal energy of the oscillator.

Now the average energy of an oscillator is given by

$$\langle \varepsilon \rangle = \frac{\sum_{n=0}^{\infty} \varepsilon_n e^{-\varepsilon_n/kT}}{\sum_{n=0}^{\infty} e^{-\varepsilon_n/kT}} \tag{6.65}$$

$$= \frac{\sum_{n=0}^{\infty} \left(n + \frac{1}{2}\right) \hbar\omega_0 \exp\left[-\left(n + \frac{1}{2}\right) \frac{\hbar\omega_0}{kT}\right]}{\sum_{n=0}^{\infty} \exp\left[-\left(n + \frac{1}{2}\right) \frac{\hbar\omega_0}{kT}\right]}$$

$$= \frac{\hbar\omega_0 \sum_{n=0}^{\infty} \left(n + \frac{1}{2}\right) e^{\left(n+\frac{1}{2}\right)x}}{\sum_{n=0}^{\infty} e^{\left(n+\frac{1}{2}\right)x}}, \tag{6.66}$$

where $x = -\frac{\hbar\omega_0}{kT}$.

$$\therefore \langle \varepsilon \rangle = \frac{\hbar\omega_0 \left(\frac{1}{2} e^{x/2} + \frac{3}{2} e^{3x/2} + \frac{5}{2} e^{5x/2} + \cdots\right)}{\left(e^{x/2} + e^{3x/2} + e^{5x/2} + \cdots\right)} = \hbar\omega_0 \frac{d}{dx} \ln\left[e^{x/2} + e^{3x/2} + e^{5x/2} + \cdots\right]$$

$$= \hbar\omega_0 \frac{d}{dx} \ln\left[e^{x/2}\left(1 + e^{x} + e^{2x} + \cdots\right)\right].$$

But

$$\ln\left(1 + e^{x} + e^{2x} + \cdots\right) = \ln\left(1 - e^{-x}\right).$$

$$\therefore \langle \varepsilon \rangle = \hbar\omega_0 \frac{d}{dx}\left[\frac{x}{2} - \ln\left(1 - e^{-x}\right)\right] = \hbar\omega_0 \left[\frac{1}{2} + \frac{e^{x}}{1 - e^{x}}\right] = \hbar\omega_0 \left[\frac{1}{2} + \frac{1}{e^{-x} - 1}\right]$$

$$= \hbar\omega_0 \left[\frac{1}{2} + \frac{1}{e^{\hbar\omega_0/kT} - 1}\right] = \frac{1}{2}\hbar\omega_0 + \frac{\hbar\omega_0}{e^{\hbar\omega_0/kT} - 1}. \tag{6.67}$$

Again, if we use $\varepsilon_n = n\hbar\omega_0$ in eqn (6.65), we get

$$\langle \varepsilon \rangle = \frac{\hbar\omega_0}{e^{\hbar\omega_0/kT} - 1}. \tag{6.68}$$

From eqn (6.67) it is clear that for $T \to 0$, $\langle \varepsilon \rangle = \frac{1}{2}\hbar\omega_0$. This means that atoms are having vibrational energy $\frac{1}{2}\hbar\omega_0$ even at $T = 0$. Since $\frac{1}{2}\hbar\omega_0$ is temperature independent, it has no contribution to specific heat of solids.

Thus, the internal energy is

$$E = 3N\langle \varepsilon \rangle = \frac{3}{2} N\hbar\omega_0 + \frac{3N\hbar\omega_0}{e^{\hbar\omega_0/kT} - 1}. \tag{6.69}$$

∴ *molar specific heat* of the solid is

$$C_v = \left(\frac{dE}{dT}\right)_V = 3Nk\left(\frac{\hbar\omega_0}{kT}\right)^2 \frac{e^{\hbar\omega_0/kT}}{\left(e^{\hbar\omega_0/kT} - 1\right)^2}. \tag{6.70}$$

Now the *characteristic temperature* $\left(\theta_c = \frac{\hbar\omega_0}{k}\right)$, also known as the *Einstein's temperature* (θ_E) is given by

$$\theta_E = \frac{\hbar\omega_0}{k}. \tag{6.71}$$

Eqn (6.70) then becomes

$$C_v = 3Nk\left(\frac{\theta_E}{T}\right)^2 \frac{e^{\theta_E/T}}{\left(e^{\theta_E/T}-1\right)^2}. \tag{6.72}$$

From eqn (6.72) it is evident that C_v is a function of temperature (T).

Special cases

1. **When the temperature T is large**: For $T \to$ large value,

$$kT \gg \hbar\omega_0 \text{ or, } T \gg \frac{\hbar\omega_0}{k}, \text{ or, } T \gg \theta_E.$$

We get

$$e^{\hbar\omega_0/kT} - 1 = 1 + \frac{\hbar\omega_0}{kT} + \cdots - 1 \simeq \frac{\hbar\omega_0}{kT}. \tag{6.73}$$

Eqn (6.67) then reduces to

$$\langle\varepsilon\rangle = \tfrac{1}{2}\hbar\omega_0 + kT \simeq kT.$$

Thus, *at high temperature the average energy $\langle\varepsilon\rangle$ is the same as that obtained from classical theory.*

Now from eqn (6.70),

$$C_v = \left(\frac{dE}{dT}\right)_V = 3Nk\left(\frac{\hbar\omega_0}{kT}\right)^2 \frac{\left(1+\frac{\hbar\omega_0}{kT}\right)}{\left(\frac{\hbar\omega_0}{kT}\right)^2} \quad \left[\because e^{\hbar\omega_0/kT} - 1 \simeq \frac{\hbar\omega_0}{kT}\right]$$

$$= 3Nk\left(1 + \frac{\hbar\omega_0}{kT}\right). \tag{6.74}$$

As $T \to \infty$, we get

$$C_v \to 3Nk = 3R.$$

This is *Dulong and Petit's law* as obtained from *classical theory.*

2. **When the temperature T is low**: When $T \to 0$,

$$kT \ll \hbar\omega_0 \text{ or, } T \ll \frac{\hbar\omega_0}{R} \text{ or, } T \ll \theta_E.$$

$$\therefore e^{\hbar\omega_0/kT} - 1 \simeq e^{\hbar\omega_0/kT}.$$

From eqn (6.66),

$$\langle\varepsilon\rangle = \tfrac{1}{2}\hbar\omega_0 + \hbar\omega_0 e^{-\hbar\omega_0/kT}. \tag{6.75}$$

This shows that *the average energy decreases exponentially with temperature in the low temperature region.*

$$\therefore C_v = \left(\frac{dE}{dT}\right)_V = 3Nk\left(\frac{\hbar\omega_0}{kT}\right)^2 e^{-\hbar\omega_0/kT} \tag{6.76}$$

$$\text{or, } C_v = 3Nk\left(\frac{\theta_E}{T}\right)^2 e^{-\theta_E/T} \tag{6.77}$$

Hence, for $T \ll \theta_E$,

$$C_v \propto e^{-\theta_E/T}.$$

This shows *rapid decrease* in C_v with the decrease of T. For most of the solids C_v is found to vary as T^3.
Fig. 6.12 shows a theoretical plot of C_v against T/θ_E using eqn (6.77). Though Einstein's theory agrees well with experiment over a wide range of temperature, it shows a more rapid decrease of C_v in the two temperature regions (for liquid HeT), $C_v \propto T^3$.

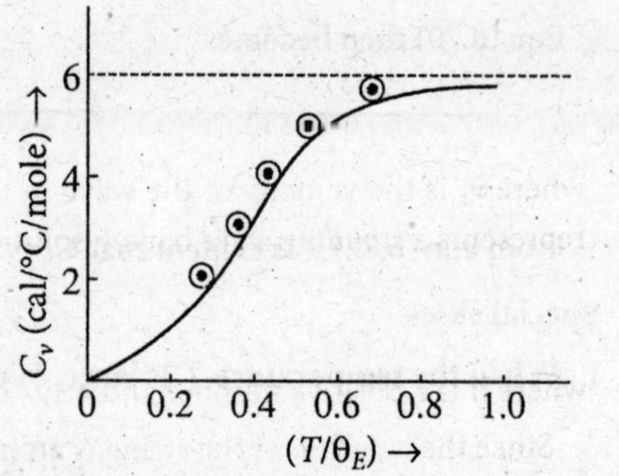

Fig. 6.12: Plot of C_v against T/Θ_E according to Einstein's theory

Drawback of Einstein's theory

Einstein assumed that each atom of the solid vibrates as a single unit independent of its neighbours. The vibrational motion of the crystal as a whole was considered to be the same as the vibrational motion of a single atom. For this reason all the atoms of the solid were assumed to vibrate with a common frequency ω, the natural frequency of vibration of single atom. This assumption is over-simplified since the atoms in a crystal are bound together to form a system of *couple oscillators* which cannot vibrate independently. Actually a solid is capable of vibrating elastically in many different modes.

6.11 Debye's Theory of Specific Heat of Solids

Debye, in 1912, modified *Einstein's theory* of specific heat of solids on the basis of the following assumptions:

1. Any solid is capable of vibrating elastically in many different modes, the frequency varying from one mode to another. The number of modes of vibration of the solid are, however, limited to $3N$, where N is the number of atoms in the solid. This highest frequency propagating through the crystal is called the *Debye frequency* $\nu_D = 3N$.
2. The crystal as a whole behaves as a system of *coupled harmonic oscillators* which can propagate a range of frequency rather than a single frequency.
3. The internal energy of a solid resides on the *elastic standing waves* instead of individual atoms. The energy of those elastic standing waves is *quantized* and the quantum of vibrational energy is called a *phonon* having energy $h\nu$, where ν is the frequency and h is the *Planck's constant.* A phonon travels with the speed of sound wave in the solid.
4. If a continuum solid is subjected to elastic vibration, two types of vibrations are produced, viz., *longitudinal vibrations* and *transverse vibrations.* The number of modes of *longitudinal* and *transverse vibrations* per unit volume in the frequency range ν to $\nu + d\nu$ are respectively

$$\frac{4\pi\nu^2 d\nu}{v_l^3} \quad \text{and} \quad \frac{2 \times 4\pi\nu^2 d\nu}{v_t^3},$$

where v_l and v_t are the velocities of the longitudinal and transverse waves respectively. The factor 2 before $\frac{4\pi\nu^2 d\nu}{v_t^3}$ is due to the fact that the transverse vibration has *two* states of polarization at right angles to each other.

6.11.1 Density of Modes

We consider a one-dimensional continuous string of length L, fixed at both ends. If $u(x, t)$ be the displacement of a point of the string at a distance x and time t, then the equation of motion of the string is

$$\frac{\partial^2 u}{\partial x^2} = \frac{1}{v_s^2}\frac{\partial^2 u}{\partial t^2}, \tag{6.78}$$

where v_s is the velocity of the wave in the string. Since the string is fixed at $x = 0$ and $x = L$, eqn (6.78) represents a *standing wave* having solution of the type

$$u(x, t) = A\sin\left(\frac{n\pi x}{L}\right)\cos 2\pi\nu_n t, \tag{6.79}$$

where n is a positive number and $\geqslant 1$.

Since the length L of the string is an integral multiple of half wavelengths

$$L = n\frac{\lambda_n}{2} \tag{6.80}$$

and the frequency

$$\nu_n = \frac{v}{\lambda_n} = \frac{nv}{2L}. \tag{6.81}$$

Thus, the frequency ν has discrete values, i.e.,

$$\nu_1 = \frac{v}{2L},\ \nu_2 = \frac{v}{L},\ \nu_3 = \frac{3v}{2L} = \cdots.$$

This shows that the frequency spectrum of a continuous string is *discrete* containing an infinite number of parallel lines equidistant from each other [Fig. 6.13].

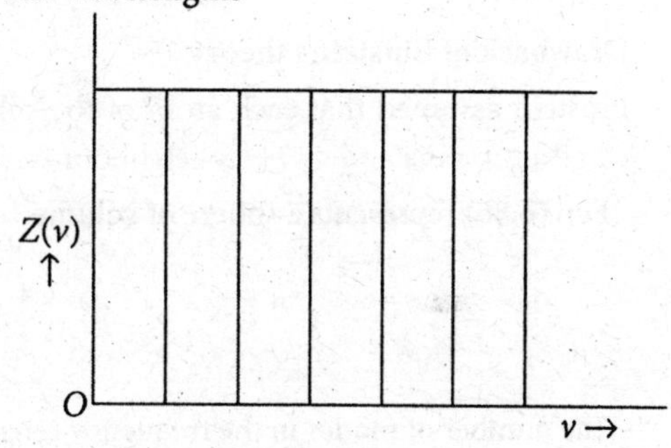

Fig. 6.13: Frequency spectrum of a continuous string

From eqn (6.81),

$$n = \frac{2L}{v_s}\nu_n,\ \therefore dn = \frac{2L}{v_s}d\nu. \tag{6.82}$$

$d\nu$ represents the possible mode of vibration in the frequency interval $d\nu$.

Three-dimensional case

In three dimensions, the wave equation is

$$\frac{\partial^2 u}{\partial x^2} + \frac{\partial^2 u}{\partial y^2} + \frac{\partial^2 u}{\partial z^2} = \frac{1}{v_s^2}\frac{\partial^2 u}{\partial t^2}. \tag{6.83}$$

Assuming the three-dimensional continuous medium as a cube of side L, with fixed faces, the standing wave solution of eqn (6.83) is

$$u(x, y, z, t) = A\sin\left(\frac{n_x\pi x}{L}\right)\sin\left(\frac{n_y\pi y}{L}\right)\sin\left(\frac{n_z\pi z}{L}\right)\cos(2\pi\nu_n t), \tag{6.84}$$

where n_x, n_y, n_z are positive integers ≥ 1.

Differentiating eqn (6.84) twice w.r.t. x, y and z, we get

$$\frac{\partial^2 u}{\partial x^2} = -\left(\frac{n_x\pi}{L}\right)^2 u(x, y, z, t)$$

$$\frac{\partial^2 u}{\partial y^2} = -\left(\frac{n_y\pi}{L}\right)^2 u(x, y, z, t)$$

and $$\frac{\partial^2 u}{\partial z^2} = -\left(\frac{n_z\pi}{L}\right)^2 u(x, y, z, t).$$

Substituting these values in eqn (6.83), we get

$$\left(\frac{n_x\pi}{L}\right)^2+\left(\frac{n_y\pi}{L}\right)^2+\left(\frac{n_z\pi}{L}\right)^2=\frac{1}{v_s^2}4\pi^2\nu^2=\frac{4\pi^2}{\lambda^2}\text{ or, }n_x^2+n_y^2+n_z^2=\frac{4\pi^2}{\lambda^2}\times\frac{L^2}{\pi^2}=\frac{4L^2\nu^2}{v_s^2}. \tag{6.85}$$

Eqn (6.85) gives the possible modes of vibration, where n_x, n_y, n_z determine possible frequencies or wavelengths.

We now consider a network of points in the space defined by three positive integral coordinates n_x, n_y and n_z. The radius vector R of any point from the origin is

$$R^2=n_x^2+n_y^2+n_z^2=\frac{4L^2\nu^2}{v_s^2}. \tag{6.86}$$

Eqn (6.86) represents a sphere of volume

$$V'=\frac{4}{3}\pi R^3,\ \therefore dV'=4\pi R^2dR.$$

The number of modes in the frequency range ν to $\nu+d\nu$ is equal to the number of points in the spherical shell between volumes V' and $V'+dV'$ or in the range R to $R+dR$ is

$$dn=4\pi R^2dR.$$

Since a mode of vibration is determined by the numbers n_x, n_y, n_z which are $\geqslant 1$, we shall consider the points lying in the *first octant* only. Hence, the number of possible mode of vibration $Z(\nu)\,d\nu$ in the frequency interval ν to $\nu+d\nu$ is given by

$$Z(\nu)\,d\nu=\frac{1}{8}\times 4\pi R^2dR=\frac{1}{8}\times 4\pi\frac{4L^2\nu^2}{v_s^2}\frac{2L}{v_s}\,d\nu=\frac{4\pi L^3\nu^2}{v_s^3}\,d\nu=\left(\frac{4\pi V}{v_s^2}\right)\nu^2d\nu, \tag{6.87}$$

where $V=L^3$, the volume of the cube.

But the elastic waves propagating in solid are of two types, viz., *transverse waves* and *longitudinal waves* with velocity of propagation v_l and v_t respectively.

$$\therefore Z(\nu)\,d\nu=4\pi V\left(\frac{2}{v_t^3}+\frac{1}{v_l^3}\right)\nu^2d\nu. \tag{6.88}$$

6.11.2 Debye Approximation

Since the interatomic distance is much smaller than the wavelength of the propagating elastic wave, the solid may be considered as a *continuum*. If the solid contains N (finite) number of atoms, the total number of vibrational mode is limited to $3N$. This is the upper limits of the frequency of the elastic wave which can propagate through the crystal and is called the *Debye frequency* $\nu_D(=3N)$. Hence, the frequency spectrum of a continuous medium is cut off at ν_D [Fig. 6.13].

Thus, *Debye frequency* can be determined by equating the total number of possible frequency to 3*N*.

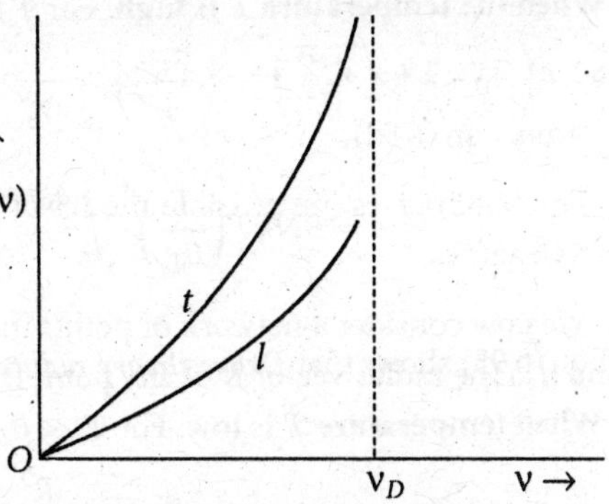

$$\therefore \int_0^{\nu_D} Z(\nu)\,d\nu = 3N \text{ or, } \int_0^{\nu_D} 4\pi V\left(\frac{2}{v_t^3}+\frac{1}{v_l}\right)\nu^2 d\nu = 3N$$

$$\text{or, } 4\pi V\left(\frac{2}{v_t^3}+\frac{1}{v_l}\right)\frac{\nu_D^3}{3} = 3N$$

$$\text{or, } \nu_D^3 = \frac{9N}{4\pi V}\left(\frac{2}{v_t^3}+\frac{1}{v_l^3}\right)^{-1}. \tag{6.89}$$

In view of assumption (iii), the average energy of an oscillator obeying *Bose-Einstein statistics* in the frequency range ν to $\nu + d\nu$ is

$$\langle\varepsilon\rangle = \frac{h\nu}{e^{h\nu/kT}-1}, \tag{6.90}$$

Fig. 6.14: Frequency spectrum of transverse (*t*) and longitudinal (*l*) modes in a continuum showing cut-off at ν_D

where *T* is the temperature.

The vibrational energy of the crystal is

$$E = \int_0^{\nu_D}\langle\varepsilon\rangle Z(\nu)\,d\nu = \int_0^{\nu_D} 4\pi V\left(\frac{2}{v_t^3}+\frac{1}{v_l^3}\right)\frac{h\nu^3}{e^{h\nu/kT}-1} \text{ [using eqns (6.88) and (6.90)]}$$

$$= 4\pi v h\left(\frac{2}{v_t^3}+\frac{1}{v_l}\right)\int_0^{\nu_D}\frac{\nu^3 d\nu}{e^{h\nu/kT}-1}$$

$$\text{or, } E = \frac{9Nh}{\nu_D^3}\int_0^{\nu_D}\frac{\nu^3 d\nu}{e^{h\nu/kT}-1} \text{ [from eqn (6.89)]} \tag{6.91}$$

Now putting

$$\frac{h\nu}{kT} = x \text{ and } \frac{h\nu_D}{kT} = x_m,$$

$$\nu = \frac{xkT}{h} \text{ and } d\nu = \frac{kT}{h}dx.$$

$\therefore$ from eqn (6.91),

$$E = \frac{9Nh}{\nu_D^3}\left(\frac{kT}{h}\right)^4\int_0^{x_m}\frac{x^3 dx}{e^x-1} = 9N\left(\frac{kT}{h\nu_D}\right)^3 kT\int_0^{x_m}\frac{x^3 dx}{e^x-1} = 9NkT\left(\frac{T}{\theta_D}\right)^3\int_0^{\theta_D/T}\frac{x^3 dx}{e^x-1}, \tag{6.92}$$

where $\theta_D = \frac{h\nu_D}{k}$ is called the *Debye temperature.*

Now the specific heat at constant volume is

$$C_v = \left(\frac{dE}{dT}\right)_V = 9Nk\left(\frac{T}{\theta_D}\right)^3\int_0^{\theta_D/T}\frac{e^x x^4}{(e^x-1)^2}dx = 3R\left(\frac{\theta_D}{T}\right)F_D, \text{ [for } N = N_a\text{]}, \tag{6.93}$$

where

$$F_D = 3\left(\frac{T}{\theta_D}\right)^4\int_0^{\theta_D/T}\frac{e^x x^4}{(e^x-1)^2}dx \tag{6.94}$$

is the *Debye function.*

Special cases

1. **When the temperature T is high:** For $T \gg \theta_D, x = \frac{\theta_D}{T}$ is small compared to unity.

So, $e^x - 1 \simeq 1 + x + \frac{x^2}{2} + \cdots - 1 \simeq x$.

$\therefore$ from eqn (6.92),

$$E = 9NkT\left(\frac{T}{\theta_D}\right)^3 \int_0^{\theta_D/T} x^2 dx = 3NkT \text{ and } C_v = \left(\frac{dE}{dT}\right)_V = 3Nk = 3R. \quad (6.95)$$

[for $N = N_a$, the Avogadro number]

Eqn (6.95) shows that *Debye theory reduces to classical Dulong and Petit's law at high temperature.*

2. **When temperature T is low:** For $T \ll \theta_D$,

$$x_m = \frac{\theta_D}{T} \to \infty.$$

Then eqn (6.92) reduces to

$$E = 9NkT\left(\frac{T}{\theta_D}\right)\int_0^{\infty} \frac{x^3 dx}{e^x - 1}$$

But $\int_0^{\infty} \frac{x^3 dx}{e^x - 1} = \frac{\pi^4}{15}$.

$$\therefore E = 9NkT\left(\frac{T}{\theta_D}\right)^3 \frac{\pi^4}{15} = \frac{3}{5}\pi^4 Nk\frac{T^4}{\theta_D^3}. \quad (6.96)$$

Eqn (6.96) shows that *the vibrational energy of the crystal is proportional to T^4 which is analogous to Stefan's law.*

Now

$$C_v = \left(\frac{dE}{dT}\right)_v = \frac{12}{5}\pi^4 Nk\left(\frac{T}{\theta_D}\right)^3 = \frac{12}{5}\pi^4 R\left(\frac{T}{\theta_D}\right)^3$$

[for $N = N_a$, the Avogadro number]

Hence, at low temperature, $C_v \propto T^3$. *This is Debye T^3-law.*

Debye law is able to explain the experimental results with excellent agreement. Fig. 6.15 shows the experimental result for silver at $\theta_D = 225$ K.

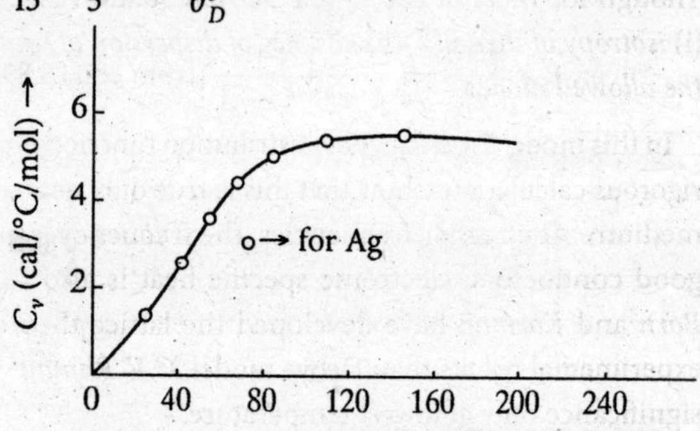

Fig. 6.15: Plot of C_v against T according to Debye theory

6.12 Comparison between Einstein's Equation and Debye Equation

1. According to Einstein all oscillators in the solid oscillate with a single frequency $\nu_0 = \omega_0/2\pi$, which means all the atomic oscillators are excited in unison. But according to Debye, there is a spectrum of frequency from $\nu = 0$ to $\nu = \nu_D$ (the maximum value of ν) as the atoms are not excited to the same extent. That is why in Einstein's theory there is *no upper limit of frequency* but in Debye theory there is an upper limit of frequency, called *cut-off frequency*.

2. At high temperature both Einstein and Debye modes give the same result $C_v = 3R$, which is the classical *Dulong-Petit's law.* But at low temperature Einstein's theory proposes that $C_v \propto e^{-h\nu/kT}$ whereas Debye theory predicts $C_v \propto T^3$, which agrees more closely with experimental results.
3. According to Einstein's theory the atoms in the solid vibrate independently with a constant frequency corresponding to *optical branch.* But according to Debye couple vibration is set up in the solid and the frequency of the wave travelling through it ranges from that of sound wave to infrared part of electronic wave.
4. Fig. 6.16 shows the temperature variation of C_v for Diamond (θ_E = 1326 K) and Al (θ_D = 396 K). The curves of Fig. 6.16 show that Debye theory is much more closer to experimental result than Einstein's theory.

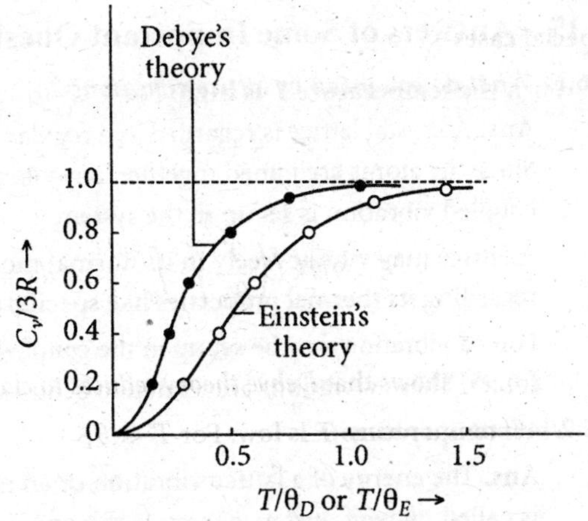

Fig. 6.16: Shows the difference of Einstein's and Debye's results on C_v

6.13 Discussion on Debye Equation

Though for most of the solids Debye equation gives satisfactory result, it is based on the assumptions: (i) *isotropy of the solid,* (ii) *absence of dispersion of sound waves in the crystal medium* and (iii) *degeneracy of the allowed modes.*

In this model the frequency distribution function is proportional to ν^2 with a sharp cut-off at $\nu = \nu_D$. More rigorous calculations show that this is true only near $\nu = 0$, where the solid behaves as an elastic continuum medium. At all other frequencies, the frequency spectrum differs widely from Debye's assumptions. For good conductors, electronic specific heat is also to be considered along with that of lattice vibration. *Born* and *Karman* have developed the lattice theory of atomic heat capacity which agrees better with experimental results than Debye model. *C.V. Raman* on the basis of X-ray reflection has shown that ν_D has significance only at lowest temperature.

6.14 Drawbacks of the Debye Theory

Though Debye theory explains the results of many solids, it suffers from the following limitations:

(i) The theory cannot be applied to crystals consisting more than one type of atoms.

(ii) In Debye's theory the cut-off frequency is assumed to be the same both for longitudinal and transverse vibrations which is not true. Also the total number of vibrational modes is assumed to be $3N$ but a continuous solid should have infinite frequencies.

(iii) The theory assumes that θ_D is independent of temperature but practically it is found to vary up to 10 % with temperature.

(iv) Debye's model is valid for long wavelengths only which indicates the presence of low frequencies only.

(v) The theory does not take into account of the interaction among the atoms and the contribution of *electronic specific heat.*

6.15 Answers of Some Important Questions

6.1. *What do you mean by lattice vibration?*

Ans. A crystal lattice is regarded as a regular arrangement of atoms joined together by elastic spring. Since the atoms are joined together, any vibration of an atom is communicated to its neighbour and coupled vibration is set up in the system.

A lattice may vibrate freely in its normal mode due to its internal energy and provide information regarding its thermal properties like specific heat and thermal conductivity.

Forced vibration may be set up in the coupled system due to external mechanical or electromagnetic forces. These vibrations are related to acoustical and optical properties of the crystal.

6.2. *What is a phonon?*

Ans. The energy of a lattice vibration or an elastic wave is quantized and a quantum of this energy is called *phonon,* just as *photon* is the quantum of electromagnetic energy. A phonon travels in a medium with the velocity of sound (v_s) and its energy is $\varepsilon = h\nu = \hbar\omega$, where $\hbar = h/2\pi$, h is the *Planck's constant* and ω is its angular frequency.

Just like a photon, a phonon has zero spin and obeys *Bose-Einstein statistics* and *Bose-Einstein's distribution law*:

$$n = \frac{1}{e^{h\nu/kT} - 1}.$$

6.3. *What are N-process and U-process?*

Ans. *N-process or Normal process*: An X-ray photon may be *elastically* scattered by a crystal producing a scattered photon.

When an incident photon is elastically scattered by a crystal lattice (Bragg's diffraction), the conservation law of momentum yields $\hbar\vec{k}' = \hbar\vec{k} + \hbar\vec{G}$ and $\hbar\omega_{ph} = \hbar\omega'_{ph}$, where $\vec{k}$ and $\vec{k}'$ are the wave vectors of the incident and scattered photons and $\vec{G}$ is the reciprocal lattice vector. It is evident from the second equation that $\omega_{ph} = \omega'_{ph}$. *Such a process of scattering in which the frequency of the incident photon* (ω_{ph}) *is equal to the frequency of the scattered photon* (ω'_{ph}) *is called the N-process.*

In an *inelastic scattering* of an X-ray photon by a crystal a phonon of wave vector $\vec{K}$ may be *emitted.* The wave vector conservation law is $\vec{k}' + \vec{K} = \vec{k} + \vec{G}$ and the *energy momentum conservation laws* are $\hbar\vec{k}' + \hbar\vec{K} = \hbar\vec{k} + \hbar\vec{G}$ and $\hbar'\omega'_{ph} + \hbar\omega = \hbar\omega_{ph}$, ω is the frequency of the generated phonon.

If a phonon is *absorbed* in the scattering process, the conservation law becomes $\vec{k}' = \vec{k} + \vec{K} + \vec{G}$ and the momentum and energy conservation laws are $\hbar\vec{k}' = \hbar\vec{k} + \hbar\vec{K} + \hbar\vec{G}$ and $\hbar\omega'_{ph} = \hbar\omega_{ph} + \hbar\omega$.

Thus, $\omega'_{ph} \neq \omega_{ph}$.

Such a process in which the frequency of the incident photon is not equal to the frequency of the scattered photon is called the U-process.

6.4. *How phonons can be generated? Does the observed shift in frequency agree with theoretical value?*

Ans. If visible light from an intense laser source is scattered by quartz or saphire crystals, phonons are generated in the microwave frequency range. The observed shift of frequency between the incident and scattered photon agrees well with theoretical values.

6.5. *Name the different branches of the dispersion relation curve. What is the difference between the two branches?*

[LU 1996; PU 1995]

Ans. If we plot ω against k, we get two branches of the curve. One branch for which $\omega = \omega_+$ is called the *optical branch* and the other branch corresponding to $\omega = \omega_-$ is called the *acoustical branch* as shown in Fig. 6.6.

Difference between two branches

(i) For *optical branch* where $k \to 0$, $\cos ka \to 1$, we get $A/B = -\frac{M}{m}$.

This shows that the two atoms of masses M and m move in opposite directions with their amplitudes inversely proportional to their masses.

For $m = M$, $A/B = -1$, i.e., if the two atoms be of equal mass, then also move in opposite directions.

For acoustical branch as $k \to 0$,

$$\cos ka = 1 - \frac{k^2 a^2}{2} + \cdots.$$

In such case, we get $A/B = +1$.

This shows that in acoustical branch the two atoms of different masses move in the same direction with equal amplitude.

Hence, *if the centre of mass of two atoms does not move, then in optical branch the two atoms move in opposite directions and in the acoustical branch they move in the same direction.*

(ii) The vibration of the optical branch can be excited by a force due to electric field of light waves from optical radiations whereas acoustical vibration can be excited by a force which makes the atoms in a crystal move in the same direction such as sound wave.

(iii) Acoustical vibration is exhibited by monatomic crystals whereas diatomic crystals exhibit optical vibrations.

6.6. *What is forbidden bond in lattice vibration?*

Ans. Fig. 6.6 shows that wave-like solution does not exist in the frequency range $\sqrt{2\beta/m}$ and $\sqrt{2\beta/M}$ and as such there is a band of frequencies between two branches which cannot propagate. This band of frequency is called *forbidden frequency band.*

The width of the forbidden frequency band depends on M/m in the sense longer the value of M/m, wider is the band. The optical and acoustical frequencies coincide for $k = \pm\frac{\pi}{2a}$, when the forbidden band vanishes.

The forbidden zone is defined by k-values as $-\pi/2a < k < \pi/2a$.

For $k = \pi/2a$, the wavelength is the smallest at the zone boundary of the crystal.

6.7. *What do you understand by lattice heat capacity?*

Ans. The internal energy in solids increases in the following two ways:

(i) The atoms assumed to be free vibrators vigorously vibrate about their equilibrium position. The energy of this vibration converts into heat and causes rise of temperature.

(ii) The free electrons in metals and semiconductors are thermally excited to higher energy states causing increase in internal energy.

If the contribution towards increase in internal energy (U) due to lattice vibration and thermal excitation of electrons be represented by U_{lattice} and U_{electron} respectively, then

$$U = U_{\text{lattice}} + U_{\text{electron}} \tag{6.97}$$

$$\text{and } C_v = \left(\frac{dU}{dT}\right)_V = \frac{d}{dt}(U_{\text{lattice}}) + \frac{d}{dt}(U_{\text{electron}}) = C_{\text{lattice}} + C_{\text{electron}}. \tag{6.98}$$

But $C_{\text{electron}} \ll C_{\text{lattice}}$.

$\therefore C_v \simeq C_{\text{lattice}}$, where C_{lattice} is the lattice specific heat of the solid.

6.8. *Give a comparative study of Einstein's and Debye's theory of specific heat of solid.*

Ans. The following table gives a comparative study of Einstein's and Debye's theory of specific heat of solid.

Table 6.1

Einstein's Theory	*Debye's Theory*
1. Einstein assumed that the atoms in the solid are independent simple harmonic oscillators.	1. Debye took into account the coupling between the adjacent atoms and the continuous elastic nature of the solid.
2. All the atoms vibrate with a single frequency which is the natural frequency of the system.	2. The frequency of the propagating waves ranges from some low value of $3N$, where N is the number.
3. The atomic oscillators are distinguishable and obey Maxwell-Boltzmann law of distribution of energy.	3. The atomic oscillators are distinguishable and obey Maxwell-Boltzmann law of energy distribution.
4. The vibrational energy of the solid is quantized and a single quantized unit is called a *phonon.*	4. The vibrational energy of the solid is quantized in terms of phonon.
5. According to Einstein's theory at high temperature, $C_v = 3R$.	5. According to Debye's theory at high temperature, $C_v = 3R$.
6. At low temperature Einstein's theory predicts that $C_v \propto e^{-h\nu/kt}$.	6. At low temperature Debye's theory predicts that $C_v \propto T^3$.
7. This theory gives a lower value of C_v than Debye's theory.	7. This theory gives higher value of C_v than Einstein's theory.

6.9. *Write down the Einstein's and Debye's expressions for the specific heat of solids.*

Ans. According to Einstein the atoms in a crystal are identical and independent three-dimensional quantum mechanical oscillators, all vibrating with a common frequency ω. The expression for molar specific heat of solid obtained on the basis of above conception is

$$C_v = 3R\left(\frac{\theta_E}{T}\right)^2 \frac{e^{\theta_E/T}}{\left(e^{\theta_E/T}-1\right)^2},$$

where $\theta_E = \frac{\hbar\omega}{T}$ is the Einstein's temperature. Einstein's theory explains the fact that $C_v \to 0$ as $T \to 0$, but fails to account the temperature variation of C_v with satisfactions.

Debye assumed that the atoms of a crystal are coupled harmonic oscillator and obtained an expression of molar specific heat as below:

$$C_v = 9R\left(\frac{T}{\theta_D}\right)^3 \int_0^{\theta_D/T} \frac{e^x x^4}{(e^x-1)^2}\,dx,$$

where

$$x = \frac{h\nu}{kT} \text{ and } \theta_D = \frac{h\nu_D}{k}.$$

θ_D is the *Debye temperature* and ν the frequency of an oscillator. At low temperature this model shows that $C_v \propto T^3$. This is in good agreement with experimental observations.

6.16 Solved Problems

6.1. (a) *Calculate the mean velocity of sound in Diamond from the following data: Debye temperature of diamond* = 2000 K, density of diamond = 3500 kg-m^3, atomic mass of diamond = 12 amu.

(b) *If the interatomic spacing be* 1.54 Å, what is the frequency of the dominant mode of lattice vibration?

Solution: If v_s, ρ, M and N_a be the velocity of sound wave, density, atomic mass and the Avogadro number respectively, then the Debye temperature for Diamond is given by

$$\theta_D = \frac{hv_s}{k}\left(\frac{3N_a\rho}{4\pi M}\right)^{1/3}.$$

$$\therefore v_s = \frac{k\theta_D}{h}\left(\frac{4\pi M}{3N_a\rho}\right)^{1/3} = \frac{(1.38\times10^{-23})\times 2000}{6.626\times10^{-34}}\left[\frac{4\pi\times12}{3\times6.023\times10^{26}\times3500}\right]^{1/3} = \mathbf{1.2\times10^4\ m/s}.$$

The order of magnitude of dominant phonon wavelength at $T(T \ll \theta_D)$ is given by

$$\lambda \simeq \left(\frac{\theta_D}{T}\right)a,$$

where a is the interatomic spacing.

$$\therefore \lambda \simeq \left(\frac{2000}{298}\right)\times1.54\times10^{-10} = 1.03\times10^{-9}\text{ m}.$$

The required frequency of vibration of the Diamond lattice is

$$\nu = v_s/\lambda = \frac{1.2\times10^4}{1.03\times10^{-9}} = \mathbf{1.16\times10^{13}\ Hz}.$$

6.2. *The elastic stiffness constant for sodium chloride is* 5×10^{10} N/m^2. *What is the value of the force constant and the restrahlen frequency for the crystal?*

Given: Lattice constant for NaCl crystal = 5.5 Å, mass of Sodium atom = 23 amu, mass of Chlorine atom = 35.5 amu and 1 amu = 1.66×10^{-27} kg.

Solution: Here the force constant of the crystal is $\beta = a \times C = (5.5\times10^{-10})\times(5\times10^{10}) = 27.5$ N/m, where a and C are respectively lattice spacing and stiffness constant.

The frequency is given by

$$\omega^2 = 2\beta\left(\frac{1}{m}+\frac{1}{M}\right) = \frac{2\times27.5\times58.5}{15.5\times23.0\times1.66\times10^{-27}}$$

$$\text{or, } \omega = \sqrt{\frac{2\times27.5\times58.5}{15.5\times23.0\times1.66\times10^{-27}}} = \mathbf{4.872\times10^{13}\ rad/s}.$$

Now the restrahlen frequency ν is

$$\nu = \frac{\omega}{2\pi} = \frac{4.872 \times 10^{13}}{2\pi} = \mathbf{7.754 \times 10^{12}/s.}$$

6.3. *Determine the phase velocity and group velocity of the wave motion along a one-dimensional lattice. What happens to the group velocity when $ka = \pm\pi$?*

Solution: The phase velocity of the wave motion along a one-dimensional lattice is

$$v_p = \frac{\omega}{k} = \frac{\omega_{max}}{q}\left|\sin\frac{ka}{2}\right| = v_0\left|\frac{\sin\frac{ka}{2}}{\frac{qa}{2}}\right|,$$

where $v_0 = \left(\frac{a}{2}\right)\omega_{max}$.

The group velocity,

$$v_g = \frac{d\omega}{dk} = v_0\left|\cos\frac{ka}{2}\right|.$$

Here v_0 is the wavelength limit both for v_p and v_g.

If $ka = \pm\pi$, $v_g = 0$, then the neighbouring atoms differ in phase by π radians corresponding to standing waves.

6.4. *From the data provided, calculate the wavelength of the electromagnetic wave for strong reflection by* NaCl *crystal. Assume that extension in* [100] *direction produces negligible contraction in the perpendicular directions.*

Radii *of* Na^+ *and* Cl^- ions = 0.98 Å and 1.81 Å respectively; Young's modulus of NaCl in [100] direction = 5×10^{10} N/m^2; Atomic masses of Na and Cl = 23 and 35.5 respectively.

Solution: If m and M be the masses of the ions and β the force constant, then the frequency of radiation strongly reflected by the ionic crystal is

$$\omega^2 = 2\beta\left(\frac{1}{m} + \frac{1}{M}\right).$$

As the extension in [100] direction produces no contraction in perpendicular direction,

$\therefore \beta = aY$, where Y and a are Young's modulus and interatomic distance respectively.

$$\therefore \omega^2 = 2aY\left(\frac{1}{m} + \frac{1}{M}\right).$$

The interatomic distance along [100] direction is $a = (0.98 + 1.810) \times 10^{-10}$ m.

$$\therefore \omega^2 = 2(0.98 + 1.81) \times 10^{-10} \times (5 \times 10^{10}) \times \left(\frac{1}{23} + \frac{1}{35.5}\right) \times \frac{1}{1.67 \times 10^{-27}}$$

or, $\omega = \mathbf{3.46 \times 10^{13}}$ **rad/s.**

Now the strongly reflected electromagnetic radiation by NaCl crystal will have wavelength

$$\lambda = \frac{c}{\nu} = \frac{2\pi c}{\omega} = \frac{2\pi(3 \times 10^{10})}{3.46 \times 10^{13}} = \mathbf{5.45 \times 10^{-3}\ m.}$$

6.5. *If Einstein's temperature of a material is* 157 K, find the value of C_v for the material at 100 K in cal/mol/K using Einstein's formula. Also calculate Einstein's frequency. **[NBU 2001]**

Solution: We know that the specific heat of the solid according to Einstein is

$$C_v = 3R\frac{x^2e^x}{(e^x-1)^2},$$

where $x = \theta_E/T$; $\theta_E = 157$ K and $T = 100$ K.

$$\therefore x = \frac{157}{100} = \mathbf{1.57}.$$

So

$$C_v = \frac{3\times1.99\times(1.57)^2e^{1.57}}{(e^{1.57}-1)^3}, \quad [\because R = 1.99 \text{ cal/mol/K}]$$

$$= \mathbf{4.88\ cal/mol/K.}$$

The Einstein's frequency is

$$\nu_E = \frac{k\theta_E}{h} = \frac{(1.38\times10^{-23})\times157}{6.6\times10^{-34}} = \mathbf{3.28\times10^{12}\ Hz.}$$

6.6. *Calculate the Debye frequency and Debye temperature for aluminium from the following data: Density of atoms in aluminium* = $6.02\times10^{28}/\text{m}^3$, *Velocity of longitudinal wave* $v_l = 6374$ m/s, *Velocity of transverse wave* $v_t = 3111$ m/s.

Solution: The Debye frequency is given by

$$\omega_D^3 = \frac{18N\pi^2}{V}\Big/\left(\frac{2}{v_t^3}+\frac{1}{v_l^3}\right).$$

Putting $N/V = 6.02\times10^3$, $v_l = 6374$ m/s, $v_t = 3111$ m/s, we have

$$\omega_D^3 = 18\times(6.02\times10^3)\times(3.1416)^2\Big/\left[\frac{2}{(3111)^3}+\frac{1}{(6374)^3}\right] = 1.5215\times10^{41}(\text{rad/s})^3.$$

$$\therefore \omega_D = \sqrt[3]{1.5215\times10^{41}} = 5.338\times10^{13} \text{ rad/s or, } \nu_D = \frac{\omega_D}{2\pi} = \mathbf{8.4956\times10^{12}\ Hz.}$$

The Debye temperature is

$$\theta_D = \frac{\hbar\omega_D}{2\pi} = \frac{(1.054\times10^{-34})\times(5.338\times10^{13})}{1.38\times10^{-23}} = \mathbf{407.6\ K.}$$

6.7. *If gold and copper both have the same structure, calculate the Debye temperature of copper if the Debye temperature of gold is* 170 K.

Given: Velocity of sound in Gold = 2100 m/s, Velocity of sound in Copper = 3800 m/s, Density of Gold = 1.93×10^4 kg/m^3, Density of Copper = 8.96×10^3 kg/m^3, Atomic of weight of Gold = 1970 amu and Atomic weight of Copper = 63.54 amu.

Solution: If ν_D be the Debye frequency, then the Debye temperature is

$$\theta_D = \frac{h\nu_D}{k} \text{ and } \nu_D^3 = \frac{9N}{4\pi V}\Big/\left[\frac{1}{v_l^3}+\frac{2}{v_t^3}\right],$$

where N is the number of atoms in volume V.

Assuming $v_l = v_t = v$, we get

$$v_D^3 = \frac{9N}{4\pi V}\frac{v^3}{3} = \frac{3N}{4\pi V}v^3 \text{ or, } v_D = v\left(\frac{3N}{4\pi V}\right)^{1/3} \quad \therefore \theta_D = \frac{h}{k}v\left(\frac{3N}{4\pi V}\right)^{1/3}.$$

If M_{Au} be the atomic weight of gold and N_A the Avogadro's number, then $N/V = (N_A/M_{Au}) \times 1.39 \times 10^4$.

For copper,

$$N/V = \left(\frac{N_A}{M_{Cu}}\right) \times 8.96 \times 10^3,$$

where M_{Cu} is the atomic weight of copper.

Now

$$\frac{\theta_D(Cu)}{\theta_D(Au)} = \frac{\frac{h}{k}v_{Cu}\left[\frac{3}{4\pi} \times \frac{N_A}{M_{Cu}} \times 8.96 \times 10^3\right]^{1/3}}{\frac{h}{k}v_{Au}\left[\frac{3}{4\pi} \times \frac{N_A}{M_{Au}} \times 1.93 \times 10^4\right]^{1/3}}$$

$$\text{or, } \theta_D(Cu) = \theta_D(Au) \times \left(\frac{v_{Cu}}{v_{Au}}\right) \times \left(\frac{8.96 \times 10^3}{19.3 \times 10^3}\right)^{1/3} \times \left(\frac{197}{63.54}\right)^{1/3}$$

$$= 170 \times \frac{3800}{2100} \times \left(\frac{8.96 \times 10^3}{19.3 \times 10^3}\right)^{1/3} \times \left(\frac{197}{63.54}\right)^{1/3}.$$

$\therefore$ Debye temperature for copper is $\theta_D(Cu) = \mathbf{347.7\ K}$.

6.8. *Show that in the Debye approximation, the total zero-point energy per* gm-mol *of a solid is given by* $\frac{9}{8}R\theta_D$.

Solution: If $D(\omega)\,d\omega$ be the number of modes of vibrations between frequency interval ω and $\omega + d\omega$, then the vibrational energy of the crystal, for these modes is $dE = \langle\varepsilon(\omega)\rangle D(\omega)\,d\omega$ and the total energy of the crystal is

$$E = \int_0^{\omega_m} < \langle\varepsilon(\omega)\rangle D(\omega)\,d\omega. \tag{1}$$

Now the possible energy level of the harmonic oscillator is $\varepsilon_n = \left(n + \frac{1}{2}\right)\hbar\omega$.

The average value of ε_n at thermal equilibrium is

$$\langle\varepsilon\rangle = h\nu\left[\frac{1}{2} + \frac{1}{e^{h\nu/kT} - 1}\right]. \tag{2}$$

Also

$$D(\omega) = \frac{3V\omega^2}{2\pi^2 v_0^3}, \tag{3}$$

where v_0 is the phase velocity of the wave and is a constant. v_0 may be taken to be equal to the velocity of longitudinal elastic waves propagating through the crystal.

Using eqns (2) and (3) in eqn (1), we get

$$E = \frac{3\hbar V\omega_m^4}{16\pi^2 v_0^3} + \frac{3\hbar V}{2\pi^2 v_0^3}\int_0^{\omega_m}\frac{\omega^2}{e^{\hbar\omega/kT} - 1}\,d\omega. \tag{4}$$

Putting $x = \frac{\hbar\omega}{kT}$,

$$x_m = \frac{\hbar\omega_m}{kT} = \frac{\hbar\nu_0}{kT}\left(\frac{6N\pi^2}{V}\right)^{1/3} \tag{5}$$

$$\text{and } \theta_D = \frac{\hbar\omega_m}{R} = \frac{\hbar\nu_0}{k}\left(\frac{6N\pi^2}{V}\right)^{1/3} \tag{6}$$

and using the values of x_m and θ_D in eqn (4), we get

$$E = \frac{9}{8}Nk\theta_D + 9NkT\left(\frac{T}{\theta_D}\right)^3 \int_0^{x_m} \frac{x^3}{e^x - 1}\,dx \left[\because \omega_m = \nu_0\left(\frac{6\pi^2 N}{V}\right)^{1/3}\right]. \tag{7}$$

Now as $T \to 0,\ x_m \to \infty$,

$$\int_0^{x_m} \frac{x^3}{e^x - 1}\,dx = \int_0^{\infty} \frac{x^3}{e^x - 1}\,dx = \frac{\pi^4}{15}.$$

$$\therefore E = \frac{9}{8}Nk\theta_D + 9NkT\left(\frac{T}{\theta_D}\right)^3 \times \frac{\pi^4}{15} = \frac{9}{8}Nk\theta_D + \frac{3}{5}\pi^4 NkT\left(\frac{T}{\theta_D}\right)^3. \tag{8}$$

The zero-point energy is obtained by putting $T = 0$ in eqn (8).

$$\therefore \boldsymbol{E = \frac{9}{8}Nk\theta_D = \frac{9}{8}R\theta_D}. \tag{9}$$

6.9. *How will specific heat get modified if Planck constant were increased* 5-fold.

Solution: We have the Debye temperature,

$$T_D = \frac{h\nu_D}{k}.$$

Let $h' = 5h$.

$$\therefore T_D' = \frac{h'\nu_D}{k} = \frac{5h\nu_D}{k} = 5T_D$$

$$C_v = \frac{12\pi^4 Nk}{5}\left(\frac{T}{T_D}\right)^3 \left[\because C_v = 77.9 \times 3R\left(\frac{T}{T_D}\right)^3\right]$$

$$C_v' = \frac{12\pi^4 Nk}{5}\left(\frac{T}{T_D}\right)^3 = \frac{12\pi^4 Nk}{5}\left(\frac{T}{5T_D}\right)^3 = \frac{1}{5^3}\frac{12^4 Nk}{5}\left(\frac{T}{T_D}\right)^3 = \frac{\boldsymbol{C_v}}{\mathbf{125}}.$$

$\therefore C_v$ **would diminish 125 times.**

6.10. *At a very low temperature the specific heat of rock salt is given by*

$$C_v = A\left(\frac{T}{T_D}\right)^3$$

as per Debye T^3-law. Find the quantity of heat required to raise the temperature of 3 moles of rock salt from 10 K to 50 K. [*Take* T_D = 281 K, A = 464 cal/mole/K].

Solution: We have specific heat

$$C_v = A\left(\frac{T}{T_D}\right)^3,$$

A = 464 cal/mole/K.

Now quantity of heat required to raise temperature of 3 moles of rock salt from 10 K to 50 K is

$$H = \int_{10k}^{50k} (3\text{ mole}) C_v \, dT = \int_{10k}^{25k} (3\text{ mole}) \frac{4T^3}{T_D^3} dT$$

$$= (3\text{ mole}) \frac{A}{T_D^3} \int_{10K}^{25K} T^3 \, dT = \frac{(3\text{ mole})(464\text{ cal/mol/K}) T^4}{(281)^3} \Bigg|_{0k}^{25k}$$

$$= \frac{(3\text{ mole})\,(464\text{ cal mol}^{-1}\text{ K}^{-1})}{(281)^3} \cdot \frac{(50k)^4 - (10k)^4}{4} = \mathbf{97.9\ cal.}$$

This is the required heat.

6.11. *Find the gradient of optical branch of dispersion curve at the maximum frequency.*

Solution: We know that the maximum frequency in optical branch occurs at $k = 0$ and is given by

$$\omega_+ = \sqrt{2\beta\left(\frac{1}{m} + \frac{1}{M}\right)} = \text{constant.}$$

6.12. *In a diatomic linear lattice,* $\frac{m}{M} = \alpha \ll L$, *then find the ratio of the square of the widths of the optical and acoustic branches.*

Solution: We have

$$\frac{\text{Optical branch width}}{\text{Acoustic branch width}} = \frac{\sqrt{2\beta\left(\frac{1}{m} + \frac{1}{M}\right)} - \sqrt{\frac{2\beta}{m}}}{\sqrt{\frac{2\beta}{m}} - 0} = \frac{\sqrt{\frac{1}{m} + \frac{1}{M}} - \sqrt{\frac{1}{m}}}{\sqrt{\frac{1}{M}}}$$

$$= \frac{\sqrt{\frac{1}{m}\left(1 + \frac{m}{M}\right)} - \sqrt{\frac{1}{m}}}{\sqrt{\frac{1}{M}}} = \frac{\sqrt{\frac{1}{m}}\left(1 + \frac{m}{M}\right)^{1/2}\sqrt{\frac{1}{m}}}{\sqrt{\frac{1}{M}}}.$$

As $\frac{m}{M} \ll 1$, binomial theorem can be applied.

$$\text{Width ratio} = \sqrt{\frac{M}{m}}\left(1 + \frac{m}{2M}\right) - \sqrt{\frac{M}{m}} = \sqrt{\frac{M}{m}}\left(1 + \frac{m}{2M} - 1\right) = \sqrt{\frac{M}{m}} \cdot \frac{m}{2M} = \frac{1}{2}\sqrt{\frac{m}{M}}$$

$$= \frac{\sqrt{\alpha}}{2}.$$

$\therefore$ width ratio squared $= \frac{\alpha}{4}$.

6.13. *If Debye characteristic temperature for* Al and Cu be 398 K and 315 K respectively, find the ratio of their specific heats at $T = 1$ K.

Solution: According to Debye theory, the specific heat is given by

$$\text{Cu} = \frac{12\pi^4 R}{5T^3} T^3 \text{ [the letters having their usual meanings].}$$

$$\therefore \frac{C_v^{Cu}}{C_v^{Al}} = \frac{12\pi^4 TR^3}{5\left(T_D^{Cu}\right)^3} \times \frac{5(T_D)^3}{12\pi^4 RT^3} = \left(\frac{T_D^{Al}}{T_D^{Cu}}\right)^3 = \left(\frac{398\text{ K}}{315\text{ K}}\right)^3 = \mathbf{2.}$$

6.14. *If Einstein's frequency of copper is* **7100 GHz, find Einstein characteristic temperature. Find molar specific heats at 100 K and 800 K. When do you get Dulong-Petit result?**

Solution: We have Einstein frequency ν_E = 7100 GHz = 7100 × 10^9 Hz.

$$T_E = \frac{h\nu_E}{k} = \frac{(6.626 \times 10^{-34}\,\text{Js})\,(7100 \times 10^9\,\text{Hz})}{1.38 \times 10^{-23}\,\text{JK}^{-1}} = 341\,\text{K}.$$

Now

$$C_v = 3R\left(\frac{T_E}{T}\right)^2 \frac{e^{T_E/T}}{\left(e^{T_E/T}-1\right)^2} = 3\left(8.3\,\text{JK}^{-1}\,\text{mol}^{-1}\right)\left(\frac{341\,\text{K}}{T\text{K}}\right)\frac{e^{341\,\text{K}/T\text{K}}}{\left(e^{341\,\text{K}/T\text{K}}-1\right)^2}$$

$$\text{or, } C_v = 2.895 \times 10^6 \frac{e^{341/T}}{T^2\left(e^{341/T}-1\right)^2}\,\text{JK}^{-1}\,\text{mol}^{-1}.$$

For T = 100 K,

$$C_v = 2.895 \times 10^6 \frac{e^{341/100}}{(100)^2\left(e^{341/100}-1\right)^2} = 10.23\,\text{JK}^{-1}\,\text{mol}^{-1}.$$

For T = 300 K,

$$C_v = 2.895 \times 10^6 \frac{e^{341/300}}{(300)^2\left(e^{341/300}-1\right)^2} = 22.38\,\text{JK}^{-1}\,\text{mol}^{-1}.$$

For T = 900 K,

$$C_v = 2.895 \times 10^6 \frac{e^{341/900}}{(900)^2\left(e^{341/300}-1\right)^2} = 25\,\text{JK}^{-1}\,\text{mol}^{-1}.$$

Hence, Dulong-Petit result C_v = **25 JK^{-1} mol^{-1}** is obtained at high temperature.

6.15. *The Debye temperature of diamond is* **2000 K. Find the highest possible vibrational frequency (the Debye frequency). What is the mean velocity of sound in diamond? What is the mean frequency of lattice vibration.**

Also find the molar heat capacity of diamond at **10 K. [Take density of diamond 3.5 gm/cm^3, interatomic spacing 1.5 Å]**

Solution: We have Debye temperature T_D = 2000 K.

Debye frequency

$$\nu_D = \frac{kT_D}{h} = \frac{(1.38 \times 10^{23}\,\text{J/K})\,(200)}{6.626 \times 10^{-34}\,\text{Js}}.$$

Again,

$$\nu_D^3 = \frac{9N}{4\pi V}\left(\frac{1}{v_l^3} + \frac{1}{v_t^3}\right)^{-1},$$

where the solid has N atomic oscillators and has volume V. v_l, v_t are longitudinal and transverse speeds respectively.

Assuming $v_l = v_t = v$, mean velocity of sound in diamond.

$$\nu_D^3 = \frac{9N}{4\pi V}\left(\frac{1}{v^3} + \frac{2}{v^3}\right)^{-1} = \frac{9N}{4\pi V}\left(\frac{3}{v^3}\right)^{-1} = \frac{3nv^3}{4\pi V}.$$

$$\therefore \nu_D = \left(\frac{3N}{4\pi V}\right)^{1/3} v.$$

$$\therefore \nu_D = \frac{kT_D}{h} = \left(\frac{3N}{4\pi V}\right)^{1/3} v \text{ or, } v = \frac{kT_D}{h}\left(\frac{4\pi V}{3N}\right)^{1/3}.$$

For 1 mole solid, $N = N_A = 6.023 \times 10^{23}$ (Avo. No.).

Also $V = M/\rho$, M = molecular weight, ρ = density.

$$\therefore \nu_D = \left(\frac{3N_A}{4\pi V}\right)^{1/3} v = \left(\frac{3\rho N_A}{4\pi M}\right)^{1/3} v.$$

Also

$$v = \frac{kT_D}{h}\left(\frac{4\pi V}{3N_A}\right)^{1/3} = \frac{kT_D}{h}\left(\frac{4\pi M}{3\rho N_A}\right)^{1/3}.$$

$$\therefore T_D = \frac{hv}{k}\left(\frac{3\rho N_A}{4\pi M}\right)^{1/3}.$$

For diamond, $M = 12\text{ g} = 12 \times 10^{-3}$ kg, $\rho = 3.5$ gm-cm^{-3} = 3500 kg-m^{-3} ($\because$ 1 gm-cm^{-3} = 10^3 kg-m^{-3}), $T_D = 2000$ K.

$$\therefore v = \frac{(1.38 \times 10^{-23}\text{ J/K})\,(2000\text{ K})}{6.626 \times 10^{-34}\text{ Js}}\left[\frac{4\pi}{3500\text{ kg/m}^3} \cdot \frac{12 \times 10^{-3}\text{ kg}}{3(6.023 \times 10^{23})}\right]^{1/3} \text{ or, } v = \mathbf{1.2 \times 10^4\ m/s}.$$

Frequency of lattice vibration $\nu = v/\lambda$.

λ = wavelength of lattice vibration = $\dfrac{T_D}{T}d$; d = interatomic spacing 1.5 Å = 1.5×10^{-10} m; T = room temperature = 300 K.

$$\therefore \nu = \frac{v}{\lambda} = \frac{vT}{T_D d} = \frac{(1.2 \times 10^4\text{ ms}^{-1})\,(300\text{ K})}{(2000\text{ K})\,(1.5 \times 10^{-10}\text{ m})} = \mathbf{1.2 \times 10^{13}\ Hz}.$$

Specific heat,

$$C_v = \frac{12\pi^4 R}{5T_D^3}T^3 = \frac{12\pi^4(8.31\text{ JK}^{-1}\text{ mole}^{-1})}{5}\left(\frac{10\text{ K}}{2000\text{ K}}\right) = \mathbf{0.243 \times 10^{-3}\ JK^{-1}\ mole^{-1}}.$$

6.16. ***Evaluate the frequency of phonon considering inelastic scattering of phonon of frequency ν_{ph} by phonon in absence of Bragg diffraction. Take crystal to be a continuum of refractive index μ.***

Solution: We have phonon frequency,

$$\nu_{ph} = \frac{\omega_{ph}}{2\pi}.$$

Photon speed in medium

$$= \frac{c}{\mu} = \nu_{ph}\lambda_{ph} \text{ or, } \lambda_{ph} = \frac{c}{\mu\nu_{ph}} = \frac{2\pi c}{\mu\omega_{ph}}.$$

Wave constant of photon,

$$k_{ph} = \frac{2\pi}{\lambda_{ph}} = 2\pi\frac{\mu\omega_{ph}}{2\pi c} = \frac{\mu}{c}\omega_{ph}.$$

This photon is scattered inelastically as a result of which its wave vector changes from $\vec{k}_{ph}$ to $\vec{k}'_{ph}$; frequency changes from ω_{ph} to ω'_{ph} and a phonon of frequency ω, wave vector $\vec{k}$ is created but in absence of Bragg diffraction.

The conservation of wave vector and momentum are $\vec{k}'_{ph} + \vec{k} = \vec{k}_{ph}$ or, $\hbar\vec{k}_{ph} + \hbar\vec{k} = \hbar\vec{k}_{ph}$ and conservation of energy is $\hbar\omega_{ph} = \hbar\omega'_{ph} + \hbar\omega$, where $\omega = v_s k$, v_s = speed of phonon $\ll c$.

So, phonon can carry off a small fraction of the incident photon energy. So, it is justified to take $\omega < \omega_{ph}$; $\omega_{ph} = \omega'_{ph}$, $k_{ph} = k'_{ph}$.

So, we get an isosceles triangle formed by $\vec{k}'_{ph}$, $\vec{k}_{ph}$ and $\vec{k}$.

Now from the conservation of wave vector,

$$\vec{k}'_{ph}\vec{k} = \vec{k}_{ph} \text{ or, } \vec{k} = \vec{k}_{ph} - \vec{k}'_{ph}$$

$$\text{or, } k^2 = k_{ph}^2 + \vec{k}'^2{}_{ph} - 2\vec{k}_{ph}\cdot\vec{k}'_{ph} = k_{ph}^2 + \vec{k}'^2_{ph} - 2\vec{k}_{ph}\vec{k}'_{ph}\cos\theta = k_{ph}^2 + k_{ph}^2 - 2k_{ph}^2\cos\theta$$

$$= 2k_{ph}^2(1-\cos\phi) = 2k_{ph}^2\cdot 2\sin^2\frac{\phi}{2} = 4k_{ph}^2\sin^2\frac{\phi}{2}$$

$$\text{or, } k = 2k_{ph}\sin\frac{\phi}{2} \text{ or, } \frac{\omega}{v_s} = 2\left(\frac{\mu\omega_{ph}}{c}\right)\sin\frac{\phi}{2}$$

$$\text{or, } \omega = \frac{2v_s\mu\omega_{ph}}{c} = \text{frequency of photon.}$$

EXERCISE

6(A): Objective (MCQ) Type Questions

Select the correct answer.

6.1. The vibration of a crystal lattice is

(a) simple harmonic (b) coupled vibration (c) mixture of (a) and (b) (d) none of these.

6.2. The free vibration of a lattice yields

(a) thermal properties (b) optical properties
(c) electrical properties (d) elastic properties of the solid.

6.3. The forced vibration of a lattice yields

(a) electrical properties (b) elastic properties (c) thermal properties (d) optical properties.

6.4. The atomic nature of the lattice becomes unimportant in the

(a) short wavelength side (b) long wavelength side
(c) intermediate wavelength of the lattice wave (d) none of these.

6.5. If v_p and v_g be the phase velocity and group velocity of the lattice wave, then in the long wavelength side

(a) $v_g > v_p$ (b) $v_g = v_p$ (c) $v_g < v_p$ (d) $v_g = v_p = \infty$.

6.6. The lattice wave behaves like a standing wave without any transfer of energy at frequency

(a) $\omega > \sqrt{2\beta/m}$ (b) $\omega = \sqrt{2\beta/m}$ (c) $\omega < \sqrt{2\beta/m}$ (d) none of these.

6.7. The forbidden frequency band of solids disappears at $k = \pm\pi/2a$, if

(a) $m > M$ (b) $m < M$ (c) $m = M$ (d) $mM = 1$,

where m and M are the masses of the light and heavy atoms respectively.

6.8. For optical branch, the light and heavy atoms move in opposite directions if their respective amplitudes A and B are related by

(a) $A/B = 1$ (b) $A/B = -1$ (c) $AB = 1$ (d) none of these.

6.9. In polyatomic lattices having N atoms per primitive cell, the number of frequency bands is

(a) $N - 1$ (b) N (c) $N + 1$ (d) $2N$.

6.10. If the light mass m and heavy mass M be equal, then the frequency range in both the monatomic and diatomic lattices is the same and the forbidden band

(a) does not exist (b) exists (c) exists in a few cases (d) none of these.

6.11. If the heavier mass $M \to \infty$, then

(a) the acoustical branch disappears (b) the optical branch disappears

(c) both the above branches disappear (d) none of these.

6.12. If the light mass $m \to 0$, then

(a) acoustical branch disappears (b) optical branch disappears

(c) both acoustical and optical branches disappear (d) none of these.

6.13. At $k = \pi/2a$, the natural frequency of the higher atom is

(a) $\sqrt{(2\beta/M)}$ (b) $\sqrt{2\beta/M}$ (c) $\sqrt{\beta/2m}$ (d) $\sqrt{2\beta m}$.

6.14. In a solid the molar specific heat $C_v = C_{\text{lattice}}$, because $C_{\text{electron}} = 0$

(a) due to absence of free electrons in solids

(b) since free electrons do not vibrate

(c) since free electrons transition from lower to higher energy state is forbidden

(d) none of these.

6.15. If γ be the volume expansion coefficient of metals, then the relation of C_v with T below 20 K is

(a) $C_v = \alpha T^3$ (b) $C_v = \alpha T$ (c) $C_v = \alpha T^3 + \gamma T$ (d) $C_v = \alpha T^2$,

where α is a constant.

6.16. Phonons are quantized energy units and obey

(a) MB-statistics (a) BE-statistics (a) FD-statistics (a) none of these.

6.17. In inelastic scattering of a photon with lattice a phonon is

(a) created (b) absorbed (c) created or absorbed (d) none of these.

6.18. In Einstein's theory of lattice specific heat, solids vibrate with the

(a) same frequency (b) band of frequency (c) discrete set of frequency (d) none of these.

6.19. In Einstein's theory of lattice specific heat, the average energy $\langle\varepsilon\rangle$ of an oscillator is

(a) $\frac{h\nu}{e^{h\nu/kT}}$ (b) $\frac{h\nu}{e^{h\nu/kT}-1}$ (c) $\frac{h\nu}{e^{h\nu/kT}+1}$ (d) $\frac{h\nu}{e^{-h\nu/kT}}$.

6.20. For $T \ll \theta_E$, where θ_E is the Einstein's temperature,

(a) $C_v \propto e^{-\theta_E/kT}$ (b) $C_v \propto e^{\theta_E/kT}$ (c) $C_v \propto e^{kT/\theta_E}$ (d) $C_v = 3R$.

6.21. Debye's theory of lattice specific heat assumes that the vibration of the atomic oscillators is
(a) independent and free (b) coupled (c) damped (d) none of these.

6.22. Debye frequency of a solid having N number of atoms is
(a) N (b) $2N$ (c) $3N$ (d) none of these.

6.23. In Debye model the atomic oscillators vibrate with
(a) a frequency ν_D (b) a range of frequency $\nu = a$ minimum to $\nu = \nu_D$
(c) $\nu > \nu_D$ (d) $\nu = 0$ to $3N$,
where θ_D is the Debye temperature.

6.24. At low temperature Debye's theory predicts that
(a) $C_v = 3R$ (b) $C_v = 3R\left(\frac{\theta_E}{T}\right)^2 e^{-\theta_E/T}$ (c) $C_v = 324R(T/\theta_D)^3$ (d) $C_v \propto T^3$,
where θ_D is the Debye temperature.

6.25. Debye theory holds for
(a) $T > 300$ K (b) $T < 300$ K (c) $T = 100$ K (d) $T > 100$ K.

6.26. Which of the acoustical branch and optical branch disappears if lighter mass $m \to 0$ in a diatomic linear lattice?
(a) acoustical branch (b) optical branch (c) both (d) none of these.

6.27. A crystal can be treated as a continuous medium (i.e., atomic nature or lattice nature becomes unimportant) if the wavelength, in comparison to the lattice constant is
(a) small (b) large (c) equal (d) no such relation.

6.28. Phonons are quantised energy units
(a) having spin 1/2 (b) obeying BE-statistics
(c) having momentum $\frac{h\nu}{c}$, ($c = 3 \times 10^8$ m/s) (d) representing atoms in the state of vibration.

6.29. Phonon is
(a) elastic quanta (b) Planck quanta (c) electromagnetic quanta (d) photon quanta.

6.30. Which of the following laws fails at low temperature?
(a) Dulong-Petit law, Debye law (b) Debye law, Einstein law
(c) Einstein law, Dulong-Petit law (d) Debye law.

6.31. Einstein law and Debye law of specific heat merge with Dulong-Petit law at
(a) high temperature (b) low temperature (c) all temperature (d) $T \to 0$ K.

ANSWERS

6.1. (b) **6.2.** (a) **6.3.** (d) **6.4.** (b) **6.5.** (b) **6.6.** (b) **6.7.** (c) **6.8.** (b) **6.9.** (b) **6.10.** (a) **6.11.** (a) **6.12.** (b) **6.13.** (b) **6.14.** (a) **6.15.** (c) **6.16.** (b) **6.17.** (c) **6.18.** (a) **6.19.** (b) **6.20.** (a) **6.21.** (b) **6.22.** (c) **6.23.** (b) **6.24.** (d) **6.25.** (b) **6.26.** (b) **6.27.** (b) **6.28.** (b) **6.29.** (a) **6.30.** (c) **6.31.** (a).

6(B): Short Answer Type Questions

6.1. What do you mean by lattice vibration and phonon ?

6.2. What are N-process and U-process as applied to lattice vibration?

6.3. Explain how a phonon is emitted or absorbed in an inelastic collision of an X-ray photon with the lattice of a solid.

6.4. Give dispersion relation for one-dimensional monatomic lattice.

6.5. What is forbidden frequency band in a linear diatomic crystal? Is it constant?

6.6. Explain the origin of acoustical and optical branch in a linear diatomic lattice. Why are these branches named so?

6.7. What are the fundamental assumptions in Einstein's theory of lattice specific heat of solid?

6.8. Discuss the limitations of Einstein's theory of lattice specific heat.

6.9. Show that for one-dimensional monatomic lattice, the phase velocity is equal to the group velocity at low temperature.

6.10. Discuss the concept of Brillouin zones as applicable to one-dimensional monatomic lattice.

6.11. Discuss the movements of atoms in the optical and acoustical modes of wave propagation for a linear diatomic crystal.

6.12. *In the long wavelength limit, the velocity of sound is independent of the frequency.* — How?

6.13. Clearly explain the fundamental assumptions in Debye model of lattice heat capacity.

6.14. In which respect Debye theory is superior to Einstein's theory of lattice specific heat? Explain.

6.15. Discuss qualitatively the concept of phonon for elastic vibration in a solid. [CU(H) 1997]

6.16. Give a comparative study of Einstein's theory and Debye theory of specific heat of solids.

6(C): Numerical Problems

6.1. Show that in the Debye approximation the total zero point energy per gm-mol of a solid is given by $(9/8)R\theta_D$. [BU(H) 1999]

6.2. Estimate the Debye temperature (θ_D) for iron in which the propagation velocities of longitudinal and transverse vibrations are equal to 5850 m/s and 3230 m/s respectively. [BU(H) 1999]

[*Hints.* Planck's Constant $h = 6.626 \times 10^{-34}$ J-s; Boltzmann's Constant $k = 1.38 \times 10^{-23}$ J/K; Atomic weight of iron = 55.847; Density of iron = 7870 kg/m^3.]

6.3. Compute the cut-off frequency for a linear monatomic lattice if the velocity of sound and the inter-atomic spacing in the lattice be 3×10^3 m/s and 3×10^{-10} m respectively. [*Ans.* 2×10^{13} Hz]

6.4. NaCl has the same structure as KCl. The Debye temperatures of NaCl and KCl are 281 K and 230 K respectively. If the lattice heat capacity of NaCl at 5 K is 1.6×10^{-2} J/mol/K, estimate the heat capacities of KCl at 5 K and 3 K. [*Ans.* 2.9×10^{-2} J/mol/K, 6.26×10^{-3} J/mol/K]

6.5. The Debye temperature of silver is 210 K. Assuming Debye's T^3-law, find the molar specific heat of silver at constant volume at 20 K. Taking the Einstein temperature the same as Debye temperature, calculate also the molar specific heat at 20 K using Einstein's formula.

Comment on the result by comparing them with experimental value of 1.67 J/mol/K.

[*Ans.* 1.68 J/mol/K; 0.076 J/mol/K]

6.6. Visible light of wavelength 4000 Å undergoes scattering from a diamond crystal of refractive index 2.42. Calculate the maximum frequency of the phonon generated and the fractional change in the frequency of the incident radiation [The velocity of sound in diamond is 1.2×10^4 m/s].

[*Ans.* 9.12×10^{11} rad/s, 1.9×10^{-4}]

6.7. What are photons? The photon dispersion relation for a vibrating diatomic chain in which alternate atoms are M_1 and M_2 is given by

$$\omega^2 = K_1\left(\frac{1}{M_1}+\frac{1}{M_2}\right) \pm K_1\left[\left(\frac{1}{M_1}+\frac{1}{M_2}\right)^2 - \frac{4\sin^2 ka}{M_1 M_2}\right]^{1/2},$$

where K_1 = force constant and $k = \frac{2\pi}{\lambda}$. Obtain minimum-maximum angular frequencies of the acoustical and optical branch. **[CU(H) 2016]**

6.8. The Debye temperature of diamond is 2000 K. Calculate the mean velocity of sound in diamond. [Given, the density of diamond = 3500 kg/m^3] **[BU(H) 2013]**

6(D): Broad Answer Type Questions

6.1. (a) Obtain dispersion relation for one-dimensional atomic crystal and discuss the nature of acoustic and optical models.

(b) Show that the group velocity vanishes at the zone-boundary. Give physical interpretation of the result.

6.2. (a) Derive an expression for the frequency of vibration of an infinite chain of identical atoms as a function of the wave vector.

(b) Show that the chain behaves as a low pass filter.

(c) Show that the number of possible modes of vibration in a linear chain of atoms is equal to the number of vibrating atoms.

6.3. (a) Derive vibrational modes of a diatomic linear lattice.

(b) Name the different branches of the dispersion relation curve. What is the difference between the two branches? **[LU 1996, 1995]**

6.4. (a) What is a phonon? Give an evidence for the existence of phonons.

(b) Obtain an expression for the momentum and energy of phonon. Explain *N-process* and *U-process*.

(c) Discuss the inelastic scattering of photons by phonons. **[HPU 1997, 1994, 1992]**

6.5. (a) Derive an expression for lattice specific heat capacity following Einstein model clearly explaining the underlying assumptions.

(b) Discuss the temperature dependence of lattice heat capacity and compare with experimental observations. **[GNDU 1997; PU 1997; KU 1996, BU 1995; HPU 1995]]**

6.6. (a) Describe Einstein model of lattice heat capacity.

(b) Discuss the success and failure of this model.

6.7. (a) Write down the assumptions of Debye model of lattice specific heat of solid.

(b) Discuss the Debye model of lattice heat capacity. What is Debye T^3-law?

6.8. Give the Debye's theory of specific heat of a crystalline solid and show that in suitable limits, it gives Dulong and Petit's law and the T^3-law.

[You may use the sum: $\sum_{n=1}^{\infty} \frac{1}{n^4} = \frac{\pi^4}{90}$.] **[BU(H) 1997]]**

6.9. (a) Why is Debye theory of specific heat more acceptable than the Einstein theory?

(b) Obtain low temperature behaviour of the lattice contribution to the specific heat on the basis of Debye's theory.

(c) What is Debye Temperature? **[CU(H) 2004, 2002]**

6.10. (a) In what respect Einstein's theory of lattice specific heat is superior to the classical theory?

(b) Obtain the values of molar specific heat at (i) $T \gg \theta_E$ and (ii) $T \ll \theta_E$, θ_E being the Einstein's temperature.

(c) What are the shortcomings of Einstein's theory?

6.11. (a) Discuss the temperature dependence of a molar specific heat of solids according to Debye.

(b) What are the drawbacks of the above theory?

(c) At a very low temperature, the specific heat of rock salt is given by $C_v = A\left(\frac{T}{\theta_D}\right)^3$ according to Debye's T^3-law. Find the quantity of heat required to raise the temperature of 3 moles of rock salt from 10 to 50 K. [$A = 464$ cal/mol/K, $\theta_D = 281$ K] [*Ans.* 97.9 cal]

[*Hints.* The required heat $\theta = \frac{3A}{\theta_D^3}\int_{10}^{50} T^3 dT$]

6.12. (a) What is the difference between Einstein's and Debye's theories of specific heats of solids.

(b) Explain why Debye's theory is more appropriate in explaining low temperature behaviour of specific heats of solids. **[CU(H) 2014]**

6.13. Derive the number of vibrational modes of a crystalline solid in the frequency range ν to $\nu + d\nu$. Hence, obtain an expression for Debye temperature and explain the significance of cut-off frequency. Also calculate the specific heat at very low temperatures. Interpret the result. **[BU(H) 2013]**

6.14. (a) Derive an expression for the specific heat capacity of a solid on the basis of Einstein's theory.

(b) How far do the results from this theory agree with experimental results? **[BU(H) 2012]**

6.15. (a) State the differences between Einstein model and Debye model in explaining the specific heat of solid.

(b) Calculate the number of vibrational modes of a monatomic solid in the frequency range ω to $\omega + d\omega$. Hence, obtain an expression for Debye temperature and explain the significance of Debye cut-off frequency.

(c) Show that in the Debye approximation the total zero point energy of one gram-mole of a solid is $(9/8)R\theta_D$ [where θ_D is the Debye temperature].

(d) Explain the variation of specific heat with temperature according to Debye model. **[BU(H) 2010]**

6.16. (a) Obtain an expression for specific heat capacity of a solid using Debye model.

[Mahatma Gandhi U 2012]

(b) Discuss the failure of classical theory in explaining the observed temperature dependence of specific heat of a solid.

6.17. What are phonons? The phonon dispersion relation for a vibrating diatomic chain in which alternate atoms are M_1 and M_2 is given by

$$\omega^2 = K_1\left(\frac{1}{M_1} + \frac{1}{M_2}\right) \pm K_1\left[\left(\frac{1}{M_1} + \frac{1}{M_2}\right)^2 - \frac{4\sin^2 ka}{M_1 M_2}\right]^{1/2},$$

where K_1 = force constant and $k = \frac{2\pi}{\lambda}$. Obtain minimum and maximum angular frequencies of the acoustical and optical branch. **[CU (H) 2016]**

6.18. The Debye temperature of diamond is 2000 K. Calculate the mean velocity of sound in diamond. [Given, the density of diamond = 3500 kg/m^3.] **[BU (H) 2013]**

Chapter 7

Free Electron Theories of Metals

7.1 Introduction

The valence electrons of an atom in metal are loosely bound to the nucleus and at ordinary temperature become free of the atomic system. Such electrons in a metal move about freely through the volume of the metal and constitute what is called *free electron gas.* The loosely bound valence electrons are called *Conduction electrons.* Many important physical properties like electrical, thermal and magnetic properties of solids can be explained with the help of this *free electron gas model.* The *free electron* theory of metals has been developed in three different stages as follows:

1. *The Classical Free Electron Theory*: This theory was developed in 1900 by *Drude* and *Lorentz* according to which the free electrons move at *random chaotic motion* within the volume of the metal just like the ideal gas molecules in a chamber. These free electrons obey the *laws of classical mechanics.*

2. *The Quantum Free Electron Theory*: This theory was developed by *Sommerfeld* in 1928 which assumes that the free electrons obey the *laws* of *quantum mechanics.*

3. *The Zone Theory*: This is also known as the *Band Theory of Solids* and was formulated by *Bloch* in 1928.

7.2 Classical Free Electron Theory of Metals

The theory assumes that

(i) In atoms the valence electrons are free to undergo *random chaotic motion* within the volume of the metal very much like the ideal gas molecules in a container. The collection of such electrons forms what is called *electron gas.*

(ii) The motion of these free electrons obeys *classical Maxwell-Boltzmann velocity distribution law* and *the laws of kinetic theory of gases.*

(iii) The free electrons move at random and collide *elastically* with each other or positive ions fixed to the lattice.

(iv) The electrons being negatively charged move in opposite direction to an externally applied electric field.

(v) The free electrons are *non-interacting* and obey *Pauli's exclusion principle.*

Electrical Conductivity

Assuming the free electrons to obey *Kinetic theory of gases,* the *root mean square* (rms) *velocity* of an electron at temperature T is given by

$$C = \sqrt{\frac{3kT}{m}}, \text{ where } k \text{ is the Boltzmann constant. Thus, } C \propto \sqrt{T}. \quad (7.1)$$

The *electrical conductivity* in metals arise due to the motion of the conduction electrons under the influence of applied electric field. If E be the electric field and $-e$ the charge of an electron, then the force exerted on the electron is $F = -eE$.

The *accelerated electrons* collide with the rest of the medium and comes to rest. But due to the action of E, the electrons again are accelerated. Due to this collision we can attribute a *frictional force* acting on the electron.

If v be the velocity of the electron and τ, the time between two consecutive collisions, then the frictional force is $-mv/\tau$, where m is the *effective mass* of the electron. The equation of motion of the electron is

$$m\frac{dv}{dt} = -eE - m\frac{v}{t}. \tag{7.2}$$

At steady state condition,

$$\frac{dv}{dt} = 0.$$

$$\therefore m\frac{v}{\tau} = -eE \text{ or, } v = \frac{-e\tau}{m}E. \tag{7.3}$$

v is called the *steady state velocity* of the electron.

Drift Velocity and Charge Mobility: In absence of an electric field the electrons move randomly in all possible direction and the corresponding velocity is called *random velocity.*

But in presence of an electric field, in addition to random velocity, there is an additional net velocity associated with the electrons and is called the *drift velocity* (v_d). The drift velocity of an electron of mass m in electric field is given by

$$v_d = -\frac{e\tau}{m}E \tag{7.4}$$

$$\text{or, } v_d = -\mu E, \tag{7.5}$$

where $\mu = \frac{eE}{m}$ is called the *mobility* of the electron and is defined as *the average drift velocity in unit electric field.*

Now if n be the number of conduction electrons per unit volume, then the charge in unit volume is $(-ne)$. The amount of charge crossing unit area per unit time, i.e., *current density* is

$$J = (-ne)v_d = -ne\left(-\frac{e\tau}{m}E\right) = \frac{ne^2\tau}{m}E. \tag{7.6}$$

But

$$J = \sigma\vec{E} \tag{7.7}$$

is the *vector form of Ohm's law,* where σ is the *electrical conductivity.*

Comparing eqns (7.6) and (7.7), we get

$$\sigma = \frac{ne^2\tau}{m}. \tag{7.8}$$

Thus, the *electrical conductivity* of a metal increases with the increase of free electron density (n).

Mean Free Lifetime: The average lifetime between two consecutive collisions is called the *mean free lifetime.*

Mean Free Path: The average distance travelled by an electron between two consecutive collisions in presence of the applied electric field is called the *mean free path* (λ).

If $\lambda_1, \lambda_2, \lambda_3, \ldots$ be the successive free paths of the electron between two consecutive collisions, then its mean free path is given by

$$\lambda = \frac{\lambda_1 + \lambda_2 + \lambda_3 + \cdots + \lambda_n}{n} = \frac{1}{n}\sum_{i=1}^{n}\lambda_i. \quad (7.9)$$

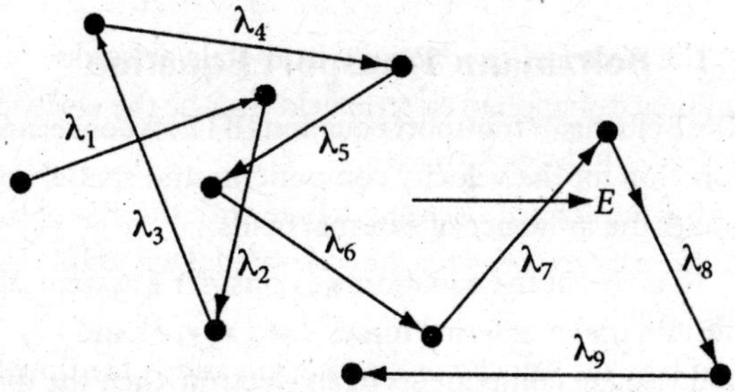

Fig. 7.1: Free paths of an electron in metal in presence of E

Relaxation Time: Let the applied electric field E be cut off after the drift velocity of the electron has reached a steady value. Putting $E = 0$ in eqn (7.2), we get

$$m\frac{dv_d}{dt} = -m\frac{v_d}{\tau} \text{ or, } \frac{dv_d}{v_d} = -\frac{dt}{\tau} \text{ or, } v_d(t) = v_d(0)\, e^{-t/\tau}, \quad (7.10)$$

where $v_d(0)$ is the drift velocity of the electron at the instant when the field is cut off. The variation of v_d with t is shown in Fig. 7.2.

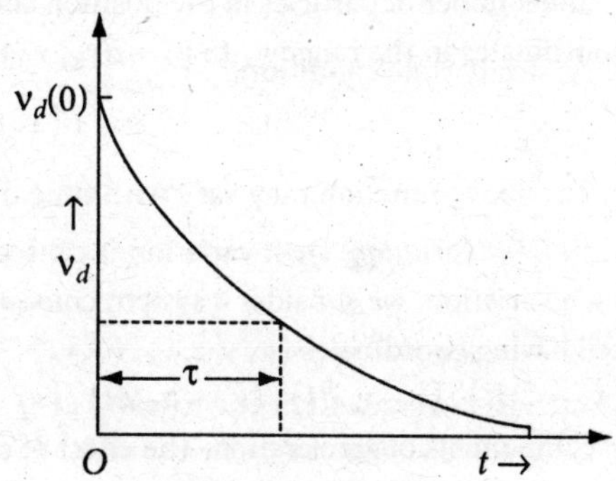

Fig. 7.2: Variation of v_d with t

To define τ, let us put $t = \tau$ in eqn (7.10),

$$v_d(t) = v_d(0)/e. \quad (7.11)$$

Hence, *relaxation time* τ is defined as *the time at which the drift velocity decays to* $\left(\frac{1}{e}\right)$ *times its initial value.*

Again, the electron mobility is the steady state drift velocity per unit electric field.

∴ electron mobility

$$\mu = \frac{\langle v_d \rangle}{E} = \frac{e\tau}{m}, \quad (7.12)$$

where $\langle v_d \rangle$ is the *average drift velocity* of the electron.

∴ from eqn (7.3),

$$\sigma = \frac{ne^2\tau}{m} = ne \cdot (e\tau/m) \text{ or, } \sigma = ne\mu. \quad (7.13)$$

The *resistivity of the material* of the conductor is

$$\rho = \frac{1}{\sigma} = \frac{1}{ne\mu} = \frac{m}{ne^2\tau}. \quad (7.14)$$

Eqn (7.13) shows that σ depends on n and μ, both of which depend on temperature of the conductor. In *metals,* μ decreases slightly with temperature and n is practically constant. So, *the electrical conductivity of metals decreases with the rise of temperature.*

For *semiconductor, n* increases exponentially with temperature causing an increase of conductivity.

In insulators, n remains constant and above certain temperature μ increases exponentially leading to *dielectric breakdown.*

7.3 Boltzmann Transport Equation

The Boltzmann transport equation (BTE) is concerned with the determination of the steady state distribution function for the velocity components and spatial coordinates in the transport process occurring in metal under the influence of external fields.

To arrive at the equation we consider a system of particles (e.g., electrons) which are in dynamic equilibrium under external forces. Let (x, y, z) and (v_x, v_y, v_z) represent respectively the position coordinates and velocity components of an electron, then the distribution function is given by

$$f(x, y, z, v_x, v_y, v_z). \tag{7.15}$$

The number of particles in the position coordinate range x to $x + dx$, y to $y + dy$, z to $z + dz$ and velocity coordinates in the range v_x to $v_x + dv_x$, v_y to $v_y + dv_y$, v_z to $v_z + dv_z$ is represented as

$$f(x, y, z, v_x, v_y, v_z)\, dx\, dy\, dz. \tag{7.16}$$

The above function may vary with time due to followings:

(i) *Drift variation*: Drift variation is caused due to external field and takes place continuously. To evaluate such variation, we consider a system consisting of particles at an instant $t + dt$, which is drifted to a phase cell having coordinates (x, y, z, v_x, v_y, v_z). The number of particles being the same as were at $(x - v_x\, dt)$, $(y - v_y\, dt)$, $(z - v_z\, dt)$; $(v_x - \alpha_x\, dt)$, $(v_y - \alpha_y\, dt)$, $(v_z - \alpha_z\, dt)$ at time t, α_x, α_y and α_z being the x-, y-, z-components of acceleration. The effect of collision has been neglected due to small interval of time so that the variations in velocity and accelerations are small, i.e., these are practically constant.

The change in the number of particles due to drift having coordinates (x, y, z, v_x, v_y, v_z) in time dt is

$$(\Delta f)_d = f(x - v_x\, dt, y - v_y\, dt, z - v_z\, dt, v_x - \alpha_x\, dt, v_y - \alpha_y\, dt, v_z - \alpha_z\, dt). \tag{7.17}$$

Expanding in *Taylor's theorem* and retaining only the first order terms we get in the limit $\Delta t \to 0$,

$$(\Delta f)_d = \left(-v_x \frac{\partial f}{dx} - v_y \frac{\partial f}{dy} - v_z \frac{\partial f}{dz} - \alpha_x \frac{\partial f}{\partial v_x} - \alpha_y \frac{\partial f}{\partial v_y} - \alpha_z \frac{\partial f}{\partial v_z}\right) dt \tag{7.18}$$

$$\left(\frac{df}{dt}\right)_{\text{drift}} = -v_x \frac{\partial f}{dx} - v_y \frac{\partial f}{dy} - v_z \frac{\partial f}{dz} - \alpha_x \frac{\partial f}{\partial v_x} - \alpha_y \frac{\partial f}{\partial v_y} - \alpha_z \frac{\partial f}{\partial v_z}. \tag{7.19}$$

(ii) *Collision variation*: Let the probability that a particle changes its velocity from v_x, v_y, v_z to $v'_x + dv'_x$, $v'_y + dv'_y$, $v'_z + dv'_z$ in unit time be represented as

$$\theta(v_x, v_y, v_z;\ v'_x, v'_y, v'_z)\, dv'_x\, dv'_y\, dv'_z. \tag{7.20}$$

Then the total number of particles undergoing such change in velocity is

$$a = f(x, y, z, v_x, v_y, v_z) \int \theta(v_x, v_y, v_z, v'_x, v'_y, v'_z)\, dv'_x\, dv'_y\, dv'_z. \tag{7.21}$$

Similarly, the number of particles whose velocity changes from v_x, v_y, v_z to v''_x, v''_y, v''_z is

$$b = \int f(v''_x, v''_y, v''_z)\, \theta(v''_x, v''_y, v''_z,\ v_x, v_y, v_z)\, dv''_x\, dv''_y\, dv''_z. \tag{7.22}$$

From eqns (7.21) and (7.22), we get the rate of change of f due to collision

$$\left(\frac{df}{dt}\right)_{\text{coll}} = b - a. \tag{7.23}$$

Under condition of equilibrium

$$\left(\frac{df}{dt}\right)_{\text{drift}} + \left(\frac{df}{dt}\right)_{\text{coll}} = 0 \tag{7.24}$$

$$\text{or, } -v_x\frac{\partial f}{\partial x} - v_y\frac{\partial f}{\partial y} - v_z\frac{\partial f}{\partial z} - \alpha_x\frac{\partial f}{\partial x} - \alpha_y\frac{\partial f}{\partial y} - \alpha_z\frac{\partial f}{\partial z} + (b-a) = 0$$

$$\text{or, } v_x\frac{\partial f}{\partial x} + v_y\frac{\partial f}{\partial y} + v_z\frac{\partial f}{\partial z} + \alpha_x\frac{\partial f}{\partial x} + \alpha_y\frac{\partial f}{\partial y} + \alpha_z\frac{\partial f}{\partial z} = (b-a) \tag{7.25}$$

$$\text{or, } \vec{v}\cdot\nabla f + \vec{\alpha}\cdot\nabla' f = (b-a). \tag{7.26}$$

The eqns (7.25) and (7.26) are *Boltzmann transport equations* (BTE).

7.3.1 Boltzmann Transport Equation for Electrons

If an electron of charge e and mass m moves in homogeneous electric field $\vec{E}$ and magnetic field $\vec{H}$ at right angles to each other, then the *Lorentz force* on it is given by

$$\vec{F} = m\vec{\alpha} = -e\left[\vec{E} + \frac{1}{c}\vec{v}\times\vec{H}\right], \tag{7.27}$$

here $\vec{\alpha}$ is the acceleration of the electron.

If $\vec{E}$ be in the xy-plane and $\vec{H}$ in the z-direction, then

$$m\alpha_x = -e\left[E_x + \frac{1}{c}v_yH_z\right] \tag{7.28}$$

$$m\alpha_y = -e\left[E_y + \frac{1}{c}v_zH_z\right] \tag{7.29}$$

$$\text{and } m\alpha_z = 0. \tag{7.30}$$

The Boltzmann transport eqn (7.25) then becomes

$$v_x\frac{\partial f}{\partial x} + v_y\frac{\partial f}{\partial y} + v_z\frac{\partial f}{\partial z} - \left[\frac{eE_x}{m} + \frac{e}{c}\frac{v_y}{m}H_z\right]\frac{\partial f}{\partial v_x} - \left[\frac{eE_y}{m} - \frac{e}{c}\frac{v_x}{m}H_z\right]\frac{\partial f}{\partial v_y} + 0 = \left(\frac{\partial f}{\partial t}\right)_{\text{coll}} \tag{7.31}$$

Eqn (7.31) is the *Boltzmann transport equation for the electron.*

7.3.2 Lorentz Solution of Boltzmann Transport Equation for the Electron

Lorentz assumed that the electric field $\vec{E}$ is along x-direction and there is no magnetic field. Then the function f is independent of t, y and z. Under this condition eqn (7.31) becomes

$$\left(\frac{\partial f}{\partial t}\right)_{\text{coll}} = v_x\frac{\partial f}{\partial x} - \frac{eE_x}{m}\frac{\partial f}{\partial v_x}. \tag{7.32}$$

Lorentz assumed that:

(a) *the collisions of electron is elastic,*

(b) *the scattering of electron is isotropic,*

(c) *the equilibrium distribution function f does not change due to electron scattering* and

(d) *the difference of equilibrium function f from the equilibrium forms f_0 so that*

$$\left(\frac{\partial f}{\partial t}\right)_{\text{coll}} = \frac{f - f_0}{\tau}, \tag{7.33}$$

where τ is called the *relaxation time.*

Eqn (7.33) represents the rate at which the distribution function approaches the condition of equilibrium when $\vec{E}$ or the temperature gradient is absent.

The solution of eqn (7.33) is then

$$(f - f_0) = Axe^{-t/x}, \tag{7.34}$$

where A is a constant. Eqn (7.34) shows that the decay of $(f - f_0)$ is exponential.

From eqns (7.33) and (7.34), we get

$$\left(\frac{\partial f}{\partial t}\right)_{\text{coll}} = v_x \frac{\partial f}{\partial x} - \frac{eE_x}{m}\frac{\partial f}{\partial v_x} = -\frac{f - f_0}{\tau}. \tag{7.35}$$

Replacing f by f_0 in the right-hand side of eqn (7.35), we get

$$v_x \frac{\partial f_0}{\partial x} - \frac{eE_x}{m}\frac{\partial f_0}{\partial v_x} = -\frac{(f - f_0)}{\tau} \tag{7.36}$$

$$\text{or, } f - f_0 = -\tau\left[v_x \frac{\partial f_0}{\partial x} - \frac{eE_x}{m}\frac{\partial f_0}{\partial v_x}\right]$$

$$\text{or, } f = f_0 - \tau\left[v_x \frac{\partial f_0}{\partial x} - \frac{eE_x}{m}\frac{\partial f_0}{\partial v_x}\right]. \tag{7.37}$$

Eqn (7.37) is the solution of the *Boltzmann transport equation* for the electron. By the application of this solution we can obtain more refined results for electrical and thermal conductivities.

7.4 Sommerfeld's Theory of Electrical Conductivity

Sommerfeld in 1928 gave the theory of electrical conduction in metal using *Fermi-Dirac* (FD) *statistics* instead of *Maxwell-Boltzmann* (MB) statistics as the latter was unable to explain the results of *Wiedemann-Franz law.*

In presence of electric field $\vec{E}$,

$$v_x \frac{\partial f_0}{\partial x} - \frac{eE_x}{m}\frac{\partial f_0}{\partial v_x} = -\frac{f}{\tau}. \tag{7.38}$$

When steady state is reached,

$$-\frac{eE_x}{m}\frac{\partial f}{\partial v_x} = \frac{f}{\tau}. \tag{7.39}$$

Again,

$$\frac{1}{m}\frac{\partial f_0}{\partial v_x} = \frac{1}{m}\frac{\partial f_0}{\partial E}\frac{\partial E}{\partial v_x} = \frac{\partial f_0}{\partial E}\left(\frac{1}{m}\frac{\partial E}{\partial v_x}\right) = \frac{\partial f_0}{\partial E} v_x.$$

$$\therefore f = \tau e E_x v_x \left(\frac{1}{m}\frac{\partial f_0}{\partial v_x}\right) = \tau e E_x \frac{\partial f_0}{\partial E} v_x = \tau e E_x v_x \left(\frac{\partial f_0}{\partial E}\right). \tag{7.40}$$

Eqn (7.40) represents the *distribution function of electrons.*

To calculate the number of electronic states per unit volume associated with element $dp_x\, dp_y\, dp_z$ in momentum space is

$$g(p) = \frac{1}{h^3}\, dp_x\, dp_y\, dp_z. \tag{7.41}$$

Considering electron spin,

$$g(p) = \frac{2}{h^3}\, dp_x\, dp_y\, dp_z. \tag{7.42}$$

The average number of occupied electron states in *absence of field at thermal* equilibrium is

$$n_0 = g(p)\, f_0(p) = \frac{2}{h^3}\, f_0(p)\, dp_x\, dp_y\, dp_z, \tag{7.43}$$

where $f_0(p)$ is the *Fermi-distribution function* in terms of total momentum p.

If now an electric field E_x be applied along x-direction, then *in the steady state* the average number of electrons per unit volume in the range p_x to $p_x + dp_x$, p_y to $p_y + dp_y$, p_z to $p_z + dp_z$ is

$$n = \frac{2}{h^3} f(p)\, dp_x\, dp_y\, dp_z. \tag{7.44}$$

Therefore, the current density is

$$J_x = \sum -Nev_x = -\frac{2e}{h^3} \iiint v_x f(p)\, dp_x\, dp_y\, dp_z. \tag{7.45}$$

Substituting the value of $f(p)$ from eqn (7.40) in eqn (7.45), we get

$$J_x = -\frac{2e}{h^3} \iiint v_x \tau e E_x v_x \left(\frac{\partial f_0}{\partial E}\right) dp_x\, dp_y\, dp_z = -\frac{2e^2}{h^3} \iiint v_x^2 \tau E_x \left(\frac{\partial f_0}{\partial E}\right) dp_x\, dp_y\, dp_z \tag{7.46}$$

$$= -\frac{2e^2 E_x}{h^3} \int_0^\infty \frac{v^2}{3} \tau \left(\frac{\partial f_0}{\partial E}\right) 4\pi p^2\, dp. \tag{7.47}$$

$$\because dp_x\, dp_y\, dp_z = 4\pi p^2\, dp \text{ and } \langle v_x^2 \rangle = v^{2/3}.$$

Again,

$$\because E = \frac{1}{2} mv^2,\; v^2 = \frac{2E}{m}. \tag{7.48}$$

Also

$$E = \frac{p^2}{2m},\; p = \sqrt{2mE}.$$

$$\therefore dE = \frac{2p\, dp}{2m} \text{ or, } p\, dp = m\, dE. \tag{7.49}$$

Substituting the value of v^2 and $p\,dp$ in eqn (7.48), we get

$$J_x = -\frac{2e^2E_x}{h^3}\int_0^\infty \frac{1}{3}\left(\frac{2E}{m}\right)\tau\frac{\partial f_0}{\partial E}4\pi\sqrt{2mE}\,m\,dE = -\frac{16\pi e^2}{3h^3}(2m)^{1/2}E_x\int_0^\infty \tau E^{1/2}\frac{\partial f_0}{\partial E}\,dE$$

$$= \frac{16\pi e^2}{3h^3}(2m)^{1/2}E_x\tau E^{3/2}\left[\because \int_0^\infty \frac{\partial f_0}{\partial E}dE = -1\right]. \quad (7.50)$$

The Fermi energy at 0 K is

$$E_{F(0)} = \frac{h^2}{2m}\left(3\pi^2N\right)^{2/3}.$$

$$\therefore J_x = \frac{16\pi e^2(2m)^{1/2}}{3h^3}E_x\tau\left(\frac{h^2}{2m}\right)^{3/2}\left(2\pi^2N\right). \quad (7.51)$$

Now the *electrical conductivity,*

$$\sigma = \frac{J_x}{E_x} = \frac{Ne^2\tau_F}{m}, \quad (7.52)$$

where N is the *electron density.*

7.5 Thermal Conductivity

Thermal conductivity is defined as the ratio of thermal energy flux to the temperature gradient giving rise to energy flow.

If H be the thermal energy carried by the electron per unit area per second and $\frac{dT}{dx}$, the temperature gradient in the x-direction, then *thermal conductivity,*

$$K = \frac{H}{dT/dx}. \quad (7.53)$$

To calculate the flux H across a plane (S) perpendicular to the x-axis, we assume that, the average component of electron velocity in the x-direction is zero at the scattering centre but the velocity of the electron becomes v_{rms} after it has traversed a distance λ, the mean free path, at which it is scattered again by an ion of the crystal lattice.

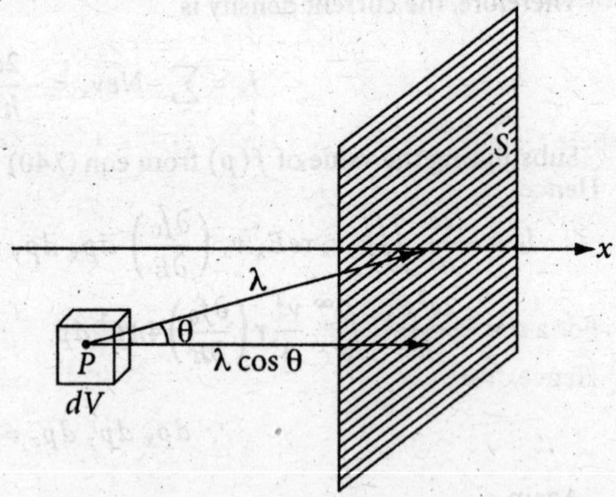

Fig. 7.3: Flux H crossing plane S

We consider an element of volume dV containing the scattering centre at P. The amount of flux crossing the plane between solid angle Ω and $\Omega + d\Omega$ is given by

$$dH = Uf(\Omega)\,d\Omega, \quad (7.54)$$

where U represents the energy carried by a single electron. $f(\Omega)\,d\Omega$ is the number of electron per unit area per unit time in the incremental solid angle $d\Omega$.

If the velocity distribution in dV be equally probable in all possible directions, then the tip of the velocity vector will lie on the surface of the sphere with centre at P. The number of electron crossing the plane S per unit area per unit time along the normal direction is $mv_{\text{rms}} \cos\theta$, where n is the number of electrons per unit volume. Then

$$f(\Omega)\, d\Omega = nv_{\text{rms}} \cos\theta \frac{d\Omega}{4\pi} = \frac{1}{2} nv_{\text{rms}} \cos\theta \sin\theta\, d\theta. \tag{7.55}$$

Now

$$U = U(0) + \lambda\cos\theta \left(\frac{\partial U}{\partial x}\right)_{x=0} + \cdots \text{ [by Taylor's theorem]}. \tag{7.56}$$

If $\frac{\partial U}{\partial x}$ is small over a distance comparable with λ, then the series in eqn (7.56) will be convergent and we may neglect the terms after the second. So

$$U = U(0) + \lambda\cos\theta \left(\frac{\partial U}{\partial x}\right)_{x=0}. \tag{7.57}$$

Now substituting the value of $f(\Omega)\, d\Omega$ and U from eqns (7.56) and (7.57) in eqn (7.54) and then integrating, we get

$$\begin{aligned} H = \int dH &= \int_0^\pi \left[U(0) + \lambda\cos\theta \left(\frac{\partial U}{\partial x}\right)_{x=0}\right] \times \left[\frac{1}{2} nv_{\text{rms}} \cos\theta \sin\theta\right] d\theta \\ &= \frac{1}{2} nv_{\text{rms}} \lambda \left(\frac{\partial U}{\partial x}\right)_{x=0} \int_0^\pi \cos^2\theta \sin\theta\, d\theta \\ &= \frac{1}{3} nv_{\text{rms}} \lambda \left(\frac{\partial U}{\pi x}\right)_{x=0} \left[\because \int_0^\pi \cos^2\theta \sin\theta\, d\theta = \frac{2}{3}\right]. \end{aligned} \tag{7.58}$$

Due to the presence of temperature gradient thermal diffusion of electrons along x-direction takes place causing the variation of electron energy U.

$$\therefore n\left(\frac{dU}{\partial x}\right)_{x=0} = n\frac{\partial U}{\partial T}\frac{\partial T}{\partial x} = C_v \frac{\partial T}{\partial x}, \tag{7.59}$$

where $C_v = n\frac{\partial U}{\partial T} = \frac{\partial}{\partial T}(nU)$ is the *specific heat of electron gas.*

Hence,

$$H = \frac{1}{3} v_{\text{rms}} \lambda C_v \frac{dT}{dx}. \tag{7.60}$$

For a *classical ideal gas*, $C_v = \frac{3}{2}nk$, where k is the *Boltzmann constant.*

Hence, from eqn (7.53), *thermal conductivity*

$$K = \frac{H}{\left(\frac{dT}{dx}\right)} = \frac{1}{3} v_{\text{rms}} \lambda \times \frac{3}{2} nk \text{ or, } K = \frac{1}{2} nkv_{\text{rms}} \lambda. \tag{7.61}$$

7.6 Wiedemann-Franz Law

Wiedemann and Franz in 1853 found that good conductors of electricity are also the good conductors of heat because both of these phenomena owe their origin to the common cause, viz., the electrons.

The states that, *at a particular temperature the ratio of thermal conductivity K to the electrical conductivity σ is the same for all metals, i.e.,*

$$\frac{K}{\sigma T} = \text{constant}. \tag{7.62}$$

This is known as *Wiedemann-Franz law.*

Now to obtain the relation, we know that the *electrical conductivity* of a metal at temperature T is given by

$$\sigma = \frac{ne^2\lambda}{\sqrt{3mkT}}, \tag{7.63}$$

where e and m are respectively the charge and mass of an electron, λ the mean free path, n the density of electrons and k, the Boltzmann constant.

Again, the *thermal conductivity* K is given by

$$K = \tfrac{1}{3}C_v\lambda v_{\text{rms}}, \tag{7.64}$$

where $C_v = \frac{3}{2}Kn$, the *specific heat of electron gas.*

Hence, the ratio of K to σ is

$$\frac{K}{\sigma} = \frac{1}{3}C_v\lambda v_{\text{rms}} \times \frac{\sqrt{3mkT}}{ne^2\lambda} = \frac{1}{3}\left(\frac{3}{2}kn\right)\sqrt{\frac{3kT}{m}}\frac{\sqrt{3mkT}}{ne^2} = \left(\frac{3}{2}\frac{k^2}{n^2}\right)T \tag{7.65}$$

$$\text{or,}\ \frac{K}{\sigma} = \left(\frac{3}{2}\frac{k^2}{e^2}\right)T = LT\ \left[\because C_v = \frac{3}{2}kn;\ v_{\text{rms}} = \sqrt{\frac{3kT}{m}}\right] \tag{7.66}$$

where $L = \left(\frac{3}{2}\frac{k^2}{e^2}\right) = 1.12 \times 10^{-8}\ \text{W}\Omega/\text{K}^2$.

The above value of L largely deviates from the experimental value. Sommerfeld using *FD-statistics* to the free electron gas calculated the value of C_v as

$$C_v = \frac{n\pi^2k^2T}{2E_F}, \tag{7.67}$$

where

$$E_F = \tfrac{1}{2}V_F^2. \tag{7.68}$$

Then,

$$K = \frac{1}{6}\frac{n\pi^2k^2T\lambda V_F}{E_F}. \tag{7.69}$$

So

$$\frac{K}{\sigma} = \frac{1}{6}\left[\frac{n\pi^2k^2\lambda V_F}{\frac{1}{2}mV_F^2}\right]T\left[\frac{e^2}{3\pi^2}\frac{3\pi^2n\lambda}{mv_F}\right] \text{ or, } \frac{K}{\sigma T} = \frac{1}{3}\left(\frac{\pi^2k^2}{e^2}\right) = L = 2.45 \times 10^{-8}\ \text{W}\Omega/\text{K}^2, \tag{7.70}$$

where $L = \frac{1}{3}\frac{\pi^2k^2}{e^2}$ is known as *Lorentz number.*

Eqn (7.70) is in good agreement with the experimental results. Table 7.1 shows $\frac{K}{\sigma T}$ values for some of the metals.

Table 7.1

Metals	K(W/cm-K)	σ(℧/cm)	$\frac{K}{\sigma T} \times 10^8$ (WΩ/K²)
Li	0.72	1.17	2.22
Be	2.3	3.57	2.36
Fe	0.80	1.12	2.61
Ag	4.18	6.62	2.31
Cu	3.85	6.41	2.20

7.7 Merits and Demerits of Free Electron Theory

Merits

1. The theory proves the validity of *Ohm's law.*
2. It can explain the phenomena of *electrical and thermal conduction* in metals and verifies *Wiedemann-Franz law.*
3. It explains the high *lustre* and complete *opacity* of metals.

Demerits

1. Electrical conductivity of semiconductor or insulator could not be explained by this theory.
2. $\frac{k}{\sigma T}$ is constant in view of classical free electron theory but it is not so at low temperature.
3. The theory predicts that the resistivity ρ is proportional to $\sqrt{T}$ but actually ρ varies as T.
4. The theory fails to explain *heat capacity* and *paramagnetic susceptibility* of the conduction electrons.
5. The *Classical theory fails to account for the long mean free path* (more than one centimetre) at low temperature.

Temperature Dependence of Resistivity: If an electric field E is applied to the metal the electron will acquire an acceleration, $a = \frac{eE}{m}$, where e and m are the charge and mass of the electron. If C be the random velocity (rms velocity) and λ the mean free path, then the velocity acquired by the electron just before the next collision is

$$u = at = \frac{eE}{m}\left(\frac{\lambda}{C}\right),$$

where $t = \lambda/C$.

The *average drift velocity* of the electron is

$$v_d = \frac{u}{2} = \frac{eE\lambda}{2mC} = \frac{eE\lambda}{2m \times \sqrt{\frac{3kT}{m}}} \quad \left[\therefore C = \sqrt{\frac{3kT}{m}}\right]$$

$$= \frac{eE\lambda}{\sqrt{12mkT}}. \qquad (7.71)$$

Using this value of v_d in eqn (7.14), the resistivity

$$\rho = \frac{1}{ne\mu} = \frac{1}{ne\left(\frac{v_d}{E}\right)} = \frac{E}{ne \times \frac{eE\lambda}{\sqrt{12mkT}}} \text{ or, } \rho = \frac{\sqrt{12mkT}}{ne^2\lambda}. \qquad (7.72)$$

Since m, k, n, e, λ are constant, $\rho \propto \sqrt{T}$. Thus, the *resistivity of a conductor* is proportional to the square of the absolute temperature (T). But experimental result shows that $\rho \propto T$.

Molar Specific Heat: The KE of one gm-mol of monovalent metal is

$$U = \frac{1}{2}mC^2N_A = \frac{1}{2}mN_A\left(\frac{3kT}{m}\right) = \frac{3}{2}RT \; [\because N_Ak = R],$$

where N_A is the *Avogadro number* and R, the *universal gas constant.*

$\therefore$ the molar specific heat of the electron is

$$C_v = \frac{dU}{dT} = 1.5R \text{ J/mol/K}. \qquad (7.73)$$

The value of $C_v = 1.15R$ is about *ten times* the experimental value. *This disagreement is due to the wrong assumption that all the free electrons contribute to the specific heat.*

7.8 Quantum Theory of Free Electrons

We consider that the free electrons move freely in a solid in a constant potential V. Assuming $V = 0$, the *Schrödinger equation* for a free electron of mass m is

$$-\frac{\hbar^2}{2m}\nabla^2\psi = E\psi,$$

where E is the total energy of the electron.

The plane wave solution of the above equation is

$$\psi = \psi_0\, e^{i\vec{k}\cdot\vec{r}}, \tag{7.74}$$

where $k = \frac{2\pi}{\lambda}$ is the *wave vector* and $r^2 = x^2 + y^2 + z^2$.

$$\therefore \frac{\partial^2\psi}{\partial x^2} = -k_x^2\psi, \quad \frac{\partial^2\psi}{\partial y^2} = -k_y^2\psi, \quad \frac{\partial^2\psi}{\partial z^2} = -k_z^2\psi.$$

So

$$\nabla^2\psi = \frac{\partial^2\psi}{\partial x^2} + \frac{\partial^2\psi}{\partial y^2} + \frac{\partial^2\psi}{\partial z^2} = -\left(k_x^2 + k_y^2 + k_z^2\right)\psi = -k^2\psi. \quad \left[\because k_x^2 + k_y^2 + k_z^2 = k^2\right]. \tag{7.75}$$

$$\therefore -\frac{\hbar^2}{2m}\nabla^2\psi = -\frac{\hbar^2}{2m}(-k^2\psi) = E\psi, \quad \left[\because \frac{\hbar^2k^2}{2m} = E\right]$$

$$\text{or, } E = \frac{\hbar^2k^2}{2m} = \left(\frac{h}{2\pi}\right)^2\left(\frac{2\pi}{\lambda}\right)^2\cdot\frac{1}{2m} = \frac{h^2}{2m\lambda^2} = \frac{h^2p^2}{2mh^2}, \quad \left[\because \lambda = \frac{h}{p}\right]$$

$$\therefore E = \frac{p^2}{2m}. \tag{7.76}$$

Eqn (7.76) is the energy of a free particle. It is evident from Fig. (7.4) that the energy E is continuous. If L be the length of the solid, then applying the boundary condition $k = \frac{2\pi n}{L}$, we get

$$E_n = \frac{\hbar^2k^2}{2m} = \frac{n^2h^2}{8mL^2}. \tag{7.77}$$

For $L = 1$ cm, the permitted energy levels differ by about 10^{-19} eV which is very small, the energy levels are, therefore, said to form a *quasi-continuous band.*

Electrical Conductivity

If an external field E is applied to the metal, the force exerted on a free electron is $-eE$. If $\frac{dp}{dt}$ represents the rate of change of momentum, then

$$-eE = \frac{dp}{dt}.$$

But according to *de Broglie's hypothesis,*

$$p = \frac{h}{\lambda} = \frac{h}{2\pi}\frac{2\pi}{\lambda} = \hbar k.$$

$$\therefore -eE = \frac{d}{dt}(\hbar k) = \hbar\frac{dk}{dt} \text{ or, } dk = -\frac{eE}{\hbar}dt. \tag{7.78}$$

Eqn (7.78) shows that, the origin of k-space moves through a distance dk in time dt in presence of the external field E. Due to collisions of the free electrons with the imperfections, the displacement of k-space becomes steady ($= \Delta k$) and then τ is the *average collision time.*

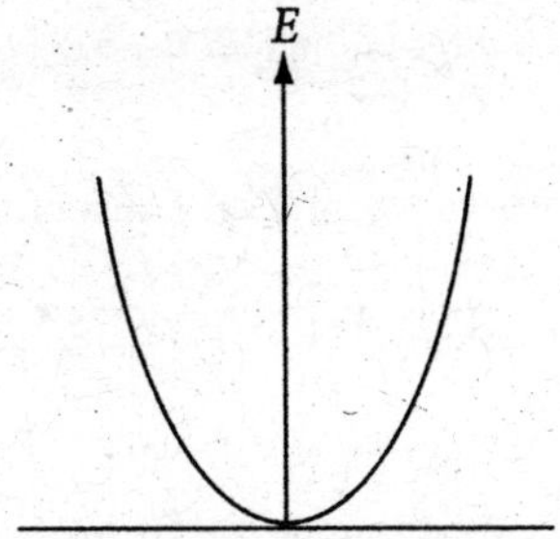

Fig. 7.4: E as a function of k

$$\therefore \Delta k = -\frac{eE\tau}{\hbar}.$$

Now the change in velocity is

$$\Delta v = \frac{\hbar}{m}\Delta k = -\left(\frac{\hbar}{m}\right)\left(\frac{eE\tau}{\hbar}\right) = -\frac{eE\tau}{m}. \tag{7.79}$$

Now the *current density,*

$$J = n(-e)\Delta v, \ (n = \text{electron density})$$

$$= \frac{ne^2E\tau}{m} = \sigma E. \tag{7.80}$$

Eqn (7.80) is the *Ohm's law,* where

$$\sigma = \frac{ne^2\tau}{m} \tag{7.81}$$

is the *electrical conductivity* of the metal.

7.9 Free Electron Gas in One-dimensional Box

We consider an electron of mass m confined in a one-dimensional line in a crystal of length L bounded by infinite potential barriers at $x = 0$ and $x = L$ as shown in Fig. 7.5. The problem is equivalent to an electron moving in one-dimensional potential box, the PE within the box being zero.

So

$$V(x) = \begin{cases} 0 & \text{for } 0 < x < L \\ \infty & \text{for } x \le 0 \text{ and } x \ge L. \end{cases} \tag{7.82}$$

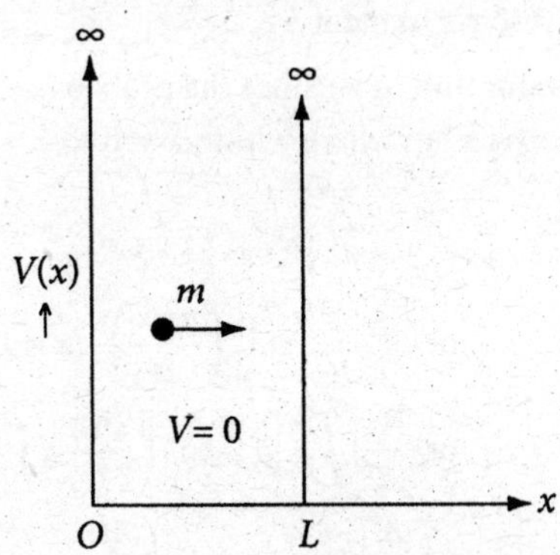

Fig. 7.5: An electron in one-dimensional potential box

Wave Function: The *Schrödinger time-independent equation* for the electron is

$$\frac{d^2\psi_n}{dx^2} + \frac{2m}{\hbar^2}(E_n - V)\psi_n = 0, \tag{7.83}$$

where ψ_n is the *wave function* for the electron in the nth state of energy E_n. Since $V = 0$,

$$\frac{d^2\psi_n}{dx^2} + \frac{2mE_n}{\hbar^2}\psi_n = 0. \tag{7.84}$$

The general solution of eqn (7.84) is

$$\psi_n(x) = A\sin kx + B\cos kx, \tag{7.85}$$

where A and B are constants and $k = \sqrt{\dfrac{2mE_n}{\hbar^2}}$.

Applying the *boundary condition*, $\psi_n(0) = 0$ and $\psi_n(L) = 0$, we get $B = 0$.

$$\therefore \psi_n(x) = A\sin kx. \tag{7.86}$$

Again, since $\psi_n(L) = 0$,

$$\therefore \sin kL = 0, \ (\because A \text{ cannot be zero})$$

$$\text{or, } k = \frac{n\pi}{L}, \tag{7.87}$$

where $n = 1, 2, 3, \ldots$.

Therefore, the allowed wave function is

$$\psi_n(x) = A\sin\left(\frac{n\pi}{L}\right)x. \tag{7.88}$$

From eqn (7.87),

$$\frac{2mE_n}{\hbar^2} = \frac{n^2\pi^2}{L^2} \text{ or, } E_n = \frac{\hbar^2}{2m}\frac{n^2\pi^2}{L^2} = \frac{n^2h^2}{8mL^2}. \tag{7.89}$$

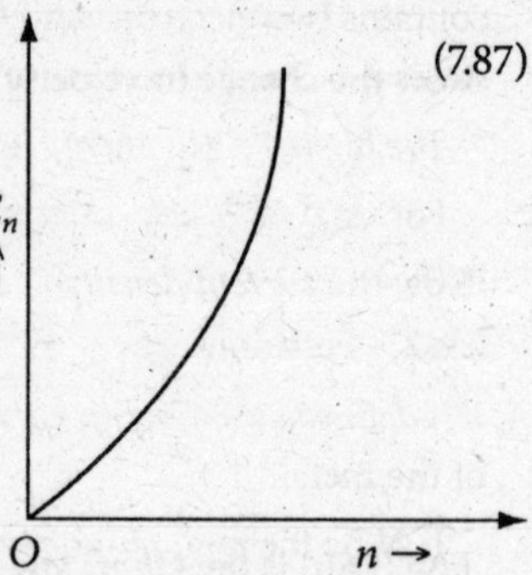

Fig. 7.6: Variation of E_n with n

Thus, $E_n \propto n^2$. The plot of E_n against n is shown in Fig. 7.6.

The allowed wave function $\psi_n(x)$ and energy E_n exist for integral values of n, called the *quantum number*. Thus, the energy spectrum of the electron is *discrete* in nature.

Evaluation of A: Since the probability of finding the electron anywhere within $x = 0$ to $x = L$ is unity,

$$\therefore \int_0^L \psi_n^*(x)\,\psi_n(x)\,dx = 1$$

$$\text{or, } A^2\int_0^L \sin^2\left(\frac{n\pi x}{L}\right)dx = 1$$

$$\text{or, } \frac{A^2}{2}\int_0^L \left[1-\cos\left(\frac{2n\pi}{L}x\right)\right]dx = 1$$

$$\text{or, } \frac{A^2}{2}\int_0^L dx = 1 \text{ or, } A = \sqrt{2/L}.$$

$$\therefore \psi_n(x) = \sqrt{2/L}\sin\left(\frac{n\pi x}{L}\right). \tag{7.90}$$

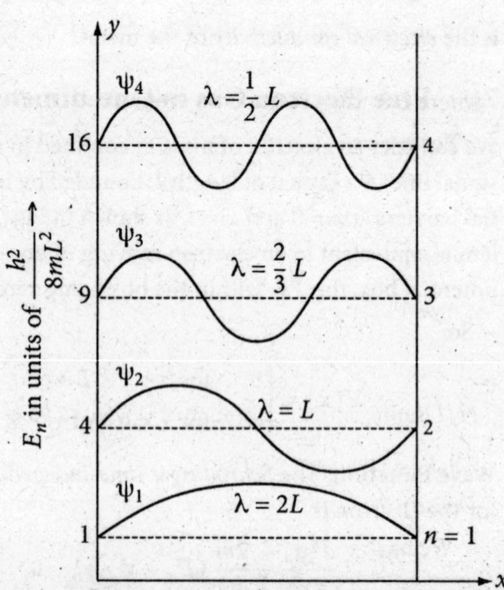

Fig. 7.7: ψ_n and E_n values of an electron in one-dimensional box

The energy levels and the wave functions of the electron for $n = 1, 2, 3, 4, \ldots$ are shown in Fig. 7.7.

7.9.1 Fermi Level

Suppose, we have N electrons to be accommodated on the line following *Pauli's exclusion principle.* According to this principle each quantum state can be occupied by at the most one electron.

In a solid a conduction electron is characterised by *two quantum numbers, n and m_s*, where n is the *principal quantum number* and m_s *is the magnetic spin quantum number.* Now for each value of n, m_s can have two possible values $+\frac{1}{2}$ and $-\frac{1}{2}$. Hence, each energy level can accommodate two electrons with $m_s = \pm\frac{1}{2}$, one with spin up (↑) and other spin down (↓).

If there are seven electrons, then the energy state $n = 1, 2, 3$ each contains two electrons and the state $n = 4$ has only one electron. The states $n = 5$ and above are empty as shown in Fig. 7.8.

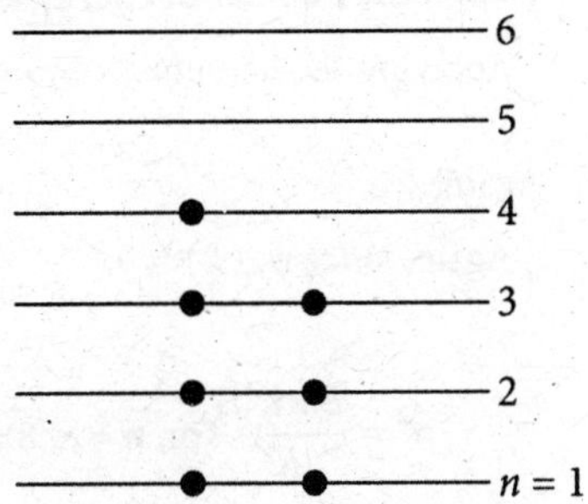

Fig. 7.8: Electron distribution in energy levels

The highest filled energy level at 0 K is called the Fermi level.

For total of N non-interacting electrons, the number of filled levels is $N/2$.

7.9.2 Fermi Energy

It is defined as the energy corresponding to the highest filled level. Fermi energy E_f depends on the temperature of the metal.

If N be the total number of electrons to be accommodated on a line, then for even N, $2n_f = N$, where n_f is the *principal quantum number* of the Fermi level.

Now for a free particle confined in a one-dimensional box, the energy in the nth quantum state is

$$E_n = \frac{\hbar^2}{2m}\left(\frac{n\pi}{L}\right)^2,$$

where L is the length of one side of the box.

For $n = n_f$,

$$E_f = \frac{\hbar^2}{2m}\left(\frac{n_f\pi}{L}\right)^2 = \frac{\hbar^2}{2m}\left(\frac{N\pi}{2L}\right)^2 \quad [\because n_f = N/2]$$

$$= \frac{1}{4\pi^2}\frac{h^2}{2m}\left(\frac{N}{2L}\right)^2\pi^2$$

$$\text{or, } E_f = \frac{h^2}{2m}\left(\frac{N}{4L}\right)^2. \tag{7.91}$$

N/L represents the number of electrons per unit length.

7.9.3 Density of States

Density of states is the number of electronic states per unit energy range and is represented by $Z(E)$.

We have energy of the electron in the nth level

$$E = \frac{\hbar^2}{2m}\left(\frac{n\pi}{L}\right)^2 = \frac{\hbar^2}{2m}\left(\frac{\pi}{L}\right)^2 n^2.$$

$$\therefore \frac{dE}{dn} = \frac{\hbar^2}{2m}\left(\frac{\pi}{L}\right)^2 2n = \frac{h^2}{4\pi^2}\frac{1}{2m}\frac{\pi^2}{L^2}2n = \frac{h^2}{m}\frac{n}{4L^2}.$$

$$\therefore \frac{dn}{dE} = \frac{4m\,L^2}{h^2\;n}. \qquad (7.92)$$

Eqn (7.92) gives *the number of energy levels in the energy range* E and $E + dE$. But there are two energy states corresponding to each level for spin up and spin down.

Hence, the number of energy levels in the energy range E and $E + dE$ is

$$Z(E) = 2\frac{dn}{dE} = \frac{8mL^2}{h^2}\cdot\frac{1}{n^2}. \qquad (7.93)$$

Again,

$$\because E = \frac{h^2}{8m}\frac{n^2}{L^2},$$

$$\therefore n^2 = \frac{8mL^2E}{h^2} \text{ or, } n = \sqrt{8mE}\frac{L}{h} \text{ or, } \frac{1}{n} = \frac{h}{L\sqrt{8mE}}. \qquad (7.94)$$

From eqns (7.93) and (7.94),

$$Z(E) = \frac{8mL^2}{h^2}\times\frac{h}{L\sqrt{8mE}} = \frac{4L}{h}\left(\frac{m}{2E}\right)^{1/2}. \qquad (7.95)$$

Eqn (7.95) represents the *density of states* in one-dimensional case. Fig. 7.9 shows that levels are filled from $E = 0$ to $E = E_f$ and beyond E_f all the levels are vacant. Thus, *Fermi level* divides the filled and unfilled levels in the metals.

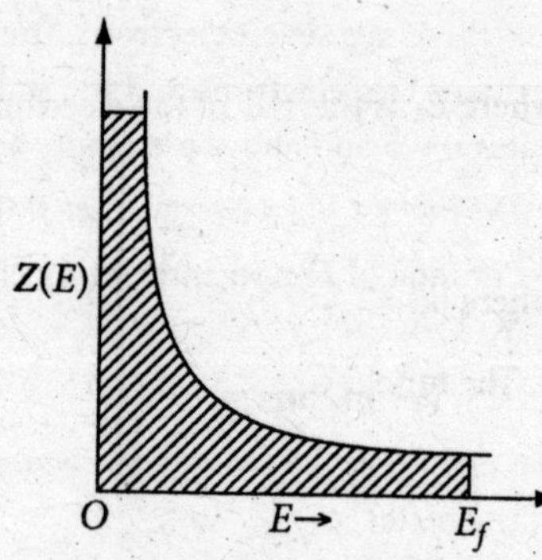

Fig. 7.9: Plot of $Z(E)$ versus E_f

7.9.4 Average KE in the Ground State

The total energy E_0 of all the N electrons in ground state is given by

$$E_0 = 2\sum_{n=1}^{N/2} E_n.$$

The factor 2 is due to the fact that each level contains 2 electrons and for N electrons the number of levels is $N/2$.

$$\text{or, } E_0 = 2\frac{\hbar^2}{2m}\left(\frac{\pi}{L}\right)^2\sum_{n=1}^{N/2} n^2. \qquad (7.96)$$

But

$$\sum_{n=1}^{s} n^2 = \frac{1}{6}s\left(2s^2 + 3s + 1\right) \simeq \frac{1}{3}s^3, \text{ for } s \geq 1.$$

$$\therefore \sum_{n=1}^{N/2} n^2 \simeq \frac{1}{3}\left(\frac{N}{2}\right)^3. \qquad (7.97)$$

$\therefore$ eqn (7.96) becomes

$$E_0 = 2\frac{\hbar^2}{2m}\left(\frac{\pi}{L}\right)^2\times\frac{1}{3}\left(\frac{N}{2}\right)^3 = \frac{1}{3}\frac{\hbar^2}{2m}\left(\frac{N\pi}{2L}\right)^2 N$$

$$= \frac{1}{3}NE_f \left[\because E_f = \frac{\hbar^2}{2m}\left(\frac{N\pi}{2L}\right)^2\right]. \qquad (7.98)$$

∴ the average KE in the ground state of the electron confined in a one-dimensional potential box is

$$\langle E_0 \rangle = \frac{E_0}{N} = \frac{1}{3} E_f. \tag{7.99}$$

7.10 Free Electron Gas in Three Dimensions

Considering the metal as a cubical electron gas container of side L, the *three-dimensional Schrödinger equation* for the electron of mass m is

$$\nabla^2 \psi_k(\vec{r}) + \frac{2m}{\hbar^2} E_k \psi_k(\vec{r}) = 0, \ [\because \text{potential energy } V = 0] \tag{7.100}$$

where E_k is the KE of the electron in the k state. The solution of eqn (7.100) is

$$\psi_k(\vec{r}) = A e^{i\vec{k}\cdot\vec{r}}, \tag{7.101}$$

where $|\vec{k}| = \sqrt{k_x^2 + k_y^2 + k_z^2}$ and $|\vec{r}| = \sqrt{x^2 + y^2 + z^2}$.

The function $\psi_k(\vec{r})$ in eqn (7.101) must be periodic in x, y, z with period L such that

$$\left.\begin{aligned} \psi(x+L, y, z) &= \psi(x, y, z) \\ \psi(x, y+L, z) &= \psi(x, y, z) \\ \psi(x, y, z+L) &= \psi(x, y, z) \end{aligned}\right\}. \tag{7.102}$$

Applying the first boundary condition to the wave function $\psi_k(\vec{r})$, we get

$$e^{i[k_x(x+L)+k_y y+k_z z]} = e^{i[k_x x+k_y y+k_z z]}.$$

$$\therefore e^{iK_x L} = 1 \text{ or, } k_x = 0, \ \pm\frac{2\pi}{L}, \ \pm\frac{4\pi}{L}, \ldots, \frac{2n\pi}{L}. \tag{7.103}$$

Similar results follow for k_y and k_z. If m_s be the *spin magnetic quantum number* of the electron, then its state can be represented by a set of *four* quantum numbers k_x, k_y, k_z and m_s.

The value of A in eqn (7.101) can be obtained from the condition of normalization, i.e.,

$$\int_0^V \psi_k^*(\vec{r})\, \psi_k(\vec{r})\, dV = 1 \text{ or, } \int_0^L \int_0^L \int_0^L A^2 e^{-i(\vec{k}^*\cdot\vec{r})} \cdot e^{i(\vec{k}\cdot\vec{r})}\, dx\, dy\, dz = 1 \text{ or, } A^2 L^3 = 1.$$

$$\therefore A = \left(\frac{1}{L^3}\right)^{1/2} = \left(\frac{1}{V}\right)^{1/2}.$$

∴ the normalised wave function for the electron in the kth state is

$$\psi_k(\vec{r}) = \left(\frac{1}{V}\right)^{1/2} e^{i(\vec{k}\cdot\vec{r})}. \tag{7.104}$$

7.10.1 Fermi Energy and Fermi Sphere

We have

$$k^2 = \frac{2mE}{\hbar^2} \text{ or, } E = E_f = \frac{\hbar^2 k^2}{2m}. \qquad (7.105)$$

E_f is the *Fermi energy* of the electron.

Fermi Sphere: A sphere of radius k_f in the k-space containing points representing the occupied states of a system of N free electrons in the ground state is called a *Fermi Sphere.*

The energy at the surface of the sphere is Fermi energy E_f given by

$$E_f = \frac{\hbar^2 k_f^2}{2m}.$$

The volume of the Fermi sphere is $\frac{4}{3}\pi k_f^3$ and the total number of electron state is

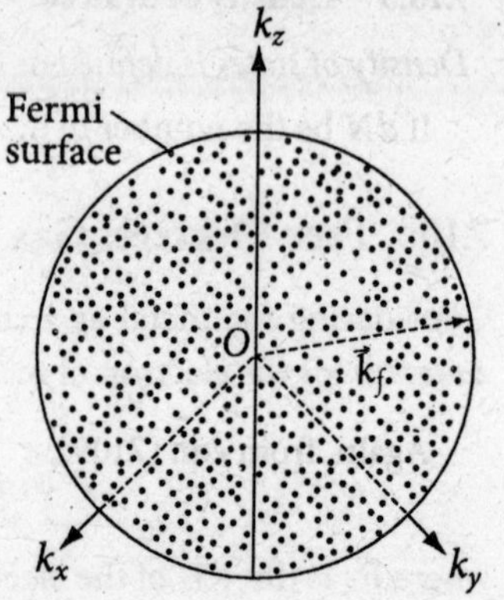

Fig. 7.10: The Fermi sphere

$$2\frac{\frac{4\pi}{3}k_f^3}{\left(\frac{2\pi}{L}\right)^3} = \frac{L^3 k_f^3}{3\pi^2} = \frac{V k_f^3}{3\pi^2} = N. \qquad (7.106)$$

The factor 2 comes because there are two allowed values of m_s for each value of k.

From eqn (7.106),

$$k_f = \left(\frac{3\pi^2 N}{V}\right)^{1/3}. \qquad (7.107)$$

So, k_f depends on the concentration (N/V) of the electron.

Then the *Fermi energy* E_{f_0} at $T = 0$ is expressed as

$$E_{f_0} = \frac{\hbar^2}{2m}k_f^2 = \frac{\hbar^2}{2m}\left(\frac{3\pi^2 N}{V}\right)^{2/3} = \frac{\hbar^2}{2m}(3\pi^2 n)^{2/3}, \qquad (7.108)$$

where $N/V = n$.

For $kT \ll E_f$, the variation of E_f with T is given by *Sommerfeld equation*

$$E_f(T) = E_{f_0}\left[1 - \frac{\pi^2}{12}\frac{(kT)^2}{E_{f_0}^2}\right]. \qquad (7.109)$$

Thus, E_{f_0} depends on electron concentration $(N/V) = n$ as well as mass m of the electron, Therefore, E_{f_0} remains the same if two identical metals are joined together. Further, the electrons have some definite energy even at $T = 0$ as $E_{f_0} \neq 0$.

7.10.2 Electron Velocity at Fermi Level

The velocity of an electron at the Fermi surface is given by

$$mv_f = \hbar k_f.$$

$$\therefore v_f = \frac{\hbar k_f}{m} = \frac{\hbar}{m}\left(\frac{3\pi^2 N}{V}\right)^{1/3} \quad \text{[using eqn (7.107)].} \qquad (7.110)$$

7.10.3 Density of States

Density of states is defined as the number of available energy states per unit energy range.

If dN be the number of available energy states between energy E to $E + dE$, then the density of states is

$$D(E) = \frac{dN}{dE}. \tag{7.111}$$

$$\therefore D(E)\,dE = dN \text{ and } \int_0^{E_f} D(E)\,dE = N. \tag{7.112}$$

Again, from eqn (7.109),

$$E_f = \frac{\hbar^2}{2m}\left(3\pi^2\frac{N}{V}\right)^{2/3}.$$

$$\therefore N^{2/3} = \left(\frac{V}{3\pi^2}\right)^{2/3}\frac{2m}{\hbar^2}E_f \text{ or, } N = \left(\frac{V}{3\pi^2}\right)\left(\frac{2m}{\hbar^2}\right)^{3/2}E_f^{3/2}. \tag{7.113}$$

From eqns (7.112) and (7.113), we get

$$\int_0^{E_f} D(E)\,dE = \frac{V}{3\pi^2}\left(\frac{2mE_f}{\hbar^2}\right)^{3/2}.$$

Expressing the above integral in indefinite form,

$$\int D(E)\,dE = \frac{V}{3\pi^2}\left(\frac{2mE}{\hbar^2}\right)^{3/2} = \int \frac{V}{3\pi^2}\left(\frac{2m}{\hbar^2}\right)^{3/2}E^{1/2}\,dE$$

$$\text{or, } D(E) = \frac{V}{3\pi^2}\left(\frac{2m}{\hbar^2}\right)^{3/2}E^{1/2}. \tag{7.114}$$

The plot of $D(E)$ against E is a parabola as shown in Fig. 7.11.

The density of states at the *Fermi surface* is obtained by putting $E = E_f$ in eqn (7.114), i.e.,

$$D(E_f) = \frac{V}{3\pi^2}\left(\frac{2m}{\hbar^2}\right)^{3/2}E_f^{1/2}. \tag{7.115}$$

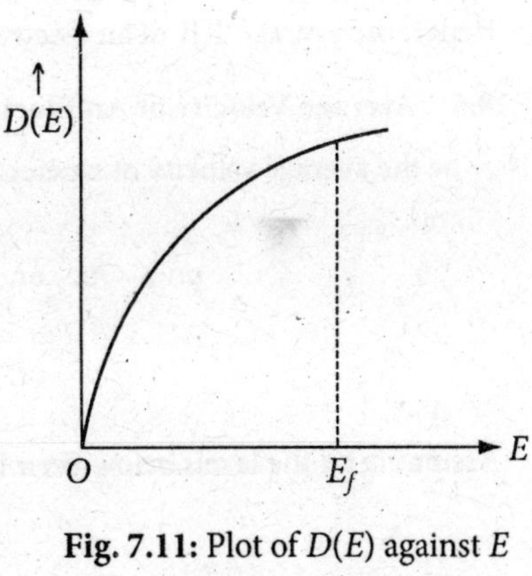

Fig. 7.11: Plot of $D(E)$ against E

Eqn (7.115) is valid at 0 K.

If $f(E)$ be the *Fermi function*, then the density of filled energy states is given by $N(E) = D(E)\,f(E)$.

Now

$$f(E) = \frac{1}{e^{(E-E_f)/kT}+1},$$

where k is the *Boltzmann constant*.

At $T = 0$, $f(E) = 1$ for $E < E_f$, this implies that all energy levels for $E < E_f$ are occupied.

For $E > E_f$, $f(E) = 0$, i.e., all energy levels for $E > E_f$ are vacant.

$\therefore$ the number of filled states between E and $E + dE$ at temperature T is

$$N(E)\,dE = D(E)\,f(E)\,dE = \left(\frac{V}{3\pi^2}\right)\left(\frac{2m}{\hbar^2}\right)^{3/2}\frac{E^{1/2}\,dE}{e^{(E-E_f)/kT}+1}.$$

For $E > E_f$, $(E - E_f) \gg kT$, i.e., $(E - E_f)/kT \gg 1$.

$$\therefore N(E)\,dE = \frac{V}{3\pi^2}\left(\frac{2m}{\hbar^2}\right)^{3/2} \frac{E^{1/2}\,dE}{e^{(E-E_f)/kT}} \tag{7.116}$$

$$= \frac{\pi V}{3}\left(\frac{8m}{h^2}\right) \frac{E^{1/2}\,dE}{e^{(E-E_f)/kT}} \left[\because \hbar = \frac{h}{2\pi}\right]. \tag{7.117}$$

7.10.4 Average KE at $T = 0$

The average KE of an electron at absolute zero is given by

$$\langle E_0 \rangle = \frac{1}{N}\int_0^\infty E D(E) f(E)\,dE = \frac{V}{3\pi^2}\left(\frac{2m}{\hbar^2}\right)^{3/2} \frac{1}{N}\left[\int_0^{E_{f_0}} E E^{1/2} f(E)\,dE + \int_{E_{f_0}}^\infty E E^{1/2} f(E)\,dE\right]$$

[using the value of $D(E)$ from eqn (7.114)]

$$\text{or, } \langle E_0 \rangle = \frac{V}{3\pi^2}\left(\frac{2m}{\hbar^2}\right)^{3/2} \frac{1}{N}\int_0^{E_{f_0}} E^{3/2}\,dE = \frac{2}{5}\frac{V}{3\pi^2}\left(\frac{2m}{\hbar^2}\right)^{3/2} \frac{1}{N} E_{f_0}^{5/2}. \tag{7.118}$$

Using the value of N from eqn (7.113) and remembering that $E_f = E_{f_0}$ at $T = 0$, we get

$$\langle E_0 \rangle = \frac{3}{5} E_{f_0}. \tag{7.119}$$

Hence, the average KE of an electron at $T = 0$ is $(3/5)\,E_{f_0}$.

7.10.5 Average Velocity of An Electron

If v_f be the average velocity of an electron at the *Fermi surface*, then

$$m v_f = \hbar k_f \text{ or, } v_f = \frac{\hbar}{m} k_f = \frac{\hbar}{m}\left(3\pi^2 \frac{N}{V}\right)^{1/3} \left[\therefore k_f = 3\pi^2 \frac{N}{V}\right]$$

$$\text{or, } v_f^3 = \frac{\hbar^3}{m^3} 3\pi^2 \frac{N}{V} \text{ or, } N = \frac{V}{3\pi^2}\left(\frac{m}{\hbar}\right)^3 v_f^3.$$

Assuming all the levels below Fermi level to be occupied we can in general write,

$$N = \frac{V}{3\pi^2}\left(\frac{m}{\hbar}\right)^3 v^3. \tag{7.120}$$

$\therefore$ the number of energy states between v and $v + dv$ is

$$dN = \frac{V}{3\pi^2}\left(\frac{m}{\hbar}\right)^3 3v^2 dv.$$

So, the average speed of the electron at 0 K is given by

$$\langle v \rangle = \frac{1}{N}\int_0^{v_f} v\,dN = \frac{1}{N}\frac{V}{3\pi^2}\left(\frac{m}{\hbar}\right)^3 \int_0^{v_f} 3v^2 v\,dv = \frac{1}{N}\frac{V}{3\pi^2}\left(\frac{m}{\hbar}\right)^3 \left[\frac{3}{4}v^4\right]_0^{v_f} = \frac{1}{N}\frac{V}{3\pi^2}\left(\frac{m}{\hbar}\right)^3 \frac{3}{4} v_f^4$$

$$\text{or, } \langle v \rangle = \frac{3}{4} v_f, \left[\because N = \frac{V}{3\pi^2}\left(\frac{m}{\hbar}\right)^3 v_f^3\right]. \tag{7.121}$$

7.10.6 Electronic Specific Heat

If E and E_0 be the internal energy of the solid at any temperature T and at absolute zero ($T = 0$), then it can be shown that

$$E = E_0 + \frac{\pi^2}{6}(kT)^2 g\left[E_{f_0}\right], \tag{7.122}$$

where

$$g\left[E_{f_0}\right] = \frac{3N}{2E_{f_0}} = \frac{3N}{2kT_f}, \tag{7.123}$$

where T_f is the *Fermi temperature* given by $kT_f = E_f(0)$.

Now substituting eqn (7.122) in (7.123), we get

$$E = E_0 + \frac{\pi^2}{4}\frac{NkT^2}{T_f}. \tag{7.124}$$

∴ the electronic specific heat at constant volume is given by

$$C_v = \frac{dE}{dT} = \frac{\pi^2}{2}Nk\left(T/T_f\right). \tag{7.125}$$

If $E_{f_0} = 5$ eV and $T = 300$ K, then from eqn (7.125), $C_v = Nk/40$, which is a small fraction of the classical value $1.5Nk$. The reason is that, *the thermal excitation of the free electron is possible if these find unoccupied states within a few times kT in energy. Electrons close to the Fermi level E_f can only be excited to such levels, whereas the electrons having much less energy than E_f, cannot be thermally excited. Hence, a small fraction of the total number of electrons can contribute to C_v. But in classical theory all the electrons are assumed to contribute to C_v.* This explains the small value of $C_v(= Nk/40)$ against $1.5\,Nk$ obtained from classical theory.

Eqn (7.125) shows that $C_v \propto T$. Also at low temperature *Debye theory* predicts that $C_v \propto T^3$. So, the *total specific heat* of metal at low temperature can be expressed as

$$C_v = AT + BT^3, \tag{7.126}$$

where A and B are constants

$$\text{or,}\quad \frac{C_v}{T} = A + BT^2. \tag{7.127}$$

The variation of C_v/T against T^2 is shown in Fig. 7.12. This variation C_v with temperature is in agreement with experimental observation.

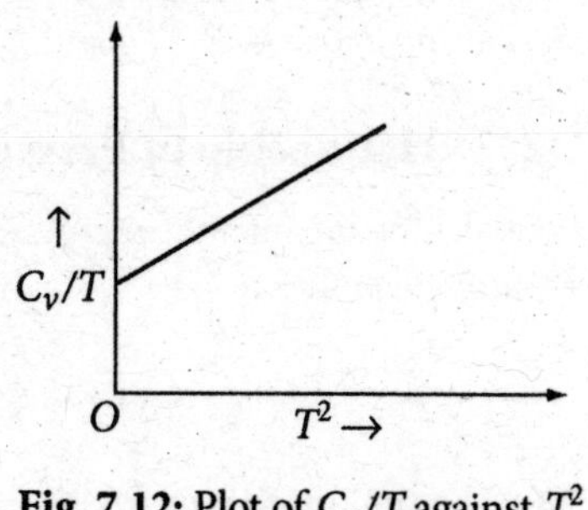

Fig. 7.12: Plot of C_v/T against T^2

7.11 Fermi Distribution Function

To know how the electrons are distributed among different energy states, we consider the electrons in the metal as *electron gas* obeying Fermi Dirac (FD)-statics. According to FD-statistics the probability that an electron will occupy an energy level E is given by

$$f(E) = \frac{1}{e^{(E-E_f)/kT} + 1}, \tag{7.128}$$

where E_f is the *Fermi level* which is assumed to be constant for a given metal and $F(E)$ is the *Fermi function*. From eqn (7.128),

$$\left.\begin{aligned} \text{at } T = 0\,\text{K for } E < E_f,\ f(E) &= 1 \\ E > E_f,\ f(E) &= 0 \end{aligned}\right\} \tag{7.129}$$

This implies that at 0 K all the levels below E_f are filled and those above E_f are vacant as shown in Fig. 7.13. As T increases Fermi function plot shows a deviation as shown in Fig. 7.14. At $T > 0$, if $E = E_f$, then from eqn (7.128),

$$f(E) = \tfrac{1}{2}. \tag{7.130}$$

Thus, *Fermi level is that energy state at which the probability of electron occupation is 1/2 at any temperature* $T > 0$ K. It is also the level of maximum energy of the filled states at 0 K.

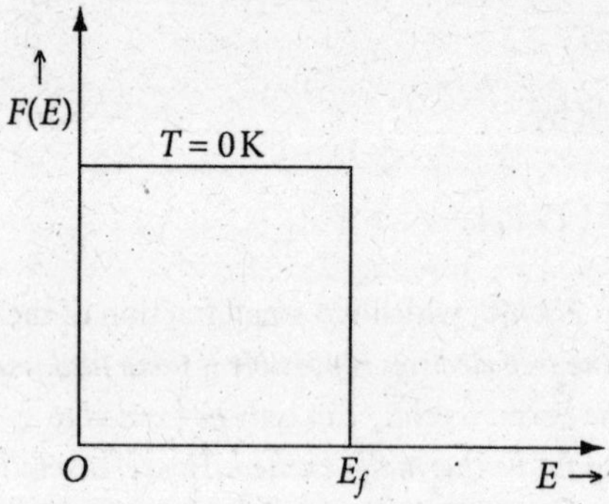

Fig. 7.13: Fermi distribution function at $T = 0$ K

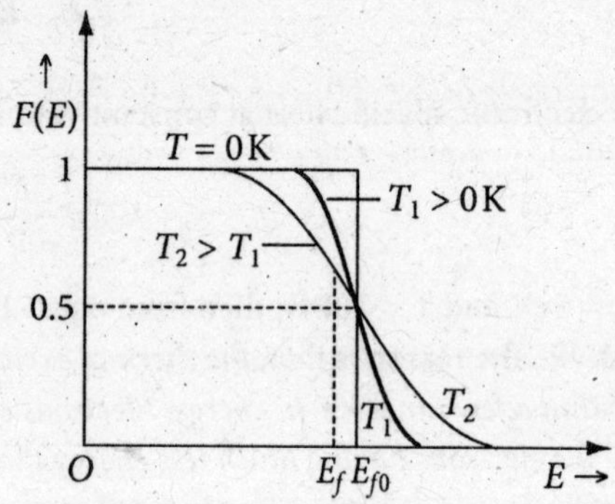

Fig. 7.14: Fermi distribution function at $T > 0$ K

Fig. 7.14 shows that the position of Fermi level is not fixed but changes with temperature. An approximate relationship between $E_f(T)$ and $E_f(0)$ is given by Eqn (7.109).

7.12 Relationship of Pressure and Volume of the Electron Gas

If E and V be the internal energy and volume of a system of particles at pressure P, then from the *principle of thermodynamics,*

$$P = -\frac{\partial E}{\partial V}.$$

If N be the number of free electrons with average energy $\langle E_0 \rangle$, then the total internal energy of the electron gas is $N\langle E_0 \rangle$.

So

$$P = -N\frac{\partial}{\partial V}\langle E_0 \rangle.$$

But from eqn (7.119),

$$P = -\frac{3}{5}N\frac{\partial E_{f_0}}{\partial V}.$$

Again,

$$E_{f_0} = \frac{\hbar^2}{2m}\left(\frac{3\pi^2 N}{V}\right)^{2/3}.$$

$$\therefore \frac{\partial E_{f_0}}{\partial V} = -\frac{2}{3}\frac{\hbar^2}{2m}(3\pi^2 N)^{2/3}\left(\frac{1}{V}\right)^{5/3}.$$

$$\therefore P = \frac{2}{5}N\frac{\hbar^2}{2m}(3\pi^2 N)^{2/3}\left(\frac{1}{V}\right)^{5/3} \text{ or, } P = \frac{2}{5}\frac{NE_{f_0}}{V}. \tag{7.131}$$

This gives the relation between pressure and volume of a free electron gas.

7.13 de Broglie Wavelength of an Electron having Energy E_f

If v_f and E_f be the velocity of an electron of mass m and Fermi energy respectively, then

$$\tfrac{1}{2}mv_f^2 = E_f \text{ or, } m^2v_f^2 = 2mE_f$$

or the momentum of the electron $p = mv_f = \sqrt{2mE_f}$.

According to *de Broglie*, the wavelength of an electron moving with Fermi energy E_f is

$$\lambda_f = \frac{h}{mv_f} = \frac{h}{\sqrt{2mE_f}} = \left(\frac{h}{\sqrt{2m}}\right)\left(\frac{2m}{\hbar^2}\right)^{1/2}\left(\frac{8m}{3n}\right)^{1/3} \quad \text{[putting the value of } E_f\text{]}$$

$$\text{or, } \lambda_f = 2(\pi/3n)^{1/3}. \tag{7.132}$$

7.14 Answers of Some Important Questions

7.1. *Explain the terms 'free electrons' and 'free electron gas'.*

Ans. *Free Electrons*: A solid metal is composed of a number of atoms in which the outermost electrons (i.e., the valence electrons) are very loosely bound to the nucleus. At room temperature these electrons get free from the influence of the parent atom and move from atom to atom as well as in the volume of the metal at *random chaotic motion*, just like the ideal gas molecules in a container. Thus, the *valence electrons* of the metal become the conduction electrons. Since these electrons are free from mutual interaction, the PE is zero and whole energy is kinetic. With the application of an external electric field, these electrons move in a direction opposite to the field and give rise to electrical conduction.

Free Electron Gas: The free electron Fermi gas is a gas of free and non-interacting electrons obeying *Pauli's exclusion principle*. The free electron gas is negatively charged.

The free electrons move inside the metal very much like atoms or molecules of an ideal gas in a container. These electrons are referred to as a *free electron gas*. The free electron gas differs from an ordinary gas in the following respects:

(a) the free electron gas is negatively charged but the charge on ideal gas is zero

(b) the concentration of an electron gas is quite large compared to that of a perfect gas.

7.2. *State the postulates of classical free electron theory of metals.*

Ans. See Art. 7.2.

7.3. *State some important phenomena which could be understood with the help of classical free electron theory of metal.*

Ans. The classical free electron theory of metals successfully explains the following phenomena:

(a) The theory verifies Ohm's law in electrodynamics.

(b) It explains the electric and thermal conductivities of metals.

(c) It derives the relation between electric and thermal conductivities (i.e., Wiedemann-Franz law).

(d) It explains the optical properties like *complete opacity* and *high lustre* of metals.

7.4. *Point out some of the failures of classical free electron theory of metals.*

Ans. Some of the failures of classical free electron theory are as follows:

(a) Though classical theory correctly predicts the room temperature variation of electrical resistivity (ρ), it fails to account the variation ρ with temperature at low temperature. The theory predicts that at low temperature $\rho \propto T^{-1}$ but actually $\rho \propto T$.

(b) The theory fails to explain *electronic heat capacity* of conduction electrons.

(c) It fails to explain the paramagnetic susceptibility of conduction electrons.

(d) $K/\sigma T$ is constant in view of classical electron theory but it is not so at low temperatures.

(e) The theory is unable to explain the existence of *long mean free path* (~ 10^8 to 10^9 interatomic spacings which is more than one centimetre) at low temperatures.

7.5. *Indicate some of the successes of quantum theory of free electrons.*

Ans. Following are some of the successes of *quantum theory of free electrons.*

(a) The theory can explain the electrical and thermal conductivities in metal and semiconductors.

(b) It can explain the paramagnetic susceptibility and heat capacity due to free electrons in metals.

(c) It can explain the existence of long mean path at low temperatures.

(d) The theory can explain the various electron emission processes from metal like *Richardson's thermionic emission*, etc.

(e) It can explain the variation of $K/\sigma T$ at low temperature.

7.6. *How can the difference between electronic specific heat of metal according to classical theory and quantum theory of free electron be explained? Which value is correct?*

Ans. The molar electronic specific heat of metal according to classical theory is $C_v = \frac{3}{2}kT$, whereas that according to quantum theory is $C_v = \frac{3}{2}Nk\left(T/T_f\right) = (0.01)\frac{3}{2}kT$.

$\therefore$ for $T \to 0$, $C_v \to 0$.

The difference between the two values of C_v is due to the fact that *classical theory assumes, all the electrons contribute to specific heat but according to quantum theory only a small fraction contributes to it.*

Quantum theory suggests that only those electrons which are within an energy range kT below Fermi level contribute to C_v. This is because if the electron gas is heated at temperature T, these electrons gain energy ~ kT and jump to the higher energy states. As the unfilled energy levels are not available to deep lying electrons, they cannot do so and cannot contribute to C_v.

The number of electrons which can contribute to C_v is of the order of

$$N\left(\frac{kT}{E_f}\right) = N\left(\frac{T}{T_f}\right).$$

$$\therefore C_v = \frac{3}{2}Nk = \frac{3}{2}\left(N\frac{T}{T_f}\right)k = \frac{3}{2}Nk\left(\frac{T}{T_f}\right).$$

For $T \to 0, C_v \to 0$.

Again, for $T = 100$ K and $T_f = 1000$ K, $T/T_f = 0.1$.

$$\therefore C_v \simeq (0.01)\tfrac{3}{2}Nk.$$

This value of C_v obtained from quantum mechanical consideration agrees with experimental results and is a correct value of C_v.

7.7. *What are Fermi sphere and Fermi surface?*

Ans. The occupied states in the ground state of a system of N particles may be represented by points inside a sphere in the k-space [Fig. 7.10]. The energy corresponding to the surface of the sphere is the *Fermi energy* and the sphere is called the *Fermi sphere* and the surface of the sphere the *Fermi surface*. The volume of the Fermi sphere is $\left(\frac{4}{3}\pi\right)k_f^3$ and the total number of electrons in it is given by

$$2 \times \frac{\frac{4}{3}\pi k_f^3}{\left(\frac{2\pi}{L}\right)^3} = \frac{L^3}{3\pi^2}k_f^3 = \frac{V}{3\pi^2}k_f^3 = N \text{ (say)}.$$

The factor 2 is due to the fact that each state can accommodate two electrons with parallel and anti-parallel spin.

$$\therefore k_f = \left(\frac{3\pi^2 N}{V}\right)^{1/3}.$$

Thus, k_f depends on the concentration of electrons (N/V).

The Fermi energy is

$$E_f = \frac{\hbar^2}{2m}\left(3\pi^2\frac{N}{V}\right)^{2/3}.$$

So, $E_f \propto (N/V)$. The Fermi energy does not change if identical metals are joined together.

7.8. *Define density of states and write down its expression.*

Ans. The number of energy levels (or orbitals) between unit energy interval is called the *density of states*. If dN be the number of energy levels in the energy interval E and $E + dE$, then density of states is given by

$$D(E) = \frac{dN}{dE}.$$

The expression for the density of states is given by

$$D(E) = \frac{V}{2\pi^2}\left(\frac{2m}{\hbar^2}\right)^{3/2} E^{1/2},$$

where V is the volume and m is the mass of an electron.

7.9. *Find an expression for the bulk modulus of elasticity of an electron gas.*

Ans. If an increase in pressure dP decreases the volume of the electron gas by an amount dV, then bulk modulus of elasticity of the gas is given by

$$B = -V\frac{dP}{dV}.$$

Since the pressure of the electron gas is given by

$$P = \frac{2}{5}N\frac{\hbar^2}{2m}\left(3\pi^2 N\right)^{2/3}\left(\frac{1}{V}\right)^{5/2}, \text{ [cf. (eqn 7.131)]}$$

$$\therefore \frac{dP}{dV} = -\frac{5}{2}\left[\frac{2}{5}\frac{N\hbar^2}{2m}\left(3\pi^2 N\right)^{2/3}\right]\left(\frac{1}{V}\right)^{7/3} = -\frac{5}{3}\frac{P}{V}.$$

$$\therefore B = \frac{5}{3}P = \frac{2}{3}\frac{NE_{f_0}}{V} = \frac{10}{9}\frac{N\langle E_0\rangle}{V} \text{ or, } B = \frac{10}{9}\frac{E}{V}\left[\because \langle E_0\rangle = \frac{3}{5}E_{f_0} \text{ and } P = \frac{2}{5}\frac{NE_{f_0}}{V}\right].$$

7.15 Solved Problems

7.1. *Assuming that each silver atom provides one conduction electron, calculate* (i) *the density of electrons* (ii) *the mobility of electrons in silver.* [*Given: The density of silver* = 10.5×10^3 kg/m^3. *The atomic weight of silver* = 107.9. *The conductivity of silver at* 20 °C *is* $6.8\times 10^7\ \Omega^{-1}\,\text{m}^{-1}$].

Solution: (i) Volume of 1 gm-mole of silver = $\dfrac{M_A}{\rho_s}$.

$\therefore \left(\dfrac{M_A}{\rho_s}\right)$ m^3 contains N_A electrons [N_A = Avogadro number].

Hence, 1 m^3 contains $\left[\dfrac{\rho_s N_A}{M_A}\right]$ electrons, i.e.,

$$n = \frac{\rho_s N_A}{M_A} = \frac{\left(10.5\times 10^3\right)\times\left(6.02\times 10^{26}\right)}{107.9} = \mathbf{5.86\times 10^{28}}.$$

(ii) We have electronic mobility $\mu = \frac{\sigma}{ne}$, where $\sigma = 6.8\times 10^7\ \Omega^{-1}\,\text{m}^{-1}$; $n = 5.86\times 10^{28}$; $e = 1.6\times 10^{-19}$ C.

$$\therefore \mu = \frac{6.8\times 10^7}{\left(5.86\times 10^{28}\right)\times\left(1.6\times 10^{-19}\right)} = \mathbf{0.725\times 10^{-2}\ m^2V^{-1}s^{-1}}.$$

7.2. *Find the relaxation time and mean free path of conduction electron in copper.* [*Given: For copper density of free electrons,* $n = 8.5\times 10^{28}/\text{m}^3$. *Resistivity,* $\rho = 1.69\times 10^{-8}\,\Omega$-m *and average velocity of electrons* = 1.154×10^5 m/s.

Solution: If n and τ be the density of conduction electrons and relaxation time of an electron of mass m and charge e, then its resistivity is

$$\rho = \frac{m}{ne^2\tau} \text{ or, } \tau = \frac{m}{ne^2\rho} \text{ or, } \tau = \frac{9.1\times 10^{-31}}{\left(8.5\times 10^{28}\right)\times\left(1.6\times 10^{-19}\right)^2\times\left(1.69\times 10^{-8}\right)} = \mathbf{2.47\times 10^{-14}\ sec}.$$

Mean free path $\lambda = \tau\bar{c} = \left(2.47\times 10^{-14}\right)\times\left(1.154\times 10^5\right) = \mathbf{2.85\times 10^{-9}\ m}.$

7.3. *Assuming that each aluminium atom contributes* 3 *free electrons for conduction, calculate the free electron concentration, mobility and drift velocity of electrons in aluminium wire of length* 5 m *and resistance* 0.06 Ω *carrying current of* 15 A. [*Given: Resistivity of aluminium* $\rho = 2.7\times 10^{-8}\ \Omega$ cm. *Atomic weight* $M_A = 26.98$; *Density* $\rho_s = 2.7\times 10^3$ kg/m^2; *Avogadro number* $N_A = 6.025\times 10^{23}$.]

Solution: Molar volume of aluminium is $\frac{M_A}{\rho_s}$. $\therefore$ $\frac{M_A}{\rho_s}$ m^3 contains N_A atoms.
Since each atom contributes 3 free electrons,
$\therefore$ 1 m^3 contains $\left(\frac{\rho_s N_A}{M_A}\right) \times 3$ electrons

$$\text{i.e., } n = \frac{3\rho_s N_A}{M_A} = \frac{3 \times (2.7 \times 10^3) \times (6.025 \times 10^{26})}{26.98} = \mathbf{1.8088 \times 10^{29}\ electrons/m^3}.$$

Again, $\frac{1}{\rho} = ne\mu$

$$\text{or, } \mu = \frac{1}{ne\rho} = \frac{1}{(1.8088 \times 10^{29}) \times (1.602 \times 10^{-19}) \times (2.7 \times 10^{-8})} = \mathbf{1.278 \times 10^{-3}\ m/s/V}.$$

But $\mu = v_d/E$.
$\therefore$ drift velocity $v_d = \mu E$, where

$$E = \frac{V}{D} = \frac{IR}{D} = \frac{(15) \times (0.06)}{5} = 0.18\ \text{V/m}.$$

$$\therefore v_d = \mu E = (1.278 \times 10^{-3}) \times (0.18) = \mathbf{0.230 \times 10^{-3}\ m/s}.$$

7.4. *Calculate the number of energy states for the free electrons in a cubical box of side* 0.02 m *lying between an energy of* 0 *to* 3 eV. [*Fermi energy of silver is* 5 eV].

Solution: The number of energy states between E to $E + dE$ is

$$D(E)dE = \frac{\pi V}{2}\left[\frac{8m}{h^2}\right]^{3/2} E^{1/2} dE.$$

$\therefore$ the number of energy states between 0 and 3 eV is

$$n = \int D(E)\,dE = \frac{\pi V}{2}\left(\frac{8m}{h^2}\right)^{3/2} \int_0^{3/2} E^{1/2} dE = \frac{\pi V}{3}\left(\frac{8m}{h^2}\right)^{3/2} \left[E^{3/2}\right]_0^3$$

$$= \frac{\pi \times (0.02)^3}{3}\left[\frac{72.8 \times 10^{37}}{6.62 \times 6.62}\right]^{3/2}_{3/2} \times \left[4.8 \times 10^{-19}\right]^{3/2}$$

$$= 8.4 \times 10^{-6} \times 6.677 \times 10^{55} \times 3.3 \times 10^{-28} = \mathbf{1.89 \times 10^{21}}.$$

7.5. *Consider silver in the metallic state with one free electron per atom. Calculate the Fermi energy.* [*Given: density of silver is* 10.5 gm/cm^3 *and atomic weight* 108.] **[CU(H) 1998, 2009; BU 1992; HPU 1996; VU(H) 2008]**

Solution: Volume occupied by 1 gm-mole of silver = $\frac{108\ \text{gm}}{10.5\ \text{gm/cm}^3}$.

No. of atoms in 1 gm-atom = Avogadro number $N_A = 6.02 \times 10^{23}$.

$$\therefore \text{electron density } N/V = \frac{6.02 \times 10^{23} \times 10.5}{108} \times 1 = 5.85 \times 10^{22}\ \text{free electrons/cm}^3$$

$$= \mathbf{5.85 \times 10^{28}\ electrons/m^3}$$

Now Fermi energy,

$$E_f = \frac{\hbar^2}{2m}\left(3\pi^2 \frac{N}{V}\right)^{2/3} = \frac{h^2}{8m}\left(\frac{3}{\pi}\frac{N}{V}\right)^{2/3} = \frac{(6.625)^2 \times 10^{-68}}{8 \times 9.1 \times 10^{-31}} \times \left[\frac{3 \times 5.85 \times 10^{28}}{3.142}\right]^{2/3}$$

$$= 0.6 \times 10^{-37} \times 14.6 \times 10^{18} = 8.77 \times 10^{-19}\ \text{J} = \frac{8.77 \times 10^{-19}}{1.6 \times 10^{-19}} = \mathbf{5.48\ eV}.$$

7.6. *Calculate the Fermi energy at absolute zero of sodium metal if its atomic radius is 1.86 Å.*

Solution: If the number of free electrons present in volume V be N, then the Fermi energy of sodium at $T = 0$ is $E_{f_0} = \frac{\hbar^2}{2m}\left(\frac{3\pi^2 N}{V}\right)^{2/3}$.

Now Sodium has BCC structure.

$\therefore$ volume of the unit cell of sodium is $V = a^3 = \left(\frac{4r}{\sqrt{3}}\right)^3$.

$\because$ interatomic distance of sodium is $a = \left(\frac{4r}{\sqrt{3}}\right)$ and r is the atomic radius of sodium.

$$\therefore V = \left(\frac{4 \times 1.86 \times 10^{-10}}{\sqrt{3}}\right)^3 = 7.93 \times 10^{-29}\ \text{m}^3.$$

Since there are two atoms/unit cell of sodium the number of free electron in a unit cell is

$$N/V = \frac{2}{7.93 \times 10^{-29}} = 2.52 \times 10^{28}.$$

$$\therefore E_{f_0} = \frac{\left(1.05 \times 10^{-34}\right)^2}{2 \times 9.1 \times 10^{-31}} \times \left(3\pi^2 \times 2.52 \times 10^{28}\right)^{2/3} = 4.98 \times 10^{-19}\ \text{J} = \frac{4.98 \times 10^{-19}}{1.6 \times 10^{-19}}\ \text{eV} = \mathbf{3.11\ eV}.$$

7.7. (a) *Calculate the internal energy of electron gas per unit volume at* 0 K. (b) *Also calculate the molar specific heat due to free electrons at constant volume in metallic silver at* 100 K. [*Given: The density of silver* = 10.5 gm/cm^3, *atomic weight* = 108, *density of electron gas* $N/V = 5.9 \times 10^{28}/\text{m}^3$ *and* E_{f_0} = 5.5 eV.]

Solution: (a) The internal energy of the electron gas per unit volume is

$$U = \frac{3}{5}\left(\frac{N}{V}\right) E_{f_0} = \frac{3}{5} \times \left(5.9 \times 10^{28}\right) \times \left(8.8 \times 10^{-19}\right)\ \text{J/m}^3 = \mathbf{3.11 \times 10^{10}\ J/m^3}.$$

(b) The electron molar specific heat is

$$C_v = \frac{\pi^2}{2} Nk\left(T/T_f\right) = \frac{\pi^2}{2} \times \left(6.02 \times 10^{23}\right) \times \left(1.38 \times 10^{-23}\right) \times \left(\frac{100}{63768}\right) = \mathbf{0.064\ J/K/mol.}$$

Using $N/V = 6.02 \times 10^{23}$; $k = 1.38 \times 10^{-23}$ J/K;

$$T_f = \frac{E_f(0)}{k} = \frac{8.8 \times 10^{-19}\ \text{J}}{1.38 \times 10^{-23}\ \text{J/K}} = \mathbf{63768\ K.}$$

7.8. *Aluminium metal crystallizes in FCC structure. If each atom contributes single electron as free electron and the lattice constant a is* 4 Å, *calculate, treating conduction electrons as free electron Fermi gas* (a) *Fermi energy E_f and Fermi vector* (k_f) (b) *Total* KE *of free electron gas per unit volume at* 0 K. [$\hbar = 1.054 \times 10^{-27}$ erg-sec; *electron rest mass* $= 9.11 \times 10^{-28}$ gm] **[PU 1991]**

Solution: The number of atoms per unit cell in FCC structure is 4.

Since each atom contributes one free electron, the number of electron per unit cell is 4.

Volume of the unit cell $V = a^3 = \left(4 \times 10^{-8}\right)^3 = 64 \times 10^{-24}$ cc.

$\therefore$ free electron density $N/V = \frac{4}{64 \times 10^{-24}} = \mathbf{6.25 \times 10^{12}}$.

Now Fermi energy,

$$E_f = \frac{\hbar^2}{2m}\left(3\pi^2\frac{N}{V}\right)^{2/3} = \frac{\left(1.054\times10^{-27}\right)^2}{2\times9.11\times10^{-28}}\left(3\pi^2\times6.25\times10^{22}\right)^{2/3}$$

$$= 9.191\times10^{-12}\text{ erg-s} = \frac{9.191\times10^{-12}}{1.6\times10^{-12}} = \mathbf{5.74\,eV}.$$

Fermi vector, $k_f = \left(3\pi^2\frac{N}{V}\right)^{1/3} = \left(3\pi^2\times6.25\times10^{22}\right)^{1/3} = \mathbf{1.227\times10^8\ cm^{-1}}$.

Now at 0 K, the total KE of free electron gas per unit volume

$$= \langle E_0\rangle\times\frac{N}{V} = \frac{3}{5}E_f\times\frac{N}{V} = \frac{3}{5}\times5.74\times6.25\times10^{22} = \mathbf{21.52\times10^{22}\ eV}.$$

7.9. *A copper wire of length 0.5 m and diameter 0.3 mm has a resistance of 0.12 Ω at 20 °C. If the thermal conductivity of copper at 20 °C is 390 Wb/m^{-1} K^{-1}, calculate Lorentz number. Compare this value with the value predicted by free electron theory.*

Solution: We have resistance $R = \rho\frac{l}{a}$, where ρ is the resistivity of a wire of length l and diameter a.

$\therefore \rho = \frac{aR}{l}$ and conductivity

$$\sigma = \frac{l}{(\pi r^2)\,R} = \frac{0.5}{\pi\times\left(0.15\times10^{-3}\right)^2\times0.12} = 5.89\times10^7\ \Omega^{-1}\text{m}^{-1}.$$

Now

$$\frac{\sigma_T}{\sigma} = L = 1.5\times\left(\frac{k}{e}\right)^2 = \frac{1.5\times\left(1.38\times10^{-23}\right)^2}{\left(1.6\times10^{-19}\right)^2} = \mathbf{1.116\times10^{-8}\ W\Omega\,K^2}.$$

The experimental value of L is obtained as

$$L = \frac{\sigma_T}{\sigma\times T} = \frac{390}{5.89\times10^7\times293} = \mathbf{2.26\times10^{-8}\ W\Omega\,K^{-2}}.$$

7.10. *A current of 2A exists in a wire of cross section area 1 mm^2. If each cubic metre of the wire contains 6×10^{28} free electrons find the drift speed.* **[KU(H) 2011]**

Solution: The current densitv in the wire is $\vec{J} = -n|e|\vec{v}_d$

$$\text{or, } v_d = \frac{J}{n|e|} = \left(\frac{2A}{1\,\text{mm}^2}\right)\frac{1}{(6\times10^{28}\,\text{m}^{-1})\,(1.6\times10^{-19}\,\text{C})}$$

$$= \frac{2A}{(10^{-3}\,\text{m})^2\,(6\times10^{28}\,\text{m}^{-3})\,(1.6\times10^{-19}\,\text{C})}$$

$$\text{or, } v_d = \mathbf{2\times10^{-4}\ ms^{-1}}.$$

7.11. *Calculate the relaxation time of copper if conductivity is 59.17×10^6 ℧m^{-1} and free electron concentration is $n = 8.5\times10^{28}$ m^{-3}.*

Solution: The relaxation time

$$\tau = \frac{\sigma m}{n|e|} = \frac{(59.17\times10^6\,℧\text{m}^{-1})\,(9.1\times10^{-31}\,\text{kg})}{(8.5\times10^{28}\,\text{m}^{-3})\,(1.6\times10^{-19}\,\text{C})^2} = \mathbf{2.47\times10^{-14}\ s}.$$

7.12. *Find the heat capacity of electron gas in copper at room temperature assuming one free electron per atom if Fermi energy of copper is 7 eV and T = room temperature.*

Solution: Heat capacity of electron gas is given by

$$C_v^e = \frac{\pi^2}{2} R \frac{T}{T_F} = \frac{\pi^2}{2}(N_A k)\frac{kT}{E_F} = \frac{\pi^2}{2} N_A k^2 \frac{T}{E_F} \quad [\because E_F = kT_F]$$

$$= \frac{\pi^2}{2}(6.023 \times 10^{23}\,\text{mol}^{-1})\,(1.38 \times 10^{-23}\,\text{JK}^{-1})\frac{300\,\text{K}}{7 \times 1.6 \times 10^{-19}}$$

$$= \mathbf{0.152\,JK^{-1}\,mol^{-1}}.$$

7.13. *Calculate the number of energy states available for free electrons in a cubical box of side* 0.01 m *lying below an energy of* 1 eV.

Solution: The number of energy states is given by

$$z(E)\,dE = \frac{V}{2\pi^2}\left(\frac{2m}{\hbar^2}\right)^{3/2} E^{1/2}\,dE.$$

∴ energy state available between 0 to E is

$$\int_0^{1\,\text{eV}} z(E)\,dE = \frac{V}{2\pi^2}\left(\frac{2m}{\hbar^2}\right)^{3/2}\int_0^E E^{1/2}\,dE = \frac{V}{2\pi^2}\left(\frac{2m}{\hbar^2}\right)^{3/2}\frac{E^{1/2}}{3/2} = \frac{V}{3\pi^2}\left(\frac{2m}{\hbar^2}\right)^{3/2} E^{1/2}.$$

For $E = 1$ eV, number of energy states available is

$$\frac{(0.01\,m)^3}{3\pi^2}\left[\frac{2 \times 9.1 \times 10^{-31}\,\text{kg}}{\left(\dfrac{6.626 \times 10^{-34}\,\text{Js}}{2\pi}\right)^2}\right]^{3/2}(1.6 \times 10^{-19})^{3/2} = \mathbf{4.5 \times 10^{21}}.$$

7.14. *A metallic conductor having area of cross section* 1 cm² *carries a current of* 1 A. *The specific resistance of the conductor is* 1.7×10^{-6} Ω-cm. *Calculate the electric field inside the conductor.*

Solution: The current density in the conductor is given by

$$J = \frac{i}{A} = \frac{1\text{A}}{1\,\text{cm}^2} = \frac{1\text{A}}{(10^{-2}\,\text{m})^2} = 10^4\,\text{Am}^{-2}.$$

If E be the electric field inside the conductor and ρ, the resistivity, then

$$\rho = E/J$$

$$\text{or, } E = \rho J = (1.7 \times 10^{-6}\,\Omega\,\text{cm})\,(10^4\,\text{Am}^{-2}) = (1.7 \times 10^{6}\,\Omega \cdot 10^{-2}\,\text{m})$$

$$= (1.7 \times 10^{-8}\,\Omega\,\text{m})\,(10^4\,\text{Am}^{-2}) = 1.7 \times 10^{-4}\,\Omega\text{Am}^{-1}.$$

But $\Omega\text{Am}^{-1} = \frac{\text{V}}{\text{A}}\,\text{Am}^{-1} = \text{Vm}^{-1}$.

$$\therefore \mathbf{E = 1.7 \times 10^{-4}\,Vm^{-1}}$$

is the required electric field inside the conductor.

7.15. *Fermi energy for* Li *is* 4.72 eV *and density of electron is* 4.6×10^{28} m⁻¹. *Find the electron density for a metal with Fermi energy* 2.35 eV.

Solution: Let Fermi energy for Li is given by

$$E_{F_1} = \frac{\hbar^2}{2m}(2\pi^2 n_1)^{2/3}$$

and that for the metal be

$$E_{F_2} = \frac{\hbar^2}{2m}(3\pi^2 n_2)^{2/3}.$$

$$\therefore \frac{E_{F_1}}{E_{F_2}} = \left(\frac{n_1}{n_2}\right)^{2/3} \quad \text{or,} \quad \frac{4.72\text{ eV}}{2.35\text{ eV}} = \left(\frac{4.6\times10^{28}\text{ m}^{-3}}{n_2}\right)^{2/3}$$

$$\text{or, } n_2^{2/3} = (4.6\times10^{28})^{2/3}\frac{2.35}{4.72}$$

$$\text{or, } n_2 = (6.39\times10^{18})^{3/2} = \mathbf{1.6\times10^{28}\ m^{-3}}.$$

7.16. *The atomic radius of sodium is 1.86 Å. Calculate Fermi energy of sodium at absolute zero.*

Solution: Here, $r = 1.86$ Å, Na is BCC, $r/a = \sqrt{3}/4$.

Volume of unit cell of Na is $V = a^3 = (4.295\text{ Å})^3$.

No. of atoms per unit cell = 2.

Each atom contributes 1 free electron.

∴ no. of free electrons contributed per unit cell = $N = 2$.

$$\therefore n = \frac{N}{V} = \frac{2}{(4.295\times10^{-10}\text{ m})^3} = 2.52\times10^{28}\text{ m}^{-3}.$$

$$\therefore \text{Fermi energy } E_F = \frac{\hbar^2}{2m}(3\pi^2 n)^{2/3} = \left(\frac{6.626\times10^{-34}\text{ Js}}{2\pi}\right)^2 \times \frac{(3\pi^2\times2.52\times10^{28}\text{ m}^{-3})^{2/3}}{2(9.1\times10^{-31}\text{ kg})}$$

$$= 5.026\times10^{-19}\text{ J} = \frac{5.026\times10^{-19}}{1.6\times10^{-19}}\text{ eV} = \mathbf{3.14\ eV.}$$

EXERCISE

7(A): Objective (MCQ) Type Questions

Select the correct answer.

7.1. Free electrons in metal are
(a) K-shell electrons (b) valence electron (c) L-shell electrons (d) none of these.

7.2. Free electron gas is similar in many respect to ideal gas except that it is
(a) positively charged (b) negatively charged (c) both (a) and (b) (d) none of these.

7.3. The ratio of the density of ideal gas to that of free electron gas is
(a) > 1 (b) < 1 (c) = 1 (d) none of these.

7.4. Classical free electron theory was developed by
(a) Sommerfeld (b) Bloch (c) Dude and Lorentz (d) Einstein.

7.5. Quantum theory of free electron was discovered by
(a) Drude and Lorentz (b) Bloch (c) Bohr (d) Sommerfeld.

7.6. The zone theory of free electrons was developed by
(a) Sommerfeld (b) Bloch (c) Drude (d) Lorentz.

7.7. The free electrons are non-interacting and obey
(a) Pauli's exclusion principle (b) principle of superposition
(c) principle of least action (d) none of these.

7.8. Free electrons in presence of an external electric field E, move in a direction
(a) parallel to E (b) opposite to E (c) perpendicular to E (d) none of these.

7.9. If K and σ be the thermal and electrical conductivities of a metal at temperature T, then

(a) $\frac{KT}{\sigma}$ = constant (b) $\frac{K\sigma}{T}$ = constant (c) $\frac{\sigma}{kT}$ = constant (d) $\frac{K}{\sigma T}$ = constant.

7.10. If n, v_d and e be the electron density, electron drift velocity and electronic charge respectively, then current density is given by

(a) $J = nv_d$ (b) $J = nev_d$ (c) $J = \frac{ne}{v_d}$ (d) $J = \frac{1}{nev_d}$.

7.11. Drift velocity for unit electric field is

(a) electrical conductivity (b) thermal conductivity

(c) electric polarization (d) mobility of charge carriers.

7.12. Drift velocity of an electron in metal is given by

(a) $v_d = \frac{mE}{e\tau}$ (b) $v_d = \frac{e\tau}{m}E$ (c) $v_d = -\frac{e\tau}{m}E$ (d) $v_d = -\frac{e\tau}{mE}$.

7.13. Electrical conductivity of a metal is

(a) $\sigma = \frac{ne^2}{\tau m}$ (b) $\sigma = \frac{ne^2\tau}{m}$ (c) $\sigma = \frac{m}{ne^2\tau}$ (d) $\sigma = \frac{\tau m}{ne^2}$.

7.14. The electron mobility is given by

(a) $\mu = m/e\tau$ (b) $\mu = \frac{e}{\tau m}$ (c) $\mu = \frac{m\tau}{e}$ (d) $\mu = \frac{e\tau}{m}$.

7.15. Resistivity of a metal of the conductor is given by

(a) $\rho = \frac{m}{ne^2\tau}$ (b) $\rho = \frac{ne^2\tau}{m}$ (c) $\rho = \frac{n}{me^2\tau}$ (d) $\rho = \frac{me^2\tau}{n}$.

7.16. The relation between resistivity and absolute temperature of a metal is

(a) $\rho \propto T$ (b) $\rho \propto \sqrt{T}$ (c) $\rho \propto T^2$ (d) $\rho \propto \frac{1}{T}$.

7.17. According to quantum theory, the free electron gas obeys

(a) MB-statistics (b) BE-statistics (c) FD-statistics (d) none of these.

7.18. At any temperature other than zero degree kelvin, the probability of finding an electron at Fermi level is

(a) 1 (b) 0 (c) 0.5 (d) any positive value.

7.19. The average KE in the ground state of the electron confined in one-dimensional potential is

(a) $3E_f$ (b) $\frac{1}{3}E_f$ (c) $\frac{3}{E_f}$ (d) $\frac{3}{5}E_{f_0}$.

7.20. If N be the number of free electrons in volume V, then the Fermi velocity

(a) $v_f \propto (N/V)$ (b) $v_f \propto (N/V)^{1/3}$ (c) $v_f \propto (N/V)^{1/2}$ (d) $v_f = N/V$.

7.21. In three-dimensional case, the average KE of an electron at 0 K is

(a) $\langle E_0 \rangle = 5/3E_{f_0}$ (b) $\langle E_0 \rangle = \frac{5E_{f_0}}{4}$ (c) $\langle E_0 \rangle = \frac{3}{5}E_{f_0}$ (d) $\langle E_0 \rangle = E_{f_0}$.

7.22. The variation of molar specific heat of free electrons at temperature T is given by

(a) $C_v/T = A + BT^3$ (b) $C_v/T = BT^3$ (c) $C_v/T = A + BT^2$ (d) $C_v/T = A$.

where A and B are constants.

7.23. At Fermi level, the probability of electron occupation at $T > 0$ K is

(a) 0 (b) 1 (c) $\frac{1}{2}$ (d) $\frac{1}{4}$.

7.24. The pressure-volume relation of an electron gas is

(a) $P = \frac{1}{5}\frac{NE_{f_0}}{V}$ (b) $P = \frac{2}{5}\frac{NE_{f_0}}{V}$ (c) $P = \frac{NE_{f_0}}{V}$ (d) $P = \frac{3}{5}\frac{NE_{f_0}}{V}$.

7.25. The de Broglie wavelength of an electron at the Fermi surface is

(a) $\lambda_f = 2\left(\frac{\pi}{3n}\right)^{1/3}$ (b) $\lambda_f = \left(\frac{\pi}{3n}\right)^{2/3}$ (c) $\lambda_f = 2\left(\frac{\pi}{3n}\right)^{1/2}$ (d) $\lambda_f = 2\left(\frac{\pi}{3n}\right)^{2/3}$.

7.26. According to quantum theory of free electrons the molar specific heat of free electrons is

(a) $C_v = \frac{3}{2}Nk$ (b) $C_v = (0.01)Nk$ (c) $C_v = (0.01)\frac{3}{2}Nk$ (d) $C_v = (0.001)\frac{3}{2}Nk$.

7.27. The density of states between energies E and $E + dE$ is

(a) $\frac{4\pi}{h^3}(2m)^{3/2}E^{1/2}\,dE$. (b) $\frac{4\pi}{h^3}m^3 EdE$ (c) $\frac{4\pi}{h^3}(2m)^{2/3}E^{1/2}\,dE$ (d) none of these.

7.28. The Bulk modulus of elasticity for a free electron gas at 0 K is

(a) $B = \frac{9}{10}(E/V)$ (b) $B = \frac{10}{9}(E/V)$ (c) $B = \frac{10}{9}(V/E)$ (d) $B = \frac{10}{9}EV$.

ANSWERS

7.1. (b) **7.2.** (b) **7.3.** (b) **7.4.** (c) **7.5.** (d) **7.6.** (b) **7.7.** (a) **7.8.** (b) **7.9.** (d) **7.10.** (b) **7.11.** (d) **7.12.** (c) **7.13.** (b) **7.14.** (d) **7.15.** (a) **7.16.** (b) **7.17.** (c) **7.18.** (c) **7.19.** (b) **7.20.** (b) **7.21.** (c) **7.22.** (c) **7.23.** (c) **7.24.** (b) **7.25.** (a) **7.26.** (c) **7.27.** (a) **7.28.** (b).

7(B): Short Answer Type Questions

7.1. Explain the terms *free electron* and *free electron* gas.

7.2. Explain the salient features of quantum theory of free electrons.

7.3. Define the terms: *Drift velocity* and *mean free path* of free electron in a metal. What is relaxation time?

7.4. What are the merits and demerits of free electron theory of metals after Drude?

7.5. Show that electrical resistivity of a metal is given by $\rho = \frac{m}{ne^2\tau}$, the letters have their usual significances.

7.6. Show that the energy levels of a free electron can be considered to from a quasi-continuous band.

7.7. Find out an expression for the total energy of an electron gas in the ground state.

7.8. Discuss the effect of temperature on the occupancy of energy levels.

7.9. What do you mean by *Fermi sphere* and *Fermi surface*? What is its volume?

7.10. Define Fermi energy, Fermi temperature and Fermi velocity of a free electron in a metal.

7.11. According to classical and quantum theories of free electrons the molar specific heats are $C_v = \frac{3}{2}Nk$ and $C_v = \frac{3}{2}Nk\left(T/T_f\right)$. How can the difference of the two values be accounted for?

7.12. Explain complete *opacity* and *high lustre* on the basis of the Drude-Lorentz free electron theory of metals.

7.13. Define density of states. Find out its expression for three-dimensional motion of free electrons.

7.14. Show that the total molar specific heat of free electrons at low temperature is given by $C_v = AT + BT^3$, where A and B are constants.

7.15. Show that the number of electrons in the Fermi surface is given by $\frac{V}{3\pi^2}\left(\frac{2m}{\hbar^2}\right)^{3/2}E_{f_0}^{3/2}$, the different letters have their usual significances.

7.16. Derive Wiedemann-Franz law from free electron theory. Discuss its validity. **[CU(H) 2010]**

7.17. The free electron model describes electrons in a metal kept at absolute zero as those occupying states inside the Fermi sphere in the k-space. With this model explain conduction of electricity in metals and an expression for conductivity. **[CU(H) 2006]**

7.18. Derive Wiedemann-Franz law from free electron theory of metal. **[CU(H) 2014]**

7.19. Obtain expression for pressure of a free electron gas in terms of its Fermi energy. **[CU(H) 2014]**

7(C): Numerical Problems

7.1. The resistivity of aluminium at room temperature is $2.62 \times 10^{-8}\,\Omega$-m. Calculate (a) the drift velocity (b) their mobility (c) their relaxation time (d) mean free path on the basis of classical free electron theory. [*Ans.* (a) 0.396 m/s; (b) 3.96×10^{-3} m^2/V/s; (c) 2.3×10^{-14} s (d) 2.65 nm]

7.2. Show that if the mean free path is independent of the velocity, the electrical conductivity of Maxwell-Boltzmann free electron gas may be expressed as

$$\sigma = \frac{4}{3}\frac{nc^2\lambda}{(2\pi kTm)^{1/2}}.$$

7.3. The Fermi energy of silver is 5.51 eV. What is the average energy of free electron at 0 K? At what temperature will the average molecular energy in an ideal gas have this value? **[BU(H) 1998]**

7.4. Show that the average energy of an electron in a metal at $T = 0$ K is $\frac{3}{5}E_f$, where E_f is the Fermi energy. **[CU(H) 1997, 2004]**

7.5. There are 2.5×10^{28} free electrons per cubic metre of sodium. Calculate Fermi energy and Fermi velocity. [*Ans.* $E_f = 3.1$ eV; $v_f = 1.048 \times 10^6$ m/s]

7.6. If Fermi temperature is of the order of 5×10^4 K provide a rough estimate of electronic specific heat at 100 K. [*Ans.* 0.025 J]

7.7. A particle is confined in an one-dimensional infinite potential well of width 0.2 mm. If it is found that when the energy of the particle is 230 eV, its eigenfunction has five antinodes. Find the mass of the particle and show that it can never have energy equal to 1 keV. **[BU (H) 2011]**

7.8. A uniform copper wire of length 0.5 m and diameter 0.3 mm has a resistance of 0.12 Ω at 293 K. If the thermal conductivity of the specimen at the same temperature is 390 Wm^{-1}K^{-1}, calculate the Lorentz number. Compare this value with theoretical value based on free electron theory. **[CU (H) 2017]**

7.9. (a) A thin film of dimension 2 mm × 2 mm × 0.1 mm contains 4×10^{12} electrons. Find the magnitude of the Fermi wave vector of the system in the free electron approximation.

(b) The Fermi energy of Al is 12 eV and its electrical resistivity is $3 \times 10^{-8}\,\Omega$m. Calculate the mean free path of the conduction electrons and their mean drift velocity in a field 1000 Vm^{-1}. [Given, the atomic weight of Al is 27 and density 2700 kg m^{-3}]. **[CU (H) 2016]**

7.10. A metal has a static conductivity of 4×10^7 ℧/m. Assuming that the true charge carriers are free electrons and there are 2×10^{28} electrons per metre cube calculate relaxation time. **[CU (H) 2012]**

7.11. Calculate the Fermi energy at 0 K of metallic silver containing one free electron per atom. The density and atomic weight of silver are 10.5 gm/m^3 and 108 respectively.
$h = 6.6 \times 10^{-34}$ J.sec; $m_e = 9 \times 10^{-34}$ kg.

7(D): Broad Answer Type Questions

7.1. (a) Describe the free electron gas model of metals.

(b) What is free electron approximation? In this approximation find the radius of the Fermi sphere for a crystal with Bravais lattice having Z electrons per atoms and n unit cells per unit volume.

(c) Find the wavelength of an electron at the Fermi level. **[BU(H) 1996]**

7.2. (a) Deduce an expression for the Fermi energy of a free electron gas in terms of the electron concentration n. What is the value of the Fermi energy in a typical metal (electron density = 10^{28} m^3)?

(b) Show that the conductivity of free electron gas can be expressed in the form $\sigma = \frac{ne^2\tau}{m}$, where the symbols have their customary connotations.

Explain: (a) Why only electrons near the Fermi surface contribute to the electrical conductivity of free electron metal;

(b) Why the resistance of a metal increases with temperature? **[BU(H) 1995]**

7.3. (a) What are the major difficulties in the classical theory of the free electron gas in metal ? Discuss how were these removed.

(b) Explain the significance of the Fermi energy of a metal.

(c) Prove that the average KE per election in a metal is equal to $\frac{3}{5}E_f$. **[BU(H) 1994]**

7.4. (a) Obtain expressions for the electrical and thermal conductivities of metal in free electron approximation and hence, establish the Wiedemann-Franz law. **[BU(H) 1993]**

(b) Show that the average KE of a conduction electron at 0 K is (3/5) E_f, where E_f is the Fermi energy and that the average speed is $\left(\frac{3}{4}\right) v_f$, where v_f is the velocity at the Fermi surface. **[BU(H) 2003]**

7.5. (a) Find the radius of the Fermi sphere and the wavelength of an electron at the Fermi level for a crystal having one electron per atom per unit cell and N unit cells per unit volume. **[BU(H) 1998]**

(b) What is meant by Fermi energy E_f? Obtain an expression for Fermi energy for free electron gas in one dimension at absolute zero. Hence, show that the average KE in the ground state is one third of Fermi energy. **[BU(H) 2000]**

7.6. (a) What do you understand by Fermi gas? Can a metal be associated with two Fermi temperature? **[PU 1991]**

[*Hints.* $T_f = \frac{E_f}{k}$. Since there is only one volume of E_f at 0 K, there is only one value of T_f.]

(b) Does the Fermi energy of a metal depend upon temperature and total number of electrons in the sample? **[PU 1995, 1992]**

7.7. (a) What are the main drawbacks of classical free electron theory?

(b) What modification was introduced by Sommerfeld into the classical theory?

(c) Quantum mechanically obtain an expression for the eigenfunction and energy eigenvalues of a free electron confined in a cube of side L.

(d) Show that the introduction of boundary conditions leads to the quantisation of energy levels.

(e) Find out the number of electrons per unit volume of $\bar{k}$-space. ($\bar{k} \rightarrow$ electron wave vector). **[BU(H) 2015]**

7.8. (a) An otherwise free electron is trapped in a one-dimensional box of length L having infinitely hard walls. Obtain an expression by considering its wave nature.

(b) Solve the Schrödinger equation for this electron with appropriate boundary conditions to obtain the energy eigenvalues and eigenfunctions.

(c) Show that the average position of the electron is the middle of the box for all quantum states. **[BU(H) 2013]**

7.9. What is meant by the free electron gas model of metals? How does a free electron gas differ from an ordinary gas? Derive the wave functions and the energy eigenvalues of an electron using the Sommerfeld's free electron theory in one dimension and hence, obtain an expression for the Fermi energy. [BU(H) 2011]

7.10. (a) State clearly the limitations of classical free electron model.

(b) State the four quantum numbers of electron in Sommerfeld's free electron theory of metals.

(c) Briefly discuss the advantages of this model over classical free electron model. [BU(H) 2008]

7.11. (a) Explain the origin of non-zero value of average energy of degenerate free electrons at $T = 0$ K.

(b) What is meant by mean free path of free electrons in metal? Calculate the electrical conductivity with mean free path λ for a metal with 6×10^{22} conduction electrons per cc in unit of λ ohm^{-1} cm^{-1}. [CU(H) 2017]

7.12. (a) Derive Wiedemann-Franz law from free electron theory. What is the remarkable feature of this law? [CU(H) 2016]

(b) Using free electron theory obtain an expression of resistivity of a metal in terms of Fermi velocity and mean free path. [CU(H) 2015, 2009]

7.13. (a) Discuss showing the changes in Fermi-Dirac distribution function as temperature is changed from 0 K.

(b) What is Fermi energy?

(c) Derive an expression for Fermi energy of a free electron gas. [CU(H) 2012]

7.14. Using free electron theory, establish a relation between electrical conductivity of a metal and mean free path. [CU(H) 2011]

7.15. (a) What are the main assumptions in Debye model? How does it differ from Einstein's model?

(b) Find the expression for number of models of vibrations in the range ν to $\nu + d\nu$ in a cubic solid having N-atoms.

(c) Sketch the Fermi distribution function for three temperatures $T_1 > T_2 > T_3$ on the same graph. [CU(H) 2017]

7.16. (a) Assuming that in a solid of volume V the number of modes of vibration having angular frequency ω and $\omega + d\omega$ is given by

$$dN = \frac{V\omega^2 d\omega}{2\pi^2 c^3},$$

where c is the velocity of sound in solids.

(b) Calculate Debye cut-off frequency and write the physical significance of it. From this obtain the temperature variation of the specific heat for very low temperature. [CU(H) 2015]

Chapter 8

Band Theory of Solids

8.1 Introduction

In a solid, the potential due to the fixed nuclei and the average potential due to all other electrons form a *periodic potential.* The motion of an electron in this periodic potential can be described by *Schrödinger equation,* whose solution gives a set of energy states. These states may be occupied by electrons according to *Pauli's exclusion principle.* All states are not permitted to electrons but bands of possible energies separated by *forbidden* energies are allowed.

8.2 Formation of Energy Bands

When two atoms are brought close to each other a single sharp energy level splits up into a pair of levels.

When two atoms are brought together their wave functions ψ_1 and ψ_2 overlap, the resultant wave function being either $(\psi_1 + \psi_2)$ or $(\psi_1 - \psi_2)$ as shown in Figs 8.1 (a) and (b).

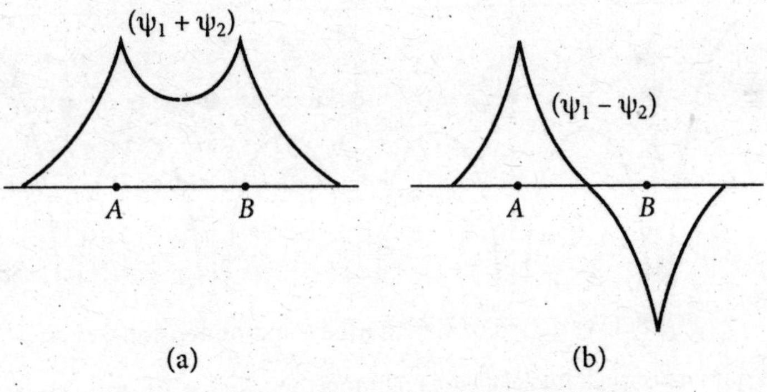

Fig. 8.1: Resultant wave functions of nuclei *A* and *B*

The electron has a finite probability of existing midway between the nuclei for $(\psi_1 + \psi_2)$. Here the binding energy increases due to the binding force of both the electrons. But for $(\psi_1 - \psi_2)$ the probability of existence of the electron is zero, and the extra contribution of energy disappears. So, there is a difference in energy between the states having wave functions $(\psi_1 + \psi_2)$ and $(\psi_1 - \psi_2)$. Hence, when two atoms are brought close together each energy state splits up into two energy states. When *N* number of atoms are brought together each energy state splits into *N* energy states whose separation is very small if *N* is very large. These energy states, then form a *quasi-continuous* energy band. In other words, each energy level may split into a band of energy level when atoms are brought together.

The formation of band out of energy levels due to the decrease of separation between the two atoms is shown in Fig. 8.2. The *strength of interaction* and the *extent of overlap* determine the width of the band.

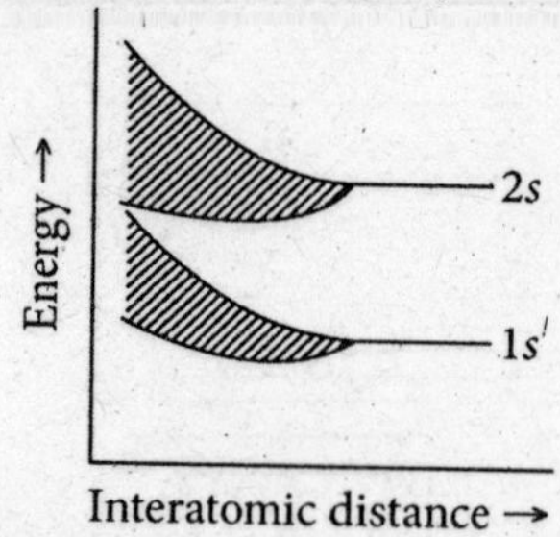

Fig. 8.2: Band formation of 1*s* and 2*s* states

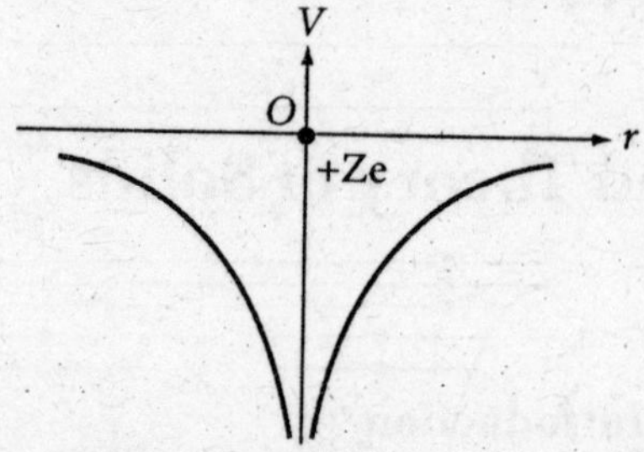

Fig. 8.3: Variation of *V* with *r* of an isolated atom

8.3 Periodic Potential in a Crystalline Solid

The PE of an electron at a distance *r* from a nucleus of charge Ze is given by

$$V = \frac{Ze^2}{4\pi\varepsilon_0 r}, \tag{8.1}$$

where ε_0 is the *permittivity of the free space.* The graphical variation of *V* with *r* is shown in Fig. 8.3.

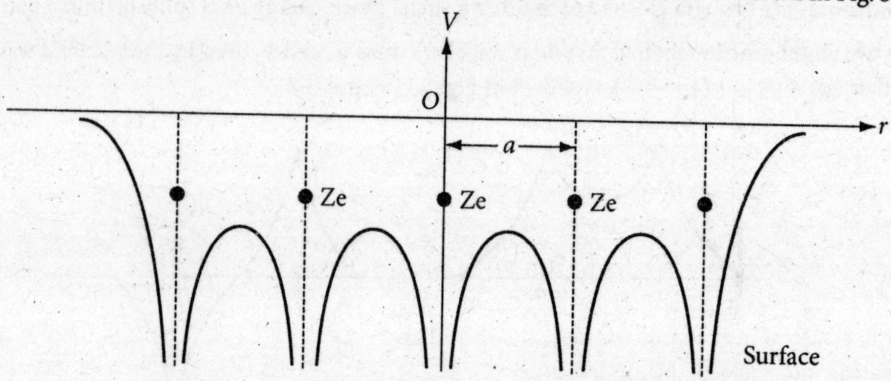

Fig. 8.4: Variation of *V* with *r* for a one-dimensional crystal

In a solid, a large number of such nuclei are brought together. So, the PE of an electron is the sum total of all the potential energies due to individual nuclei. Under such condition the PE will be a *periodic function of r* in an infinite one-dimensional lattice as shown in Fig. 8.4.

8.3.1 Development of Energy Bands in Sodium Crystals

Fig. 8.5 shows the PE profile for the linear array of sodium atoms in a crystal with r_0 as the equilibrium spacing between two neighbouring atoms.

Though outside the crystal the PE rises towards $E = 0$ asymptotically, the every *E* is considerable below $E = 0$ level between neighbouring atoms due to mutual interaction. The 2*s* and 2*p* levels of atoms are separated by a potential hump but the band formed by 3*s* electrons are only half-filled. The electrons in this band have energy higher than the potential hump and behave as free electrons and as such sodium metal is a good conductor of electricity.

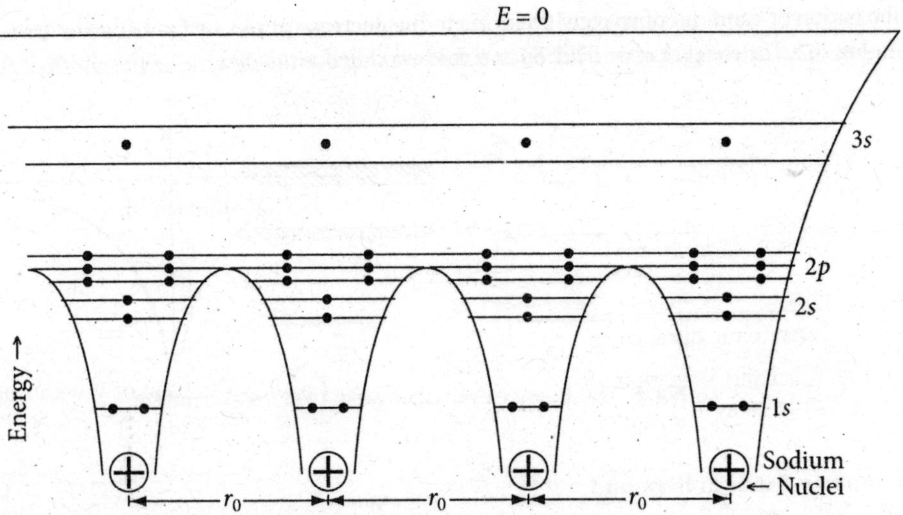

Fig. 8.5: Potential energy profiles and energy levels along an array of atoms in a sodium crystal

The mechanism of development of energy bands in sodium crystal is represented in Fig. 8.6. When the atoms of sodium are far apart, the allowed states of the system are just the states of single atom due to the absence of interaction between them. With the decrease of interatomic spacing (r), the valence orbitals of the atoms begin to overlap and the energy of the 3*s* level splits into closely spaced *N* levels where *N* is the number of atoms. These closely spaced *N* energy levels form an energy band consisting of a virtually continuous spread of permitted energies known as *quasi-continuous energy band.*

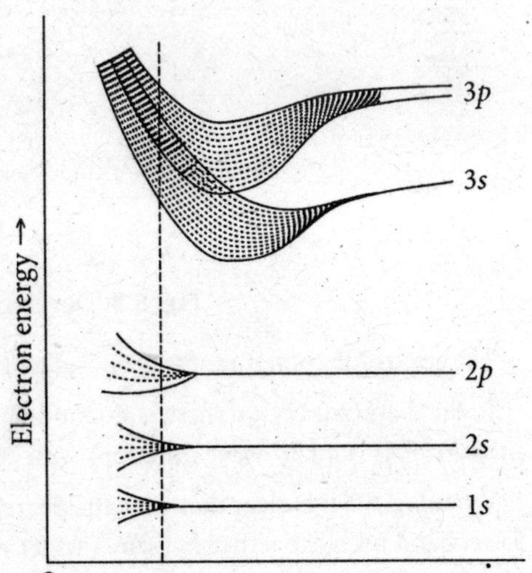

Fig. 8.6: Energy bands in sodium crystal

The empty 3p level also spreads into a band. The band formed out of 3*s* and 3p levels undergoes broadening with the further decrease of interatomic spacing. The 2*s* and 2p levels also spread into energy bandwidth the decrease of *r*. If *r* is reduced to the equilibrium separation r_0 (0.367 nm for sodium) the energy bands due to 3*s* and 3p states overlap. With the decrease of *r*, the 3*s* level is the first occupied level which undergoes broadening and due to the effect of attractive forces between the atoms the average energy of the 3*s* level decreases. Hence, *we see that in a real solid, consisting of large number of atoms, splitting of energy levels into energy bands occurs with the decease of r due to the interaction between the atoms.*

In metals, the energy bands so formed overlap but in other materials they may be separated by a gap called *band gap* (E_g). The energy of the highest filled level is called the *Fermi energy* (E_F). All energy levels below E_F are filled at 0 K but those above E_F are empty. According to *Pauli's principle* each *s*-band of a crystal having *N* atoms can accommodate 2*N* electrons.

If the highest s-band is completely filled, then the electron cannot drift within the crystal due to the application of an external electric field. Such solids are called *insulators.*

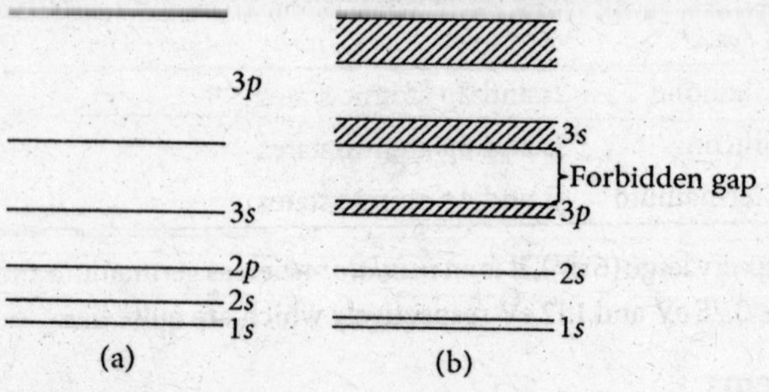

Fig. 8.7 (a) Energy levels of an atom (b) Energy bands of a solid

8.3.2 Energy Bands in Diamond

The development of energy band in diamond is shown in Fig. 8.8. Diamond is the crystalline form of carbon.

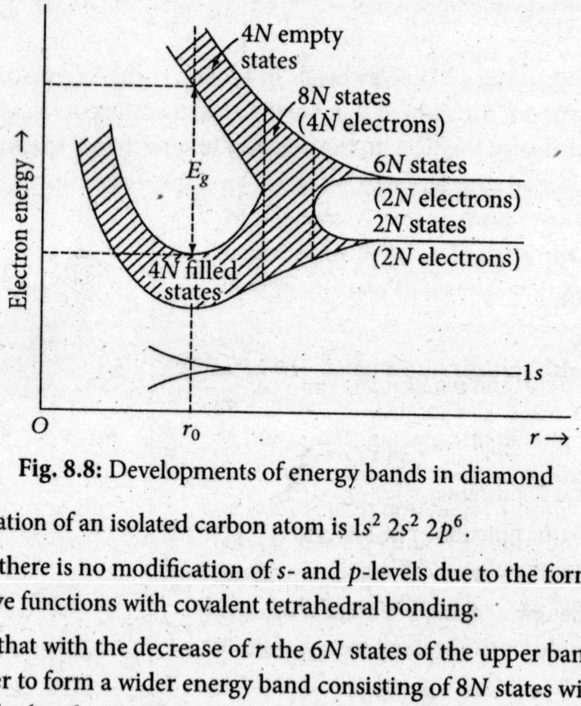

Fig. 8.8: Developments of energy bands in diamond

The electronic configuration of an isolated carbon atom is $1s^2\ 2s^2\ 2p^6$.

In the diamond crystal there is no modification of s- and p-levels due to the formation of sp^3 hybrid, a mixture of $2s$- and $2p$-wave functions with covalent tetrahedral bonding.

From Fig. 8.8 it is clear that with the decrease of r the $6N$ states of the upper band and $2N$ states of the lower band merge together to form a wider energy band consisting of $8N$ states with $4N$ electrons. If the interatomic distance r be further decreased to r_0 (the equilibrium separation), the $2N$ states in $2s$ sub-shell and the filled $2N$ states in $2p$ sub-shell approach each other and spread into a band of $4N$ states which are all filled with electrons. The remaining $4N$ states in $2p$ sub-shell which are all empty spread into another band being separated from the lower filled band by a forbidden energy gap (E_g). The completely filled lower band is called the *valence band* and the upper empty band is called the *conduction band.* If an electron from the lower valence band transits to the upper conduction band by receiving energy E_g (= 6 eV for diamond) electrical conduction takes place.

The formation of sp^3-hybrid stem for different crystals is shown in Table 8.1.

Table 8.1

Crystal	*sp^3 hybrid stem from*	*Energy gap E_g in eV*
Diamond	$2s$ and $2p$ atomic states	6
Silicon	$3s$ and $3p$ atomic states	0.78
Germanium	$4s$ and $4p$ atomic states	1.2

E_g for diamond being very large (6 eV), it is an *insulator* whereas germanium and silicon are *semiconductor* as E_g values for them is 0.78 eV and 1.12 eV respectively which are quite low.

8.4 Bloch Theorem

The *Schrödinger equation* for an electron moving in a one-dimensional periodic potential is given by

$$\frac{d^2\psi}{dx^2} + \frac{2m}{\hbar^2}[E - V(x)]\psi = 0 \tag{8.2}$$

where $V(x) = V(x + a)$ and a is the period equal to *lattice constant.*

Bloch theorem states that the solution of the Schrödinger equation for a periodic potential must be of the form:

$$\psi(x) = e^{\pm ikx}u_k(x), \tag{8.3}$$

where

$$u_k(x) = u_k(x + a). \tag{8.4}$$

Thus, the solutions are plane waves of the type $e^{\pm ikx}$ modulated by the function $u_k(x)$ having the same periodicity as the lattice.

The wave function $\psi(x) = e^{\pm ikx}u_k(x)$ is called *Bloch function* and the wave vector $\vec{k}$ represents the direction of the *Bloch wave.*

Proof of Bloch theorem

Let $f(x)$ and $g(x)$ be two *real* and *independent functions* satisfying eqn (8.2). Then the general solution of eqn (8.2) is

$$\psi(x) = Af(x) + Bg(x), \tag{8.5}$$

A and B being two arbitrary constants.

Since $V(x) = V(x+a)$, the functions $f(x+a)$ and $g(x+a)$ must also be solutions of eqn (8.2). According to the theory of differential equation, a differential equation of second order can have only two independent solutions. So, $f(x + a)$ and $g(x + a)$ can be written as

$$f(x + a) = \alpha_1 f(x) + \alpha_2 g(x) \tag{8.6}$$

$$g(x + a) = \beta_1 f(x) + \beta_2 g(x), \tag{8.7}$$

where $\alpha_1, \alpha_2, \beta_1, \beta_2$ are functions of E.

Now the general solution of *Schrödinger equation* is $\psi(x) = Af(x) + Bg(x)$ and

$$\begin{aligned}\psi(x + a) &= Af(x + a) + Bg(x + a)\\ &= A\alpha_1 f(x) + A\alpha_2 g(x) + B\beta_1 f(x) + B\beta_2 g(x) \quad \text{[using eqns (8.6) and (8.7)]}\\ &= (A\alpha_1 + B\beta_1) f(x) + (A\alpha_2 + B\beta_2) g(x).\end{aligned} \tag{8.8}$$

Now let

$$A\alpha_1 + B\beta_1 = \lambda A \tag{8.9}$$

$$\text{and } A\alpha_2 + B\beta_2 = \lambda B, \tag{8.10}$$

where λ is a constant.

$$\therefore \psi(x+a) = \lambda Af(x) + \lambda Bg(x) = \lambda[Af(x) + Bg(x)] = \lambda\psi(x). \tag{8.11}$$

A and B will have non-zero values if the determinant of their coefficient vanishes, i.e.,

$$\begin{vmatrix} \alpha_1 - \lambda & \beta_1 \\ \alpha_2 & \beta_2 - \lambda \end{vmatrix} = 0 \tag{8.12}$$

$$\text{or, } \lambda^2 - (\alpha_1 + \beta_2)\,\lambda + (\alpha_1\beta_2 - \alpha_2\beta_1) = 0. \tag{8.13}$$

To prove that $\alpha_1\beta_2 - \alpha_2\beta_1 = 1$, we have

$$\frac{d^2 f(x)}{dx^2} + \frac{2m}{\hbar^2}[E - V(x)]\,f(x) = 0 \tag{8.14}$$

$$\text{and } \frac{d^2 g(x)}{dx^2} + \frac{2m}{\hbar^2}[E - V(x)]\,g(x) = 0. \tag{8.15}$$

Multiplying eqn (8.14) by $g(x)$ and eqn (8.15) by $f(x)$ and then subtracting, we get

$$f(x)\frac{d^2 g(x)}{dx^2} - g(x)\frac{d^2 f(x)}{dx^2} = 0$$

$$\text{or, } f(x)\frac{dg(x)}{dx} - g(x)\frac{df(x)}{dx} = \text{a constant}$$

$$f(x)\frac{dg(x)}{dx} - g(x)\frac{df(x)}{dx} = W(x)$$

is the *Wronskian* of the solution and is constant in this case.

Now using eqns (8.6) and (8.7) in the above equation, we have

$$\begin{aligned}
W(x+a) &= f(x+a)\frac{dg(x+a)}{dx} - g(x+a)\frac{df(x+a)}{dx} \\
&= f(x+a)\left[\beta_1\frac{df(x)}{dx} + \beta_2\frac{dg(x)}{dx}\right] - g(x+a)\left[\alpha_1\frac{df(x)}{dx} + \alpha_2\frac{dg(x)}{dx}\right] \\
&= [\alpha_1 f(x) + \alpha_2 g(x)]\left[\beta_1\frac{df(x)}{dx} + \beta_2\frac{dg(x)}{dx}\right] - [\beta_1 f(x) + \beta_2 g(x)]\left[\alpha_1\frac{df(x)}{dx} + \alpha_2\frac{dg(x)}{dx}\right] \\
&= (\alpha_1\beta_2 - \alpha_2\beta_1)\left[f(x)\frac{dg(x)}{dx} - g(x)\frac{df(x)}{dx}\right] \\
&= (\alpha_1\beta_2 - \alpha_2\beta_1)\,W(x).
\end{aligned} \tag{8.16}$$

$$\therefore W(x+a) = (\alpha_1\beta_2 - \alpha_2\beta_1)\,W(x).$$

But $W(x+a) = W(x) =$ a constant.

Hence, $\alpha_1\beta_2 - \alpha_2\beta_1 = 1$.

Now eqn (8.13) reduces to

$$\lambda^2 - (\alpha_1 + \beta_2)\,\lambda + 1 = 0. \tag{8.17}$$

The quantity $(\alpha_1 + \beta_2)$ is a real function of energy E. Since eqn (8.17) is quadratic, it will have two roots and hence, two wave functions $\psi_1(x)$ and $\psi_2(x)$ having the property $\psi_1(x+a) = \lambda\psi(x)$. The product of the roots, i.e., $\lambda_1\lambda_2 = 1$.

Special cases

Case I. *For energy ranges such that* $(\alpha_1 + \beta_2)^2 < 4$.

In this case eqn (8.17) will have *a pair of complex roots.* As $\lambda_1\lambda_2 = 1$, these complex roots will be complex conjugate of each other.

$$\left.\begin{aligned} \therefore \lambda_1 &= e^{iKa} \\ \text{and } \lambda_2 &= e^{-iKa} \end{aligned}\right\} \tag{8.18}$$

where k is real.

The corresponding functions $\psi_1(x)$ and $\psi_2(x)$ will satisfy the relation:

$$\left.\begin{aligned} \psi_1(x+a) &= e^{iKa}\psi_1(x) \\ \text{and } \psi_2(x+a) &= e^{-iKa}\psi_2(x) \end{aligned}\right\} \tag{8.19}$$

or, in general,

$$\psi(x+a) = e^{\pm ika}\psi(x). \tag{8.20}$$

The function of eqn (8.20) can be shown to be *Bloch function* of the type $\psi(x) = e^{\pm ikx}\,u_k(x)$.

Replacing x by $(x+a)$ in eqn (8.3), we get

$$\psi(x+a) = e^{\pm ik(x+a)}\,u_k(x+a) = e^{\pm ika}\,e^{\pm ikx}\,u_k(x+a) = e^{\pm ika}\psi(x) = \lambda\psi(x), \tag{8.21}$$

where

$$\lambda = e^{\pm ika}. \tag{8.22}$$

Hence, eqns (8.20) and (8.21) are the same and thus *Bloch theorem is proved.*

Case II. *For the energy range such that* $(\alpha_1 + \beta_2)^2 > 1$.

In this case eqn (8.17) will have real roots λ_1 and λ_2 and may be written as

$$\lambda_1 = e^{+\mu a} \text{ and } \lambda_2 = e^{-\mu a}, \tag{8.23}$$

where μ is real. Then the corresponding solutions to the *Schrödinger equation* are

$$\psi_1(x) = e^{\mu x}\,u(x) \text{ and } \psi_2(x) = e^{-\mu x}\,u(x). \tag{8.24}$$

Since

$$\frac{\psi(x+na)}{\psi(x)} \to \infty, \text{ as } n \to \infty,$$

these solutions are not acceptable. This is contrary to the requirement because *the wave function should remain finite at all points in space due to their probabilistic interpretation.* For Case I, however,

$$|\psi(x+na)| = |\psi(x)|$$

and the ratio

$$\frac{\psi(x+na)}{\psi(x)} \text{ is finite}$$

The roots $\lambda_1 = e^{iKa}$ and $\lambda_2 = e^{-iKa}$ are *allowed* roots whereas $\lambda_1 = e^{\mu a}$ and $\lambda_2 = e^{-\mu a}$ are *forbidden* roots though all are functions of $(\alpha_1 + \beta_2)$ in energy E. The allowed roots are related to allowed energy regions and the forbidden roots to forbidden energy regions.

Hence, *the energy spectrum of an electron moving in a periodic potential consists of allowed and forbidden energy regions (or bands).*

8.5 Kronig-Penney Model

The solution of the *Schrödinger equation* with periodic potential [Fig. 8.4] is not very easy. *Kronig* and *Penney,* therefore, introduced a simpler model for the shape of the potential. They assumed that the *potential inside the crystal is rectangular in shape as shown in* Fig. 8.9.

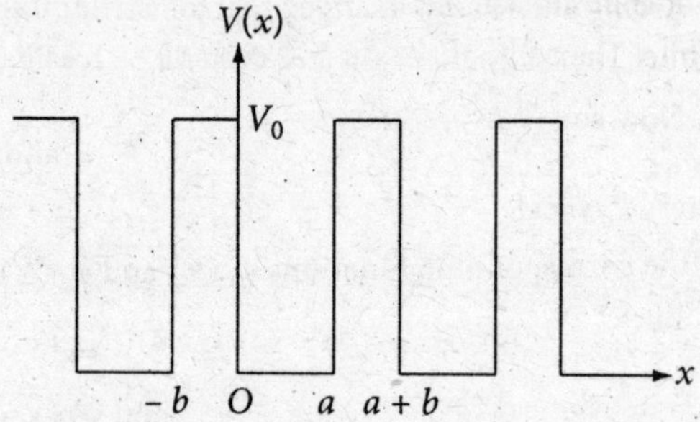

Fig. 8.9: Kronig-Penny model of potential

The potential is defined as

$$V = 0, \qquad 0 < x < a$$

$$\text{and } V = V_0, \quad -b < x < 0.$$

The moving electrons in the lattice are not actually free but influenced to some extent by the periodic potential. The wave functions relating to this model may be obtained by solving *Schrödinger equation* for the following two regions:

$$\frac{d^2\psi}{dx^2} + \left(\frac{2m}{\hbar^2}\right) E\psi = 0, \qquad \text{for } 0 < x < a \text{ with } V(x) = 0$$

$$\text{and } \frac{d^2\psi}{dx^2} + \left(\frac{2m}{\hbar^2}\right)(E - V_0)\,\psi = 0, \text{ for } -b < x < 0 \text{ with } V(x) = V_0.$$

Putting $\alpha^2 = \frac{2mE}{\hbar^2}$ and $\beta^2 = \frac{2m}{\hbar^2}(V_0 - E)$, we get

$$\frac{d^2\psi}{dx^2} + \alpha^2\psi = 0 \tag{8.25}$$

$$\text{and } \frac{d^2\psi}{dx^2} + \beta^2\psi = 0. \tag{8.26}$$

According to *Bloch,* the solutions are of the form:

$$\psi(x) = u_k(x)\, e^{iKx}, \tag{8.27}$$

where $u_k(x)$ is a *periodic function* of x with periodicity of $(a + b)$, i.e.,

$$u_k(x) = u_k(x + a) = u_k(x + na),$$

where n is an integer and k the *wave number,* where

$$\frac{2\pi}{\lambda} = \frac{p}{\hbar}.$$

The solution applicable to eqns (8.25) and (8.26) as shown by *Bloch* is

$$\psi(x) = u_k(x)\, e^{iKx},$$

where $u_k(x) = u_k(x+a) = u_k(x+na)$, n being an integer.

After a long mathematical calculations *Kronig* and *Penney* have shown that only those values of k are allowed which satisfy the following equation:

$$\left(\frac{\beta^2-\alpha^2}{2\alpha\beta}\right)\sinh \beta b \sin \alpha a + \cosh \beta b \cos \beta a = \cos k(a+b). \tag{8.28}$$

Kronig and *Penney* assumed that for a fruitful solution, $V_0 \to \infty$ and $b \to 0$ in such a way that $V_0 b$ remains finite. Then $\sinh \beta b \to \beta b$ and $\cosh \beta b \to 1$, as $b \to 0$.

Now eqn (8.28) reduces to

$$\beta b\left(\frac{\beta^2-\alpha^2}{2\alpha\beta}\right)\sin \alpha a + \cos \alpha a = \cos ka. \tag{8.29}$$

But

$$\beta^2 - \alpha^2 = \frac{2m}{\hbar^2}\left[(V_0 - E) + E\right].$$

$$\therefore \frac{2mV_0}{2\hbar^2}(ab)\frac{\sin \alpha a}{\alpha a} + \cos \alpha a = \cos ka \tag{8.30}$$

$$\text{or, } \left(\frac{mV_0 ab}{\hbar^2}\right)\left(\frac{\sin \alpha a}{\alpha a}\right) + \cos \alpha a = \cos ka \text{ or, } P\frac{\sin \alpha a}{\alpha a} + \cos \alpha a = \cos ka, \tag{8.31}$$

where $P = \frac{mV_0ab}{\hbar^2}$.

The quantity $V_0 \to \infty$, therefore, P is a measure of $V_0 b$, the *area of the potential barrier.*

The physical significance of P is that, as P increases the area $V_0 b$ of the potential barrier increases. *This means the electron is bound more strongly to a particular potential well. The potential barrier becomes very weak as P → 0 and the electron becomes free.*

Eqn (8.31) will be satisfied for those values of αa for which $\left(P\frac{\sin \alpha a}{\alpha a} + \cos \alpha a\right)$ will have values between +1 and −1. Only such values of αa will give a solution of the eqn (8.31).

Fig. 8.10 shows the plot of

$$\left(P\frac{\sin \alpha a}{\alpha a} + \cos \alpha a\right)$$

against αa for $P = \frac{3\pi}{2}$. The part of the vertical axis lying between −1 and +1, as indicated by the horizontal lines, represents the values acceptable to

$$\left(P\frac{\sin \alpha a}{\alpha a} + \cos \alpha a\right).$$

The abscissa is a measure of energy because

$$\alpha^2 = \frac{2mE}{\hbar^2}.$$

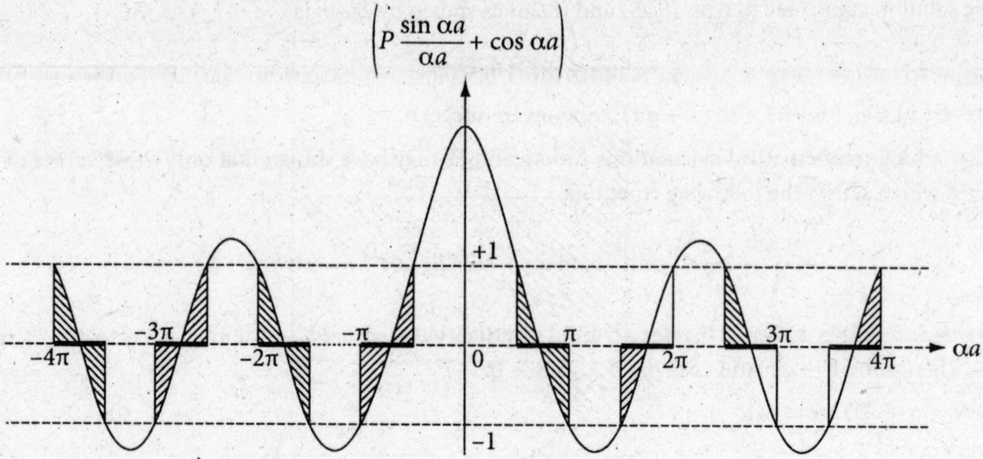

Fig. 8.10: Plot of $\left(P\dfrac{\sin \alpha a}{\alpha a}+\cos \alpha a\right)$ versus αa using $P = 3\pi/2$

We can draw the following conclusions from Fig. 8.10:

(i) The energy spectrum of the electron consists of alternate regions of *allowed energy bands* (solid lines on abscissa) and *forbidden energy bands* (broken lines). The boundaries of allowed ranges of αa corresponds to the values of $\cos ka = \pm 1$ or, $ka = n\pi$ or, $k = \frac{n\pi}{a}$.

(ii) The width of the allowed band increases with αa but the width of the forbidden band decreases.

(iii) P is a measure of the *barrier potential strength.* If P is large, the allowed energy bands are compressed into energy levels and a line spectrum is obtained. This is because for $P \to \infty$, the function $\left(P\frac{\sin \alpha a}{\alpha a} + \cos \alpha a\right)$ crosses +1 and −1 regions at a steeper angle. When $\alpha a = \pm n\pi$, $\cos \alpha a = \cos ka$ with $ka = \pm n\pi$ or, $k = \pm\frac{n\pi}{a}$. These values of k represent the *discontinuity* in the $(E\text{-}k)$ curve for electron.

(iv) The energy spectrum in two extreme cases using eqn (8.31):

(a) When $P \to \infty$, eqn (8.31) has a solution only for $\sin \alpha a = 0$ or, $\alpha a = \pm n\pi$, where n is an integer.

Now

$$\alpha = \pm\frac{n\pi}{a} \text{ or, } \alpha^2 = \frac{n^2\pi^2}{a^2} = \frac{2mE}{\hbar^2} \text{ or, } E = \frac{n^2\pi^2\hbar^2}{2ma^2} \text{ or, } E = \frac{n^2h^2}{8ma^2}. \tag{8.32}$$

Thus, the energy is independent of k. *The electron is completely bound and the energy levels are discrete. The electron in this case will be confined to the potential well and moves only in the cell of width a.*

(b) When $P \to 0$,

$$\cos ka = \cos \alpha a \text{ or, } \alpha^2 = k^2 = \frac{2mE}{\hbar^2} \text{ or, } E = \frac{\hbar^2 k^2}{2m} = \frac{h^2}{8\pi^2 m}\left(\frac{2\pi}{\lambda}\right)^2. \tag{8.33}$$

$$\therefore E = \left(\frac{h^2}{2m}\right)\frac{1}{\lambda^2} = \frac{h^2}{2m}\left(\frac{p^2}{h^2}\right) = \frac{p^2}{2m} = \frac{1}{2}mv^2. \tag{8.34}$$

The energy E in this case corresponds to that of a *free particle.* All energies are allowed to the electrons as there is no energy level exists.

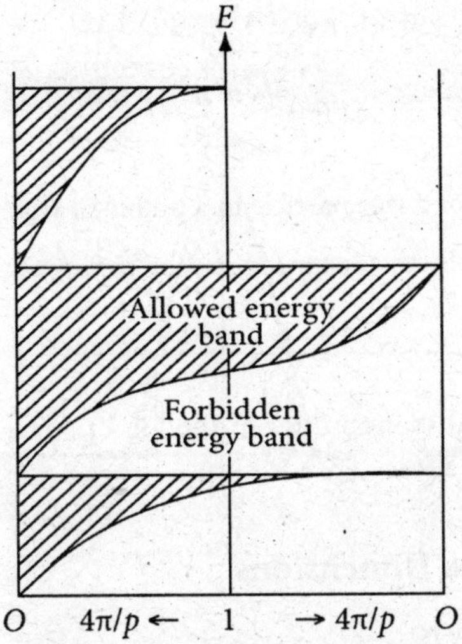

Fig. 8.11: Allowed and forbidden energy band as a function of P

Fig. 8.11 shows the energy spectrum of P. At the extreme left for $P = 0$, the energy spectrum is *quasi-continuous* and on the extreme right for $P \to \infty$, a line spectrum is resulted. For $0 < P < \infty$, the position and width of allowed and forbidden energy bands can be shown by drawing a vertical line.

8.6 Energy (E) versus Wave Vector (k) Relationship

Eqn (8.33) gives a parabolic curve when E is plotted against k as shown by dotted lines in Fig. 8.12. An electron being acted by the lattice potential has a (E-k) curve as shown by solid lines. Near the energy gap at $ka = \pm n\pi$, the solid line deviates from the parabola considerably.

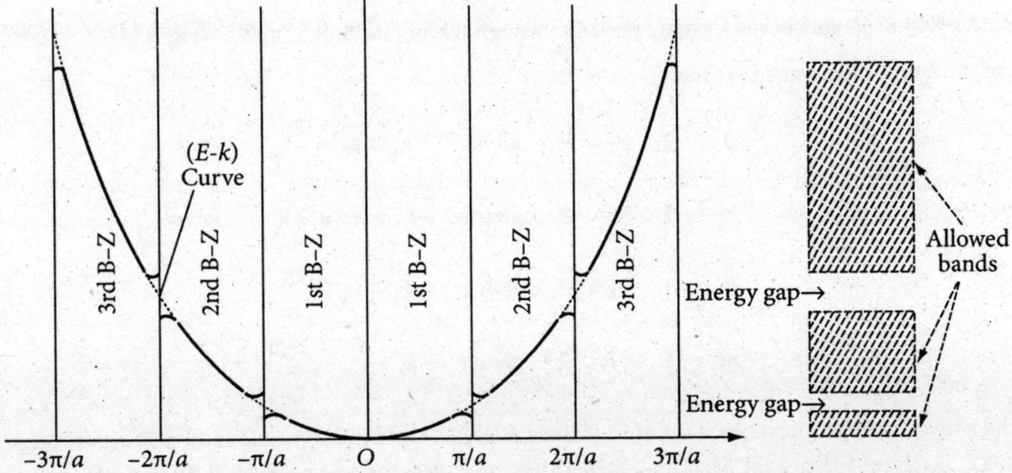

Fig. 8.12: The (E-k) curve for nearly free electron

When $k = \pm\frac{n\pi}{a}$, the free electron energy is given by eqn (8.34)

$$\text{i.e., } E = \frac{\hbar^2 n^2 \pi^2}{2ma^2} = \frac{n^2 h^2}{8ma^2}. \tag{8.35}$$

This is the same energy as that for a free particle in a potential well of width a.

If the size of the well is replaced by the distance between the atoms, say 0.3 nm, then from eqn (8.35),

$$E = 4n^2 \text{ eV}. \tag{8.36}$$

This shows that beyond a few discontinuities in the $(E\text{-}k)$ curve, $E \to \infty$. The regions beyond the discontinuities are called *Brillouin zones.*

8.7 Brillouin Zones in Two Dimensions

The condition for any discontinuity is

$$k = \pm\frac{n\pi}{a}. \tag{8.37}$$

For *two-dimensional case,* the corresponding conditions is

$$k_x n_1 + k_y n_2 = \frac{\pi}{2}\left(n_1^2 + n_2^2\right),$$

where n_1 and n_2 are integers relating to x- and y-axis respectively. To sketch the first zone n_1 and n_2 are made in turn to ±1 or 0. The borders of the first zone are

$$n_1 = +1, \quad n_2 = 0, \quad \text{so,} \quad k_x = \pm\frac{\pi}{a}$$

$$n_1 = 0, \quad n_2 = \pm 1, \quad \text{so,} \quad k_y = \pm\frac{\pi}{a}.$$

So, the *first Brillouin zone* is a square passing through the points A, B, C and D [Fig. 8.13].

For the 2*nd Brillouin zone,* we have

$$n_1 = +1, \quad n_2 = +1, \quad \text{giving} \quad k_x + k_y = \frac{2\pi}{a}$$

$$n_1 = -1 \quad n_2 = +1, \quad \text{giving} \quad -k_x + k_y = \frac{2\pi}{a}$$

$$n_1 = +1, \quad n_2 = -1, \quad \text{giving} \quad k_x - k_y = \frac{2\pi}{a}$$

$$n_1 = -1, \quad n_2 = -1, \quad \text{giving} \quad -k_x - k_y = \frac{2\pi}{a}.$$

The above equations describe a set of four lines at 45° to the k_x and k_y axes and passing through the points P, Q, R and S [Fig. 8.13]. Thus, the *second Brillouin zone* is the area between P, Q, R, S and A, B, C, D. Putting n_1 and n_2 = 0, ±1 and ±2, the *third Brillouin zone* may be obtained [Fig. 8.13].

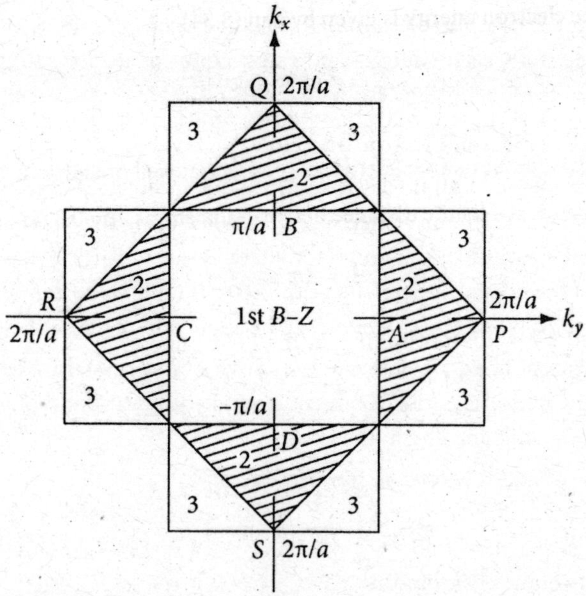

Fig. 8.13: Brillouin zones for a square lattice

8.7.1 Number of Wave Functions in a Band

We have seen that in an infinitely long one-dimensional crystal there are certain allowed energy ranges which have a continuous distribution of energy.

For a linear crystals of length L, the periodic boundary condition is $\psi(x + L) = \psi(x)$.

As the wave function are *Bloch functions,* we have $e^{iK(x+L)}\, u_k(x + L) = e^{iKx}\, u_k(x)$.

Since $u_k(x + L) = u_k(x)$.

$\therefore e^{iK(x+L)} = e^{iKx}$ or, $e^{iKL} = 1$.

$$\therefore k = \frac{2\pi n}{L} \tag{8.38}$$

with $n = \pm 1, \pm 2, \ldots$

$$\text{or, } k = \pm\frac{2\pi}{L}, \pm\frac{4\pi}{L}, \ldots. \tag{8.39}$$

So, the number of possible wave function in range dk is

$$dk = \frac{2\pi}{L}\, dn, \text{ [from eqn (8.38)]}$$

$$\text{or, } dn = \frac{L}{2\pi}\, dk. \tag{8.40}$$

In the first Brillouin zone, the maximum value of k is π/a, where a is the length of the *primitive cell.* Again, if in a crystal of length L, the number of primitive cells is N, then $a = L/N$. So, the maximum value of k in the *first Brillouin zone* is $N\pi/L$. This means that the series in eqn (8.39) terminates at $N\pi/L$. Thus, the total number of allowed k-values in the *first Brillouin zone* is N. *Hence, the total number of possible states in any energy band is equal to the number of primitive unit cells* N.

Since an electron has *two spin states* and *Pauli's exclusion principle* allows at best two electrons with opposite spins in a single state, each energy band can accommodate a maximum of $2N$ electrons. For distinguishing metals, semiconductors and insulators, this resultant plays an important role.

8.7.2 Concept of Effective Mass of Electron and Hole

In a crystal the electrons interact with the lattice and are not completely free. The wave-particle motion, therefore, differs from that of an electron in free space. So, while using the equations of electrodynamics for charge carriers in a crystal, we must keep in mind the altered value of particle mass called the *effective mass* (m^*) due to the effect of the crystal lattice. Then the *electrons* (and *holes*) can be regarded as almost free charge carriers and *Newton's second law of motion* can be applied to describe their dynamic behaviour.

We consider an electron of charge e and mass m moving in a crystal lattice of electric field E. The acceleration $a = \frac{eE}{m}$ in the periodic lattice is not constant. Let the variation of acceleration is due to the variation of electron's mass while moving in the lattice.

$\therefore$ the acceleration $a = \frac{eE}{m^*}$ and the force on the electron is

$$F = m^* a. \tag{8.41}$$

Now the group velocity of the electron wave is

$$v_g = \frac{d\omega}{dk} = 2\pi\frac{dv}{dk} = \frac{2\pi}{h}\frac{dE}{dk} = \frac{1}{\hbar}\frac{dE}{dk}, \tag{8.42}$$

where $E = h\nu$ and $\hbar = h/2\pi$.

Acceleration

$$a = \frac{dv_g}{dt} = \frac{1}{\hbar}\frac{d^2E}{dk\,dt} = \frac{1}{\hbar}\frac{d^2E}{dk^2}\frac{dk}{dt}.$$

But $\hbar k = p$ and $\frac{dp}{dt} = F$.

$$\therefore \frac{dk}{dt} = \frac{F}{\hbar}.$$

So

$$a = \frac{1}{\hbar^2}\frac{d^2E}{dk^2}F \text{ or, } F = \frac{\hbar^2}{\frac{d^2E}{dt^2}}a. \tag{8.43}$$

Comparing eqns (8.41) and (8.43), we get

$$m^* = \frac{\hbar^2}{\frac{d^2E}{dk^2}}. \tag{8.44}$$

Thus, determining the value of $\frac{d^2E}{dk^2}$, we can calculate the value of the *effective mass* m^*.

(i) Variation of E with k: The variation of E with k for the first allowed band is shown in Fig. 8.14 (a). Using this type of variation of E with k, the group velocity $v_g = v$ can be calculated using eqn (8.42).

(ii) Variation of v with k: Fig. 8.14 (b) shows the variation of v with k. For $k = 0$, $v = 0$ and with the increase of k (i.e., energy), v increases and reaches a maximum value at $k = k_0$, where k_0 is the value of k at the point of inflection of the $(E\text{-}k)$ curve. Beyond k_0, v begins to decrease and finally assumes the zero value at $k = \pi/a$. The negative value of k shows a similar behaviour.

(iii) The variation of m^* with k: Fig. 8.14 (c) shows the variation of m^* with k. Around $k = 0$, m^* increases, reaches its maximum value at the point of inflection of the $(E\text{-}k)$ curve. Above the point of inflection, m^* is *negative* and as $k \to \pi/a$, m^* reduces to a small negative value.

The negative mass is due to the force on the electron (caused by Bragg reflection) in one direction leading to a gain in momentum in opposite direction. This results in negative effective mass.

The equation $E = \frac{\hbar^2 k^2}{2m}$ is satisfied near the bottom of the band and hence, m^* has a constant value. But with the increase of k, m^* increases. Fig. 8.14 (b) shows that beyond the inflection point k_0, velocity decreases due to the decrease of acceleration. This implies that in this region of k the lattice exerts a large retarding force on the electron and it behaves as a positively charged particle called *hole.*

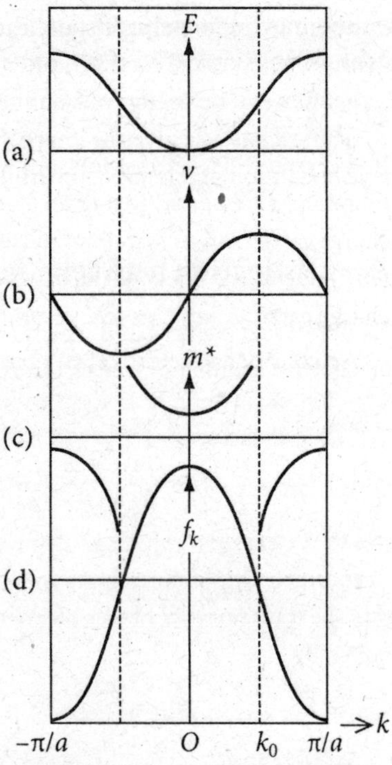

Fig. 8.14: E, m^* and f_k as a function of k

(iv) Variation of f_k with k: The degree of freedom of an electron is defined as

$$f_k = \frac{m}{m^*} = \frac{m}{\hbar^2}\left(\frac{d^2E}{dk^2}\right). \tag{8.45}$$

f_k gives a measure of extent to which the electron is free. When m^* is large, f_k is small, the particle then behaves as a *heavy* particle. *If $f_k = 1$, the electron behaves as a free electron.*

8.8 Crystal Momentum

The momentum associated with the motion of a free electron is $\hbar\vec{k}$, where $\vec{k}$ is the wave vector. This momentum of free electron is called the *true momentum* which helps to understand the dynamical behaviour of it. For example, the energy of the free electron is

$$E = \frac{\hbar k}{2m} = \left(\frac{1}{2m}\right)\left(\frac{h^2}{4\pi^2}\right)\left(\frac{4\pi^2}{\lambda^2}\right) = \left(\frac{1}{2m}\right)\left(\frac{h^2}{\lambda^2}\right) = \frac{1}{2m}p^2. \tag{8.46}$$

This $\vec{k}$ was first introduced by *Sommerfeld* while formulating the theory of free electrons.

However, when an electron moves in a *periodic potential,* $\hbar\vec{k}$ does not represent the true momentum as the energy does not vary with $\vec{k}$ in the manner as it does in case of a free electron and as such is known as the *crystal momentum.*

For an electron moving in a periodic potential, $\hbar\vec{k}$ is not the true electron momentum. This is also confirmed by the fact that, the Bloch wave functions which are eigenstate of the Hamiltonian, are not simultaneously eigenstates of the momentum operator

$$\vec{p}\psi_k(\vec{r}) = \frac{\hbar}{i}\nabla\psi_k(\vec{r}) = \frac{\hbar}{i}\nabla\left[e^{i\vec{k}\cdot\vec{r}}\,u_k(\vec{r})\right] = \hbar k\psi_k(\vec{r}) + e^{i\vec{k}\cdot\vec{r}}\frac{\hbar}{i}\nabla u_k(\vec{r}). \tag{8.47}$$

$\therefore\ \vec{p}\psi_k(\vec{r})$ is not simply $\hbar k\psi_k(\vec{r})$, still $\hbar\vec{k}$ is an extension of momentum $\vec{p}$ to the periodic potential of the crystal. This is why $\hbar\vec{k}$ for a lattice is known as the *crystal momentum,* $\vec{k}$ being the wave vector introduced in *Bloch theorem.*

It is to be remembered that, the free electron momentum $\vec{p} = \hbar\vec{k}$ is a constant of motion but for an electron moving in periodic potential $\hbar\vec{k}$ is not constant of motion and cannot be calculated directly by quantum mechanical method, only its average value can be calculated. On the other hand, the crystal momentum is a perfectly well-defined constant for a state of given energy just as the true momentum of a dynamical particle gives its energy. For this reason, *in dealing with interaction of electron with lattice, we use conservation of crystal momentum instead of true electron momentum.*

8.9 Distinction between Metals, Insulators and Semiconductors

We consider an energy band which is filled with electrons up to a value k_1 ($k_1 < \pi/a$) as shown in Fig. 8.15. The effective number of free electrons in the band is responsible to the conductivity of this band.

Now

$$f_k = \frac{m}{\hbar^2}\left(\frac{d^2E}{dk^2}\right)$$

gives the extent up to which an electron in the k state is free which would participate in electron conduction. The total number of free electrons in the band is given by

$$N_{eff} = \sum_k f_k, \tag{8.48}$$

$\sum_k$ corresponds to various occupied energy states in the band.

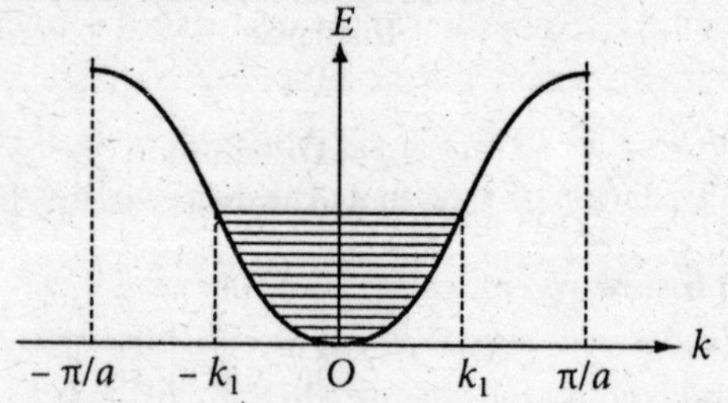

Fig. 8.15: Energy band filled up to the state k_1

Now the number of possible states in k to $k + dk$ for a one-dimensional lattice of length L is

$$dn = \frac{L}{2\pi}\,dk.$$

∴ the total number of free electrons present within the limit $-k_1$ to $+k_1$ is

$$N_{eff} = 2 \times \frac{1}{2\pi}\int_{-k_1}^{+k_1} f_k dk = \frac{L}{\pi}\int_{-k_1}^{k_1} f_k dk. \tag{8.49}$$

The factor 2 on the right side of eqn (8.49) is due to the fact that each state can occupy a maximum of two electrons with opposite spin.

$$\therefore N_{eff} = \frac{L}{\pi}\int_{-k_1}^{k_1}\frac{m}{\hbar^2}\left(\frac{d^2E}{dk^2}\right)dk \quad \left[\because f_k = \frac{m}{\hbar^2}\left(\frac{d^2E}{dk^2}\right)\right]$$

$$= \frac{2L}{\pi}\frac{m}{\hbar^2}\int_0^{k_1}\frac{d^2E}{dk^2}dk = \frac{2Lm}{\pi\hbar^2}\left(\frac{dE}{dk}\right)_{k=k_1}. \tag{8.50}$$

The behaviour of solids as *insulator, semiconductor* and *conductor* can be explained on the basis of *Band Theory* as follows:

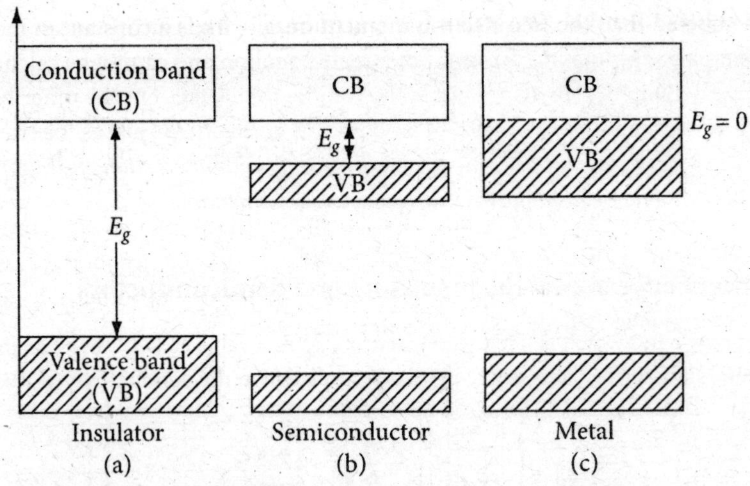

Fig. 8.16: Distribution of electrons in insulator, semiconductor and metal. The shaded areas are occupied by electrons

(a) Insulators: From the $(E\text{-}k)$ curve, it is clear that, at the top (and bottom) of the band $\frac{dE}{dk} = 0$ as $k = 0$, π/a. So from eqn (8.50), $N_{eff} = 0$. Thus, when the valence band is completely filled (and the conduction band is completely empty) the solid behaves as a *perfect insulator* [Fig. 8.16(a)]. There is a *forbidden energy gap or band gap* (E_g) between the valence band and conduction band. For example, diamond is a good insulator having band gap $E_g \sim 6$ eV. As the electrons in the valence band ordinarily do not possess this amount of energy, they cannot transit to the conduction band and participate in the electrical conduction process. As a result, the solid behaves as an insulator. *All solids with large band gap* (E_g) *are good electrical insulators.*

(b) Semiconductors: There is a class of solids like Germanium (Ge) and Silicon (Si), etc. for which the band gap E_g is relatively small ~ 1 eV. It is, therefore, possible for some electrons at room temperature to have thermal energy to jump across the forbidden gap into the empty conduction band. In presence of an electric field, these electrons constitute a limited current. The conductivity of such materials lies between good conductor and insulator [Fig. 8.16(b)].

(c) Conductors (Metals): Since the value of $\frac{dE}{dk}$ is the maximum at the point of inflection of $(E\text{-}k)$ curve (Fig. 8.15), the effective number of electrons N_{eff} is the maximum. So, in a material with completely filled valence band, there is a large number of the electrons available for conduction. Such materials are conductors like all metals. The valence band and conduction band in conductors partially overlap so that the band gap $E_g = 0$, Fig. 8.16 (c).

Effect of Temperature

In case of *good conductors* (metals) with the increase in temperature, more and more phonons are excited which scatter electrons and reduce their mobility. This, in turns, leads to the increase in resistivity and decrease the conductivity of metals. But in case of semiconductors, increase in temperature causes more and more electrons in the valence band to have sufficient energy to jump into the conduction band. Hence, *with the rise of temperature the conductivity of the semiconductors increases. This is an important distinction between metals and semiconductors.*

8.10 Direct Experimental Evidence for Band Structure in Solids

The essential requirement for the direct evidence of the band structure of the energy levels in solids are:

(i) The existence of energy gap ΔE between allowed bands should be experimentally confirmed.

(ii) There should be experimental confirmation that allowed bands are partly filled in metals and completely filled in insulators.

The existence ΔE between allowed bands is experimentally confirmed from the study of variation of resistivity with temperature, light absorption and photo-conductivity in semiconductors and insulators. The confirmation of the fact that, allowed bands are partly filled in metals and completely filled in insulators comes from the experimental study of soft X-ray spectroscopy.

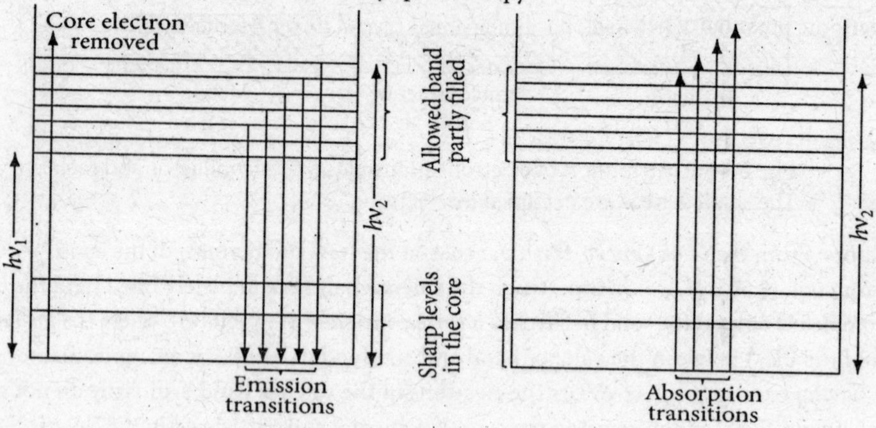

Fig. 8.17: Various transitions in soft X-ray spectrocopy

In Fig. 8.17, A represents the energy level of a core electron, say, 1s (K-shell) state. This type of core levels are sharp. Let there be bombardment of a suitable cathode ray electron to the target atom, the core electron is removed and a vacancy is created there. This vacancy in the 1s state can be filled by the transition of an electron in the higher allowed band which is partly filled [Fig. 8.17 (a)].

Since there are no energy levels in higher band from where transition would take place to ν_2, there will be no emission, if $\nu > \nu_2$. Sudden emissions of all transitions will occur behaving as emission edge within $\nu_1 < \nu < \nu_2$. For $\nu < \nu_1$, no emission will occur due to the absence of transition levels. The observed emission bandwidth is a measure of the extent up to which the allowed band is filled.

Fig. 8.17 (a) shows the various absorption transitions of X-rays in the same specimen.

For $\nu < \nu_2$, there is no absorption of energy but for $\nu = \nu_2$, there is an absorption edge. For $\nu > \nu_2$, an absorption band arises. Hence, the existence of the unfilled levels in the allowed band is evidenced by the absorption of energy. In case of metals, the absorption and emission edges coincide in frequency.

The emission bands for non-metals are different from those of metals and have no sharp edges. In case of metals, the transitions from filled to unfilled levels occur over a region proportional to kT at the Fermi levels and hence, are not perfectly sharp. This is why the width of the emission edge is temperature dependent.

8.11 Limiting Cases of Periodic Potential

In the foregoing discussion of *Kronig-Penney model,* Sommerfeld's free electron theory is used which is poorly related to the actual form of potential. The effect of ion cores cannot be totally ignored. Therefore, two limiting cases arise where the periodic potential is either very weak or very strong. In case of very strong periodic potential each electron is almost bound to the minimum of the potential and the rest of the lattice is a perturbation to this minimum. This is *tight binding approximation.*

In case of solids where the above methods are inadequate, we use more complex methods like *orthogonalised plane wave* (OPW) or *augmented plane wave* (APW) methods.

8.11.1 Nearly Free Electron (NFE) Approximation

Here the crystal potential is assumed to be very weak in comparison to the electron KE E, and the effect of the periodic potential as a *perturbation.* Under these conditions the allowed bands will be broad and the forbidden gaps quite narrow.

Let the periodic potential $V(x)$ in a one-dimensional crystal be represented as

$$-\frac{2mV(x)}{\hbar^2} = \gamma f(x). \tag{8.51}$$

$f(x)$ represents the periodicity of lattice. Then the *one-dimensional Schrödinger's equation* may be written as

$$\frac{\partial^2\psi}{\partial x^2} + \left[\frac{2m}{\hbar^2}E + \gamma f(x)\right]\psi(x) = 0 \text{ or, } \frac{\partial^2\psi}{\partial x^2} + \left[K_0^2 + \gamma f(x)\right]\psi(x) = 0. \tag{8.52}$$

As $f(x)$ is a periodic function of x, we can express it as a Fourier series, i.e.,

$$f(x) = \sum_{n=-\infty}^{\infty} C_n\, e^{-\frac{2\pi inx}{a}} \tag{8.53}$$

with

$$C_n = \frac{1}{a}\int_0^a f(x)\, e^{\frac{2\pi inx}{a}}\, dx, \tag{8.54}$$

where a is the lattice constant of the crystal.

If V_n be the Fourier expansion coefficient of $V(x)$, then according to eqn (8.51)

$$V_n = -\frac{\hbar^2}{2m}\gamma C_n, \tag{8.55}$$

The wave function $\psi(x)$ must have the Bloch form:

$$\psi(x) = e^{iKx}\, U(x) = e^{iKx} \sum_{n=-\infty}^{\infty} b_n\, e^{-\frac{2\pi inx}{a}}, \tag{8.56}$$

where $U(x)$ is a periodic function of a, it can be expressed as a Fourier series as

$$U_k(x) = \sum_{-\alpha}^{\infty} b_n\, e^{-\frac{2\pi inx}{a}}. \tag{8.57}$$

In case the electron is completely free, $\gamma = 0$ and $n = 0$, when

$$\psi(x) = b_0\, e^{iKx}. \tag{8.58}$$

In the limit, $U_k(x) \to b_0$ and $K \to K_0$. So, as $x \to 0$ all b_n (except b_0) approach zero. Then the approximate expression for the wave function is of the form:

$$\psi(x) = b_0\, e^{iKx} + \gamma\left[e^{iKx} \sum_{n\neq 0} b_n\, e^{-\frac{2\pi nx}{a}}\right]. \tag{8.59}$$

Eqn (8.59) is valid for small values of γ

$$\text{or, } \psi(x) = \left[b_0 + \gamma \sum_{n \neq 0} b_n e^{-\frac{2\pi n x}{a}}\right] e^{iKx} = U(x)\, e^{iKx}, \tag{8.60}$$

where

$$U(x) = b_0 + \gamma \sum_{n \neq 0} b_n e^{-\frac{2\pi n x}{a}}.$$

Substituting eqn (8.59) in eqn (8.52), we get

$$b_0\left(K_a^2 - K^2\right) e^{iKx} + \gamma \sum_{n \neq 0}\left[\left(K_0^2 - K_a^2\right) b_n + b_0 c_n\right] e^{iK_n x} + \gamma^2 \sum_{n \neq 0} \sum_{n' \neq 0} b'_n C_n e^{i\left(K - \frac{2\pi n}{a} - \frac{2\pi n'}{a}\right)x} = 0, \tag{8.61}$$

where

$$K_n = -K - \frac{2\pi n}{a}. \tag{8.62}$$

As a first approximation, we neglect γ^2 term in eqn (8.61). Then multiplying the equation by $e^{-iK_m x}$ and integrating over unit cell from $x = 0$ to $x = a$, we have

$$b_0\left(K_0^2 - K^2\right) \int_0^a e^{\frac{2\pi i m x}{a}} dx + \gamma \sum_{n \neq 0}\left[\left\{\left(K_0^2 - K_n^2\right) b_n + b_0 C_n\right\} \times \int_0^{a_0} e^{\frac{2\pi i (m-n) x}{a}} dx\right] = 0. \tag{8.63}$$

For $m = 0$,

$$\int_0^a e^{\frac{2\pi i (m-n) x}{a}} dx = 0,$$

for all values of n in the summation. Then

$$b_0\left(K_0^2 - K^2\right) = 0, \text{ or, } K = K_a. \tag{8.64}$$

Again, for $m \neq 0$,

$$\int_0^a e^{\frac{2\pi i m x}{a}} dx = 0$$

and the second integral in eqn (8.61) gives zero except $m = n$.

So

$$\gamma\left[\left(K_0^2 - K_m^2\right) b_m + b_0 C_m\right] a = 0 \text{ or, } b_m = \frac{b_0 C_m}{K_m^2 - K_0^2} = \frac{b_0 C_m}{K_m^2 - K^2}. \tag{8.65}$$

$\therefore$ from eqn (8.64),

$$K^2 = K_0^2, \text{ or, } \frac{2mE}{\hbar^2} = K_0^2 \text{ or, } E = \frac{K_0^2 \hbar^2}{2m}. \tag{8.66}$$

The relation is the same as that for a free particle.

Now substituting the value of b_m from eqn (8.65) into eqn (8.60), we get

$$\psi(x) = b_0 e^{iKx}\left[1 + \gamma \sum_{n \neq 0} \frac{C_n}{K_n^2 - K^2} e^{-\frac{2\pi i n x}{a}}\right]. \tag{8.67}$$

Eqn (8.64) shows that the contribution of the first order correction to the energy of a free particle due to the periodic potential is zero. To obtain second order energy correction, we multiply eqn (8.61) by e^{-iKx}

and then integrate from $x = 0$ to $x = a$,

$$ab_0\left(K_0^2 - K^2\right) + \gamma \sum_{n\neq 0}\left[\left(K_0^2 - K_n^2\right)b_n + b_0 C_n\right]\int_0^a e^{-\frac{2\pi i n x}{a}}\,dx + \gamma^2 \sum_{n\neq 0}\sum_{n'\neq 0} b'_{n'} C_n \int_0^a e^{-\frac{2\pi i(n+n')x}{a}}\,dx = 0. \tag{8.68}$$

Now

$$\int_0^a e^{-\frac{2\pi i n x}{a}}\,dx = 0,$$

for all allowed values of n and

$$\int_0^a e^{-\frac{2\pi i(n+n')x}{a}}\,dx = 0, \text{ if } n \neq -n'.$$

Thus, eqn (8.68) reduces to

$$b_0\left(K_0^2 - K^2\right)a + \gamma^2 \sum_{n'\neq 0} b'_n C'_n a = 0. \tag{8.69}$$

Again, replacing n by $-n$ in eqn (8.53), we get

$$f(x) = \sum_{n=-\infty}^{\infty} C_{-n}\, e^{\frac{2\pi n i x}{a}}. \tag{8.70}$$

The complex conjugate of eqn (8.53) is

$$f^*(x) = \sum_{n=-\infty}^{\infty} C_n^*\, e^{\frac{2\pi i n x}{a}} \text{ or, } f(x) = \sum_{n=-\infty}^{n=\infty} C_n^*\, e^{\frac{2\pi n i x}{a}}. \tag{8.71}$$

Since $f^*(x) = f(x)$, comparing eqns (8.70) and (8.71), we get

$$C_{-n} = C_n^*. \tag{8.72}$$

Now using eqns (8.65) and (8.72) in eqn (8.69), we get

$$K_0^2 = K^2 + \gamma^2 \sum_{n\neq 0} \frac{C_n^* C_n}{K^2 - K_n^2} = K^2 + \gamma^2 \sum_{n\neq 0} \frac{C^* C_n}{K^2 - \left(K - \frac{2\pi n}{a}\right)^2} \quad \left[\because K_n = K - \frac{2\pi n}{a}\right]. \tag{8.73}$$

The above relation can be expressed in terms of $E\left(= \frac{\hbar^2 K^2}{2m}\right)$ as follows:

$$E = \frac{\hbar^2 K^2}{2m} + \sum_{n\neq 0} \frac{|V_n|^2}{\left(\frac{\hbar^2 K^2}{2m}\right) - \frac{\hbar^2}{2m}\left(K - \frac{2\pi n}{a}\right)^2}. \tag{8.74}$$

If $K^2 \approx K_n^2$ for a particular value of n, then one of the quantities

$$\left(K^2 - K_n^2\right) \text{ in } \sum_{n\neq 0} \frac{C_n}{K_n^2 - K^2} e^{-\frac{2\pi i n x}{a}}$$

of eqn (8.67) tends to zero and the corresponding term will tend to infinity despite γ is very small. Then the eqn (8.59) is not a suitable form of the wave function $\psi(x)$. Further, if $K^2 \approx K_n^2$, then

$$K = \pm K_n = \pm K \mp \frac{2\pi n}{a}. \tag{8.75}$$

Since, for the upper sign, this equation gives $n = 0$, hence excluded from eqn (8.67). For the lower sign, we get

$$K = \frac{n\pi}{a}. \tag{8.76}$$

Eqn (8.62) refers to all the band edge points as n may assume both positive as well as negative values.

Now let $K \approx n\pi/a$, whence $K_n = K - 2\pi n/a \simeq -\frac{n\pi}{a}$. Then $K_n^2 \simeq K^2$ and b_n will not be very large. Under this condition γb_n will not be very small. The wave function in this case can be written as

$$\psi(x) = b_0\, e^{iKx} + \gamma b_n\, e^{iKx}\, e^{-\frac{2\pi i n x}{a}} = b_0\, e^{iKx} + \gamma b_n\, e^{iK_n x}. \tag{8.77}$$

As $K_n = -K$ at the band edge, $\psi(x)$ in eqn (8.77) in the result of superposition of a wave travelling along the positive x-direction and the other along the negative x-direction, i.e., $\psi(x)$ represents a standing wave.

Now putting

$$f(x) = \sum_{n=-\infty}^{+\infty} C_n\, e^{-\frac{2\pi i n x}{a}}$$

and $\psi(x)$ as given by eqn (8.77) in eqn (8.52), we get

$$b_0^2\left(K_0^2 - K^2\right) e^{iKx} + \gamma b_n \left(K_0^2 - K^2\right) e^{i\left(K - \frac{2\pi n}{a}\right)x} + \gamma b_0 \sum_{n' \neq 0} C_n^* e^{i(K - 2\pi n'/a)x}$$

$$+ \gamma^2 b_n \sum_{n' \neq 0} C_0^* e^{i\left(K - \frac{2\pi n}{a} - \frac{2\pi n'}{a}\right)x} = 0. \tag{8.78}$$

Multiplying eqn (8.78) by e^{-iKx} on both sides and then integrating within limits, $x = 0$ to $x = a$, we find that the contributions of the second and third terms become zero and that from the fourth term is obtained only for $n' = n$. Finally,

$$\left(K_0^2 - K^2\right) b_0 + \gamma^2 C_n^* b_n = 0. \tag{8.79}$$

Similarly, multiplying both sides of eqn (8.78) by

$$e^{-iK_n x} = e^{-i(k - 2\pi n/a)x}$$

and then integrating from $x = 0$ to $x = a$, we get

$$C_n b_0 + \left(K_0^2 - K_n^2\right) b_n = 0. \tag{8.80}$$

The eqns (8.79) and (8.80) are a set of homogeneous equations whose solutions other than $b_0 = b_n = 0$ exist, if

$$\begin{vmatrix} \left(K_0^2 - K^2\right) & \gamma^2 C_n^* \\ C_n & K_0^2 - K_n^2 \end{vmatrix} = 0 \text{ or, } \left(K_0^2 - K^2\right)\left(K_0^2 - K_n^2\right) - \gamma^2 C_n C_n^* = 0. \tag{8.81}$$

This is a quadratic equation having solution

$$K_0^2 = \tfrac{1}{2}\left[\left(K^2 + K_n^2\right) \pm \sqrt{\left(K^2 - K_n^2\right) + 4\gamma^2 C_n^* C_n}\,\right]. \tag{8.82}$$

Using the relations, $E = \frac{\hbar^2 K_0^2}{2m}$.

Eqn (8.82) can be expressed as

$$E(k) = \frac{\hbar^2}{4m}\left[K + \left(K - \frac{2\pi n}{a}\right)^2 \pm \sqrt{\left\{K^2 - \left(K - \frac{2\pi n}{a}\right)^2\right\}^2 + \left\{\frac{4m|V_n|^2}{\hbar^2}\right\}}\,\right]. \tag{8.83}$$

At the band edge, $K - K_n = \frac{n\pi}{a}$, where eqn (8.83) reduces to

$$E = E_n \pm |V_n| \tag{8.84}$$

with

$$E_n = \frac{\hbar^2}{2m}\left(\frac{n\pi}{a}\right)^2 \tag{8.85}$$

is the energy of a free particle at the band edge points.

The results of the foregoing discussion can be summarized as below:

(i) There exists an energy gap or discontinuity in the E versus K curve due to the internal Bragg reflection at the band edge points $K = \pm\frac{n\pi}{a}$. Eqn (8.81) shows that the energy gap for the forbidden band is of width $2|V_n|$, V_n being the nth Fourier coefficient in the Fourier expansion of the lattice potential.

(ii) For $|K| > \frac{n\pi}{a}$, $E(K)$ is greater than the value given by eqn (8.64) where $(K^2 - K_n^2)$ is sufficient to reduce γb_n sufficiently small. Similarly, for $|K| < \frac{n\pi}{a}$, $E(K)$ must be less than the value given by eqn (8.84). These conditions will be satisfied, if the positive sign and negative sign be chosen for $|K| > \frac{n\pi}{a}$ and $|K| < \frac{n\pi}{a}$ respectively in eqn (8.84).

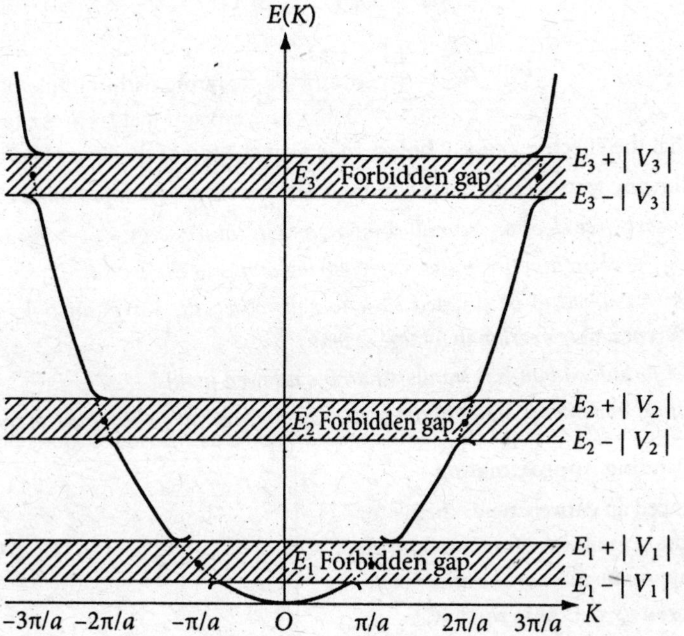

Fig. 8.18: E versus K curve for free electron approximation

(iii) The schematic representation of variation of $E(K)$ with K for an electron is shown in Fig. 8.18 as obtained from eqns (8.71) and (8.80). Fig. 8.18 shows that the E versus K curves are nearly parabolic near the band edge points.

Effective Mass

The effective mass of the electron (m_e^*) can be calculated by taking

$$K = \frac{n\pi}{a} + K',$$

where $K' \ll \pi/a$ in eqn (8.83).

$$\therefore E(K) = \frac{\hbar^2}{2m}\left[\left(\frac{n\pi}{a}\right)^2 + K'^2 + m\frac{\Delta E}{\hbar^2}\sqrt{1 - 4K'^2\left(\frac{n\pi}{a}\right)^2\left(\frac{\hbar^2}{m\Delta E}\right)^2}\,\right], \tag{8.86}$$

where

$$\Delta E = 2|V_n|. \tag{8.87}$$

For points near the bottom of the upper band K' is very small. From eqn (8.86) by expanding in *binomial theorem* and neglecting higher order terms we get

$$E(K) = E_n + \frac{1}{2}\Delta E + \frac{\hbar^2 K'^2}{2m}\left(1 + \frac{4E_n}{\Delta E}\right). \tag{8.88}$$

Differentiating twice, we get

$$\frac{d^2E}{d^2k'} = \frac{\hbar^2}{m}\left(1 + \frac{\Delta E_n}{\Delta E_n}\right).$$

$$\therefore m_e^* = \frac{\hbar^2}{\frac{d^2E}{dK^2}} = \frac{\hbar^2}{\frac{d^2E}{dK'^2}} = \frac{m}{1 + \frac{4E_n}{\Delta E}}. \tag{8.89}$$

It can be shown that the effective mass of holes (m_h^*) near the top of the lower band is the same as m_e^*.

In conclusion, following facts were observed in the *nearly free electron approximation method:*

(i) *The PE of the nearly free electron is small compared to its total energy.*

(ii) *There is a strong interaction between the neighbouring atoms in the crystal.*

(iii) *As the atoms are assumed to be situated close to each other, the wave functions of the neighbouring electrons in different atoms overlap to a large extent.*

(iv) *The width of the forbidden bands is smaller than the allowed bands.*

(v) *The wave functions and energy states of the crystal do not resemble with those of individual atoms.*

8.11.2 The Tight Binding Approximation

Bloch in 1928 introduced an entirely new scheme for the construction of the crystal wave function known as the *tight binding approximation* or *the linear combinations of atomic orbitals* (LCAO) method.

This approximation deals with the situations in which the overlap of atomic wave function is enough to require corrections to the pictures of the isolated atoms. In this method, one starts from the wave functions for an electron in a free atom and then constructs a crystal orbital which just equals the LACO's and describes the electron in the periodic field of crystal as a whole.

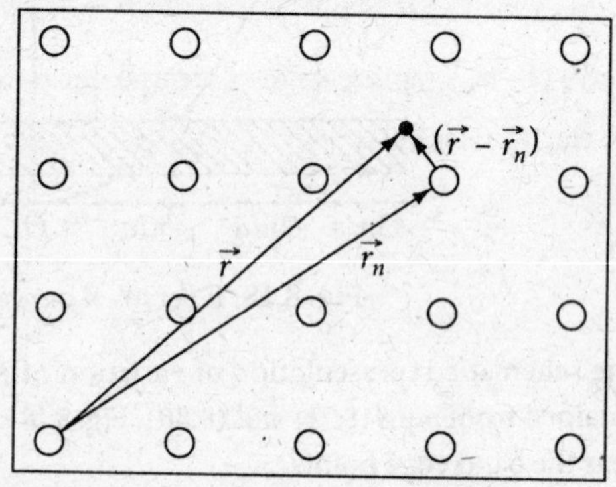

Fig. 8.19: Two-dimensional square lattice showing the positions of the *n*th atom and an electron

We consider a two-dimensional square array of atoms in a crystal where $\vec{r}$ and $\vec{r}_n$ represent respectively the position vector of the electron and *n*th atom respectively as shown in Fig. 8.19.

Fig. 8.20 shows the potentials of an electron in a free atom denoted by $V(\vec{r})$ and that in a crystal by $V_0(\vec{r}-\vec{r}_n)$.

In this method, the electron wave function is assumed to be influenced by the nearest atoms and not by the other atoms in the crystal. If $\psi_k(\vec{r}-\vec{r}_n)$ be the wave function for an electron in a free atom and $\psi_k(\vec{r})$ the wave function of the electron in a crystal, then $\psi_k(\vec{r})$ may be expressed as a linear combination as

$$\psi_k(\vec{r}) = \sum_n e^{\left(i\vec{k}\cdot\vec{r}_n\right)}\psi_0(\vec{r}-\vec{r}_n). \tag{8.90}$$

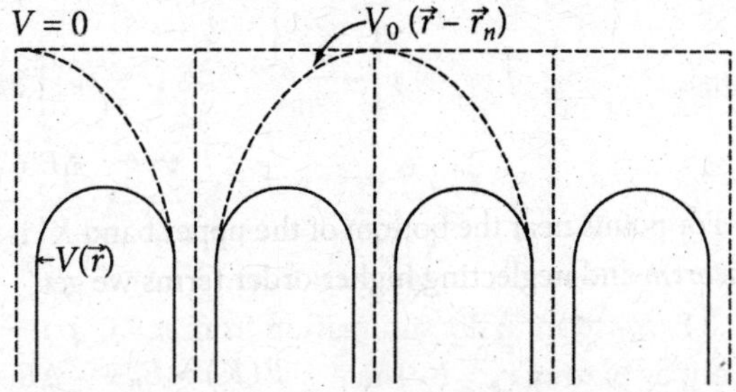

Fig. 8.20: Potential of an electron in an atom $[V(\vec{r})]$ and in a crystal $[V_0(\vec{r}-\vec{r}_n)]$

The summation extends over all the atoms in the crystal. It can be shown, the wave function $\psi_k(\vec{r})$ is a *Bloch function,* since

$$\psi_k(\vec{r}+\vec{r}_m) = \sum_m e^{i\left(\vec{k}\cdot\vec{r}_n\right)}\psi_0(\vec{r}+\vec{r}_m-\vec{r}_n) = \sum_n e^{i\left(\vec{k}\cdot\vec{r}_n\right)}\psi_0(\vec{r}-\vec{r}_n+\vec{r}_m)$$
$$= e^{i\left(\vec{k}\cdot\vec{r}_m\right)}\sum_n e^{i\left[\vec{k}\cdot(\vec{r}_n-\vec{r}_m)\right]}\psi_0[\vec{r}-(\vec{r}_n-\vec{r}_m)] = e^{i\left(\vec{k}\cdot\vec{r}_m\right)}\psi_k(\vec{r}). \tag{8.91}$$

The crystal *Hamiltonian H* can be written as

$$H = H_0 + H' \tag{8.92}$$

with

$$H_0 = -\frac{\hbar^2}{2m}\nabla^2 + V_0(\vec{r}-\vec{r}_n) \tag{8.93}$$

and

$$H' = V(\vec{r}) - V_0(\vec{r}-\vec{r}_n), \tag{8.94}$$

where $V_0(\vec{r}-\vec{r}_n)$ represents the potential due to the nth ground state atom on the electron at $\vec{r}$.

The electron energy is

$$\vec{E}(k) = \frac{\int \psi_k^*(\vec{r})\,[H_0+H']\,\psi_k(\vec{r})\,d\vec{r}}{\int \psi_k^*(\vec{r})\,\psi_k(\vec{r})\,d\vec{r}}. \tag{8.95}$$

Now
$$\int \psi_k^*(\vec{r})\,\psi_k(\vec{r})\,d\vec{r} = \sum_m\sum_n e^{i\left[\vec{k}\cdot(\vec{r}-\vec{r}_m)\right]} \times \int \psi_0^*(\vec{r}-\vec{r}_m)\,\psi_0(\vec{r}-\vec{r}_n)\,d\vec{r}. \tag{8.96}$$

If the overlap of the electron wave function between the neighbouring atoms be neglected, then

$$\int \psi_0^*(\vec{r}-\vec{r}_m)\,\psi_{0(\vec{r}-\vec{r}_n)}d\vec{r} = \delta_{mn} \tag{8.97}$$

and

$$\int \psi_k^*(\vec{r})\,\psi_k(\vec{r})\,d\vec{r} = N, \tag{8.98}$$

N is the total number of atoms in the crystal.

$$\therefore E(\vec{k}) = \frac{1}{N}\int \psi_k^*(\vec{r})\,(H_0 + H')\,\psi_k(\vec{r})\,d\vec{r} \tag{8.99}$$

$$= \frac{1}{N}\int \psi_k^*(\vec{r})\left[-\frac{\hbar^2}{2m}\nabla^2 + V_0(\vec{r}-\vec{r}_n)\right]\psi_k(\vec{r})\,d\vec{r} + \frac{1}{N}\int \psi_k^*(\vec{r})\left[V(\vec{r}) - V_0(\vec{r}-\vec{r}_n)\right]\psi_k(\vec{r})\,d\vec{r}$$

$$= \frac{E_0}{N}\int \psi_k^*(\vec{r})\,\psi_k(\vec{r})\,d\vec{r} + \frac{1}{N}\sum_m\sum_n e^{i[\vec{k}\cdot(\vec{r}_n-\vec{r}_m)]} \times \int \psi_0^*(\vec{r}-\vec{r}_m)\left[V(\vec{r}) - V_0(\vec{r}-\vec{r}_n)\right]\psi_0(\vec{r}-\vec{r}_n)\,d\vec{r}. \tag{8.100}$$

Every term in the summation from $n = 0$ to $n = N-1$ contains N identical terms which may be found out by putting $r_n = 0$. Then

$$E(\vec{k}) = E_0 + \sum_m e^{-i(\vec{k}\cdot\vec{r}_m)} \int \psi_0^*(\vec{r}-\vec{r}_m)\left[V(\vec{r}) - V_0(\vec{r})\right]\psi_0(\vec{r})\,d\vec{r}. \tag{8.101}$$

If ψ_0 is spherically symmetric (like the ground state of alkali metal), then the contribution due to all the nearest neighbour may be assumed to be the same. The above integral for $\vec{r}_m = 0$ is

$$\int \psi_0^*(\vec{r})\left[V(\vec{r}) - V_0(\vec{r})\right]\psi(\vec{r})\,d\vec{r} = -\alpha \tag{8.102}$$

and for the nearest neighbour atoms

$$\int \psi_0^*(\vec{r}-\vec{r}_n)\left[V(\vec{r}) - V_0(\vec{r})\right]\psi_0(\vec{r})\,d\vec{r} = -\beta, \tag{8.103}$$

where $\vec{r}_m$ is a vector connecting the atom at the origin with a nearest neighbour. α and β are constants called the *overlap integrals.* Thus, eqn (8.101) reduces to the form,

$$E(\vec{k}) = E_0 - \alpha - \beta\sum_m e^{i(\vec{k}\cdot\vec{r}_m)}. \tag{8.104}$$

Eqn (8.104) shows that the energy of the electron in the crystal differs from that of the free atom by a constant α plus a term which depends on the wave vector $\vec{k}$. *It is thus the last part which transforms the discrete atomic levels into an energy band in the solid.*

The summation in eqn (8.104) extends only over the nearest neighbour. Since free ion potential $V_0(\vec{r})$ is more negative than the lattice potential $V(\vec{r})$, $[V(\vec{r}) - V_0(\vec{r})]$ is a negative quantity and as such α and β are always positive. Thus, $E < E_0$.

8.11.3 Application of Tight Binding Approximation to Cubic Bravais Lattices

(a) The Simple Cubic (SC) Lattice: In this case, there are six neighbours for any given ion at

$$\vec{r}_m = (\pm a, 0, 0),\ (0, \pm a, 0),\ (0, 0, \pm a), \tag{8.105}$$

where a is the lattice constant.

Therefore,

$$E = (E_0 - \alpha) - 2\beta\left(\cos k_x a + \cos k_y a + \cos k_z a\right). \tag{8.106}$$

(b) The Body-Centred Cubic (BCC) Lattice: The eight nearest neighbours are at

$$r_m = (\pm a/2,\ \pm a/2,\ \pm a/2).$$

Therefore,

$$E = (E_0 - \alpha) - 8\beta \cos\left(\frac{k_x a}{2}\right)\cos\left(\frac{k_y a}{2}\right)\cos\left(\frac{k_z a}{2}\right). \tag{8.107}$$

(c) **The Face-Centred Cubic (FCC) Lattice**: Here the twelve nearest neighbours are at

$$r_m = (0, \pm a, \pm a),\ (\pm a, 0, \pm a),\ (\pm a, \pm a, 0) \tag{8.108}$$

and

$$E = (E_0 - \alpha) - 4\beta\left[\cos\left(\frac{k_y a}{2}\right)\cos\left(\frac{k_z a}{2}\right) + \cos\left(\frac{k_z a}{2}\right)\cos\left(\frac{k_x a}{2}\right) + \cos\left(\frac{k_x a}{2}\right)\cos\left(\frac{k_y a}{2}\right)\right]. \tag{8.109}$$

Thus, the s-band is described by a constant energy term $(E_0 - \alpha)$ and a term depending on the electron wave vector $\vec{k}$.

From eqn (8.106), E is minimum for $k_x = k_y = k_z = 0$.

Thus, the bottom of the energy band is given by

$$E_{\text{bottom}} = E_0 - \alpha - 6\beta. \tag{8.110}$$

The energy maximum corresponds to $k_x = k_y = k_z = \pm\pi/a$ and top of the energy band is given by

$$E_{\text{top}} = E_0 - \alpha + 6\beta\ (\text{as } k_x = k_y = k_z = \pm\pi/a). \tag{8.111}$$

Thus, the width of the energy band is

$$E_{\text{top}} - E_{\text{bottom}} = (E_0 - \alpha + 6\beta) - (E_0 - \alpha - 6\beta) = 12\beta. \tag{8.112}$$

This corresponds to the s state of the isolated atom. Eqn (8.112) shows that, the width of energy band increases with β which represents the overlap of the wave functions on the neighbouring atoms.

The inner electronic levels of free atoms gives rise to *narrow bands* in the solid whereas the outer shell electrons give rise to *wider bands*.

Similarly, from eqn (8.107) for the BCC lattice, $E_{\text{bottom}} = E_0 - \alpha - 8\beta$ and $E_{\text{top}} = E_0 - \alpha + 8\beta$.

$\therefore$ energy bandwidth is

$$E_{\text{top}} - E_{\text{bottom}} = 16\beta \tag{8.113}$$

and for FCC lattice from eqn (8.109), $E_{\text{bottom}} = E_0 - \alpha - 12\beta$ and $E_{\text{top}} = E_0 - \alpha + 12\beta$.

$\therefore$ bandwidth

$$E_{\text{top}} - E_{\text{bottom}} = 24\beta. \tag{8.114}$$

The variation of energy bandwidth in SC, BCC, FCC lattice is shown in Fig. 8.21.

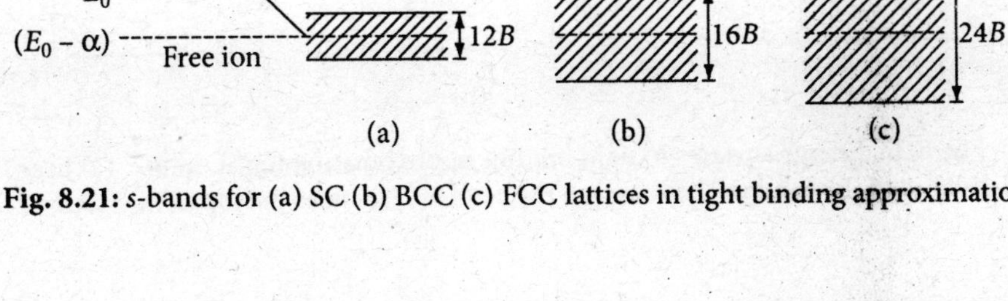

Fig. 8.21: s-bands for (a) SC (b) BCC (c) FCC lattices in tight binding approximation

8.11.4 Energy Surfaces

In tight binding approximation, the PE of the valence electrons in the field of their parent and neighbouring ions is larger than their K.E. This indicates that the separation between the ions is small compared with the electron in de Broglie wavelength, i.e., $|\vec{k}|\, a \ll 1$.

$$\therefore \cos ka \simeq 1 - \frac{k^2 a^2}{2}. \tag{8.115}$$

Then

$$E(\vec{k}) \simeq (E_0 - \alpha) - 2\beta\left[\left(1 - \frac{k_x^2 a^2}{2}\right) + \left(1 - \frac{k_y^2 a^2}{2}\right) + \left(1 - \frac{k_z^2 a^2}{2}\right)\right]$$
$$= (E_0 - \alpha) - 6\beta + \beta a^2 \left(k_x^2 + k_y^2 + k_y^2\right)$$
$$\text{or, } E(\vec{k}) = (E_0 - \alpha) - 6\beta + \beta a^2 k^2 + \cdots \text{ (for SC lattice).} \tag{8.116}$$

Similarly, for BCC lattice,

$$E(\vec{k}) = (E_0 - \alpha) - 8\beta + \beta a^2 k^2 + \cdots \tag{8.117}$$

and for FCC lattice,

$$E(\vec{k}) = (E_0 - \alpha) - 12\beta + \beta a^2 k^2 + \cdots. \tag{8.118}$$

8.11.5 Effective Mass of the Electron

The effective mass of the electron is

$$m^* = \frac{\hbar^2}{\partial^2 E/\partial k^2} = \frac{\hbar^2}{2\beta a^2} \left[\because \frac{\partial^2 E}{\partial k^2} = 2\beta a^2 \text{ (for SC lattice)}\right]. \tag{8.119}$$

Thus, $m^* \propto 1/\beta$, this implies that *for a narrow energy band the electron behaves as a very heavy particle in* SC lattice.

From eqn (8.113), the width of the energy band $\left(E_{\text{top}} - E_{\text{bottom}}\right) = 12\beta$, which refers to the bottom region of the band and corresponds to the constant spherical energy surfaces (i.e., $k = 0$) as shown in Fig. 8.22 (a). This means that the lowest energy state is at the centre of the zone and the higher energy states are at the corners [Fig. 8.22 (c)]. With the increase of energy $(k > 0)$, the energy surfaces bulge more and more towards the centre of the zone faces, until they intersect the zone boundary. Since the normal derivative of $E(\vec{k})$ is zero, the energy surfaces intersect the zone boundary orthogonally.

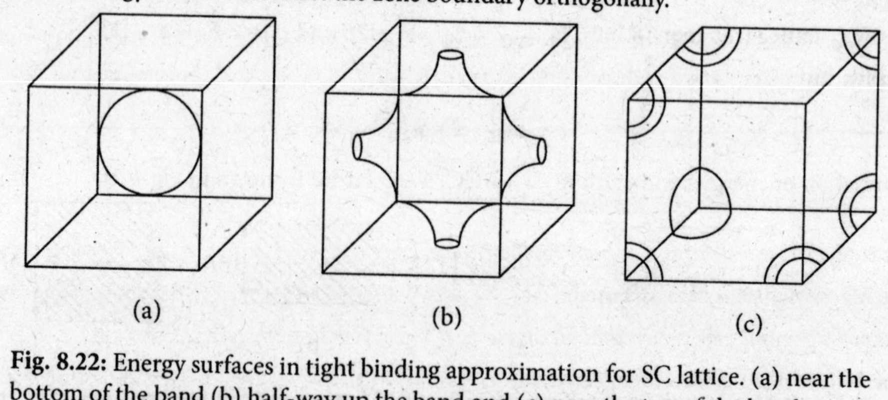

(a) (b) (c)

Fig. 8.22: Energy surfaces in tight binding approximation for SC lattice. (a) near the bottom of the band (b) half-way up the band and (c) near the top of the band.

A cross section through the energy surfaces at equal intervals is depicted in Fig. 8.23. Fig. 8.23(a) shows that for small k the energy surfaces are spherical around the zone centre. The energy surfaces are also spherical with corners at the covers of the zone. Fig. 8.23 (b) shows that the energy surfaces are spherical up to large value of k (i.e., nearly free electron approximation).

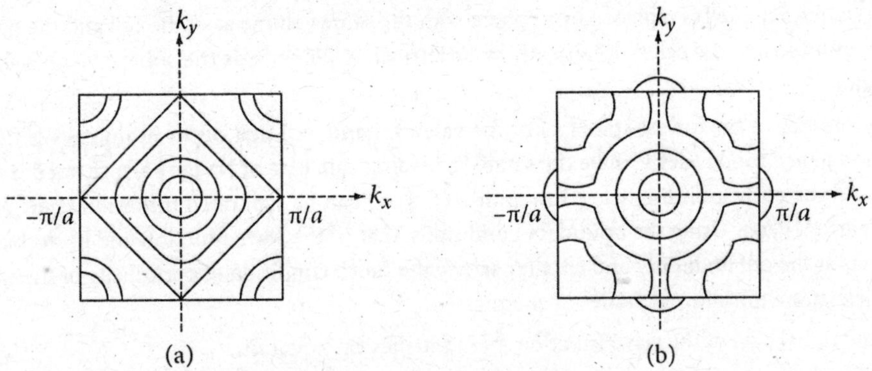

Fig. 8.23: Electron energy contours for SC lattice. (a) for tight binding approximation (b) for nearly free electron approximation.

Fundamental difference in the assumptions of nearly free electron approximation and tight binding approximation

Nearly free electron approximation (NFA)	*Tight binding approximation* (TBA)
1. The PE of the electron is small compared to its total energy.	1. The PE of the electron is nearly equal to its total energy.
2. The width of the forbidden bands is smaller compared to its allowed bands.	2. The width of the forbidden bands is larger compared to allowed energy bands.
3. As the atoms of the crystal are assumed to be closed spaced, the electron wave functions of neighbouring atoms overlap.	3. The atoms of the crystal are widely separated so that the electron wave functions of neighbouring atoms do not overlap to appreciable extent.
4. The interaction between the neighbouring atoms is strong.	4. The interaction between the neighbouring atoms is relatively weak.
5. The wave functions and the energy states in the crystal do not resemble with those of an individual atom.	5. The wave functions and energy as a whole are closely related to those of an individual atom.

8.12 Wigner-Seitz Cellular Method

Slater in 1934 introduced the *cellular method* to calculate band structure using *Wigner-Seitz's method.*

The following points are assumed:

1. *The solid is monatomic such that all atoms in it are translationally equivalent.*
2. *The Wigner-Seitz cell* (cellular polyhedron) is chosen as the unit cell so that each atom lies at the centre of the cell and hence, shows the symmetry properties of the crystal.

3. As the electron wave functions obey Bloch theorem it is necessary to find out the wave functions in one Wigner-Seitz cell of the structure.
4. The method aims to make maximum use of spherical symmetry of the atom and the translational symmetry of the crystal in the integration of the wave function within each *Wigner-Seitz cell.*
5. The Wigner-Seitz cell is replaced by a sphere with the same volume as of the cell and the potential $V(r)$ at any point inside the cell is spherically symmetrical, while outside the sphere $V(r)$ is assumed to be negligible.

The symmetry of the lowest state $(\vec{k})$ of the valence band and that of the atomic wave function from which the valence band is derived are the same. For example, in case of Na the valence band is derived from 3*s* state and the corresponding wave function $\psi(\vec{r}, \theta, \phi)$ can be separated into radial part $R(\vec{r})$ and the angular part $Y(\theta, \phi)$. Using the boundary conditions, that (i) the wave function and its derivative must be continuous at the cell boundary and (ii) they satisfy the Bloch condition of periodicity of the wave function we can determine the allowed values of energies E_0.

Now the radial part of the wave function $R(\vec{r})$ satisfies the equation,

$$\frac{1}{r^2}\frac{d}{dr}\left(r^2\frac{dR}{dr}\right) + \frac{2m}{\hbar^2}(E_0 - V)R = 0. \tag{8.120}$$

According to the condition stated above,

$$\left(\frac{dR}{dr}\right)_{r_s} = 0. \tag{8.121}$$

As $R(r)$ must resemble 3*s* state near the ion, it must have two spherical nodes. The allowed values of E_0 corresponding to different values of r_s may be calculated. The energy obtained from the condition

$$\left(\frac{dR}{dr}\right)_{r_s} = 0$$

is found to be much smaller than that corresponding to a free atom and this leads to main source of binding energy in metals like Na.

According to standard perturbation theory, in a cubic crystal, the energies of higher bands having no degeneracy at $\vec{k} = 0$ are given by

$$E_k = E_0 + \frac{\hbar^2 k^2}{2m^*}, \tag{8.122}$$

where m^* is the effective mass and

$$m^* = \frac{1}{m}\left[1 + \frac{2}{m}\sum_\alpha \frac{|\langle 0|p_x|\alpha\rangle|^2}{E_0 - E_\alpha}\right]. \tag{8.123}$$

In eqn (8.123), $\langle 0|p_x|\alpha\rangle$ represents the matrix element of the x-component of momentum operator between the states $\vec{k} = 0$ in the band O and another band α having energy E_α.

If E_k be the energy of an electron in the field of its own ion core, then the crystal energy is ΣE_k. The PE of all other electrons inside the cell is neglected as there is only one electron inside the cell at any moment. As the electron-ion, electron-electron and ion-ion interactions cancel in this approximation, the average energy of an electron is given by

$$E = E_0 + E(B), \tag{8.124}$$

where $E(B)$ is the average KE of the electron and $E(B) = \frac{3}{5}E_F$; E_F, the Fermi energy.

Then the *binding energy* is given by

$$E_0 = E(A) - [E_0 + E(B)] = [E(A) - E_0] - E(B), \tag{8.125}$$

where $E(A)$ is the energy of a free atom.

The difference $[E(A) - E_0]$ is due to the boundary conditions

$$\left(\frac{d\psi_0}{dr}\right)_{r=\infty} = 0$$

in the free atom and

$$\left(\frac{d\psi}{dr}\right)_{r=r_s} = 0$$

in the metal. Fig. 8.24 shows this difference. As the electron density is more near the region of the nucleus, $E_0 < E(A)$ in the solid and as such the PE is lowered.

In case of Na, $E(A) - E_0 \approx 3\,\text{eV}$ and $E(B) \sim 2\,\text{eV}$ for a half-filled band.

So, the binding energy

$$E_0 = 3\,\text{eV} - 2\,\text{eV} = 1\,\text{eV} \text{ [from Fig. 8.24]}.$$

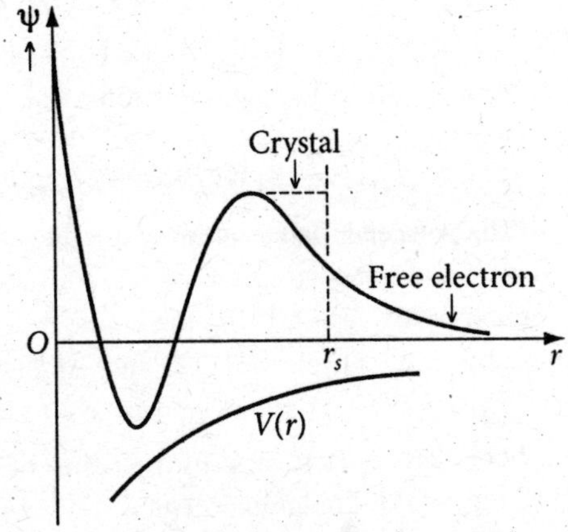

Fig. 8.24: Renormalisation of atomic wave function in Wigner-Seitz model

It is to be remembered that the Wigner-Seitz method does not depend upon the structure of the crystal but depends on the atomic volume.

In conclusion, the cellular method is greatly over-simplified and nowadays it has little use. The main disadvantage lies in the fact that when we replace the Wigner-Seitz cell by a sphere, we ignore the crystal structure and all anisotropic effects are completely masked out.

8.13 The Augmented Plane Wave (APW) Method

Slater in 1937 used the results of cellular method to overcome the following two major difficulties, viz.,

(i) The potential used in cellular method has a discontinuous derivative whenever the boundary between two cells is crossed [Fig. 8.25], but actually the potential is quite flat in these regions.

(ii) The method requires lot of numerical calculations to match the wave functions at the boundary of the cells (specially for $\vec{k} = 0$ states).

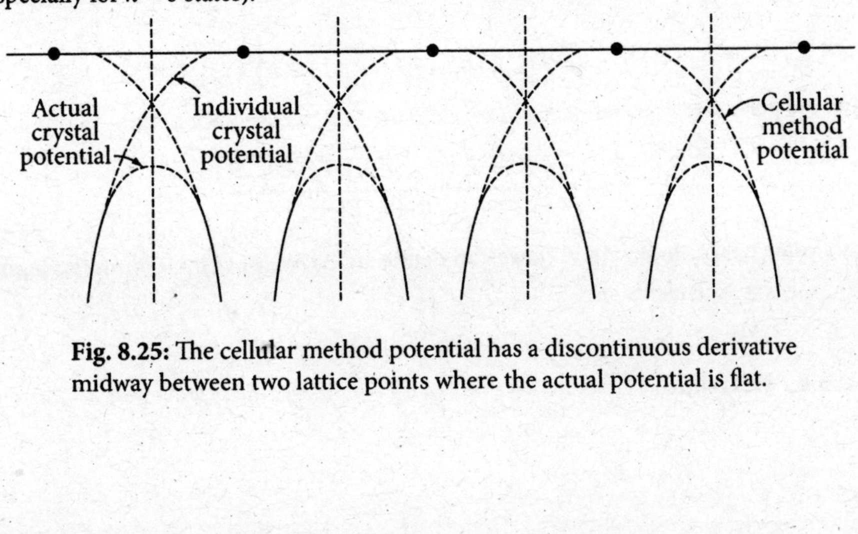

Fig. 8.25: The cellular method potential has a discontinuous derivative midway between two lattice points where the actual potential is flat.

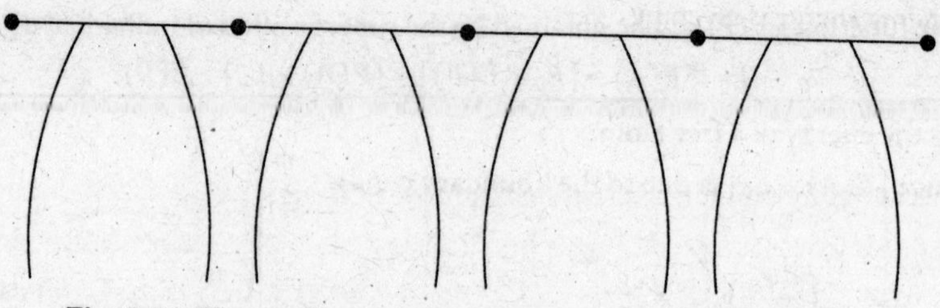

Fig. 8.26: The muffle-tin potential plotted along a line of ions

The potential due to an ion at the centre of the unit cell is given by

$$V(\vec{r}) = \begin{cases} V_i(\vec{r}) - V_0 & \text{for } r \le r_i \\ V_0 & \text{for } r_i < r < r_s \\ 0 & \text{for } r \ge r_s \end{cases} \tag{8.126}$$

where V_0 is constant and r is the radius of the sphere inscribed in the Wigner-Seitz cell. V_0 is actually the average of the Coulomb potential $-e^2/r$ between the spheres of radii r_i and r_s, i.e.,

$$V_0 = -\frac{e^2 \int_{r_i}^{r_s} \frac{4\pi r^2}{r} dr}{\int_{r_i}^{r_s} 4\pi r^2 dr} = \frac{3e^2 \left(r_s^2 - r_i^2\right)}{2\left(r_s^3 - r_i^3\right)}.$$

From the two solutions in the region $r > r_i$ and $r < r_i$, a solution of the *Schrödinger's equation* can be written as

$$\sum_{l,m} C_{l,m} R_l(\vec{r}, E) Y_{l,m}(\theta, \phi), \; r < r_i$$

$$\phi_{k,g(r)} = C_0 e^{i(\vec{k}+\vec{g})\cdot\vec{r}}. \tag{8.127}$$

Here $R_l(\vec{r}, E)$ is a solution of the radial equation

$$-\frac{1}{r^2}\frac{\partial}{\partial r}\left(r^2 \frac{\partial R_l}{\partial r}\right) + \left[\frac{l(l+1)}{r^2} + \frac{2mV(r)}{\hbar^2}\right] R_l = \left(\frac{2mE}{\hbar^2}\right) R_l \tag{8.128}$$

at a given energy E. $C_{l,m}$, C_0 are constants so that the two solutions for $r > r_i$ and $r < r_i$ are continuous at r_i. $g(\vec{r})$ is a *reciprocal lattice vector* and $Y(\theta, \phi)$ is a *spherical harmonic*.

Now

$$e^{i(\vec{k}+\vec{g})\cdot\vec{r}} = 4\pi \sum i^l j_l\left(\left|\vec{k}+\vec{g}\right| \vec{r}\right) \times Y_{lm}(\theta, \phi) Y_{lm}^*(\theta', \phi'). \tag{8.129}$$

θ', ϕ' define the polar directions of $\left(\vec{k}+\vec{g}\right)$, we get

$$C_{lm} = \frac{4\pi C_0 i^l j_l\left(\left|\vec{k}+\vec{g}\right| r_i\right) Y_{lm}^*(\theta', \phi')}{R_l(\vec{r}_i, E)}. \tag{8.130}$$

Using this value of C_{lm} in eqn (8.127), we can define an APW, $\phi_k(\vec{r}) = e^{i\vec{k}\cdot\vec{r}}$ in the interstitial space satisfying

$$\phi_k(\vec{r}) = e^{i\vec{k}\cdot\vec{l}} \phi_k\left(\vec{r} - \vec{l}\right) \tag{8.131}$$

inside the sphere with centre at $\vec{l}$.

The exact wave function $\psi_k(\vec{r})$ can be obtained by the superposition of individual $\phi_k(\vec{r})$, i.e.,

$$\psi_k(\vec{r}) = \sum_g \alpha_k - g\phi_k - g(\vec{r}). \tag{8.132}$$

Using $\psi_k(\vec{r})$ as a trial function in a variational estimate of energy, we get

$$\left[\left(\vec{k}-\vec{g}\right)^2 - E\right]\alpha_{k-g} + \sum \beta_{gg'}\alpha_{k-g'} = 0, \tag{8.133}$$

where

$$\beta_{gg'} = \frac{4\pi r_i^2}{\Omega_0}\left\{-\left[\left(\vec{k}-\vec{g}\right)\cdot\left(\vec{k}-\vec{g}\right)-E\right]\frac{j_l\left(|\vec{g}-\vec{g}'|r_i\right)}{|\vec{g}-\vec{g}'|}\right.$$
$$\left.+\sum_{l=0}^{\infty}(2l+1)P_l\left(\cos\theta_{gg'}\right)j_l\left(|\vec{k}-\vec{g}|r_i\right)j_l\left(|\vec{k}-\vec{g}'|r_i\right)\times\frac{R_l'(r_i,E)}{R_l(r_i,E)}\right\}, \tag{8.134}$$

where $\theta_{gg'}$ is the angle between $\left(\vec{k}-\vec{g}\right)$ and $\left(\vec{k}-\vec{g}'\right)$, $R_l'(r_i,E)$ is the derivative of $R_l(r_i,E)$ and Ω_0, the volume of the unit cell.

Now equating the secular determinant of the coefficient to zero, the band structure can be determined, e.g.,

$$\left|(k-g)^2 - E(k)\,\delta_{gg'} + \beta_{gg'}\right| = 0. \tag{8.135}$$

To have more accurate $E(k)$ value, a larger secular determinant is needed. APW method is good for calculating the *band structure in metals.*

8.14 Orthogonalised Plane Wave (OPW) and Pseudopotential Method

Herring in 1940 introduced the OPW method to overcome the limitations in the NFE and LCAO methods in the calculations of energy band in solids.

The core electron wave functions are localised about the lattice sites and the valence electron wave functions have a rapid oscillatory behaviour having a tail. The OPW method uses the fact that the valence wave functions are orthogonal to the core wave functions. Hence, instead of expanding the wave functions in terms of simple waves, *Herring* used an expansion of the wave functions in terms of waves which are already orthogonal to core wave functions. As a result of this orthogonalisation, *wiggles* are introduced in the plane waves in the neigbourhood of ion-sites.

An OPW of wave number $\vec{k}$ may be expressed as

$$|\text{OPW}_k\rangle = (1-P)\,|\vec{k}\rangle \tag{8.136}$$

with

$$|P\rangle = \sum_\alpha |\alpha\rangle\langle\alpha| \tag{8.137}$$

as the *projection operator* and $|\alpha\rangle = \psi_a(r)$ is the core electron wave function. In eqn (8.137), $|\vec{k}\rangle$ and $P|\vec{k}\rangle$ represent respectively as a *plane wave* and a *core function* so that, OPW = plane wave − core function as represented in Fig. 8.27.

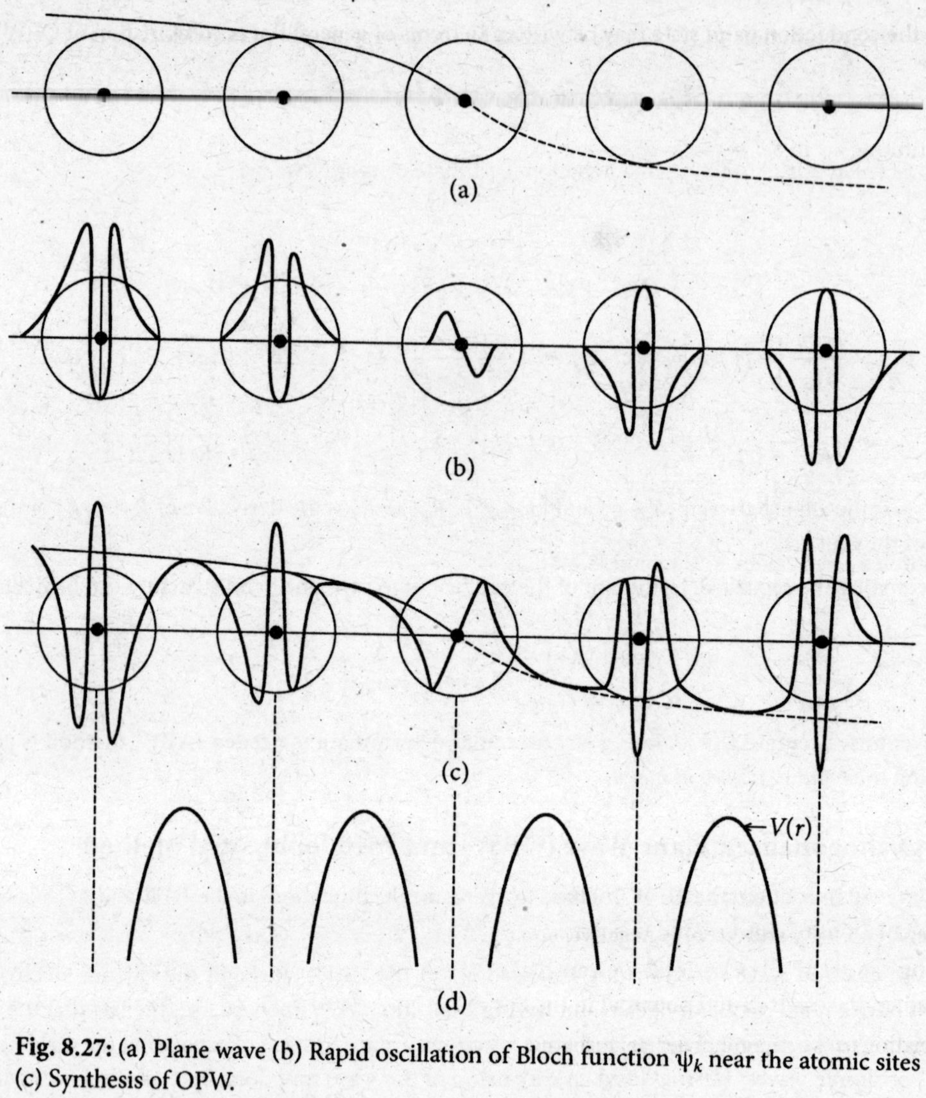

Fig. 8.27: (a) Plane wave (b) Rapid oscillation of Bloch function ψ_k near the atomic sites (c) Synthesis of OPW.

Supposing $|\beta\rangle$ to be one of the core states

$$\langle\beta|\text{OPW}_k\rangle = \langle\beta|(1-P)|\vec{k}\rangle = \langle\beta|\vec{k}\rangle - \langle\beta|P|\vec{k}\rangle$$
$$= \langle\beta|\vec{k}\rangle - \sum_{\alpha}\langle\beta|\alpha\rangle\langle\alpha|\vec{k}\rangle = \langle\beta|\vec{k}\rangle - \sum_{\alpha}\langle\alpha|\vec{k}\rangle\,\delta_{\alpha,\beta} = 0. \tag{8.138}$$

Here the *orthogonalised character of the core states* has been used;

$$\langle\beta|\alpha\rangle = \delta_{\alpha,\beta}. \tag{8.139}$$

Assuming T to be the KE operator and $V(\vec{r})$, the electron-ion interaction, $\vec{r}$ be the position of the electron relative to the ion, then the *Hamiltonian* of the system is

$$H = T + V(\vec{r}) = -\frac{\hbar^2}{2m}\nabla^2 + V(\vec{r}). \tag{8.140}$$

Now the conduction band state may be written in terms of general linear combination of OPW's as

$$\psi_k = \sum_q a_q(\vec{k})\,(1-P)\,|\vec{k}+\vec{q}\rangle. \tag{8.141}$$

Substituting ψ_k in *Schrödinger equation,*

$$H\psi_k = E(\vec{k})\,\psi_k, \tag{8.142}$$

we get

$$\sum_q a_q(\vec{k})\,H(1-P)\,|\vec{k}+\vec{q}\rangle = E(\vec{k})\sum_q a_q(\vec{k})\,(1-P)\,|\vec{k}+\vec{q}\rangle. \tag{8.143}$$

Let

$$\psi_k = \sum_q a_q(\vec{k})\,|\vec{k}+\vec{q}\rangle, \tag{8.144}$$

then

$$\psi_k = (1-P)\phi_k. \tag{8.145}$$

Substituting this value of ψ_k in eqn (8.143), we get

$$\begin{aligned}[T+V(r)]\phi_k - HP\psi_k &= E(\vec{k})\,\phi_k - E(\vec{k})\,PQ_k\\ \text{or, } T\phi_k + \left[V(r) - HP + E(\vec{k})\,P\right]\phi_k &= E(\vec{k})\,\phi_k\\ \text{or, } T\phi_k + W(\vec{r})\,\phi_k &= E(\vec{k})\,\phi_k,\end{aligned} \tag{8.146}$$

where

$$W(\vec{r}) = V(\vec{r}) + \left[E(\vec{k}) - H\right]P = V(\vec{r}) + \sum_\alpha\left[E(\vec{k}) - E_\alpha\right]|\alpha\rangle\langle\alpha|. \tag{8.147}$$

$W(\vec{r})$ is known as the pseudopotential and $\phi_k(\vec{r})$ is the pseudo-wave function. Here $H|\alpha\rangle = E_\alpha|\alpha\rangle$.

Eqn (8.146) shows that ψ_k and $W(\vec{r})$ lead to the same set of values as their real parts.

Since, $E(\vec{k}) > E_\alpha$ and $V(\vec{r})$ is negative, eqn (8.147) shows that $|W(\vec{r})| < V(\vec{r})$, this means by properly choosing the term $[E(k) - H]P$ we can make $W(\vec{r})$ as small as we please. Thus, for calculating the eigenvalues, the electron-ion potential in the low order perturbation theory can be successfully used.

According to the second order perturbation theory,

$$E(\vec{k}) = \frac{\hbar^2 k^2}{2m} + \langle\vec{k}|W|\vec{k}\rangle + \sum_q{}' \frac{|\langle\vec{k}+\vec{q}|W|\vec{k}\rangle|^2}{\frac{\hbar^2}{2m}\left[k^2 - |\vec{k}+\vec{q}|^2\right]}. \tag{8.148}$$

Here prime indicates that $q \neq 0$.

The term

$$E_{bs}(\vec{k}) = \sum_q{}' \frac{|\langle\vec{k}+\vec{q}|W|\vec{k}\rangle|^2}{\frac{\hbar^2}{2m}\left[k^2 - |\vec{k}+\vec{q}|^2\right]} \tag{8.149}$$

in eqn (8.148) represents the *band structure energy* of the $\vec{k}$ state.

If $\vec{r}$ be the position of an electron and $\vec{R}_j$ that of an ion, then the individual ionic potential is given by $w(|\vec{r}-\vec{R}|)$.

Thus,

$$W(\vec{r}) = \sum_j w\left(\left|\vec{r} - \vec{R}\right|\right) \tag{8.150}$$

$$\therefore \langle \vec{k} + \vec{q}\,|W|\vec{k}\rangle = S(\vec{q})\,\langle \vec{k} + \vec{q}\,|w|\vec{k}\rangle, \tag{8.151}$$

where

$$S(\vec{q}) = \frac{1}{N}\sum_j e^{-i\vec{q}\cdot\vec{R}_j} \tag{8.152}$$

is the *structure factor.* When the ions occupy precisely the lattice position at equilibrium, $S(\vec{q})$ becomes unity. This method has been successfully used to calculate the band structure of metals (Be, Na, K) and semiconductors (Gi, Si, etc.).

8.15 Answers of Some Important Questions

8.1. *Discuss how energy bands are formed?*

Ans. See Art. 8.2.

8.2. *State Bloch Theorem. What is the outcome of the theory?*

Ans. Bloch theorem states that the solution of the Schrödinger equation for a periodic potential must be of the form:

$$\psi(x) = e^{\pm ikx} u_k(x),$$

where $u_k(x) = u_k(x + a)$, a being the period. Thus, the solutions are plane waves of the type $e^{\pm ikx}$ modulated by the function $u_k(x)$ having the same periodicity as the lattice.

The *outcome of Bloch theorem* is that the energy spectrum of an electron moving in a periodic potential consists of allowed and forbidden energy regions (or bands).

8.3. *What is the concept of effective mass of the electron?*

Ans. In a crystal, the electrons interact with the lattice and are not completely free. The wave-particle motion, therefore, differs from that of an electron in free space. So, while using the equations of electrodynamics for charge carriers in a crystal we must keep in mind the altered value of the particle mass called the *effective mass* (m^*) due to the effect of the crystal lattice. Then the *electrons* (and *holes*) can be regarded as free charge carriers and *Newton's second law of motion* can be applied to describe their dynamic behaviour.

Statement: *When an electron in a periodic potential is accelerated by an electric field or magnetic field, the mass of the electron varies and this mass is called the effective mass of the electron.*

If E be the externally applied electric field accelerating the electron in the lattice, then the effective mass of the electron is given by

$$m^* = \frac{\hbar^2}{(d^2E/dk^2)},$$

where $\hbar = h/2\pi$ and k is the wave vector.

8.4. *What do you mean by valence band?*

Ans. In a solid, the highest filled energy band which includes electrons shared in *covalent bands* or electrons transferred in *ionic bonds* is known as *valence band.*

Valence band includes electrons which are transferred in case of *ionic bond* or shared in case of covalent bonds. These electrons are not free to gain energy from an external electric field and hence, are unable to move through the solid. The sharp atomic levels are spread out in solids and all the energy states up to a forbidden energy region are occupied (at least at 0 K) and there is no more free electrons.

8.5. *What is a conduction band?*

Ans. The band of energy of valence electrons consisting of closely spaced energy levels is called conduction band. The bottom of the conduction band has energy E_c.

The electrons in the valence orbit of atoms are loosely bound to the nucleus and are free to move in the solid forming *free electron gas.* There is a change in energy of these electrons due to mutual interaction. The value of energy of these electrons varies from zero to a maximum value called the *Fermi energy* E_f. These electrons participate in the electric conduction process in presence of an external electric field. The energy band of these electrons is called the *conduction band.*

8.6. *What is forbidden gap?*

Ans. The energy gap between the valence band and conduction band is called *forbidden energy gap* or *band gap.* Electrons cannot have any energy in the forbidden energy gap. The energy of the forbidden energy is called *Band Gap* (E_g). If E_v and E_c represent respectively the energy of the topmost level of the valence band and bottom level of conduction band, then $E_g = E_c - E_v$.

If E_g is very small, thermal excitation may excite the electrons in the valence band to the conduction band, so that the solid, although an insulator at 0 K, may become a conductor at temperature $T > 0$ K.

8.7. *What are the sources of electrical resistance in solids?*

Ans. The electrical conductivity depends mainly on two factors, viz., (a) temperature and (b) defects (e.g., impurities).

According to *Mathiessen's rule,*

$$\rho_{\text{total}} = \rho_{\text{pure}} + \rho_{\text{impurity}},$$

where ρ_{pure} is the temperature dependent of resistivity and ρ_{impurity} is the resistivity due to scattering of electrons by impurity atoms.

The *temperature dependence* of resistivity arises due to the scattering effect of thermal vibrations of the atoms on the electrons. *If A_0 be the amplitude of the thermal vibrations,* then the *mean free path* λ of the electron is

$$\lambda \propto 1/A_0^2.$$

Again, $A_0^2 \propto kT$. $\therefore$ $\lambda_0^2 \propto 1/T$.

$$\therefore \rho \propto 1/\lambda \propto T,$$

i.e., *the resistivity is directly proportional to the temperature T of the solid.*

The *defect dependence* of resistivity is due to the additional scattering of electrons by the *defect or impurities.* Assigning two relaxation time, we can write,

$$\frac{1}{\tau} = \frac{1}{\tau_0} + \frac{1}{\tau'},$$

where τ_0 is independent of temperature while τ' depends on temperature.

$$\therefore \rho = \frac{m}{ne^2\tau} = \frac{m}{ne^2}\left(\frac{1}{\tau_0} + \frac{1}{\tau'}\right) = \frac{m}{ne^2\tau_0} + \frac{m}{ne^2\tau'} = \rho_0 + \rho_{pure}(T) \text{ [Mathiessen's rule].}$$

In copper, Nickel, etc. both the contributions have been observed.

8.8. *Give a comparative study of conductor, semiconductor and insulator.*

Ans. The following table gives a comparative study of conductor, semiconductor and insulator.

Table 8.1

Conductor	*Semiconductor*	*Insulator*
1. The valence band (VB) and conduction band (CB) overlap.	The VB and CB are separated by a forbidden gap.	The VB and CB are separated by larger forbidden gap.
2. The band gap $E_g = 0$.	E_g ranges between 0.2 to 2.5 eV.	$E_g \sim 6$ eV (Diamond).
3. Electrical conductivity σ decreases with the increase of temperature (T).	σ increases with the increase of T.	$\sigma = 0$ and is independent of T.
4. Resistivity $\rho \propto T$.	Resistivity decreases with the increase of temperature.	Resistivity $\rho = \infty$ at all temperatures.

8.16 Solved Problems

8.1. *An insulator has an optical absorption which occurs for all wavelengths shorter than 1800 Å. Find the width of the forbidden energy band for this insulator.* **[LU 1994]**

Solution: Here the minimum energy required by an electron from conduction band to valence band across the forbidden energy gap is

$$E_g = h\nu = \frac{hc}{\lambda} = \frac{(6.626\times10^{-34})\times(3\times10^8)}{1800\times10^{-10}}\,\text{J} = \frac{(6.625\times10^{-34})\times(3\times10^8)}{(1800\times10^{-10})\times(1.6\times10^{-19})}\,\text{eV} = \mathbf{6.9\,eV.}$$

8.2. *Find the lowest energy band using Kronig-Penney model for $P \ll 1$.*

Solution: We have from the equation

$$P\frac{\sin\alpha a}{\alpha a} + \cos\alpha a = \cos ka,$$

the energy of the lowest band corresponds to $k = \pm\frac{\pi}{a}$.

$$\therefore P\frac{\sin\alpha a}{\alpha a} + \cos\alpha a = \pm1, \text{ (for lowest energy).}$$

Considering the magnitude of ±1, we get

$$\frac{P}{\alpha a}\sin\alpha a = 1-\cos\alpha a \text{ or, } \frac{P}{\alpha a}\left(2\sin\frac{\alpha a}{2}\cos\frac{\alpha a}{2}\right) = 2\sin^2\frac{\alpha a}{2} \text{ or, } \frac{P}{\alpha a} = \frac{2\sin^2\frac{\alpha a}{2}}{2\sin\frac{\alpha a}{2}\cos\frac{\alpha a}{2}} = \tan\frac{\alpha a}{2}.$$

When $P \ll 1$,

$$\frac{P}{\alpha a} = \frac{\alpha a}{2} \text{ or, } \alpha^2 = \frac{2P}{a^2} = \frac{2mE}{\hbar^2} \text{ or, } E = \frac{\hbar^2}{ma}P = \frac{h^2}{4\pi^2 ma}P, \text{ i.e., } \mathbf{E = \frac{\hbar^2}{ma}P.}$$

8.3. *The energy wave vector dispersion relation for a one-dimensional crystal of lattice constant a is given by $E(k) = E_0 - \alpha - 2\beta \cos ka$, where E_0, α, β are constants. Obtain the effective mass of the electron at the bottom and at the top of the band.* **[CU(H) 2003]**

Solution: We have the effective mass of the electron m^* is given by

$$m^* = \frac{\hbar^2}{d^2E/dk^2} = \frac{\hbar^2}{2\beta a^2 \cos ka}.$$

At the bottom of the band $\cos ka = 1$.

$$\therefore m^* = \frac{\hbar^2}{2\beta a^2}.$$

At the top of the band $\cos ka = -1$.

$$\therefore m^* = -\frac{\hbar^2}{2\beta a^2}.$$

8.4. *The energy near a valence band edge is given by $E(k) = -1 \times 10^{-26} k^2$ ergs. An electron is removed from the orbital $k = 1 \times 10^7 k_x$ cm^{-1}. The band is otherwise full. Give the sign and magnitude of the effective mass of the hole.*

Solution: The energy of the electron is given by

$$E = \frac{\hbar^2}{2m^*} k^2.$$

Here m^* is the effective mass of the electron.

But

$$E_k = -1 \times 10^{-26} k^2 \text{ ergs.} \therefore \frac{\hbar^2}{2m^*} = -1 \times 10^{-26}.$$

The effective mass of electron is

$$m^* = -\frac{\hbar^2}{2 \times 10^{-26}} \text{ gm.}$$

The effective mass of a hole is opposite to the effective mass of an electron at the same point on the energy band.

$\therefore m_h = -m_e$.

$$\therefore m_h^* = -m_e^* = \frac{\hbar^2}{2 \times 10^{-26}} = \frac{(1.05 \times 10^{-27}) \times (1.05 \times 10^{-27})}{2 \times 10^{-26}} = \mathbf{0.55 \times 10^{-28}} \textbf{ gm.}$$

8.5. *If the energy near the valence band of a crystal is given by $E = -Dk^2$ $(D = 10^{-39} \text{ Jm}^2)$. An electron with wave vector $\vec{k} = 10^{10} \hat{k}_x$ m^{-1} is removed from an orbital in the completely filled valence band. If the effective mass of the hole is $m_h^* = 5.5 \times 10^{30}$ kg, determine the velocity, momentum and energy of the hole.*

Solution: The momentum of the hole is

$$p_h = \hbar k_h = -(1.05 \times 10^{-34}) \times 10^{10} \hat{k}_x = -1.05 \times 10^{-24} \hat{k}_x \text{ J-s/m.}$$

The velocity of the hole is

$$v_h = \frac{p_h}{m_h^*} = -\frac{1.05 \times 10^{-24} \hat{k}_x}{5.5 \times 10^{-30}} = \mathbf{-1.9 \times 10^5 \hat{k}_x \text{ m/s.}}$$

The energy of the electron with wave vector $\vec{k}_e$ is

$$E_e = -Dk_e^2 = -10^{-39} \times \left(10^{10}\,\hat{k}_x\right)^2 = -10^{-19}\text{ J}.$$

Now the energy of the hole with zero at the top of the valence band is $E_h = -E_e = \mathbf{10^{-19}}$ **J.**

8.6. *An alloy of metal is found to have a resistivity of the order of* 10^{-6} Ω m *at* 27 °C. *When it is heated to a temperature of* 1000 K, *the resistivity increases by* 6.5 %. *Using Mattheissen rule, determine the resistivity of the alloy.*

Solution: We have the total resistivity of the alloy $\rho = \rho_0 + \alpha T_1$, at temperature T_1 or, $\rho + 0.065\rho = \rho_0 + \alpha T_2$, at temperature $T_2 > T_1$ and

$$0.065\rho = \alpha\left(T_2 - T_1\right) \text{ or, } \alpha = \frac{0.065\rho}{(T_2 - T_1)} = \frac{0.065\rho}{700} = 0.9 \times 10^{-10}\,\Omega\,\text{mK}^{-1}.$$

$$\therefore \rho_0 = \rho - \alpha T_1 = 10^{-6} - 0.9 \times 10^{-10} \times 300 = 10^{-6} - 2.7 \times 10^{-8} = 10^{-6}(1 - 0.0027)$$

or, $\rho_0 = \mathbf{0.9793 \times 10^{-6}}$ **Ω-m.**

8.7. *The ratio of hole mobility to electron mobility in a semiconductor is b. If the Hall coefficient is zero, what fraction of the total current is carried by electrons?*

Solution: We have the Hall coefficient

$$R_H = \frac{P\mu_s^2 - n\mu_e^2}{|e|(p\mu_p + n\mu_e)^2} = 0 \text{ [the different letters have their usual meaning]}$$

$$\text{or, } p\mu_p^2 = n\mu_e^2 \text{ or, } \frac{p}{n} = \frac{\mu_e^2}{\mu_p^2}.$$

Also

$$\frac{\mu_p}{\mu_e} = b \text{ or, } \mu_p = b\mu_e \text{ or, } \frac{p}{n} = \frac{1}{b^2} \text{ or, } n = pb^2.$$

$$\therefore \frac{\text{Electron current}}{\text{Total current}} = \frac{J_{xn}}{J_x} = \frac{|e|E_x n\mu_e}{|e|E_x(p\mu_p + n\mu_e)} = \frac{n\mu_e}{p\mu_p + n\mu_e}$$

$$= \frac{(bp^2)\mu_e}{p(b\mu_e) + (pb^2)\mu_e} = \frac{b^2}{b + b^2} = \frac{b}{1+b}.$$

∴ the fraction of total current carried by electrons is $\left(\frac{b}{1+b}\right)$.

8.8. (a) *At* 300 K *the intrinsic carrier concentration in silicon is* 3×10^{16} m^{-3}. *If the electron and hole mobilities are* 0.14 *and* $0.06\frac{m^2}{V_s}$ *respectively, determine the intrinsic resistivity of silicon at* 300 K.

[cf. KU(H) 2004]

(b) *The sample is then doped with* 10^{23} *phosphorous atoms*-m^{-3}. *Determine the equilibrium hole concentration. Also find conductivity.*

Solution:

(a) We have conductivity

$$\sigma = n_i|e|(\mu_e + \mu_p) = \frac{1}{\rho} \text{ [the different letters having their usual meaning]}$$

$$\text{or, } \sigma = (3 \times 10^{16}\,\text{m}^{-3})\,(1.6 \times 10^{-19}\,\text{C})\,(0.14 \times 0.06)\frac{m^2}{V_s} = 9.6 \times 10^{-4}\,\mho\text{m}^{-1}.$$

$$\therefore \rho = \frac{1}{\sigma} = \frac{1}{9.6 \times 10^{-4}\, \mho\text{m}^{-1}} = \mathbf{1.04 \times 10^3\ \Omega m.}$$

(b) Assuming full donor ionisation: $n_c = N_d^+ = 10^{23}$ atoms-m^{-3}.

$$n_v = \frac{n_i^2}{n_c} = \frac{(3 \times 10^6\ \text{m}^{-3})^2}{10^{23}\ \text{m}^{-3}} = 9 \times 10^9\ \text{m}^{-3}.$$

$$\therefore \sigma = n_c|e|\mu_e = (10^{23}10^{-3})\,(1.6 \times 10^{-19})C\left(0.14\frac{m^2}{V_s}\right) = \mathbf{2.24 \times 10^3\ \mho m^{-1}}.$$

8.9. *The energy gap in* Ge *is* 0.72 eV. *Find the maximum wavelength of electromagnetic radiation which can cause transition from the valence band to the conduction band.* **[BU(H) 1997]**

Solution: The energy gap is given by

$$E_g = h\nu = \frac{hc}{\lambda} \text{ or, } \lambda = \frac{hc}{E_g}.$$

∴ the required wavelength of electromagnetic radiation causing transition from valence band to the conduction band is

$$\lambda = \frac{(6.626 \times 10^{-34}\ \text{Js})\ (3 \times 10^8\ \text{ms}^{-1})}{0.72 \times 1.6 \times 10^{-19}\ \text{J}} = \mathbf{1.73 \times 10^{-6}\ m.}$$

8.10. *Mobilities of electron and holes in a sample of intrinsic semiconductor at* 300 K *are* 0.36 m^{2}V^{-1}s^{-1} *and* 0.17 m^{2}V^{-1}s^{-1} *respectively. If the conductivity of the specimen is* 2.12 Ω^{-1}, *compute the forbidden energy gap.* [Take $m_e^* = m_p^* = m$]

Solution: We have the conductivity

$$\sigma = 2\left(\frac{2\pi kT}{h^2}\right)(m_e^* m_p^*)\,e^{-E_g/2kT}|e|\,(\mu_e + \mu_p),$$

where the different symbols have their usual meaning

$$\text{or, } \sigma = e^{E_g/2kT} = \frac{2}{\sigma}\left(\frac{2\pi kT}{h^2}\right)(m_e^* m_p^*)^{3/4}\,|e|\,(m_e + \mu_p)$$

$$\text{or, } E_g/2kT = \ln\left[\frac{2}{\sigma}\left(\frac{2\pi kT}{h^2}\right)^{3/2}(m_e^* m_p^*)\,|e|(\mu_e + \mu_p)\right]$$

$$= \ln\left[\frac{2}{2.12\,\Omega^{-1}\text{m}^{-1}}\left\{\frac{2\pi(1.38 \times 10^{-23}\ \text{JK}^{-1})(300\ \text{K})}{(6.626 \times 10^{-34}\ \text{Js})^2}\right\}^{3/2}\right.$$

$$\left.\times\,(9.1 \times 10^{-31}\ \text{kg})^{2\times3/4}\ (1.6 \times 10^{-19}\ \text{C})\{(0.36 + 0.17)\text{m}^2V^{-1}s^{-1}\}\right].$$

$$\frac{E_g}{2kT} = 13.82 \text{ or, } E_g = 2kT(13.82).$$

$$\therefore E_g = 2\left(1.38 \times 10^{-23}\ \text{JK}^{-1}\right)(300\ \text{K})(13.82) = 1.14 \times 10^{-19}\ \text{J}$$

$$= \frac{1.14 \times 10^{-19}}{1.6 \times 10^{-19}}\ \text{eV} = \mathbf{0.7125\ eV.}$$

8.11. *The effective mass of a hole is equal to twice the effective mass of an electron, find the distance (in* eV*) of the Fermi level in an intrinsic semiconductor from the centre of the forbidden band at room temperature.*

Solution: We have the Fermi energy

$$E_f = \frac{E_c + E_v}{2} + \frac{3}{4}kT\ln\frac{m_p^*}{m_e^*} = \frac{E_c + E_v}{2} + \frac{3}{4}(1.38\times10^{-23}\ \text{JK}^{-1})(300\ \text{K})\ln\frac{2m_p^*}{m_e^*}$$

$$= \frac{E_c + E_v}{2} + \frac{3}{4}\frac{(1.38\times10^{-23})(300)}{1.6\times10^{-19}}\ln 2\ \text{eV} = \frac{E_c + E_v}{2} + 0.0135\ \text{eV}.$$

Hence, the distance (in eV) of Fermi level from the centre of forbidden band at room temperature is **0.0135 eV.**

8.12. *A rectangular bar of silicon is 2 cm long and has a cross-sectional area of 1 cm. The donor concentration is 10^{15} cm^{-3}. The resistance measured between the two ends of the bar is 10 Ω. Calculate the electron mobility.*

Solution: The conductivity is given by

$$\sigma = N_d^+|e|\mu_e = N_d|e|\mu_e \text{ [assuming full donor ionisation]}$$

$$\text{or, } \mu_e = \frac{\sigma}{N_d|e|}.$$

Resistance of the bar $R = \frac{1}{\sigma}\frac{l}{A}$ or, $\sigma = \frac{l}{kA}$.

[l = length, A = cross-sectional area of the rectangular bar].

∴ electron mobility

$$\mu_e = \frac{l}{RAN_d|e|} = \frac{2\ \text{cm}}{(10\ \Omega)(1\ \text{cm}^2)(10^{15}\ \text{cm}^{-3})(1.6\times10^{-19}\ \text{C})}$$

$$= \frac{2\times10^{-2}\ \text{m}}{(10\ \Omega)(10^{-2}\ \text{cm})[10^{15}\times(10^{-2}\ \text{m})^{-3}](1.6\times10^{-19}\ \text{C})}$$

$$= 0.125\frac{\text{m}}{\Omega\text{m}^{-1}\text{C}} = 0.125\frac{\text{m}^2}{\text{VA}^{-1}\text{C}}$$

$$= \mathbf{0.125\ m^2V^{-1}s^{-1}}\ [\because A = CT^{-1}].$$

8.13. *The conductivity of* Ge *is* 0.7 Ω^{-1}m^{-1} *at* 0 °C *and* 2 Ω^{-1}m^{-1} *at* 20 °C. *What is the band gap of* Ge?

[BU(H) 1993]

Solution: Conductivity $\sigma = k''e^{-E_g/2kT}$, k'' = constant.

$$\therefore \frac{\sigma_1}{\sigma_2} = \frac{e^{-E_g/kT_1}}{e^{-E_g/kT_2}} = \exp\left[\frac{E_g}{2k}\left(\frac{1}{T_2} - \frac{1}{T_1}\right)\right]$$

$$\text{or, } \ln\frac{\sigma_1}{\sigma_2} = \frac{E_g}{2k}\left(\frac{1}{T_2 - \frac{1}{T_1}}\right)$$

$$\text{or, } E_g = \frac{2k\ln\frac{\sigma_1}{\sigma_2}}{\frac{1}{T_2} - \frac{1}{T_1}} = \frac{2(1.38\times 10\ \text{JK}^{-1})\ln\frac{0.7}{2}}{\frac{1}{20+273} - \frac{1}{273}} = 1.1588\times10^{-19}\ \text{J}$$

$$= \frac{1.1588\times10^{-19}}{1.6\times10^{-19}}\ \text{eV} = \mathbf{0.72\ eV}.$$

8.14. *A pure semiconductor has an energy gap of 1 eV. For temperatures 0 K and 290 K respectively, calculate the probability of an electron occupying a state near the bottom of conduction band.*

Solution: The probability of an electron occupying an energy state is given by the Fermi distribution function

$$f(E) = \frac{1}{e^{(E-E_F)/kT}}.$$

$$\therefore f(E_c) = \frac{1}{e^{(E_c-E_F)/kT}+1} = \frac{1}{e^{E_g/2kT}}.$$

$\therefore E_c - E_v = E_g, E_c - E_T = \frac{E_g}{2}$, assuming Fermi level to be midway in the bandgap.

At $T = 0\,\text{K}, f(E_c) = \frac{1}{e^{\infty}+1} = 0.$

At $T = 290\,\text{K}, f(E_c) = \dfrac{1}{e\left[\dfrac{1\times 1.6\times 10^{-19}}{2(1.38\times 10^{-23}\,\text{JK}^{-1})(290\,\text{K})}\right]} + 1 = \mathbf{2\times 10^{-9}}.$

8.15. *Assuming $m_e^* = m_p^*$, indicate the position of the Fermi level in band diagram at 300 K in an intrinsic Germanium.* **[KU(H) 1999]**

Solution: We have Fermi energy

$$E_F = \frac{E_c + E_v}{2} + \frac{3}{2}kT\ln\frac{m_p^*}{m_e^*}.$$

$$\therefore m_p^* = m_e^*, \ln\left(\frac{m_p^*}{m_e^*}\right) = \ln 1 = 0.$$

$$\therefore E_F = \frac{E_C + E_T}{2}.$$

Fermi level would then lie halfway between E_c and E_v, i.e., at the middle of the band gap E_g.

EXERCISE

8(A): Objective (MCQ) Type Questions

Select the correct answer.

8.1. According to the band theory of solids, an electron can have

(a) all values of energy (b) only the energy corresponding to energy bands

(c) the energy in the forbidden gap (d) none of these.

8.2. The width of energy band depends on

(a) the extent of overlap of wave functions ψ_1 and ψ_2 corresponding to the two atoms

(b) the interaction between the two atoms

(c) both (a) and (b)

(d) none of these.

8.3. Bloch theorem is applicable to

(a) periodic potential (b) constant potential (c) infinite potential (d) none of these.

8.4. The energy spectrum of an electron moving in a periodic potential consists of

(a) continuous energy regions (b) forbidden energy regions

(c) for allowed and forbidden energy regions (d) zero energy regions.

8.5. The number of wave functions in an energy band in dk of an electron moving in a one-dimensional potential well is

(a) $dn = \frac{2\pi}{L} dk$ (b) $dn = \frac{L}{2\pi} dk$ (c) $dn = 2\pi L dk$ (d) $dn = \frac{2\pi L}{dk}$.

8.6. In an $(E\text{-}k)$ curve, the condition for energy discontinuity is

(a) $k = \pm\frac{n\pi}{a}$ (b) $k = \pm\frac{a}{n\pi}$ (c) $k = \frac{2a}{n\pi}$ (d) $k = \frac{n\pi}{a}$.

8.7. The K-electron of copper atom is so lightly bound to the nucleus that its effective mass is

(a) $m_e^* = 0$ (b) $m_e^* = \infty$ (c) $m_e^* = m_e$ (d) $0 < m_e^* < \infty$.

8.8. The effective mass of an electron moving in a periodic lattice is

(a) $m_e^* = \hbar^2 \Big/ \frac{d^2E}{dk^2}$ (b) $m_e^* = \hbar^2 \frac{d^2E}{dk^2}$ (c) $m_e^* = \frac{\hbar^2 d^2k}{dE^2}$ (d) $m^* = m_e$.

where E and k are the field and wave vectors relating to the motion of the electron.

8.9. The extent to which an electron is free is given by f_k, where f_k is equal to

(a) $\frac{m}{\hbar^2}\left(\frac{d^2E}{dk^2}\right)$ (b) $m\hbar^2\left(\frac{d^2E}{dk^2}\right)$ (c) $\frac{\hbar^2}{m}\left(\frac{d^2E}{dk^2}\right)$ (d) $\frac{m}{\hbar^2}\left(\frac{d^2k}{dE^2}\right)$.

8.10. If temperature is increased, then the electrical conductivity of a semiconductor

(a) decreases (b) remains constant (c) increases (d) reduces to zero.

8.11. A semiconductor behaves at 0 K as

(a) an insulator (b) a good conductor (c) a semiconductor (d) none of these.

8.12. The solution of Schrödinger equation for an electron moving in a periodic potential is of the form:

(a) $\psi_k(\vec{r}) = e^{i\vec{k}\cdot\vec{r}}$ (b) $\psi_k(\vec{r}) = u_k(\vec{r})$ (c) $\psi_k(\vec{r}) = u_k(\vec{r})\, e^{i\vec{k}\cdot\vec{r}}$ (d) $\psi_k(\vec{r}) = u_k(\vec{r}) + Ae^{i\vec{k}\cdot\vec{r}}$.

8.13. In a crystal having N primitive cells, the maximum number of electrons per band is

(a) N (b) $2N$ (c) ∞ (d) 0.

8.14. For a one-dimensional lattice of length L, the total effective number of free electrons in an energy band filled with electrons up to a certain volume k_1 $\left(k_1 < \frac{\pi}{a}\right)$ is

(a) $N_{\text{eff}} = \frac{\pi\hbar^2}{2Lm}\left(\frac{dE}{dk}\right)_{k=k_1}$ (b) $N_{\text{eff}} = \frac{2Lm}{\pi\hbar^2}\left(\frac{dE}{dk}\right)_{k=k_1}$

(c) $N_{\text{eff}} = \frac{Lm}{2\pi\hbar^2}\left(\frac{dE}{dk}\right)_{k=k_1}$ (d) $N_{\text{eff}} = \frac{2L}{\pi\hbar^2}\left(\frac{dE}{dk}\right)_{k=0}$.

8.15. The velocity of an electron from $(E\text{-}k)$ curve is

(a) $v = \frac{1}{\hbar}\frac{dE}{dk}$ (b) $v = \hbar\frac{dE}{dk}$ (c) $v = \frac{1}{\hbar}\frac{d^2E}{dk^2}$ (d) $v = \hbar \Big/ \left(\frac{dE}{dk}\right)$.

ANSWERS

8.1 (b) **8.2** (c) **8.3** (a) **8.4** (c) **8.5** (b) **8.6** (a) **8.7** (b) **8.8** (a) **8.9** (a) **8.10** (c) **8.11** (a) **8.12** (c) **8.13** (b) **8.14** (b) **8.15** (a).

8(B): Short Answer Type Questions

8.1. Discuss how energy bands are formed.

8.2. Compare the salient features of band theory and free electron theory. Discuss the merits and demerits of these theories.

8.3. State Bloch theorem. What is Bloch function?

8.4. Explain the concept of conduction band, valence band and forbidden energy gap.

8.5. Distinguish metal, semiconductor and insulator in view of band theory of solids. [NBU 2001, CU 1983]

8.6. Explain the concept of effective mass of electron. [CU 1997]

8.7. What is negative effective mass? Explain with reference to the band theory of solids. [PU 2003]

8.8. Explain why the conductivity of metal decreases while that of the semiconductor increases with the rise of temperature.

8.9. Why does a completely filled band contribute nothing to an electric current in presence of an electric field?

8.10. Explain the concept of *holes*.

8.11. Show graphically the variation of energy with the wave number k for the first allowed band in a solid. Hence, explain the variation in sign of the effective mass of an electron in the band. What is a hole? [CU 2000]

8.12. Discuss how the effective mass (m^*) varies with wave vector $(\vec{k})$.

8.13. What are *Brillouin zones*? Discuss the formation of Brillouin zones for a linear lattice. [IIPU 1991; PU 1997]

8.14. Prove that the total number of possible energy states in an energy band of a finite crystal is equal to the number of unit cells.

8.15. Discuss the origin of electrical resistance in a metal. What is *Metheissen's rule* regarding temperature variation of resistivity?

8.16. Starting from dispersion relation

$$\omega = \omega_0 \sin\left(\frac{ka}{2}\right)$$

for linear monatomic chain of length (Na, N is the number of atoms), obtain and sketch the density of states as a function of ω. [CU(H) 2017]

8.17. The dispersion relation in $3d$ lattice is given by $\epsilon_k = \alpha \cos k_x a + \beta \cos k_y a + \gamma \cos k_z a$, where a is the lattice constant and α, β, γ are lattice constants. Find the effective mass tensor at the corner of the first Brillouin zone $(\pi/a, \pi/a, \pi/a)$. [CU(H) 2016]

8.18. Obtain the dispersion relation for one-dimensional monatomic lattice. [CU(H) 2015]

8.19. In the Kronig-Penney model, the following equation is obtained after simplification:

$$P \sin \alpha a + \cos \alpha a = \cos ka, \text{ where } \alpha = \sqrt{\frac{2mE}{\hbar^2}}.$$

(a) What is the physical significance of P?

(b) By plotting the above equation as a function of αa, discuss the allowed and forbidden regions in the plot. (Assume a fixed value of P). [CU(H) 2015]

8.20. Discuss the formation of allowed and forbidden energy bands on the basis of Kronig-Penney model. Set up the appropriate Schrödinger equation and boundary conditions. Use block function as the wave function. [CU(H) 2006]

8.21. What are phonons? Describe the characteristics of acoustical and optical phonons with reference to dispersion curves of linear diatomic lattice. **[Delhi U 2014]**

8.22. How are the materials classified into conductors, semiconductors and insulators on the basis of E vs K graph? **[Delhi U 2014]**

8.23. Discuss the Kronig-Penney model for the motion of an electron in periodic potential. Obtain an expression for effective mass of an electron in a metal. Discuss the significance of Brillouin Zone. **[Delhi U 2014]**

8.24. Distinguish between metals, semiconductors and insulators on the basis of band theory of solids. **[Pune U 2014]**

8.25. Define block function and write a note on it. **[Pune U (MSc) 2008]**

8(C): Numerical Problems

8.1. Assuming a periodic potential of one-dimensional lattice of spacing a to be a box shaped waveform which has a value at each atom $-V$ and zero at $0.1a$ on either side of each lattice point, calculate the width of the first energy gap in the electron energy system. [*Ans.* 0.37 V]

8.2. If the origin be at the bottom of the conduction band, calculate the crystal momentum for a free electron of energy 0.02. [Effective mass $m_e^* = 0.2m_0$] [*Ans.* 3.416×10^{-26} kg-m/s]

8.3. An electric field of strength $E = E_0 \cos \omega t$ is applied along the x-direction of a crystal lattice. Find expressions for current density and electrical conductivity at frequency ω.

[*Ans.* $J_x = \sigma_s (E_0) \left[\frac{\cos \omega t}{(1+\omega^2\tau^2)} + \frac{\omega\tau \sin \omega\tau}{(1+\omega^2\tau^2)}\right]$ and $\sigma_\omega = \frac{j_x}{E_0 \cos \omega t}$]

8.4. Assuming the copper to be highly pure, find the resistivity of copper at 700 K.
[Given, the resistivity of copper at 300 K = $1.7 \times 10^{-8}\,\Omega$-m]. [*Ans.* $\rho = 552 \times 10^{-8}\,\Omega$-m]

8.5. Estimate the band gap of an insulator if it absorbs electromagnetic radiation of all wavelength below 3000 Å. [*Ans.* 4.14 eV]

8.6. Show that for a two-dimensional square lattice, the kinetic energy of a free electron at a corner of the first Brillouin zone is higher than that of an electron at midpoint of a side face of zone by a factor of 2. **[Pune U (MSc) 2008]**

8(D): Broad Answer Type Questions

8.1. (a) Discuss briefly the origin of energy bands in solids.

(b) How would you distinguish between a semiconductor from an insulator on the basis of band theory? **[BU(H) 1995]**

(c) Prove that the motion of an electron through the periodic potential of solids gives rise to band structure. **[PU 1992]**

8.2. (a) Using the Kronig-Penney model, show that the energy spectrum of an electron consists of a number of allowed energy bands separated by forbidden regions.

(b) Discuss the nature of energy spectrum in the limiting cases of very weak and very strong potentials. **[BU(H) 2004]**

(c) How does the resistivity of a metal and a semiconductor vary with temperature? Explain qualitatively the temperature variation of resistivity in the two cases. **[BU(H) 1999]**

8.3. (a) Compare the salient features of the band theory and free electron theory. Discuss the merits and demerits of these theories. **[BU(H) 2001]**

(b) What are valence band and conduction band? What do you mean by forbidden band?

(c) Distinguish amongst metals, semiconductors and insulators. Why does a semiconductor acts as an insulators at 0 K? **[BU(H) 2002]**

8.4. (a) Discuss the periodic character of potential in a crystal.

(b) State and prove Bloch theorem in periodic potential in a crystal. **[HPU 1996; LU 1996]**

8.5. (a) What do you mean by Brillouin zones?

(b) Discuss the formation of Brillouin zones for a (i) linear lattice (ii) two-dimensional lattice. **[HPU 1991; LU 1994; PU 1997]**

8.6. (a) What do you mean by effective mass of an electron? What is its significance?

(b) Show that the effective mass of an electron is inversely proportional to the second derivative of the $(E\text{-}k)$ curve.

(c) Discuss the condition when the effective mass of an electron becomes positive, negative and infinity.

8.7. (a) The band theory for a cubic crystal leads to discontinuities in the $E(k)$ surfaces whenever k satisfies the condition $nk = n^2\pi/a$. Show that this condition is equivalent with that for Bragg reflection of the electrons by the set of planes with miller indices (hkl).

(b) Explain the origin of the forbidden energy gap in crystalline solids and the concept of effective mass of an electron in a lattice.

8.8. (a) The solution of Schrödinger equation for a one-dimensional periodic lattice is given by

$$P\frac{\sin \alpha a}{\alpha a} + \cos \alpha a = \cos ka$$

with $\alpha^2 = (2mE/\hbar^2)$. The symbols have their usual significances. Discuss the formation of energy bands in a solid.

(b) Prove that the effective mass of an electron in any energy band is given by

$$\frac{1}{m^*} = \frac{1}{\hbar^2}\frac{\partial^2 E}{\partial k^2}.$$

8.9. (a) Describe band theory of solids on the basis of Kronig-Penney model. Distinguish among conductors, semiconductors and insulators.

(b) Discuss the temperature dependence of resistivity in metal and semiconductor.

(c) What is the concept of *hole*.

8.10. (a) Show that when an electron moves through a crystal lattice the effective mass of the electron can be expressed as

$$m^* = \hbar^2 \Big/ \frac{d^2E}{dk^2},$$

where the symbols have their usual significances.

(b) According to simple model, the energy of a free electron in a metal is given by the following function of the wave number $E = A - B\cos ka$, where A and B are constants and a is the distance between adjacent atoms. Draw the $(E\text{-}k)$ curve. What value of effective mass can you deduce from the above formula? With a spacing $a = 2.0$ Å what is the range of the electron wavelengths in the first Brillouin zone. [CU(H) 2007]

8.11. What is Bloch theorem? Starting from the relation

$$P\frac{\sin \alpha a}{\alpha a} + \cos \alpha a = \cos ka,$$

discuss the motion of electrons in one dimension and the variation of energy, velocity and effective mass as a function of wave vector.

[Here $\alpha = \sqrt{\frac{2mE}{\hbar^2}}$, $P = \frac{mV_0 ba}{\hbar^2}$

$V_0 b$ = area of the potential barrier.] [VU(H) 2009]

8.12. Discuss the formation of allowed and forbidden energy bands on the basis of Kronig-Penney model. Set up the appropriate Schrödinger equation and boundary conditions. Use Bloch function as the wave function. [CU(H) 2006]

8.13. (a) State Bloch's theorem in periodic crystals.

(b) In the Kronig-Penney model, the following equation is obtained after simplification:

$$\frac{P}{\alpha a}\sin \alpha a + \cos \alpha a = \cos ka,\ \alpha = \frac{2mE}{\hbar^2},\ P = \frac{mV_0 ab}{\hbar^2}.$$

(i) Obtain the energy band gap at $k = \pi/a$ in the limit $V_0 ab \ll \frac{\hbar^2}{m}$.

(ii) What is the energy of the lowest band at $k = 0$ in the limit $P \ll 1$?

(c) Find the expression for effective mass of an electron in a lattice. What happens to this mass close to the edges of the Brillouin zones and why? [CU(H) 2017]

8.14. (a) What do you mean by *effective mass* of an electron in a lattice? Calculate the effective mass m^* as a function of the crystal of lattice constant a having dispersion relation $E(k) = E_0 - \alpha - 4\beta \cos(ka)$, where E_0, α, β are constants.

(b) Find the width of the energy band in the above question. Also find the value of k at which the velocity of the electron is maximum.

(c) Argue that the effective mass of the hole is opposite to that of an electron. [CU(H) 2013]

8.15. The energy wave vector dispersion relation for one-dimensional crystal of lattice constant a is given by $E(k) = E_0 - \alpha - 2\beta \cos ka$, where E_0, α, β are constants.

(a) Find the value of k at which the velocity of an electron is maximum.

(b) Find the difference between the top and bottom of the energy band.

(c) Obtain expression for the effective mass m^* of the electron as a function of k. Sketch m^* $V_s k$.

[CU(H) 2003]

8.16. (a) State and prove Bloch's theorem. (b) Discuss in detail the Kronig-Penney model. [Bhabnagar U 2017]

8.17. (a) Discuss the nearly free electron model and explain how it leads to the formation of forbidden gap and band structure.

(b) Explain the concept of reduced, extended and periodic zone schemes used for the representation of energy bands with neat diagram. [Pune U 2008]

Chapter 9

Theory of Semiconductors

9.1 Introduction

In view of electrical conductivity, materials are classified as *conductors, semiconductors* and *insulators.* Semiconductors have electrical conductivity intermediate between conductors and insulators.

The resistivity of some typical materials at 300 K is shown in Table 9.1.

Table 9.1

Conductor *Metal*	*Resistivity* (Ω-m)	*Semiconductor*	*Resistivity* (Ω-m)	*Insulator*	*Resistivity* (Ω-m)
Cu	1.7×10^{-8}	Pure Ge	0.47	Glass	10^{10} to 10^{11}
Al	2.8×10^{-8}	Pure Si	3000	Diamond	10^{14}
Fe	10×10^{-8}	Sn (Grey)	2×10^{-6}	Mica	9×10^{14}

Some important characteristics of semiconductors are as follows:

(i) Semiconductors have *negative* temperature coefficient unlike metals which have *positive* temperature coefficient of resistance.

(ii) Semiconductors have *small band gap* which varies from $E_g = 0.2$ to 2.5 eV, whereas for insulators, like Diamond, $E_g = 6$ eV.

This property determines which wavelength can be emitted or absorbed by the semiconductor and helps to construct *light emitting diodes* (LEDs).

(iii) Addition of a suitable chemical impurity may change the electrical properties of semiconductors manyfold. The process of addition of impurity is called *doping* which enhances the conductivity of the material.

Depending on the nature of the *dopant* semiconductors are classified as follows:

(a) *Pure or intrinsic semiconductors.* In this type of semiconductors no impurity atoms are added.

(b) *Impurity or extrinsic semiconductors.* Suitable impurity atoms are added to the semiconductors.

9.2 Intrinsic Semiconductors

Pure *germanium* (Ge) and pure *silicon* (Si) are important examples of intrinsic semiconductors, where electrons and *holes* are produced by *phonons* or *photons* of suitable frequency. Both Ge and Si are group IV elements of the periodic table and each has four valence electrons. Fig. 9.1 (a) shows a covalent Ge crystal in two dimensions.

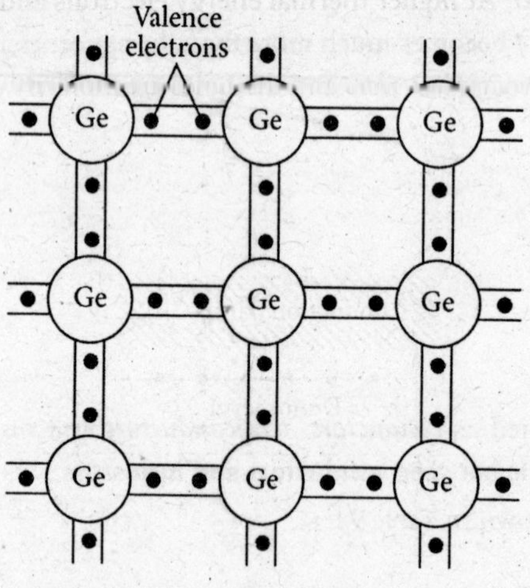

Fig. 9.1: (a) Two-dimensional Ge crystal

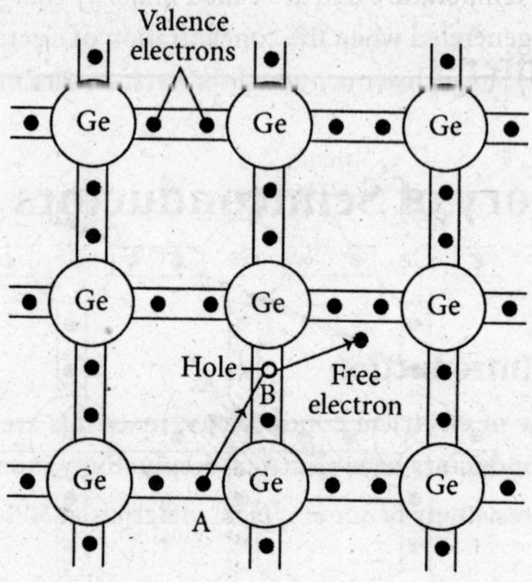

Fig. 9.1: (b) Production of electron and hole

A valence electron breaks its *covalent bond* if it gets sufficient thermal energy and becomes free. The vacancy left behind at *B* serves as a *hole*. Thus, a free electron and a hole are created as shown in Fig. 9.1 (b). A valence electron in the neighbourhood at *A* after getting sufficient thermal energy jumps to the position of hole at *B* and the bond is reconstructed. In this process a *hole* is created at *A* and the *hole* moves from *B* to *A*. In this way the *electron* and *hole* move in the crystal in opposite direction to each other and conduction of electricity takes place. As the charge carriers are created due to the breaking of the *covalent bond*, the concentration of electrons (n_i) becomes equal to the concentration of holes (p_i), i.e., $n_i = p_i$.

Due to thermal agitation fresh electron-hole pair is generated continuously while electron-hole pair disappears by *recombination*. A free electron is converted into a bound electron creating a *covalent bond*. The Fermi level (E_f) in intrinsic semiconductor lies midway in the forbidden gap. The electron and hole so generated are called *intrinsic charge carriers*.

9.3 Extrinsic Semiconductor

If a small amount of selected *impurity* be added to the semiconductor, then free electrons and holes may be generated. Such materials are called *impurity* (or *extrinsic*) *semiconductor*. The impurity increases the carrier concentration and hence, the electrical conductivity of the semiconductor. Impurity semiconductors are of the following two types:

(a) *n*-type semiconductors — The elements of group *V* such as *Phosphorous* (P), *Arsenic* (As) or *Antimony* (Sb) have five electrons in their outermost orbits. If any one of such element be added to *tetravalent Germanium* (Ge) or *Silicon* (Si) during crystal growing, four electrons of the impurity atom form covalent bonding with the four neighbouring semiconducting atom and the fifth electron is released with lesser thermal energy and the parent atom becomes *positively charged* [Fig. 9.2 (a)]. No hole is created during this process. As the pentavalent impurity atoms donate negative charges (i.e., electrons), so the semiconductors doped with pentavalent impurity is called *n-type semiconductor*.

The energy level of the fifth electron lies close to the conduction band [Fig. 9.2 (b)] and is called the *donor level*. That is why most of the electrons in the donor level are easily excited to the conduction band at room temperature and are called *majority charge carrier.* At higher thermal energy electrons and holes are also generated when the concentration of electrons (n) becomes much more than the concentration of holes (h). In such semiconductors electrons are *majority charge carriers* and the holes are *minority charge carriers.*

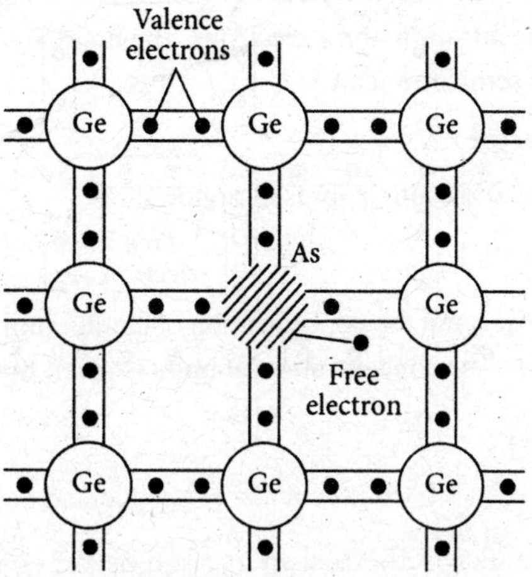

Fig. 9.2: (a) A pentavalent *As* in Ge crystal

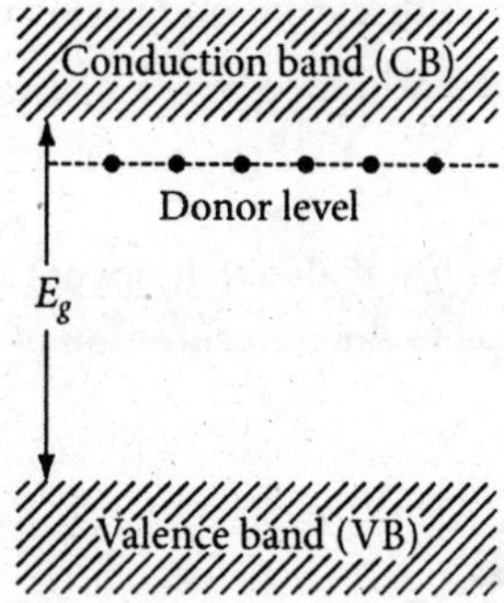

Fig. 9.2: (b) Donar level in energy diagram

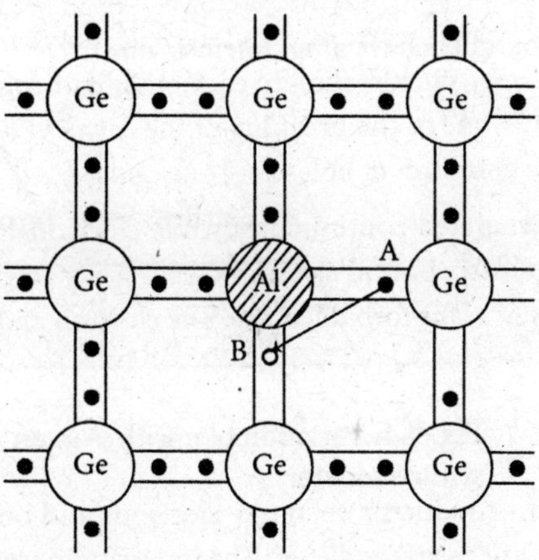

Fig. 9.3: (a) Trivalent impurity atom (Al) in a Ge crystal

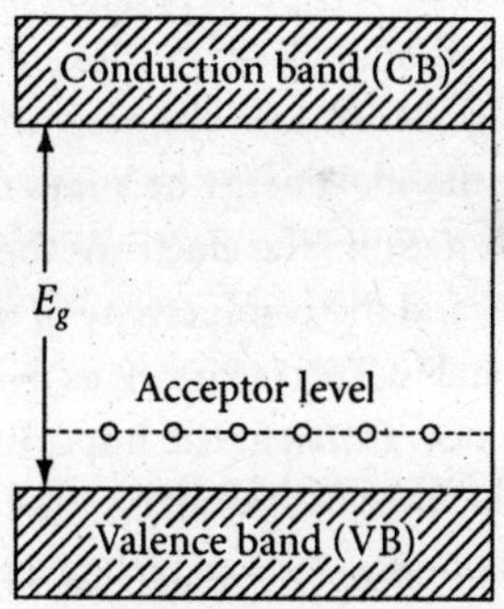

Fig. 9.3: (b) Acceptor level in energy diagram

(b) *p*-type semiconductor — If a trivalent impurity atom of group III of the periodic table, such as *Aluminium* (Al), *Gallium* (Ga) or *Indium* (In) be added in small amounts to *Germanium* or *silicon,* a p-type semiconductor is formed [Fig. 9.3 (a)]. These trivalent atoms have three electrons in their outermost orbits. When a small amount of Al is added to a Ge crystal during its growing, all the three electrons of Al form covalent bond with three neighbouring Ge atoms leaving a vacancy at B. To complete the bond it needs one more electron from the neighbouring atoms. This electron may be supplied by Ge, thereby creating a *vacancy* or *hole* at A of the Ge atom. As Al accept one extra electron, the corresponding energy level is called the *acceptor level* which lies just above the *valence band* as shown in Fig. 9.3 (b). The trivalent impurity (Al) is called the *acceptor impurity* and the semiconductor is called the *p-type semiconductor.*

These acceptor atoms, even at low temperatures, get ionized taking electrons from the valence band and thereby leaving holes in the valence band. It is to be kept in mind that during this process of ionization only holes and no electrons are generated. If, however, the temperature of the semiconductor is sufficiently high, then along with this process, electron-hole pairs are generated by the breakage of covalent bonds. That is why hole concentration (p) is much higher than the concentration of the electrons. The holes are the *major* and electrons, the *minor* carriers of electricity in *p-type* semiconductor.

9.3.1 Effect of Temperature on Extrinsic Semiconductor

If D and D^+ represent respectively a neutral donor atom and a singly ionised donor, then

$$e^- + D^+ \rightleftharpoons D + \Delta E_d. \tag{9.1}$$

At high temperature all donor atoms are thermally ionised and the concentration of conduction electrons (n) becomes equal to the concentration of donors (N_D). The concentration of holes is given by

$$np = n_i^2, \tag{9.2}$$

where

$$n_i = \sqrt{C}T^{3/2}\, e^{-E_g/2kT},$$

where E_g represents the binding energy of an electron to the semiconductor atom in the crystal, i.e., $e^- + e^+ \rightleftharpoons E_g$.

At higher temperature, $n_i > N_D$ and the semiconductor will behave as an intrinsic one.

The range, where $n = N_D$, independent of T is called *extrinsic*. The variation of electron concentration with T is shown in Fig. 9.4. At low temperature a point is reached at which thermal energy becomes too small to cause electron excitation. The electrons then fall from CB to donor level and the conductivity of the material decreases considerably. This is known as *freeze-out*, i.e., the electrons are now *frozen* at the impurity sites. The temperature at which the freeze-out occurs can be evaluated from the relation $E_D \simeq kT$, which gives $T \simeq 100$ K.

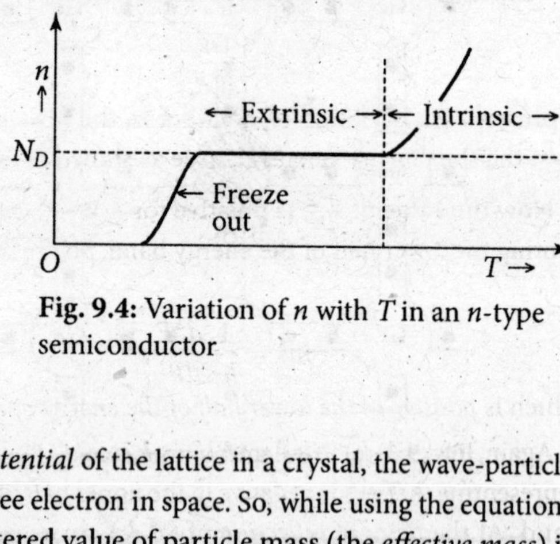

Fig. 9.4: Variation of n with T in an n-type semiconductor

9.3.2 Effective Mass of the Electron

Due to the interaction of the electron with the *periodic potential* of the lattice in a crystal, the wave-particle motion of an electron differs considerably from that of a free electron in space. So, while using the equations of electrodynamics for charge carriers in a crystal, the altered value of particle mass (the *effective mass*) is to be considered, so that the particle may be considered as a *free particle*. The dynamical behaviour of the particle (the charge carrier) then can be described by the *laws of classical physics*.

Let an electron at state k in a lattice be subjected to an external electric field X, the force acting on the electron being eX. If dx be the displacement of the electron in time dt, then the work done is equal to the energy gained by the electron, i.e., $dE = eX\,dx$.

The velocity of the electron is

$$v = \frac{dx}{dt} \text{ or, } dx = v\,dt. \therefore\ dE = eXv\,dt.$$

But

$$v = \frac{1}{\hbar}\frac{dE}{dk}.\ \therefore\ dE = eX\frac{1}{\hbar}\frac{dE}{dk}dt \text{ or, } \frac{dk}{dt} = \frac{eX}{\hbar}.$$

$$\because \hbar k = p \text{ and } \hbar\frac{dk}{dt} = \frac{dp}{dt} = F,\ \therefore \hbar\frac{dk}{dt} = eX = F.$$

Also

$$\frac{dv}{dt} = \frac{1}{\hbar}\frac{d^2E}{dk\,dt} = \frac{1}{\hbar}\frac{d^2E}{dk^2}\frac{dk}{dt} = \frac{1}{\hbar}\frac{d^2E}{dk^2}\frac{F}{\hbar}$$

or acceleration

$$a = \frac{1}{\hbar}\frac{d^2E}{dk^2}\frac{F}{\hbar} \text{ or, } \frac{a}{F} = \frac{1}{\hbar^2}\frac{d^2E}{dk^2}.$$

Putting $\frac{a}{F} = m^*$, the *effective mass* of the electron

$$m^* = \frac{1}{\hbar^2}\frac{d^2E}{dk^2}. \tag{9.3}$$

9.3.3 Variation of m^* with k

Since the *effective mass* of an electron moving in the periodic potential of a crystal lattice is

$$m^* = \frac{1}{\hbar^2}\frac{d^2E}{dk^2},$$

m^* depends on $\frac{d^2E}{dk^2}$ and is a function of k.

$$v = \frac{1}{\hbar}\frac{dE}{dk}$$

represents the variation of E with k in the $(E\text{-}k)$ curve and $\frac{d^2E}{dk^2}$, the slope of $\frac{dE}{dk}$ versus k curve as shown in Fig. 9.5 (a).

Now the value of $\frac{d^2E}{dk^2}$ is positive for $k = -k_0$ to $+k_0$ representing the lower half of the energy band. So, the value of m^* is

$$m^* = \frac{1}{\hbar^2}\frac{d^2E}{dk^2},$$

which is *positive at the lower half of the energy band.*

Again, for $k = +k_0$ to $+\frac{\pi}{a}$ and $k = -k_0$ to $-\frac{\pi}{a}$, $\frac{d^2E}{dk^2}$ is *negative* representing that m^* is negative in the upper half of the energy band. At the *point of inflection* of $(E\text{-}k)$ curve m^* becomes *infinite.* The variation of m^* with k is shown in Fig. 9.5 (c).

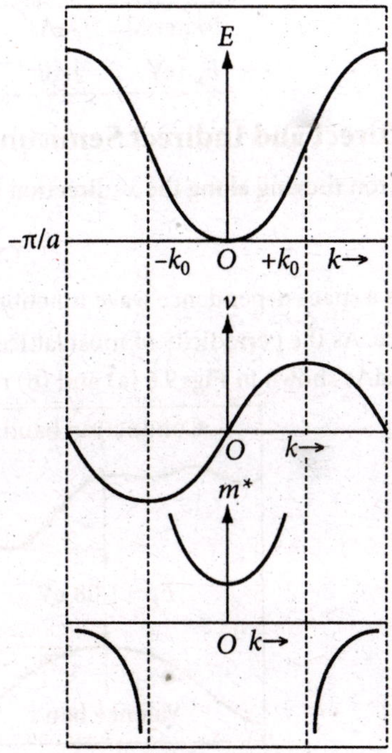

Fig. 9.5: (a) Variation of E with k
(b) Variation of v with k
(c) Variation of m^* with k

The large effective mass can be due to *strong binding force between the electron and the lattice at specified region of k.* The negative m^* may be interpreted *due to Bragg reflection for* $k \to \pm\pi/a, \pm 2\pi/a$ *for which a force in one direction leads to a gain in momentum in opposite direction.*

9.4 Compound Semiconductors

Semiconductor elements are covalent having diamond like structure and belong to Group IV of the periodic table. Zinc blend lattice and wurtzite lattice which occur in binary semiconducting compounds are closely related to them. The semiconducting compounds are strongly covalent and made up of elements placed symmetrically about Group IV in the periodic table.

The compound semiconductors like InSb, GaAs, InP, etc., the inter-metallic compounds of Groups III and V have energy gap between a few tenths to several electron volts and as such find important applications in electronic devices at wider ranges of temperature than Ge and Si.

The compounds such as PbS, PBTe and PbSe are formed between Groups IV and VI and CdS, ZnS and CdTe between Groups II and VI belong to other group of compound semiconductors. The crystalline defects or derivation from *stoichiometry* are responsible for conductivity in these materials. If there is an excess of cations or deficiency of anions, the *non-stoichiometric* compounds behave as an n-type semiconductors. But the presence of excess anions or deficiency of cations leads to a p-type semiconductor.

The energy gap E_g of some compound semiconductors at $T = 300$ K is shown in Table 9.2.

Table 9.2

Material	GaAs	GaP	InS	InAs	InP	PbS	GaSb
E_g (eV)	1.40	2.24	0.16	0.33	1.29	0.40	0.77

9.5 Direct and Indirect Semiconductors

An electron moving along the x-direction in a crystal lattice may be represented by the wave function

$$\psi_k(x) = V\left(\vec{k}, x\right) e^{iK_x x}.$$

This is a space dependence wave function and $V\left(\vec{k}, x\right)$ modulates $\psi_k(x)$ according to the periodicity of the lattice. As the periodicity of most lattices varies with direction we consider the $E\left(\vec{k}\right)$ versus $\vec{k}$ curve for Si and GaAs shown in Figs 9.6 (a) and (b) respectively.

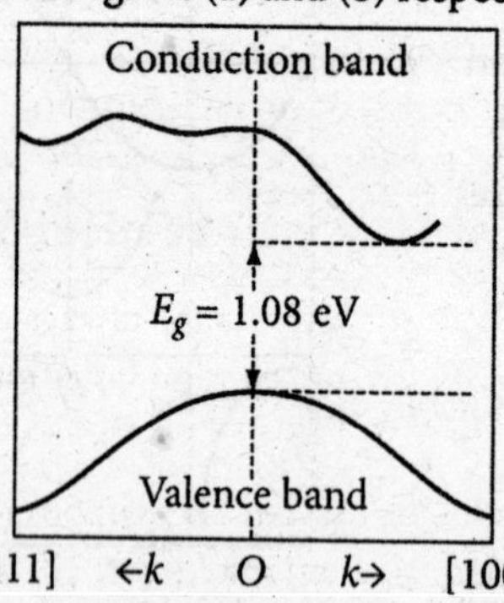

Fig. 9.6 (a) Si

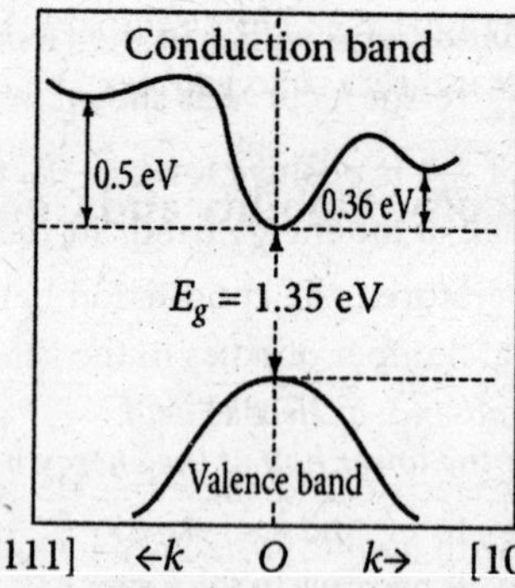

Fig. 9.6 (b) GaAs

Dispersion curves bounding the conduction and the valence bands in Si and GaAs

Figs 9.6 show that the maxima of the dispersion curves for the valence bands of Si and GaAs lie in the middle of the Brillouin zone, resembling a parabola with the vertex facing upwards. In Si, the minima of the curve for the conduction band lie at a distance $k < \pi/a$ from the middle of the zone in the [100] direction. There are two maxima of this type of GaAs, one in the middle of the Brillouin zone and the other at a distance $0.75\pi/a$ from the middle in the direction [100]. The minima and the maxima of the dispersion curve are referred to as the *bottom* and *top* of the energy band. For $\vec{k} = 0$, there is a minimum in the conduction band and a maximum in the valence band of GaAs. But for Si, the valence band maximum is at a different value of $\vec{k}$ than its conduction band minimum. Hence, in GaAs, an electron can make a smallest energy transition from conduction band to valence band without a change in $\vec{k}$-value. But in Si, an electron can make a transition from conduction band minimum to valence band maximum with a change in $\vec{k}$-value. That is why GaAs and Si are referred to as *direct* and *indirect* band gap semiconductors [Fig. 9.7 (a)].

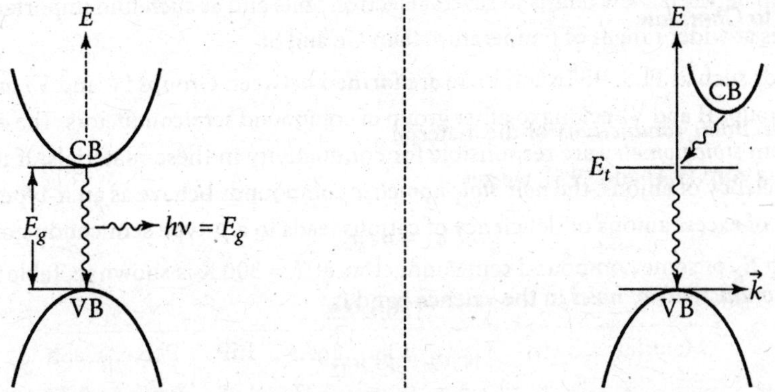

Fig. 9.7: (a) Direct BG semiconductor. Direct transition with a photon emission

Fig. 9.7: (b) Indirect BG semiconductor. Indirect transition via a defect level.

In a direct band gap semiconductor, an electron in the conduction band can transit to an empty state in the valence, thereby giving rise to photon of energy $E_g = h\nu$. But in case of an indirect band semiconductor, an electron in the conduction band minimum cannot transit directly to the valence band instead undergoes a change in momentum as well as energy involving a defect state (or photon) [Fig. 9.7 (b)]. In this type of transition of a conduction electron, $E = E_g + \hbar\omega$, where $\hbar\omega$ is the energy of the emitted photon. The difference in energy appears as heat in the lattice. This is important in deciding which type of semiconductor (direct band gap or indirect band gap) is to be used in devices requiring light output, e.g., LED or Lasers.

9.6 Drift Velocity, Mobility and Conductivity of Intrinsic Semiconductors

At ordinary temperatures, the conduction band of intrinsic semiconductor always contains some free electrons and equal number of holes in the valence band. These electrons and holes move at random and give *no current in absence of electric field.*

In presence of an electric field, these free electrons and holes are accelerated in opposite directions to each other and collide with *atomic nuclei, phonons,* etc. present in the semiconductor. The electrons and holes acquire an *average velocity* which reduces to zero in the subsequent collision. This extra velocity gained by the charge carriers in presence of an electric field E is called the *drift velocity* (v_d). The drift velocity is proportional to the applied field strength E, i.e.,

$$v_d \propto E \text{ or, } v_d = \mu E, \tag{9.4}$$

where μ is a constant of proportionality and is called the *mobility. The mobility of the charge carrier is defined as the drift velocity gained by a carrier for unit electric field strength.*

In intrinsic semiconductor as the electrons move in *nearly empty conduction* band and the holes move in *nearly full valence band,* the *mobility, conductivity,* etc. of electrons and holes are, in general, different.

If v_{de}, μ_e and n denote the drift velocity, mobility and concentration of free electrons in the conduction band respectively, then the current density

$$J_e = nev_{de} \text{ or, } J_e = ne\mu_e E \; [\because v_{de} = \mu_e E]. \tag{9.5}$$

But according to *Ohm's law*,

$$J_e = \sigma_e E, \tag{9.6}$$

where σ_e is the *electronic conductivity* of the material.

Now comparing eqns (9.4) and (9.5), we get

$$\sigma_e = ne\mu_e. \tag{9.7}$$

Similarly, the *conductivity of holes* in the valence band is

$$\sigma_h = pe\mu_h, \tag{9.8}$$

where p and μ_h represent respectively the *concentration* and *mobility of holes.*

$\therefore$ the total conductivity is

$$\sigma = \sigma_e + \sigma_h = e\,(n\mu_e + p\mu_h). \tag{9.9}$$

For *intrinsic semiconductor,*

$$n = p = n_i \text{ (say)}.$$

$$\therefore \sigma = en_i\,(\mu_e + \mu_h). \tag{9.10}$$

The flow of current in semiconductors arises due to

(i) the drift velocity of the carriers in presence of the applied field and the current is called the *drift current* and

(ii) diffusion of carriers due to *concentration gradient* of the dopant, the corresponding current is called the *diffusion current.*

In semiconductors with uniform distribution of impurities the diffusion current is nil. So, the drift current has been taken into account.

Temperature Dependence of Mobility

The electron and hole mobilities are greatly influenced by *lattice scattering* and *impurity scattering.* The effect of dislocation is, however, very small. With the rise of temperature the frequency of lattice scattering increases causing a decrease in mobility. On the other hand, at low temperature the effect is small as the thermal motions of the carriers is slower.

If μ_L and μ_I be the mobility of a carrier in presence of *lattice scattering* and *impurity scattering* respectively, then it has been found theoretically that

$$\mu_L = AT^{-3/2} \text{ and } \mu_I = BT^{3/2},$$

where A and B are constants for a given sample.

The resultant resistivity is given by

$$\rho = \frac{1}{\sigma} = \rho_L + \rho_I$$

$$\text{or, } \frac{1}{\sigma} = \frac{1}{\sigma_L} + \frac{1}{\sigma_I}$$

$$\text{or, } \frac{1}{ne\mu} = \frac{1}{ne\mu_L} + \frac{1}{ne\mu_I}$$

$$\text{or, } \frac{1}{\mu} = \frac{1}{\mu_L} + \frac{1}{\mu_I}$$

$$\text{or, } \frac{1}{\mu} = \frac{T^{3/2}}{A} + \frac{T^{-3/2}}{B}.$$

This gives the temperature dependence of carrier mobility.

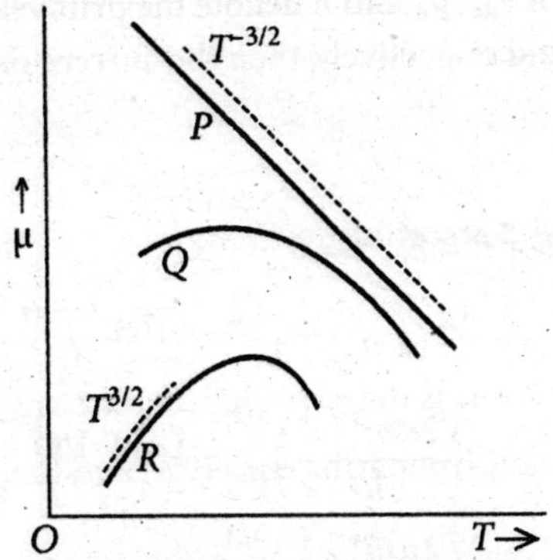

Fig. 9.8: Temperature dependence of μ for pure (*P*), less pure (*Q*) and least pure (*R*) materials

9.6.1 Electron Concentration of Intrinsic Semiconductor in the Valence Band

If $D(E)$ be the *density of states* and $f(E)$, *the Fermi-distribution function*, then the free electron density in an energy interval E and $E + dE$ is

$$dn = D(E)\,f(E)\,dE. \tag{9.11}$$

But

$$f(E) = \frac{1}{e^{(E-E_f)/kT}} \tag{9.12}$$

and

$$D(E) = \frac{4\pi}{h^3}(2m)^{3/2}E^{1/2}. \tag{9.13}$$

Using eqn (9.13) in eqn (9.11), we get

$$dn = \frac{4\pi}{h^3}(2m)^{3/2}E^{1/2}f(E)\,dE. \tag{9.14}$$

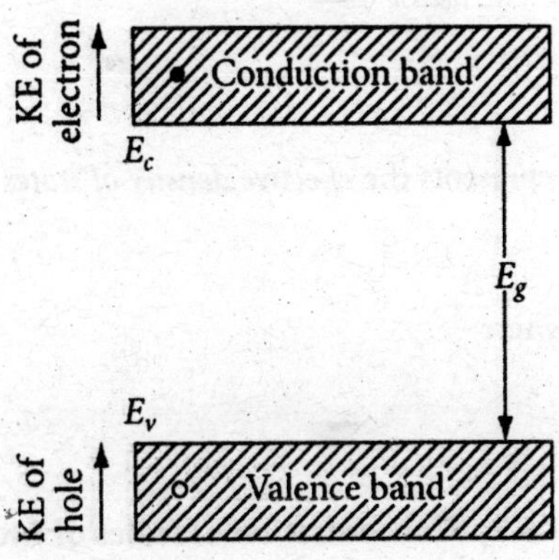

Fig. 9.9: Band model of an intrinsic semiconductor

From Fig. 9.9 it is clear that an electron of energy E in the conduction band can have energy $(E - E_c)$.
$\therefore$ from eqn (9.14),

$$dn = \frac{4\pi}{h^3}(2m_e^*)^{3/2}(E - E_c)^{1/2}\frac{1}{e^{(E-E_c)/kT}+1}dE, \tag{9.15}$$

where m_e^* is the *effective mass of an electron* in the conduction band.

Therefore, the *electron concentration* in the conduction band is given by

$$n = \frac{4\pi}{h^3}(2m_e^*)^{3/2}\int_{E_c}^{\infty}\frac{(E - E_c)^{1/2}\,dE}{e^{(E-E_c)/kT}+1}. \tag{9.16}$$

At room temperature, $kT \simeq 0.026$ eV.

$$\therefore \text{ for } E > E_c,\ e^{(E-E_c)/kT} + 1 \simeq e^{(E-E_c)/kT}.$$

So

$$n = \frac{4\pi}{h^3}(2m_e^*)^{3/2}e^{(E_f-E_c)/kT}\int_{E_c}^{\infty}(E - E_c)^{1/2}e^{-(E-E_c)/kT}\,dE.$$

Let

$$\frac{E - E_c}{kT} = x. \therefore dE = kT\,dx.$$

For $E = E_c$, $x = 0$.

$$\therefore n = \frac{4\pi}{h^3}(2m_e^*)^{3/2} e^{(E_f - E_c)/kT} \int_0^\infty x^{1/2}(kT)^{1/2} e^{-x} kT\,dx = \frac{4\pi}{h^3}(2m_e^* kT)^{3/2} e^{(E_f - E_c)/kT} \int_0^\infty x^{1/2} e^{-x} dx$$

$$= 2\left(\frac{2\pi m_e^* kT}{h^2}\right)^{3/2} e^{-(E_c - E_f)/kT} \left[\text{Using} \int_0^\infty x^{1/2} e^{-x} dx = (\pi/4)^{1/2}\right] \quad (9.17)$$

$$\text{or, } n = 2\left(\frac{2\pi m_e^* kT}{h^2}\right)^{3/2} f(E_c), \quad (9.18)$$

where

$$f(E_c) = \frac{1}{e^{(E_c - E_f)/kT} + 1} \simeq e^{-(E_c - E_f)/kT}.$$

The factor

$$2\left(\frac{2\pi m_e^* kT}{h^2}\right)^{3/2}$$

represents the *effective density of states of electrons* at the conduction band edge.

$$\therefore n = N_c\, e^{-(E_c - E_f)/kT}, \quad (9.19)$$

where

$$N_c = 2\left(\frac{2\pi m_e^* kT}{h^2}\right)^{3/2}. \quad (9.20)$$

9.6.2 Concentration of Holes of Intrinsic Semiconductor in the Valence Band

The number of holes per unit volume in the energy range E to $E + dE$ is

$$dp = D(E)\,[1 - f(E)]\,dE, \quad (9.21)$$

where $[1 - f(E)]$ is the *probability* of finding a hole in the energy state E.

Now

$$1 - f(E) = 1 - \frac{1}{e^{(E - E_f)/kT}} = \frac{e^{(E - E_f)/kT}}{e^{(E - E_f)/kT} + 1}.$$

In the valence band, since $E < E_f$, $e^{(E - E_f)/kT}$ in the denominator may be neglected.

$$\therefore 1 - f(E) \simeq e^{(E - E_f)/kT}. \quad (9.22)$$

Therefore, the density of holes in the valence band is

$$D(E) = \frac{4\pi}{h^3}(2m_h^*)^{3/2}(E_v - E)^{1/2}, \quad (9.23)$$

where m_h^* is the *effective mass of the hole* in the valence band.

Substituting eqns (9.22) and (9.23) in eqn (9.21), we get

$$p = \frac{4\pi}{h^3}(2m_h^*)^{3/2} \int_{-\infty}^{E_v} (E_v - E)^{1/2}\, e^{(E-E_f)/kT}\, dE$$

$$= \frac{4\pi}{h^3}(2m_h^*)^{3/2}\, e^{(E_v-E_f)/kT} \int_{-\infty}^{E_v} (E_v - E)^{1/2}\, e^{(E-E_v)/kT}\, dE.$$

Let

$$\frac{(E_v - E)}{kT} = x. \;\therefore\; dE = -kT\, dx.$$

For $E = E_v$, $x = 0$.

$$\therefore p = \frac{4\pi}{h^3}(2m_h^*)^{3/2} e^{(E_v-E_f)/kT} \int_0^{\infty} x^{1/2}(kT)^{1/2}\, e^{-x}(-kT)dx$$

$$= \frac{4\pi}{h^3}(2m_h^*)^{3/2} e^{(E_v-E_f)/kT}(kT)^{3/2} \int_0^{\infty} x^{1/2}\, e^{-x}\, dx$$

$$= \frac{4\pi}{h^3}(2m_h^*\, kT)^{3/2}\, e^{(E_v-E_f)/kT}(\pi/4)^{1/2} \left[\because \int_0^{\infty} x^{1/2}\, e^{-x} dx = (\pi/4)^{1/2}\right]$$

$$= 2\left(\frac{2\pi m_h^*\, kT}{h^2}\right)^{3/2} e^{-(E_f-E_v)/kT} \tag{9.24}$$

$$\text{or, } p = N_v\, e^{-(E_f-E_v)/kT}, \tag{9.25}$$

where

$$N_v = 2\left(\frac{2\pi m_h^* kT}{h^2}\right)^{3/2}, \tag{9.26}$$

N_v is the *effective density of holes in the valence band edge.*

9.6.3 Fermi Level

Since $n = p = n_i$ for an *intrinsic semiconductor,* we have from eqns (9.19) and (9.25)

$$N_c\, e^{-(E_c-E_f)/kT} = N_v\, e^{-(E_f-E_v)/kT}$$

$$\text{or, } e^{(2E_f-E_c-E_v)/kT} = \frac{N_v}{N_c}$$

$$\text{or, } (2E_f - E_c - E_v)/kT = \ln(N_v/N_c)$$

$$\text{or, } E_f = \frac{E_c + E_v}{2} + \frac{1}{2}kT\ln(N_v/N_c). \tag{9.27}$$

Using the values of N_c and N_v from eqns (9.20) and (9.26) in eqn (9.27), we get

$$E_f = \frac{E_c + E_v}{2} + \frac{3}{4}kT\ln(\mu_h^*/m_e^*). \tag{9.28}$$

At $T = 0$ K,

$$E_f = \frac{E_c + E_v}{2}. \tag{9.29}$$

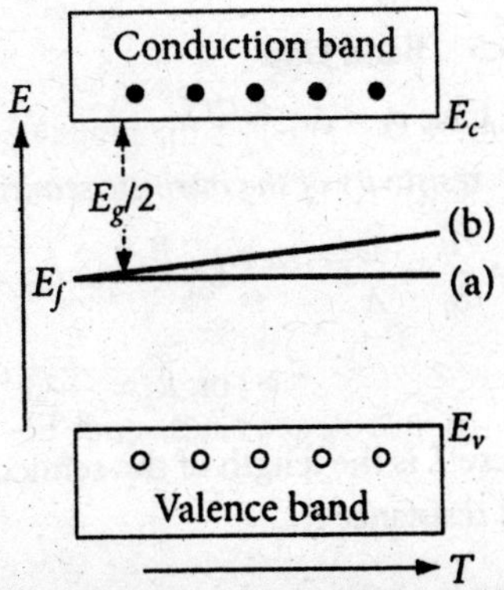

Fig. 9.10: Position of Fermi level in intrinsic semiconductor (a) at $T = 0$ K (b) at $T > 0$ K

Eqn (9.29) shows that *Fermi level is located half-way between the valence and conduction bands.* If $\mu_h^* = m_e^*$, the position of the Fermi level is independent of temperature. In general, $\mu_h^* > m_e^*$ and the Fermi level is just above the middle and rises slightly with the increase of temperature.

9.6.4 Electrical Conductivity of Intrinsic Semiconductors

For an *intrinsic semiconductor,* the electrical conductivity σ is given by $\sigma = (ne\mu_e + pe\mu_h)$, where μ_e and μ_h denote the *mobilities of electrons and holes respectively.*

$$\because n = p = n_i.$$

$\therefore$ from eqns (9.17) and (9.24),

$$n_i^2 = np = 4\left(\frac{2\pi kT}{h^2}\right)^3 (m_e^* \mu_h^*)^{3/2} e^{(E_v - E_c)/kT}$$

$$= 4\left(\frac{2\pi kT}{h^2}\right)^3 (m_e^* \mu_h^*)^{3/2} e^{-E_g/kT} \left[\because (E_c - E_v) = E_g\right].$$

Therefore,

$$n_i = 2\left(\frac{2\pi kT}{h^2}\right)^{3/2} (m_e^* m_h^*)^{3/4} e^{-E_g/2kT}. \tag{9.30}$$

$$\therefore \sigma = (\mu_e + \mu_h)\, 2e\left(\frac{2\pi kT}{h^2}\right)^{3/2} (m_e^* m_h^*)^{3/4} e^{-E_g/2kT} \tag{9.31}$$

$$\text{or, } \sigma_i = Ae^{-E_g/2kT}, \tag{9.32}$$

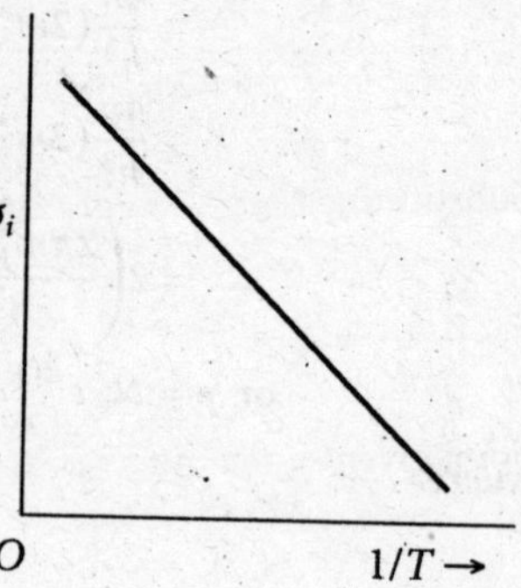

Fig. 9.11: Variation of σ_i with T

σ_i is the *electrical conductivity of the intrinsic semiconductor, where*

$$A = (\mu_e + \mu_h)\, 2e\left(\frac{2\pi kT}{h^2}\right)^{3/2} (m_e^* \mu_h^*)^{3/4}.$$

Taking logarithm on both sides of eqn (9.32), we get

$$\ln \sigma_i = \ln A - E_g/2kT. \tag{9.33}$$

The plot of $\ln \sigma_i$ against $1/T$ is shown in Fig. 9.11 which shows that σ_i increases with the decrease of temperature.

9.6.5 Band Gap

We have $\sigma_i = Ae^{-E_g/2kT}$.

$\therefore$ *resistivity of the intrinsic semiconductor* is

$$\rho_i = \frac{1}{\sigma_i} = \frac{1}{A} e^{E_g/2kT} \text{ or, } \frac{R_i a}{L} = (1/A)\, e^{E_g/2kT}, \left[\because \rho_i = \frac{R_i a}{L}\right]$$

$$\text{or, } R_i = \frac{L}{aA} e^{E_g/2kt} = Ce^{E_g/2kT} \left[\text{where } C = \frac{L}{aA}\right],$$

where L is the length of the semiconductor of cross-sectional area a and resistance R_i.

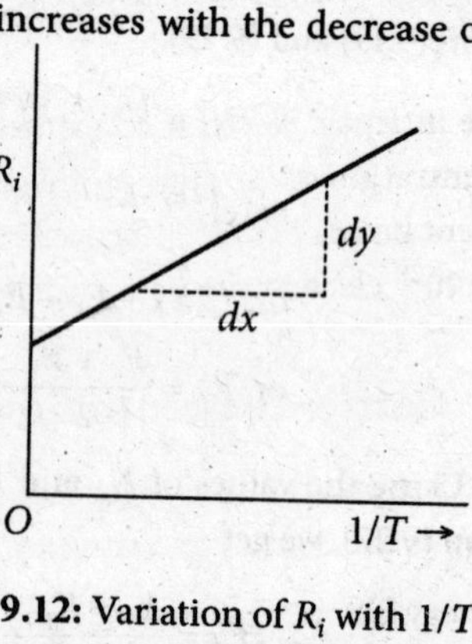

Fig. 9.12: Variation of R_i with $1/T$

Taking logarithm on both sides,

$$\ln R_i = \ln C + E_g/2kT. \tag{9.34}$$

Plotting $\ln R_i$ against $\frac{1}{T}$, we get a straight line as shown in Fig. 9.12.

The slope of the curve is $E_g/2kT$. From Fig. 9.12,

$$\text{slope}, \frac{dy}{dx} = \frac{E_g}{2k}. \therefore E_g = 2k\left(\frac{dy}{dx}\right), \tag{9.35}$$

where k is the *Boltzmann constant.*

9.6.6 Law of Mass Action

Statement: *For a given semiconductor, the product of electron and hole concentrations is a constant at a given temperature and is equal to the square of the intrinsic carrier concentration.*

The *law of mass action* holds for both *intrinsic* and *extrinsic semiconductors.* Due to the addition of impurity atoms to the semiconductor n increases and corresponding p decreases so that $np = n_i^2$.

But

$$np = n_i^2 = N_c N_v e^{-(E_c-E_v)/kT} = N_c N_v e^{-E_g/kT} \; [\because E_c - E_v = E_g]. \tag{9.36}$$

Substituting the values of N_c and N_v from eqns (9.20) and (9.26) in eqn (9.36), we get

$$np = n_i^2 = 4\left(\frac{2\pi kT}{h^2}\right)^3 (m_e^* \mu_h^*)^{3/2} e^{-E_g/kT}.$$

For a given semiconductor at a particular temperature,

$$np = n_i^2 = \text{constant}. \tag{9.37}$$

This proves the *laws of mass action.*

The *intrinsic carrier concentration* can be obtained from eqns (9.17) and (9.24) as below:

$$n_i = (N_c N_v)^{1/2} e^{-E_g/2kT} = 2\left(\frac{2\pi\, kT}{h^2}\right)^{3/2} (m_e^* \mu_h^*)^{3/4} e^{-E_g/2kT}. \tag{9.38}$$

The intrinsic electron concentration in pure Ge at $T = 300$ K is about $n_i = 2.4 \times 10^{19}/\text{m}^3$, when the concentration of Ge atom is $4.4 \times 10^{28}/\text{m}^3$. This indicates that at ordinary temperatures only, about five covalent bonds per 10^{10} Ge atoms are broken contributing to intrinsic conduction. But for metals like Cu about 10^{28} electrons/m^3 are available for conduction.

9.7 Carrier Concentration in *N*-type Semiconductors

Fig. 9.13 shows the energy level diagram of an n-type semiconductor, where E_d is the donor energy level and N_d is the donor concentration of the material.

At very low temperature all *donor levels* are filled with electrons and with the increase in temperature more and more donor atoms get ionized. The released electrons from the donors increase the concentration of the electrons in the conduction band.

Neglecting the concentration of thermally generated holes in the valence band $n = N_d^+$, where N_d^+ is the *concentration of donors.*

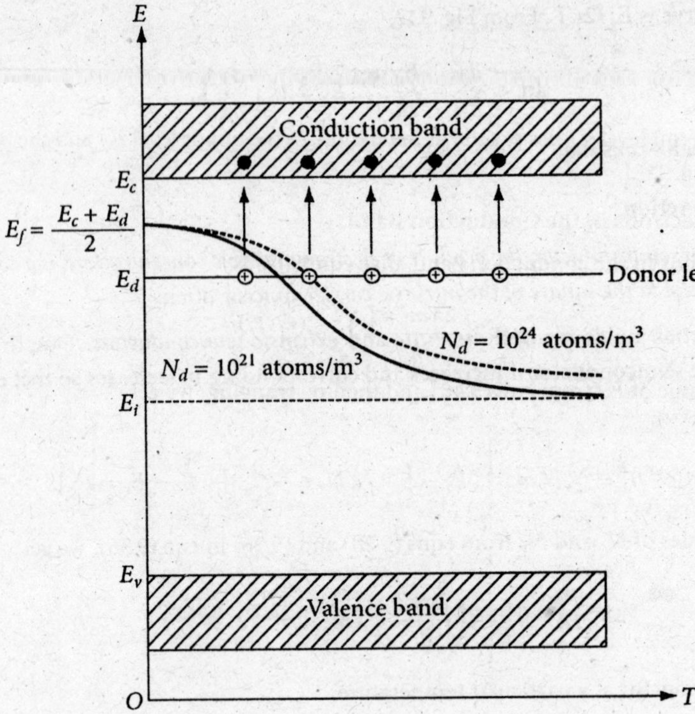

Fig. 9.13: Variation of Fermi level with temperature in *n*-type semiconductor

The density of electrons in the conduction is given by eqn (9.17)

$$n = 2\left(\frac{2\pi m_e^* kT}{h^2}\right)^{3/2} e^{(E_f - E_c)/kT}.$$

If the Fermi level E_f lies more than a few kT above the donor level E_d, then the density of ionized donors is given by

$$N_d\,[1 - F(E_d)] \simeq N_d\, e^{(E_d - E_f)/kT}. \tag{9.39}$$

At very low temperatures, the electron-hole pairs are not generated due to the breakage of covalent bonds. Then the number of electrons in the conduction band must be equal to the number of ionized donors in donor level E_d, i.e., $n = N_d^+$, N_d^+ is the *concentration of ionized donors*.

$$\therefore 2\left(\frac{2\pi m_e^* kT}{h^2}\right)^{3/2} e^{(E_f - E_c)/kT} = N_d\, e^{(E_d - E_f)/kT}.$$

Taking logarithm on both sides of the above equation we get after rearranging

$$\frac{(E_f - E_c)}{kT} - \frac{(E_d - E_f)}{kT} = \ln N_d - \ln 2\left(\frac{2\pi m_e^* kT}{h^2}\right)^{3/2}$$

$$\text{or, } 2E_f - (E_d + E_c) = kT \ln\left[\frac{N_d}{2\left(\frac{2\pi m_e^* kT}{h^2}\right)^{3/2}}\right] \quad \text{or, } E_f = \left(\frac{E_d + E_c}{2}\right) + \frac{kT}{2}\ln\left[\frac{N_d}{2\left(\frac{2\pi m_e^* kT}{h^2}\right)^{3/2}}\right]. \tag{9.40}$$

At $T = 0$ K,

$$E_f = \frac{E_d + E_c}{2}. \tag{9.41}$$

Thus, at 0 K, the Fermi level E_f lies exactly at the middle of the donor level E_d and the bottom level of the conduction band [Fig. 9.13].

9.7.1 Density of Electrons in the Conduction Band

The density of electrons in the conduction band after eqn (9.17) is

$$n = 2\left(\frac{2\pi m_e^* kT}{h^2}\right)^{3/2} e^{(E_f - E_c)/kT}.$$

Substituting the value of E_f from eqn (9.40) and then rearranging, we get

$$e^{(E_f - E_c)/kT} = \exp\left[(E_d - E_c)/2kT + \frac{1}{2}\ln\frac{Nd}{2\left(\frac{2\pi m_e^* kT}{h^2}\right)^{3/2}}\right]$$

$$= \exp\left[(E_d - E_c)/2kT + \ln\frac{N_d^{1/2}}{\left\{2\left(\frac{2\pi m_e^* kT}{h^2}\right)^{3/2}\right\}^{1/2}}\right]$$

$$= \exp\left(\frac{E_d - E_c}{2kT}\right)\frac{N_d^{1/2}}{\left\{2\left(\frac{2\pi m_e^* kT}{h^2}\right)^{3/2}\right\}^{1/2}}$$

$$\therefore n = 2\left(\frac{2\pi m_e^* kT}{h^2}\right)^{3/2}\frac{N_d^{1/2}}{\left\{2\left(\frac{2\pi m_e^* kT}{h^2}\right)^{3/2}\right\}^{1/2}} e^{(E_d - E_c)/kT}$$

$$\text{or, } n = (2N_d)^{1/2}\left(\frac{2\pi m_e^* kT}{h^2}\right)^{3/4} e^{(E_d - E_c)/kT}. \tag{9.42}$$

Thus, density of electrons in the conduction band of an n-type semiconductor is proportional to $N_d^{1/2}$ at moderately low temperatures. At higher temperatures, the intrinsic behaviour dominates over donor concentration. Also, n increases exponentially as E_f moves away from E_i towards the conduction band.

9.7.2 Variation of Fermi Level with Temperature

Eqn (9.40) shows that E_f increases slightly with the increase of temperature (T), when T is not high. As temperature increases more and more, donor atoms become ionized more and more and at a particular temperature all the donor atoms get completely ionized. If temperature still goes on increasing, the breaking of covalent bond takes place and the material behaves as an intrinsic semiconductor, when $n_i \gg N_d$. At this condition, the Fermi level gradually decreases and approaches the intrinsic Fermi level E_i as shown in Fig. 9.13.

The rate of lowering of Fermi level from $E_f = \frac{E_d + E_c}{2}$ to E_i with rise of temperature is more for lower donor concentration than the higher one. So, the highly doped semiconductor behaves intrinsically after all the donor atoms are ionized [Fig. 9.13].

9.7.3 Temperature Dependence of Electrical Conductivity of *n*-Type Semiconductor

The electrical conductivity of an *n*-type semiconductor is given by

$$\sigma = en\mu_e = eN_d^+\mu_e, \quad (9.43)$$

assuming the electrons are mainly responsible for conduction. The value of the electron concentration n and mobility μ_e can be determined at particular temperature and σ can be calculated using eqn (9.43).

The variation of σ with T for an *n-type* semiconductor is represented in Fig. 9.14.

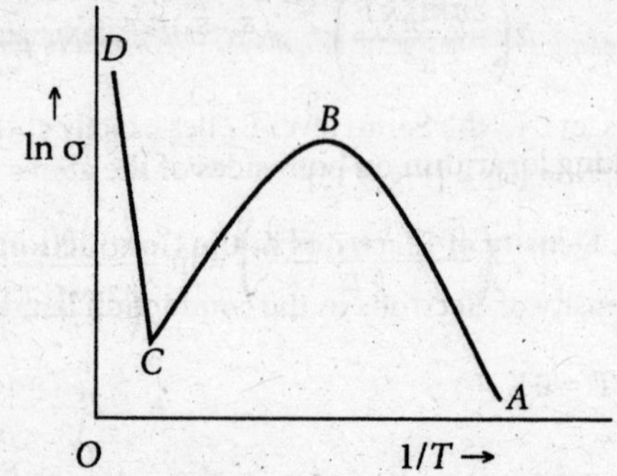

Fig. 9.14: Variation of σ with *T* for an *n*-type semiconductor

The following points are apparent from Fig. 9.14:

(i) From low temperature at *A* the conductivity increases and becomes maximum at *B*, where all the donor atoms are ionized.

(ii) With further increase of temperature, σ decreases and becomes minimum at *C*. This is due to *the decrease of mobility with the increase of temperature.*

(iii) From *C* to *D*, σ increases sharply. This is due to *the increase in electron concentration due to breaking of covalent bonds at high temperature dominating the decrease in mobility.*

9.8 Carrier Concentration in *p*-type Semiconductor

Fig. 9.15 shows the energy level diagram of a *p*-type semiconductor, where E_a and N_a represents respectively the acceptor energy level and concentration of acceptor atoms in the *p*-type semiconductor.

At very low temperatures all the acceptor levels are empty and at $T > 0$ K with the increase of temperature a part of acceptors is ionized by acquiring electrons from the valence band. Thus, holes are created in the valence band which causes *p*-type conduction. Neglecting the holes created due to thermal effect, the hole concentration (p) becomes equal to the concentration of the ionized acceptors (N_a^-).

Thus,

$$p = N_a^-. \quad (9.44)$$

Now the *density of holes* in the valence band is given by

$$p = 2\left(\frac{2\pi m_h^* kT}{h^2}\right)e^{\frac{(E_v-E_f)}{kT}}. \quad (9.45)$$

The Fermi level E_f lies below E_a, the density of ionized acceptors is given by

$$N_a F(E_a) = N_a\, e^{(E_f-E_a)/kT}.$$

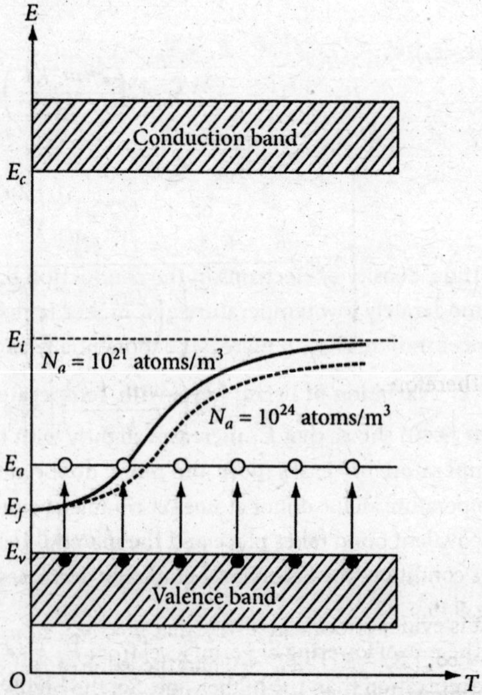

Fig. 9.15: Variation of Fermi level position with temperature

Substituting the values of p and N_a^- in eqn (9.44), we get

$$2\left(\frac{2\pi m_h^* kT}{h^2}\right)^{3/2} e^{(E_v-E_f)/kT} = N_a\, e^{(E_f-E_a)/kT}, \text{ i.e., } e^{(E_v+E_a-2E_f)/kT} = \frac{N_a}{2\left(\frac{2\pi m_h^* kT}{h^2}\right)^{3/2}}.$$

Taking logarithm on both sides of the above equation, we get

$$\left(\frac{E_v+E_a-2E_f}{kT}\right) = \ln\frac{N_a}{2\left(\frac{2\pi m_h^* kT}{h^2}\right)^{3/2}} \text{ or, } E_f = \frac{E_v+E_a}{2} - \frac{kT}{2}\ln\frac{N_a}{2\left(\frac{2\pi m_h^* kT}{h^2}\right)^{3/2}}. \tag{9.46}$$

At $T = 0$ K,

$$E_f = \frac{E_v+E_a}{2}. \tag{9.47}$$

Hence, *at $T = 0$ K, the Fermi level lies exactly at the middle of the acceptor level and top of the valence band as shown in* Fig. 9.15.

9.8.1 Density of Holes in the Valence Band of a *p*-Type Semiconductor

The density of holes in the valence band from eqn (9.24) is

$$p = 2\left(\frac{2\pi m_h^* kT}{h^2}\right)^{3/2} e^{(E_v-E_f)/kT}.$$

Substituting the value of E_f from eqn (9.46), we get

$$e^{(E_v-E_f)/kT} = \exp\left\{\frac{E_v}{kT} - \frac{E_v+E_a}{2kT} + \frac{1}{2}\ln\frac{N_a}{2\left(\frac{2\pi m_h^* kT}{h^2}\right)^{3/2}}\right\} = \exp\left\{\frac{E_v-E_a}{2kT} + \ln\frac{N_a^{1/2}}{\left[2\left(\frac{2\pi m_h^* kT}{h^2}\right)^{3/2}\right]^{1/2}}\right\}$$

$$= \left[\exp\frac{(E_v-E_a)}{2kT}\,\frac{N_a^{1/2}}{\left\{2\left(\frac{2\pi m_h^* kT}{h^2}\right)^{3/2}\right\}^{1/2}}\right].$$

Therefore,

$$p = 2\left(\frac{2\pi m_h^* kT}{h^2}\right)\left[\exp\frac{(E_v-E_a)}{2kT}\right]\frac{N_a^{1/2}}{\left[2\left(\frac{2\pi m_h^* kT}{h^2}\right)^{3/2}\right]^{1/2}}$$

$$\text{or, } p = (2N_a)^{1/2}\left(\frac{2\pi m_h^* kT}{h^2}\right)^{3/4} \exp\left(E_v-E_a\right)/kT. \tag{9.48}$$

It is evident from eqn (9.48), that $p \propto N_a^{1/2}$ at moderately low temperature. At higher temperature, i.e., at $T \to \infty$, $(E_v - E_f)/kT \to 0$ and the value of p approaches the intrinsic concentration n_i. Also p increases exponentially as E_f moves away from E_i towards the valence band.

9.8.2 Temperature Dependence of Electrical Conductivity of *p*-Type Semiconductor

If μ_h be the mobility and p the concentration of holes, then the electrical conductivity is

$$\sigma_p = ep\mu_h = eN_a^- \mu_h. \tag{9.49}$$

The conductivity of the minority carriers has been neglected. The temperature variation of σ_p is similar to that of n-type semiconductor.

9.8.3 Variation of Fermi Level with Temperature of *p*-Type Semiconductor

Eqn (9.46) shows that, the Fermi level E_f slightly decreases with temperature at low temperature. But with the increase of temperature more and more acceptor atoms are ionized. Again, electron-hole pair is created with the rise of temperature due to breaking of covalent bonds when the material behaves as an intrinsic semiconductor. The Fermi level E_f moves gradually towards the intrinsic Fermi level E_i as shown in Fig. 9.15.

The variation of E_f with acceptor concentration is also shown in Fig. 9.15.

9.9 Charge Neutrality Equation

We consider a mixed semiconductor containing both donor and acceptor impurities. We assume that at surrounding temperature the impurity atoms are ionized and no charge carriers are generated due to breaking of covalent bonds.

If n, p be the density of electrons and holes and N_d, N_a the density of donor and acceptor atoms respectively, then under the above condition $n = N_d$ and $p = N_a$.

Then, for overall charge neutrality of the material,

$$p + N_d = n + N_a. \tag{9.50}$$

According to the *law of mass action*, in any semiconductor *at thermal equilibrium* $np = n_i^2$, where n_i is the intrinsic carrier concentration.

$$\therefore n = \frac{n_i^2}{p} \text{ and } p = \frac{n_i^2}{n}.$$

From eqn (9.50),

$$p = n + N_a - N_d = \frac{n_i^2}{p} + N_a - N_d \text{ or, } p^2 - p(N_a - N_d) - n_i^2 = 0. \tag{9.51}$$

Eqn (9.51) is a quadratic equation having roots

$$p = \frac{N_a - N_d}{2} \pm \left[\frac{(N_a - N_d)^2}{4} + n_i^2\right]^{1/2}. \tag{9.52}$$

Similarly,

$$n = \frac{N_d - N_a}{2} \pm \left[\frac{(N_d - N_a)^2}{4} + n_i^2\right]^{1/2}. \tag{9.53}$$

Special Cases

Case I For intrinsic semiconductor,

$$N_a = N_d = 0, \therefore n = p = n_i. \tag{9.54}$$

Case II For n-type semiconductors, $N_a = 0$

$$\therefore p = -\frac{N_d}{2} \pm \left[\frac{(-N_d)^2}{4} + n_i^2\right]^{1/2} = \frac{-N_d \pm \left(N_d + 4n_i^2\right)^{1/2}}{2} = \frac{-N_d \pm N_d\left[1 + 4\frac{n_i^2}{N_d^2}\right]^{1/2}}{2}$$

$$\text{or, } p = \frac{-N_d + N_d\left(1 + 4\frac{n_i^2}{N_d^2}\right)^{1/2}}{2}.$$

$\because$ p cannot be negative.

Now

$$\left(1 + 4\frac{n_i^2}{N_d^2}\right)^{1/2} \simeq 1 + \frac{2n_i^2}{N_d^2} \text{ (neglecting the higher order terms).}$$

$$\therefore p = -\frac{N_d}{2} + \frac{N_d}{2}\left(1 + \frac{2n_i^2}{N_d^2}\right) \text{ or, } p = \frac{n_i^2}{N_d}. \tag{9.55}$$

Similarly,

$$n = \frac{N_d}{2} + \frac{N_d}{2}\left(1 + \frac{2n_i^2}{N_d^2}\right) = N_d + \frac{n_i^2}{N_d}.$$

At $T \to 0$, $n_i \simeq 0$.

$$\therefore n = N_d. \tag{9.56}$$

Case III For p-type semiconductor, $N_d = 0$.

$\therefore$ we can show that

$$n \simeq \frac{n_i^2}{N_a} \tag{9.57}$$

$$\text{and } p = N_a. \tag{9.58}$$

Hence, at ambient temperatures, the density of majority carriers is approximately equal to the density of impurity atoms. This means, at low temperatures for large impurity density the charge carriers are totally due to ionization of impurity atoms. On the other hand, if impurity density is low and temperature is high, then the contribution due to impurity atoms is less. Under such condition, charge carriers are generated due to breaking of covalent bonds, i.e., the material behaves as an intrinsic semiconductor.

9.10 Carrier Transport in Semiconductors

(a) Drift Current: In presence of an electric field E, the free electrons and holes acquire certain drift velocity, $v_d = \mu E$, μ is the mobility of charge carrier. This movement of the charge carriers constitutes a current, called the *drift current*.

For the *holes*, the drift current density is

$$J_{hd} = pev_{hd} = pe\mu_h E, \tag{9.59}$$

where v_{hd} and μ_h are the drift velocity and mobility of holes respectively.

Similarly, the drift current density for *electrons* is

$$J_{ed} = nev_{ed} = ne\mu_e E, \tag{9.60}$$

where v_{ed} and μ_e represent respectively the drift velocity and mobility of the electron.

$\therefore$ the total drift current density is

$$J_d = J_{hd} + J_{ed} = ne\mu_e E + pe\mu_h E = (ne\mu_e + pe\mu_h)\, E \tag{9.61}$$

or electrical conductivity

$$\sigma = \frac{J_d}{E} = ne\mu_e + pe\mu_h. \tag{9.62}$$

For intrinsic semiconductor,

$$\sigma_i = n_i\,(\mu_e + \mu_h)\, e, \tag{9.63}$$

$\because n = p = n_i$ for intrinsic semiconductor.

(b) Diffusion current: In a semiconductor, in addition to conduction current, transport of charge carriers also occurs by a process known as *diffusion* leading to *diffusion current*.

Diffusion is a process in which the particles (electrons or holes) move from a region of higher concentration to the region of lower concentration. The resulting current due to concentration gradient of the charge carriers is called *diffusion current.*

The electron flux at any point due to diffusion is proportional to concentration gradient at that point, i.e., diffusion of electron flux

$$N \propto \left.\frac{dn}{dx}\right|_{x=x_1} \quad \text{or, } N = -D_e \frac{dn}{dx},$$

where D_e is a constant of proportionality and is called *diffusion constant* for electrons.

$\therefore$ diffusion current density of electrons is

$$(J_e)_x = eD_e \frac{dn}{dx}. \tag{9.64}$$

Similarly, the diffusion current due to holes is

$$(J_h)_x = -eD_h \frac{dn}{dx}, \tag{9.65}$$

where D_h is the *diffusion coefficient for the holes.*

Therefore, the total current density is

$$J_x = (J_e)_x + (J_h)_x = e\left[\left(n\mu_e E_x + D_e \frac{dn}{dx}\right) + \left(p\mu_h E_x - D_h \frac{dp}{dx}\right)\right]. \tag{9.66}$$

9.10.1 Einstein's Equation

Mobility and diffusion constant for a particular type of charge carrier are intimately related. The higher the value of mobility, greater will be the tendency to diffusion. In the steady state the conduction current density (J_c) must be equal and opposite to the diffusion current, i.e.,

$$J_c = -J_{eD} \text{ or, } \sigma E = -eD_e \frac{dn}{dx}. \tag{9.67}$$

The presence of electric field produces a potential difference $V(x)$ over a distance x. The variation of electron concentration (n) with x is given by *Boltzmann relation*:

$$n = Ce^{-\frac{eV}{kt}} = Ce^{-\frac{eEx}{kT}}.$$

$$\therefore \frac{dn}{dx} = Ce^{-\frac{eEx}{kT}}\left(-\frac{eE}{kT}\right) = -\frac{neE}{kT} \text{ or, } \frac{dn}{dx} = -\frac{neE}{kT}. \tag{9.68}$$

From eqns (9.67) and (9.68), we get

$$ne\mu_e E = \left(\frac{eD_e}{kT}\right)(neE) \text{ or, } D_e = \frac{\mu_e kT}{e} \quad [\because \sigma = ne\mu_e]. \tag{9.69}$$

This is known as *Einstein's equation.*

For hole diffusion,

$$D_h = \frac{\mu_h kT}{e}. \tag{9.70}$$

Hence,

$$\frac{D_e}{D_h} = \frac{\mu_e}{\mu_h}. \tag{9.71}$$

The calculated values of D for Si are

$$D_e = 3.4 \times 10^{-3} \text{ m}^2/\text{s for electrons}$$
$$D_h = 10^{-3} \text{ m}^2/\text{s for holes}$$
$$\text{and for Ge, } D_e = 9.9 \times 10^{-3} \text{ m}^2/\text{s for electrons}$$
$$D_h = 10^{-3} \text{ m}^2/\text{s for holes.}$$

9.10.2 Lifetime and Diffusion Length

Though the electron and hole concentrations in a semiconductor under thermal equilibrium are steady, the equilibrium itself is dynamic. That is, at any instant of time, the rate of generation of electron-hole pair is exactly equal to the rate of recombination of such pair. Such equilibrium concentration changes by several processes.

The carriers do not exist for indefinite period of time but recombine to form neutral particles. It will be shown that the excess carrier density decreases exponentially with time, the carrier density at time τ being $\propto e^{-t/\tau}$, where τ is the *lifetime of the carriers*. So, *the lifetime τ can be defined as the time at which carrier density falls to $1/e$ of its initial value.*

The average distance L that a carrier diffuses before recombination is called the ***diffusion length.***

The diffusion length L and lifetime τ of carriers are related as follows:

$$L_n = \sqrt{D_e \tau_e} \text{ (for electrons)} \tag{9.72}$$

$$L_h = \sqrt{D_h \tau_h} \text{ (for holes).} \tag{9.73}$$

9.11 Theory of Generation and Recombination of Charge Carriers

If a semiconductor be supplied with sufficient energy, the covalent tetrahedral bonds are broken and free electrons are generated. These free electrons are raised to the conduction band subsequently and a conducting hole is left behind in the valence band. This process is called the *electron-hole pair production* or generation as shown in Fig. 9.16.

In n-type semiconductor the number of electrons is much more than the number of holes. During pair production the number *minority carriers* (holes) increases but the increase of *majority carriers* (electrons) is negligible above the equilibrium number already existing in the material.

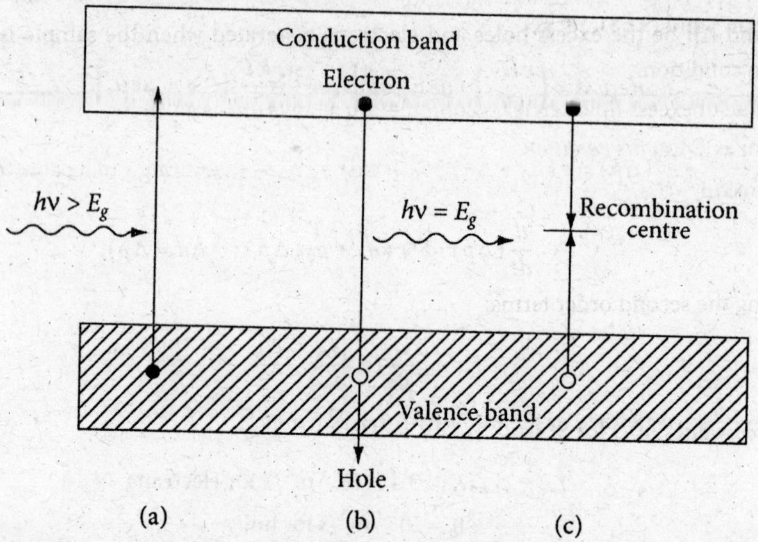

Fig. 9.16: (a) Generation of an electron-hole pair (b) Direct recombination of an electron and a hole with the emission of a photon (c) An electron recombines with a hole via a recombination centre

If the semiconductor be irradiated with light of frequency ν, such that $h\nu > E_g$, where E_g is the *forbidden energy gap*, electron-hole pair is created. When the light is cut off, the excess carriers disappear through electron-hole recombination and the material returns to thermal equilibrium. The recombination takes place when a conduction band electron is bound again with a hole in the valence band [Fig. 9.16 (b)]. The process is called *band to band recombination* which results with the emission of photon such that energy $h\nu = E_g$.

In some semiconductors (e.g., Ge and Si) the electron-hole recombination takes place via a recombination centre which is an energy level at about the centre of the gap. An electron from the conduction band and a hole from the valence band jump to the recombination centre and *annihilate* [Fig. 9.16 (c)].

Minority carrier lifetime: Under the thermal equilibrium condition, the densities of electrons (n_0) and holes (p_0) in a semiconductor remain constant in spite of the generation and recombination process taking place constantly. In other words, on an average, for every electron excited to conduction band, one electron recombines or drops to the valence band.

Let G and R denote respectively the rate of generation and recombination at equilibrium.

Then

$$R = G \text{ or, } R - G = 0.$$

For band to band recombination, R is proportional to the product of n_0 and p_0, i.e.,

$$R \propto n_0 p_0 \text{ or, } R = r n_0 p_0,$$

where r is the constant of proportionality.

$$\therefore r n_0 p_0 - G = 0. \tag{9.74}$$

Let Δp and Δn be the excess holes and electrons generated when the sample is disturbed from the equilibrium condition.

The net rate of excess hole carrier recombination at any time is given by

$$-\frac{d}{dt}(\Delta p) = r(n_0 + \Delta n)(p_0 + \Delta p) - rn_0 p_0 \text{ (assuming } r \text{ to be constant).} \tag{9.75}$$

$$\therefore \frac{d}{dt}(\Delta p) = -r(n_0 + p_0)\Delta p, \ (\because \Delta n = \Delta p).$$

Neglecting the second order terms,

$$\frac{d}{dt}(\Delta p) = -\frac{\Delta p}{\tau}, \tag{9.76}$$

where $\tau = \frac{1}{r(n_0+p_0)}$ is defined as the *minority carrier lifetime* τ.

$$\text{or, } \frac{d}{dp}(\Delta p) = -\frac{dt}{\tau}. \tag{9.77}$$

Integrating,

$$\Delta p = Ae^{-\frac{t}{\tau}}. \tag{9.78}$$

At $t = 0$, $\Delta p = (\Delta p)_0$,

$$\therefore (\Delta p)_0 = A.$$

So

$$\Delta p = (\Delta p)_0 e^{-\frac{t}{\tau}}. \tag{9.79}$$

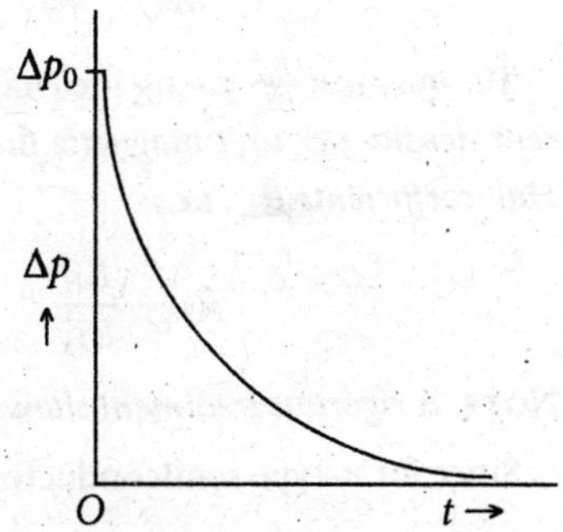

Fig. 9.17: Variation of Δp with t

The equation indicates Δp *(or* Δn*) disappears exponentially with time* as shown in Fig. 9.17.

From eqn (9.79), at $t = \tau$.

$$\Delta p = (\Delta p)_0 e^{-1} \text{ or, } \frac{\Delta p}{(\Delta p)_0} = \frac{1}{e}. \tag{9.80}$$

Thus, *the minority carrier lifetime is defined as the time taken for the excess charge carriers to reduce to* $\frac{1}{e}$ *times its initial value, once the source generating these excess charge is cut off.*

9.12 Hall Effect

When a piece of conductor (metal or semiconductor) carrying a current is placed in a transverse magnetic field, an electric field is produced inside the conductor in a direction normal to both the current and magnetic field. This phenomenon is known as Hall effect and the generated voltage is known as Hall voltage.

Let an n-type semiconductor carrying a current flow consisting entirely of electrons be moving from right to left just opposite to the conventional flow of current. The semiconductor is placed in a magnetic field B in a direction perpendicular to the velocity (v) of the electrons. The *Lorentz force* acting on the electrons is $F = BeV$, which causes the electron to be deflected in the downward direction. As a result, the negative charges will be accumulated on the bottom face of the semiconductor slab. The top surface will develop positive charge. A potential difference will, therefore, be established between the top and bottom surface with the bottom surface negative. The potential difference causes a field E_H, called the *Hall field*, in the negative y-direction as shown in Fig. 9.18 and exerts a force eE_H on the electron in the upward y-direction.

In equilibrium condition the Lorentz force and the force due to E_H will balance each other when no further accumulation of electrons takes place and E_H attains a steady value.

In case of a p-type semiconductor, the charge carriers are holes and the bottom surface will be positively charged relative to the top surface due to the accumulation of holes.

At equilibrium, $eE_H = Bev$ or, $E_H = Bv$.

If J_x be the current density in the x-direction, then

$$J_x = nev,$$

where n is the concentration of the current carriers.

$$\therefore E_H = \frac{BJ_x}{ne} \text{ or, } \frac{E_H}{BJ_x} = \frac{1}{ne}. \qquad (9.81)$$

The quantity $\frac{E_H}{BJ_x}$ giving the *Hall field per unit current density per unit magnetic flux density is called Hall coefficients* R_H, i.e.,

$$R_H = \frac{E_H}{BJ_x}. \qquad (9.82)$$

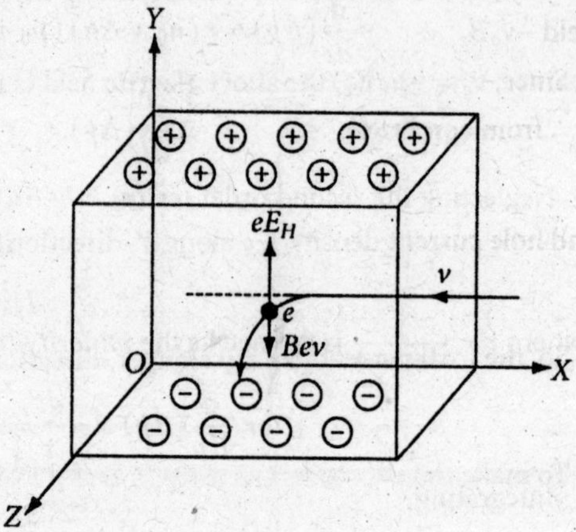

Fig. 9.18: Hall effect and motion of electrons in an n-type semiconductor

NOTE. *A rigorous treatment shows that* $R_H = \frac{r}{ne}$, *where r is a numerical constant.*

Since for n-type semiconductor the carriers are negatively charged electrons,

$$\therefore R_H = -\frac{1}{ne}. \qquad (9.83)$$

For p-type semiconductor the carriers are positively charged.

$$\therefore R_H = \frac{1}{pe}, \qquad (9.84)$$

where p is the density of holes. Eqns (9.83) and (9.84) show that *the Hall coefficients R_H are of opposite sign for n-type and p-type semiconductors.*

When the contribution of both the electrons and holes to the current are appreciable, Hall coefficient may be proved to be

$$R_H = \frac{p\mu_h^2 - n\mu_n^2}{e(n\mu_e + p\mu_h)}, \qquad (9.85)$$

since $\sigma = ne\mu_e + pe\mu_h$ and μ_e, μ_h are the mobilities of electrons and holes respectively.

To prove eqn (9.85), the electron and hole current densities along x-direction can be expressed as

$$\left.\begin{aligned} J_{xe} &= ne\mu_e E_x \\ J_{xh} &= pe\mu_h E_x \end{aligned}\right\}, \qquad (9.86)$$

since the magnitude of charge of both electron and hole is e. Then the total current density J_x is

$$J_x = J_{xe} + J_{xh} = e(n\mu_e + p\mu_h)E_x, \qquad (9.87)$$

the changes in μ_e and μ_h due to magnet resistance have been neglected.

Now the Lorentz force acting on the electron along the y-direction is

$$\vec{F} = e\left(\vec{v}_e \times \vec{B}\right) = ev_e B_z \hat{j}, \tag{9.88}$$

where v_e is the electron drift velocity along the y-direction. This Lorentz force $\vec{F}$ is equivalent to an electric field $-v_e B_z$.

Since, $v_e = -\mu_e E_x$, the above electric field is equal to $B_z \mu_e E_x$.

$\therefore$ from eqn (9.86),

$$J_{ye} = ne\mu_e\left(B_z\mu_e E_x\right) = ne\mu_e^2 B_z E_x \tag{9.89}$$

and hole current density J_{yh} along Y-direction is

$$J_{yh} = pe\mu_p^2 B_z E_x. \tag{9.90}$$

So, the total current density along Y-direction is

$$J_y = J_{ye} + J_{yh} = e\left(n\mu_e^2 - p\mu_h^2\right) B_z E_x. \tag{9.91}$$

To make the net current density J_y zero a Hall field E_y is required.

$$\therefore \frac{E_y}{E_x} = \frac{J_y}{J_x} \text{ or, } J_x E_y = J_y E_x \text{ or, } e\left(n\mu_e + p\mu_h\right) E_y = -e\left(n\mu_e^2 - p\mu_h^2\right) B_z E_x$$

$$\text{or, } E_y = \frac{p\mu_h^2 - n\mu_e^2}{n\mu_e + p\mu_h} B_z E_x. \tag{9.92}$$

The Hall coefficient R_H is given by

$$R_H = \frac{E_y}{B_z J_x} = \frac{p\mu_h^2 - n\mu_e^2}{e(n\mu_e + p\mu_h)^2}. \tag{9.93}$$

9.12.1 Measurement of the Hall Coefficient

A rectangular wafer of semiconductor having thickness t and width w is taken, through which a current I is passed along the x-direction. The wafer is placed between the pole pieces of an electromagnet so that the magnetic field B acts along the z-direction as shown in Fig. 9.19.

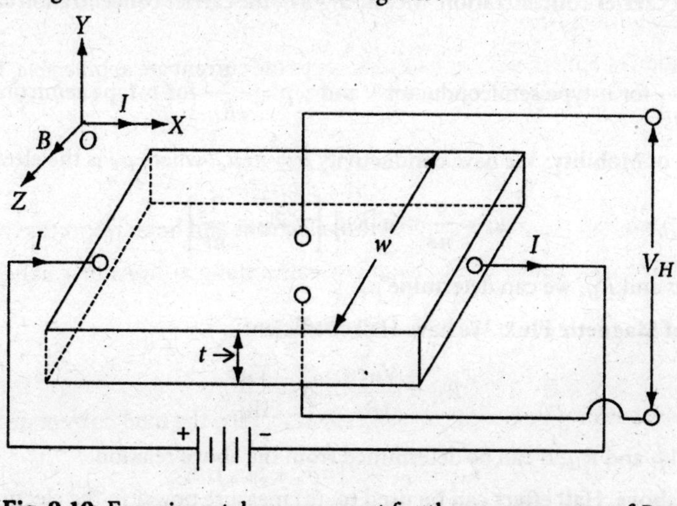

Fig. 9.19: Experimental arrangement for the measurement of R_H

The Hall voltage V_H is measured by placing two probes at the top and bottom faces of the sample.

If E_H be the Hall field, then $E_H = \frac{V_H}{t}$.
Hence, $R_H = \frac{E_H}{BJ_x} = \frac{V_H}{BJ_x w}$. $\therefore$ $V_H = R_H J_x Bw$, [*w is the width of the wafer*].
Now $J_x = \frac{I}{\alpha} = \frac{I}{wt}$ [t = thickness of the wafer].

$$\therefore V_H = R_H B J_x w = R_H I B w / wt \text{ or, } R_H = \frac{V_H t}{IB}. \tag{9.94}$$

If B is in Wb/m^2 and V_H is in volt, then R_H is in m^3/coulomb. The polarity of V_H will be *opposite for n-type and p-type materials.* It has been assumed in the above analysis that *the charge carriers move in the energy bands with a steady drift velocity without any attractive forces.*

Hall angle (θ_H) : The Hall angle θ_H is defined as

$$\tan\theta_H = \frac{E_H}{E_x}.$$

We have from eqn (9.81),

$$E_H = \frac{J_x B}{ne} = \frac{(nev_x)B}{ne} \; [\because J_x = nev_x]$$

$$\text{or, } E_H = \frac{v_x B E_x}{E_x}.$$

$$\therefore \frac{E_H}{E_x} = \tan\theta = \frac{v_x B}{E_x} = \mu B \; [\because v_x = \mu E_x] \tag{9.95}$$

μ = mobility of the charge carriers.

9.12.2 Applications of Hall Effect

(i) Determination of semiconductor type: Since R_H *is positive for p-type* and *negative for n-type semiconductor,* it can be used to determine the type of semiconductor.

(ii) Measurement of carrier concentration: Measuring R_H, the carrier concentration can be obtained from the relations:

$$n = \frac{1}{e|R_H|} \text{ for } n\text{-type semiconductor} \quad \text{and} \quad p = \frac{1}{eR_H} \text{ for } p\text{-type semiconductor.}$$

(iii) Determination of Mobility: We have conductivity $\sigma = ne\mu_e$, where μ_e is the electron mobility, then

$$\mu_e = \frac{\sigma}{ne} = \sigma|R_H| \left[\because R_H = \frac{1}{ne}\right].$$

Thus, measuring σ and R_H, we can determine μ_e.

(iv) Measurement of Magnetic Flux: We have Hall coefficient

$$R_H = \frac{V_H t}{IB}, \therefore B = \frac{V_H t}{IR_H}.$$

Thus, measuring V_H and R_H, B can be determined from the above relation.

(v) Apart from the above, Hall effect can be used to: (a) measure power in the electromagnetic wave and (b) determine Hall effect multiplier.

9.13 Cyclotron Resonance

If a constant magnetic field $\vec{B}$ be applied to a semiconductor, then the electrons in the conduction band and the holes in the valence band describe spiral (circular) orbits about $\vec{B}$. The angular frequency of rotation is given by

$$m^* \frac{v^2}{r} = \pm eVB \qquad (9.96)$$

$$\text{or, } \omega_c = \pm \frac{eB}{m^*}. \qquad (9.97)$$

The ± sign is due to the fact that the electrons and holes rotate in opposite sense. .

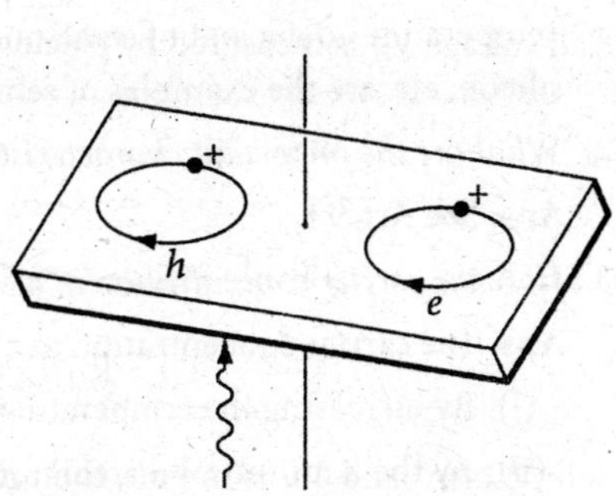

Fig. 9.20: Cyclotron motion of *e* and *h* in magnetic field *B*

If now a radio frequency electric field is applied perpendicular to static magnetic field $\vec{B}$, then absorption of energy may take place under resonance condition,

$$\frac{d\vec{v}}{dt} = \frac{e}{m^*} \vec{v} \times \vec{B}. \qquad (9.98)$$

The time period of rotation is

$$T = \oint dt = \int \frac{dr}{v} \qquad (9.99)$$

$$\vec{v} = \frac{\partial E}{\partial p} = \frac{1}{\hbar}\frac{\partial E}{\partial k} = \frac{1}{\hbar} = \frac{dE}{dk_\perp}, \qquad (9.100)$$

where $dk_\perp$ is normal distance in $\vec{k}$-space projected on a plane perpendicular to $\vec{B}$.

But

$$\hbar \frac{dk}{dt} = e\vec{v} \times \vec{B} \text{ or, } \vec{k} = \frac{e}{\hbar} \vec{r} \times \vec{B}. \qquad (9.101)$$

$$\therefore T = \oint \frac{dr}{v} = \hbar \oint \frac{d\vec{r} \cdot dk_\perp}{dE} = \frac{\hbar^2}{eB} \frac{|d\vec{k} \times d\vec{k}_\perp|}{dE} \qquad (9.102)$$

$$\text{or, } T = \frac{\hbar^2}{eB}\frac{dA}{dE} \quad [A = \text{area of the orbit in } k\text{-space}] \qquad (9.103)$$

$$\text{or, } \omega_c = \frac{eB}{m^*} = \frac{2\pi}{T} = \frac{2\pi eB}{\hbar^2}\frac{dE}{dA} \qquad (9.104)$$

$$\text{or, } m^* = \frac{\hbar^2}{2\pi}\frac{dA}{dE}. \qquad (9.105)$$

Thus, measuring ω_c, in different direction of $\vec{B}$, we can measure the effective mass m^* in different directions. This technique is also used to distinguish between electron and hole. This is a standard technique. The cyclotron resonance experiments are performed at low temperatures (~ $4k$) using relatively pure samples by which the collision time τ is lengthened and $\omega_c\tau$ becomes larger.

9.14 Answers of Some Important Questions

9.1. *What do you understand by the term semiconductor?*

Ans. Semiconductors are materials having electrical conductivity intermediate between conductors and insulators. The resistivity of semiconductors varies between 10^{-5} to $10^{-4}\,\Omega$-m, whereas those for conductors and insulators vary from 10^{-8} to $10^{-6}\,\Omega$-m and 10^7 to $10^8\,\Omega$-m respectively. The band gap for semiconductors varies from 0.2 to 2.5 eV as compared to 6 eV for insulators (diamond).

Temperature coefficient of semiconductors is negative, whereas for metals it is positive. Germanium, silicon, etc. are the examples of semiconductors which are widely used.

9.2. *What are the important characteristics of semiconductors?*

Ans. See Art. 9.1.

9.3. *How the carrier concentration in a semiconductor can be increased?*

Ans. The carrier concentration in a semiconductor can be increased by following two processes:

(i) By increasing the temperature of the semiconductor and

(ii) By the addition of a small amount of suitable impurity atoms to the semiconductor during growing.

9.4. *Explain the concept of hole.*

Ans. If an electron of an atom is removed, then a vacancy (or *hole*) is left behind which has a tendency to catch an electron due to its electropositive nature. By thermal excitation or by adding *dopants* to the semiconductor, an electron can be removed from the completely filled valence band of a semiconductor leaving behind a vacancy called *hole* in that band. The hole in the valence behaves as a free positive electronic charge.

When a vacancy is created in the valence band, a lower energy electron in the valence band can be excited to the higher energy vacant level by applying an electric field when the hole moves down to the lower energy level. The movement of the hole in the electric field is just opposite to the direction of movement of the electron suggesting that the hole is a *positively charged particle.* The energy of a hole is increased when it moves from higher energy level to the lower energy level. Thus, the movement of a hole in an applied electric field causes electrical conduction in the material. The conductivity in semiconductor arises due to the movement of electrons in conduction band and holes in valence band.

9.5. *What is doping in semiconductors?*

Ans. *Doping* in semiconductors is a process in which a small quantity of suitable impurity atom is added to the material during crystal growth. The dopants are added in the ratio 1 in 10^6 to 10^8 atoms which increases the electrical conductivity to a large extent. Doping does not change the structure of the semiconductor as the impurity atoms replace the regular atoms in the crystal. When a crystal is doped, the semiconductor is known as *extrinsic semiconductor.* Extrinsic semiconductors are of the following two types:

(a) *n-type* extrinsic semiconductor:

Example: Ge or Si doped with pentavalent As or Sb.

(b) *p-type* extrinsic semiconductor:

Example: Ge or Si doped with trivalent Al or Ga.

9.6. *How the charge neutrality is maintained in an intrinsic semiconductor even on the addition of neutral pentavalent and trivalent atoms to it?*

Ans. Intrinsic semiconductors like Ge or Si are tetravalent elements having four electrons in the outermost orbit. If a pentavalent element like As or Sb having five electrons in the valence band is added to it, the four electrons of As are used to form covalent bonds with four electrons of Ge and the fifth electron is free making the semiconductor n-type.

If a trivalent impurity like Al is added to Ge, three electrons of it form covalent bond with three valence orbit electrons of Al. But the fourth electron cannot find an electron to form a covalent bond. This gives rise to a hole which behaves like a positive electron and the material becomes a p-type semiconductor.

No charge is added or taken out of the material and hence, the overall charge neutrality is maintained. Thus, the semiconductor even on the addition of pentavalent or trivalent impurity remains electrically neutral.

9.7. *Discuss the effect of temperature on metal and semiconductor.*

Ans. The effect of temperature creates an important distinction between metals (conductors) and semiconductors. In case of a metal, the conductivity decreases with the rise of temperature as the temperature coefficient of resistance is *positive.* But the temperature coefficient of resistance for semiconductor is *negative* which causes an increase of conductivity with temperature. Further, as the temperature increases, more and more *phonons* are excited which scatter electrons and reduce their mobility. In semiconductor, the decrease in mobility is more compensated by the increased number of charge carriers. The electrons in the valence orbit get sufficient energy with the increase of temperature and jump to the conduction band, thereby increasing the electrical conductivity of the semiconductor.

9.8. *What do you mean by effective mass of an electron? How does it vary with the wave vector $\vec{k}$?*

Ans. The wave-particle motion of an electron is different from that of a free electron as the electron interacts with the *periodic potential* of the crystal lattice. For an electron with energy E and mass m_0 moving in free space,

$$m_0 = \frac{\hbar^2 k^2}{2E}.$$

But the *effective mass* of the electron moving in a periodic potential is given by

$$m^* = \frac{1}{\hbar^2}\frac{d^2E}{dk^2}.$$

Hence, the dynamical behaviour of the electron is the same as that of a free electron of mass m^*. Hence, using the effective mass m^*, we can describe the motion of the electron in the crystal lattice by applying *principles of classical physics.*

Variation of m^* with $\vec{k}$: See last part of Art. 9.3.2.

9.9. *Why does a semiconductor behave as an insulator at* 0 K?

Ans. The electrons at 0 K in the valence band do not have sufficient energy to be excited from valence band to the conduction band across the forbidden gap of energy (E_g). The conduction band under such condition is empty and electrical conduction ceases to take place. That is why at 0 K, a semiconductor behaves as an insulator.

9.10. *Why do all insulators behave as semiconductor at temperatures sufficiently higher than* 0 K?

Ans. Insulators are characterized by large band gap (e.g., $E_g \sim 6$ V for Diamond) and at ~ 0 K the electrons in the filled valence band do not get sufficient energy (i.e., $kT \gg E_g$) to jump across the forbidden gap to the conduction band. But if temperature is sufficiently large so that $kT \gg E_g$, then the jump of the electrons from valence band to the conduction band is possible and electrical conduction can take place. Hence, all insulators may behave as semiconductor, if $T \gg 0$ K.

9.11. *Discuss the variation of position of Fermi level in extrinsic semiconductor with temperature.*

Ans. At low temperature, the Fermi level E_f lies half-way between E_j and E_a in n-type semiconductor, whereas in p-type semiconductor E_f lies half-way between E_a and E_v. At high temperature, E_f coincides with the *Fermi level* of intrinsic semiconductor $\left(E_f\right)$.

9.12. *State the law of mass action in semiconductor. Discuss the effect of addition of impurity atoms to the material.*

Ans. The law of mass action states that, for a given semiconductor, the product of electron and hole concentrations is constant at a given temperature and is equal to the square of the intrinsic carrier concentration.

If n and p represent respectively the concentration of electrons and holes, then at a particular temperature, $np = n_i^2$, where n_i is the concentration of the intrinsic semiconductor.

If the addition of impurity atoms increases the value of n, then there will be a corresponding decrease in p, so that np = constant.

9.13. *Write down Einstein's equation and show that*

$$\frac{D_e}{D_p} = \frac{\mu_e}{\mu_h},$$

where the symbols have their usual meanings.

Ans. If μ_e be the electron mobility of an electron of charge e at temperature T, then its *diffusion length* is given by $D_e = \mu_e\, kT/e$, where k is the *Boltzmann constant.*

For p-type semiconductor, $D_p = \mu_h\, kT/e$.

$$\therefore \frac{D_e}{D_h} = \frac{\mu_e\, kT/e}{\mu_h kT/e} = \mu_e/\mu_h,$$

where μ_h is the mobility of holes.

9.14. *Write down the expression for Hall coefficient. What is its unit in* SI *system?*

Ans. If a wafer of semiconductor of thickness t be placed in a magnetic field B along Z-direction carrying a current I along y-direction, then the *Hall coefficient* is given by

$$R_H = \frac{V_H t}{IB},$$

where V_H is the *Hall voltage* developed between the faces perpendicular to y-axis.

The unit of R_H in SI system is m^3/coulomb.

9.15. *Mention some important applications of Hall effect.*

Ans. See Art. 9.12.2.

9.15 Solved Problems

9.1. *Calculate the resistivity of* Ge *at room temperature from the following data: Intrinsic carrier density of* Ge *at RT* = $2.37 \times 10^{19}/m^3$. *The electron mobility* μ_e = $0.38\, m^2/V/s$. *The hole mobility* μ_h = $0.18\, m^2/V/s$.

Solution: The intrinsic conductivity of Ge is given by $\sigma_i = n_i e\,(\mu_e + \mu_h)$ and resistivity $\rho = 1/\sigma$.

$$\therefore \sigma_i = \left(2.37 \times 10^{19}\right) \times \left(1.6 \times 10^{-19}\right) \times (0.38 + 0.18) = 2.1235\Omega\text{-m}.$$

$$\therefore \rho = \frac{1}{\sigma_i} = \frac{1}{2.1235} = \mathbf{0.471\Omega\text{-m}.}$$

9.2. *Calculate the intrinsic carrier density and conductivity at 300 K in Germanium. Given: Atomic weight of Ge = 72.6, Density of Ge = 5400 kg/m³, Mobility of electrons μ_e = 0.4 m²/V/s, Mobility of holes μ_h = 0.2 m²/V/s and Band Gap of Ge E_g = 0.7 eV.*

Solution: We have the intrinsic carrier density

$$n_i = 2\left[\frac{2\pi mkT}{h^2}\right]^{3/2}\left[e^{-E_g/2kT}\right] = 4.83 \times 10^{21} \times (300)^{3/2} \times e^{-\frac{0.7}{0.052}} = 3.5 \times 10^{19}/\text{m}^3.$$

Conductivity $\sigma = n_i e\,(\mu_e + \mu_h) = \left(3.5 \times 10^{19}\right) \times \left(1.6 \times 10^{-19}\right) \times 0.6 = \mathbf{3.36\ \Omega^{-1}m^{-1}}$.

9.3. *Calculate the density of donor atoms that has to be added to an intrinsic germanium semiconductor of conductivity 4 mho/cm. Given: the electron mobility of the n-type semiconductor is 3900 cm²/V/s.*

Solution: If N_d be the donor concentration, then assuming the concentration of donor atoms equal to the concentration of free electron n, we have

$$n = N_d = \frac{\sigma_n}{e\mu_e} = \frac{4}{\left(1.6 \times 10^{-19}\right) \times 3900} = \mathbf{0.64 \times 10^{16}/cm^3}.$$

9.4. *The resistivity of an intrinsic semiconductor is 4.5 Ω-m at 20 °C and 2.0 Ω-m at 32 °C. Find the energy band gap.*

Solution: The plot of $\ln \rho_i$ against $\frac{1}{T}$ is shown in the adjoining figure.

The band gap E_g is given by

$$E_g = 2k\left(\frac{dy}{dx}\right).$$

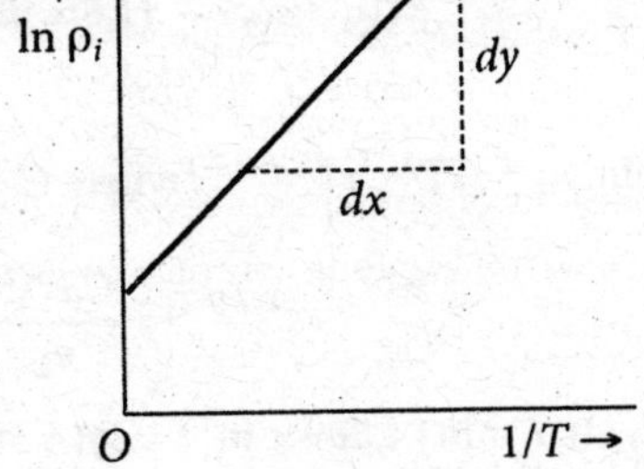

Fig. 9.21

Now

$$dy = \ln \rho_2 - \ln \rho_1 = 0.6532 - 0.3010 = 0.3522$$

$$dx = \frac{1}{T_2} - \frac{1}{T_1} = \frac{1}{(273 + 20)} - \frac{1}{(273 + 32)}$$

$$= \frac{1}{293} - \frac{1}{305} = 0.134 \times 10^{-3}.$$

$$\therefore \frac{dy}{dx} = \frac{0.3522}{0.134 \times 10^{-3}} = 2.628 \times 10^3.$$

Hence, $E_g = 2 \times \left(8.616 \times 10^{-5}\right) \times \left(2.628 \times 10^3\right) = \mathbf{0.453\ eV}$.

9.5. *Show that the ratio of maximum resistivity to intrinsic resistivity is expressed as*

$$\frac{\rho_m}{\rho_i} = \frac{(\mu_e + \mu_h)}{2\sqrt{\mu_e \mu_h}},$$

where μ_e and μ_h denote the mobilities of electron and hole respectively.

Solution: We have the intrinsic resistivity $\sigma_i = n_i e(\mu_e + \mu_h)$ and conductivity at any other dropping level is $\sigma = e(n_e\mu_e + n_h\mu_h)$, the different symbols having their usual meanings.

$$\text{or, } \sigma = e\left[n_e\mu_e + \frac{n_i^2\mu_h}{n_e}\right] \quad [\because n_e n_h = n_i^2].$$

The condition for the resistivity to be maximum the conductivity is to be minimum, i.e.,

$$\frac{d\sigma}{dn_e} = 0 \text{ or, } e\left[\mu_e - \frac{n_i^2\mu_h}{n_e^2}\right] = 0 \text{ or, } \mu_e = \frac{n_i^2\mu_h}{n_e^2} \text{ or, } n_e = n_i\sqrt{(\mu_h/\mu_e)} \text{ and } n_h = n_i\sqrt{\frac{\mu_e}{\mu_h}}.$$

$$\therefore \sigma_m = e\left[n_i(\mu_e\mu_h)^{1/2} + \frac{n_i^2\mu_h}{n_i}\sqrt{\frac{\mu_e}{\mu_h}}\right] = n_i e\left[(\mu_e\mu_h)^{1/2} + (\mu_e\mu_h)^{1/2}\right] = n_i e\left[2(\mu_e\mu_h)^{1/2}\right].$$

$$\therefore \frac{\sigma_i}{\sigma_m} = \frac{\rho_m}{\rho_i} = \frac{n_i e(\mu_e + \mu_h)}{2n_i e(\mu_e\mu_h)^{1/2}} \text{ or, } \frac{\rho_m}{\rho_i} = \frac{(\mu_e + \mu_h)}{2\sqrt{\mu_e\mu_h}} \text{ (Proved).}$$

9.6. *Holes were injected into a bar of germanium. 12 microsecond later the density decreased to* $10^{18}/\text{m}^3$. *The lifetime of hole in the bar is 4.5 μs. Compute the density of holes injected initially.*

Solution: If $(\Delta p)_0$ and (Δp) be the change in density of holes at $t = 0$ and $t = t$, then

$$\Delta p = (\Delta p)_0 e^{-t/\tau} \text{ or, } 10^{18} = (\Delta p)_0 e^{-\frac{12}{4.5}} = (\Delta p)_0 e^{-2.7}.$$

$$\therefore (\Delta p)_0 = 10^{18} \times e^{+2.7} = \mathbf{1.4 \times 10^{19}}.$$

9.7. *Compute the forbidden energy gap in germanium from the following data: Mobilities of electrons and holes in intrinsic germanium at 300 K are respectively* $0.36\,\text{m}^2/\text{V/s}$ *and* $0.17\,\text{m}^2/\text{V/s}$; *Resistivity of Germanium at 300 K = 2.12 Ω-m.*

Solution: We have the intrinsic conductivity of Ge, $\sigma_i = n_i e(\mu_e + \mu_h)$.

$$\therefore n_i = \frac{\sigma_i}{e(\mu_e + \mu_h)} = \frac{0.472}{(1.6\times10^{-19})\times(0.17+0.36)} \quad \left[\because \sigma = \frac{1}{\rho} = \frac{1}{2.12} = 0.472\,\Omega^{-1}\,\text{m}^{-1}\right]$$

or, $n_i = 0.5566 \times 10^{19}/\text{m}^3$.

Again, $n_i = CT^{3/2}e^{-E_g/2kT}$, where $C = 2\left[\frac{2\pi mk}{h^2}\right] = 4.83 \times 10^{21}$ and $2kT = 0.052$ at $T = 300$ K.

$$\therefore e^{E_g/2kT} = \frac{CT^{3/2}}{n_i} = \frac{(4.83\times10^{21})\times(300)^{3/2}}{0.5566\times10^{19}} = 4.509\times10^6.$$

$\therefore E_g/2kT = \ln(4.509\times10^6) = \ln(4.509) + 6\ln 10 = 15.322$ or, $E_g = 2kT \times 15.322 = 0.052 \times 15.322$.

$\therefore E_g = \mathbf{0.797\ eV}$.

9.8. *The electron and hole mobilities of* Si *sample are* $0.135\,\text{m}^2/\text{V/s}$ *and* $0.048\,\text{m}^2/\text{V/s}$ *respectively. Determine the conductivity of intrinsic* Si *at 300 K if the intrinsic carrier concentration is* 1.5×10^{16} atoms/m³. *The sample is then doped with* 10^{23} atoms/m³ *of phosphorus. Determine the equilibrium hole concentration, conductivity and position of the Fermi level relative to the intrinsic level.*

Solution: The conductivity of Si is given by $\sigma = n_i e(\mu_e + \mu_h)$.

Here $n_i = 1.5\times10^{16}$ atoms/m³, $\mu_e = 0.135\,\text{m}^2/\text{V/s}$, $\mu_h = 0.048\,\text{m}^2/\text{V/s}$.

$\therefore \sigma = (1.5\times10^{16})\times(1.6\times10^{-19})\times(0.135+0.048) = 1.5\times1.6\times0.183\times10^{-3} = \mathbf{0.4392\times10^{-3}\,\Omega\text{-m}}.$

Now the sample is doped with 10^{23} phosphorus atoms/m³. $\therefore N_d = 10^{23}/\text{m}^3$.

So, the hole concentration is

$$p = \frac{n_i^2}{N_d} = \frac{(1.5 \times 10^{16})^2}{10^{23}} = \mathbf{1.5 \times 10^9/m^3}.$$

Conductivity $\sigma_{ex} = N_d e \mu_e = 10^{23} \times (1.6 \times 10^{-19}) \times 0.135 = \mathbf{2.16 \times 10^3/\Omega\text{-m}}.$

$$E_f = \frac{E_g}{2} + \frac{3kT}{4}\left(\ln\frac{m_h^*}{m_e^*}\right) = \frac{E_g}{2} + \frac{3 \times (1.38 \times 10^{-28}) \times 300}{4 \times 1.6 \times 10^{-19}}\left(\ln\frac{\mu_e}{\mu_h}\right)\left[\because \mu \propto \frac{1}{m}\right]$$

$$= \frac{E_g}{2} + 0.01935 \times \ln\left(\frac{0.135}{0.048}\right) = \frac{E_g}{2} + 0.01935 \times 1.034 = \frac{E_g}{2} + 0.02.$$

$\therefore$ the Fermi level will be **0.02 eV** above the intrinsic level.

9.9. *The intrinsic concentration of a semiconductor is* $3.1 \times 10^{13}/\text{cm}^3$ *at room temperature. It is doped with* 10^{14} *donor* atoms/cm^3. *Assuming that the donors are all ionized, calculate the electron and hole concentrations and the percentage increase in conductivity of the doped material over the intrinsic values. Given: electron mobility* = 3800 cm^2/V/s *and hole mobility* = 1800 cm^2/V/s.

Solution: If n, p and N_d be the concentrations of electron, hole and donor, then from the condition of charge neutrality $n - p = N_d$.

Also, if n_i be the intrinsic concentration, then $np = n_i^2$.

$$\therefore n - \frac{n_i^2}{n} = N_d \text{ or, } n^2 - nN_d - n_i^2 = 0.$$

$$\therefore n = \frac{N_d \pm \sqrt{N_d^2 + 4n_i^2}}{2} \text{ or, } n = N_d + \sqrt{\left(\frac{N_d}{2}\right)^2 + n_i^2}.$$

Substituting

$$n_i = 3.1 \times 10^{13}/\text{cm}^3$$

$$N_d = 10^{14}/\text{cm}^3,$$

we get

$$n = \mathbf{1.09 \times 10^{14}/cm^3}.$$

$$\therefore p = \frac{n_i^2}{n} = \frac{(3.1 \times 10^{13})^2}{1.09 \times 10^{14}} = \mathbf{0.88 \times 10^{13}/cm^3}.$$

Now the total conductivity,

$$\sigma = e(n\mu_e + p\mu_h) = (1.6 \times 10^{-19})\left[(1.09 \times 10^{14}) \times 3800 + (0.88 \times 10^{13}) \times 1800\right]$$

$$= 0.068\ \Omega^{-1}\,\text{cm}^{-1}.$$

The intrinsic conductivity

$$\sigma_i = en_i(\mu_e + \mu_h) = (1.6 \times 10^{-19}) \times (3.1 \times 10^{13})\,[3800 + 1800] = 0.028\ \Omega^{-1}\,\text{cm}^{-1}.$$

$\therefore$ % increase in σ over the intrinsic value is

$$\frac{\sigma - \sigma_i}{\sigma} \times 100 = \frac{0.068 - 0.028}{0.028} \times 100 = \mathbf{143\,\%}.$$

9.10. *A semiconducting crystal* 12 mm *long,* 5 mm *wide and* 1 mm *thick has a magnetic flux density of* 0.5 Wb/m^2 *applied from front to back perpendicular to the largest faces. When a current of* 20 mA *flows lengthwise through the specimen, the voltage measured across its width is found to be* 37 μV. *Find the Hall coefficient of the conductor.*

Solution: If J_x, E_y and B_z be respectively the current density along x-direction, electric field along y-direction and magnetic field along z-direction, then Hall coefficient,

$$R_H = \frac{E_y}{J_x B_z} = \frac{1}{ne},$$

where n and e are the charge concentration and electronic charge respectively.

$$\text{or, } R_H = \frac{V_y}{bJ_x B_z} \quad \left[\because E_y = \frac{V_y}{b}\right]$$

$$J_x = \frac{I_x}{bt} \quad [b = \text{breadth and } t = \text{thickness of the sample}].$$

$$\therefore R_H = \frac{V_y\, bt}{bI_x\, B_z} = \frac{V_y t}{I_x B_z}.$$

Here

$$V_y = 37 \times 10^{-16}\ \text{V}$$

$$t = 10^{-3}\ \text{mm}$$

$$B_z = 0.5\ \text{Wb/m}^2$$

$$I_x = 20 \times 10^{-3}\ \text{A}.$$

$$\therefore R_H = \frac{(37 \times 10^{-6}) \times (10^{-3})}{(20 \times 10^{-3}) \times (0.5)} = \mathbf{3.7 \times 10^{-6}\ m^3/C}.$$

9.11. *A current of* 50 A *is established in a slab of copper* 0.5 cm *thick and* 2 cm *wide. The slab is placed in a magnetic field* $B = 1.5$ T. *The magnetic field is perpendicular to the plain of the slab and to the current. The free electron concentration in copper is* 8.48×10^{28} electrons/m^3. *What is is the magnitude of Hall voltage across the width of the slab?*

Solution: The Hall coefficient is given by

$$R_H = \frac{1}{ne} \text{ and } V_H = \frac{R_H IB}{t}.$$

$$\therefore V_H = \frac{IB}{net} \text{ is the Hall voltage.}$$

Here

current, $I = 50$ A

Magnetic field, $B = 1.5$ T

Concentration of electrons, $n = 8.4 \times 10^{28}/\text{m}^3$.

Thickness of the sample, $t = 0.5\ \text{cm} = 0.5 \times 10^{-2}\ \vec{m}$.

$$\therefore V_H = \frac{50 \times 1.5}{(8.4 \times 10^{28}) \times (1.6 \times 10^{-19}) \times (0.5 \times 10^{-2})}$$

$$= 11.161 \times 10^{-7}\ \text{V} = 1.116 \times 10^{-6}\ \text{V}$$

$$= \mathbf{1.116\ \mu V}.$$

9.12. *A current of* 1 mA *flows through an n-type* Ge *strip of* 1 mm *thick and* 1 mm *wide placed in a magnetic field B. If the Hall voltage produced inside the strip be* 1 mV, *what is the value of B?* [*Given*: $R_H = 10^{-2}$ m^3/coulomb.]

Solution: If w be the width and t be the thickness of the Ge strip, then Hall coefficient,

$$R_H = \frac{V_H w}{IB} \text{ or, } B = \frac{V_H w}{IR_H} = \frac{(1\times10^{-3})\times(1\times10^{-3})}{(10^{-2})\times(1\times10^{-3})} = \mathbf{0.1\,T}.$$

EXERCISE

9(A): Objective (MCQ) Type Questions

Select the correct answer.

9.1. The temperature coefficient of semiconductor is

(a) positive (b) negative (c) 0 (d) none of these.

9.2. The band gap of semiconductor is of the order of

(a) 0.1 eV (b) 10 eV (c) 50 eV (d) 100 eV.

9.3. When a semiconductor is doped with suitable impurity, its conductivity

(a) remains constant (b) decreases (c) increases (d) becomes zero.

9.4. In n-type Ge (or Si) semiconductor, the impurity atom is

(a) trivalent (b) tetravalent (c) pentavalent (d) hexavalent.

9.5. In p-type Ge (or Si) semiconductor, the impurity atom is

(a) trivalent (b) tetravalent (c) pentavalent (d) none of these.

9.6. In intrinsic semiconductor, the carrier concentration varies as

(a) $1/T$ (b) $T^{3/2}$ (c) T^2 (d) T.

9.7. At 0 K pure Ge is an

(a) extrinsic semiconductor (b) insulator (c) intrinsic semiconductor (d) a superconductor.

9.8. The electrical conductivity in a pure semiconductor, when an external electric field is applied, is due to

(a) holes (b) electrons (c) ions (d) electrons and holes.

9.9. The majority carriers in p-type Ge are

(a) free electrons (b) holes (c) ions (d) conduction electrons.

9.10. In a p-type semiconductor, the concentration of holes is proportional to the square root of the concentration of

(a) donor (b) acceptor atom (c) intrinsic impurity atom (d) none of these.

9.11. The value of $\Delta T/T$ necessary to double the number of carriers (or conductivity) is

(a) $\frac{kT}{E_g}$ (b) $kT\ln 2$ (c) $\frac{2kT}{E_g}\ln T$ (d) $\frac{E_g}{2kT}$.

9.12. At 0 K the Fermi level in an intrinsic semiconductor lies at

(a) the centre of the top of the valence band and bottom of the conduction band

(b) the centre of the donor level and intrinsic Fermi level

(c) the centre of donor level and bottom of the conduction band

(d) the centre of the acceptor level and top of the valence band.

9.13. At 0 K, the resistivity of pure Si is 300 Ω-m. If $\mu_e = 0.14$ and $\mu_h = 0.05\,\text{m}^2\,\text{V}^{-1}\,\text{s}^{-1}$, then intrinsic carrier concentration per cubic metre is

(a) zero (b) 10^{17} (c) 2.19×10^{17} (d) 1.095×10^{17}.

9.14. The density of a charge carrier in a pure semiconductor is proportional to

(a) $e^{-E_g/kT}$ (b) $e^{-2E_g/kT}$ (c) e^{-E_g/kT^2} (d) $e^{-E_g/2kT}$.

9.15. At low temperatures and higher concentration of acceptor impurities, the density of holes (p) varies as

(a) N_a (b) $\frac{1}{N_a}$ (c) $\sqrt{N_a}$ (d) N_a^2,

N_a being the concentration of donor atoms.

9.16. With increase of temperature the Fermi level E_f in n-type semiconductor moves towards the

(a) intrinsic Fermi level (E_i) (b) bottom of conduction band (E_c)

(c) top of valence band E_v (d) none of these.

9.17. With the increase of temperature the Fermi level E_f in a p-type semiconductor moves gradually towards the

(a) intrinsic Fermi level E_i (b) bottom of the conduction band E_c

(c) top of the valence band (E_v) (d) none of these.

9.18. The average distance an electron diffuse before recombination is related to the diffusion coefficient and lifetime by the equation

(a) $L_n = \sqrt{D_e \tau_n}$ (b) $L_n = \frac{1}{D_e \tau_n}$ (c) $L_n = D_e \tau_n$ (d) $l_n = \frac{D_e}{\tau_n}$.

9.19. The mean lifetime of the carrier injected into a semiconductor is the time in which the carrier value falls to

(a) 10 % of its initial value (b) 27 % of its initial value

(c) 37 % of its initial value (d) 17 % of its initial value.

9.20. Hall coefficient of a n-type semiconductor is

(a) $R_H = \frac{1}{ne}$ (b) $R_H = -\frac{1}{ne}$ (c) $_H = \frac{r}{ne}$ (d) $R_H = rne$.

9.21. When both electrons and holes to the current is appreciable, Hall coefficient is given by

(a) $R_H = \frac{r}{\pm ne}$ (b) $R_H = \frac{p\mu_p^2 - n\mu_n^2}{e(n\mu_e + p\mu_p)}$ (c) $R_H = \frac{1}{ne}$ (d) $R_H = \frac{e(n\mu_e + p\mu_p)}{p\mu_p^2 - n\mu_n^2}$.

9.22. Which quantity cannot be calculated by measuring Hall coefficient?

(a) mobility of charge carriers (b) energy band gap of the semiconductor

(c) type of semiconductor (d) concentration of charge carriers.

9.23. The effective mass of the electron is given by

(a) $m^* = \frac{\hbar^2 E}{2k^2}$ (b) $m^* = \frac{2E}{\hbar^2 k^2}$ (c) $m^* = \frac{\hbar^2 k^2}{2E}$ (d) $m^* = \frac{2\hbar^2 k^2}{E}$.

9.24. The law of mass action is

(a) $np = n_i^2$ (b) $nn_i^2 = p$ (c) $\frac{n}{p} = n_i^2$ (d) $n = n_i^2 p$.

9.25. If $m_p^* > m_e^*$, then

(a) $\mu_e < \mu_p$ (b) $\mu_e = \mu_p$ (c) $\mu_e > \mu_p$ (d) $\mu_e = \frac{1}{\mu_p}$.

ANSWERS

9.1 (b) **9.2** (a) **9.3** (c) **9.4** (c) **9.5** (a) **9.6** (b) **9.7** (b) **9.8** (d) **9.9** (b) **9.10** (b) **9.11** (c) **9.12** (a) **9.13** (d) **9.14** (d) **9.15** (c) **9.16** (a) **9.17** (a) **9.18** (a) **9.19** (c) **9.20** (b) **9.21** (b) **9.22** (b) **9.23** (c) **9.24** (a) **9.25** (c).

9(B): Short Answer Type Questions

9.1. What is a semiconductor? How does it differ from metal and insulator?

9.2. What is intrinsic semiconductor? Show that in an intrinsic semiconductor both the electrons and holes contribute towards electrical conductivity. [LU 1996; PU 1997]

9.3. Distinguish between intrinsic and extrinsic semiconductors.

9.4. What is the concept of *hole*? How does it differ from a free electron?

9.5. What do you understand by *donor* and *acceptor*? Show their energy level in the energy band diagram. Give two examples of each.

9.6. What is effective mass of electron. Show that it depends on $\frac{d^2E}{dk^2}$ and is a function of the wave vector k.

9.7. Can a semiconductor containing both n-type and p-type impurities behave as an intrinsic semiconductor? Explain with reason.

9.8. Derive an expression for electron concentration in an intrinsic semiconductor.

9.9. Find out an expression for the hole concentration in an intrinsic semiconductor.

9.10. Discuss the effects of doping As and Al atoms separately in Ge crystal.

9.11. What is law of mass action? What is its significance?

9.12. Discuss the variation of Fermi level in n-type and p-type semiconductors with temperature.

9.13. Define (a) mean lifetime (b) diffusion length and write the relation between the two.

9.14. Describe the basic principle of Hall effect in semiconductors. [CU(H) 2004, '01, '96]

9.15. Show that the Hall coefficient for an intrinsic semiconductor is given by

$$R_H = -\frac{1}{ne}\left(\frac{\mu_e - \mu_h}{\mu_e + \mu_h}\right),$$

where μ_e, μ_h are the mobilities of electrons and holes and n is the intrinsic carrier concentration.

9.16. Show that at a given temperature a semiconductor has a maximum conductivity σ_m given by

$$\sigma_m = 2en_i\sqrt{\mu_e - \mu_h},$$

where μ_e, μ_h are the mobilities of electrons and holes and n_i is the intrinsic carrier concentration.

9.17. What are mobility and conductivity? Obtain an expression for the conductivity of a doped semiconductor. [NBU(H) 2008]

9.18. What is Hall effect? Find an expression for Hall voltage and Hall coefficient. State the importance of this effect. [NBU(H) 2009]

9.19. Find an expression for (i) mobility of electrons and (ii) electrical conductivity on the basis of free electron theory of metals. [VU(H) 2009]

9.20. Qualitatively discuss the formation of energy bands for electrons in solids, and distinguish between metals and semiconductors (Use necessary diagram). [VU(H) 2002, '01]

9.21. What is Hall effect? Obtain an expression for Hall coefficient in a metal when the carriers are only electrons. Why is the Hall coefficient positive for some metals? [VU(H) 2007, '06, '05]

9.22. Show that the effective number of carriers is maximum when the band is half-filled and zero when filled or empty (Use effective mass). [CU(H) 2009]

9.23. Show that when an electron moves through a crystal, the effective mass of the electron can be expressed as

$$m^* = \frac{\hbar^2}{\left(\frac{d^2E}{dk^2}\right)},$$

where the symbols have their usual significances. **[CU(H) 2007]**

9.24. (a) Find the difference between the top and bottom of the energy band.

(b) Obtain an expression for the effective mass m^* of the electron as a function of k. Sketch m^* vs. k.

[CU(H) 2003]

9.25. Show graphically the variation of energy with the wave vector k for the first allowed band in a solid. Hence, explain the variation in sign of the effective mass for an electron in the band. What is a hole?

[CU(H) 2000]

9.26. Derive an expression for Hall coefficient. What information do you obtain from Hall effect experiment?

[VU(H) 1995]

9(C): Numerical Problems

9.1. A wafer of Ge crystal of area 1 cm^2 and thickness 0.3 mm is placed in a potential difference of 2 volts. Calculate the current produced.

[Given, concentration of free electrons in Ge = 2×10^{19}/m^3, $\mu_e = 0.36\,\text{m}^2\text{V}^{-1}\text{s}^{-1}$ and $\mu_h = 0.17\,\text{m}^2\text{V}^{-1}\text{s}^{-1}$] [*Ans.* 1.13 A]

9.2. Estimate the temperature at which diamond has the same conductivity as Si at 27 °C.

[Given, band gaps of diamond and silicon are 5.4 eV and 1.1 eV respectively). [*Ans.* 1200 °C]

9.3. For intrinsic germanium at 300 K, calculate the resistivity of the sample from the given data: $n_i = 2.4 \times 10^{19}$/m^3; $\mu_e = 0.39\,\text{m}^2\text{V}^{-1}\text{s}^{-1}$, $\mu_h = 0.19\,\text{m}^2\text{V}^{-1}\text{s}^{-1}$. [*Ans.* 0.449 Ω-m]

9.4. Find the resistance of an intrinsic Ge rod 1 cm long, 1 mm wide and 1 mm thick at 300 K.

[Given, $n_i = 2.5 \times 10^{19}$/m^3; $\mu_e = 0.39\,\text{m}^2\text{V}^{-1}\text{s}^{-1}$ and $\mu_h = 0.19\,\text{m}^2\text{V}^{-1}\text{s}^{-1}$ at 300 K] [*Ans.* 4.31×10^3 Ω]

9.5. If the effective mass of holes in a material is 4 times that of electrons, find the temperature at which the Fermi level will be shifted by 10 % from the middle of the forbidden energy gap.

[Given, E_g = 1 eV.] [*Ans.* 1116 K]

9.6. In an n-type semiconductor, the Fermi level lies 0.5 eV below the conduction band. If the concentration of donor atoms is trippled, find the new position of the Fermi level [kT = 0.03 eV].

[*Ans.* 0.48 below the conduction band]

9.7. The Fermi energy of lithium at 0 K is 4.72 eV. Calculate the number of conduction electrons per unit volume in lithium. **[CU(H) 1997]**

[*Ans.* 4.66×10^{22}/cm^3]

9.8. An n-type Ge sample is 3.5 mm thick and is placed in a magnetic field B = 0.5 T. If the donor density is 10^{21}/m^3 and the current density J = 500 A/m^2, find the Hall voltage. [*Ans.* 5.47 mV]

9.9. The R_H of a sample is $3.66 \times 10^{-4}\,\text{m}^3\text{C}^{-1}$ and its resistivity is 8.93×10^{-3} Ω-m. Find μ and n.

[*Ans.* $\mu = 0.041\,\text{m}^2\text{V}^{-1}\text{s}^{-1}$, $n = 1.708 \times 10^{22}$ per m^3.]

9.10. Find the ratio of diffusion coefficient of electrons and holes of a silicon single crystal at 27 °C. [Mobilities of electrons and holes at 27 °C are 0.17 $m^2V^{-1}s^{-1}$ and 0.025 $m^2V^{-1}s^{-1}$ respectively]

[*Ans.* $D_e/D_p = 0.68$]

9.11. If the band gap in Ge is 0.47 eV, compare the intrinsic conductivities of Ge at 27 °C and 47 °C.

[*Ans.* 2.27]

9.12. Calculate the position of Fermi level E_f and the conductivity at 300 K for Ge crystal containing 5×10^{22} arsenic atoms/m^3. Also calculate conductivity if the mobility of electron is 0.39 $m^2V^{-1}s^{-1}$.

[*Ans.* $E_f = 0.16$ eV below E_c, $\sigma = 3120\ \Omega^{-1}\,m^{-1}$]

9.13. Determine the average band gap of Si if the conductivity of intrinsic Si is 4.17×10^{-5} and 4×10^{-4} $(\Omega\text{-m})^{-1}$ at 0 °C and 27 °C respectively. [*Ans.* 1.11 eV]

9.14. Calculate the band gap of an intrinsic semiconductor having resistivity 4.5 Ω-m at 20 °C and 2 Ω-m at 32 °C. [*Ans.* 0.96 eV]

9.15. A copper strip 2.0 cm wide and 1.0 mm thick is placed in a magnetic field of 1.5 Wb/m^2 perpendicular to the strip. If a current of 200 A is set up along the length of the strip, what Hall voltage appears across the strip? [The free electron concentration of copper is 8.48×10^{28} electrons/m^3.] **[BU(H) 2003]**

[*Hints.* $V_H = \frac{IB}{net}$, $I = 200$ A, $B = 1.5$ Wb/m^2, $n = 8.48 \times 10^{28}$ electrons/m^3, $t = 1$ mm $= 10^{-3}$ m, $e = 1.6 \times 10^{-19}$ C.

$$\therefore V_H = \frac{200 \times 1.5}{(4.48 \times 10^{28}) \times (1.6 \times 10^{-19}\,c) \times (0.5 \times 10^{-2})} = 7.81 \times 10^{-6}\text{ V} = 7.81\,\mu\text{V}.]$$

9.16. In a metallic sodium there are 2.6×10^{22} free electrons per cm^3. Compute the Fermi energy (in eV) in sodium. Find out E_f. **[BU(H) 2006]**

[*Hints.* Fermi energy, $E_f = \frac{h^2}{8m}\left(\frac{3n}{\pi}\right)^{2/3}$.

Put $h = 6.625 \times 10^{-34}$ J-s; $m = 9.1 \times 10^{-31}$ kg; $n = 2.6 \times 10^{22}$ electron/cm^3 $= 2.6 \times 10^{28}$ electron/m^3.]

9.17. In a Hall effect experiment on Zn, a potential of 4.5 μV is developed across a foil of thickness 0.02 mm when a current of 1.5 A is passed in a direction perpendicular to a magnetic field 2.0 T. Calculate (i) Hall coefficient of Zn and (ii) Electron density. **[Pune U 2013; Mumbai U 2013]**

9(D): Broad Answer Type Questions

9.1. (a) What is intrinsic semiconductor? Explain why impure semiconductors exhibits intrinsic behaviour at high temperatures.

(b) Deduce expressions for the number density of electrons in the conduction band and holes in the valence band of an intrinsic semiconductor. Show that the product of these two quantities does not depend on E_f. Write down expression for conductivity of an intrinsic semiconductor.

(c) Deduce an expression for the Hall coefficient of a solid with one type of carrier only. **[BU(H) 1994]**

9.2. (a) What are *p*- and *n*-type semiconductors? Where is the Fermi level in the two cases compared to that in an intrinsic semiconductor?

(b) Write the expression for the electrical conductivity in an intrinsic semiconductor and discuss its temperature dependence.

(c) Define Hall constant. Derive its expression for a solid with only holes as carriers. **[BU(H) 1996]**

9.3. (a) What are intrinsic and extrinsic semiconductors? How can the band gap of an intrinsic semiconductors be determined from conductivity measurement?

(b) Define Hall constant. Find its expression for a solid with only one electron as carrier. [BU(H) 2000]

(c) Discuss briefly how energy bands are formed in solid. [BU(H) 1997]

9.4. (a) Deduce expression for the concentration of electrons in conduction band and holes in valence band of an intrinsic semiconductor at temperature T. Show that the product np is independent of the Fermi level. What is the implication of this result?

[You may assume the formula for the density of states of free electrons namely,

$$g(E) = \frac{1}{2\pi^2}\left(\frac{2m}{\hbar^2}\right)^{3/2} E^{1/2}].$$

(b) On the basis of deductions made in part (a) of the question, find the location of the Fermi level in an intrinsic semiconductor. Draw neat diagrams showing the location of the Fermi level in n- and p-type semiconductors.

(c) How will you experimentally determine whether a given semiconductor is n- or p-type? [BU(H) 1999]

9.5. (a) Show that the product of concentration of electrons in the conduction band and that of holes in the valence band of a semiconductor depends upon the energy gap.

(b) Deduce an expression for the density of holes in the valence band of a p-type semiconductor. Show the variation of position of the Fermi level. [BU(H) 2002]

9.6. (a) What is Hall effect?

(b) Find Hall coefficient in a metal, where the carriers are only electrons. Why is the Hall coefficient positive in some metals? [CU(H) 1992]

(c) Mention some important applications of Hall effect.

9.7. (a) What is the concept of effective mass? On what factors does it depend?

(b) State and establish the law of mass action in semiconductors.

(c) Discuss the effect of temperature and impurity towards electrical conductivity of semiconductors.

9.8. (a) Describe the drift and diffusion currents in a semiconductor. Derive their expressions.

(b) Deduce Einstein relation.

(c) Explain with necessary theory the concept of minority carrier lifetime in the extrinsic semiconductor.

9.9. (a) What do you understand by intrinsic and extrinsic semiconductors?

(b) Derive the expression:

$$R = \frac{1}{|e|}\frac{(n_h\mu_h^2 - n_e\mu_e^2)}{(n_h\mu_h + n_e\mu_e)^2}$$

for Hall constant when n and μ are carrier concentration and mobility of charge carriers. Subscripts e and h refer to electron and hole. [RU(H) 1986]

9.10. (a) Explain the Hall effect. Derive an expression for the Hall coefficient of semiconductor on two-band model of carriers in semiconductors.

(b) Describe with theory an experimental arrangement for the measurement of Hall coefficient.

(c) How the magnetic field can be determined by Hall effect experiment?

9.11. What is meant by Hall effect? Discuss the experiment with suitable diagram for finding Hall coefficient and Hall voltage. Mention two applications of Hall effect. **[BU(H) 2008]**

9.12. What do you mean by the effective mass of an electron? What can it mean physically for an electron to have zero or negative effective mass? Explain how it gives rise to the concept of hole conduction in solids. **[VU(H) 1995]**

9.13. Explain Hall effect and derive an expression for Hall coefficient for a semiconductor based on two-band model of charge carriers. **[Loyala College, Chennai 2016]**

9.14. (a) What is effective mass of electron. Find out its expression. Give its physical significance.

(b) What do you mean by direct and indirect semiconductors? Discuss charge neutrality.

(c) Obtain Einstein's equation,

$$D_e = \frac{\mu_e kT}{e},$$

the different letters having their usual meanings.

Chapter 10

Semiconductor Devices

10.1 Introduction

The *p-type* and *n-type* extrinsic semiconductors are prepared by adding suitable impurities to intrinsic semiconductors. The p-type and n-type extrinsic semiconductors show drastic increase in their conductivity. Though the use of individual p-type or n-type semiconductors find little use, their combination forming a pn-junction, having non-linear characteristics, is used in a larger group of semiconductor devices. Simply by joining a p-type material to an n-type a pn-junction may be obtained, but it will have no useful electrical properties as in this case the lattice structure remains discontinuous.

10.2 Techniques for Formation of *pn*-Junction

There are *four* principal techniques for the formation of pn-junction: (i) *Grown junction technique* (ii) *Alloy junction technique* (iii) *Epitaxial technique* and (iv) *Diffusion technique.*

In grown junction technique, a *seed* is dipped and then withdrawn from a crucible of molten semiconductor material (e.g., Ge or Si) and then the molten material is allowed to cool and solidify on the seed. More and more material solidifies and accumulates forming a crystal. For producing a p-type, one kind of impurity is first added to the melt and when enough is grown, impurities of other type is added to form n-type. The pn-junction so formed is called *linearly graded junction* where the impurity density varies linearly with distance from the junction. In diffusion technique, semiconductor materials available in the market are purified by *multiple zone refining technique.* The n-type silicon is cut into thin slices called *wafers.* An aluminium film is placed on a wafer of n-type silicon and the combination is placed in an oven at 600 °C. After several minutes, the aluminium film melts and diffuses into the surface of the material, thereby producing a p-type semiconductor immediately below the surface. The pn-junction formed in this way is called a *step-graded junction* where the acceptor or the donor impurity in the semiconductor is constant up to the junction.

10.3 Formation of Depletion Region in *pn*-Junction

We know that in an n-type semiconductor the concentration of free electrons is greater than that of holes whereas in a p-type semiconductor the concentration of holes is greater than that of electrons. For this reason, when a pn-junction is formed, some electrons from n-side diffuse into the p-region and some holes diffuse into the n-region across the pn-junction.

The diffusion currents for electron and hole diffusion are given by

$$I_{nd} = qAD_n \frac{dn}{dx} \text{ and } I_{pd} = -qAD_p \frac{dp}{dx}$$

respectively, where d, n and p stand for diffusion, electron and hole respectively. The negative sign in the expression for I_{pd} represents that the holes move in the direction opposite to that of electrons.

The free electrons after crossing the junction from n-side recombine with holes in the p-side. In a similar way, the holes diffusing from p-side recombine with electrons in the n-side. As a result, we are left with ionised acceptor atoms on p-side and ionised donor atoms on the n-side near the junction. These ions in the neighbourhood of the junction are known as *uncovered charges. The region near the junction which is depleted of mobile charges and contains only immobile ionic charges is called 'depletion region'* (or 'the space charge region' or 'the transition region'), the thickness being 0.5 μm. The variation of concentration of acceptors and donors, space charges, and uncovered charges are shown in Figs 10.1(b), (c) and (d) respectively.

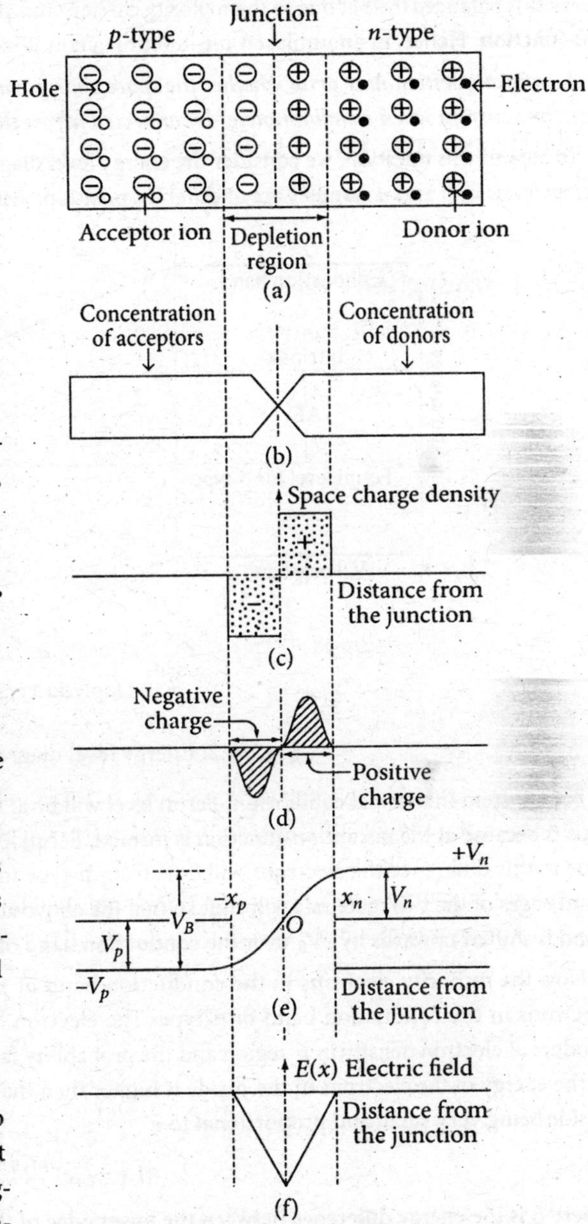

Fig. 10.1: (a) pn-junction, (b) Concentration of acceptors and donors, (c) Space charge density (d) Uncovered charges, (e) Potential barrier, (f) Variation of electric field (E).

The *immobile* positive and negative charges, which are uncovered, produce an electric field across the junction directed from the n-side to the p-side and is termed as *barrier field* (F_B). This field opposes the movement of the charge carriers across the junction. For a particular value of barrier field, equilibrium is established when the movement of carriers across the junction is stopped. Due to the barrier field (F_B) across the junction some expenditure of energy (E_B) is required for the transfer of electrons from the n-side to the p-side. The resulting barrier potential V_B is then given by

$$E_B = eV_B \text{ or, } V_B = \frac{E_B}{e}.$$

For the transfer of a hole from the p-region to n-region across the junction the same amount of energy (E_B) will also be required. The magnitude of E_B depends on the width of the uncovered region.

In a pn-junction, the p-region contains minority electrons and the n-region, the minority holes. Under *unbiased condition,* the barrier field helps the minority carriers to cross the junction easily giving rise to a current flow. The majority carriers gain energy from the thermal sources and constitutes a current flow by crossing the junction in the opposite direction to that due to minority carriers. In equilibrium the currents due to minority carriers is balanced by that due to the majority carriers and yield zero current due to electrons in an unbiased *pn*-junction. Hence, in an unbiased *pn*-junction a condition of *dynamical equilibrium* is established.

A natural question may arise whether the space charge equivalent battery (PD $V_B = E_B/e$) will be able to send any current when a milliammeter is connected across the pn-junction.

To answer this question, we consider the energy level diagram of a *pn*-junction as shown in Fig. 10.2. The *Fermi levels* in *p* and *n* samples are at different positions with reference to the band edges.

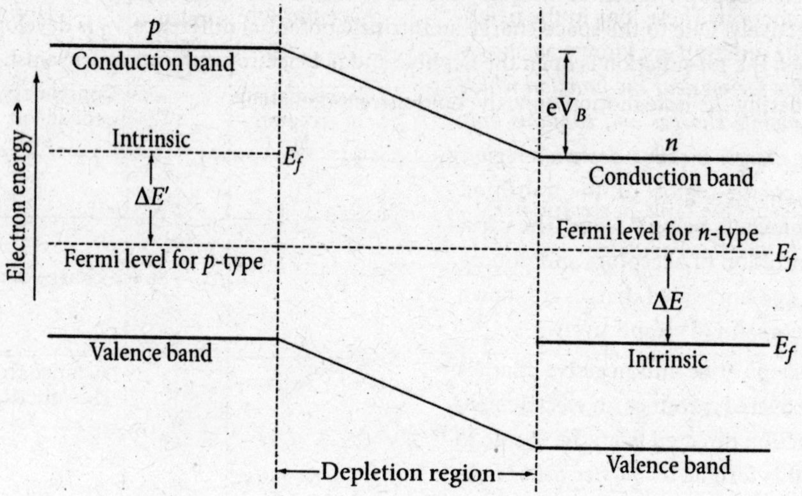

Fig. 10.2: Energy level diagram for a *pn*-junction

For a system in thermal equilibrium Fermi level will be at the same level when the *pn*-junction is formed. This is because at the instant *pn*-junction is formed, Fermi level in *p*-type is at a higher level than in *n*-type. Due to this difference the electrons will flow from *p*-type to *n*-type until they attain a common level. The band edges of the two materials will shift so that the above alignment be possible. In *p*-type the conduction band is shifted upwards by eV_B over the conduction band of *n*-type, V_B being the *potential barrier.*

Now the minority electrons in the conduction band of *p*-type are at higher energy than the majority electrons in the conduction band of *n*-type. The electron current I_e from *p* to *n* is proportional to the product of electron density in *p*-region and the probability factor. The probability factor in this case is unity as the energy of the electrons in the *p*-side is higher than that in the *n*-side. The number of electrons in the *p*-side being very small and proportional to $e^{-E/kT}$,

$$|\vec{I}_e| = Ce^{-E/kT}, \tag{10.1}$$

where *E* is the energy difference between the lower edge of the conduction band of *p*-type and Fermi level.

Again, the electron density in *n*-type is proportional to

$$e^{(E_f-E_c)/kT} = e^{-(E_c-E_f)/kT}.$$

The probability that the electron will have sufficient energy to overcome the energy $\left[E-(E_c-E_f)\right]$ is proportional to $\exp\left[-(E-E_c+E_f)/kT\right]$. Hence, the electron current from *n*-side to *p*-side is

$$\left|\overleftarrow{I_e}\right| = Ce^{[-(E_c-E_f)/kT]} \cdot e^{[-(E-E_c+E_f)/kT]} = Ce^{-E/kT}. \tag{10.2}$$

From eqns (10.1) and (10.2),

$$\left|\vec{I}_e\right| = \left|\overleftarrow{I}_e\right|. \tag{10.3}$$

The same argument holds good for holes too. Hence, *we conclude that no current flow will occur across the ammeter due to barrier voltage V_B.*

10.4 The Potential Barrier Across *pn*-Junction

We consider an abrupt pn-junction whose p-side and n-side are uniformly doped with concentration N_a and N_n respectively. Due to the space charge an intrinsic potential difference V_0 is developed across the junction. When the pn-junction is open the net hole and net electron currents must vanish separately.

Now considering the hole motion only, the hole current density is

$$J_p = ep\mu_p E - eD_p \frac{dp}{dx}, \tag{10.4}$$

where E represents the built-in electric field.

According to *Einstein's equation,*

$$\frac{D_p}{\mu_p} = \frac{kT}{e}.$$

$\therefore$ from eqn (10.4),

$$\frac{dp}{p} = \frac{e}{kT} E\, dx.$$

$$\because E = -\frac{dV}{dx}, \therefore \frac{dp}{p} = -\frac{e}{kT} dV. \tag{10.5}$$

To have the potential rise from p- to n-region we integrate the above equation from $-x_p$ to $+x_n$.

$$\therefore \ln \frac{p_n}{p_p} = -\frac{e}{kT}\left(V_n + V_p\right) = -\frac{e}{kT} V_B, \tag{10.6}$$

where $V_B = V_n + V_p$ is the *contact potential difference* or *potential barrier.* p_p and p_n represent respectively the hole concentration in the p- and n-regions.

From eqn (10.6),

$$p_n = p_p e^{-eV_B/kT}. \tag{10.7}$$

As all the impurity atoms practically get ionised at moderate temperature, therefore, $p_p = p_{p_0} = N_a$. Similarly, in n-side, $n_n = n_{n_0} = N_d$.

Now by the *law of mass-action,* the concentration of minority holes in n-side is

$$p_n = \frac{n_i^2}{n_n} = \frac{n_i^2}{N_d}.$$

From eqn (10.6), the height of the potential barrier at the pn-junction is

$$V_B = \frac{kT}{e} \ln \frac{p_p}{p_n} = \frac{kT}{e} \ln \frac{N_a N_d}{n_i^2}. \tag{10.8}$$

Eqn (10.8) indicates that the value of V_B depends on the concentration of donor and acceptor impurity.

If the p- and n-regions of a typical Ge pn-junction be doped with 10^{22} impurity atoms/m^3, the value of V_B at room temperature becomes 0.3 V.

Is this barrier potential V_B directly measurable?

To answer the above question we consider a pn-junction short circuited through a voltmeter [Fig. 10.3]. To show the reading of the voltmeter a small current must flow through the circuit which would cause Joule heating in the circuit. This heat must be supplied by the pn-junction as there is no external source of emf and as such the pn-junction must cool simultaneously. But according to the *second law of thermodynamics,* it is not possible to derive work by cooling a body below its equilibrium temperature. Hence, no current is possible to flow through the circuit and the voltmeter will show zero reading. Actually, the V_B is balanced by the contact potential between the metal to semiconductor and the net potential becomes zero.

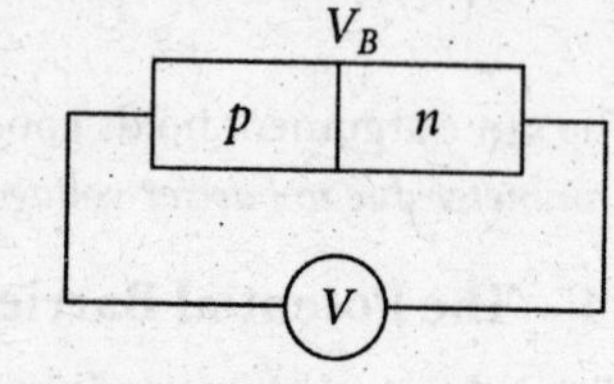

Fig. 10.3: A pn-junction is short-circuted through a voltmeter showing zero reading.

10.5 Width of Depletion Region of the pn-Junction

The width of the depletion region depends on the impurity concentration and the voltage. We consider a pn-junction [Fig. 10.1(c)], where the impurity concentration changes abruptly from acceptors to donors at $x = 0$. The space charge densities in the p- and n-regions are eN_a and eN_d respectively. If V be the potential at a distance x from the junction, then from *Poisson's equation in one dimension,*

$$\frac{d^2V}{dx^2} = \frac{eN_a}{\epsilon}, \text{ for } x < 0 \tag{10.9}$$

$$\text{and } \frac{d^2V}{dx^2} = -\frac{eN_d}{\epsilon}, \text{ for } x > 0, \tag{10.10}$$

where ϵ is the absolute permittivity of the medium.

Integrating eqns (10.9) and (10.10), we get

$$\frac{dV}{dx} = \frac{eN_a}{\epsilon}x + C_1 \tag{10.11}$$

$$\text{and } \frac{dV}{dx} = -\frac{eN_d}{\epsilon}x + C_2, \tag{10.12}$$

where C_1 and C_2 are constants of integration whose values may be found out from the boundary conditions $\frac{dV}{dx} = 0$ at $x = -x_p$ and $x = x_n$.

$$\therefore 0 = -\frac{eN_a}{\epsilon}x_p + C_1 \text{ or, } C_1 = \frac{eN_a}{\epsilon}x_p.$$

Similarly,

$$C_2 = \frac{eN_d}{\epsilon} = x_n.$$

$$\therefore \frac{dV}{dx} = \frac{eN_a}{\epsilon}(x_p + x) \text{ and } \frac{dV}{dx} = \frac{eN_d}{\epsilon}(x_n - x). \tag{10.13}$$

Integrating the above equations again, we get

$$V = \frac{eN_a}{\epsilon}\left(x_p x + \frac{x^2}{2}\right) + C_3$$

$$\text{and } V = \frac{eN_d}{\epsilon}\left(x_n x - \frac{x^2}{2}\right) + C_4.$$

To find the constants C_3 and C_4, we have the boundary conditions:

$$V = -V_p \text{ at } x = -x_p$$

$$\text{and } V = V_n \text{ at } x = x_n.$$

$$\therefore V = -V_p + \frac{eN_a}{2\epsilon}(x_p + x)^2 \tag{10.14}$$

$$\text{and } V = V_n - \frac{eN_d}{2\epsilon}(x_n - x)^2. \tag{10.15}$$

Since, at $x = 0$, $\frac{dV}{dx}$ in eqn (10.13) will have the same value. Therefore,

$$\frac{EN_a}{\epsilon}x_p = \frac{eN_d}{\epsilon}x_n.$$

$$\therefore eN_a x_p = eN_d x_n. \tag{10.16}$$

Eqn (10.16) represents the *condition of charge neutrality.*

From eqn (10.16), we have

$$\frac{x_p}{x_n} = \frac{N_d}{N_a} \text{ or, } \frac{x_p}{x_p + x_n} = \frac{N_d}{N_d + N_a}.$$

Similarly,

$$\frac{x_n}{x_n + x_p} = \frac{N_a}{N_a + N_d}. \tag{10.17}$$

Again, at $x = 0$, the values of V in eqns (10.14) and (10.15) are the same.

$$\therefore -V_p + \frac{eN_a}{2\epsilon}x_p^2 = V_n - \frac{eN_d}{2\epsilon}x_n^2$$

$$\text{or, } V_n + V_p = \frac{eN_a}{2\epsilon}x_p^2 + \frac{eN_d}{2\epsilon}x_n^2 = \frac{e}{2\epsilon}\left(\frac{N_a N_d}{N_a + N_d}\right)(x_p + x_n)^2. \tag{10.18}$$

Therefore, the *width of the depletion layer* is

$$W = x_p + x_n = \sqrt{\frac{2eV_B}{\epsilon}\left(\frac{N_a + N_d}{N_a N_d}\right)}, \tag{10.19}$$

where $V_B = V_n + V_p$.

Thus, knowing the values of N_a, N_d and V_B, we can determine the width (W) of the depletion region.

10.6 A *pn*-Junction under Forward and Reverse Bias

10.6.1 Forward Biased *pn*-Junction

We have seen that in an unbiased *pn*-junction a potential barrier (V_B) exists tending to oppose the movement of majority holes from the *p*-side to *n*-side and majority electrons from *n*-side to *p*-side. But if an external voltage V be applied so that the *p*-side becomes positive and the *n*-side negative, then the barrier voltage is reduced to ($V_B - V$), the junction is then said to be *forward-biased* [Fig. 10.4(a)]. This reduces the width of the depletion region. Due to the reduction of barrier potential a good number of holes from the *p*-side will be able to overcome this reduced potential barrier and diffuse into the *n*-side. Likewise, a good number of electrons will cross the reduced barrier and diffuse into the *p*-side. This flow of electrons and holes across the junction will constitute a large current, called the *forward current* (I_f). Forward current increases exponentially with the applied voltage V. As the applied voltage is increased further, the electrons and holes having lower energy will be able to cross the junction and the forward current will increase. The minority carriers flow down the potential barrier and constitute a small current in the opposite direction.

The salient features of the forward-biased *pn*-junction are as follows:

(i) the *pn*-junction offers *low forward resistance* to the flow of *forward current,*

(ii) the magnitude of the forward current depends upon the applied forward-bias,

(iii) at $V = V_B$, the potential barrier is eliminated altogether.

The mechanism of the current flow may be explained as follows:

The recombination of electron-hole near the *pn*-junction, leads to the breakage of a covalent bond near the positive terminal of the battery. This causes the liberation of an electron which enters the positive terminal. The new hole created moves towards the junction. Thus, for each electron in *n*-side combining with a hole in *p*-side, an electron enters the crystal from the negative terminal. The movement of the electrons towards the positive terminal and the holes towards the negative terminal is constant creating a high forward current. With the increase of the applied voltage (V), the electrons and holes of lower energy will be able to cross the junction leading to further increase in forward current.

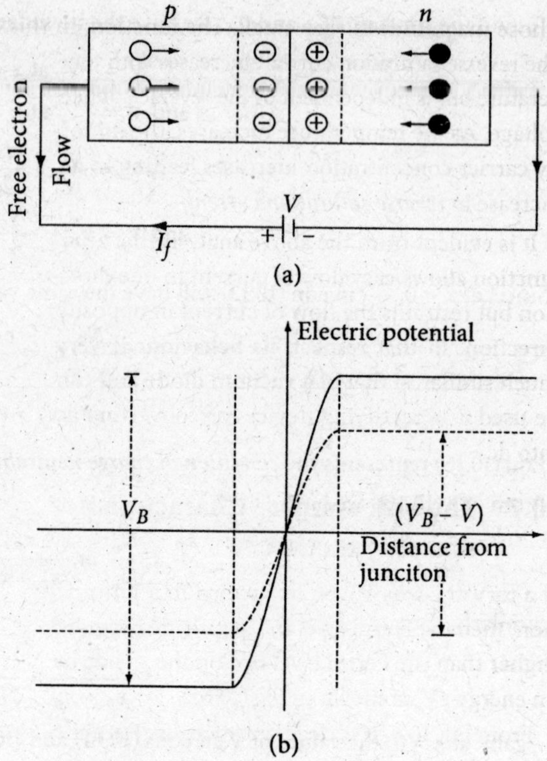

Fig. 10.4: (a) Forward biased *pn*-junction (b) Reduced potential barrier.

In forward-bias many electrons from the negative terminal of the battery enter the *n*-side. The life story of a single electron may be explained as follows:

(i) The electron after leaving the negative terminal enters the right end of the crystal.

(ii) This electron travels through the *n*-region as a free electron.

(iii) Near the junction the electron recombines and becomes a valence electron.

(iv) It travels as a valence electron through the *p*-region.

(v) It follows to the positive terminal after leaving the left end.

10.6.2 Reverse Bias

If the *p*-side is connected to the positive and the *n*-side to the negative terminals of the battery [Fig. 10.5(a)], then the *pn*-junction is said to be *reverse-biased.* The barrier height of the junction then becomes ($V_B + V$) which makes the current due to the majority carriers negligible. The width of the depletion region is increased. The minority carriers, however, move down the potential barrier and hence, their motion is not affected by the barrier height. They constitute *reverse saturation current* (I_s) in the reverse direction whose magnitude is very small. The junction, in this case, offers a *high resistance* to the flow of current. The reverse saturation current increases with temperature but is independent of the reverse applied voltage. As the temperature increases, the minority carrier concentration increases leading to an increase in *reverse saturation current.*

It is evident from the above analysis that a *pn*-junction allows easy flow of current in one direction but restrains the flow of current in opposite direction. In that respect, its behaviour is very much similar to that of a vacuum diode and can be used as a rectifying device for converting ac into dc.

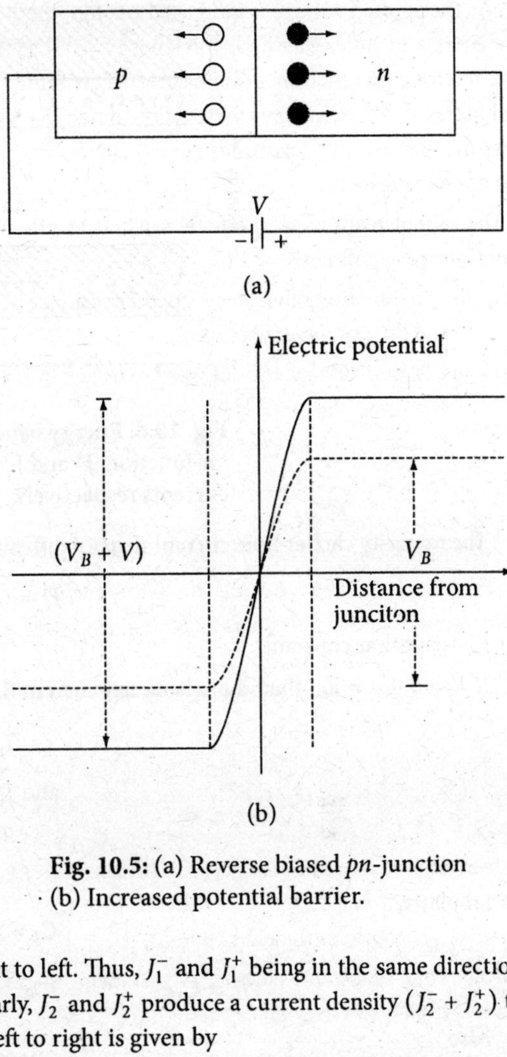

Fig. 10.5: (a) Reverse biased *pn*-junction (b) Increased potential barrier.

10.7 The Volt-Ampere Characteristics of a *pn*-Junction

If a forward-bias V volt be applied to a *pn*-junction, then the Fermi level E_{Fn} on n-side becomes higher than the Fermi level E_{Fp} on the p-side by an energy eV as shown in Fig (10.6).

From Fig. 10.6, it is clear that there are four possible current densities across the *pn*-junction, viz., $J_1^-, J_2^-, J_1^+, J_2^+$. The suffixes 1 and 2 to current density J represent respectively the majority and minority carriers. The current densities J_1^- and J_2^- due to the electrons in the conduction band produce conventional current in the directions as shown in the Fig. 10.6 as a flow of electron from left to right produces a conventional current from right to left. Thus, J_1^- and J_1^+ being in the same direction produce a current density $(J_1^- + J_1^+)$ to the right. Similarly, J_2^- and J_2^+ produce a current density $(J_2^- + J_2^+)$ to the left. Hence, the net current density flowing from left to right is given by

$$J = (J_1^- + J_1^+) - (J_2^- + J_2^+). \tag{10.20}$$

J_2^- is the minority carrier electron current density drifting from p- to n-side and is independent of the applied voltage V, if $V < V_B$. J_2^- is called the *saturation current density due to electrons.*

Similarly, J_2^+ represents the minority carrier hole current density from n-side to p-side and is also known as the *saturation current density due to holes.*

In the conduction band, J_1 is the majority carrier electron current diffusing from n-side to p-side. If J_1^- is not too large, the only restriction to the flow of this current is the barrier potential $V_T = V_B - V$. For V_T, the number of electrons which are available source of current is proportional to $e^{-eV_T/kT}$, i.e.,

$$J_1^- \propto e^{-eV_T/kT} \text{ or, } J_1^- = C_1 e^{-eV_T/kT}, \tag{10.21}$$

where C_1 is a constant.

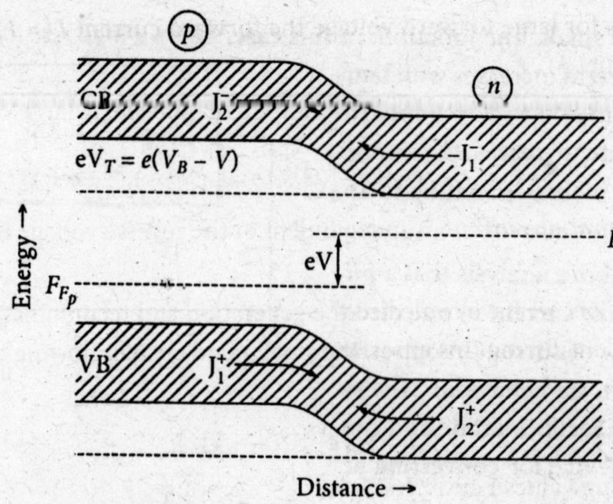

Fig. 10.6: Energy band diagram of a forward biased *pn*-junction. J^+ and J^- refer to hole and electron currents respectively.

The majority carrier hole current density diffusing from *p*- to *n*-side is given by

$$J_1^+ = C_2 e^{-eV_T/kT}. \tag{10.22}$$

C_2 is another constant.

If $V = 0$, $V_T = V_B$, then there is no net current flowing through the *pn*-junction, i.e.,

$$\left.\begin{aligned} J_1^- - J_2^- &= 0 \\ \text{and } J_1^+ - J_2^+ &= 0 \end{aligned}\right\}. \tag{10.23}$$

$$\therefore C_1 e^{-eV_B/kT} - J_2^- = 0 \text{ or, } C_1 = J_2^- e^{eV_B/kT}. \tag{10.24}$$

Similarly,

$$C_2 = J_2^+ e^{eV_B/kT}. \tag{10.25}$$

$$\therefore J_1^- = C_1 e^{-eV_T/kT} = J_2^- e^{eV_B/kT} e^{-eV_T/kT} = J_2^- e^{e(V_B - V_T)/kT}. \tag{10.26}$$

Also

$$J_1^+ = J_2^+ e^{e(V_B - V_T)/kT}. \tag{10.27}$$

Hence,

$$\begin{aligned} J &= (J_1^+ + J_1^-) - (J_2^+ + J_2^-) = (J_2^- + J_2^+)\left(e^{eV/kT} - 1\right) \\ \text{or, } J &= J_2\left(e^{eV/kT} - 1\right), \end{aligned} \tag{10.28}$$

where $J = (J_2^- + J_2^+)$.

If α be the cross-sectional area of the *pn*-junction, then the total forward current is

$$I = I_s\left(e^{eV/kT} - 1\right), \tag{10.29}$$

where $I_s = \alpha J$ and $I = \alpha J\left(e^{eV/kT} - 1\right)$. Eqn (10.29) is known as the *Shockley equation* or *rectifier equation.*

For large *forward voltage,*

$$e^{eV/kT} - 1 \simeq e^{eV/kT} \text{ and } I = I_s e^{eV/kT}. \tag{10.30}$$

Eqn (10.30) shows that, for large forward voltage the forward current $I\left(=I_f\right)$ increases exponentially with V, when T is constant.

For reverse-biased *pn*-junction, V is negative and if its magnitude is large compared to (kT/e), then from eqn (10.30),

$$I \simeq -I_s \left[e^{eV/kT} \ll 1\right], \tag{10.31}$$

where I_s is the *reverse saturation current,* is independent of the reverse voltage but increases with increase in temperature.

In the foregoing discussion, we have neglected the generation and recombination of electron and hole in the space charge region. Considering this effect, the small rated current flowing through the *pn*-junction is approximated as

$$I = I_s\left(e^{eV/\eta kT} - 1\right), \tag{10.32}$$

where η is a constant having values 1 and 2 for Ge and Si respectively.

The theoretical volt-ampere relation for a *pn*-junction is represented in Fig. 10.7.

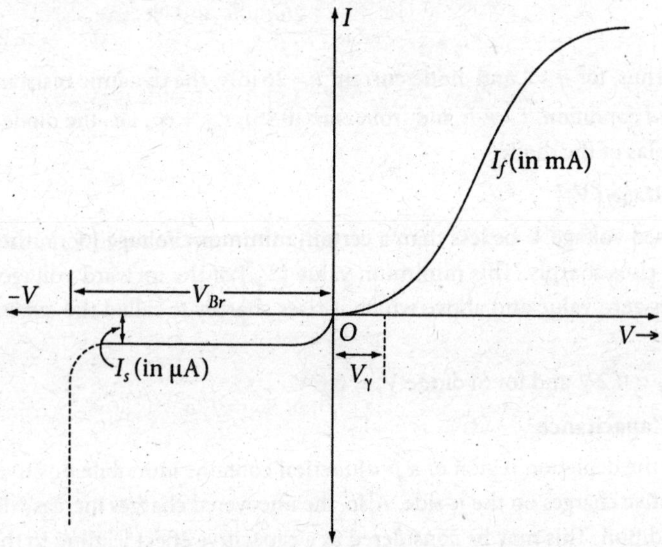

Fig.10.7: volt-ampere characteritics of a *pn*-junction

10.7.1 Junction Resistance of a *pn*-Junction: Static or dc Resistance (r_d)

It is defined as the ratio of the applied voltage V to the current I, i.e.,

$$r_d = \frac{V}{I}. \tag{10.33}$$

The dc resistance r_d is actually the resistance offered by the diode to the flow of *steady current* through it. As r_d varies inversely with I, a diode is a *non-linear element.*

10.7.2 Dynamic or Incremental Resistance (r_{ac})

When a *small signal voltage* is applied across a diode, the current (i) flowing through it varies about the dc operating current (I). The resistance offered to the flow of ac component of current is known as the *dynamic resistance* or *incremental resistance* of the diode.

The dynamic resistance of a diode is defined as *the reciprocal of the slope of the current-voltage characteristic at the operating point, i.e.,*

$$r_{ac} = \frac{\Delta V}{\Delta I} \tag{10.34}$$

If the applied voltage V be large, then the current through the diode is

$$I = I_s \left(e^{eV/\eta kT} - 1\right).$$

$$\therefore \frac{dI}{dV} = \frac{e}{\eta kT} I_s e^{eV/\eta kT} = \frac{e}{\eta kT}(I + I_s) \; \left(\because e^{eV/\eta kT} \gg 1\right).$$

$$\therefore r_{ac} = \frac{dV}{dI} = \frac{\eta kT}{e} \cdot \left(\frac{1}{I + I_s}\right). \tag{10.35}$$

In case of forward-bias, $I \gg I_0$,

$$\therefore r_{ac} = \frac{\eta kT}{e}\frac{1}{I}. \tag{10.36}$$

If $k = 1.38 \times 10^{-23}$ J/K, $e = 1.6 \times 10^{-19}$ C and $T = 300$ K, then from eqn (10.36), we get

$$r_{ac} = \frac{26\eta}{I}, \tag{10.37}$$

where I is in mA. Thus, for $\eta = 1$ and diode current $I = 26$ mA, the dynamic resistance $r_{ac} = 1\,\Omega$.

For *reverse-biased* condition, $I = -I_s$ and from eqn (10.35), $r_{ac} = \infty$, i.e., the diode resistance is very large in case of reverse-bias of the diode.

10.7.3 Cut-in Voltage (V_γ)

If the forward-biased voltage V be less than a certain minimum voltage (V_γ), then the current I is very small. If $V > V_\gamma$, I rises sharply. This minimum value (V_γ) of the forward voltage (V) below which the current approaches zero value and above which it rises sharply is called the *cut-in* or *off-set* or *threshold voltage* [Fig. 10.7].

For Ge diode, $V_\gamma \simeq 0.2$ V and for Si diode $V_\gamma \simeq 0.6$ V.

10.7.4 Junction Capacitance

We have seen that, the depletion region of a pn-junction contains immobile positive charges on the n-side and immobile negative charges on the p-side. Also, the uncovered charges increase due to the application of reverse-biased condition. This may be considered as a capacitive effect leading to the junction capacitance of a pn-junction diode.

In a pn-junction two types of capacitance effects are observed, viz., (i) *Transition capacitance* and (ii) *Diffusion capacitance.*

(i) *Transition capacitance* (C_T) : This is also known as the *depletion region capacitance, space charge region capacitance* or *barrier capacitance.*

We have mentioned that, the width of the depletion region increases with the increase of reverse-biased voltage. This increase in width of the depletion region causes an increase in the amount of uncovered charges leading to the capacitive effect of the pn-junction.

If dQ be the change in charge due to a change dV in voltage, then the *incremental capacitance* C_T is given by

$$C_T = \left|\frac{dQ}{dV}\right|. \tag{10.38}$$

The value of C_T is dependent on the nature of the pn-junction.

10.7.5 Abrupt or Step-graded *pn*-Junction

If there is an abrupt change from *p*-type to *n*-type impurity in a *pn*-junction, then it is called an *abrupt* or *step-graded junction.*

The nature of an abrupt *pn*-junction is shown in Fig. 10.8, where the space charge density is a function of the distance from the junction. In such an asymmetric junction, the acceptor impurity concentration (N_a) is assumed to be greater than the donor impurity concentration (N_d).

From the condition of charge neutrality,

$$eN_a x_p = eN_d x_n. \tag{10.39}$$

If $N_a \gg N_d$, then $x_p \ll x_n$.

Now the width of the depletion region of junction is

$$W = x_p + x_n \approx x_n \; [\because x_p \ll x_n].$$

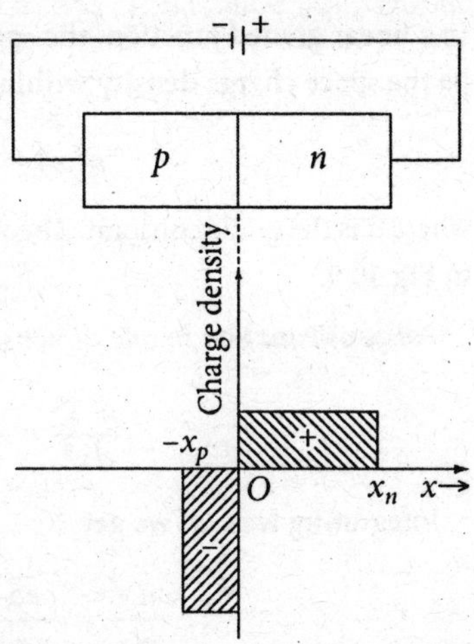

Fig. 10.8: Charge density variation in an abrupt *p-n* junction

Now *Poisson's equation* in one dimension is

$$\frac{d^2V}{dx^2} = -\frac{eN_d}{\epsilon}, \tag{10.40}$$

ϵ being the absolute permittivity of the medium

$$\text{or,}\; \frac{dV}{dx} = -\frac{eN_d}{\epsilon}x + C_1. \tag{10.41}$$

Assuming the electric field outside the depletion region to be zero, i.e., $\frac{dV}{dx} = 0$ at $x = x_n \simeq W$, we have

$$C_1 = \frac{eN_d}{\epsilon}W. \; \therefore \; \frac{dV}{dx} = \frac{eN_d}{\epsilon}(W - x) \text{ or, } dV = \frac{eN_d}{\epsilon}(W - x)\,dx.$$

Integrating w.r.t. x we get the effective barrier voltage,

$$V_T = V_B + |V_a| = \frac{eN_d}{\epsilon}\left[Wx - \frac{x^2}{2}\right]_0^W = \frac{eN_d}{2\epsilon}W^2, \tag{10.42}$$

where $|V_a|$ is the magnitude of the applied reverse-bias. If α be the area of the junction, then charge in the distance $x_n \simeq W$ is

$$Q = eN_d W\alpha. \tag{10.43}$$

$$\therefore C_T = \frac{dQ}{dV_T} = eN_d\alpha\frac{dW}{dV_T} = \frac{\epsilon\alpha}{W} \text{ [using eqn (10.43)]}$$

$$\text{or, } C_T = \epsilon\alpha\sqrt{\frac{eN_d}{2\epsilon(V_B + |V_a|)}} \left[\because W = \sqrt{\frac{2\epsilon}{eN_d}\frac{1}{V_B + |V_a|}}\right] \text{ [from eqn (10.43)].} \tag{10.44}$$

Thus, $C_T \propto \frac{1}{\sqrt{V_B + |V_a|}}$. So, the capacity is voltage sensitive and is used as a *varactor* or *varicaps* (i.e., voltage variable capacitor).

10.7.6 Linear Graded *pn*-Junction

In a linear graded junction, the space charge density varies almost linearly in the transition region. If $\rho(x)$ be the space charge density within $-W/2 < x < W/2$, then

$$\rho(x) = eax, \tag{10.45}$$

where a is the grade constant. The variation of $\rho(x)$ with x is shown in Fig. 10.9.

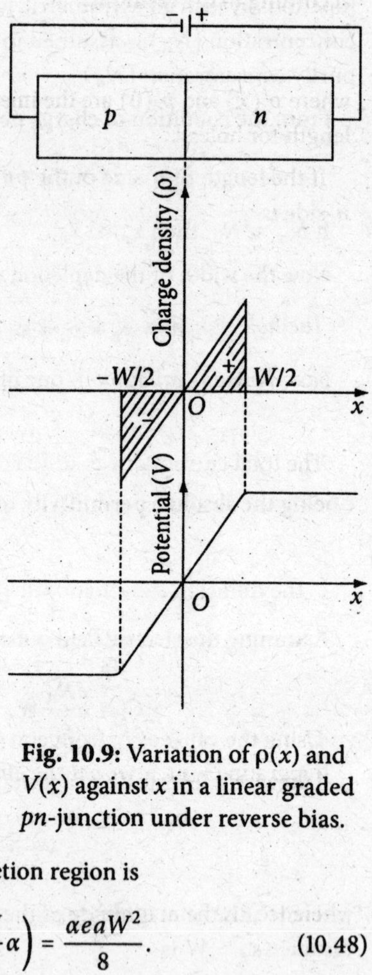

Fig. 10.9: Variation of $\rho(x)$ and $V(x)$ against x in a linear graded *pn*-junction under reverse bias.

Poisson's equation in one dimension is

$$\frac{d^2V}{dx^2} = -\frac{eax}{\epsilon}. \tag{10.46}$$

Integrating w.r.t. x, we get

$$\frac{dV}{dx} = -\frac{ea}{\epsilon}\frac{x^2}{2} + C_1.$$

Since, the electric field $\frac{dV}{dx} = 0$ at $x = \pm W/2$,

$$\therefore C_1 = \frac{eaW^2}{8\epsilon}.$$

So

$$\frac{dV}{dx} = \frac{ea}{2\epsilon}\left(\frac{W^2}{4} - x^2\right).$$

Integrating again,

$$V = V_T = \frac{ea}{2\epsilon}\left[\frac{W^2}{4}x - \frac{x^3}{3}\right]_{-W/2}^{+W/2}$$

$$\text{or, } V_T = V_B + |V_a| = \frac{eaW^3}{12\epsilon}. \tag{10.47}$$

If α be the area of the junction, then charge on one side of the depletion region is

$$Q = \text{average density} \times \text{volume} = \frac{1}{2}\left(ea\frac{W}{2}\right)\cdot\left(\frac{W}{2}\alpha\right) = \frac{\alpha eaW^2}{8}. \tag{10.48}$$

Hence, the *transition capacity* is

$$C_T = \frac{dQ}{dV_T} = \frac{\alpha eaW}{4}\frac{dW}{dV_T} = \frac{\epsilon\alpha}{W}\left[\text{by eqn (10.48), } \frac{dW}{dV_T} = \frac{4\epsilon}{eaW^2}\right]$$

$$\text{or, } C_T = \epsilon\alpha\left(\frac{ea}{12\epsilon V_T}\right)^{1/3}\left[\text{From eqn (10.47), } W = \left(\frac{12\epsilon V_T}{ea}\right)^{1/3}\right]. \tag{10.49}$$

$\therefore$ for a *linear graded pn-junction,* $C_T \propto \frac{1}{V_T^{\frac{1}{3}}}$.

Actually, C_T lies between 5 to 20 μμF.

10.7.7 Diffusion capacitance

A variation of forward-biased voltage leads to the changes in the concentration of injected carriers near a *pn*-junction. The rate of change of injected charges with the applied voltage is defined as the *diffusion* or *storage* capacitance (C_D). This capacitance arises in addition to transition capacitance under reverse-bias.

We consider a forward-biased *pn*-junction. The injected hole concentration decreases with distance according to the relation:

$$p'(x) = p'(0)\, e^{-x/L_p}, \tag{10.50}$$

where $p'(x)$ and $p'(0)$ are the injected hole concentrations at $x = x$ on *n*-side and $x = 0$. L_p is the diffusion length for holes.

If the length of *n*-side of the *pn*-junction be much greater than L_p, then the excess minority charge on *n*-side is

$$Q = \int_0^{\alpha} Aep'(x)\, dx = AeL_p p'(0). \tag{10.51}$$

The *hole diffusion current density* at $x = 0$ is

$$J_p = -eD_p \left[\frac{dp'(x)}{dx}\right]_{x=0} = \frac{eD_p p'(0)}{L_p}.$$

The total current due to diffusion of hole is

$$I_p = AJ_p = \frac{AeD_p}{L_p} p'(0). \tag{10.52}$$

∴ the diffusion capacitance due to holes is

$$C_{Dp} = \frac{dQ}{dV} = AeL_p \frac{dp'(0)}{dV} = \frac{L_p^2}{D_p}\frac{dI_p}{dV} = g_p \frac{L_p^2}{D_p} \text{ [using eqn (10.51)]}. \tag{10.53}$$

Using the value of I_p from eqn (10.52), $g_p = \frac{dI_p}{dV}$ is the *hole-conductivity.*
Likewise, the diffusion capacitance for the electrons is given by

$$C_{Dn} = g_n \frac{L_n^2}{D_n}. \tag{10.54}$$

Hence, the total diffusion capacitance is

$$C_D = C_{Dp} + C_{Dn} = g_p \frac{L_p^2}{D_p} + g_n \frac{L_n^2}{D_n}. \tag{10.55}$$

In a *pn*-junction diode, $g \propto I$, i.e., diffusion capacitance $C_D \propto I$ and has a large value in *forward-biased condition.*

10.8 Diode as a Circuit Element

A *pn*-junction diode is represented by a circuit symbol as shown in Fig. 10.10(a). When the junction is forward-biased, the arrowhead represents the direction of current flow. Fig 10.10(b) shows a simple circuit consisting of a diode in series with a load resistance R_L.

When the current I passes through the diode a voltage $V = IR_L$ is dropped across R_L and the voltage across the diode becomes

$$V = V_i - IR_L,$$

where V_i is the applied voltage. The I–V characteristic of a diode is shown in Fig. 10.7.

Fig. 10.10(a): Circuit symbol for a diode.

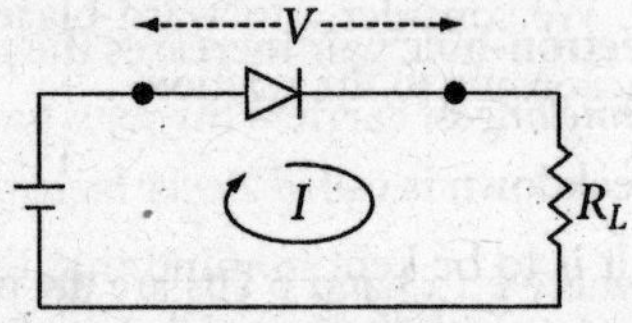

Fig. 10.10(b): A diode in series with a load resistace R_L.

For the analysis of the diode circuit, various techniques are applied, out of which *the graphical method* and *piecewise linear equivalent circuit method* are most popular. The former is based on complicated nonlinear mathematical techniques and the latter assumes that a diode characteristic is a straight-line segments as shown in Fig. 10.11(a).

Here the forward-biased voltage $V > V_\gamma$, the cut-in voltage, and the diode is replaced by a circuit shown in Fig. 10.10(b). If, however, the diode is reverse-biased or the forward-biased voltage $V < V_\gamma$, then the diode is replaced by an open circuit [Fig. 10.11(c)]. The circuit is linear and can be easily analysed by the known laws of circuit analysis.

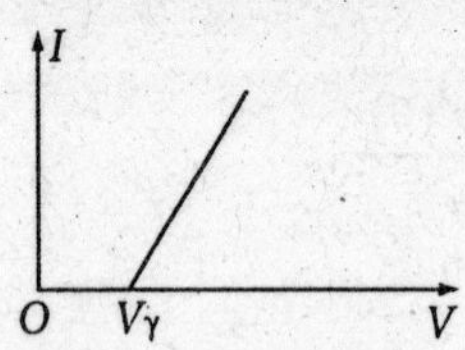

Fig. 10.11(a): Piecewise linear characteristic of a diode.

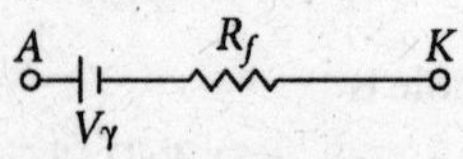

Fig. 10.11(b): Model of diode in ON state with cut-in voltage V_γ.

Fig. 10.11(c): Model of diode in OFF state

10.9 Some Special *pn*-Junction Diodes

A *pn*-junction, when operated in the reverse-biased condition, a small saturation current (I_s) flows, which is almost independent of reverse-biased voltage. If, however, the reverse-bias exceeds a certain critical value, the value of current suddenly approaches a very large value and the situation is known as *breakdown* in diodes. The critical voltage at which this breakdown occurs is known as *breakdown voltage* (V_{Br}) which may damage the diode, if not specially designed.

A diode having adequate power dissipation capabilities to operate in the breakdown region is called a *breakdown diode* or a *Zener diode*.

The mechanisms responsible for the breakdown are as below:

(i) *Avalanche Breakdown*: Under reverse-biased condition, a small saturation current flows down the junction barrier and the involved minority carriers acquire energy from the applied voltage. These carriers are colliding with the crystal ions continuously. When the applied voltage is large, the carriers acquire sufficient energy from the field and some of the collisions may become so violent that electrons are knocked out of the covalent bonds of the crystal atoms producing new electron-hole pairs. The new carriers acquire sufficient energy from the applied field and collide with other immobile ions generating further electron-hole pairs. This cumulative process of electron-hole pair creation is known as *avalanche multiplication.* As a result of avalanche breakdown a large current flows at a particular voltage (V_{Br}) as shown by dotted line in Fig. 10.7.

(ii) *Zener breakdown*: If the two sides of a *pn*-junction are heavily doped, the width of the depletion region (W) becomes very thin, and the electric field at the junction may be very large with a relatively small reverse voltage (V), [i.e., $E = V/W \rightarrow$ large value for $W \rightarrow$ small value] under this condition chance of creation of new carriers through collision becomes extremely small, but due to the large force (eE) of the carriers on the bound electrons, generation of new carriers takes place due to the rupture of covalent bonds. The new electron-hole pair increases the reverse current. This phenomenon arises due to the *quantum mechanical tunnelling* of carriers through barrier at the junction and was first investigated by *Zener*. That is why this breakdown is called *Zener breakdown.*

It is to be kept in mind that the *generation of carriers by Zener breakdown is not due to the collision of carriers with the immobile ions.* The diodes working on the principle of the above two breakdown processes are generally called *Zener diodes.*

10.9.1 The Circuit Symbol and Characteristics of a Zener Diode

Figs 10.12(a) and (b) show the circuit symbol and typical characteristic curves of a *Zener diode* respectively. It is evident from curve of Fig. 10.12(b) that, in the reverse-biased condition, there is a region in the characteristic, where the voltage is almost independent of the diode current. This property of the Zener diode is exploited in a *voltage regulator device.* The voltage across the Zener diode may also be used as a reference voltage and the corresponding diode is called a *reference diode.*

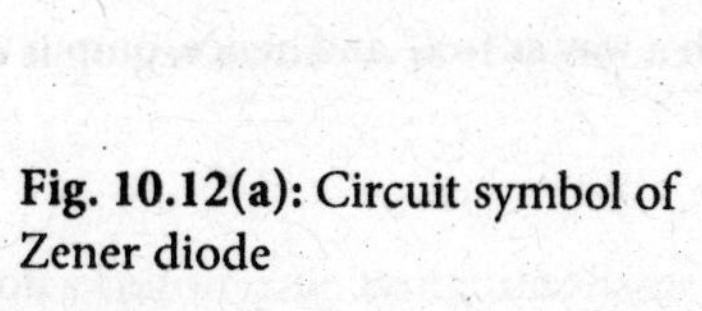

Fig. 10.12(a): Circuit symbol of Zener diode

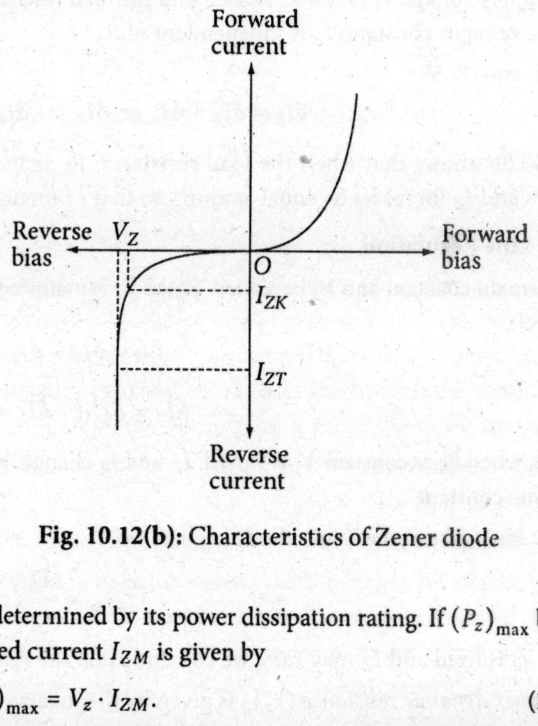

Fig. 10.12(b): Characteristics of Zener diode

The upper limit of the diode current may be determined by its power dissipation rating. If $(P_z)_{max}$ be the wattage of the Zener diode, then maximum rated current I_{ZM} is given by

$$(P_z)_{max} = V_z \cdot I_{ZM}. \tag{10.56}$$

The diode will be destroyed if the diode current I_z exceeds I_{ZM} (i.e., $I_Z > I_{ZM}$). I_{ZK} is known as the *knee current* corresponding to the knee of the reverse characteristic curves which is 5 to 10 % of I_{ZM}.

The test current I_{ZT} of the diode which corresponding to 25 % of $(P_Z)_{max}$ is given by

$$V_Z \cdot I_{ZT} = \tfrac{1}{4}(P_Z)_{max}. \tag{10.57}$$

$$\therefore I_{ZT} = \frac{\frac{1}{4}(P_Z)_{max}}{V_Z} = \frac{\frac{1}{4}(P_Z)_{max}}{(P_Z)_{max}} I_{ZM} = \tfrac{1}{4} I_{ZM}. \tag{10.58}$$

10.9.2 Application of Zener Diode: Simple Zener Regulator

The circuit diagram of a typical application of Zener diode is shown in Fig. 10.13. From Fig. 10.12(b) it is clear that the voltage across the Zener diode is almost independent of the current through it, which makes it useful as a voltage regulator. The output voltage $V_O = V_z$ in a Zener diode is independent of the variations of input voltage $V_i\ (> V_z)$ and remains constant so long as the diode works in the breakdown region.

Now from the circuit in Fig. 10.13,

$$I_S = I_Z + I_L, \tag{10.59}$$

where

$$I_S = \frac{V_i - V_z}{R_s} \text{ and } I_Z = \frac{V_O}{R_L}. \tag{10.60}$$

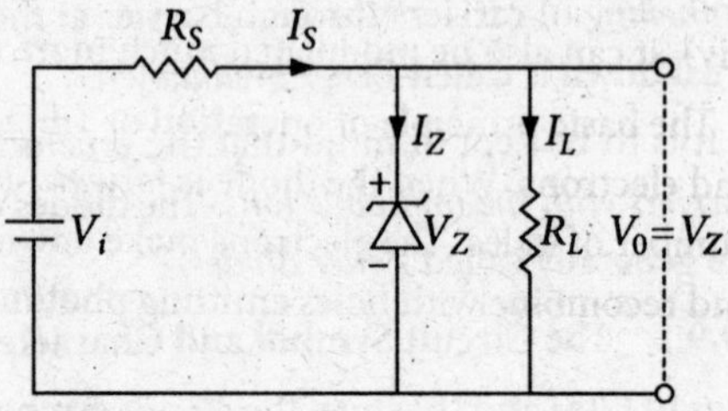

Fig. 10.13: A simple Zener regulator circuit diagram

10.9.3 Load Regulation

Let the supply voltage V_i remain constant and the load resistance be varied. As the Zener voltage V_z across the diode remains constant, I_S is independent of R_L.

$\therefore$ from eqn (10.59),

$$\delta I_S = \delta I_Z + \delta I_L \text{ or, } \delta I_Z = -\delta I_L \ [\because \delta I_S = 0]. \tag{10.61}$$

Eqn (10.61) shows that when the load resistance R_L is increased keeping the supply voltage fixed, I_L decreases and I_Z increases by equal amount, so that I remains constant.

10.9.4 Line Regulation

Let R_L remain constant and V_i be varied. Since V_O remains constant at V_Z, I_L remains constant. Then from eqn (10.59),

$$\delta I_s = \delta I_Z + \delta I_L.$$

$$\therefore \delta I_Z = \delta I_s \ (\because \delta I_L = 0). \tag{10.62}$$

Hence, when R_L is constant V_i is varied, I_Z and I_S change in such a way as to I_L and hence, output voltage V_O become constant.

The *dc resistance* of the Zener diode is given by

$$(r_z)_{dc} = \frac{V_O}{I_Z}.$$

Since V_O is fixed and I_Z may vary, we conclude that the Zener dc resistance $(r_Z)_{dc}$ is not a constant.

The *Zener dynamic resistance* $(r_z)_d$ is given by the inverse of the slope of the characteristic, i.e.,

$$(r_z)_d = \frac{\delta V_O}{\delta I_z}.$$

Since the Zener characteristic is almost parallel to the current axis, $\delta V_O = 0$. So, $(r_z)_d = 0$.

Practically, $(r_z)_d$ has a finite value which is very small for a good Zener diode.

10.9.5 Light Emitting Diode

A semiconductor *pn*-junction when properly biased can serve as a light emitter, i.e., a *semiconductor lamp.* A light emitting diode (LED) has several advantages over conventional lamps, viz.,

(i) an LED has greater emission power per unit volume,
(ii) it has greater ruggedness,
(iii) it can be built at considerably low cost,
(iv) it can also be modulated much more rapidly than the conventional lamps.

The basic principle of operation of LED is the emission of radiation due to the recombination of holes and electrons. When the diode is *forward-biased,* the electrons move into the *p*-side and encounter a large number of holes. The electrons make downward transition from the conduction band to the valence band and recombine with holes emitting photons with energy $h\nu = E_g$, i.e., $e + h \rightarrow h\nu$

$$h\nu = E_g \text{ or, } \frac{hc}{\lambda} = E_g \text{ or, } \lambda = \frac{hc}{E_g}, \tag{10.63}$$

where λ is the wavelength of the light emitted, c is the velocity of light and E_g is the band gap. Gallium arsenide phosphide ($GaAs_xP_{1-x}$) and Gallium phosphite (GaP) are used to construct such devices.

The colour of the emitted light depends on the forbidden energy gap (E_g) [cf. eqn (10.63)]. Table (10.1) shows the colour of light emitted from LED with different semiconductors:

Table 10.1

LED *made of*	*Colour of radiation*
GaP with an addition of Cd	Red and green
SiC	Yellow
GaAs	Infrared

Under certain condition the emitted light from LED is coherent. Such diodes are referred to as *injection-junction laser.*

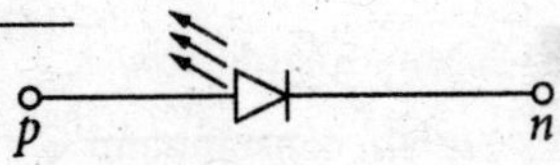

Fig. 10.14: Circuit symbol of an LED

The circuit symbol of an LED is shown in Fig. 10.14 and the circuit using it is shown in Fig. 10.15. The current through the LED is

$$I = \frac{V_i - V_D}{R_S}, \tag{10.64}$$

where R_s is the resistor which limits the current to safety range, V_i and V_D are the source voltage and the voltage drop across the LED respectively. The brightness of the LED depends on the current I flowing through it.

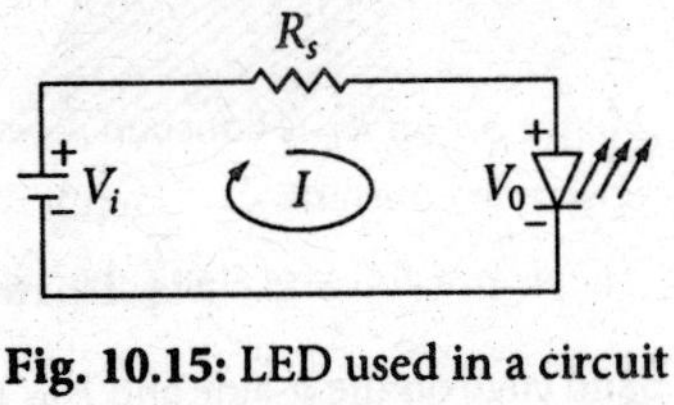

Fig. 10.15: LED used in a circuit

10.9.6 Use of LED

LEDs have the following uses:

(a) Digital display system of calculators, watches and many other micro-electronic circuits.
(b) The infrared LED finds application in burglar alarm system, CD players, optical fibre communication and systems which require visible light.
(c) LED is used as an indicator of power line voltage in different instruments.

10.9.7 Tunnel Diode

The tunnel diode is a *pn*-junction working on the principle of *quantum mechanical tunnelling.* It was discovered by *Esaki* in 1958 and he received Nobel prize [Ref. I.L. Esaki, *New Phenomena In Narrow Ge pn-junctions*—Phys. Rev. **109**, 603, (1958).]

It is a *pn*-junction device operating in a certain region of the $I-V$ characteristic by the *quantum mechanical tunnelling of electrons through the potential barrier of the junction.* The Zener effect is also a tunnelling effect for reverse current but the tunnelling process in tunnel diode needs a very negligible reverse-bias to start with. The concentration of impurity atoms in tunnel diode is very large (~ 10^{18} or $10^{19}/\text{cm}^3$) in both *pn*-regions, the barrier width of such a junction is very small (~ 100 Å).

10.9.8 Operation of the Tunnel Diode

Fig. 10.16(a) shows the energy band structure of a *pn*-junction between two degenerate semiconductors. In equilibrium condition the Fermi level is constant throughout the junction. Fig. 10.16(a) shows that E_{Fp} lies below the valence [Fig. 10.16]. *The principle of operation of a tunnel diode.* (a) zero bias condition (no tunnelling) (b) small reverse-bias (electron tunnelling from *p* to *n*) (c) small forward-bias (electron tunnelling from *n* to *p*) (d) increased forward-bias (electron tunnelling from *n* to *p* decreases as bands pass by each other).

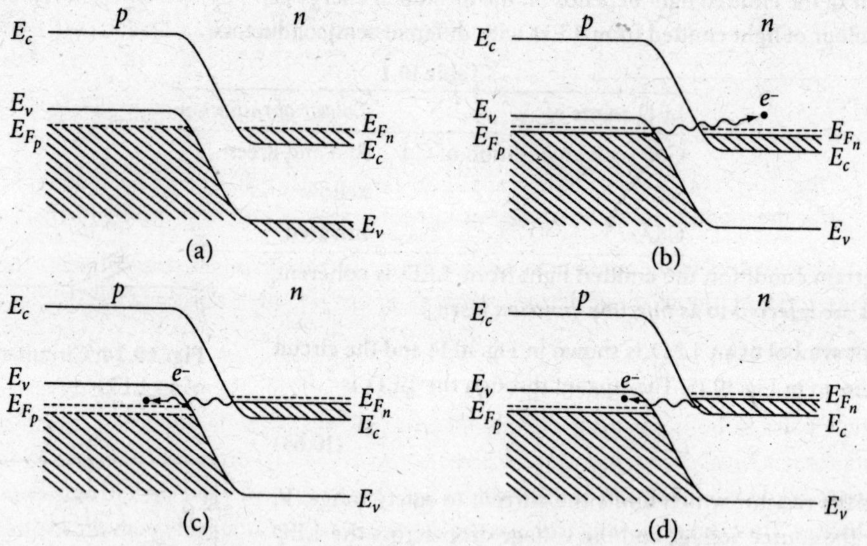

Fig. 10.16: The principle of operation of the tunnel diode.

Band edge on the *p*-side and E_{Fn} lies above the conduction band edge on the *n*-side, i.e., the bands overlap so that E_F is constant.

Let now a small reverse-bias be applied to the *pn*-junction. Then the tunnelling of electrons occurs from the filled valence band states below E_{Fp} to the empty conduction band states above E_{Fn}. The electron tunnelling increases with the increase of reverse-biased voltage [Fig. 10.16(b)].

With the application of a small forward-bias (V), E_{Fn} moves up in energy w.r.t. E_{Fp} by an amount eV. As a result, the electrons below E_{Fn} on the *n*-side are placed opposite empty states above E_{Fp} on the *p*-side [Fig. 10.16(c)]. Then electron tunnelling occurs from *n*- to *p*-side as shown by the part *OA* of the $I-V$

characteristic curve in Fig. 10.17(a). With the increase of the forward voltage, the forward tunnelling current continues to increase as more filled states are placed opposite empty states [Fig. 10.15(a)]. As E_{Fp} continues to move up w.r.t. E_{Fn}, a point is reached when the bands begin to pass each other. The number of filled states opposite empty states decreases and the tunnelling current decreases as shown by the part *AB* of the curve in Fig. 10.17(a). Thus, the portion *AB* of the *I–V* characteristic curve shows a decrease of tunnelling current with the increase of forward-bias. It thus produces a *region of negative slope* where the resistance of the junction is negative ($\because \frac{dV}{dI}$ is negative). This negative resistance region has a good number of important applications. If the forward-bias is increased beyond the negative resistance region, the current *I* begins to increase again as the bands pass each other [Portion *BC* of the *I–V* characteristic curve of Fig. 10.17(a)].

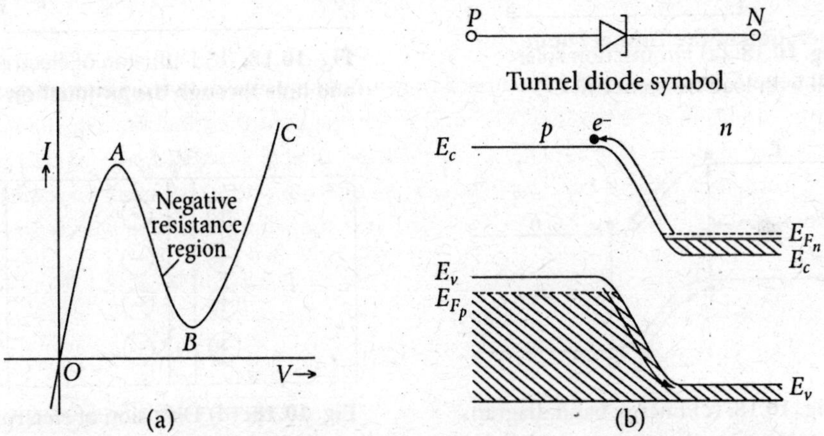

Fig. 10.17: (a) Tunnel diode symbol and characteristic (b) Energy band diagram for the diode beyond the tunnelling condition [BC of the curve in Fig. 10.17(a)].

Tunnel diodes are used in the construction of amplifiers, oscillators or switching devices. The tunnelling process being very fast, tunnel diode devices can work at very high frequencies (~ 10 GHz).

10.9.9 Photo Diodes

We consider a reverse-biased *pn*-junction on which light is incident. The incident light creates additional electron-hole pairs in both *p*- and *n*-regions. Because of this a very large change in minority carrier concentration and a small change in majority carrier concentration are produced. These additional minority carriers produced due to the action of the incident light enhance the reverse current as they fall down the potential barrier. The current flowing through the diode varies almost linearly with the incident light flux. The diodes working on the basis of the above principle are called *photo diodes.*

Photo diodes find important applications in light detection, light operated switches, reading of computer punched cards and tapes, etc.

10.9.10 Solar Cell

A solar cell is a pn-junction diode which converts sunlight directly to electricity with large conversion efficiency.

The principle of operation of a solar cell is as follows:

When a *pn*-junction diode is exposed to sunlight, the incident photons creates electron-hole pairs in both *p*- and *n*-regions, of the junction as shown in Fig. 10.17(a). The generated electron-hole pairs over a small distance from the junction move to the space charge region *X* through diffusion [Fig. 10.17(b)].

Then the strong electric field existing in the X-region separates the electron-hole pairs. The electrons in the p-region slide down the barrier potential to move to the n-region. Similarly, the holes in the n-region move towards the p-region, as shown in Fig. 10.18(c).

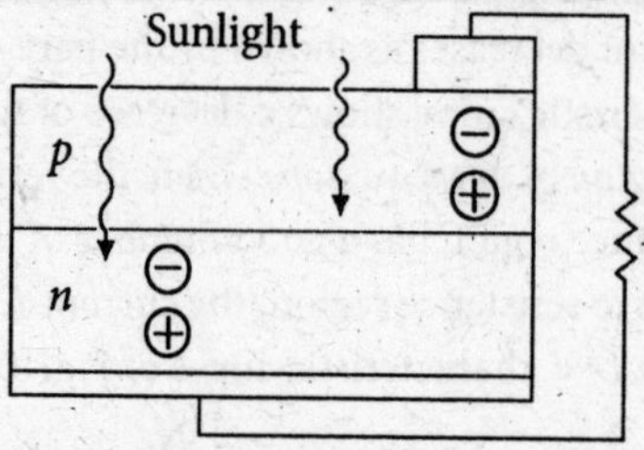

Fig. 10.18: (a) *pn*-junction solar cell with load resistance R_L

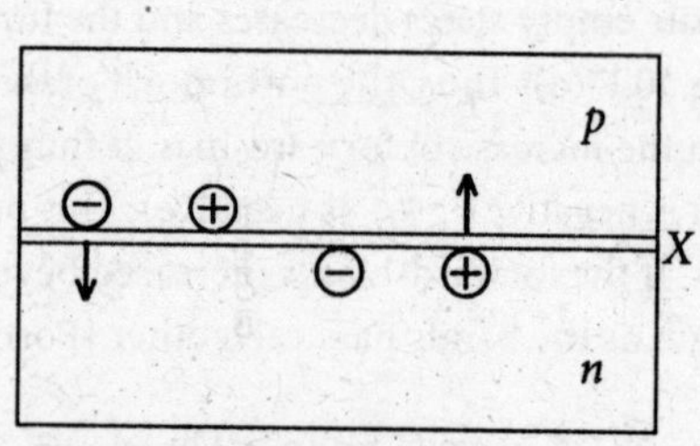

Fig. 10.18: (b) Diffusion of electron and hole through the *pn*-junction

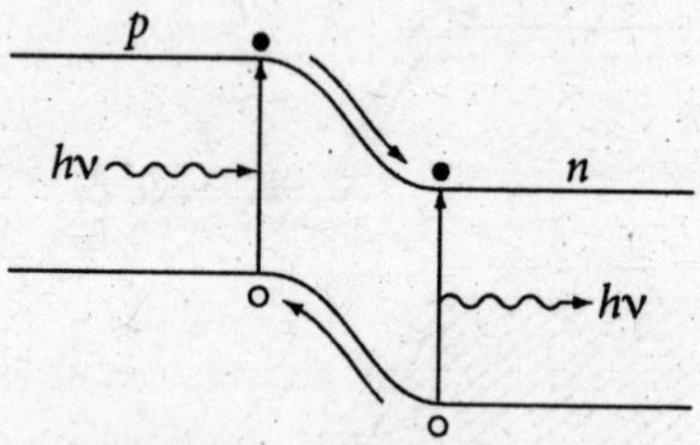

Fig. 10.18: (c) Energy band diagram for diffusion of electrons and holes

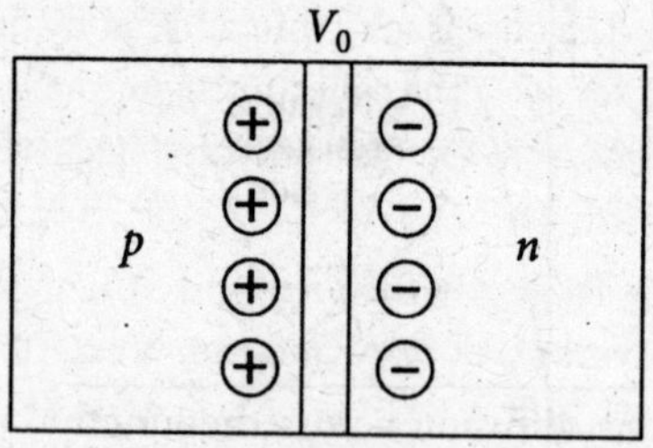

Fig. 10.18: (d) Diffusion of electron and hole through the *pn*-junction

In open circuit condition, the accumulated electrons and holes in the two sides of the pn-junction diode produce an open circuit voltage V_0. If a load resistance R_L is connected across the diode, a current will flow through the load, Fig. 10.18(a). The short circuit current (the maximum current) will be obtained by making $R_L = 0$ so long as the junction is exposed to sunlight. The magnitude of current through R_L is directly proportional to the light intensity.

10.9.11 Use

Solar cells are used in: (i) satellites and space vehicles as most important long duration power supply, (ii) domestic power supply to fans, lights, etc. (iii) power supply to motors and pumps in irrigation purpose, etc.

10.10 Bipolar Junction Transistor (BJT)

John Bardeen and *W. Brattain* demonstrated the amplifying action of the first transistor at the *Bell Telephone Laboratory* in America. After that *Shockley* in 1951 invented the first junction transistor. From the integrated circuits to electronic devices and modern computers all are due to the discovery of transistor. Nowadays, the impact of transistor on electronic industry is enormous.

10.10.1 Construction of Transistor

A transistor is a three-layer semiconductor device in which there are either two p- and one n-type layer of materials or two n- and one p-type layer of materials. The former is called an pnp- and the latter npn-transistors as shown in Fig. 10.19.

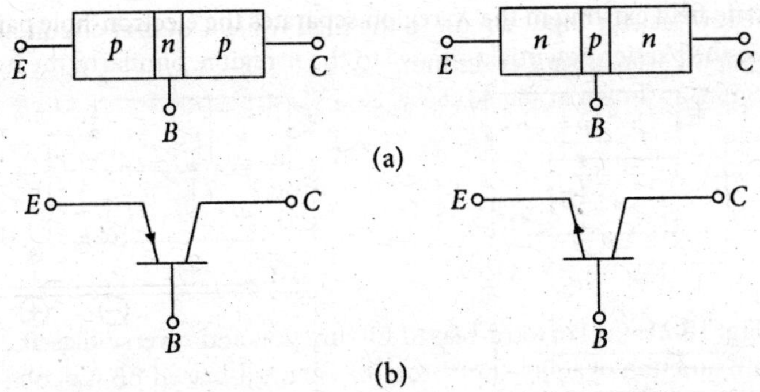

Fig. 16.19: (a) Two types of BJT transistors and (b) the circuit symbols of *pnp*- and *npn*-transistors

In case of *npn*-transistor, the emitter (E) is to emit or inject electrons into the base (B) and hence, it is heavily doped. The base is very thin and lightly doped as it has to pass most of the emitter ejected electrons on to the collector (C). The collector is intermediately doped between the emitter and the base. The collector collects electrons from the base and dissipates more heat than the emitter and the base and as such it is the largest in volume.

The majority carriers in the *pnp*- and *npn*-transistors are holes and electrons respectively. The arrowhead in the circuit symbols of *pnp*- and *npŋ*-transistors indicates the direction of conventional current.

The electrons diffusion across the junction produces *two depletion regions*. The barrier potential at ~ 25 °C for Si and Ge are approximately 0.7 V and 0.3 V respectively. As the doping levels of the three regions are different, the width of the emitter depletion region is smaller compared to collector depletion region as shown in Fig. 10.20.

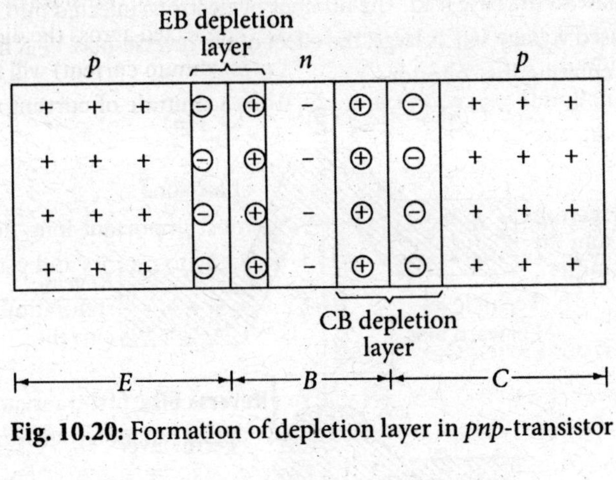

Fig. 10.20: Formation of depletion layer in *pnp*-transistor

10.10.2 Transistor Biasing

In normal operation emitter-base junction is forward-biased and the collector-base junction is reverse-biased [Fig. 10.21]. The emitter in *pnp*-transistor injects holes into the base which are finally collected by the collector. The *EB*-junction being forward-biased its dynamic resistance is small whereas the dynamic resistance of *CB*-junction is large as it is reverse-biased. Because of the transistor action, almost the same current flows from the low resistor input circuit to the high resistor output circuit and hence, the name *transistor* is derived from the words *transfer resistor*.

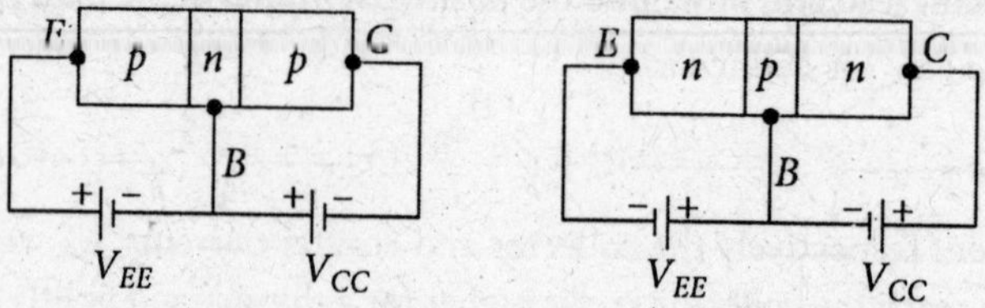

Fig. 10.21: (a) Forward-biased EB junction and reverse-biased CB junction of a *pnp*-transistor, (b) Forward-biased EB junction and reverse-biased CB junction of an *npn*-transistor.

The abbreviation BJT is used for the junction transistor and the term *bipolar* is used as both electrons and holes taking part in the injection process in the oppositely polarized materials.

10.10.3 Transistor Operation

Operation of an *npn*-Transistor

Under normal operation the emitter-base junction (J_E) is forward-biased and the collector-base junction (J_C) is reverse-biased [Fig. 10.21(b)]. Due to forward-bias of J_E, the emitter diode allows some free electrons to flow from the emitter (E) to base (B). The width of the base being smaller than the diffusion length of the electrons and due to higher density of electrons (i.e., the minority carriers), the electron-hole recombination in base region is almost negligible. Most of the electrons (almost 95 %) diffuse to the collector depletion layer. This decreases the collector energy hill as shown in Fig. 10.22. Only 5 % of the electrons injected by the emitter fall along the recombination path. The recombining electrons become valence electrons and flow through the hole in the base to the base lead. The number of electrons injected into the base by the emitter is large if the forward-biased voltage V_{BE} is large, the effect of the reverse-bias V_{CB}, however, is negligible to the number of electrons entering the collector.

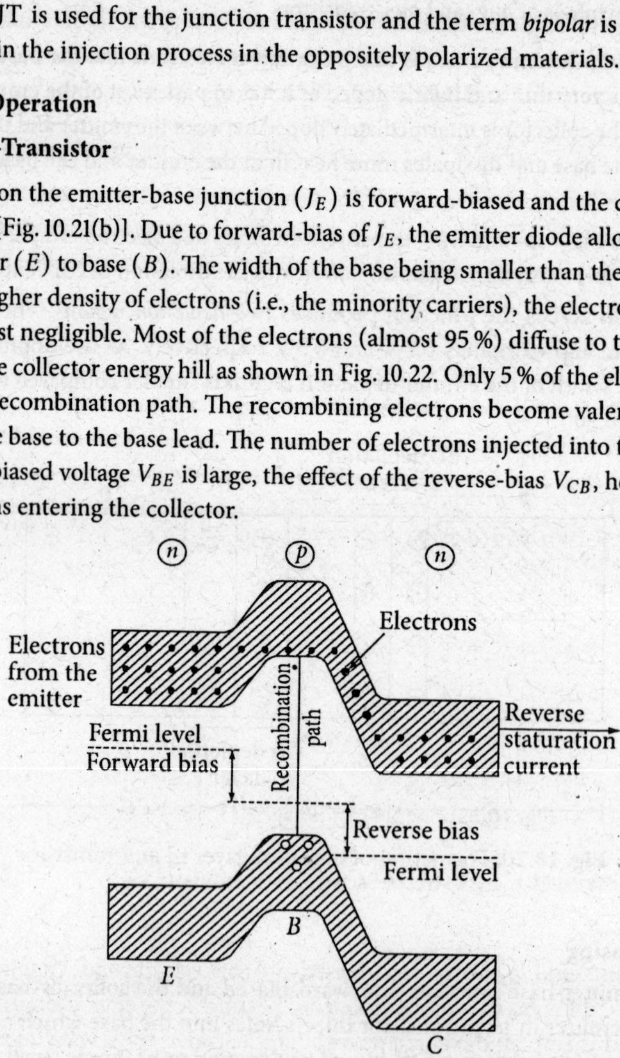

Fig. 10.22: Energy level diagram of a forward bias *npn*-junction

A steady stream of electrons is pushed into the collector region by the depletion layer field of J_C, which enters the external collector lead and flows into the positive terminal of the battery.

The *dc transistor alpha* (α_{dc}) is defined as

$$\alpha_{dc} = \frac{I_C}{I_E}, \tag{10.65}$$

where I_C and I_E represent respectively the collector and emitter currents, α_{dc} represents the fraction of emitter current that can reach the collector. α_{dc} is higher for a thinner and lightly doped base.

Operation of a *pnp*-Transistor

In case of a *pnp*-transistor, the emitter is of high conductivity *p*-type material and the base is of low conductivity material. When the *pnp*-transistor is forward-biased [Fig. 10.21(a)], the emitter current (I_E) consisting of holes flows through the thin layer of base and reaches the reverse-biased *CB*-junction. The holes passing across the *CB*-junction are collected by the collector and constitute the collector current (I_C). The upward direction of potential in Fig. 10.23 represents a down hill of potential as the holes are positively charged.

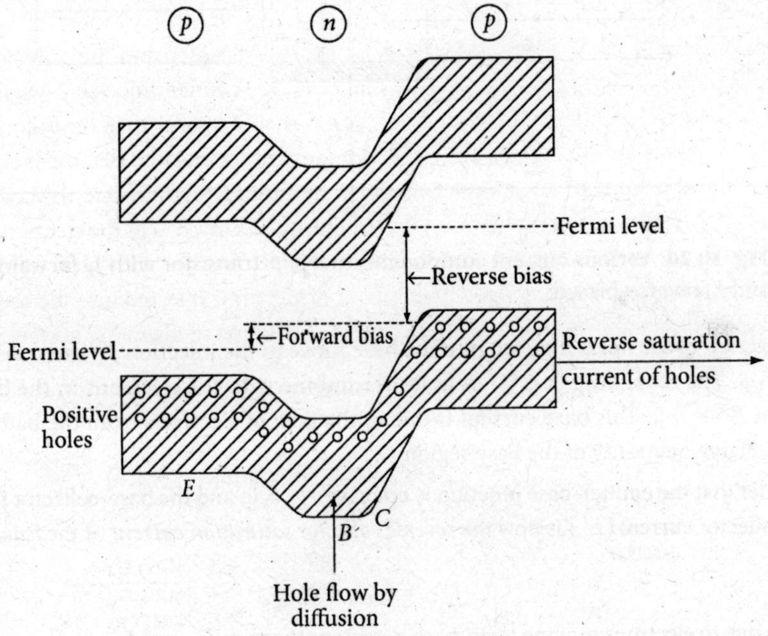

Fig. 10.23: Energy band diagram of *pnp*-transistor

10.11 Different Current Components in a Transistor

10.11.1 *pnp*-Transistor

Let the emitter-base junction (J_E) be forward-biased and the collector-base junction (J_C) reverse-biased of a *pnp*-transistor. The different current components existing in the transistor are as shown in Fig. 10.24. When the external bias is zero all the currents through the transistor is zero due to the development of intrinsic potential barrier across the *pn*-junction. The forward-bias V_{EE} reduces the emitter-base (J_E) potential barrier and the reverse collector bias V_{CC} increases the potential barrier across J_C. The reduction in potential barrier at J_E helps the injection of holes from p- to n-region across J_E and the current I_{EP} results. Simultaneously, the injection of electrons from the base to emitter constitutes the current I_{En}. Hence, the emitter current I_E is given by

$$I_E = I_{Ep} + I_{En}. \tag{10.66}$$

In commercial transistor, the conductivity of the emitter is made much larger than that of the base and as such in a pnp-transistor $I_{Ep} \gg I_{En}$, i.e.,

$$I_E \simeq I_{E_p}. \tag{10.67}$$

This is purposely done as I_{En} does not contribute to the collector current (I_C).

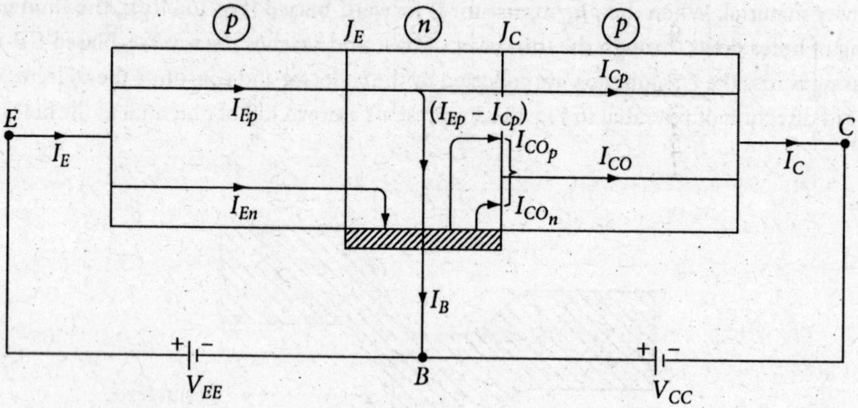

Fig. 10.24: Various current components in a pnp-transistor with J_E forward-biased and J_C reverse-biased.

A major fraction of the holes injected into the base move to the injection J_C and constitute the current component I_{CP}. The remaining (1–3 %) of holes recombine with the electrons in the base giving rise to a base current $I_{Ep} - I_{CP}$. This base current is due to the flow of electrons from the battery to the base to maintain the charge neutrality of the base region.

Now consider that the emitter-base junction is open, i.e., $I_E = 0$, and the base-collector junction is reverse-biased. The collector current (I_C) is now the *reverse collector saturation current* or the *leakage current* (I_{CO}) where

$$I_{CO} = I_{COn} + I_{COp},$$

where I_{COn} is due to electrons moving from p- to n-regions through J_C, and I_{COp} is due to the holes moving from p to n through J_C as shown in Fig. 10.24. The component I_{CO} is too much temperature sensitive and is responsible for the change of transistor characteristics with the change of temperature.

Thus, we have

$$I_C = I_{CO} + I_{Cp} \tag{10.68}$$

$$\text{and } I_E = I_B + I_C. \tag{10.69}$$

Practically, I_{CO} is in microampere and I_C is in milliampere, i.e., $I_C \gg I_{CO}$, so

$$I_C \simeq I_{Cp}. \tag{10.70}$$

It is to be noted that for a pnp-transistor I_E is positive, but both I_C and I_{CO} are negative.

10.11.2 *npn*-Transistor

In the operation of an *npn*-transistor, the roles of electron and holes are just reverse. Fig 10.25 shows the different current components of an *npn*-transistor when the emitter is forward-biased and the majority holes from the base are injected into the emitter region. As a result, the resulting emitter current I_E is given by

$$I_E = I_{En} + I_{Ep}. \tag{10.71}$$

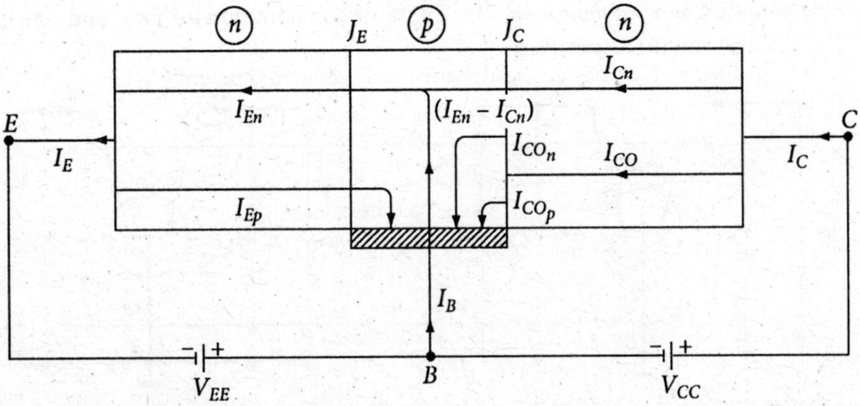

Fig. 10.25: Various current components in an *npn*-transistor with J_E forward-biased and J_C reverse-biased.

In commercial transistor the conductivity of the emitter is made much larger (by heavily doping) than that of the base region, $I_{En} \gg I_{Ep}$ and as such

$$I_E \approx E_{En}. \tag{10.72}$$

The electrons being negatively charged, the direction of the conventional current is opposite to that of electron flow.

The major part of the electrons which are injected into the base from the emitter now diffuse through J_C into the collector region. A small fraction of these electrons recombine with majority holes in the base.

The electron reaching the collector region constitutes the current component I_{Cn}. The difference $(I_{En} - I_{Cn})$ forms a part of base current. The reverse saturation current through J_C is I_{Co}, even if the emitter is open, i.e., $I_E = 0$. Now

$$I_{CO} = I_{CO_n} + I_{CO_p}, \tag{10.73}$$

where I_{CO_n} is due to the flow of minority electrons from base to collector and I_{CO_p} is due to the flow of minority holes from collector to base.

So

$$I_C = I_{CO} + I_{Cn} \tag{10.74}$$

$$\text{and } I_E = I_B + I_C. \tag{10.75}$$

Here also I_{CO} is in μA and I_C is in mA. So

$$I_C \simeq I_{Cn}. \tag{10.76}$$

10.12 Modes of Operation of a Transistor

Since a transistor has *three* terminals, viz., base (B), emitter (E) and collector (C), any one terminal may be common to both input and output circuits. As a result, a transistor may be operated in any one of the following three modes as described below.

10.12.1 Common-Base (CB) Mode

The base terminal of the transistor is made common to both the input and output circuits. This mode is also called the *grounded base configuration.* The CB mode configuration for pnp- and npn-transistors are shown in Figs 10.26(a) and (b) respectively.

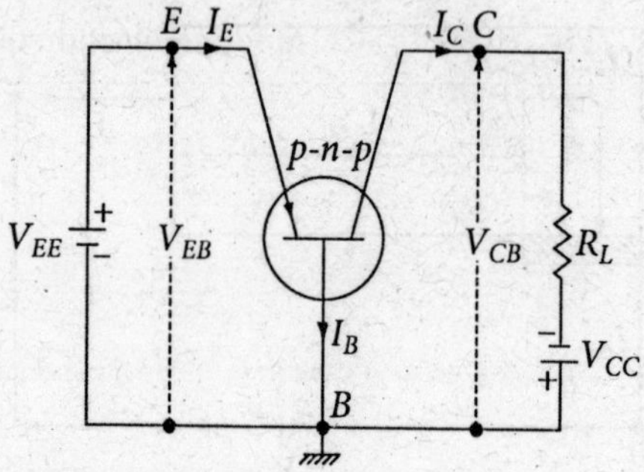

Fig. 10.26: (a) CB mode connection of a *pnp*-transistor

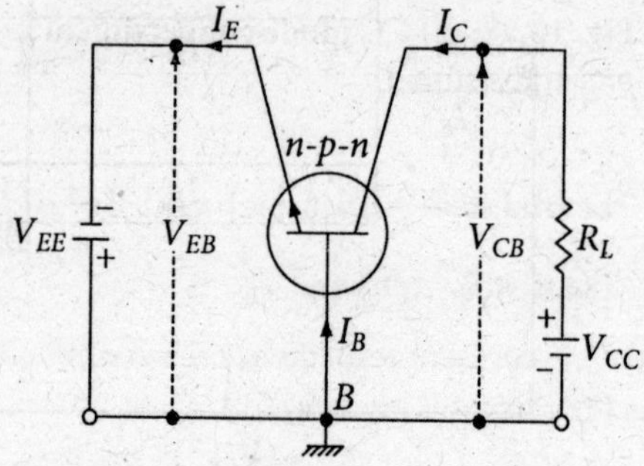

Fig. 10.26: (b) CB mode connection of a *npn*-transistor

10.12.2 Common-Emitter (CE) Mode

This mode is also known as the *grounded emitter configuration of the transistor.* Here the emitter terminal is made common to both the input and output circuits of the transistor. The CE modes of a pnp- and an npn-transistors are shown in Figs 10.27(a) and (b) respectively.

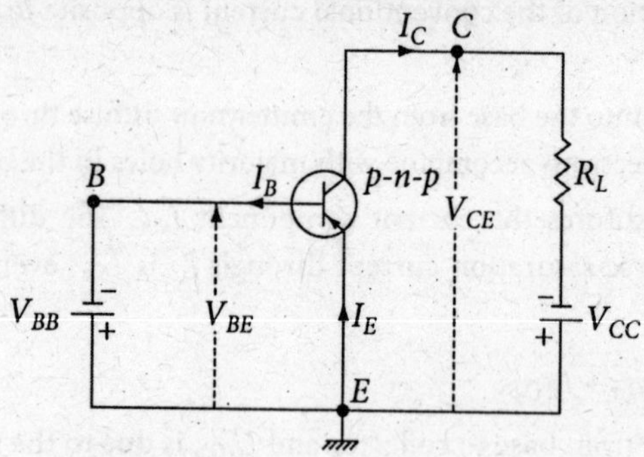

Fig. 10.27: (a) CE mode connection of a *pnp*-transistor

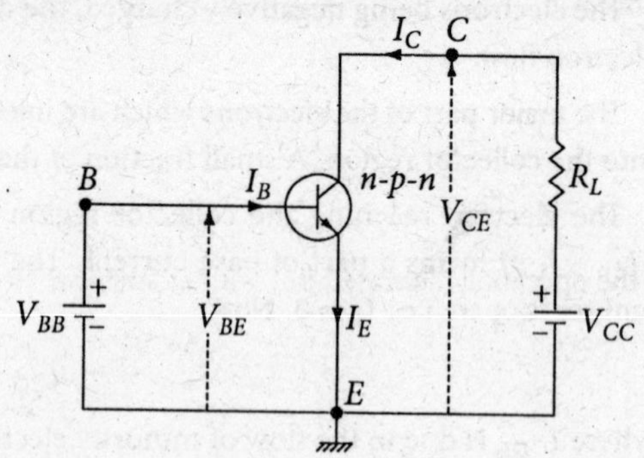

Fig. 10.27: (b) CE mode connection of an *npn*-transistor

10.12.3 Common-Collector (CC) Mode

In this mode, the collector terminal of the transistor is made common to both the input and the output circuits. The CC mode is also known as the *grounded-collector configuration.* Figs 10.28(a) and (b) show respectively the CC mode of operation of the pnp- and npn-transistors.

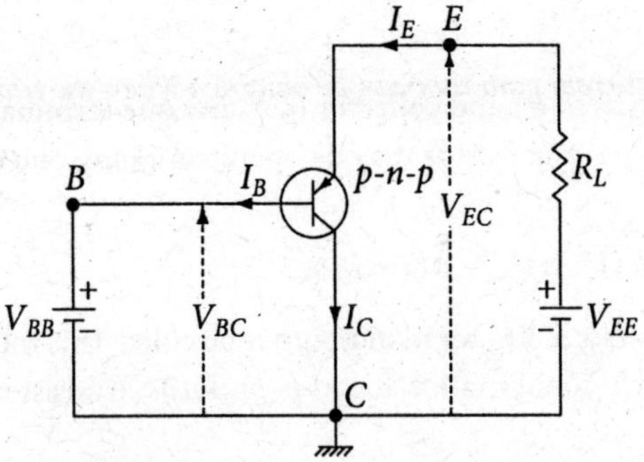

Fig. 10.28: (a) CC mode connection of a *pnp*-transistor

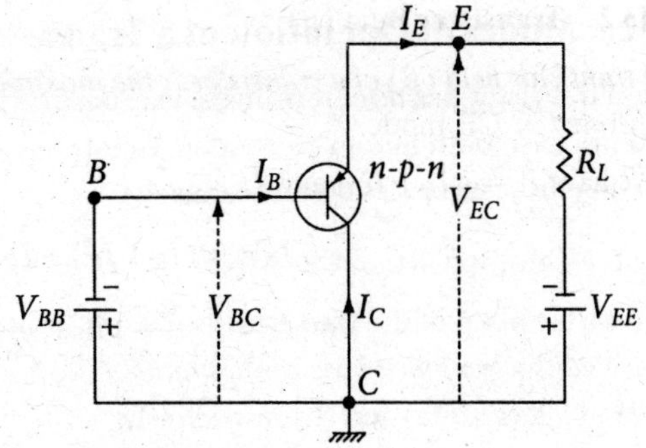

Fig. 10.28: (b) CC mode connection of an *npn*-transistor

10.13 Transistor Alpha (α) and Beta (β)

10.13.1 Transistor Alpha (α)

If I_E and I_{CO} represent respectively the *emitter current* and *reverse saturation current* of a transistor, then the collector current is given by

$$I_C = I_{CO} + \alpha I_E,$$

where α is called the *transistor alpha.* It gives the fraction of I_E that reaches the collector.

In case of dc operation, the transistor alpha is given by

$$\alpha_{dc} = \frac{I_C}{I_E}. \tag{10.77}$$

Since

$$I_{CO} \ll I_C,\ I_C \simeq \alpha I_E \text{ or, } \alpha \simeq \frac{I_C}{I_E}. \tag{10.78}$$

$$\therefore \alpha \simeq \alpha_{dc}.$$

If the operation is carried out with ac, then we define *small-signal short circuit current gain by* α_{ac}, where

$$\alpha_{ac} = \frac{\text{change in the collector current}}{\text{change in the emitter current}}$$

at constant collector to base voltage (V_{CB})

$$\text{or, } \alpha_{ac} = -\frac{\Delta I_C}{\Delta I_E}\bigg|_{V_{CB}=\text{constant}}. \tag{10.79}$$

The negative sign in eqn (10.79) represents that as α and α_{ac} are positive, I_C and I_E are of opposite sign for a particular transistor. For a good transistor, $\alpha \simeq \alpha_{dc} \simeq \alpha_{ac}$ and the value of α for such a transistor lies between 0.98 to 0.988, though α is not constant. α actually varies with I_E, V_{CB} and temperature.

10.13.2 Transistor Beta (β)

The transistor beta (β) of a transistor is the maximum current gain that can be obtained when the transistor is operated in CE mode.

We have $I_C = \alpha I_E + I_{CO}$ and $I_E = I_B + I_C$.

$$\therefore I_C = \alpha(I_B + I_C) + I_{CO} \text{ or, } (1-\alpha)I_C = \alpha I_B + I_{CO}$$

$$\text{or, } I_C = \left(\frac{\alpha}{1-\alpha}\right) I_B + \frac{1}{(1-\alpha)} I_{CO} = \beta I_B + (1+\beta) I_{CO}, \tag{10.80}$$

where

$$\beta = \frac{\alpha}{1-\alpha}. \tag{10.81}$$

β is called the *transistor* β and is related to *transistor* α by eqn (10.81).

In dc mode of operation, dc β is defined as

$$\beta_{dc} = \frac{\text{collector current}}{\text{base current}} = \frac{I_C}{I_B} \text{ (in CE mode)}. \tag{10.82}$$

β_{dc} is also denoted by h_{FE}. So

$$\beta_{dc} = h_{FE} = \frac{I_C}{I_B}. \tag{10.83}$$

Usually, $I_B \gg I_{CO}$. $\therefore I_C \simeq \beta I_B$.

Hence, $\beta_{dc} = \beta$.

Usually in commercial transistor β_{dc} varies from 50 to 400 and h_{FE} depends on I_C.

For operation with ac we define *small signal short circuit* CE current gain by β_{ac} or h_{fe}, where

$$\beta_{ac} = h_{fe} = \frac{\text{change in collector current}}{\text{change in base current}} \text{ at fixed collector to emitter voltage } (V_{CE})$$

$$= \left.\frac{\Delta I_C}{\Delta I_B}\right|_{V_{CE}=\text{constant}}. \tag{10.84}$$

Generally, for small I_C, $h_{fe} > h_{FE}$ and for large I_C, $h_{fe} < h_{FE}$, but $(h_{fe} \sim h_{FE}) < 20\,\%$.

10.13.3 Relation between α and β

According to definition,

$$\alpha_{dc} = \frac{I_C}{I_E} \text{ and } \beta_{dc} = \frac{I_C}{I_B}.$$

Again, $I_E = I_B + I_C$.

$$\therefore \frac{I_E}{I_C} = \frac{I_B}{I_C} + 1 \text{ or, } \frac{1}{\alpha_{dc}} = \frac{1}{\beta_{dc}} + 1 \text{ or, } \beta_{dc} = \frac{\alpha_{dc}}{1-\alpha_{dc}}. \tag{10.85}$$

Since $\alpha_{dc} \simeq \alpha$ and $\beta_{dc} \simeq \beta$,

$$\therefore \beta = \frac{\alpha}{1-\alpha}. \tag{10.86}$$

10.14 Basic Common-Emitter Transistor Amplifier

The circuit diagram of a *basic common-emitter transistor amplifier* is shown in Fig. 10.29.

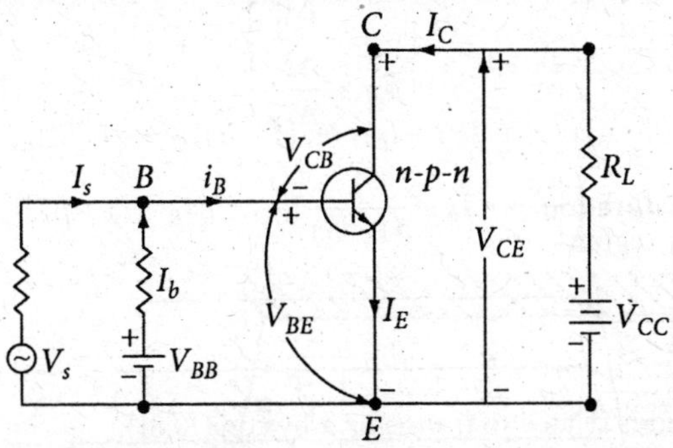

Fig. 10.29: Basic common-emitter (CE) amplifier using an *npn*-transistor

In the above figure,

R_L is the load resistance against which output voltage is developed.

V_{CC} is the supply voltage to the collector (usually 12 to 20 V).

V_{CE} is the collector to emitter voltage.

V_{BE} is the base to emitter voltage.

V_{BB} is the supply voltage to the base and corresponds to the value at the centre of its characteristics [i.e., active region, Fig. 10.30].

v_s is the signal voltage.

Under the zero signal condition (i.e., $v_s = 0$), V_{BB} supplies a dc current I_B into the base and hence, the dc collector current I_C adds to I_B to give the emitter current I_E, i.e.,

$$I_E = I_B + I_C. \tag{10.87}$$

When a signal voltage v_s is applied, an ac current i_s flows which makes the base current i_B given by

$$i_B = I_B + i_s. \tag{10.88}$$

As a result, the ac collector current becomes

$$i_C = I_C + i_c, \tag{10.89}$$

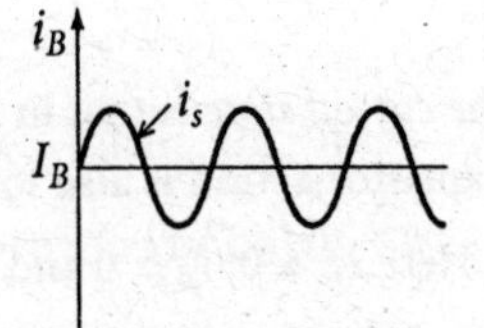

Fig. 10.30: Total current through the base when a signal voltage is applied

where i_c is the ac component of collector current. Since a small base current i_B controls a large collector current i_c, the transistor has a *current gain* A_i. Again, since $V_{BE} < V_{CE}$, there will be a *voltage gain* A_v.

Hence, the power gain of the circuit is

$$A_p = A_v \times A_i. \tag{10.90}$$

The *quiescent point* must lie well within the active region of the transistor characteristic [Fig. 10.31] so that the amplification be faithful. The curve *P* in this figure represents the maximum power the transistor can handle. The values of α_{ac} and β_{ac} are given by

$$\left.\begin{aligned} \alpha_{ac} &= \frac{\Delta I_C}{\Delta I_E} \\ \beta_{ac} &= \frac{\Delta I_C}{\Delta I_B} \end{aligned}\right\}. \tag{10.91}$$

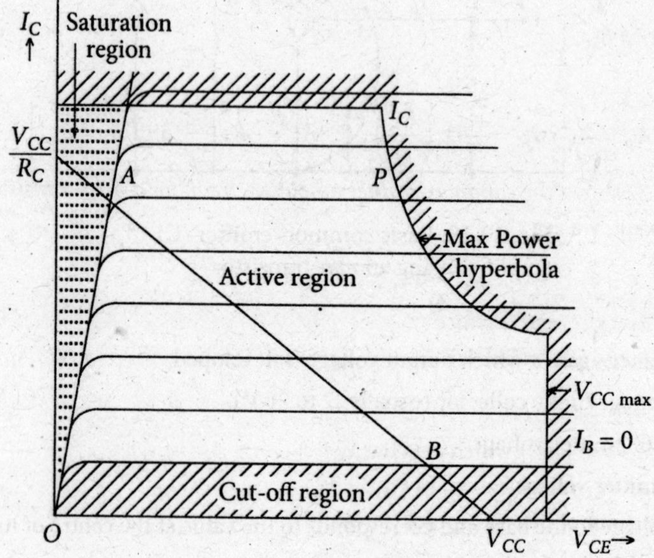

Fig. 10.31: The active region, cut-off region and saturation region

Saturation State: *The saturation state* of the transistor is that in which both the base and the collector are forward-biased. The normal transition work is lost in this state. In this state,

$$(I_C)_{sat} = \frac{V_{CC}}{R_L} \text{ and } (I_B)_{sat} = \frac{(I_C)_{sat}}{\beta_{ac}}. \tag{10.92}$$

The cut-off state is that in which the emitter-base junction (J_E) is reverse-biased. In this state, the normal transistor action is also lost.

Here $I_C \simeq 0, I_B \simeq 0$ and $V_{CE} \simeq V_{CC}$.

Amplification Action of the Transistor: We have

$$V_{CC} = V_{CE} + I_C R_L \text{ or, } \delta V_{CC} = -R_L \delta I_C \; (\because V_{CC} = \text{constant}). \tag{10.93}$$

Again,

$$v_B - \delta V_{BE} = r_i \delta I_B \left[\because r_i = \frac{\delta V_{BE}}{\delta I_B}\right]. \tag{10.94}$$

∴ voltage gain,

$$A_v = \frac{v_0}{v_s} = \frac{\delta V_{CF}}{\delta V_{BE}} = -\frac{R_L\, \delta I_C}{r_i\, \delta I_B} = -R_L \frac{\beta_{ac}}{r_i}. \tag{10.95}$$

Here β_{ac} is large and r_i is small.

$\therefore |A_v| > 1$ for $R_L > 1$.

Again,

$$\beta_{ac} = g_m \times r_i. \tag{10.96}$$

$$\therefore A_v = -g_m R_L. \tag{10.97}$$

Also current gain, $A_i = \beta_{ac}$.

$\therefore$ power gain,

$$A_p = A_v \times A_i = -\beta_{ac}^2\left(\frac{R_L}{r_i}\right). \left[\because A_v = -R_L\frac{\beta_{ac}}{r_i} \text{ and } A_i = \beta_{ac}\right] \tag{10.98}$$

The Load Line: Fig. 10.32 shows the *common-emitter amplifier circuit for a pnp-transistor.* Using *Kirchhoff's law* to the collector circuit, we get

$$V_{CE} = V_{CC} - I_C R_L. \tag{10.99}$$

Eqn (10.99) represents a curve of the form:

$$y = mx + c. \tag{10.100}$$

By plotting eqn (10.99) on the collector characteristic curves, we get the *dc load line* (AB) which cuts the V_{CE} axis at V_{CC} and I_C axis at V_{CC}/R_L as shown in Fig. 10.33. The coordinates on the point of intersection of the load line and the characteristic curve corresponding to the fixed value of I_B [Fig. 10.33] gives the actual dc value of I_C flowing through the transistor and V_{CE} (dc) operating across the transistor. The point is called the *quiescent operating point* or Q-*point*.

The Q-point is usually selected at about the centre of the load line for low amplitude sinusoidally varying voltage or currents applied at the input of the amplifier. Then the output signal will be a faithful reproduction of the small amplitude input signal.

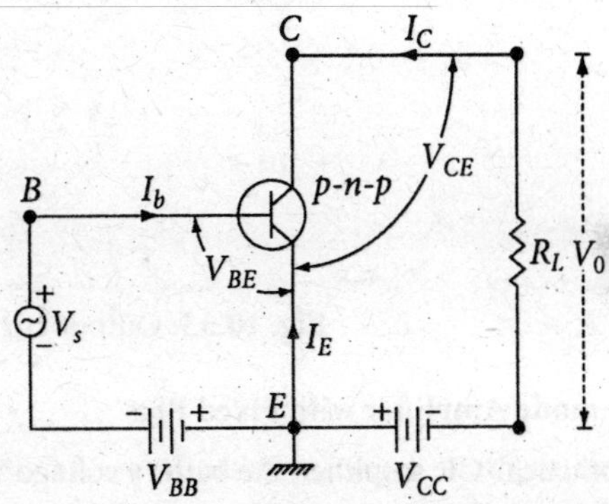

Fig. 10.32: CE amplifier using a *pnp* transistor

To analyse the amplifying action of the CE-mode *pnp*-transistor, let a sinusoidal current of amplitude 100 μA be impressed on the dc base current $I_B = -300\ \mu$A. The quiescent point (when the signal is not applied and $I_B = -300\ \mu$A) is represented by the point Q on the dc load line AB. The point Q will move from A to Q down to B and then from B to A via Q. Due to the variation of base current, the collector current i_C will undergo a variation with time as shown on the left side of I_C-axis. The output current waveform is a sine curve as shown in Fig. 10.33.

Here the change in collector current i_c is in mA though the change in the base current i_b is in μA. Thus, the circuit produces amplification. If the ac components of base current $i_b = 100\ \mu$A and the ac component of collector current $i_c = 9$ mA, then the current amplification is

$$A = \frac{i_c}{i_b} = \frac{9 \times 10^{-3}\text{ A}}{100 \times 10^{-6}\text{ A}} = 90.$$

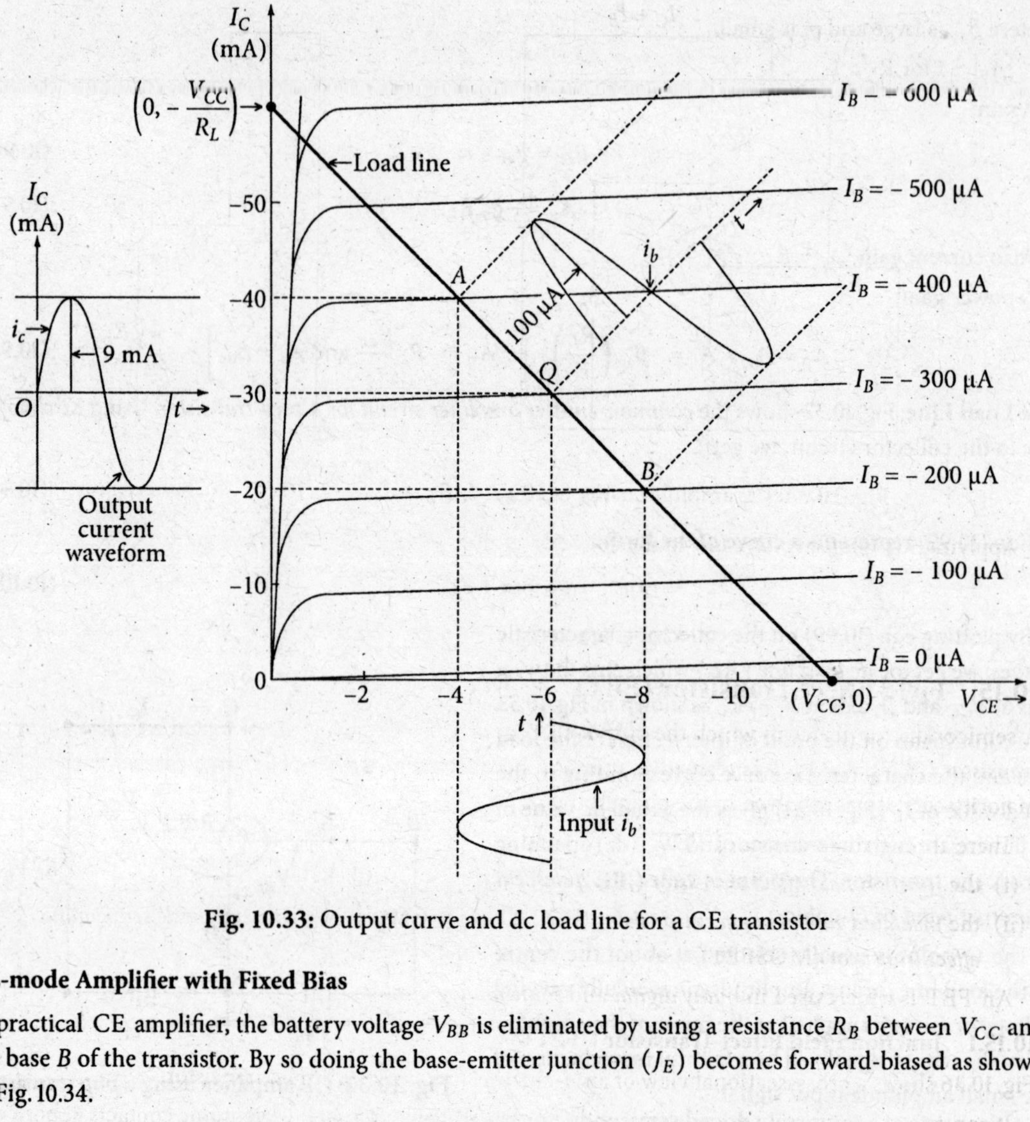

Fig. 10.33: Output curve and dc load line for a CE transistor

CE-mode Amplifier with Fixed Bias

In practical CE amplifier, the battery voltage V_{BB} is eliminated by using a resistance R_B between V_{CC} and the base B of the transistor. By so doing the base-emitter junction (J_E) becomes forward-biased as shown in Fig. 10.34.

By applying *Kirchhoff's voltage law* to the emitter-base loop in the clockwise sense [Fig. 10.34], we get

$$-V_{CC} + I_B R_B + V_{BE} + I_E R_E = 0. \tag{10.101}$$

Again,

$$I_E = I_C + I_B = \beta I_E + I_B = (\beta + 1) I_B. \tag{10.102}$$

From eqns (10.101) and (10.102), we get

$$I_B = \frac{V_{CC} - V_{BE}}{R_B + (\beta + 1) R_E}. \tag{10.103}$$

The current I_B flows in the base circuit and is constant in magnitude.

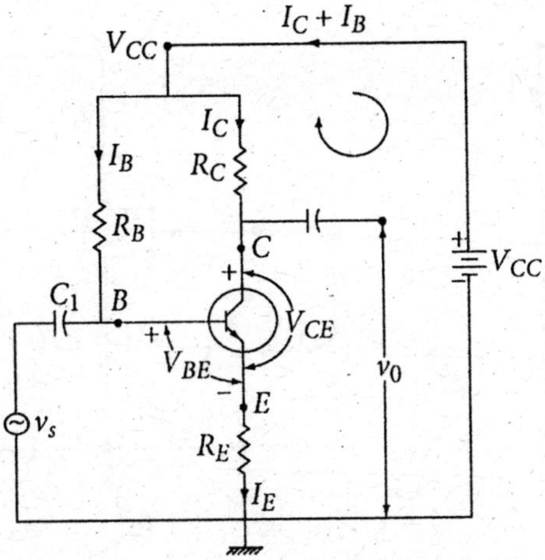

Fig. 10.34: CE amplifier using fixed bias

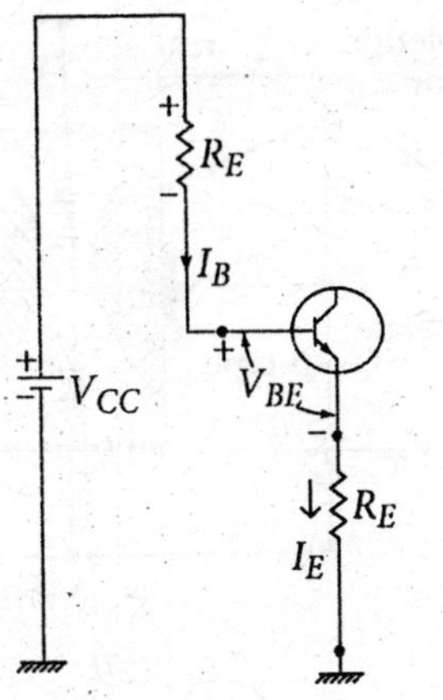

Fig. 10.35: The base circuit

Applying *Kirchhoff's voltage law* to the collector-emitter circuit [Fig. 10.35], we get

$$-I_E R_E - V_{CE} - I_C R_C + V_{CC} = 0$$

$$\text{or, } V_{CE} = V_{CC} - I_C (R_C + R_E) \quad (\because I_E \simeq I_C). \tag{10.104}$$

10.15 Field Effect Transistor (FET)

A semiconductor device in which the output current is controlled by an electric field is called a *field effect transistor* (FET). An FET is termed as unipolar because only one type of charge carriers (specially, the majority carriers) is involved in the operation of it.

There are two main classes of FET:

(i) the *junction field effect transistor* (JFET or FET),

(ii) the *insulated gate field effect transistor* (IGFET), commonly known as *metal oxide semiconductor field effect transistor* (MOSFET).

An FET is widely used in *many digital* and *some analog* circuits.

10.15.1 Junction Field Effect Transistor (JFET)

Fig. 10.36 shows a cross-sectional view of an n-channel junction field effect transistor.

It consists of a uniformly doped semiconductor bar (called a *channel*) with ohmic contacts at both ends. By applying a voltage between the end terminals a current is made to flow between the end terminals. The FET is called an *n-channel* FET or a *p-channel* FET according as the semiconductor bar is of n- or p-type. The junctions on both sides of the semiconductor bar are formed with impurities opposite to that of the channel, i.e., n-type impurities with p-type channel and p-type impurities with n-type channel. *Silicon* and *gallium arsenide* are used for basic semiconductors in FETs. The structure and the circuit symbols of an n-channel JFET are shown in Fig. 10.36. The different notations marked in the figure are:

Source: Source (S) is the terminal through which the majority carriers enter the channel region. The corresponding current is denoted by I_D.

Drain: Drain (D) is the terminal through which the majority carriers leave the bar and the conventional current is denoted by I_S.

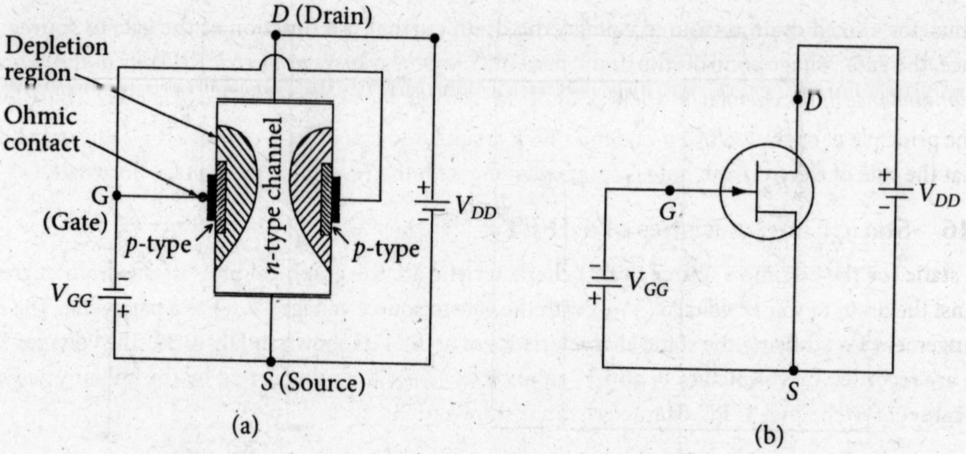

Fig. 10.36: n-channel JFET (a) Structure (b) Circuit symbol

Gate: Heavily doped regions on both sides of the doped semiconductor bar are formed by alloying, diffusion or by other techniques using impurities opposite to that of the channel. These regions are called *gate*. Hence, for n-channel, the gate is p-type.

The structure and circuit symbol of a p-channel JEFT are shown in Fig. 10.37.

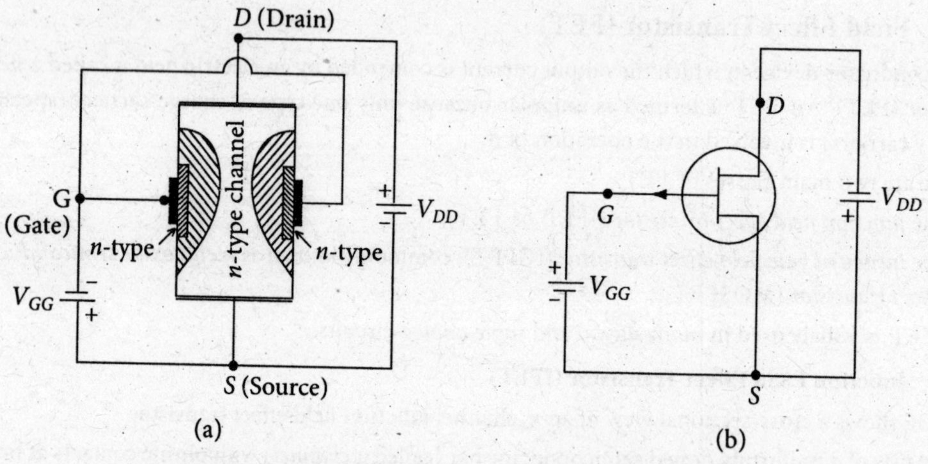

Fig. 10.37: p-channel JFET (a) Structure (b) Circuit symbol

10.15.2 Principle of Operation

Consider an n-channel JFET connected in common mode with the gate junction reverse-biased. Depletion regions are formed on both sides of the channel [Fig. 10.36]. A gradual potential drop occurs due to the passage of current between the source and the drain which makes the gate more *reverse-biased* near the drain end than the source end. *This makes the depletion region wider near the drain than that near the gate.* Then the drain current flows through an effectively wider wedge-shaped channel as the depletion region is free of mobile charges. If now the gate reverse-biased voltage be increased, the cross section of the depletion region decreases with the further extension of the depletion region into the channel. As a result, the flow of current reduces. Thus, both the drain current and the conductance decrease.

Thus, for a fixed drain to source voltage, the drain current is a function of the gate to source voltage. Hence, the gate voltage controls the drain current. *Since the transverse electric field due to the gate controls the conductance of the channel and hence, the device current, the device is called a field effect transistor* (FET).

The principle of operation of a p-channel JFET is similar to that of an n-channel JFET, the only difference is that the role of electron and hole is reversed along with the reversal of the voltage polarities.

10.16 Static Characteristics of a JFET

The static (or the common source drain) characteristic are the graphical plots of the drain current (I_D) against the drain to source voltage (V_{DS}) with the gate to source voltage (V_{GS}) as a parameter. The circuital arrangement for studying the static characteristics of a JFET is shown in Fig. 10.38. The voltages V_{GS} and V_{DS} are recorded by voltmeters V_1 and V_2 respectively and the drain current by the milliammeter mA. A typical set of n-channel JFET characteristics is shown in Fig. 10.39.

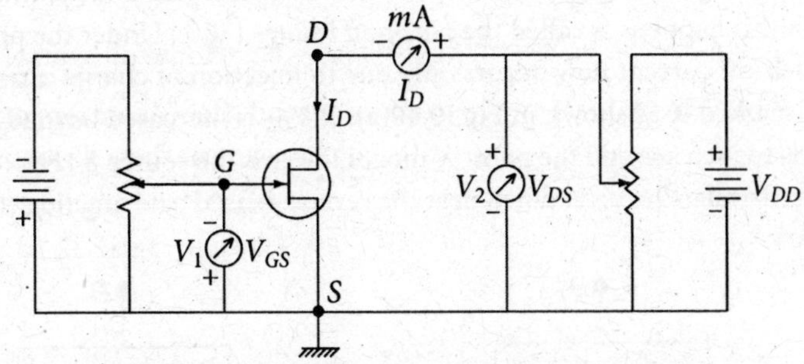

Fig. 10.38: Circuit diagram for drawing the static characteristic curves of an n-channel JFET.

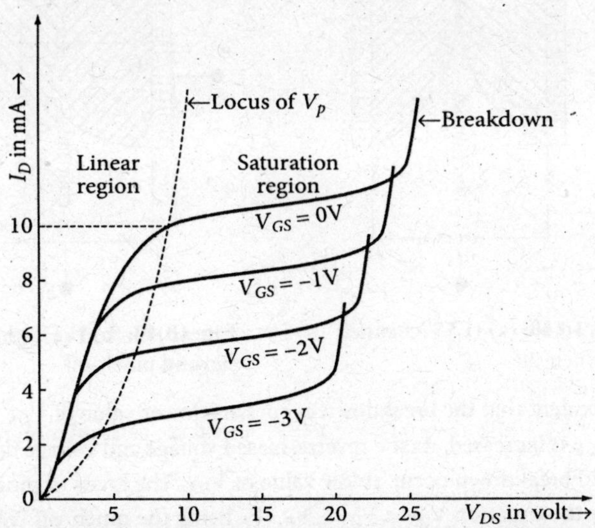

Fig. 10.39: Drain characteristics of an n-channel JFET

The static characteristic curves for an n-channel JFET are shown in Fig. 10.39. The drain characteristics can be divided into the following three regions:

(i) *The ohmic or linear region*: In this region, V_{DS} is small and I_D is approximately proportional to V_{DS}.

(ii) *The saturation region*: In this region, I_D is almost constant and is independent of V_{DS}.

(iii) *The breakdown region*: In this region, I_D rises abruptly to large value for a slight change in V_{DS} so that the characteristic curve is almost vertical to V_{DS} axis.

10.16.1 Explanation of the Nature of the Characteristics

In case of the characteristic for $V_{GS} = 0$, the n-channel bar acts as a simple resistor and I_D increases linearly with V_{DS}. With the increase of I_D, the ohmic voltage drop along the n-channel region reverse-biases the gate source junction. This decreases the effective conducting channel cross section. Due to the gradual increase in ohmic potential drop along the length of the bar the gate becomes reverse-biased more near the gate end than the source end. Hence, *the width of the depletion region increases from source to drain end* and I_D is constrained to move through a *wedge-shaped channel.* The channel becomes narrower with the increase of V_{DS}, and I_D versus V_{DS} curve becomes non-linear. If the value of V_{DS} is continued to increase, a stage will come when the depletion regions meet each other and the channel is said to be *pinched off.* The particular value of V_{DS} at which this happens is called the *pinch-off voltage* (V_p). Under the pinched-off condition I_D saturates. The pinch-off current flow occurs only due to injection of charge carriers from channel to the depletion region marked X as shown in Fig.10.40(a). If V_{DS} is increased beyond V_p, the width of the depletion region tends to increase and the point X moves towards the source S [Fig. 10.40(b)]. At X, since V_{DS} is further increased avalanche breakdown near the reverse-biased gate junction starts and I_D abruptly increases to large value.

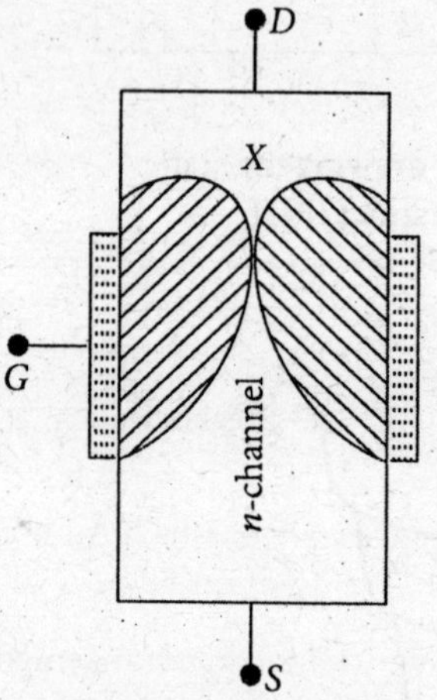

Fig. 10.40: (a) JFFT channel at pinch-off

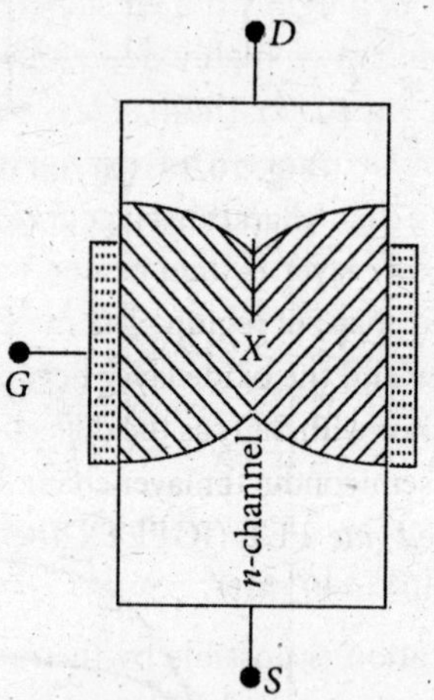

Fig. 10.40: (b) JFFT channel beyond pinch-off

From Fig. 10.40, it is evident that the breakdown occurs at a lower value of V_{DS}, if the magnitude of the reverse-biased voltage V_{GS} is increased. As the reverse-biased voltage and voltage drop due to the flow of I_D add up, the pinch-off and breakdown occur at low value of V_{DS}. The locus of pinch-off point is shown in Fig. 10.39 and is given by the equation $V_{DS} = V_{GS} - V_P$, V_P being the pinch-off voltage.

10.16.2 Transfer Characteristic Curve of a JFET

The variation of I_D with the gate to source reverse-biased voltage V_{GS} at constant value of V_{DS} $(> V_P)$ represents the *transfer characteristic of a* JFET as shown in Fig. 10.41. The upper limit of I_D is I_{DSS} for $V_{GS} = 0$.

The shape of the transfer characteristic is parabolic and is represented by

$$I_D = I_{DSS}\left(1 - \frac{V_{GS}}{V_P}\right)^2. \qquad (10.105)$$

Thus, I_D can be controlled by varying V_{GS} and hence, a JFET can be used as a *voltage controlled current source.*

For a BJT in the active region $I_C \approx \beta I_B$. Thus, BJT is a current controlled device as I_C can be varied by varying I_B.

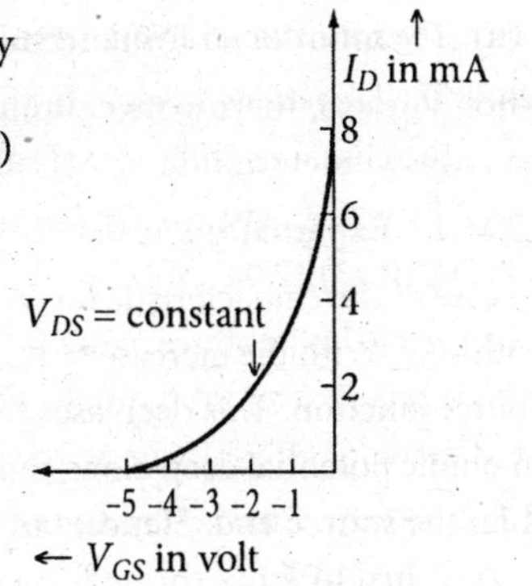

Fig. 10.41: Transfer characteristic of an *n*-channel JFET

10.17 Metal Oxide Semiconductor Field Effect Transistor (MOSFET)

MOSFET has important commercial application in digital circuit. In IC it occupies lesser space and suitable for a very large scale integration. It is also used as an important power device.

MOSFETs are of the following two types: (i) enhancement MOSFET and (ii) depletion MOSFET. Each of the two types may be of either *n*-channel (NMOS) or *p*-channel (PMOS) transistors.

(i) *Enhancement* MOSFET: An *n*-channel enhancement MOSFET consists of a lightly doped substrate of *p*-type on the two sides of which two highly doped n^+-regions are diffused as shown in Fig. 10.42. These two n^+-regions serve as the source (S) and the drain (D) respectively. The region between S and D is the channel. A thin layer of SiO_2 is grown over the surface. The oxide layer is then coated with a metal gate electrode above the channel region. The metal gate, the semiconductor substrate and the oxide layer between them form a parallel plate capacitor with SiO_2 as the dielectric. As the gate is insulated from the semiconductor layer channel by SiO_2 layer it is called an *insulated gate* FET (IGFET). The input resistance of the devices is ~ 10^4 to 10^9 MΩ.

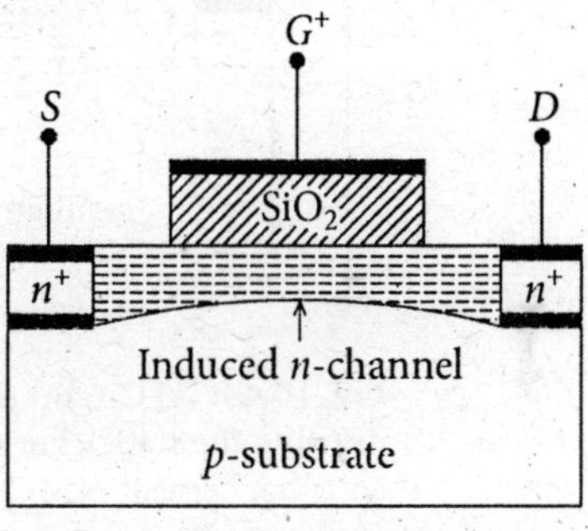

Fig. 10.42: An *n*-channel enhancement MOSFET

The device operation is possible by the metal oxide semiconductor arrangement by an electric field produced by the voltage applied to the gate and hence, the name MOSFET.

Operation

Let the source and the substrate are grounded, the drain to source voltage (V_{GS}) is zero and a positive voltage is applied to the gate of the enhancement NMOS transistor. The holes will be repelled by this positive voltage of the gate and move to the bulk from the surface of *p*-substrate. The minority electrons of the *p*-substrate are attracted by the gate and form a layer below the SiO_2, called the *inversion layer,* when $V_{GS} > V_T$, where V_T is called the *threshold voltage.* If now drain is made positive relative to the source, a current can flow between the source and the drain through the *induced n-channel* just below the SiO_2 layer. The induced negative charges in the channel increase by increasing V_{GS} which leads to the increase of drain current and hence, conductivity of the channel. The name *enhancement type* MOSFET is due to the fact that the drain current is enhanced by increasing positive gate voltage.

Circuit Symbol of an Enhancement MOSFET

When $V_{GS} = 0$, there is no conducting channel and hence, the enhancement mode of MOSFET is off. This is represented by broken channel like in the schematic circuit symbol (Fig. 10.43).

The arrow in the above figure points to the inversion layer (which acts like an *n*-channel). For a *p*-channel enhancement mode MOSFET, the arrow points outwards.

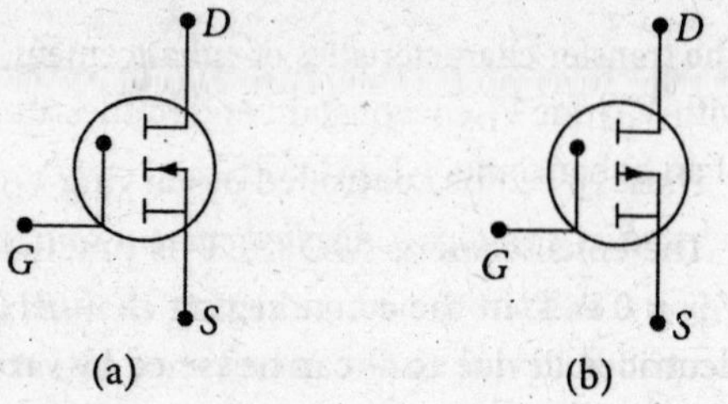

Fig. 10.43: (a) Circuit symbol for *n*-channel (b) *p*-channel enhancement mode MOSFET

10.17.1 Static Characteristics of Enhancement MOSFET

The static drain characteristics of an enhancement MOSFET are the graphical representation of the variation drain current (I_D) with the variation of source to drain voltage V_{DS} with constant gate voltage (V_{GS}) greater than V_T. A set of drain (also called volt-ampere) characteristics of an *n*-channel enhancement type MOSFET is shown in Fig. 10.44(b).

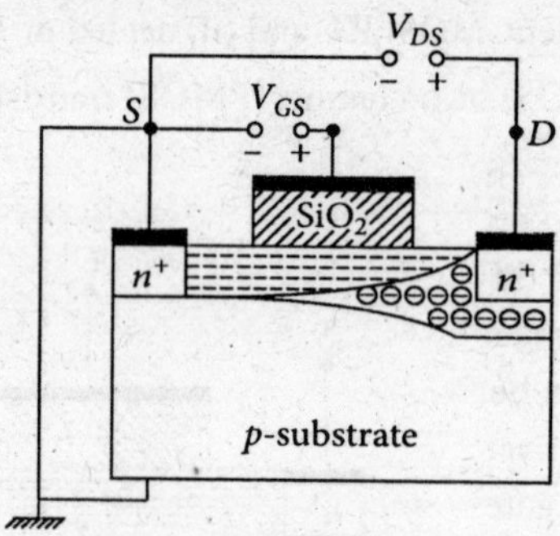

Fig. 10.44: (a) Circuit diagram for drawing the static characteristic of an enhancement NMOSFET

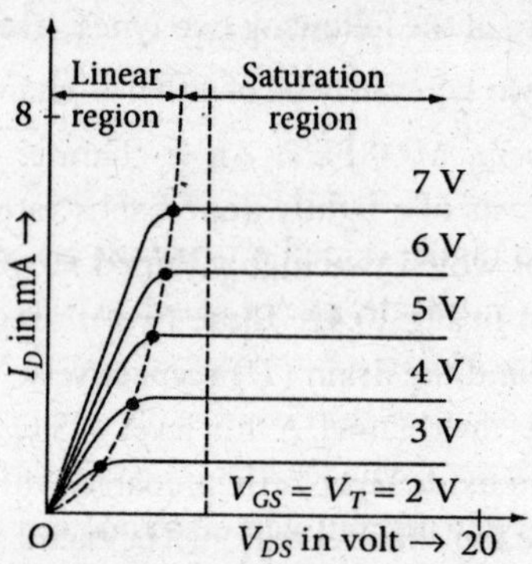

Fig. 10.44: (b) Drain characteristic of an enhancement NMOSFET

If the gate to source voltage V_{GS} is less than the threshold voltage V_T, the enhancement MOSFET is off and gate current $I_D = 0$. If V_{GS} is maintained constant such that $V_{GS} > V_T$ and V_{DS} is increased gradually from OV, then I_D increases almost linearly with V_{DS} with small values of V_{DS}. The enhancement MOSFET in this behaves as a resistor whose value is being controlled by V_{GS}. This is *linear ohmic region.*

If now V_{GS} is gradually increased, the channel conductivity increases due to the increase of electron density in the channel region and the resistance between the source and drain decreases. As V_{DS} is increased, keeping V_{GS} constant, the *pn* transition region between the substrate and the gate becomes reverse-biased but the source to substrate voltage remains zero. As a result, the width of the depletion region at the drain transition region increases and the negative space charge due to the acceptor ions increases. This helps to decrease the cross section of the effective channel with a distance from source to drain and the channel resistance is increased [Fig. 10.44(b)]. Ultimately V_{DS} assumes a critical value at which the drain end is *pinched off* and I_D becomes almost saturated.

As V_{DS} increases further the channel resistance increases and I_D increases by small amount with V_{DS} source. A sudden increase in I_D with a large value of V_{DS} may occur due to breakdown.

10.17.2 Transfer Characteristics of Enhancement MOSFET

The transfer characteristics of enhancement MOSFET is the graphical representation of the variation of I_D with V_{GS} for V_{DS} = constant and in the saturation region. Fig 10.45 shows a typical transfer characteristics of an enhancement NMOSFET.

The enhancement MOSFET is practically off (or $I_{DSS} = 0$) for $V_{GS} = 0$ as there is no conducting channel under this condition. The electric field due to the increase of V_{GS} repels the holes from SiO_2 surface into the interior of p-type substrate. The minority electrons at the same time are attracted to this region forming an induced n-channel, and a current flows.

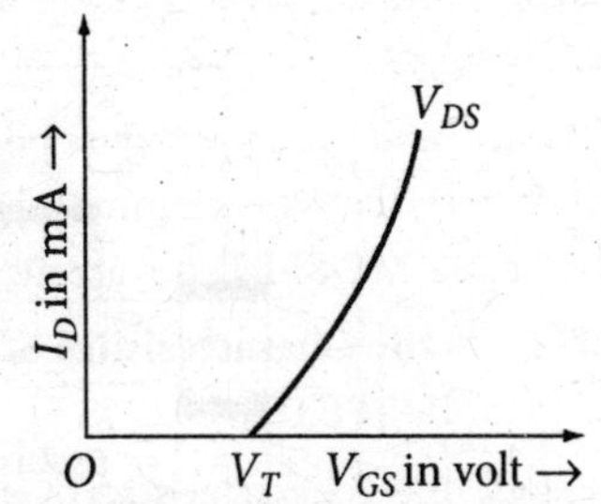

Fig. 10.45: Transfer characteristic of an enhancement NMOSFET

Now increasing V_{GS} (for $V_{GS} > V_T$), the electron density in the channel increases causing an increase in the drain current I_D in the device. The saturation level of I_D, for $V_{GS} > V_T$, is related to V_{GS} by the following non-linear relationship: $I_D = k(V_{GS} - V_T)^2$ where k is a constant which depends on the construction of the device. The value of k usually varies from 10^{-4} to 10^{-3} A/V^2.

Depletion MOSFET

Fig. 10.46 shows the structure of an n-channel depletion MOSFET. Two highly doped n^+-regions are diffused into the highly doped p-substrate, one of which acts as a source and the other as a drain. A narrow n-channel is diffused between the source and the drain. A thin layer of SiO_2 is formed on the surface of the substrate. A thin layer of SiO_2 is formed on the surface of the substrate. The metallic gate electrode is then developed on the SiO_2 surface and kept insulated from the rest of the structure.

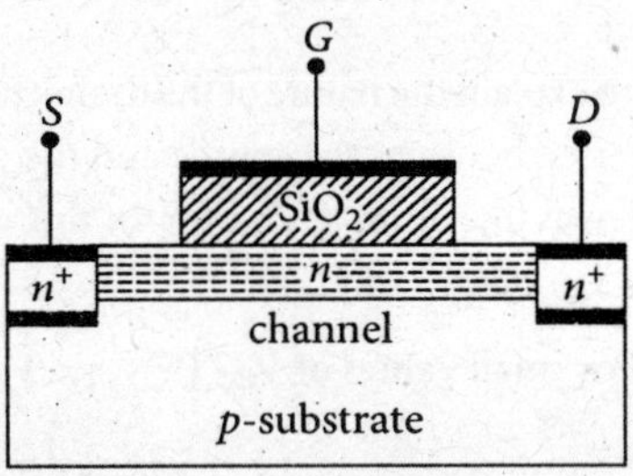

Fig. 10.46: An n-channel enhancement MOSFET

The gate metal, the semiconductor substrate and the oxide layer between them form a parallel plate capacitor with SiO_2 as the dielectric. The device current is controlled by the electric field produced by a gate voltage in the metal oxide semiconductor arrangement and hence, the name MOSFET.

10.17.3 Operation of Depletion NMOSFET

When the source and substrate are grounded and V_{GS} is zero, an appreciable drain current I_{DSS} flows with the increase of V_{GS} due to the presence of a channel. If now a negative voltage be applied to the gate, the electrons get repelled from the channel towards the interior of the p-substrate and holes are attracted towards the channel region. As a result, a depletion of majority carriers in the channel occurs due to electron-hole recombination. Hence, the device is called a *depletion* MOSFET. If the negative voltage is gradually increased, a stage will come when the channel will be depleted of majority carriers causing the current between the source and the drain zero. The minimum value of the negative voltage at which this occurs is called the *threshold voltage* (V_T).

If now a positive voltage be applied to the gate, it attracts free electrons into the channel thereby increasing the channel conductivity. The source to drain current under this condition becomes greater than I_{DSS}. This shows that a depletion MOSFET can also be operated *under enhancement* mode also.

10.17.4 Characteristic Curves of MOSFET

The circuit connection of a depletion MOSFET is shown in Fig. 10.47 and its drain characteristics are shown in Fig. 10.48.

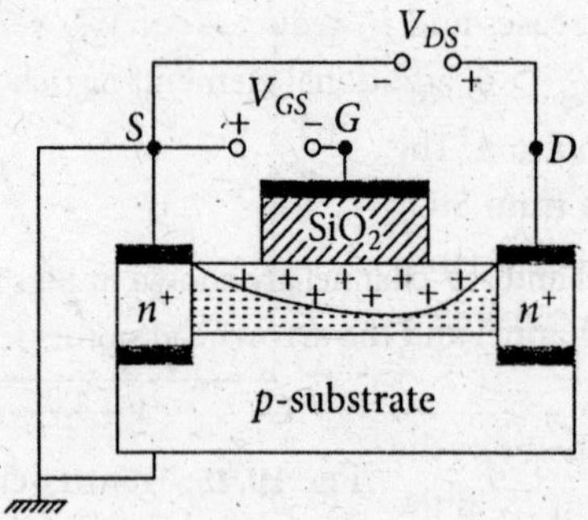

Fig. 10.47: Depletion channel MOSFET

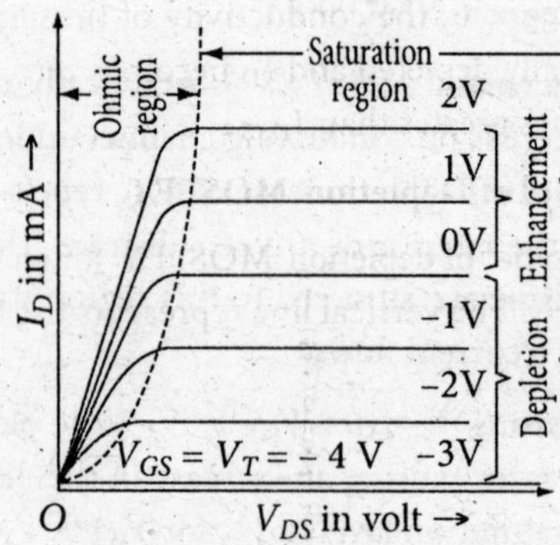

Fig. 10.48: Drain characteristics of a depletion n-channel MOSFET

To explain the nature of the drain characteristics, Fig. 10.48, let us start with $V_{GS} = 0$. As a channel is already diffused between the source and the gate, I_D first increases linearly with V_{DS}, and the depletion MOSFET behaves like a *simple resistor.* In this ohmic region I_D is given by $I_D = k\left[2\left(V_{GS} - V_T\right)V_{DS} - V_{DS}^2\right]$, where $k = I_{DSS}/V_T^2$.

For small values of $V_{DS}\ (V_{DS} \ll V_T)$,

$$I_D = 2k\left(V_{GS} - V_T\right)V_{DS}, \text{ i.e., } I_D \propto V_{DS}.$$

Application of a negative voltage to the gate causes the repulsion of electrons from the channel into the interior of the p-substrate and attraction of holes to the channel. As a result, a depletion of majority electrons in the channel occurs. For a given value of V_{DS}, I_D decreases. The MOSFET now operates in the *depletion mode.*

If V_{GS} is negative, then electrons are repelled to the interior of the p-substrate, and holes are attracted to the channel region. A depletion of majority carrier in the channel will occur and I_D reduces for a given V_{DS}. Then the MOSFET operation is said to be in the *depletion mode.*

Again, if V_{DS} is increased, keeping V_{GS} constant, I_D will saturate due to pinch-off of the channel. The channel is *more depleted near the drain than the source* [Fig 10.48] due to the voltage drop when I_D flows along the channel. I_D then flows through a narrow wedge-shaped channel, which saturates for a particular value of V_{DS}. If the gate voltage be made positive, then the electrons will be attracted to the channel region and the holes will be pushed back to the interior of the p-substrate. This enhances the conductivity of the channel and drain current $I_D > I_{DSS}$. The operation of the MOSFET is then in the *enhancement mode.*

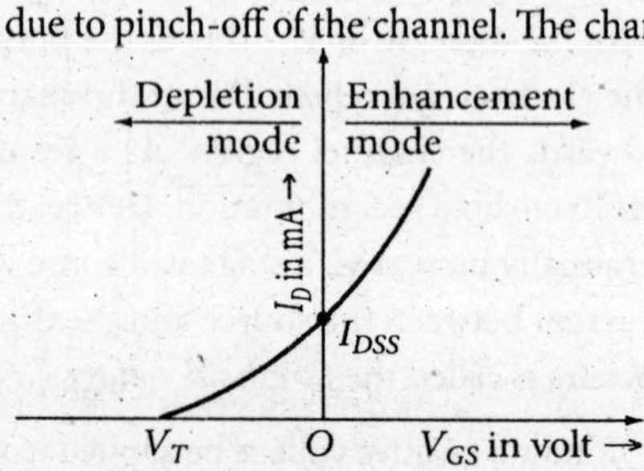

Fig. 10.49: Transfer characteristics of an n-channel MOSFET

Transfer Characteristics of the Depletion MOSFET

The transfer characteristics of an n-channel MOSFET is shown in Fig. 10.49.

If V_{DS} be in the saturation region, then for $V_{GS} = 0$, appreciable drain current I_{DSS} flows through the device. For $V_{GS} < 0$, the conductivity of the channel decreases and I_D reduces. For V_{GS} very large, the channel is totally depleted and I_D becomes zero. But for $V_{GS} > 0$, additional elements enrich the channel and I_D becomes greater than I_{DSS}.

Circuit Symbol of Depletion MOSFET

The circuit symbol of depletion MOSFET for an n-channel and a p-channel are shown in Figs 10.50(a) and (b) respectively. The vertical line represents the built-in channel and the arrowhead points to the n-type material.

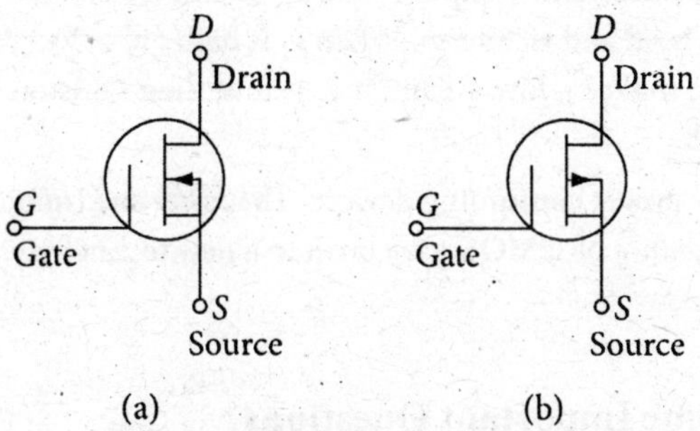

Fig. 10.50: Depletion MOSFET (a) n-channel (b) p-channel

10.18 Application of MOSFET in Digital Switching

In computer industry MOSFET finds important applications as transistors can be used as switching device. An E-MOSFET inverter with a passive load R_D is shown in Fig. 10.51. The device gives a high output at low input and low output at high input.

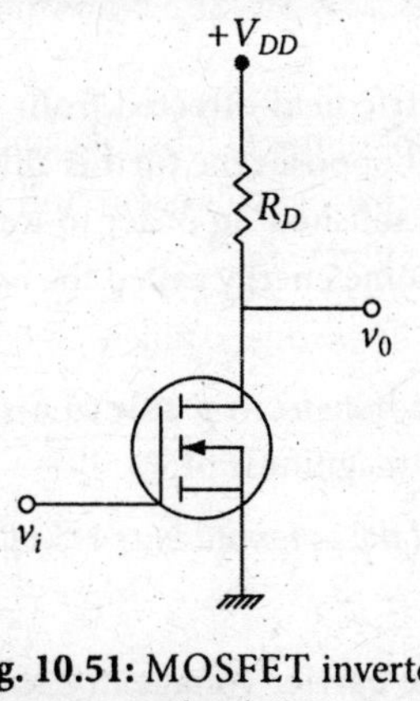

Fig. 10.51: MOSFET inverter using passive load R_D

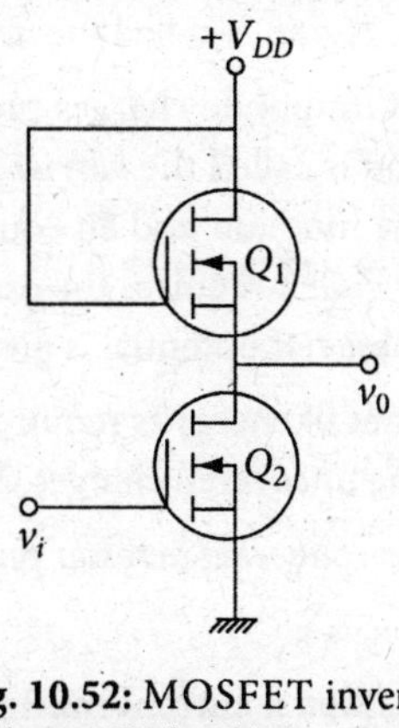

Fig. 10.52: MOSFET inverter with a MOS load

In the arrangement shown in Fig. 10.51, if the input voltage v_i is less than V_T, the MOSFET becomes off. Under this condition voltage drop against R_D is zero and the output voltage v_0 $(= V_{DD})$ is high. For $v_i > V_T$, the MOSFET saturates and v_0 becomes low.

Since passive resistors require more area than MOSFETs, IC form of passive resistors are seldom used. In Fig. 10.52, a MOSFET inverter with an active load is depicted. There are two MOSFETs Q_1 and Q_2 as shown in the figure. The MOSFET Q_1 plays the role of a larger resistor whereas the MOSFET Q_2 acts as a switch. The reduction in size of IC by the use of MOSFET makes it popular in LSI circuit connections.

10.19 The Complementary MOS (or CMOS)

An important use of CMOS is the logic inverter circuits as shown in Fig. 10.53. CMOS consists of one NMOS and one PMOSFET on the same substrate. Q_1 represents a PMOS and Q_2 is an NMOS. The source is connected to a voltage V_{SS} and Q_2 is grounded. The gates of both are connected to the input v_i and drains are connected to the output v_O. Both Q_1 and Q_2 act like switches, when Q_1 is on, Q_2 is off and vice versa. When v_i is high $(v_i > V_T)$, Q_2 is on but Q_1 is off which make v_O low. Again, if v_i is low, then Q_1 is on but Q_2 is off when v_O is high.

CMOS are extremely low power consuming devices. The *faster switching speed* and *higher input impedance* of CMOS gave birth to a new technology, called CMOS *logic*.

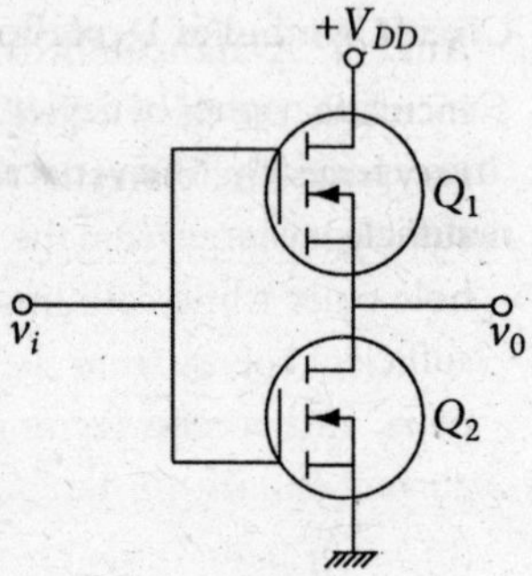

Fig. 10.53: CMOS inverter

10.20 Answers of Some Important Questions

10.1. *How does barrier potential arise in a pn-junction?*

Ans. We consider a *pn*-junction in which there are only holes in *p*-side and electrons in the *n*-side. Due to the existence of density gradient across the junction, electrons will diffuse to the *p*-side and holes to the *n*-side through the junction. After crossing the junction the electrons and holes will recombine each other and the immobile ions in the neighbourhood of the junction will be left. These are called *uncovered* charges. The vicinity of the junction is depleted of mobile electrons and holes and as such the region is called the *depletion region.*

The uncovered immobile charges produce an electric field directed from *n*-side to *p*-side across the junction and is called the *barrier field.* This field opposes the further diffusion of electrons and holes across the junction and an equilibrium is established. In order to transfer an electron from the *n*-side to *p*-side it requires the expenditure of some energy called the *barrier energy* (E_B). The corresponding barrier potential is given by $V_B = \frac{E_B}{e}$.

An equal amount of energy is required to transfer a hole from *p*-side to *n*-side across the junction. The width of the uncovered charges determines the magnitude of V_B.

10.2. *Will a voltmeter connected across a pn-junction read the potential barrier voltage* (V_B)?

Ans. The answer is *no.*

If it is possible, then a current must flow due to the barrier voltage in a short-circuit *pn*-junction diode which would heat the connecting metal wire. The heating of the wire must be due to the heat flowing from the *pn*-junction, which will be cooled simultaneously. But according to the second law of thermodynamics, *heat cannot flow out of itself from a body at lower temperature (the pn-junction) to a body at higher temperature (the connecting wire).*

Hence, we conclude that the current through the circuit must be zero. This means, the barrier voltage is balanced by the metal-to-semiconductor contact potential at the ends of the *pn*-junction diode. As the current is zero, the connecting wire may be cut without any change in the system and the voltage drop across the cut must be zero.

Thus, a voltmeter connected to the terminals of a *pn*-junction diode will show zero reading.

10.3. *What do you mean by* (a) *avalanche breakdown and* (b) *Zener breakdown? Point out the difference between them.*

Ans. (a) **Avalanche breakdown**: In case of *lightly doped pn*-junction as the applied *reverse-bias* increases, the electric field across the junction increases and at some value of the bias the field becomes very large. Therefore, the thermally generated minority carriers during traversing the junction acquire sufficient energy from the field and some of the carriers disrupt a covalent bond creating new electron-hole pairs while colliding with a neutral immobile ion. The newly created carriers again acquire sufficient energy from the field and collide with the immobile ions to generate further electron-hole pairs. This process is cumulative and is known as *avalanche multiplication*. This creates large reverse current and the *pn*-junction diode is said to be operated in the region of *avalanche breakdown.*

(b) **Zener breakdown**: In case of *heavily doped pn*-junction, the width of the depletion region is very small and the electric field there may be very high at *relatively low reverse-bias* across the junction. This field exerts a strong force on the bound electrons to tear it out from a covalent bond. Thus, new electron-hole pairs are generated and the reverse current increases. It may be noted that the carriers get negligible chance to generate new electron-hole pairs through thermal collision. Zener explained this phenomenon from the standard point of *quantum mechanical tunnelling* and hence, it is known as *Zener breakdown.*

Difference between Avalanche breakdown and Zener breakdown

Avalanche breakdown	*Zener breakdown*
1. It occurs in lightly doped *pn*-junction for reverse-bias.	1. It occurs in heavily doped *pn*-junction for reverse-bias.
2. It takes place above a reverse-biased voltage of 6 V.	2. It takes place below a reverse-biased voltage of 6 V.
3. If the junction temperature increases, vibrational motion of ions and collision frequency increases. The average distance between two successive collisions decreases and the energy gained is insufficient to start the avalanche process. Hence, with the increase of junction temperature, higher applied voltage is necessary to start the process.	3. In case of Zener breakdown, if the junction temperature increases, then the energy of the valence electron increases helping to escape from the covalent bond. This necessitates smaller reverse-biased voltage to start *Zener breakdown* with the rise of junction temperature.

10.4. *On what factors does potential barrier across a pn-junction depend?*

Ans. The potential barrier (V_B) across a *pn*-junction with concentrations of acceptor (N_a) and donor (N_d) ions at temperature T is given by

$$V_B = \frac{kT}{e} \ln \frac{N_a N_d}{n_i^2},$$

where n_i is the concentration of the intrinsic carriers. The above relation shows that the barrier voltage of a *pn*-junction depends on temperature (T) of the junction and the values of N_a, N_d and n_i.

10.5. *What is Shockley equation?*

Ans. The Shockley equation (or rectifier equation) for a *pn*-junction diode is given by

$$I = I_s\left(e^{eV/kT} - 1\right),$$

where I and I_s are the current and reverse saturation current through a *pn*-junction diode for an applied voltage V at temperature T. η is a constant related to recombination of electrons and holes in the space charge region and has value such that $1 < \eta < 2$.

10.6. *Point out the basic difference between a* BJT *and an* FET.

Ans. (i) A BJT is a current controlled device whereas an FET is a voltage controlled device. The flow of current in an FET is controlled by an applied electric field and hence, the name *field effect transistor* or FET.

(ii) BJT is a *bipolar device* in which both majority and minority charge carriers are involved in the current flow but FET is a *unipolar device* in which only majority charge carriers are responsible for the current flow.

(iii) The input impedance of an FET is very high compared to that of a bipolar transistor.

(iv) FETs occupy less space in IC circuit and are less noisy than BJTs.

(v) In an FET current, carriers move through the channel by the *drift process.* But in a BJT, carriers move through the base by *diffusion process.*

(vi) FETs are thermally more stable than the BJTs.

(vii) Since drift is a faster process than diffusion, FETs can be used at higher frequencies (~ GHz).

(viii) FETs are linear devices. Intermodulation and cross-modulation products of FETs are much smaller than those of bipolar transistors.

(ix) The gain bandwidth product for FET is low as compared to that of BJT. FETs are very sensitive to overload voltages.

10.7. *What are pinch-off and pinch-off voltage as referred to a* JFET?

Ans. In a JFET, the depletion regions meet each other at a certain value V_P of V_{DS}, the channel is then said to be pinched off. The voltage at which pinched off occurs is called the pinch-off voltage (V_P). The current (I_D) gets almost saturated at pinched off condition.

10.8. *What is a* MOSFET? *What are its different types?*

Ans. Field effect transistors are of two types: (i) the junction field effect transistor (JFET or, simply FET) and (ii) the metal oxide semiconductor field effect transistors (MOSFET).

MOSFETs are again of following two types: (a) enhancement MOSFET and (b) depletion MOSFET. Both types can be of *n*-channel or *p*-channel variety and are called NMOS and PMOS transistors respectively.

10.9. *Discuss the superiority of* NMOSFET *over that of* PMOSFET.

Ans. An NMOSFET is compact due to higher mobility of electrons and can be used in higher frequencies. For same size of the devices, the current in the NMOSFET is double that of PMOSFET. The ON resistance of NMOSFET is about one-third of ON resistance of PMOSFET. That is why

PMOSFETs have become obsolete (with the exception of CMOS technology) and NMOSFETs are widely used.

10.10. *What is depletion region in a pn-junction? On what factors does the width of the depletion region depend?*

Ans. The density of free electron in an n-type material is greater than in the p-type. The hole concentration in the p-type material, on the other hand, is much larger than that in the n-type semiconductor. When a pn-junction is formed, diffusion of electrons from n-side to p-side and diffusion of holes from the p-side to n-side take place due to the difference in concentration gradient of electrons and holes until equilibrium is reached. By this, the electrons leave behind positively charged immobile donor impurity ion on n-side and holes leave behind the negatively charged immobile acceptor impurity ion on the p-side. The electrons after diffusion into the p-side recombine with the holes. Similarly, the holes after diffusion into the n-side recombine with the electrons. Because of this diffusion and recombination positive and negative immobile ions appear in the neighbourhood of the junction. The region in the neighbourhood of the pn-junction which is depleted of mobile charges and contains only immobile ions is called the *depletion region, space charge region* or *the transition region* of the pn-junction.

If N_a and N_d be the concentration of acceptor and donor ions respectively in the depletion region of a pn-junction, then the width of the depletion region is given by

$$W = \left[\frac{2eV_B\,(N_a + N_d)}{\epsilon\, N_a\, N_d}\right]^{1/2},$$

where V_B is the potential barrier voltage.

Thus, the width of the depletion region of a pn-junction depends on V_B, and N_a and N_d of the junction.

10.11. *Write down the approximate expression for the pinch-off voltage* (V_P).

Ans. An approximate expression for the pinch-off voltage (V_P) for an *n-channel* FET is given by

$$|V_P| = \frac{eN_d}{2\epsilon}a^2,$$

where N_d = the concentration of donor in the channel, a = half the channel width, ϵ = the permittivity of the channel material and e = the electronic charge.

For p-channel,

$$|V_P| = \frac{eN_a}{2\epsilon}a^2,$$

where N_a = the concentration of acceptor in the channel.

10.21 Solved Problems

10.1. *A p-type* Ge *is made by adding acceptor atoms at a rate of one atom per* 4×10^8 Ge *atoms.* $n_i = 2.5 \times 10^{19}/\text{m}^3$ *at* 300 K *and all the acceptor atoms are ionised at* 300 K. *If the density of* Ge *atom is* 4.4×10^{28} atoms/m^3, *compare the density of electrons with intrinsic charge carriers.*

Solution: Since one acceptor atom is added for 4×10^8 Ge atoms, the number of acceptor atoms added for 4.4×10^{28} atoms is

$$\frac{4.4 \times 10^{28}}{4 \times 10^{8}} = 1.1 \times 10^{20} = N_a \text{ (say)}.$$

Now

$$n_p = \frac{n_i^2}{N_a} = \frac{(2.5 \times 10^{19})}{1.1 \times 10^{20}} = 5.7 \times 10^{18}. \therefore \frac{n_p}{n_i} = \frac{5.7 \times 10^{18}}{2.5 \times 10^{19}} = \mathbf{0.22.}$$

10.2. *The reverse saturation current for a pn-junction* Ge *diode at* 300 K *is* 5 μA. *What voltage must be applied across the junction to get a forward current of* 50 mA?

Solution: We know that the forward current is $I = I_s\left[e^{eV/kT} - 1\right]$.

Given, I = 50 mA = 50×10^{-3} A, $I_s = 5 \times 10^{-6}$ A, T = 300 K, $e = 1.6 \times 10^{-19}$ C and $k = 1.38 \times 10^{-23}$ J/K.

Now

$$e^{eV/kT} = \frac{I}{I_s} + 1 = \frac{50 \times 10^{-3}}{5 \times 10^{-6}} + 1 = 10^4 \text{ or, } V = \frac{kT}{e} \ln 10^4$$

$$\text{or, } V = \frac{\left(1.38 \times 10^{-23}\right) \times 300}{1.6 \times 10^{-19}} \times 2.304 \times 4 = \mathbf{0.238 \text{ volts.}}$$

10.3. *Calculate the diffusion constant for electron in Si, if mobility of electron at* 300 K *in Si be* 0.130 m²/V/s.

Solution: We have the Einstein's equation: $\frac{D_n}{\mu_n} = \frac{kT}{e}$.

$$\therefore D_n = \frac{kT \times \mu_n}{e} = \frac{\left(1.38 \times 10^{-23} \text{ J/K}\right) \times (300 \text{ K}) \times \left(0.130 \text{ m}^2/\text{V/s}\right)}{1.6 \times 10^{-19} \text{ C}} = \mathbf{0.0034 \text{ m}^2/\text{s}.}$$

10.4. *Find the value of Hall coefficient for a p-type Si bar, if the Hall voltage is* 50 mV, *current* = 10 μA, *and* Si *bar resistivity* $2 \times 10^3 \Omega$-m, *magnetic field* = 0.1 T *and the thickness and width of the bar each equal to* 3 mm. *What is the value of conductivity?*

Solution: The Hall coefficient,

$$R_H = \frac{V_H d}{IB} = \frac{\left(50 \times 10^{-3}\right) \times \left(3 \times 10^{-3}\right)}{(10 \times 10^{-6}) \times (0.1)} = \mathbf{150 \text{ m}^3/\text{C}.}$$

Conductivity,

$$\sigma = \frac{1}{\text{resistivity}} = \frac{1}{2 \times 10^3 \Omega\text{-m}} = \mathbf{0.5 \times 10^{-3} (\Omega\text{-m})^{-1}}.$$

10.5. *Find the barrier capacitance of a* Ge *pn-junction whose area is* 1 mm × 1 mm *having depletion region thickness* 2×10^{-3} mm.

Solution: We have the barrier capacitance

$$G = \frac{\epsilon A}{W} = \frac{k\epsilon_0 A}{W} = \frac{16 \times \left(8.85 \times 10^{-12}\right) \times 10^{-6}}{2 \times 10^{-6}} \text{ F} = \mathbf{70.8 \text{ F}} \; (\because k \text{ for Ge} = 18).$$

10.6. *The reverse saturation current in a Ge-diode increases from* 10 μA *at* 300 K *to* 105 μA *at* 330 K. *Find the band gap of the semiconductor.* **[GATE 1993]**

Solution: We have the reverse saturation current $I_s \propto e^{-E_g/kT}$, where E_g is the band gap at temperature T. When T_1 = 300 K, $I_{s_1} \propto e^{-E_g/kT_1}$ and T_2 = 330 K, $I_{s_2} \propto e^{-E_g/kT_2}$.

$$\therefore \frac{I_{s_2}}{I_{s_1}} = \exp\left[\frac{E_g}{k}\left(\frac{1}{T_1} - \frac{1}{T_2}\right)\right] \text{ or, } \ln\left(\frac{I_{s_2}}{I_{s_1}}\right) = \frac{E_g}{k}\left(\frac{1}{T_1} - \frac{1}{T_2}\right)$$

$$\text{or, } E_g = \frac{k\ln\left(\frac{I_{s_2}}{I_{s_1}}\right)}{\left(\frac{1}{T_1} - \frac{1}{T_2}\right)} = \frac{(1.38\times10^{-23})\times\ln 10.5}{\left(\frac{1}{300} - \frac{1}{300}\right)} \times \frac{1}{(1.6\times10^{-19})\text{ eV}}$$

$$\simeq \frac{(1.38\times10^{-23})\times\ln(10.5)\times300\times330}{30\times(1.6\times10^{-19})}\text{ eV} \simeq \mathbf{0.67\ eV.}$$

10.7. *Find the static and dynamic resistances of a pn-junction* Ge-diode *if the temperature is* 27 °C *and* $I_s = 1\,\mu$A *for an applied forward-bias of* 0.2 V.

Solution: If V be the forward-bias voltage and T the temperature of the junction, then the forward current $I = I_s\left(e^{eV/kT} - 1\right)$, where I_s is the reverse saturation current.

Here, $I_s = 1\times10^{-6}$ A, $V = 0.2$ V, $T = 300$ K, $e = 1.6\times10^{-19}$ C and $k = 1.38\times10^{-23}$ J/K.

$$\therefore I = \left(1\times10^{-6}\right)\exp\left[\frac{\left(1.6\times10^{-19}\right)\times0.2}{\left(1.38\times10^{-23}\right)\times300} - 1\right]\text{ A} = \mathbf{2.27\ mA.}$$

The static resistance,

$$r_{dc} = \frac{V}{I} = \frac{0.2}{2.27\times10^{-3}} = \mathbf{88\ \Omega.}$$

The dynamics resistance,

$$r_{ac} = \frac{26}{I} = \frac{26}{2.27} = \mathbf{11.4\ \Omega.}$$

10.8. *A transistor has* $\alpha = 0.99$. *If it is connected in* CE *mode what would be the change in base current for change of collector current* 0.99 mA.

Solution; We have the relation between transistor α and β as

$$\beta = \frac{\alpha}{1-\alpha} = \frac{0.99}{1-0.99} = 99.$$

Again, $\beta = \frac{\Delta I_c}{\partial I_B}$.

$$\therefore \Delta I_B = \frac{\Delta I_C}{\beta} = \frac{0.99\times10^{-3}}{99} = \mathbf{10\ \mu A.}$$

10.9. *The transistor in* Fig. 10.54 *has* $\beta = 100$ *and* $I_{CO} = 20$ nA. *Calculate* I_B, I_C, V_{CE} *and hence, decide in which region the transistor operates.* [*Given,* $V_{BE\text{ active}} = 0.7$ V, $V_{BE\text{ sat}} = 0.8$ V, $V_{CE\text{ sat}} = 0.2$ V.]

[CU(H) 1997]

Solution: Suppose, the transistor is operated in the active region.

$I_E = I_B + I_C \simeq I_B + \beta I_B = (1+\beta)I_B$.

From the base circuit applying KVL, we get $V_{BB} = I_B R_B + V_{BE} + I_E R_E$.

$$\therefore I_B = \frac{V_{BB} - V_{BE}}{R_B + (1+\beta)I_E} = \frac{5\text{ V} - 0.7\text{ V}}{50\text{ k}\Omega + 101\times2\text{ k}\Omega} = \mathbf{17.06\ \mu A.}$$

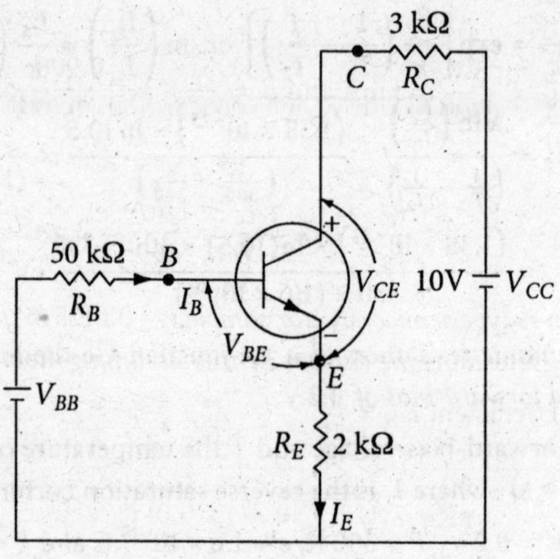

Fig. 10.54

Now $I_c = \beta I_B + (1+\beta)I_{CO} = 100 \times 17.06\,\mu\text{A} + 101 \times 20\,\text{nA} = \mathbf{1.708\,mA}$.

Applying KVL to collector circuit, $V_{CC} = I_C R_C + V_{CE} + I_E R_E$.

$$V_{CE} = V_{CC} - I_C(R_C + R_E)\ (\because I_E \simeq I_C)$$
$$= 10 - 1.708\,\text{mA} \times 5\,\text{k}\Omega = \mathbf{1.46\,V}.$$

$\therefore V_{CE} > V_{CE\,sat}$, hence the transistor is operated in the active region.

10.10. *An n-channel silicon (dielectric constant = 12) FET with a channel half-width* $a = 3 \times 10^{-6}$ m *is doped with a concentration* $N_d = 10^{21}\,\text{m}^{-3}$. *Find the pinch-off voltage.* **[GATE 2001]**

Solution: The pinch-off voltage,

$$V_P = \frac{eN_d}{2\epsilon}a^2 = \frac{(1.6 \times 10^{-19}) \times 10^{21}}{2 \times (12 \times 8.854 \times 10^{-12})} \times (3 \times 10^{-6})^2 = \mathbf{6.8\,V}.$$

10.11. *An n-channel* JFET *has* $I_{DSS} = 12$ mA. (a) *If the pinch-off voltage* $V_P = -4$ V, *find drain current for* $V_{GS} = -2$ V, (b) *If the transconductance* g_{m_0} *at* $V_{GS} = 0$ *is* 4 millimho, *find the pinch-off voltage.*

Solution: (a) The drain current,

$$I_D = I_{DSS}\left(1 - \frac{V_{GS}}{V_P}\right)^2 = 12\,\text{mA}\left(1 - \frac{-2V}{-4V}\right)^2 = \mathbf{3\,mA}.$$

$$\text{(b) } g_m = \left.\frac{\partial I_D}{\partial V_{GS}}\right|_{V_{DS}} = 2I_{DSS}\left(1 - \frac{V_{GS}}{V_P}\right)\left(-\frac{1}{V_P}\right).$$

For $V_{GS} = 0$,

$$g_{m_0} = -\frac{2I_{DSS}}{V_p} \text{ or, } V_p = -\frac{2I_{DSS}}{g_{m_0}} = -\frac{2 \times 12\,\text{mA}}{4\,\text{millimho}} = \mathbf{-6\,V}.$$

10.12. *A transistor in* CE *mode is connected with a resistance* 5 kΩ *and power supply of* 5 V *in the collector circuit. If* $\alpha = 0.998$ *and the voltage drop across the* 5 kΩ *resistance is* 5 V, *find the base current.*

[CU(H) 2000]

Solution: We have

$$I_C = \frac{5\,\text{V}}{5\,\text{k}\Omega} = 1\,\text{mA and } \beta = \frac{\alpha}{1-\alpha} = \frac{0.998}{1-0.998} = 499.$$

Since

$$\beta = \frac{I_C}{I_B}. \therefore I_B = \frac{I_C}{\beta} = \frac{1\,\text{mA}}{499} = \mathbf{2\,\mu A}.$$

10.13. *The Hall coefficient of an n-type semiconductor sample is* $-0.0125\,\text{m}^2/\text{C}$. *An electric field of* $100\,\text{V/m}$ *is applied to this sample. Determine the current density assuming mobility* $\mu = 0.36\,\text{m}^2\,\text{V}^{-1}\,\text{s}^{-1}$.

Solution: We have Hall coefficient $R_H = \frac{1}{ne}$.

$$\therefore n = \frac{1}{eR_H} = \frac{1}{(1.6\times10^{-19})\times 0.0125} = 5\times10^{20}/\text{m}^3.$$

Again, conductivity, $\sigma = ne\mu$ and

$$J = \sigma E = (ne\mu)\,E = \left(5\times10^{20}\right)\times\left(1.6\times10^{-19}\right)\times 0.36\times 100 = \mathbf{2880\,A/m^2}.$$

EXERCISE

10(A): Objective (MCQ) Type Questions

Select the correct answer.

10.1. In case of a *pn*-junction, if the reverse-bias is increased, then the width of the depletion region

(a) increases (b) decreases (c) remains constant (d) none of these.

10.2. The depletion region in a *pn*-junction has width

(a) 0.5 $\mu\mu$m (b) 0.5 cm (c) 5 cm (d) 0.5 μm.

10.3. A Si-diode has barrier potential,

(a) 26 mV (b) 2 mV (c) 0.7 V (d) 0.3 V.

10.4. The reverse saturation current of a Ge-diode operating at room temperature is of the order of

(a) A (b) mA (c) μA (d) nA.

10.5. In a *pn*-junction, the width of the depletion region

(a) increases with forward-biasing (b) decreases with reverse-biasing

(c) independent of applied bias (d) decreases with doping level.

10.6. The barrier capacitance C_T of a *pn*-junction is related to depletion region W as

(a) $C_T \propto W$ (b) $C_T \propto 1/W$ (c) $C_T \propto 1/\sqrt{W}$ (d) $C_T \propto \sqrt{W}$.

10.7. The value of diffusion capacitance is appreciable when a linear graded junction is

(a) forward-biased (b) reverse-biased (c) in open circuit (d) none of them.

10.8. A diode is

(a) a linear device (b) a non-linear device (c) unipolar (d) bilateral.

10.9. Avalanche breakdown in a *pn*-junction diode is due to

(a) collision of mobile carriers with immobile ions

(b) rupture of covalent bonds by the action of a strong electric field

(c) increase of temperature of the junction

(d) none of these.

10.10. If I be the current through a *pn*-junction, then dynamic resistance r_{dc} is proportional to

(a) I (b) $1/I$ (c) $\sqrt{I}$ (d) $1/\sqrt{I}$.

10.11. The cut-in or offset or threshold voltage (V_r) of a Ge-diode is

(a) $V_\gamma = 0.5\,\text{V}$ (b) $V_\gamma = 0.3\,\text{V}$ (c) $V_\gamma = 0.2\,\text{V}$ (d) $V_\gamma = 2\,\text{V}$.

10.12. Zener breakdown in a *pn*-junction diode is due to

(a) collision of carriers with the immobile ions

(b) strong force of the electric field on the bound electrons of immobile ions

(c) doping level in *p*-type material

(d) doping level in *n*-type material.

10.13. The relation between α and β in a transistor is

(a) $\beta = \frac{\alpha}{1-\alpha}$ (b) $\beta = \frac{1-\alpha}{1+\alpha}$ (c) $\frac{1+\alpha}{\alpha}$ (d) $\alpha(1+\alpha)$.

10.14. The value of transistor α is always

(a) 1 (b) greater than 1 (c) less than but near to 1 (d) very much greater than 1.

10.15. A transistor is a semiconductor device which is

(a) unipolar (b) bipolar (c) none of these.

10.16. If J_{EB} and J_{CB} represent respectively the emitter-base and collector-base junction of a transistor, then in normal operation

(a) J_{EB} is forward-biased (b) J_{EB} is reverse-biased

(c) J_{CB} is forward-biased (d) J_{CB} is reverse-biased.

10.17. A transistor, when used as a switch, operates between the regions

(a) saturation and cut-off (b) active and cut-off (c) saturation and active (d) none of these.

10.18. For operation in high frequency region, we prefer

(a) *npn*-transistor (b) *pnp*-transistor (c) a combination of (a) and (b) (d) none of these.

10.19. For a BJT in saturation condition,

(a) $V_{CE} = \frac{V_{CC}}{2}$ (b) $V_{CE} = V_{CC}$ (c) $V_{CE} = 1/V_{CC}$ (d) $V_{CE} \simeq 0\,\text{V}$.

10.20. If N_d be the donor ion concentration and W, the normal width of the base region, then punch through voltage (V_{PT}) of the transistor is

(a) $V_{PT} \propto W$ (b) $V_{PT} \propto W^2$ (c) $V_{PT} \propto \sqrt{W}$ (d) $V_{PT} \propto W^{3/2}$.

10.21. If the reverse collector bias is increased, then the effective width of the base

(a) decreases (b) increases (c) remains constant (d) none of these.

10.22. Carriers are injected from the emitter to base by the process of

(a) diffusion (b) drift (c) collision (d) leakage.

10.23. JFET and MOSFET devices are

(a) unipolar (b) bipolar (c) linear (d) non-linear.

10.24. A JFET compared to BJT

(a) is more noisy (b) has very low input impedance

(c) occupies less space in IC (d) is more temperature sensitive.

10.25. The SiO_2 layer in IGFET has resistance

(a) 10^5–10^8 Ω (b) 10^7–10^{12} Ω (c) 10^{10}–10^{15} Ω (d) 10^{15}–10^{18} Ω.

10.26. The threshold voltage (V_T) for an NMOSFET is about

(a) 2 V (b) 4 V (c) 7 V (d) 10 V.

10.27. For high frequency operation, NMOSFET are preferred to PMOSFET because for devices of the same size

(a) current in NMOSFET is almost double that in PMOSFET

(b) electron mobility of the former is greater than that of the latter

(c) holes are heavier than electrons

(d) none of these.

10.28. Enhancement MOSFET is widely used in

(a) discrete circuits (b) integrated circuits (c) both in (a) and (b) (d) none of these.

10.29. CMOS consists of

(a) a BJT and an NMOS (b) a JFET and a PMOS

(c) an NMOS and a PMOS (d) a JFET and an NMOS.

ANSWERS

10.1 (a) **10.2** (d) **10.3** (c) **10.4** (c) **10.5** (d) **10.6** (b) **10.7** (a) **10.8** (b) **10.9** (a) **10.10** (b) **10.11** (c) **10.12** (b) **10.13** (a) **10.14** (c) **10.15** (b) **10.16** (a,d) **10.17** (a) **10.18** (a) **10.19** (d) **10.20** (b) **10.21** (a) **10.22** (a) **10.23** (a) **10.24** (c) **10.25** (a) **10.26** (a) **10.27** (a) **10.28** (c) **10.29** (c).

10(B): Short Answer Type Questions

10.1. What are the different techniques for formation of a *pn*-junction?

10.2. Explain the formation of the barrier field and barrier voltage across the junction of an *n*-type semiconductor.

10.3. Explain the difference between a step junction and a linearly graded junction.

10.4. Explain the terms (a) barrier energy (b) barrier potential and (c) depletion region by drawing the energy-band diagram of an unbiased *pn*-junction.

10.5. Can you measure the barrier potential across a *pn*-junction by a milli-voltmeter?—Explain.

10.6. Write down Shockley equation regarding the forward current in a *pn*-junction diode. Does the reverse saturation current vary with temperature?

10.7. Define cut-in voltage of a *pn*-junction diode. Mention the values of cut-in voltage for Ge-diode and Si-diode.

10.8. What do you mean by avalanche breakdown and Zener breakdown with reference to a *pn*-junction diode? Point out the difference between them.

10.9. What is the origin of junction capacitance of a *pn*-junction diode?

10.10. Explain the operation of an LED.

10.11. What is tunnel diode? Explain its principle of working?

10.12. Explain the action of a solar cell.

10.13. What is a transistor? How does it work?

10.14. Discuss with diagram how the different current components originate in a transistor.

10.15. What are transistor's α and β. How are they related?

10.16. Show that $I_C = \beta I_B + (1+\beta) I_C$ for CE transistor in active region.

10.17. Draw a self-bias circuit of a transistor. Why is it so called?

10.18. What is a junction field effect transistor (JFET). Justify the nomenclature with suitable diagram.

10.19. What do you mean by pinch-off and pinch-off voltage? Explain.

10.20. What is a MOSFET? What are NMOS and PMOS?

10.21. Give a comparative study between an FET and a BJT.

10.22. Write the relation between drain current and gate to source voltage of an enhancement type MOSFET. How is an *n*-channel MOSFET used as a resistor? **[CU(H) 2007]**

10.23. Draw the cross-sectional view of a CMOS. **[CU(H) 2007]**

10.24. Write down the differences among BJT, JFET and MOSFET. **[VU(H) 2004]**

10.25. What is a MOSFET? What do you mean by enhancement (E)MOS, depletion (D)MOS and complementary (C)MOS? **[VU(H) 2002]**

10.26. Discuss the superiority of *n*-channel MOSFET over *p*-channel MOSFET.

10.27. Discuss the uses of MOSFET.

10(C): Numerical Problems

10.1. The reverse current in a *pn*-junction Ge-diode attains a value 90 % of its saturation value at room temperature. Find the voltage applied. [*Ans.* −60 mV]

10.2. In an *n*-type semiconductor the Fermi level lies 0.3 eV below the conduction band at 300 K. If the temperature is increased to 330 K, find the new position of the Fermi level. [*Ans.* $E_c - E_f = 0.33$ eV]

10.3. A junction has a built-in potential of 0.8 V, and a forward-bias of 0.55 V is applied. What is the potential difference across the depletion region? What is its value if the bias is reversed? [*Ans.* 0.25 V, 1.35 V]

10.4. The conductivities of *p*-region and *n*-region of a Ge *pn*-junction are 1000 $\Omega^{-1}\,m^{-1}$ and 200 $\Omega^{-1}\,m^{-1}$ respectively. The cross section of the crystal is 0.05 cm, and the relative dielectric constant for Ge is 16. Calculate the width of the depletion layer and the junction capacity when the *pn*-junction is (a) unbiased (b) when it is forward-biased by 0.1 V and (c) when it is reverse-biased by 10 V. [Assume that the junction is an abrupt one, $\mu_p = 0.19\,m^2\,V^{-1}\,s^{-1}$ and $\mu_n = 0.39\,m^2\,V^{-1}\,s^{-1}$ and $n_i = 2.15 \times 10^9/m^3$.] [*Ans.* 4.3×10^{-7} m, 165 pF, 35.1×10^{-8} m, 202 pF, 2500 nm, 28 pF]

10.5. Find the static and dynamic resistance of a *pn*-junction Ge-diode, if the temperature is 27 °C and $I_0 = 1\,\mu A$ for an applied forward-bias of 0.2 V. [*Ans.* 88Ω, 11.4Ω]

10.6. The supply voltage in a Zener diode circuit is 9 V and Zener voltage is 6 V. If the maximum Zener current that can safely flow be 20 mA, determine the value of the series resistance *R*. If a load resistance

R_L of 1 kΩ be connected across the Zener diode, calculate the load current and the Zener current. Also calculate the minimum value of R_L that can be used. [*Ans.* 150 Ω, 6 mA, 14 mA, 300 Ω]

10.7. An abrupt Si *pn*-junction is doped with $10^{21}/\text{m}^3$ boron atoms in the *p*-region and $10^{20}/\text{m}^3$ phosphorus atom in the *n*-region. Calculate at $T = 300$ K (a) the barrier potential (b) the depletion layer thickness in each side when −10 V is applied and (c) the capacitance with −10 V bias, if the area is 10^{-8} m^2. Given, $\epsilon_r = 12$ and $n_i = 1.6 \times 10^{16}/\text{m}^3$. [*Ans.* 0.514 V, 1.1×10^{-6} m, 1.1×10^{-5} m, 0.0878 pF]

10.8. The value of α of a transistor is 0.98. If it changes by 0.5 % what would be the corresponding percentage change in the value of β? [*Ans.* 33 %]

10.9. For a Si *npn*-transistor, $\beta = 100$ and $I_{CO} = 20 \times 10^{-9}$ A. It is connected in CE mode. What will be the collector current for a base current of 0.02 mA. [*Ans.* 2.002 mA]

10.10. If the values of I_C, I_{CO} and I_B for a particular transistor be 5.202 mA, 2 μA and 50 μA respectively, then find (a) the values of α, β and I_E and (b) the new value of I_B required to make $I_C = 8$ mA. [*Ans.* 0.99, 100, 5.25 mA, 77.98 μA]

10.11. For a CE configuration, it is found that for a fixed base current of 30 μA, collector current changes from 3.5 mA to 3.7 mA when collector-emitter values change from 7.5 V to 12.5 V. Calculate the output resistance and β at 12.5 V. What is the value of its α? **[CU(H) 2002]**
[*Ans.* 25 kΩ, 0.992]

10.12. Consider the circuit of Fig. 10.55 and calculate the value of V_{CE} assuming $\beta = 125$ and $V_{BE} = 0.7$ V. **[CU(H) 2002]**
[*Ans.* 16.125 V]

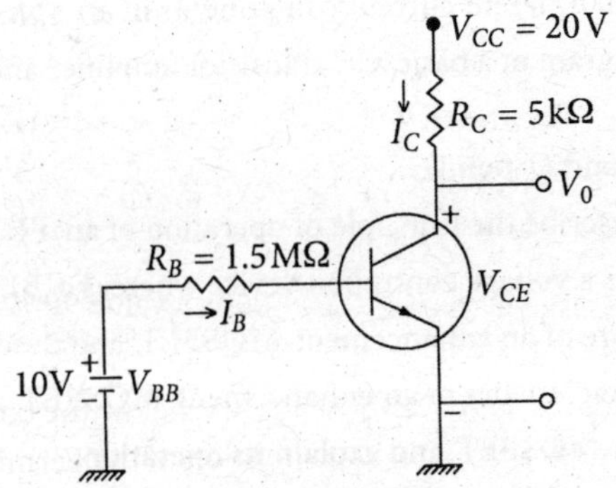

Fig. 10.55

10.13. For a constant gate to source voltage, a change in drain to source voltage of 4 volts produces a change in drain current of 40 μA. Calculate the ac drain resistance of the FET. [*Ans.* 100 kΩ]

10.14. In a JFET, $I_{DSS} = 12$ mA and $V_p = -5$ V. Calculate the drain current for a gate to source voltage of −2 V. [*Ans.* 4.32 mA]

10.15. If in a JFET, the drain current increases from 10 mA to 10.25 mA for an increase in drain to source voltage from 10 V to 15 V for a fixed V_{GS}. What is the value of drain resistance r_d? [*Ans.* 20 kΩ]

10(D): Broad Answer Type Questions

10.1. (a) Discuss with suitable diagram the formation of depletion region in a *pn*-junction and find out an expression for the potential barrier voltage developed therein.

(b) Find out an expression for the barrier voltage across a *pn*-junction.

10.2. (a) Derive an expression for the width of the depletion region across a *pn*-junction in terms of impurity concentration.

(b) Obtain Shockley equation relating current and voltage in a *pn*-junction.

10.3. (a) Draw the energy band diagram of an unbiased *pn*-junction and explain the terms *barrier potential* and *depletion region.*

(b) Find out expression for junction resistance of a *pn*-junction diode. What is cut-in voltage?

10.4. (a) How does junction capacitance arise in the *pn*-junction diode? How does it vary with (i) the width of the depletion region and (ii) the applied reverse voltage?

(b) Explain how breakdown in a *pn*-junction occurs. What is a breakdown diode and what are its applications?

10.5. (a) Distinguish between avalanche breakdown and Zener breakdown. Draw the volt-ampere characteristics of a Zener diode and explain the variation of current with applied voltage.

(b) What is a photo-diode? Explain its working principle.

10.6. (a) What is a tunnel diode? Draw the volt-ampere characteristics of a tunnel diode. Mention some of its applications.

(b) Explain the working principle of a solar cell. Indicate some of its uses.

10.7. (a) What is a transistor? Discuss the operation of an *npn*- and a *pnp*-transistors with suitable energy band diagram.

(b) Discuss the origin of different current components in an *npn*-transistor with suitable diagram.

10.8. (a) Draw the circuit diagram of a basic CE transistor amplifier and explain the amplification action of it.

(b) What is a load line and Q-point?

10.9. (a) What is a JFET? Describe the principle of operation of an FET.

(b) Show that a JFET is a voltage controlled device whereas a BJT is a current controlled device.

10.10. (a) Describe the structure of an enhancement MOSFET and describe its operation.

(b) Draw the static characteristics of an enhancement MOSFET and explain its operation.

10.11. (a) Describe a depletion MOSFET and explain its operation.

(b) Draw the static and dynamic characteristics of a depletion MOSFET and explain their nature.

10.12. (a) Discuss the application of MOSFET in digital switches.

(b) What is a CMOS? How does it work? Discuss its uses.

10.13. (a) What is a MOSFET? Point out the main physical differences between a MOSFET and JFET.

(b) Why MOSFET is commercially more important than a JFET?

10.14. (a) What is pinch-off and pinch-off voltage? Show pictorially the variation of depletion region before and after pinch-off?

(b) Show that for $V_{GS} = 0$, the MOSFET behaves like a simple resistor.

Chapter 11

Superconductivity

11.1 Introduction

H. Kamerlingh Onnes, a Dutch Physicist, while measuring electrical conductivity at low temperatures observed in 1911 that *certain metal and alloys exhibit infinite conductivity* (i.e., *almost zero resistivity*) *when they are cooled to sufficiently low temperatures.* This phenomenon is called *superconductivity* and the material is known as *superconductor.* He observed that when pure mercury was cooled down to below 4.2 K, its resistivity suddenly drops to zero as shown in Fig. 11.1. Many metals like Al, Pb, Sn, Nb, etc. show zero resistivity at around 4 K.

The temperature at which a material suddenly shows zero resistivity is known as the *transition temperature* (T_c). The material undergoes a *phase transition* from normal resistivity to zero resistivity at T_c. That is why T_c is also called *superconducting transition temperature.* The variation of resistivity (ρ) with temperature (T) for pure metal, impure metal and semiconductor is shown in Fig. 11.2, where ρ_0 is the residual resistivity due to impurity in the specimen.

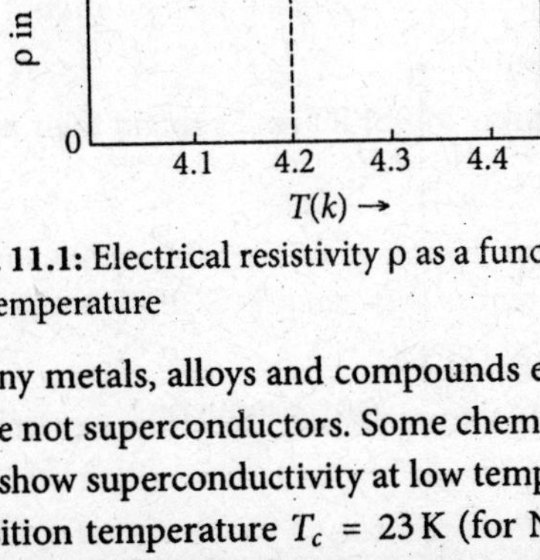

Fig. 11.1: Electrical resistivity ρ as a function of temperature

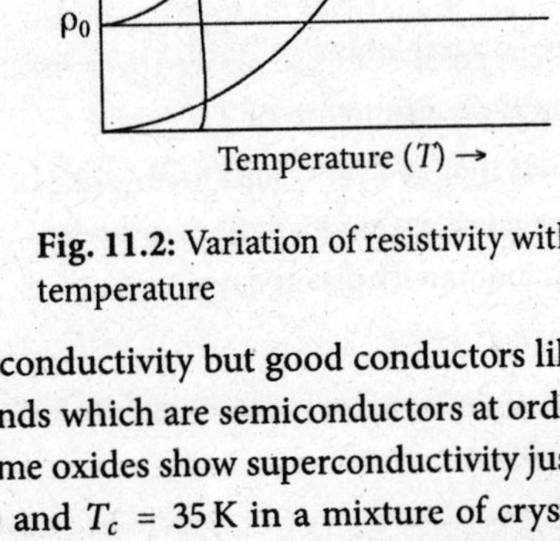

Fig. 11.2: Variation of resistivity with temperature

Many metals, alloys and compounds exhibit superconductivity but good conductors like Ag, Au and Cu are not superconductors. Some chemical compounds which are semiconductors at ordinary temperatures show superconductivity at low temperatures. Some oxides show superconductivity just below 12.5 K. Transition temperature $T_c = 23$ K (for Nb_3Ge alloy) and $T_c = 35$ K in a mixture of crystalline state in La–Ba–Cu–O has also been reported in recent years.

Superconductivity has many important applications in emerging fields such as *energy technology, telecommunication, computing, super-fast transportation, medicine, etc.*

11.2 General Properties of Superconductors

Following are some of the general properties of the superconductors:

(i) Materials having high normal resistivity exhibit superconductivity.

(ii) Superconductivity is exhibited by metallic elements in which the number of valence electrons lies between 2 and 8.

(iii) Transition temperature T_c varies from sample to sample.

(iv) The transition range for chemically pure and structurally perfect specimen is very small but for impure and structurally imperfect specimen it is broad.

(v) Transition metals having odd number of valence electrons are better than those having even number of valence electrons as regards superconductivity is concerned.

(vi) Elements with small atomic volume and atomic mass are better superconductors.

(vii) T_c varies linearly with Z^2, where Z is the number of valence electrons.

(viii) If ρ be the electrical resistivity, then the materials for which $Z\rho > 10^6$ show superconductivity.

(ix) *Ferro-magnetic* and *antiferromagnetic materials* do not show superconductivity.

(x) When a material is cooled to T_c, the following changes are found to occur:

  (a) resistivity drops to zero,

  (b) magnetic lines of flux are rejected out of superconductor (Meissner effect),

  (c) specific heat changes discontinuously,

  (d) small changes in thermal conductivity and volume of the element are found to occur.

(xi) In a superconductivity ring, current persists for a long time.

11.3 Effect of Magnetic Field

Kamerlingh Onnes found that the factors governing the superconducting state of metal are (i) *the temperature at which the specimen is kept,* (ii) *the magnetic field in which the specimen is placed.*

If the temperature T of the specimen is raised above T_c (i.e., $T > T_c$), the superconductivity disappears. It also disappears if the magnetic field H is raised above the *critical magnetic field* H_c, where H_c is a function of T.

Fig. 11.3 shows that at $T < T_c$ and in absence of magnetic field the specimen behaves as a superconductor. When $H \geq H_c$, the superconductivity disappears.

If $T = T_c$, then $H = 0$.

If $T < T_c$, H_c increases. The variation of H_c with temperature T is given by

$$H_c(T) = H_c(0)\left[1 - \left(\frac{T}{T_c}\right)^2\right], \quad (11.1)$$

where $H_c(0)$ *represents the critical magnetic field at* 0 K and T_c *is characteristic of the material.* Eqn (11.1) is known as *Tuyn's law.*

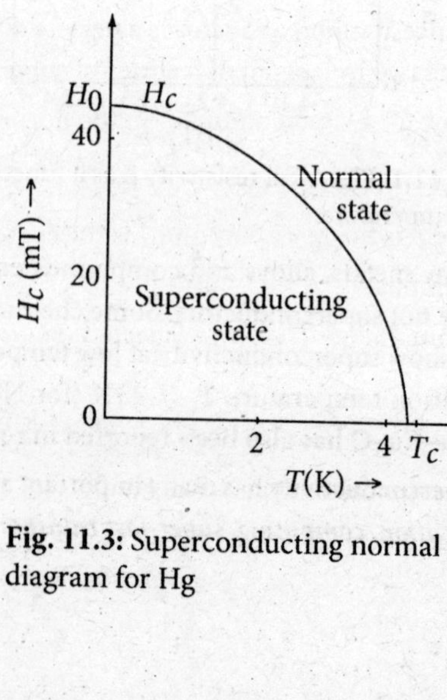

Fig. 11.3: Superconducting normal phase diagram for Hg

The Meissner Effect

If a weak magnetic field ($H < H_c$) be applied to a superconducting specimen at a temperature below the transition temperature (i.e., $T < T_c$), then no lines of magnetic induction exists inside the specimen. The specimen, therefore, pushed out the lines of magnetic induction so that $\vec{B} = 0$ inside the specimen [Fig. 11.5]. This phenomenon was discovered by *Meissner and Ochsenfeld* in 1933 and is known as *flux exclusion* or *Meissner Effect.* If, however, the temperature is raised from below T_c at $T = T_c$, the lines of flux suddenly start penetrating and the specimen returns back to the *normal state,* i.e., the effect is *reversible.*

When the specimen returns back to the normal state, the magnetic induction inside it is given by

$$\vec{B} = \mu_0 \left(\vec{H} + \vec{M}\right), \tag{11.2}$$

where $\vec{H}$ is the external magnetic field and $\vec{M}$ is the magnetization produced inside the specimen.

For superconducting material, the magnetic induction inside it is zero, i.e., $B = 0$ (according to Meissner Effect).

$$\therefore \mu_0 \left(\vec{H} + \vec{M}\right) = 0 \text{ or, } \vec{M} = -\vec{H}. \tag{11.3}$$

Eqn (11.3) shows that the specimen is perfectly *diamagnetic.* The *magnetic susceptibility* is

$$\chi = \frac{M}{H} = -1. \tag{11.4}$$

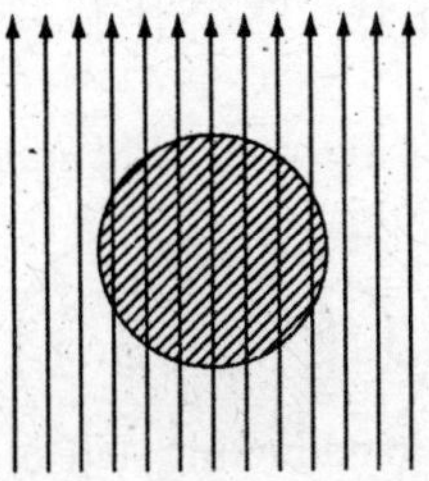

Fig. 11.4: Normal $T > T_c$ or $H > H_c$

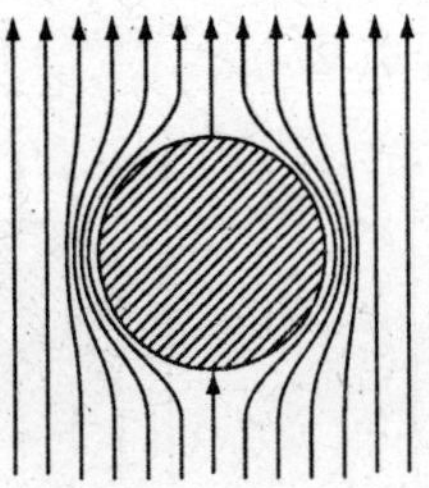

Fig. 11.5: Superconducting $T < T_c$ or $H < H_c$

Hence, a superconductor is a perfect diamagnetic, i.e., the *perfect conductivity* and *perfect diamagnetism* are the two independent features of superconductor.

Now for a superconductor in the normal state, the electric field according to *Ohm's law* is

$$\vec{E} = \rho \vec{J}, \tag{11.5}$$

where ρ is the resistivity and $\vec{J}$ is the current density. ρ tends to zero when the material is cooled to T_c.

$\therefore \vec{E}$ must be zero if $\vec{J}$ is finite.

From *Maxwell's electromagnetic equation,*

$$\nabla \times \vec{E} = -\frac{\partial \vec{B}}{\partial t}. \tag{11.6}$$

In superconducting state, $\vec{E} = 0$,

$$\therefore \frac{\partial \vec{B}}{\partial t} = 0 \text{ or, } \vec{B} = \text{constant}.$$

This implies that on cooling the specimen to transition temperature (T_c), the magnetic flux passing through the specimen should not change. The above treatment shows that if a sample at $T > T_c$ be kept in a magnetic field $\vec{B}$ and cooled below the transition temperature the magnetic flux gets trapped in the sample. On the contrary, if the sample be cooled below the transition temperature, and then magnetic field is applied the magnetic flux cannot enter the sample. That is, when a perfect conductor is cooled in the magnetic field until its resistance becomes zero, the magnetic field in the material gets *frozen-in* and cannot change subsequently irrespective of the applied field. But *Meissner effect* is quite contradictory to it. Meissner effect predicts that for defining superconducting state the essential property is perfect diamagnetism, i.e., $\vec{E} = 0, \vec{B} = 0$.

Thus, in conclusion the behaviour of a superconductor is different from that of a perfect conductor. The superconductivity state is a characteristic thermodynamic phase of a substance in which it cannot sustain steady $\vec{E}$ and steady $\vec{B}$. Thus, in the *superconducting state* of a substance $E = 0$ and $B = 0$, i.e., the substance is perfect *diamagnetic.*

The behaviour of a *perfect conductor* to cooling (below T_c) and magnetic field $\vec{B}$ is shown in Fig. 11.6.

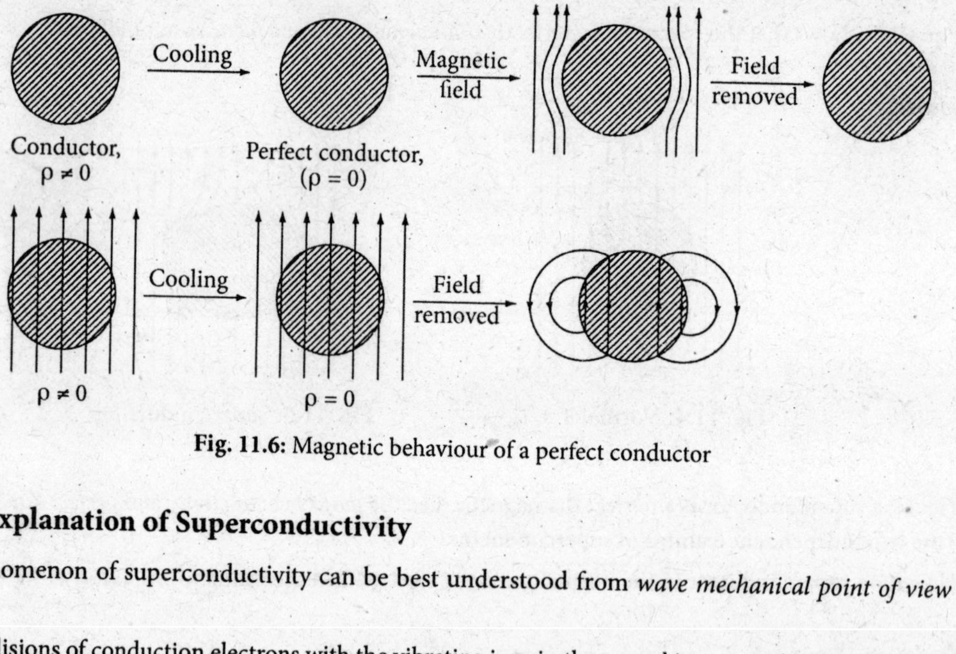

Fig. 11.6: Magnetic behaviour of a perfect conductor

11.4 Explanation of Superconductivity

The phenomenon of superconductivity can be best understood from *wave mechanical point of view* as below:

The collisions of conduction electrons with the vibrating ions in the crystal is responsible for the electrical resistance in ordinary metals. In the superconducting state *the electrons tend to scatter in pairs rather than individually. This gives rise to an attractive exchange force between the electrons and becomes very strong when the spins and momenta of the electrons are opposite.* The forces of attraction between the conduction electrons exceeds the electrostatic repulsive force, i.e., the entire system of conduction electrons then becomes a *bound system.* No energy is transferred from this system to the lattice ions. If now an electric field is applied, the electrons gain additional energy and a current flows due to their motion without any transfer of energy to the lattice. Under this condition the collisions of conduction electrons with the vibrating ions is ceased and the substance does not possess any electrical resistivity.

11.5 Types I and II Superconductors

11.5.1 Type I or Soft Superconductors

On the basis of diamagnetic response of the material, superconductors are of two types, viz., type I and type II. Type I superconductors are those for which *Meissner effect* is complete, i.e., *perfect diamagnetism.*

Below the critical magnetic field (H_c), if the magnetic field H is gradually increased from its initial value the magnetization M increases and at $H = H_c$, the diamagnetism abruptly disappear, when the sharp transition from superconducting state to normal state takes place as shown in Fig. 11.7(a). Al, Zn, Hg and Sn in pure state exhibit type I superconductivity.

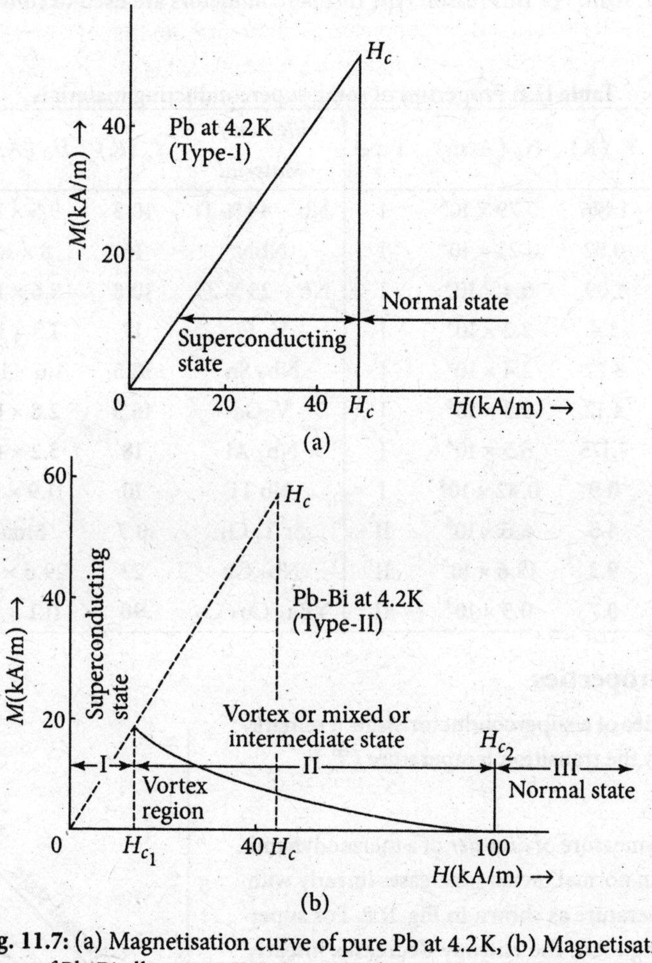

Fig. 11.7: (a) Magnetisation curve of pure Pb at 4.2 K, (b) Magnetisation curve of Pb-Bi alloy at 4.2 K. I: Superconducting state, II: Vortex or mixed or intermediate state, III: Normal state

11.5.2 Type II or Hard Superconductors

The characteristic behaviour of type II semiconductor is shown in Fig. 11.7(b). When the magnetic field is increased from $H = 0$ to $H_c = H_{c_1}$, the material behaves as a pure *superconductor* and the lines of magnetic flux are rejected. If H is increased beyond H_{c_1}, the lower critical field, the lines of flux begin penetrating and the material is in a mixed state up to H_{c_2}, the *upper critical field.* If $H > H_{c_2}$, specimen comes to normal and *Meissner effect* is incomplete in the region H_{c_1} to H_{c_2}. The region between H_{c_1} to H_{c_2} is known as the *vortex region.* Zr and Nb are some examples of type II semiconductors.

Types I and II semiconductors are respectively called the *soft superconductors and hard superconductors.* Type I superconductors give away their superconductivity at lower field strength and are referred to as soft superconductor. But Type II superconductors need relatively larger field to come back to normal state and are called *hard superconductors.*

The critical field for type I semiconductor well below T_c is about 10^6 A/m. For type II superconductors H_{c_2} is greater than 10^6 A/m. For this reason type II superconductors are used to construct *superconducting motors and magnets.*

Table 11.1: Properties of some superconducting materials

Element/ Material	T_c (K)	H_0 (A/m)	Type	*Element/ Material*	T_c (K)	H_0 (A/m)	Type
Al	1.196	0.79×10^4	I	Nb – 44 % Ti	10.5	9.5×10^6	II
Cd	0.52	0.22×10^4	I	NbN	16	8×10^6	II
Ga	1.09	0.4×10^4	I	Nb – 25 % Zr	10.8	5.6×10^6	II
In	3.4	2.2×10^4	I	V_3 Si	17	1.3×10^7	II
Sn	3.72	2.4×10^4	I	Nb_3 Sn	18.5	1.6×10^{17}	II
Hg	4.12	3.3×10^4	I	V_3Ga	16.5	2.8×10^7	II
Pb	7.175	6.5×10^4	I	Nb_3 Al	18	3.2×10^7	II
Zn	0.9	0.42×10^4	I	Nb Ti	10	11.9×10^6	II
Ta	4.5	6.6×10^6	II	Sr T_i O_3	0.7	Small	II
Nb	9.3	15.6×10^4	II	Nb_3 Ge	23	29.6×10^6	II
Os	0.7	0.5×10^4	II	YBa_2 Cu_3 O_7	90	11.1×10^7	II

11.6 Thermal Properties

The thermal properties of a superconductor show a remarkable abrupt change at the transition temperature (T_c).

11.6.1 Entropy

We know entropy is a measure of *disorder* of a thermodynamical system. Entropy, in normal metals, decreases linearly with the decrease of temperature as shown in Fig. 11.8. For superconducting metal (e.g., Al), the entropy decreases linearly up to $T = T_c$, but when cooled below T_c (i.e., $T < T_c$) the decrease in entropy with temperature is remarkable.

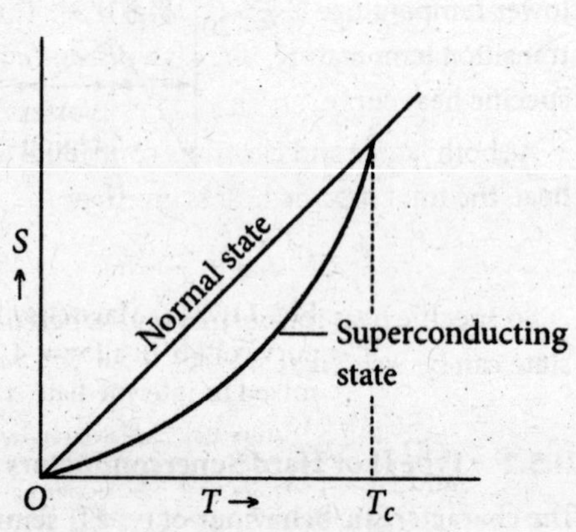

Fig. 11.8: Entropy of Al in the normal and superconducting stage

If S_N and S_S be the entropies per unit volume in the normal and superconducting phases respectively, then we can show that

$$S_N - S_S = -\mu_0 H_c \frac{dH_c}{dT}. \tag{11.7}$$

Thus, $S_N > S_S$ as $\frac{dH_c}{dT}$ is always *negative*. As the entropy decreases for superconducting electrons with the decrease of T, *we conclude that superconducting electrons are in more ordered state than in a normal state.*

At $T = 0$, $\frac{dH_c}{dT} = 0$ and $T = T_c$, $H_c = 0$. So at $T = 0$ and $T = T_c$, $S_N = S_S$. At all other points there is a definite change of entropy when transition takes place in a magnetic field. The variation of entropy with T is shown in Fig. 11.8.

11.6.2 Specific Heat

Multiplying both sides of eqn (11.7) by T, we get

$$TS_N - TS_S = -\mu_0 TH_c\frac{dH_c}{dT} \text{ or, } Q_N - Q_S = -\mu_0 TH_c\frac{dH_c}{dT}, \quad \text{i.e., } Q_{S\to N} = -\mu_0 TH_c\frac{dH_c}{dT}. \tag{11.8}$$

$\because \frac{dH_c}{dT} < 0$, the transition-heat from superconducting to normal state is positive. The heat $Q_{S\to n}$ provides the energy required to fill the superconductor with the magnetic field when transition occurs by increasing the temperature in presence of a magnetic field.

Now differentiating eqn (11.7) with T, we get

$$\frac{dS_N}{dT} - \frac{dS_s}{dT} = -\mu_0\frac{d}{dT}\left(H_c\frac{dH_C}{dT}\right) \text{ or, } T\frac{dS_N}{dT} - T\frac{dS_s}{dT} = -\mu_0 T\frac{d}{dT}\left(H_c\frac{dH_c}{dT}\right)$$

$$\text{or, } C_N - C_S = -\mu_0 T\frac{d}{dT}\left(H_c\frac{dH_c}{dT}\right) \quad \text{or, } C_S - C_N = \mu_0 T\left(\frac{dH_c}{dT}\right)^2 + \mu_0 TH_c\frac{d^2H_c}{dT^2}. \tag{11.9}$$

At $T = T_c$, $H_c = 0$ and $\frac{d^2H_c}{dT^2} = 0$.

$$\therefore (C_S - C_N)_{T=T_c} = \mu_0 T\left(\frac{dH_c}{dT}\right)^2_{T=T_c}. \tag{11.10}$$

Hence, there is an increase in *specific heat* when the material is cooled below the critical temperature (T_c). Eqn (11.10) is known as the *Rutgers equation*. Measuring C_S and C_N by calorimetric method, $\frac{dH_c}{dT}$ can be determined.

Again, from eqn (11.10),

$$C_N - C_S = -\mu_0 T_c\left(\frac{dH_c}{dT}\right)^2. \tag{11.11}$$

This means that, near $T = T_c$, $C_N < C_S$. But at lower temperature $C_N > C_s$ [Fig. 11.9]. Thus, at the transition temperature, there is a *discontinuity* in the specific heat curve.

As both lattice and electrons contribute to specific heat, the total specific heat is given by

$$C = C_{\text{lattice}} + C_{\text{electron}}.$$

So specific heat for normal and superconducting state can be written as

$$C_N = (C_{\text{lattice}})_N + (C_{\text{electron}})_N$$

$$\text{and } C_S = (C_{\text{lattice}})_S + (C_{\text{electron}})_S.$$

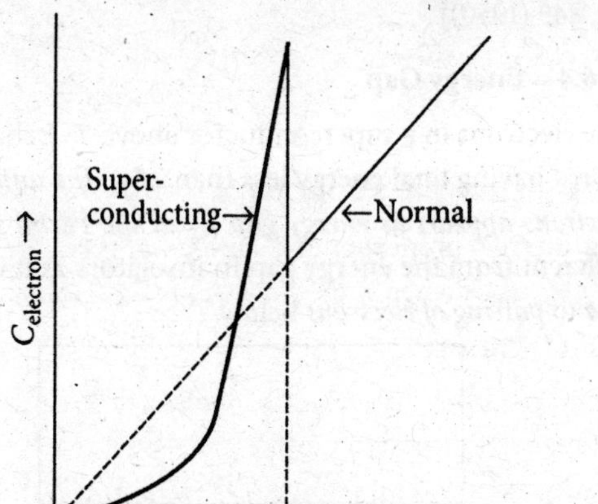

Fig. 11.9: Variation of electronic specific heat ($C_{\text{electronic}}$) with temperature (T) for normal and superconducting states of a sample.

But the properties of the lattice structure remain unchanged when a material becomes superconducting.

$$(C_{\text{lattice}})_N = (C_{\text{lattice}})_S.$$

$$\therefore C_N - C_S = (C_{\text{electron}})_N - (C_{\text{electron}})_S .$$

It has been found that $(C_S - C_N)_{T<T_c} = Ae^{-2\beta\Delta}$, where Δ is the *energy gap* in the electron energy spectra and A is a constant.

The variation of electronic specific heat (C_{electron}) with temperature (T) is shown in Fig. 11.9.

11.6.3 Isotope Effect

The transition temperature T_c of a superconductor has been found to vary with the *average isotopic mass M* of their constituents. The variation of T_c with M is found to obey the form:

$$T_c \propto \frac{1}{M^\beta} \text{ or, } T_c M^\beta = \text{a constant,} \tag{11.12}$$

where the *isotopic effect coefficient* β is given by

$$\beta = \frac{\partial (\ln T_c)}{\partial (\ln M)}.$$

$\beta \simeq \frac{1}{2}$ for all metals. For Hg, M varies from 199.5 to 203.4 atomic mass and the corresponding variation in transition temperature ranges from 4.185 K to 4.146 K.

Recent theories lead to the result $\beta = 0.5\left[1 - 0.01\{N(0)V\}^{-2}\right]$, where $N(0)$ is the density of single particle states for one spin at the *Fermi level* and V is the *model potential* between the electrons.

Since Debye temperature $\theta_D \propto$ sound velocity $\propto M^{-1/2}$, the transition temperature T_c can be related to θ_D as $T_c \propto M^{-\alpha} \propto D_\theta$ or, T_c/θ_D = constant.

The constancy of T_c/θ_D implies that, the lattice vibrations have an important role on superconductivity. It leads to the fact that *electron-photon interaction must be the basis of the phenomenon of superconductivity.* This was also suggested by *H. Fröhtich* before the publication of BCS theory [Ref. H. Fröhtich, Phys. Rev. **79**, 845 (1950)]

11.6.4 Energy Gap

The electrons in a superconductor above T_c behave like *normal electrons* but below T_c these electrons are *paired* having total energy less than $2E_f$. *The difference in energy between the normal state and paired state electrons appears as energy gap E_g at the Fermi surface.* This difference in energy in superconductors is different from the energy gap in insulators as it is due to the lattice. The energy gap in superconductor is due to *pairing of electrons* below T_c.

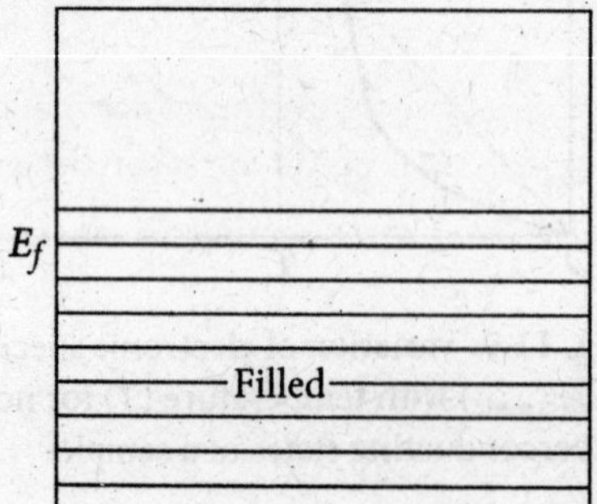

Fig. 11.10(a): Normal

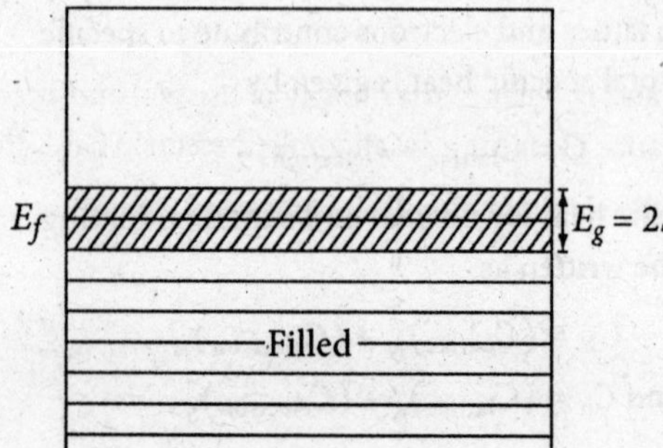

Fig. 11.10(b): Superconductor

Fig. 11.10(a) shows the conduction band in the normal state and Fig. 11.10(b) shows an energy gap equal to E_g at the Fermi level in superconductor, where $E_g = 2\Delta = 2b\,(kT_c)$ or, $E_g/kT_c = 2b$.

$2b$ is about 3.5, i.e., the gap decreases from a value ~ $3.5kT_c$ at 0 K to zero at the transition temperature. Values of E_g for some superconductors is shown in Table 11.2.

Table 11.2: Energy gap $E_g = 2\Delta$ at 0 K for some elements

Elements	2Δ (m eV)	T_c(K)	$2\Delta/kT_c$
Nb	3.05	9.50	3.8
Ta	1.40	4.48	3.6
Al	0.34	1.20	3.3
Sn	1.15	3.72	3.5
Pb	2.90	7.19	4.3
Hg	1.65	4.15	4.6

The ratio $2\Delta/kT_c$ is in good agreement with theory.

E_g is a function of temperature T. Fig. 11.11 shows the variation $[E_g(T)/E_g(0)]$ as a function of (T/T_c).

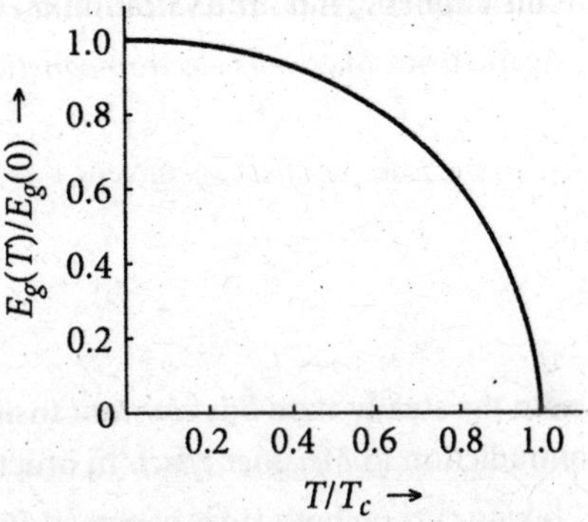

Fig. 11.11: Variation of $[E_g(T)/E_g(0)]$ with (T/T_c)

11.7 London Equations

London brothers, *F. London* and *H. London* in 1934 proposed *macroscopic theory* of *superconductivity.* They showed that the magnetic lines of flux were not completely expelled from the superconductor but remained confined in a thin surface layer which induced a flow of super-current in a closed circuit within the thin surface layer. They assumed *two fluid models*, i.e., there were *two* types of conduction electrons in a superconductor, viz., *super-electrons* and *normal electrons* at $T < T_c$. A superconductor contains only super-electrons at $T = 0$ K. But between $0 < T < T_c$, as the temperature rises the ratio of normal electrons to super-electrons increases and when $T = T_c$, all electrons are normal. Hence, at value of T between 0 and T_c the conduction electron density is

$$N = n_n + n_s, \tag{11.13}$$

where n_n and n_s refer to the density of normal and super-electrons. Since super-current does not experience any resistance even for a small applied electric field, the entire current is due to the super-electrons.

If $\vec{E}$ be the electric field, m the mass and e the charge of an electron, then assuming the super-electrons as free,

$$m\frac{d\vec{v}_s}{dt} = -e\vec{E} \tag{11.14}$$

$$\text{or,}\ \frac{d\vec{v}_s}{dt} = -\frac{e\vec{E}}{m}, \tag{11.15}$$

where $\vec{v}_s$ is the velocity of the super-electrons.

The *super-current* $\vec{J}_s$ is given by $\vec{J}_s = -n_s e\vec{v}_s$.

$$\therefore \frac{d\vec{J}_s}{dt} = -n_s e\frac{d\vec{v}_s}{dt} = \frac{n_s e^2}{m}\vec{E} \text{ [using eqn (11.15)]}$$

$$\text{i.e., } \frac{d\vec{J}_s}{dt} = \frac{n_s e^2}{m}\vec{E}. \tag{11.16}$$

Eqn (11.16) is known as the *first London Equation.*

From eqn (11.16) it is evident that $\frac{d\vec{J}_s}{dt} = 0$ for $\vec{E} = 0$.

This implies $\vec{J}_s$ *is finite and constant even in absence of* $\vec{E}$.

Again, from *Maxwell's electromagnetic equation,*

$$\nabla \times \vec{E} = -\frac{\partial \vec{B}}{\partial t}$$

$$\text{or, } \frac{\partial \vec{B}}{\partial t} = 0 \; [\because \vec{E} = 0] \tag{11.17}$$

$$\text{or, } \vec{B} = \text{constant,} \tag{11.18}$$

i.e., in the steady state $\vec{B}$ is *constant* inside the superconductor irrespective of temperature which is clear contradiction to *Meissner effect.* In order to avoid the contradiction *London* proposed some modifications.

Taking curl on both sides of eqn (11.16),

$$\nabla \times \frac{d\vec{J}_s}{dt} = \frac{n_s e^2}{m}\left(\nabla \times \vec{E}\right) = -\frac{n_s e^2}{m}\frac{d\vec{B}}{dt} \text{ [using eqn (11.17)].} \tag{11.19}$$

Integrating eqn (11.19) w.r.t. time,

$$\nabla \times \vec{J}_s = -\frac{n_s e^2}{m}\left(\vec{B} - \vec{B}_0\right). \tag{11.20}$$

The magnetic field $\vec{B}_0$ is a constant of integration and represents field at $t = 0$. Since *frozen-in magnetic field* precluded in *Meissner effect*, $\vec{B}_0 = 0$.

$\therefore$ eqn (11.20) becomes

$$\nabla \times \vec{J}_s = -\frac{n_s e^2}{m}\vec{B}. \tag{11.21}$$

This is known as *second London equation.*

Penetration Depth

We have *Maxwell's electromagnetic equation,*

$$\nabla \times \vec{B} = \mu_0 \vec{J}_s.$$

Taking curl on both sides,

$$\nabla \times \left(\nabla \times \vec{B}\right) = \mu_0\left(\nabla \times \vec{J}_s\right) \text{ or, } \nabla\left(\nabla \cdot \vec{B}\right) - \nabla^2\vec{B} = \mu_0\left(\nabla \times \vec{J}_s\right)$$

$$\text{or, } -\nabla^2\vec{B} = \mu_0\left(\nabla \times \vec{J}_s\right) \; [\because \nabla \cdot \vec{B} = 0]$$

$$\text{or, } \nabla^2\vec{B} = -\frac{\mu_0 n_s e^2}{m}\vec{B} \text{ [using eqn (11.21)]}$$

$$\text{or, } \nabla^2\vec{B} = \frac{\vec{B}}{\lambda^2}, \tag{11.22}$$

where $\lambda = \left(\frac{m}{\mu_0 n_s e^2}\right)^{1/2}$ is called *London penetration depth.*

The solution of eqn (11.22) is

$$H = H(0)\, e^{-\frac{x}{\lambda}}, \tag{11.23}$$

where $\vec{B} = \mu_0 \vec{H}$.

Eqn (11.23) shows that *H decreases exponentially with distance x inside the superconductor.* Fig. 11.11 shows the graphical variation of *H* with distance *x* inside the superconductor.

Putting $x = \lambda$, we get

$$H/H(0) = e^{-1} \text{ or, } H = \frac{H(0)}{e}. \tag{11.24}$$

Thus, *the penetration depth λ is defined as the depth from the surface at which the magnetic flux density falls to $\frac{1}{e}$ of its initial value at the surface.*

As *H* decreases exponentially, the flux inside the superconductor is zero which is in agreement with *Meissner effect.*

If $\lambda(0)$ be the penetration depth at $T = 0$ K, then at any temperature *T*,

$$\lambda(T) = \lambda(0)\left[1 - \frac{T^4}{T_c^4}\right]^{-1/2}. \tag{11.25}$$

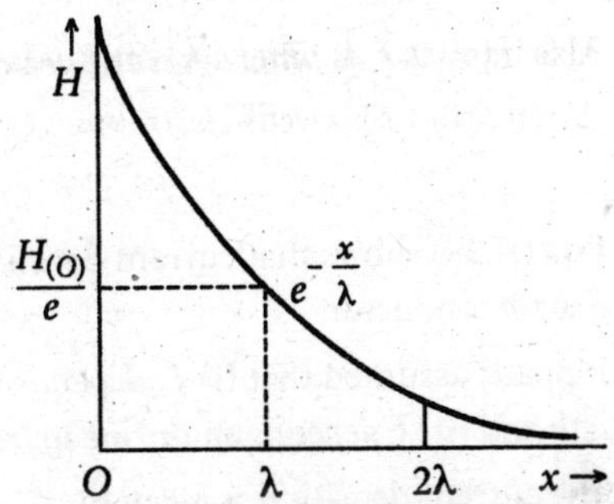

Fig. 11.12: Exponential decrease of *H* inside the superconductor

Eqn (11.25) shows that, *λ increases with the increase of T and $\lambda(T) \to \infty$ as $T \to T_c$.*

The sample changes from superconducting state to normal state at $T = T_c$ and the field *H* can penetrate the whole material. The variation of penetration depth $\lambda(\tau)$ with temperature is shown in Fig. 11.12.

The *London penetration depth λ* and the number of superelectrons n_s are inversely related to each other, and n_s is given by

$$n_s = n_0\left[1 - \left(\frac{T}{T_c}\right)^4\right] \tag{11.26}$$

$$\text{or, } \psi = \frac{n_s}{n_0} = \left[1 - \frac{T^4}{T_c^4}\right], \tag{11.27}$$

where ψ is known as the *order parameter.*

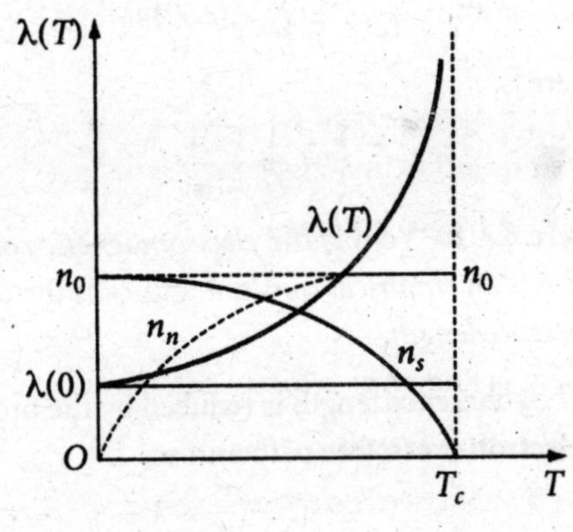

Fig. 11.13: Variation of λ, n_n and n_s with *T*

11.8 Pippard's Equation and Coherence Length

We have already seen that for a superconductor electron density n_s, the *London penetration depth* is given by

$$\lambda = \left(\frac{m}{\mu_0 n_s e^2}\right)^{1/2}, \tag{11.28}$$

where *m* and *e* are the mass and charge of the super-electron and μ_0 is the permeability of the material. Eqn (11.28) shows that λ does not depend on the impurity present in the material. But *Pippard* experimentally measured the penetration depth of tin and found that λ should change by a factor of 2 at low temperature.

This suggests the non-local modification of *London theory* and the concept of coherence length ξ depending on the mean free path of the super-electrons in the lattice.

Now we have from the first London eqn (11.16)

$$\frac{d}{dt}(\nabla \times \vec{J}_s) = \frac{n_s e^2}{m}(\nabla \times \vec{E}) = -\frac{n_s e^2}{m}\frac{d}{dt}\vec{B}\left[\because \nabla \times \vec{E} = -\frac{d\vec{B}}{dt}\right]$$

$$\text{or, } \vec{B} = -\frac{m}{n_s e^2}(\nabla \times \vec{J}_s) \text{ or, } \vec{H} = -\frac{m}{\mu_0 n_s e^2}(\nabla \times \vec{J}_s).$$

Also, $\vec{H} = \nabla \times \vec{A}$, where $\vec{A}$ is the *vector potential*.

$$\therefore \vec{A} = -\frac{m}{\mu_0 n_s^2 e^2}\vec{J}_s. \qquad (11.29)$$

Eqn (11.29) shows that current density $\vec{J}_s$ at any point depends on the vector potential $\vec{A}$ at that point of the superconductor.

Pippard assumed that (i) $\vec{J}_s$ *depends on* $\vec{A}$ *integrated over a domain of radius* ξ, *where* ξ *is the coherence length* and (ii) ξ *depends on the mean free path of the electrons to account the variation of* ξ *with impurity.*

The current density $\vec{J}_s$ is given by

$$\vec{J}_s(\vec{r}) = \int K(\vec{r}, \vec{r}')\,\vec{A}(\vec{r}')\,d\vec{r}'. \qquad (11.30)$$

The value of ξ is defined by the function, $k(\vec{r} - \vec{r}')$. According to Pippard,

$$K(\vec{r}, \vec{r}') = e^{(\vec{r}-\vec{r}')\times\frac{1}{\xi}}, \qquad (11.31)$$

where

$$\frac{1}{\xi} = \frac{1}{\xi_0} + \frac{1}{\alpha v}, \qquad (11.32)$$

where $\xi \sim 10^{-4}$ cm is the *electromagnetic coherence* and α is an *empirical constant* and ξ_0 is the *intrinsic coherence length.*

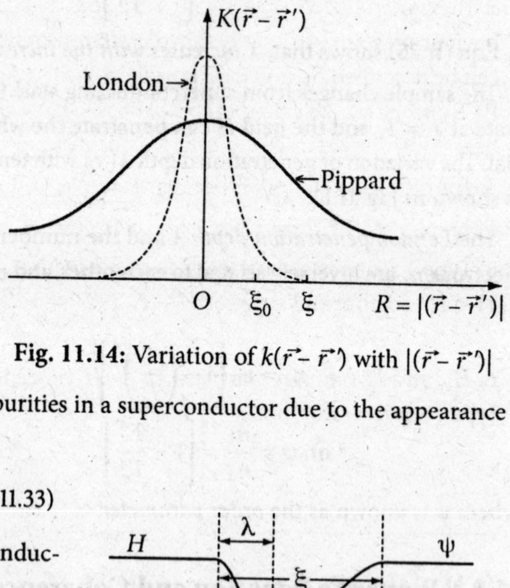

Fig. 11.14: Variation of $k(\vec{r} - \vec{r}')$ with $|(\vec{r} - \vec{r}')|$

The coherence length is reduced by the presence of impurities in a superconductor due to the appearance of electron mean free path and so,

$$k = \frac{\lambda}{\xi} \qquad (11.33)$$

is a number controlled by alloying or doping the superconductor and is known as *Ginzberg-Landau parameter.*

11.9 Variation of Magnetic Field and Order Parameter Across a Superconductor Normal Interface

The variation of the magnetic field H and order parameter ψ within the interface between normal and superconducting region is shown in Fig. 11.15.

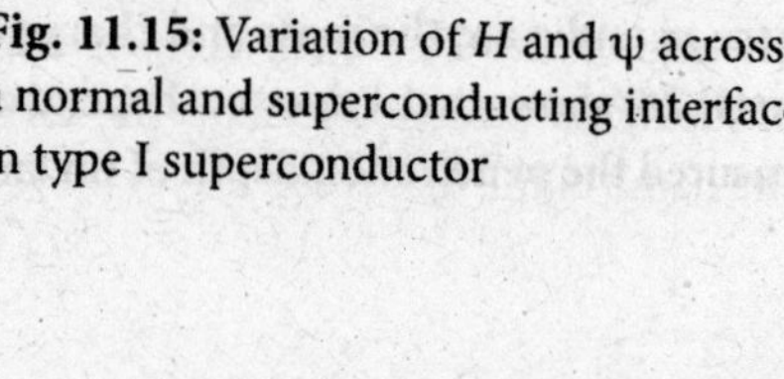

Fig. 11.15: Variation of H and ψ across a normal and superconducting interface in type I superconductor

The coherence length ξ_0 for a pure superconductor can be shown to be

$$\xi_0 = \frac{\hbar v_F}{2\Delta} \quad \left(\Delta = \tfrac{1}{2} kT_c\right), \tag{11.34}$$

where v_F is the *electron velocity at the Fermi surface.* A more accurate expression for ξ_0 is

$$\xi_0 = \frac{2\hbar v_F}{\pi \Delta}.$$

11.10 Comparison of Type I and Type II Superconductors

Type I *Superconductor*	*Type* II *Superconductor*
1. Type I superconductors strictly follow Meissner effect, i.e., the magnetic field penetrates completely at $H > H_c$, H_c is the critical field.	1. Type II superconductors do not follow Meissner effect strictly, i.e., the magnetic field does not penetrate abruptly at $H = H_c$.
2. These materials give away their superconductivity at lower value of H and hence, are called *soft superconductors.* Due to very low value of H_c these materials have very limited technical applications.	2. Relatively large field (H) is necessary to bring them to normal state (i.e., H_c is large) and hence, are called *hard superconductors.* These are used to manufacture superconducting wires used in *superconducting magnets.*
3. These are called *Pippard superconductors.* Their behaviour may not be described by *London theory* but can be described by *Pippard's non-local theory.* Also, BCS *theory* can explain their properties like specific heat, tunnelling effects and infrared absorption, etc.	3. Type II superconductors are called *London superconductors* and may be described by London theory.
4. For type I superconductors, $\frac{\lambda}{\xi} = k < \frac{1}{\sqrt{2}}$. *Examples*: Al, In, Sn and Pb.	4. For type II superconductors $\frac{\lambda}{\xi} \simeq k > \frac{1}{\sqrt{2}}$. *Examples*: Nb_3S_n and V_eGa.
5. Type I materials show *perfect diamagnetism* for $H < H_c$ and normal state for $H > H_c$.	5. *Perfect diamagnetism* is shown below $H = H_{c_1}$, where $H_{c_1} < H_c$ and normal state above $H > H_{c_2}$. The region $H_{c_1} < H < H_{c_2}$ is called the *vortex region.*
6. H_{c_1} and H_{c_2} do not exist, i.e., $H_{c_1} = H_{c_2} = H_c$ for type I superconductor.	6. For type II superconductors, $H_{c_1} = \frac{\xi}{\lambda} H_c$. *Abrikosov* showed that $H_{c_1} = \frac{\xi}{\lambda} H_c \left[\ln\left(\frac{\xi}{\lambda}\right) - 0.27\right]$. $H_{c_2} = \sqrt{2}\left(\frac{\lambda}{\xi}\right) H_c$.

11.11 Temperature Dependence of H_{c_1} and H_{c_2}

For type I superconductors, H_c undergoes variation with the increase of temperature of the material in the manner as shown in Fig. 11.16(a). The variation of H_{c_1} and H_{c_2} for type II superconductor is represented in Fig. 11.15(b).

In Fig. 11.16(b), the Meissner region represents that all the flux is expelled by the material when $H < H_{c_1}$. The region between H_{c_1} and H_{c_2} represents the penetration of flux into the material without the destruction of superconducting behaviour. The corresponding region in Fig. 11.16(b) is known as *Shubnikov region* after the name of the discoverer.

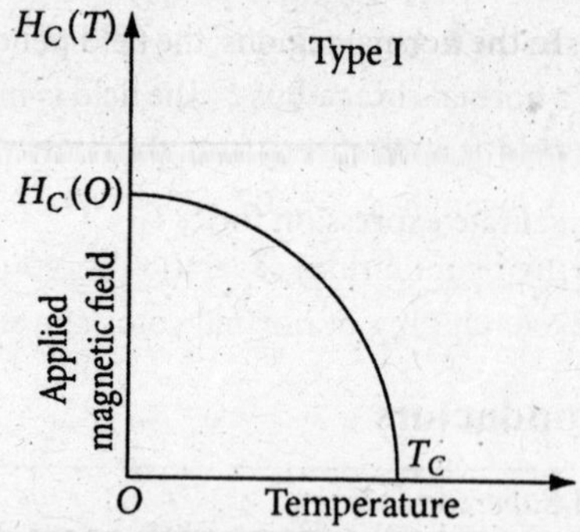

Fig. 11.16(a): Variation of H_c with the increase of temperature

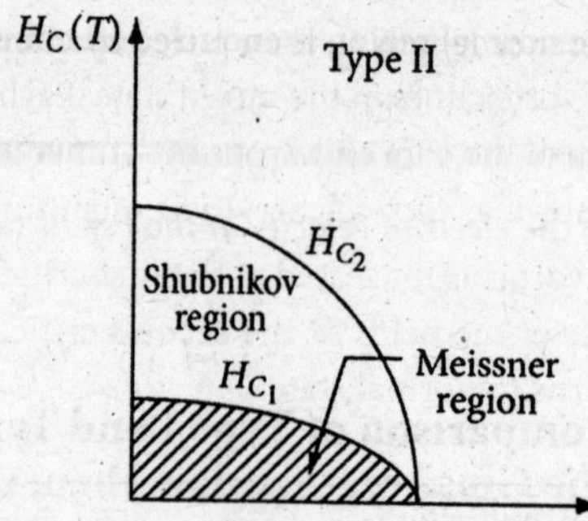

Fig. 11.16(b): Variation of H_{c_1} and H_{c_2} with temperature

11.12 Critical Currents

The critical current I_c is defined as the minimum current that can be passed in a sample without destroying its superconductivity. If a wire of radius *r* of type I superconductor carries a current I, the surface magnetic field produced is $H_I = \frac{I}{2\pi r}$ associated with *I*. If $H_I > H_c$, the material will transit to normal state. If now a transverse magnetic field *H* is applied to the wire, then the condition of transition to the normal state at the surface will be that H_c is equal to the sum of *H* and H_I. Thus, as shown in Fig. 11.17(b), $H_c = H_I + 2H$.

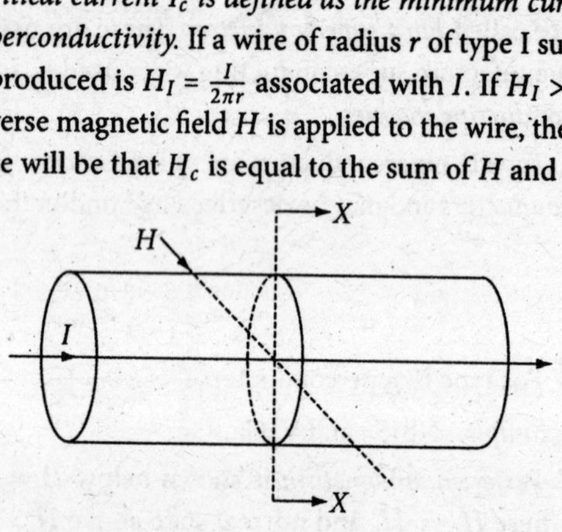

Fig. 11.17(a): Type I semiconductor carrying current *I*

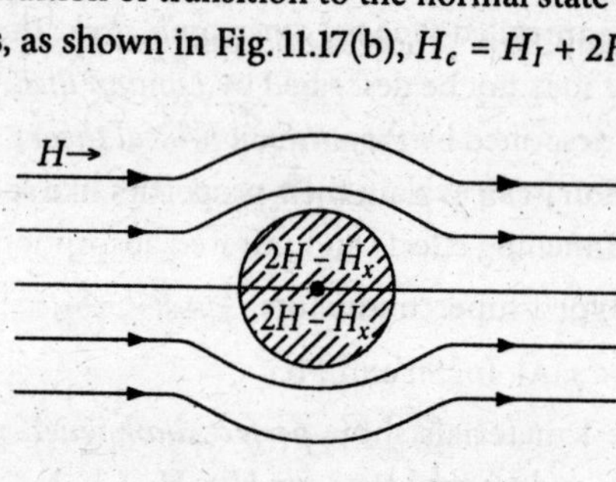

Fig. 11.17(b): A transverse magnetic field *H* is applied to the wire

But

$$H_I = \frac{I_c}{2\pi r} = H_c - 2H. \therefore I_c = 2\pi r\,(H_c - 2H). \qquad (11.35)$$

Eqn (11.35) shows that with the increase of the applied field (H), the critical current I_c decreases until reduces to zero at $H = H_c/2$. For $H = 0, I_c = 2\pi r H_c$.

For type II superconductors, the same consideration applies for $H < H_{c_1}$, when the superconductor is not in the mixed state.

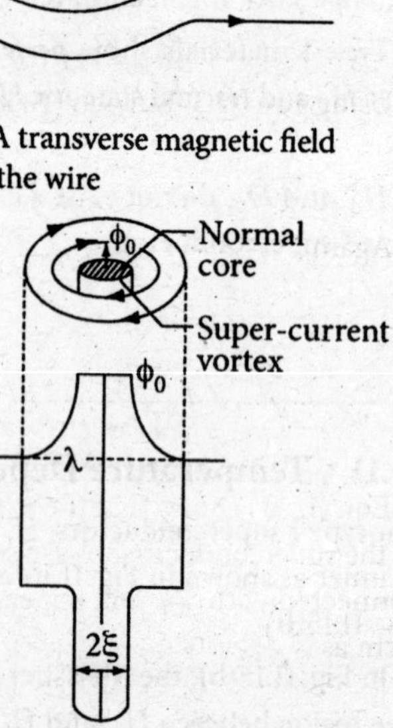

Fig. 11.18: Vortex lines in a type II mixed state

11.13 The Mixed State

There is a fine subdivision between normal and superconducting phases in the mixed state of type II superconductors. The normal core is embedded in the superconducting material as shown in Fig. 11.18. The microscopic normal region is encircled by circulating currents. In the normal regions, the field penetrates in the form of vortex lines in the mixed state. Each vortex line has a normal core radius ξ. The field is maximum at the centre of the core and drops to $\frac{1}{e}$ times of the maximum field at a distance λ from the centre. The flux vortex is known as *fluxoid* carrying a quantum of magnetic flux given by $\phi_0 = 2.07 \times 10^{-15}$ Wb.

The flux vortex is nucleated at the first critical field H_{c_1} and their equilibrium separation is reduced with the increase of the field. At the second critical field H_{c_2}, the overlapping of normal cores starts and the material turns to normal state.

11.14 The Ginzberg-Landau Theory[1]

Ginzberg and Landau asserted that the superconducting state could be characterised by a complex *order parameter* $\psi(\vec{r})$. The magnitude of $\psi(\vec{r})$ represents the *degree of superconducting order* at position $\vec{r}$ below T_c. $\psi(\vec{r})$ vanishes at $T \le T_c$.

The wave function ψ is an effective wave function of the superconducting electron, where $|\psi|^2 = n_s$, the density of superconducting electrons. Superconductivity may be considered as a *second order phase transition* with a critical temperature T_c and a positive parameter n_s. $n_s = 0$ for $T > T_c$ and n_s = finite for $T < T_c$. Then the *Gibb's free energy* $F(T, V, n_s)$ is given by

$$F(T, V, n_s) = F(T, V, 0) + \alpha n_s + \tfrac{1}{2}\beta n_s^2. \tag{11.36}$$

At equilibrium, there is no change of F with n_s, i.e.,

$$\delta F/\delta n_s = 0, \text{ i.e., } \alpha + \beta n_s = 0, \quad \text{or, } n_s = -\alpha/\beta. \tag{11.37}$$

For stability F is to be minimum. Again, for F to be minimum $\frac{\Delta^2 F}{\Delta n_s^2} > 0$ and β is positive. From eqn (11.37), α should be negative.

Therefore, for a *second order phase transition*, we assume that $\alpha(T_c) = 0$.

$$\alpha(T) = (T - T_c)\left(\frac{\partial \alpha}{\partial T}\right)_{T=T_c}. \tag{11.38}$$

Using eqn (11.37) in eqn (11.36), we get

$$F_s = F_n - \frac{\alpha^2}{2\beta}. \tag{11.39}$$

Again,

$$F_s - F_n = -\frac{H_c^2}{8\pi}. \tag{11.40}$$

$$\therefore \frac{\alpha^2}{2\beta} = \frac{H_c^2}{8\pi}. \tag{11.41}$$

Eqn (11.41) shows that the external magnetic field H contributes to the free energy by an amount $H_c^2/8\pi$ in the superconducting phase of the system. As ψ is not rigid in the presence of H an extra free energy term connected with $\nabla\psi$ will appear in addition to the term $H_c^2/8\pi$. *Ginzberg and Landau,* assumed the extra term as

$$\frac{1}{2m}\left|-i\hbar\nabla\psi + \frac{e^*}{c}\vec{A}(\vec{r})\,\psi\right|^2 \tag{11.42}$$

[1] V.L. Ginzberg and L.D. Landau, Zh. Eksp. Teor. Fiz. **20,** 1064 (1950).

so that the *Gauss's invariance* be preserved. Here $\vec{A}(\vec{r})$ is the *vector potential* and m^* is the *effective charge.*

$$\therefore F_t = F_n + \int \left[\alpha|\psi|^2 + \frac{1}{2}\beta|\psi|^4 + \frac{\mu}{8\pi} + \frac{1}{2m}\left|\left(i\hbar\nabla\psi - \frac{e^*}{c}\vec{A}(\vec{r})\,\psi\right)\right|^2 dV \right]. \tag{11.43}$$

The eqn (11.43) is invariant under the simultaneous transformation

$$\left.\begin{aligned} \psi^* &= \psi e^{\left(\frac{ie^*}{\hbar c}\right)\phi} \\ \vec{A}' &= \vec{A} + \nabla\phi \end{aligned}\right\}. \tag{11.44}$$

Now from eqn (11.43),

$$\delta F = \int \left[\alpha\psi + \beta|\psi|^2\psi + \frac{1}{2m}\left\{\left(-i\hbar\nabla\psi - \frac{e^*}{c}\psi\right)\left(i\hbar\nabla - \frac{e^*}{c}\vec{A}\right)\right\}\right] \delta\psi^* dV. \tag{11.45}$$

Now

$$\int \frac{1}{2m}\left\{\left(-i\hbar\nabla\psi - \frac{e^*}{c}\psi\right)\left(i\hbar\nabla - \frac{e^*}{c}\vec{A}\right)\right\} \delta\psi^* dV$$

$$= \frac{1}{2m}\left[\int \delta\psi^*\left(-i\hbar\nabla - \frac{e^*}{c}\vec{A}\right)\psi dV + \iint \hat{n}\cdot\left(-i\hbar\nabla - \frac{e^*}{c}\vec{A}\right)\psi\delta\psi^* da\right]. \tag{11.46}$$

Since $\delta\psi^*$ is arbitrary,

$$\iint \hat{n}\cdot\left(-i\hbar\nabla - \frac{e^*}{c}\vec{A}\right)\psi\delta\psi^* da = 0. \tag{11.47}$$

$$\therefore \delta F = \int \left[\frac{1}{2m}\left(-i\hbar\nabla - \frac{e^*}{c}\vec{A}\right)^2 \psi + \alpha\psi + \beta|\psi|^2\psi\right]\delta\psi^* dV. \tag{11.48}$$

For F to be minimum, $\delta F = 0$.

$$\therefore \frac{1}{2m}\left(-i\hbar\nabla - \frac{e^*}{c}\vec{A}\right)\psi + \alpha\psi + \beta|\psi|^2\psi = 0 \tag{11.49}$$

with the condition

$$\hat{n}\cdot\left(-i\hbar\nabla - \frac{e^*}{c}\vec{A}\right) = 0.$$

Now variation w.r.t. $\vec{A}$ gives

$$\nabla\times\nabla\times\vec{A} = -\nabla^2\vec{A} = \frac{4\pi}{c}\vec{J}, \tag{11.50}$$

where the current $\vec{J}$ due to $e^*(= -2e)$ is

$$\vec{J} = -\frac{ie^*\hbar}{2m}(\psi^*\nabla\psi - \psi\nabla\psi^*) - \frac{e^{*2}}{mc}|\psi|^2\vec{A} \tag{11.51}$$

with $\nabla\cdot\vec{A} = 0$.

Hence, the *fundamental equations of Ginzberg-Landau theory* are as follows:

$$\frac{1}{2m}\left(-i\hbar\nabla - \frac{e^*}{c}\vec{A}\right)^2 \psi + \alpha\psi + \beta|\psi|^2\psi = 0 \tag{11.52}$$

$$\text{and } \vec{J} = -i\frac{e^*\hbar}{2m}(\psi^*\nabla\psi - \psi\nabla^*) - \frac{e^{*2}}{mc}|\psi|^2 A. \tag{11.53}$$

11.15 The Bardeen-Cooper-Schrieffer (BCS) Theory of Superconductivity

In order to describe various superconducting phenomena two basic approaches are used, viz., (i) *a phenomenological theory using London equation* and (ii) *a microscopic theory proposed by Bardeen, Cooper and Schrieffer known as* BCS *theory.*[2]

The BCS theory is baed on the following:

(i) *The electrons in the superconducting state form bound pairs (the Cooper pairs). The interaction between electron and lattice, through lattice vibration, leads to electron-electron interaction forming bound electron-pairs.*

(ii) *By the exchange of virtual phonons between two electrons, a bound electron-pair is formed.*

(iii) *An electron with spin up and momentum* $(\vec{k})$ *pairs with another electron with spin down and momentum* $(-\vec{k})$ *so that the angular momentum of the system is zero.*

(iv) *The energy gap corresponds to the binding energy and the coherence length is the same as the size of the electron-pair.*

We consider two electrons 1 and 2, lying just inside the Fermi surface in a metal with conduction electrons inside the Fermi sphere. If these two electrons are close to the *Fermi surface,* they form a bound pair and their motion is correlated. The pair formation leads to *Bose-Einstein condensation* of all the electrons close to the Fermi surface, and a *sea of Cooper pair* is formed.

The electron being close to the Fermi surface moves with high velocity (v_F) in the lattice. The electron is attracted by the neighbouring positive ions forming a positive ion core as shown in Fig. 11.19. The electron charge is then greatly reduced due to the screening effect of the positive charges. The region around the electron, in equilibrium position, carries more positive charge than in the normal state as a result of which the lattice gets distorted.

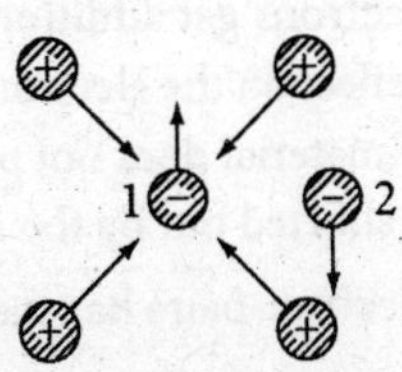

Fig. 11.19: Positive ions attracted towards an electron forming a positive ion core

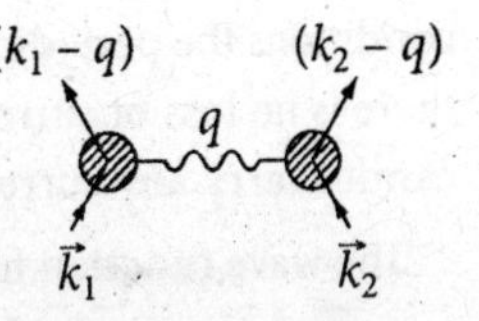

Fig. 11.20: Exchange of virtual phonon between Cooper pair

After the electron 1 leaves, the region remains positive for a time and attracts another electron 2. The electron 2 then interacts with electron 1 through the exchange of *virtual phonon* created due to lattice distortion. In other words, each electron is surrounded by a phonon cloud and electrons 1 and 2 interact by the exchange of a phonon as shown in Fig. 11.20.

In the interaction process electron 1 emits a phonon of momentum q which is absorbed by electron 2. Then the momenta of electrons 1 and 2 will be $(k_1 - q)$ and $(k_1 + q)$ respectively. The force of interaction is the maximum if electrons 1 and 2 possess opposite spin and opposite momenta, i.e., $\vec{k}_1 = \vec{k}\uparrow$ and $\vec{k}_2 = -\vec{k}\downarrow$.

Phonons are available only within a small energy range $\hbar\nu_D = k\theta_D$, where ν_D is the *Debye temperature.* The *electron-electron interaction* becomes attractive, if

$$E_{k_1} - E_{k_2} = \hbar\nu_D, \tag{11.54}$$

[2]J. Bardeen, L.N. Copper and J.R. Schrieffer, Phys. Rev, 18, 1175, (1957).

where E_{k_1} and E_{k_2} are the energy of electrons 1 and 2 respectively. The interaction process is expressed as: $k_1 - q = k_1'$ and $k_2 + q = k_2'$.

$$\therefore k_1 + k_2 = k_1' + k_2'. \tag{11.55}$$

This implies that net wave vector (momentum) of the pair is conserved. If an electron-pair has equal and opposite momenta $(\vec{k}$ and $-\vec{k})$, i.e., $\vec{k} = 0$, then in the ground state maximum overlap of phase with largest number of transitions occurs. It can be shown that for the maximum strength of the interaction the two electrons of the pair must have opposite spins, i.e., $(\vec{k}\uparrow$ and $-\vec{k}\uparrow)$.

Due to the formation of the pair, the total energy of the pair becomes less than the total energy of the free electrons. This creates a gap near the Fermi level which is approximately

$$2\Delta = 3.5kT_c, \tag{11.56}$$

where k is the *Boltzmann constant* and T_c is the *critical temperature of the material.*

Below T_c, 2Δ rises sharply according to the relation

$$2\Delta(T) \simeq 3.52kT_c\left[1 - \frac{T}{T_c}\right]^{1/2}. \tag{11.57}$$

At $T < T_c$, the *lattice-electron interaction* is stronger than *electron-electron interaction* in this temperature range. The Cooper pair moves smoothly in the lattice with constant momentum and without any transfer of energy to the lattice since the pair assembly involves a high degree of correlation in their motion. If now an electric field is applied, the electrons get additional KE and produce an electric current. Under such conditions the periodicity has no effect on the electron-pairs. The motion of the pair is not slowed down and there is no loss of current, i.e., the material does not possess any *electrical resistance* as the normal electrons cannot carry any current and are shorted out by the super-electrons.

The wave function for all the electron-pairs has the same form:

$$\psi(r', r'') = \sqrt{n_s}\, e^{iS(r)}, \tag{11.58}$$

where n_s is the number of super-electrons and $S(r)$ is a phase function. Thus, BCS theory accounts the zero resistivity at $T < T_c$ of the superconductors.

11.16 The BCS Ground State

The ground state of a Fermi gas consisting of non-interacting electrons forms the *Fermi sea* which allows arbitrarily small excitations. An excited state can be formed by removing one electron from the Fermi surface and raising it just above the Fermi surface. BCS theory leads to the fact that *an appropriate attractive interaction between the electrons forms the new ground state which is superconducting and separated by finite energy E_g from the lowest excited state.*

A Cooper pair may be formed by a weak attractive interaction between electron and phonon. The formation of a single Cooper pair leads to an energy reduction of the *Fermi sea*. The total energy reduction can be obtained by taking into account the KE of one-electron component and the energy reduction due to electron-phonon collision.

Now the KE of one-electron component is given by

$$E_k = 2\sum_k w_k \xi_k, \tag{11.59}$$

where $\xi_k = \frac{\hbar^2 k^2}{2m} - E_k^0$ and w_k is the probability that the pair state $(\vec{k}\uparrow, \vec{k}\downarrow)$ be occupied and $E_F^0 = E_F$ $(T = 0\,\text{K})$.

Again, the pair collision

$$(\vec{k}\uparrow, -\vec{k}\downarrow) \rightleftarrows (\vec{k}'\uparrow, -\vec{k}'\downarrow)$$

leads to the annihilation of a pair $(\vec{k}\uparrow, \vec{k}\downarrow)$ and simultaneous creation of a pair $(\vec{k}'\uparrow, \vec{k}'\downarrow)$. The reduction of energy associated with the process is given by

$$V'_{kk'} = \int V(\vec{r})\, e^{-(\vec{k}-\vec{k}')\cdot\vec{r}} d^3\vec{r}. \tag{11.60}$$

If states $|1\rangle_k$ and $|0\rangle_k$ represent respectively the *occupied* and *unoccupied* states of the electron-pair, then the most general state of the pair $(\vec{k}\uparrow, \vec{k}\downarrow)$ is given by

$$|\psi\rangle_k = |\psi\rangle_k = k|0\rangle_k + v_k|1\rangle_k. \tag{11.61}$$

$\therefore w_k = v_k^2$ and $1 - \omega_k = u_k^2$ are the probabilities that the pair is occupied and unoccupied respectively. The ground state of the many body system of all the *Cooper pairs* in the superconductor is

$$|\psi_{\text{BCS}}\rangle = \prod_k [u_k|0\rangle + v_k|1\rangle_k]. \tag{11.62}$$

Here we have neglected the interactions between the Cooper pairs.

For two-dimension representation,

$$|0\rangle_k = \begin{pmatrix} 0 \\ 1 \end{pmatrix}_k ; \; |1\rangle_k = \begin{pmatrix} 1 \\ 0 \end{pmatrix}. \tag{11.63}$$

Let us use *Pauli spin matrices*

$$\sigma_k^{(1)} = \begin{pmatrix} 0 & 1 \\ 1 & 0 \end{pmatrix}_k ; \; \sigma_k^{(2)} = \begin{pmatrix} 0 & -i \\ i & 0 \end{pmatrix} \tag{11.64}$$

for the description of creation and annihilation of a Cooper pair by the operators

$$\sigma_k^+ = \tfrac{1}{2}\left(\sigma_k^{(1)} + i\sigma_k^{(2)}\right); \; \sigma_k^- = \left(\sigma_k^{(1)} - i\sigma_k^{(2)}\right) \tag{11.65}$$

respectively. Here σ_k^+ transforms state $|0\rangle_k$ into $|1\rangle_k$ and σ_k^- transforms state $|1\rangle_k$ to $|0\rangle_k$.

Now

$$\begin{aligned} \sigma_k^+ |0\rangle_k &= 0, \qquad \sigma_k^+ |0\rangle_k = |1\rangle_k \\ \sigma_k^- |1\rangle_k &= |0\rangle_k, \quad \sigma_k^- |0\rangle_k = 0. \end{aligned} \tag{11.66}$$

$V_{kk'}$ is the reduction of energy in the scattering process

$$(\hat{k}\uparrow, -\vec{k}\downarrow) \rightleftarrows (\hat{k}'\uparrow, -\vec{k}'\downarrow). \tag{11.67}$$

The interaction matrix element in BCS *theory* can be assumed to be a constant. Let it be V_0/L^3. The scattering process in two dimensions is the annihilation of $\vec{k}$ and creation of $\vec{k}'$ and the operator describing the reduction in energy in the process is $-\frac{V_0}{L^3}\sigma_{k'}^+\sigma_k^-$.

Now the total reduction in energy in the collision process represented by eqn (11.67) is the *Hamiltonian* H, where

$$H = -\frac{V_0}{L^3}\sum_{kk'} \tfrac{1}{2}\left(\sigma_{k'}^+, \sigma_k^- + \sigma_k^+\sigma_{k'}^-\right) = -\frac{V_0}{L^3}\sum_{kk'} \sigma_k^+\sigma_{k'}^-. \tag{11.68}$$

The value of V_0 is restricted to the shell $\pm\hbar\omega_D$ around E_F^0 and the sum includes only the pair states.

According to *perturbation theory*, the reduction in energy due to collision is given by

$$\langle\psi_{BCS}|H|\psi_{BCS}\rangle = -\frac{V_0}{L^3}\left[\prod_x \left(U_{xx}\langle 0| + v_{xx}\langle 11|\right) \times \sum_{kk'} \sigma_k^+ \sigma_k^- \prod_y \left(u_y|0\rangle_y + v_y|1\rangle_y\right)\right]. \tag{11.69}$$

Since, the operators $\sigma_k^+\left(\sigma_k^-\right)$ operate only on $|1\rangle_k\,(|0\rangle_k)$, we have

$$\langle\psi_{BCS}|H|\psi_{BCS}\rangle = -\frac{V_0}{L^3}\sum_{kk'} v_k u_{k'} u_k u_{k'}\; \left[\because\ {}_k\langle 1|1\rangle_k = 1 = {}_k\langle 0|0\rangle_k \text{ and } {}_k\langle 1|0\rangle_k = 0\right]. \tag{11.70}$$

The total energy reduction due to collision of the *Cooper pair* is obtained by adding eqns (11.59) and (11.70) as

$$W_{BCS} = 2\sum_k u_k^2 \xi_k - \frac{V_0}{L^3}\sum_{kk'} u_k u_k u_k' u_k'. \tag{11.71}$$

The BCS ground state at $T = 0$ is represented by the minimum W_{BCS} relative to amplitude u_k and v_k as

$$\begin{aligned} v_k &= \sqrt{w_k} = \cos\theta_k \\ u_k &= \sqrt{1-w_k} = \sin\theta_k, \end{aligned} \tag{11.72}$$

i.e., $u_k^2 + v_k^2 = 1$.

So

$$\begin{aligned} W_{BCS} &= \sum_k 2\xi_k \cos^2\theta_k - \frac{V_0}{L^3}\sum_{kk'} \cos\theta_k \sin\theta_{k'} \cos\theta_{k'} \sin\theta_k \\ &= \sum_k 2\xi_k \cos^2\theta_k - \frac{1}{4}\frac{V_0}{L^3}\sum_{kk'} \sin 2\theta_k \sin 2\theta_{k'}. \end{aligned} \tag{11.73}$$

For W_{BCS} to represent the ground state it must be minimum, i.e.,

$$\frac{\partial W_{BCS}}{\partial \theta_k} = 0 \tag{11.74}$$

$$\text{or, } -2\xi_k \sin^2\theta_k - \frac{V_0}{L^3}\sum_{k'} \cos^2\theta_k \sin^2\theta_{k'} = 0$$

$$\text{or, } \xi_k \tan 2\theta_k = -\frac{1}{2}\frac{V_0}{L^3}\sum_{k'} \sin 2\theta_{k'}. \tag{11.75}$$

Now let

$$\Delta = \frac{V_0}{L^3}\sum_{k'} u_{k'} v_{k'} = \frac{V_0}{L^3}\sum_{k'} \sin 2\theta_{k'} \tag{11.76}$$

$$\text{and } E_k = \sqrt{\xi^2 + \Delta^2}. \tag{11.77}$$

$$\therefore \frac{\sin 2\theta_k}{\cos 2\theta_k} = \tan 2\theta_k = -\frac{\Delta}{\xi_k} \tag{11.78}$$

$$2u_k v_k = \sin 2\theta_k = \frac{\Delta}{\xi_k} \tag{11.79}$$

$$v_k^2 - u_k^2 = -\frac{\xi_k}{E_k}. \tag{11.80}$$

Occupation Probability of the Cooper Pair in the BCS Ground State

The probability of occupation (w_k) of the Cooper pair $(\vec{k}\uparrow, -\vec{k}\downarrow)$ in the BCS ground state is

$$w_k = v_k^2 = \frac{1}{2}\left(1 - \frac{\xi_k}{E_k}\right) = \frac{1}{2}\left(1 - \frac{\xi_k}{\sqrt{\xi_k^2 + \Delta^2}}\right). \tag{11.81}$$

The nature of variation of v_k^2 against electron energy ξ_k is shown in Fig. 11.21. It is similar to the Fermi-distribution curve at finite critical temperature T_c. The probability of occupation is unity when the energy is below E_F, but around E_F, the occupation probability is less. The energy reduction of the BCS ground state is due to the Cooper pair in a particular k-region $\pm\Delta k$ around E_F.

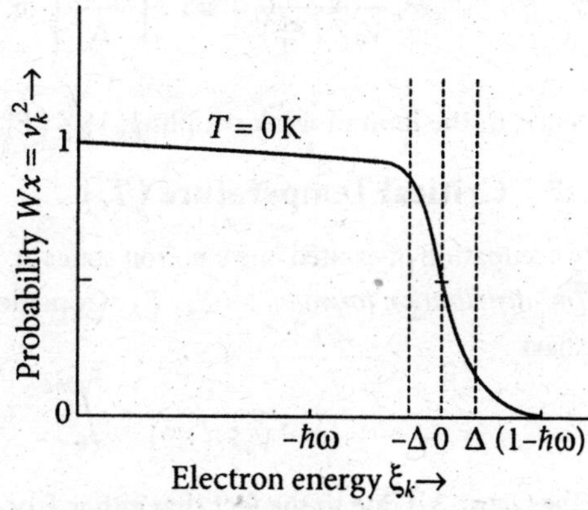

Fig. 11.21: BCS occupation probability v_k^2 for Cooper pairs in the vicinity of Fermi energy E_F

The energy of the BCS ground state $\left(W^0_{\text{BCS}}\right)$ is obtained from eqns (11.73), (11.76) and (11.81) as

$$W^0_{\text{BCS}} = \sum_k \xi_k\left(1 - \frac{\xi_k}{E_k}\right) - L^3\frac{\Delta^2}{V_0}. \tag{11.82}$$

We can show that

$$\Delta = \left(W^1_{\text{BCS}} - W^0_{\text{BCS}}\right) = 2E_{k'} = 2\sqrt{\xi_{k'}^2 + \Delta^2}. \tag{11.83}$$

Here $\xi_{k'} = \left(\frac{\hbar k'^2}{2m} - E_F^0\right)$ represents the KE of two scattered electrons from the Cooper pair and it is a very small quantity. Thus, the excitation energy of the pair is minimum but finite, i.e., 2Δ, which is also the energy needed to break up a Cooper pair.

Now from eqn (11.76),

$$\Delta = \frac{V_0}{L^3}\frac{1}{2}\sum_k 2u_k v_k \tag{11.84}$$

$$\text{or, } \Delta = \frac{1}{2}\frac{V_0}{L^3}\sum_k \frac{\Delta}{E_k} = \frac{1}{2}\frac{V_0}{L^3}\sum_k \frac{\Delta}{\sqrt{\xi_k^2 + \Delta^2}} \tag{11.85}$$

$$\text{and } I = \frac{V_0}{(2\pi)^3}\iint \frac{d\xi_k}{|\nabla_k \xi(k)|}\frac{d\xi}{\sqrt{\xi^2 + \Delta^2}}, \tag{11.86}$$

where summation

$$\frac{1}{L^3}\sum_k \frac{\Delta}{\sqrt{\xi_k^2 + \Delta^2}}$$

has been replaced by

$$\frac{1}{(2\pi)^3}\int_{S_\xi}\int_\xi \frac{dS_\xi}{|\nabla_k \xi(k)|}\frac{d\xi}{\sqrt{\xi^2 + \Delta^2}}$$

and the integral over k-space has been split into an integral over the *Fermi sphere* and the energy ξ.

$$\int_{S_\xi} \frac{dS_\xi}{|\nabla_k \xi(k)|}$$

represents the *density of states* and may be considered as a constant.

Here the *Cooper pair density* of states is

$$Z\left(E_F^0 + \xi\right) = \tfrac{1}{2}D\left(E_F^0 + \xi\right),$$

where D represents a *single particle density* of states.

$$\therefore I = \frac{V_0}{2}\int_{-\hbar\omega_D}^{+\hbar\omega_D} \frac{Z\left(E_F^0+\xi\right)}{\sqrt{\xi^2+\Delta^2}}\,d\xi \quad [\because -\hbar\omega_D \le \xi \le \hbar\omega_D]. \tag{11.87}$$

V_0 does not vanish in the region $\left(E_F^0-\hbar\omega_D\right)$ to $\left(E_F^0+\hbar\omega_D\right)$ and $Z\left(E_F^0+\xi\right)$ varies slightly. So, we can assume, $Z\left(E_F^0+\xi\right) \equiv Z\left(E_F^0\right)$; also due to symmetry,

$$\frac{1}{V_0 Z\left(E_F^0\right)} = \int_0^{\hbar\omega_D} \frac{d\xi}{\sqrt{\xi^2+\Delta^2}} \tag{11.88}$$

$$\text{or, } \frac{1}{V_0 Z\left(E_F^0\right)} = \sin^{-1}\left[\frac{\hbar\omega_D}{\Delta}\right] \quad \text{or, } \Delta_0 = \frac{\hbar\omega_D}{\sinh\left[\frac{1}{V_0 Z\left(E_F^0\right)}\right]} = h^2\hbar\omega_D \cdot e^{-1/V_0 Z\left(E_F^0\right)} \tag{11.89}$$

Since, in the limit of weak coupling, $V_0 Z\left(E_F^0\right) \ll 1$.

11.17 Critical Temperature (T_c)

The occupation of excited one electron states at T = finite is $E_k = \sqrt{\xi_k^2+\Delta^2}$, which obeys FD-*statistics* with *Fermi-distribution formula*, $F\left(E_k, T\right)$. Considering the non-occupation of the corresponding pair states, we have

$$\frac{1}{V_0 Z\left(E_F^0\right)} = \int_0^{\hbar\omega_D} \frac{d\xi}{\sqrt{\xi^2+\Delta^2}}\left[1-2F\left(E_k+E_F^0, T\right)\right]. \tag{11.90}$$

The factor 2 is due to the fact that either $\vec{k}$ or $-\vec{k}$ may be occupied.

Putting $\Delta = 0$ in eqn (11.90), we get

$$\frac{1}{V_0 Z\left(E_F^0\right)} = \int_0^{\hbar\omega_D} \frac{d\xi}{\xi}\tanh\left(\frac{\xi}{2kT_c}\right) \tag{11.91}$$

$$\text{or, } kT_c = 1.14\hbar\omega_D \exp\left[-\frac{1}{V_0 Z\left(E_F^0\right)}\right] \tag{11.92}$$

$$\text{or, } T_c = \frac{1.14\hbar\omega_D}{k}\exp\left[-\frac{1}{V_0 Z\left(E_F^0\right)}\right] \tag{11.93}$$

is the desired expression for T_c.

Now taking the ratio of eqns (11.89) and (11.93), we get

$$\frac{2\Delta_0}{kT_c} = 3.52 \text{ or, } 2\Delta_0 = 3.52kT_c. \tag{11.94}$$

So, $\frac{2\Delta_0}{kT_c}$ is a constant and is in good agreement with experimental results in case of *weak coupling* superconductors. The superconducting gap parameter $\Delta(T)$ is Δ_0 at $T = 0\,\text{K}$ which decreases with T and becomes zero for $T > T_c$.

11.18 Fluxoid

According to *Ginzberg* and *Landau*, the density of superconducting electrons $n_s \propto |\psi|^2$, where ψ is a complex wave function considered in the superconducting state. The *order parameter* ψ is a function of position at a given temperature and vanishes at $T = T_c$.

Taking the wave function

$$\psi = |\psi|\,e^{i\phi}, \tag{11.95}$$

the current density can be expressed as

$$J = -\left[\frac{2e^2}{mc}A + \frac{e\hbar}{m}\nabla\phi\right]|\psi|^2, \tag{11.96}$$

where A is the *vector potential.* Considering a closed path in a superconducting ring [Fig. 11.22] we have

$$\oint \vec{J}\cdot d\vec{l} = |\psi|^2 \oint \left[\frac{2e^2}{mc}A + \frac{e\hbar}{m}\nabla\phi\right]\cdot d\vec{l} = 0. \tag{11.97}$$

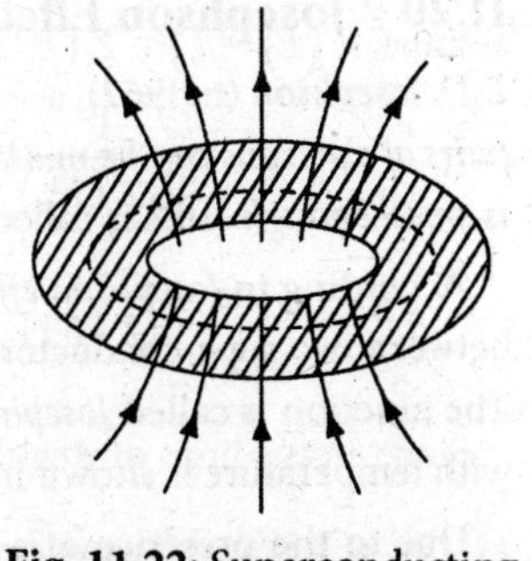

Fig. 11.22: Superconducting ring with magnetic flux through the central hole

But according to *Stokes' theorem,*

$$\oint_C \vec{J}\cdot d\vec{l} = \iint_S (\nabla\times\vec{A})\cdot d\vec{s} = \iint_S \vec{B}\cdot d\vec{s} = \phi, \tag{11.98}$$

where ϕ represents the flux enclosed by the ring.

As ψ is single valued, the phase change $d\phi$ around the closed path is $2n\pi$, where $n = 0, 1, 2, \ldots$.

$$\therefore \oint_C d\phi = \oint_C \nabla\phi\cdot d\vec{l} = 2\pi n. \tag{11.99}$$

Using eqns (11.98) and (11.99) in eqn (11.96) and solving for ϕ, we get

$$\phi = \frac{nhc}{2e} = n\phi_0, \tag{11.100}$$

where $\phi_0 = \frac{hc}{2e} = 2.07\times10^{-15}$ Wb is known as *fluxioid* (or *flux quantum*).

Let the magnetic field be switched on and the superconducting ring is in the normal state. On cooling the ring below T_c, flux is *excluded* from material of the ring but passes through the hole. If the external magnetic field now be switched off, the flux through the hole remains trapped. The super-currents produced around the ring maintain the flux through the hole of the ring.

11.19 Successes of BCS Theory

BCS theory successfully explains the following:

(i) The transition temperature of a superconductor is proportional to $\frac{1}{\sqrt{M}}$, where M is the atomic weight. If T_{c_1} and T_{c_2} be the *transition temperatures* for the atomic weights M_1 and M_2, then $\frac{T_{c_1}}{T_{c_2}} = \sqrt{\frac{M_2}{M_1}}$. This is known as the *isotope effect.*

(ii) BCS theory explains the vanishing of resistance in the superconducting state when super-current arises. The *phonons* during phonon scattering cannot supply an energy 2Δ to the *Cooper pairs* as they cannot lose KE in the process. Thus, phonon scattering is absent leading to *zero resistivity.*

(iii) Application of magnetic field $H > H_c$, a transition from superconducting state to normal state occurs. Because in presence of a magnetic field the electron-pairs move faster due to increase in KE and produce current for the expulsion of magnetic flux from the superconductor. At high KE of the electrons the phonon scattering takes place with exchange of energy and the Cooper pairs *split up* into normal electrons. The superconductor then transits to a normal conductor.

(iv) At 0 K most of the electrons are superconducting and the material shows a *sharp absorption edge* for angular frequency $\omega = 3.52\,kT_c/\hbar$ for *electromagnetic radiation.* The superconducting gap $2\Delta_0$ is experimentally established for electromagnetic radiation in the far infrared region.

The states below the superconducting gap $(2\Delta_0)$ are occupied by *paired electrons* and those above $(2\Delta_0)$ are occupied by *normal electrons.* If the material be irradiated at this state with electromagnetic radiation of appropriate energy, the super-electrons move across the gap and become normal electrons, i.e., a transition from superconducting state to normal state occurs.

11.20 Josephson Effect

B.D. Josephson (in 1962), a PG student at Cambridge, predicted that *a super-current consisting of correlated pairs of electrons can be made to flow across an insulating gap between two superconductors provided the gap is small enough.* This is called *Josephson effect* and the junction a *weak link.*

According to *Josephson effect* the *Cooper pairs* can *tunnel* through a thin layer of insulator sandwiched between two superconductors even when there is no potential difference between the two superconductors. The junction is called *Josephson junction* or *weak junction* as shown in Fig. 11.23. The variation of current with temperature is shown in Fig. 11.24.

Due to the presence of a thin insulating layer between the two superconductors, which are weakly coupled, the *tunnelling current* across the junction is very small.

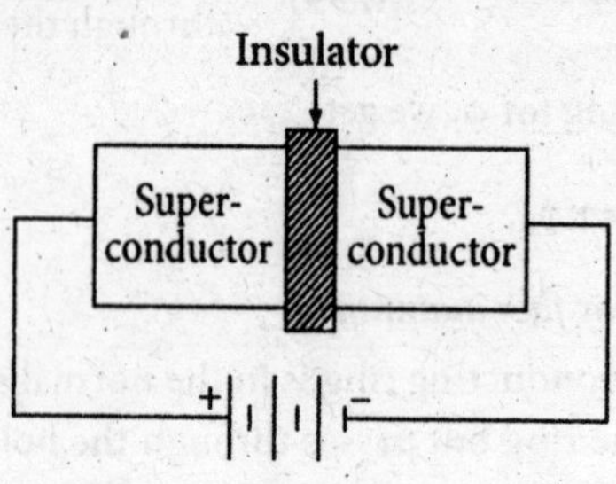

Fig. 11.23: Josephson junction of two superconductors separated by an insulator

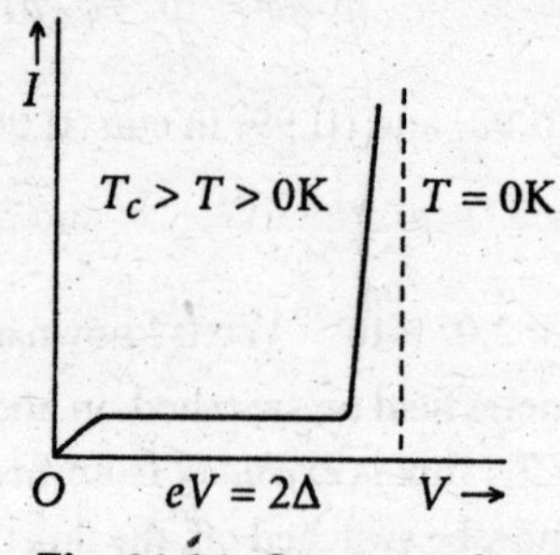

Fig. 11.24: Current-voltage relation at $T = 0\,K$ and $T_c > T > 0\,K$

The *tunnelling of Cooper pair* through the thin insulating layer between two superconductors produces the following effects:

(i) *D.C. Josephson effect*: Here DC current flows through the junction in absence of any electric and magnetic field.

(ii) *A.C. Josephson effect*: When a DC voltage is applied across the junction a RF current oscillates across the junction. Further an RF voltage applied with a DC voltage across the junctions causes a DC current through the junction.

(iii) *Macroscopic long range quantum interference*: When a DC magnetic field is applied through a superconducting circuit containing two Josephson junctions it causes the maximum super-current to show interference effects as a function of magnetic field intensity. This is utilised in *Superconducting Quantum Interference Devices* (SQUIDs) which finds important application of Josephson junction in designing sensitive magnetometers.

11.20.1 DC Josephson Effect

We consider a *superconductor-insulator-superconductor* junction as shown in Fig. 11.25.

It will be assumed that both the superconductors are *identical* and maintained at *zero potential.*

Let ψ_1 and ψ_2 be the amplitude of electron-pairs on sides (1) and (2) respectively. Then the *time-dependent Schrödinger's equation* for sides (1) and (2) respectively are

$$\left.\begin{aligned} i\hbar\frac{\partial\psi_1}{\partial t} &= \hbar T\psi_2 \\ \text{and } i\hbar\frac{\partial\psi_2}{\partial t} &= \hbar T\psi_1 \end{aligned}\right\}, \qquad (11.101)$$

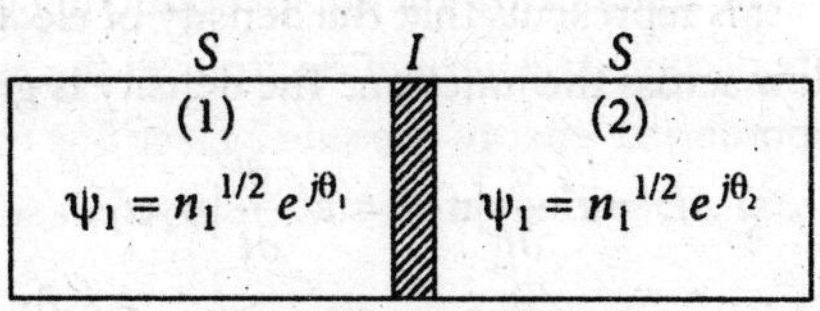

Fig. 11.25: A superconductor-Insulator-superconductor junction

where $\hbar T$ represents the effect of electron-pair coupling or transfer action across the insulator. T has the dimension of (time)$^{-1}$. T is the measure of leakage of ψ_1 into region (2) and of ψ_2 into region (1).

In case of thick insulator between the superconductors, there is no pair tunnelling through it and hence, $T_1 = 0$.

Let $\psi_1 = n_1^{1/2}\, e^{i\theta_1}$ and $\psi_2 = n_2^{1/2}\, e^{i\theta_2}$, then

$$\frac{\partial\psi_1}{\partial t} = \frac{1}{2}n_1^{-1/2}\, e^{i\theta_1}\frac{\partial n_1}{\partial t} + i\psi_1\frac{\partial\theta_1}{\partial t} = -iT\psi_2 \qquad (11.102)$$

$$\text{and } \frac{\partial\psi_2}{\partial t} = \frac{1}{2}n_2^{-1/2}\, e^{i\theta_1}\frac{\partial n_2}{\partial t} + i\psi_2\frac{\partial\theta_2}{\partial t} = -iT\psi_1. \qquad (11.103)$$

Multiplying eqn (11.102) by $n_1^{1/2}\, e^{-i\theta_1}$ and putting $\delta \equiv (\theta_2 - \theta_1)$, we get

$$\frac{1}{2}\frac{\partial n_1}{\partial t} + in_1\frac{\partial\theta_1}{\partial t} = -iT(n_1n_2)^{1/2}\, e^{i\delta}. \qquad (11.104)$$

Multiplying eqn (11.103) by $n_2^{1/2}\, e^{-i\theta_2}$, we get

$$\frac{1}{2}\frac{\partial n_2}{\partial t} + in_2\frac{\partial\theta_2}{\partial t} = -iT(n_1n_2)^{1/2}\, e^{-i\delta}. \qquad (11.105)$$

Since $e^{\pm i\delta} = \cos\delta \pm i\sin\delta$, and equating the real and imaginary parts of eqns (11.104) and (11.105), we get

$$\frac{\partial n_1}{\partial t} = 2T(n_1n_2)^{1/2}\sin\delta \qquad (11.106)$$

$$\frac{\partial n_2}{\partial t} = -2T(n_1n_2)^{1/2}\sin\delta \qquad (11.107)$$

$$\frac{\partial\theta_1}{\partial t} = -T(n_2/n_1)^{1/2}\cos\delta \qquad (11.108)$$

$$\frac{\partial\theta_2}{\partial t} = -T(n_1/n_2)^{1/2}\cos\delta. \qquad (11.109)$$

For identical superconductors (1) and (2), $n_1 \simeq n_2$.

$\therefore$ from eqns (11.108) and (11.109),

$$\frac{\partial\theta_1}{\partial t} = \frac{\partial\theta_2}{\partial t} \text{ or, } \frac{\partial}{\partial t}(\theta_2 - \theta_1) = 0. \qquad (11.110)$$

Also, from eqns (11.106) and (11.107),

$$\frac{\partial n_2}{\partial t} = -\frac{\partial n_1}{\partial t}. \qquad (11.111)$$

This represents that the density of electron-pairs changes at equal and opposite rates due to the current flow across the junction. The density is given by

$$\vec{J}_s = e^* \frac{\partial}{\partial t}|\psi_2|^2 = e^* \frac{\partial}{\partial t}|\psi_2 \psi_2^*| \qquad (11.112)$$

$$= e^* \frac{\partial}{\partial t}\left[n_2 e^{i\theta_2} e^{-i\theta_2}\right] = \left(e^* \frac{\partial n_2}{\partial t}\right) = -e^* \frac{\partial n_1}{\partial t} \qquad (11.113)$$

$$\text{or, } \vec{J}_s = -2e^* T(n_1 n_2)^{1/2} \sin(\theta_2 - \theta_1)$$

$$= -2e^* T(n_1 n_2)^{1/2} \sin\delta = \vec{J}_0 \sin\delta, \qquad (11.114)$$

where $\vec{J}_0 = -2e^* T(n_1 n_2)^{1/2}$.

In eqn (11.114), $J_0 \propto T$ and it represents the maximum zero voltage current that passes through the junction. *With zero applied voltage a DC current flows across the junction with value between J_0 and $-J_0$ depending on the value of phase difference $\delta = (\theta_2 - \theta_1)$.* This is what is called *Josephson effect.*

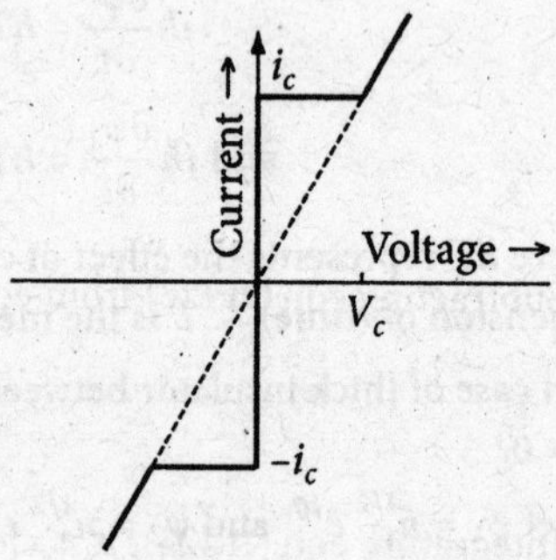

Fig. 11.26: Current-voltage relation of Josephson effect

The current-voltage characteristic of DC Josephson effect is shown in Fig. 11.26.

11.20.2 AC Josephson Effect

The Josephson junction being an insulator a potential difference (say, V) can be applied across it. An electron-pair of charge $q = -2e$ experiences a PE difference qV while passing through the junction. That is, the PE of a pair on one side is $-eV$ and that on the other side is $+eV$.

The equations of motion of the electron-pair are

$$i\hbar \frac{\partial \psi_1}{\partial t} = \hbar T \psi_2 - eV\psi_1, \quad i\hbar \frac{\partial \psi_2}{\partial t} = \hbar T \psi_1 + eV\psi_2.$$

Now $\psi_1 = n_1^{1/2} e^{i\theta_1}$ and $\psi_2 = n_2^{1/2} e^{i\theta_2}$.

$$\therefore \frac{\partial \psi_1}{\partial t} = \frac{1}{2} n_1^{-1/2} e^{i\theta_1} \frac{\partial n_1}{\partial t} + i\psi_1 \frac{\partial \theta_1}{\partial t} \text{ or, } \frac{\partial \psi_1}{\partial t} = -iT\psi_2 + i\frac{eV}{\hbar}\psi_1. \qquad (11.115)$$

Also

$$\frac{\partial \psi_2}{\partial t} = \frac{1}{2} n_2^{-1/2} e^{i\theta_2} \frac{\partial n_2}{\partial t} + i\psi_2 \frac{\partial \theta_2}{\partial t} \text{ or, } \frac{\partial \psi_2}{\partial t} = -iT\psi_1 - i\frac{eV}{\hbar}\psi_2. \qquad (11.116)$$

Multiplying both sides of eqn (11.114) by $n_1^{1/2} e^{-i\theta_1}$, we get

$$\frac{1}{2}\frac{\partial n_1}{\partial t} + i n_1 \frac{\partial \theta_1}{\partial t} = ie\frac{Vn_1}{\hbar} - iT(n_1 n_2)^{1/2} e^{i\delta}, \qquad (11.117)$$

where $\delta = (\theta_2 - \theta_1)$.

The real part of eqn (11.117) is

$$\frac{\partial n_1}{\partial t} = 2T(n_1 n_2)^{1/2} \sin\delta \qquad (11.118)$$

and the imaginary part is

$$\frac{\partial \theta_1}{\partial t} = \frac{eV}{\hbar} - T\left(\frac{n_2}{n_1}\right)^{1/2} \cos\delta. \qquad (11.119)$$

From eqn (11.116),

$$\frac{1}{2}\frac{\partial n_2}{\partial t} + in_2\frac{\partial \theta_2}{\partial t} = ie\frac{Vn_2}{\hbar} - iT(n_1n_2)^{1/2}e^{-i\delta}. \tag{11.120}$$

$$\therefore \frac{\partial n_2}{\partial t} = -2T(n_1n_2)^{1/2}\sin\delta \tag{11.121}$$

$$\text{and } \frac{\partial \theta_2}{\partial t} = -\left(\frac{eV}{\hbar}\right) - T\left(\frac{n_1}{n_2}\right)^{1/2}\cos\delta. \tag{11.122}$$

Subtracting eqn (11.122) from eqn (11.119) and putting $n_1 = n_2$, we get

$$\frac{\partial(\theta_2 - \theta_1)}{\partial t} = \frac{\partial \delta}{\partial t} = -\frac{2eV}{\hbar}. \tag{11.123}$$

Integrating eqn (11.123), it is clear that for a DC voltage across the junction, the relative phase of the probability amplitude varies as

$$\delta(t) = \delta(0) - \left(\frac{2eVt}{\hbar}\right). \tag{11.124}$$

The current density across the AC *Josephson junction* is now

$$J = J_0 \sin\delta(t) \text{ or, } J = J_0 \sin\left[\delta(0) - \frac{2eVt}{\hbar}\right]. \tag{11.125}$$

This shows the current density is *oscillatory* and the corresponding frequency of oscillation is $\omega = \frac{2eV}{\hbar}$. For a DC voltage of 1$\mu$V, the frequency produced is ~ 483.6 MHz.

Now $\because \omega = \frac{2eV}{\hbar}$, $\therefore \omega\hbar = 2eV$, i.e., a photon of energy $2eV$ is emitted or absorbed when the electron-pair crosses the barrier. The value of $e/\hbar$ can be found out by measuring V and ω.

11.21 Macroscopic Quantum Interference

We know that *Josephson effect* arises due to the *phase coherence* between two superconductors. It is, therefore, quite natural to expect that a two-junction device consisting of a pair of Josephson tunnel junctions connected by superconducting links would show interference effect, just like optical interference of *critical current* for the device.

In absence of applied magnetic field the junctions are identical and the critical current I_c from an external source divides equally in two paths so that the phase difference across the junctions is the same $(S_a = S_b)$. Under this condition, there is no voltage between P and Q and $I_c = 2I_0$ [Fig. 11.27].

If, however, an external magnetic field be applied, it perturbs the situation as shown in Fig. 11.28.

First effect is that a circulating current i flows along the inner surface of the loop which prevent any flux from passing through the loop.

$$Li = \phi_a = Ba \times \text{area of the loop} \tag{11.126}$$

in absence of weak links and L is the *self-inductance* of the circuit.

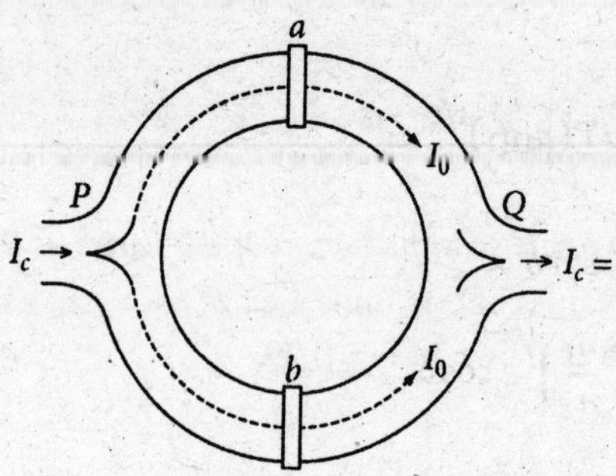

Fig. 11.27: No applied field to the superconducting ring with two weak links

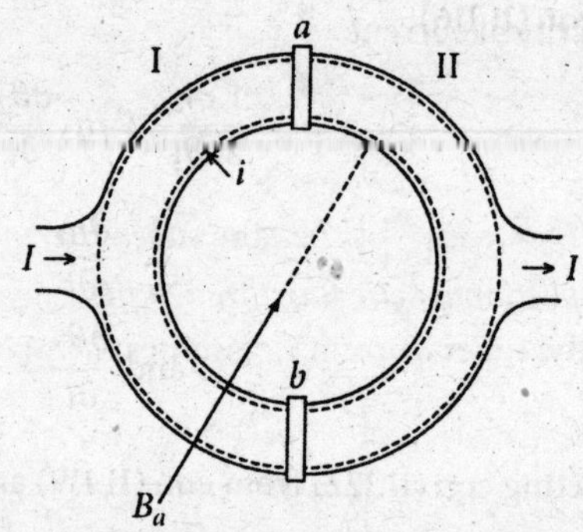

Fig. 11.28: External magnetic field B_a applied to superconductors I and II with two weak links (a) and (b)

In this case, $LI_0 \ll \phi_0$ or, $L \ll \phi_0/I_0$, where ϕ_0 is the *flux quantum*. Hence, flux (Li) through the loop is equal to the applied flux ϕ_0 [Fig. 11.27].

The *second effect* produces an *imbalance in phase difference* between the junctions of the loop.

Now the phases S_a and S_b at junctions a and b in presence of flux ϕ_a are given by

$$\Delta S = 2\pi n = \frac{2e}{\hbar}\int \vec{A}\cdot d\vec{l} + S_a + S_b$$

$$\text{or, } S_a + S_b = 2\pi n - 2\pi\frac{\phi_a}{\phi_0}. \tag{11.127}$$

Since the currents in the two junctions are $(i + 1/2)$ and $(i - 1/2)$, $S_a \neq S_b$.

$$\therefore S_a = \left(\pi n - \frac{\pi\phi_a}{\phi_0}\right) - S_0 \tag{11.128}$$

$$\text{and } S_b = \left(\pi n - \frac{\pi\phi_a}{\phi_0}\right) + S_0, \tag{11.129}$$

where $S_0 = (S_b - S_a)/2$, i.e., the total phase difference is $(S_b - S_a) = 2S_0$. Eqns (11.127) and (11.129) show that, $(S_a - S_b)$ can be controlled by varying ϕ_a. This phenomenon is equivalent to the *interference of two coherent light beams* interfering over paths a and b.

For a constant phase shift depending on barriers (a) and (b) and applied voltage, eqns (11.128) and (11.129) lead to that fact that

$$I_a \simeq I_0 \sin S_a \tag{11.130}$$

$$I_b \simeq I_0 \sin S_b. \tag{11.131}$$

$$\therefore I = I_a - I_b = I_0(\sin S_a - \sin S_b)$$

$$= I_0\left[\sin\left\{\pi\left(n - \frac{\phi_a}{\phi_0}\right) - S_0\right\} - \sin\left\{\pi\left(n - \frac{\phi_a}{\phi_0}\right) + S_0\right\}\right]$$

$$= I_0\left[2\cos\left\{\pi\left(n - \frac{\phi_a}{\phi_0}\right)\right\}\right] = 2I_0\cos\pi\left(n - \frac{\phi_a}{\phi_0}\right)\sin S_0$$

$$\text{or, } I = \left(2I_0\cos\frac{\phi_a}{\phi_0}\right)\sin S_0. \tag{11.132}$$

∴ the critical current,

$$I_c = 2I_0 \left\{ \cos\left(\frac{\phi_a}{\phi_0}\right) \right\}. \tag{11.133}$$

Thus, $I_c \propto \cos\left(\frac{\phi_a}{\phi_0}\right)$, i.e., the *super-current through two Josephson junctions a and b varies as the cosine of the applied magnetic flux through the ring.* For integral values of flux units, i.e., $\phi_a/\phi_0 = s\pi$, s being an integer, the super-current I_c assumes the maximum values as shown in Fig. 11.29.

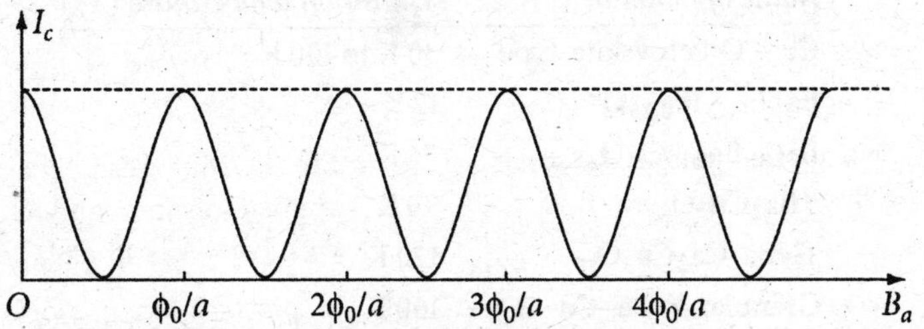

Fig. 11.29: Variation of I_c in a DC SQUID with B_a, a is the area of the loop

The result of interference between two Cooper pairs tunnelling current through two parallel Josephson junctions is exploited to construct extremely sensitive magnetometers (SQUIDs).

11.22 Applications of Josephson Effect

Some of the important applications of the Josephson effect are as follows:

(i) It is used to generate microwave frequency ω according to the relation, $\omega = \frac{2eV_0}{\hbar}$.

(ii) For switching of signals from one circuit to another circuit Josephson effect is used. The time of switching is ~ 10^{-12} s, which is extremely useful in *high speed computers.*

(iii) AC Josephson effect is used to define standard volt.

(iv) Since the frequency of the radiation emitted from the Josephson junction depends on the temperature of the source. AC Josephson junction is used to measure very low temperatures.

(v) Magnetic flux can be measured by a SQUID magnetometer (working on the principle of AC Josephson effect) with a precision ($\pi\hbar/e$).

11.23 The High Temperature Superconductivity

Generally, the superconductors with a transition temperature (T_i) above 10 K are known as the high temperature superconductors. Nowadays, superconductors with transition temperature (T_i) above 40 K or even 90 K are regarded as high temperature superconductors (HTSC).

Superconductors have large number of possible practical applications but due to the limitation of low T_c value these are not being possible. Because at low T_c the maintenance of conductivity through refrigeration requires very high cost. In recent past special efforts have been made by the workers to discover high T_c superconductors.

Bednorz and Muller in 1986 discovered a new class of oxide ceramic superconductors with T_c = 30 K. These are called high – T_c superconductors. The first group of superconductors was $La_{2-x}M_xCuO_4$ (M = Ba, Sr, Ca) with T_c = 20 K to 40 K and are referred to as '214' system having K_2NiF_4 structure with an orthorhombic distortion. Another system is having general formula $LnBa_2Cu_3O_{7-x}$ (Ln = Y, Nd, Sm, Eu, Gd, Dy, Ho, Er, Tm, Yb), where $x = 0.2$ and is called '123' system having orthorhombic structure. Non-rare earth-based copper oxide system involving Bi and Tl were found in 1988 to show superconductivity between 60 K and 125 K.

Table 11.3 shows the high temperatures superconductors and their transition temperatures.

Table 11.3

Name of material	*Transition temperature* (T_c)
Cu – O Perovskite-type	30 K to 100 K
$BaPb_{0.75}Bi_{0.25}O_3$	12 K
$La_{1.85}Ba_{0.15}CuO_4$	36 K
$YBa_2Cu_3O_7$	90 K
$Tl_2Ba_2Ca_2Cu_3O_{10}$	120 K
Granular Y–Ba–Cu–O	160 K
Lanthanum Compounds	300 K to 500 K

Mechanism for Superconductivity

The precise mechanism for the origin of high temperature superconductivity is yet to investigate. According to a group of investigators the role of oxygen is essential for high T_c oxide superconductors. A strong electronic band in the infrared spectrum and its correlation with superconductivity provides evidence that, an exciting mechanism is responsible for superconductivity of $TBa_2Cu_3O_7$ materials. According to *Anderson's resonating valence band theory* (Sr, Ba, Y, La)$_2$ Cu O_{4-y} arises due to interaction with phonons and conductivity increases by the resonance along the O–Cu″O–Cu lines of atoms. In spite of the above efforts, the actual conduction mechanism in high T_c superconductors is perhaps the most challenging problem in condensed matter physics nowadays.

11.24 Important properties of High Temperature Superconductors

11.24.1 Structural Properties

(i) The high temperature superconductors (HTSCs) have ideal *perovskite* structure either through an intergrowth phenomenon or by an ordered removal of oxygen atoms.

(ii) HTSCs have layered crystal structure of one or more CuO_2 layers in which each copper atom is strongly and covalently bonded in a nearly square-planar arrangement to four oxygen atoms.

(iii) Transfer of charge to and from CuO_2 layers takes place due to doping near the metal insulator phase boundary in all HTSCs.

11.24.2 Physical properties

(i) As the Hall coefficient R_H is positive, the charge carriers in all the oxide HTSCs are holes [except $(Ln_{1-x}M_x)_2\,CuO_{4-y}$ in which the charge carriers are electrons].

(ii) Superconductivity and magnetism coexist in rare earth compounds and the rare earth ions do not participate in the process of superconductivity.

(iii) The low temperature specific heat of HTSCs in superconducting phase is given by

$$C = \alpha T^{-2} + \gamma T + \beta T^{-3},$$

where αT^{-2} is due to *Schottky like anomaly,* γT due to *free electrons in metal* and βT^{-3} due to *Debye contribution from the lattice.* The linear term is now believed to be due to impurities.

(iv) HTSCs are all type II superconductors with Hc_1 and Hc_2 highly anisotropic.

(v) The coherence length (ξ) and penetration length (l) are highly anisotropic. The magnitude of ξ is small (~ a few lattice spacing).

(vi) In HTSCs, the transition temperature increases with the increase of pressure, whereas opposite effect is observed in conventional superconductors.

(vii) Isotope effect is almost absent in HTSCs.

(viii) The normal state transport properties are highly anisotropic and temperature dependent.

(ix) Energy gap over the entire Fermi surface below T_c and *Josephson tunnelling* exists in HTSCs.

(x) The value of $2\Delta(0)/kT$ for tunnelling of pairs perpendicular to ab-plane is 3.8 to 4, but for parallel plane $2\Delta(0)/kT$ ranges from 5.8 to 6.

11.25 Important Applications of Superconductivity

Table 11.4 shows some important applications of superconductivity.

Table 11.4: Applications of superconductivity.

Field of application	*Applications*
Magnets ($R = 0, J_c$ = high)	(i) High field magnets for various applications (ii) Nuclear Magnetic Resonance Tomography (iii) Ore refining (iv) Magnetic levitations (v) Magnetic shielding (vi) Interconnects in microelectrons
Energy related	(i) Production of magnetic fusion and magnetohydrodynamics (ii) Energy storage (iii) Electrical power transmission by cables
Transport (J_c = high at high H_{c_2})	(i) Ship drive system (ii) High speed trains
Electronics and small devices (Josephson tunnelling)	(i) In physical measurement (SQUIDs, etc.) (ii) Microwave detectors and mixer (iii) Computers (fast logic and memory circuits) (iv) Josephson devices (square law detector, paramagnetic amplifier) (v) Bolometer (vi) Electromagnetic shielding (vii) In medicine (magnetic encephalography)
Computers and information processing	(i) Active superconducting elements (A–D) converters (ii) Voltage standard (iii) Optoelectronics (iv) Matched filters

In the field of medicine, the *magnetic resonance imaging* (MRI) has gained popularity. Here intense magnetic fields are created which simulate the resonance of hydrogen atoms in human body and this in turn helps to build up a two-dimensional picture of human body.

Applications of HTSCs in magnetically elevated trains and mineral separations is highly probable.

11.26 Applications of High T_c Oxide Superconductors

Following are some of the prospecting field of application of HTSCs:

(i) The very large scale integration (VLSI) technology is used in HTSC interconnections at 77K(LNT). In near future high speed computers and telecommunication are expected to use it extensively.

(ii) Nowadays, SQUIDs working at 4.2 K (LHeT) are used in sensitive magnetometers. But SQUIDs made of oxide superconductors working at 77 K are expected to be used which would improve the ruggedness and utility of the device. These will also find important application in the field of military and medicine.

(iii) HTSCs have important applications in generation, storage, transmission and transformation of electrical power in electrical industry.

11.27 Some Applications of Superconductors

1. **Switching Devices:** If the applied magnetic field H be greater than the critical field H_c, then a change from superconducting to normal state occurs. The process is reversed if the applied field H is removed. This fact is utilised in the construction of switching element *cryotron*.
   A cryotron consists of a *tantalum* core around which is wound a *niobium* wire. Both tantalum and niobium with T_c = 4.5 K and 9.5 K respectively act as *gate* and the *control* [Fig. 11.30]. The current through the tantalum gate can be controlled by switching action. If now a control current passes through the niobium winding, a magnetic field of sufficient magnitude can be produced which would change tantalum from its superconducting to normal state. This closes the gate, and the control current cannot pass through the tantalum core. The gate reopens if the control current be made off.

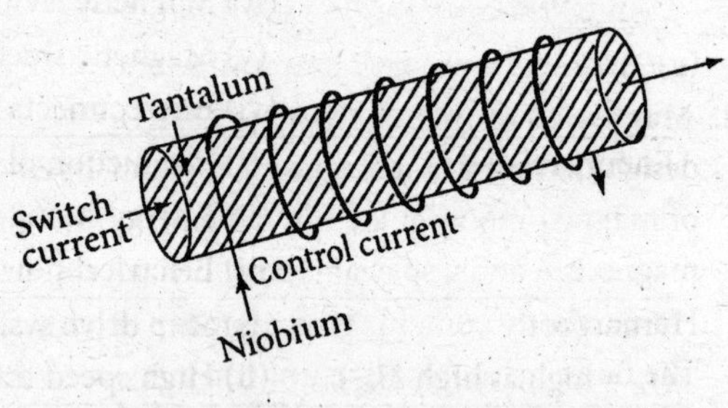

Fig. 11.30: Wire-wound cryotron

   Such superconducting elements are suitable for large scale computer applications. The low temperature environment can be maintained at low cost by using such simple and small sized basic elements.

2. **Electric Generators:** A low loss superconducting coil is rotated in an extremely large magnetic field to generate electric powers. Very high powers (~ 2500 kV) could be generated at very low voltage (450 V) by such generators. These generators are very small in size and weight, and form the basis of new generation of energy saving power system.

3. **Generation of High Magnetic Field:** High magnetic field (~ 50 T) can be produced by the application of superconducting materials. The power consumed to generate such a high field is only 10 kW, whereas in conventional generators the required power is about 3 MW. In this type of generators no cooling arrangement is required to save the copper coil from burning due to Joule heating.

4. **Low Loss Transmission Lines and Transformers:** Electric cables constructed with superconducting wires have very small power loss due to the negligible resistance of the superconductors. So transmission of powers can be done with negligible power loss if superconducting wires be used.

   If the primary and secondary winding of a transformer be made of superconducting wires, then *Joulean heating* will be minimum and loss will be negligible. Portable transformers of 2000 to 3000 mV have been designed by using superconducting wires in the windings.

5. **Magnetic Levitation:** The diamagnetic property that the magnetic lines of force are rejected by superconducting materials is used in the magnetic *levitation*. A superconducting material can be suspended in air against the repulsive force from a permanent magnet.

   In high speed transportation, the magnetic levitation effect can be used.

6. **Logic and Storage Functions in Computers:** To perform logic and storage functions in computers, superconductors are used.

   For memory elements the current-voltage relation of *Josephson junction* is highly suitable which helps to cause much faster switching time (~ 10 ps).

7. **Josephson Junction Devices:**

   (a) Josephson junction devices (using AC Josephson effect) offer a process of measuring the constant $e/\hbar$ precisely.

   (b) Voltages can also be measured accurately by devices using Josephson effect.

   (c) Magnetic field can be measured accurately by using magnetometers based on the principle of SQUID.

   (d) The magnetic structure underneath the earth's surface can be explored by using SQUIDs.

   (e) The SQUIDs are used to study tiny magnetic signals from the brain and heart.

   (f) SQUID magnetometers are used to detect paramagnetic response in the liver which gives the information regarding the iron present in the liver of the body accurately.

   (g) Sensitive ammeters and voltmeters can be constructed with SQUIDs.

8. **Magnetic Resonance Imaging (MRI):** The superconducting property has been successfully used to design the important diagnostic tool called *Magnetic Resonance Imaging* (MRI). MRI is not really a form of magnetometry at all, but depends on the property of superconductivity. In fact, MRI depends on magnetism and a special type of behaviour of atoms subjected to magnetic field.

   Human body contains large number of hydrogen atoms along with the atoms of many other elements. The nuclei of atoms (like tiny magnets) tend to line up like a compass needle in presence of an applied magnetic field. Though a fraction of nuclei tend to point out the other way, the orientation against the magnetic field is not possible energetically. The motion of these nuclei provides the signal used in MRI. When an atomic nucleus switches from being aligned against the magnetic field to being aligned along the magnetic field a small amount of electromagnetic energy in the form of the radio waves is released. The strength of the magnetic field and the kind of atom emitting the energy determine the frequency.

   The body of the patient is placed in the bore of a cylindrical magnet and a precise magnetic field is maintained through the cross section of the body. This magnetic field helps to align the nuclei of the atoms of the body and sets a frequency of the emitted electromagnetic wave. At this moment a sharp electromagnetic pulse is supplied which temporarily scrambles many of the atomic alignments with the external field. At the end of the pulse the nuclei align along the magnetic field and emit their characteristic radio signals. Then a complete picture of a magnetic slice is obtained by a sophisticated computer analysis. As two-third of water consists of hydrogen atom, it is a very useful monitor of human body's health. The water rich tissues respond strongly to MRI probes. Organic materials are also rich in hydrogen atom and as such respond nicely to MRI scans. Brain tissues hidden by the skull are possible to examine since bones appear relatively dark in MRI.

11.28 Answers of Some Important Questions

11.1. *What is superconductivity?*

Ans. Superconductivity is the phenomenon in which the electrical resistance of the material suddenly drops to zero when it is cooled to a sufficiently low temperature. The corresponding material is called the *superconductor.*

The temperature at which the resistance of a material suddenly drops to zero is called the *critical temperature* (T_c). At T_c, the material undergoes a *phase transition* from normal state to superconducting state and T_c is also called *superconducting transition temperature.*

11.2. *Mention the general properties of superconductor.*

Ans. See Art. 11.2.

11.3. *Discuss the effect of magnetic field on superconductivity.*

Ans. *Kamerlingh Onnes* found that the superconductivity of materials is destroyed by the application of a suitable magnetic field. The minimum magnetic field necessary to destroy superconductivity is called *critical field* (H_c). The critical magnetic field is found to depend on temperature of the superconductor according to relation

$$H_c = H_0\left[1-\left(\frac{T}{T_c}\right)^2\right],$$

where H_c and T_c are the *critical field at* 0 K and *the critical temperature* respectively. As the magnetic field strength is increased above H_c, the lines of magnetic flux start penetrating the material, and the *superconductivity* is destroyed.

11.4. *What is Meissner effect? Show that a superconductor behaves as a perfectly diamagnetic substance.*

Ans. See Art. 11.3 [up to eqn (11.4)]

11.5. *Give the wave mechanical explanation of superconductivity.*

Ans. See Art. 11.4.

11.6. *What are Types I and II superconductors?*

Ans. Superconductors are classified as type I (soft superconductor) and Type II (hard superconductors).

Superconductors which exhibit complete Meissner effect are called Type I *superconductors.* Al, Zn, Hg and Sn in pure state are the examples of Type I superconductor. Superconductivity of these materials is destroyed if the applied magnetic field $H > H_c$.

Type II superconductors have two values of critical fields H_{c_1} and H_{c_2}. If $H < H_c$, then the specimen is in *superconducting state* and if $H > H_c$, then it is in *normal state.* If $H_{c_1} < H < H_{c_2}$, then the material is in a *mixed state.*

11.7. *What is London penetration depth?*

Ans. The magnetic field at a depth x from the surface of a superconductor is given by $H = H_0\, e^{-x/\lambda}$, where H_0 is the magnetic field at the surface of the material. For $x = \lambda$, $H = \left(\frac{1}{e}\right) H_0$, λ is called the

London penetration depth. It is defined as *the depth from the surface of the superconductor at which the magnetic flux density falls to $\frac{1}{e}$ times the value at the surface.*

Penetration depth depends on temperature of the material according to the relation:

$$\lambda(T) = \lambda(0)\left[1 - \frac{T^4}{T_c^4}\right]^{-1/2},$$

where $\lambda(0)$ is the penetration depth at 0 K. At $T = T_c$, $\lambda(T_c) = 0$. The penetration depth of mercury at 3.5 K is about 750 Å, whereas at 0 K it is 519 Å.

11.8. *What are virtual phonons? What is a Cooper pair?*

Ans. *Virtual phonons*: When an electron moves through a lattice, the lattice is distorted and acts on the electron by virtue of electrostatic forces between them. The distortion of the lattice is oscillatory and quantized in terms of *phonons.* A constant *emission* and *re-absorption* of phonons by electrons take place as a result of interaction between the lattice and phonons. These phonons are called *virtual phonons.*

Cooper pair: According to BCS theory an attractive force between the electrons, sufficiently stronger than their mutual electrostatic repulsion, gives rise to superconductivity.

An electron pulls the surrounding ions from their equilibrium position so that the region around the electron carries more positive charge than in the normal state. After the electron leaving the region, it remains positive for a time and attracts another electron. The wave function of the latter electron is modified by the alteration of phonon number by the former electron. The attractive force between the two electrons becomes maximum when their spins are of *opposite directions.* The pair of electrons can be regarded as a single particle with zero momentum, zero spin, and charge 2*e* is called a *Cooper pair,* which obeys *Bose-Einstein statistics.* The electron interaction energy is less than that of two free electrons, i.e., negative representing the *quasi-bound sate* of the electron-pair which leads to superconductivity at low temperature.

11.9. *What is critical temperature (or transition temperature)? Explain the origin of superconductivity.*

Ans. The temperature at which the transition from normal state to superconducting state takes place on cooling in absence of magnetic field is called the *critical temperature* (T_c) or *transition temperature.* According to BCS theory at temperature $T < T_c$, the lattice-electron interaction is stronger than the electron-electron Coulombic interaction. The Cooper pair of electrons move smoothly in the lattice without any transfer of energy to the lattice as they are not scattered by the lattice points. If now an electric field is applied to the sample, the electrons gain energy and produce an electric current. Under such condition the electrons do not get slowed down and as a consequence of which the material does not possess any electrical resistivity. Thus, BCS theory accounts the zero resistivity of a superconductor.

11.10. *What is isotope effect in view of* BCS *theory?*

Ans. See Art. 11.6.3.

11.11. *What is Josephson effect? What is* SQUID *and what are its important applications?*

Ans. A super-current consisting of correlated pairs of electrons can be made to flow across an insulating gap between two superconductors provided the gap is small enough. This is known as *Josephson effect.*

The super-current arises due to the quantum mechanical tunnelling through the junction and the wave functions on both sides are highly correlated.

A SQUID is a double junction quantum interferometer in which two Josephson junctions are mounted on a superconducting ring. The working of the SQUIDs is based on flux quantization in a superconducting ring.

Application of SQUIDs:

(a) A minute magnetic signal can be detected by SQUID sensors.

(b) SQUIDs are used to detect the magnetic signals from the brain and heart.

(c) SQUID magnetometers are used to detect the *paramagnetic response* in the liver which gives the amount of iron present in the lever of the body accurately.

11.12. *What is fluxoid? Point out its one important application.*

Ans. The magnetic flux ϕ enclosed by a superconducting ring at temperature below T_c is constant and given by

$$\phi = \frac{nhc}{2e} = n\phi_0,$$

where $\phi_0 = \frac{hc}{2e} = 2.07 \times 10^{-15}$ Wb is called *fluxoid* or *flux quantum* and n = an integer.

An important application of fluxoid is in *memory element in computers* and in SQUIDs.

11.13. *Mention some important applications of superconductors.*

Ans. Some important applications of superconductors are as below:

(a) Fast electrical switching used in fast large scale computers.

(b) Logic and storage functions in computers.

(c) Electric generators.

(d) Low loss transformers and transmission lines.

(e) Magnetic levitation.

(f) Generation of high magnetic field.

(g) SQUIDs.

11.14. *What is high-T_c superconductivity? Mention two compounds exhibiting the above effect.*

Ans. Any superconductor with transition temperature $T_c > 10$ K is in general called *high-T_c superconductor. Bednorz* and *Muller, Chu* and *co-workers* developed such superconductors after 1986. The two compounds exhibiting the property of high-T_c superconductivity are:

(a) Ba–La–Cu–O, $T_c \simeq 35$ K

(b) Y–Ba–Cu–O, $T_c \simeq 90$ K.

11.15. *What are the predictions made by BCS theory?*

Ans. BCS theory makes the following predictions:

(a) There is a discontinuity in the electronic contribution to specific heat at transition temperature given by $C_s - C_n = 1.43C_n$, the subscripts s and n denote the superconducting and normal states respectively.

(b) The transition temperature T_c depends on isotopic mass through the relation $T_c = M^{-1/2}$, M is the isotopic mass.

(c) Existence of perfect diamagnetism below T_c with $\chi = -1$.

(d) The energy gap E_g is proportional to T_c (except for gapless semiconductors).

(e) The transition temperature T_c and energy gap 2Δ are given by $T_c = 1.14\theta_D e^{-1/UD(E_f)}$, where $D(E_f)$ represents the electron density of energy states at Fermi level.

$$2\Delta = E_g = 4k\theta_D e^{-1/UD(E_f)},$$

where θ_D is the *Debye temperature* and U, the *electron-electron attractive potential.*

11.29 Solved Problems

11.1. *A superconducting tin has a critical temperature of* 3.7 K *at zero magnetic field and a critical field of* 0.0306 T *at* 0 K. *What is the critical field at* 2 K.

Solution: We know that the critical field H_c is

$$H_c(T) = H_0\left[1-\left(\frac{T}{T_c}\right)^2\right].$$

Here $T_C = 3.7$ K, $H_c(0) = 0.0306$ T, $T = 2$ K.

$$\therefore H_c(2) = 0.0306\left[1-\left(\frac{2}{3.7}\right)^2\right] = \mathbf{0.02166\,T}.$$

11.2. *The critical fields for a sample are* 1.4×10^5 A/m *and* 4.2×10^5 A/m *at* $T = 14$ K *and* 13 K *respectively. What are the transition temperatures and critical field at* 0 K *and* 4.2 K?

Solution: We have

$$H_c(T_1) = H_0\left[1-\left(\frac{T_1}{T_c}\right)^2\right] \quad \text{(i)}$$

$$H_c'(T_2) = H_0\left[1-\left(\frac{T_2}{T_c}\right)^2\right]. \quad \text{(ii)}$$

$$\therefore \frac{H_c(T_1)}{H_c'(T_2)} = \frac{T_c^2 - T_1^2}{T_c^2 - T_2^2} \text{ or, } \frac{1.4}{4.2} = \frac{T_c^2 - T_1^2}{T_c^2 - T_2^2} = \frac{T_c^2 - 14^2}{T_c^2 - 13^2}$$

or, $4.2T_c^2 - 4.2 \times 14^2 = 1.4T_c^2 - 1.4 \times 13^2$ or, $T_c^2(4.2 - 1.4) = 4.2 \times 14^2 - 1.4 \times 13^2$

or, $T_c^2 = \dfrac{586.6}{28}$, $\therefore T_c = \mathbf{14.5\,K}$.

Putting $T_c = 14.5$ K and $H_c = 1.4 \times 10^5$ in eqn (i), we get

$$H_0 = \frac{H_c}{1-(14/14.5)^2}$$

with $H_c = 1.4 \times 10^5$ A/m.

$$\therefore H_0 = \frac{1.4 \times 10^5}{1 - 0.9322} \text{ or, } H_0 = \mathbf{20.7 \times 10^5\,A/m}.$$

Now

$$H_c(0) = H_0\left[1-\left(\frac{T}{T_c}\right)^2\right] = H_0[1-0] = H_0, \quad \therefore H_c(0) = \mathbf{20.7 \times 10^5\,A/m}.$$

and

$$H_c(4.2) = H_0\left[1-\left(\frac{T}{T_c}\right)^2\right] = 20.7 \times 10^5\left[1-\left(\frac{4.2}{14.5}\right)^2\right] \text{ or, } H_c(4.2) = \mathbf{18.9 \times 10^5\,A/m}.$$

11.3. *For a superconductor the critical field at 0 K is 6.39×10^4 A/m and the critical temperature for zero magnetic field is 7.18 K. What is the critical field for the material at 4 K?*

Solution: If $T_c(H)$ and $T_c(0)$ be the critical temperatures for fields H and 0 respectively, then

$$\frac{T_c(H)}{T_c(0)} = 1 - \frac{H_c(T)}{H_c(0)},$$

where $H_c(T)$ and $H_c(0)$ are the critical fields at temperatures T and 0 respectively.

In the present case, $H_c(0) = 6.39 \times 10^4$ A/m, $T_c(0) = 7.18$ K and $T_c(H) = 4$ K.

$$\therefore H_c(T) = H_c(0)\left[1 - \left(\frac{T_c(H)}{T_c(0)}\right)^2\right] = 6.39 \times 10^4\left[1 - \left(\frac{4}{7.18}\right)^2\right] = \mathbf{4.41 \times 10^4\ A/m.}$$

11.4. *The diameter of a lead wire at 4.2 K is 1 mm. If the critical temperature for lead is 7.18 K and $H_0 = 6.5 \times 10^4$ A/m, find the critical current.*

Solution: We have the critical current $I_c = 2\pi r H_c$, where H_c is the critical field given by

$$H_c = H_0\left[1 - \left(\frac{T}{T_c}\right)^2\right].$$

Here $T = 4.2$ K, $T_c = 7.18$ K and $H_0 = 6.5 \times 10^4$ A/m

$$\therefore H_c = 6.5 \times 10^4\left[1 - \left(\frac{4.2}{7.18}\right)^2\right] = 4.276 \times 10^4 \text{ A/m}.$$

Hence,

$$I_c = \pi \times (10^{-3}) \times (4.276 \times 10^4) = \mathbf{134.51\ A.}$$

11.5. *Calculate the London penetration depth for tin from the given data at 2.3 K: $T_c = 3.7$ K, density $= 7.3 \times 10^3$ kg/m^3, $M = 118.7$ and $m^* = 1.9$ m.*

Solution: We have the London penetration depth

$$\lambda(0) = \left(\frac{m^*}{\mu_0 n e^2}\right)^{1/2}.$$

$\because \frac{118.7}{7.3 \times 10^3}$ m^3 of the tin contains $2 \times 6.02 \times 10^{26}$ electrons,

$\therefore$ 1 m^3 will contain $\frac{7.3 \times 10^3 \times 12.04 \times 10^{26}}{118,7}$ electrons or, $n = 7.4 \times 10^{28}$.

So

$$\lambda(0) = \left[\frac{1.9 \times (9.1 \times 10^{-31})}{(4\pi \times 10^{-7}) \times (7.4 \times 10^{28}) \times (1.6 \times 10^{-19})^2}\right]^{1/2} \quad \text{or, } \lambda(0) = 27 \text{ nm}.$$

Again,

$$\lambda(T) = \lambda(0)\left[1 - \left(\frac{T}{T_c}\right)^4\right]^{-1/2}.$$

Putting $T = 2.3$ K and $T_c = 3.7$ K,

$$\lambda(T) = \frac{27 \times 10^{-9}}{\left[1 - \left(\frac{2.3}{3.7}\right)^4\right]^{1/2}} = \mathbf{29.1\ cm.}$$

11.6. *The critical temperature T_c for Hg with isotopic mass 199.5 is 4.18 K. Calculate its critical temperature when its isotopic mass changes to 203.4.*

Solution: If T_c and T_c' be the critical temperatures of Hg for isotopic masses 199.5 and 203.4 respectively, then

$$T_c/T_c' = (M_2/M_1)^{\beta} \text{ or, } T_c' = (M_1/M_2)^{\beta} T_c = 4.8 \times \left(\frac{199.5}{203.4}\right)^{1/2} = \mathbf{4.139\ K} \left[\because \beta = \frac{1}{2}\right].$$

11.7. *If the temperature dependence of the H_c be given by*

$$H_c(T) = H_0\left[1 - \left(\frac{T}{T_c}\right)^2\right].$$

Calculate $(C_n - C_s)/C_n$ at T_c for aluminium from the given data: H_0 = 99 gauss; T_c = 1.18 K; γ = 1.35×10^4 ergs/mole-K^2, *the specific heat constant for the normal metal.*

Solution: We have

$$C_n - C_s = -\frac{T_c}{4\pi}\left[\left(\frac{dH_c}{dt}\right)^2 + H_c\left(\frac{d^2H_c}{dT^2}\right)\right]_{T=T_c}$$

Again,

$$H_c(T) = H_0\left[1 - \left(\frac{T}{T_c}\right)^2\right], \therefore \frac{dH_c}{dt} = -\frac{2H_c}{T_c^2}T \text{ and } \frac{d^2H_c}{dT^2} = -\frac{2H_c}{T_c^2}.$$

$$\therefore C_n - C_s = \frac{H_0^2}{2\pi T_c}\left[\frac{T}{T_c} - \frac{3T^2}{T_c^2}\right] = -\frac{H_c^2}{\pi T_c}.$$

Hence,

$$\frac{C_n - C_s}{C_n} = \frac{H_0^2/\pi T_c}{\gamma T_c} = \frac{H_0^2}{\pi T_c^2 \gamma} = \frac{(99)^2}{3.4 \times (1.18)^2 \times (1.35 \times 10^4)}, \therefore \frac{C_n - C_s}{C_n} = \mathbf{0.17}.$$

11.8. The critical temperature and critical field are 7.18 K and 6.5×10^4 Am^{-1}, respectively. Find the temperature to which it has been cooled to become a superconductor in a magnetic field 2×10^4 Am^{-1}. Find the critical current density for 1 mm diameter of lead wire at 4.2 K.

Solution: Critical field is given by

$$H_c = H_0\left[1 - \left(\frac{T}{T_c}\right)^2\right].$$

$$\therefore 2 \times 10^4 \text{ Am}^{-1} = 6.5 \times 10^4 \text{ Am}^{-1}\left[1 - \frac{T}{7.18\text{ K}}\right] \text{ or, } \left(\frac{T}{7.18\text{ K}}\right)^2 = 1 - \frac{2}{6.5} \text{ or, } T = \mathbf{5.97\ K}.$$

At T = 4.2 K;

$$H_c = 6.5 \times 10^4 \text{ Am}^{-1}\left[1 - \left(\frac{4.2\text{ K}}{7.18\text{ K}}\right)^2\right] = 4.276 \times 10^4 \text{ Am}^{-1}.$$

Applying Ampere circuital law over the circle c of radius r encircling lead wire

$$\oint_c \vec{B} \cdot d\vec{r} = \mu_0 I.$$

From symmetry, $\vec{B} = B\hat{\phi}$ and B = constant on c; $d\vec{r} = \hat{\phi}dr$.

$$\therefore \oint_c B\hat{\phi} \cdot \hat{\phi}dr = \mu_0 I \text{ or, } B\oint_c d\phi = \mu_0 I \text{ or, } B2\pi r = \mu_0 I.$$

$$\therefore H = \frac{B}{\mu_0} = \frac{I}{2\pi r}.$$

At $r = R, H = \frac{I}{2\pi r}$.

Critical current,

$$I_c = \frac{I_c}{\pi R^2} = \frac{2\pi R H_c}{\pi R^2} \text{ or, } I_c = \frac{2H_c}{R} = \frac{2(4.276 \times 10^4\ \text{Am}^{-1})}{0.5 \times 10^{-3}\ \text{m}} = \mathbf{1.7 \times 10^8\ Am^{-2}}.$$

11.9. The penetration depths for lead are 396 Å and 1730 Å at 3 K and 7.1 K respectively. Find the critical temperature for lead and penetration depth at 0 K.

Solution: The penetration depth at T K is given by

$$\lambda(T) = \delta(0)\left[1 - \left(\frac{T}{T_c}\right)^4\right]^{-1/2}.$$

$$\therefore 396\ \text{Å} = \lambda(0)\left[1 - \left(\frac{3\ \text{K}}{T_c}\right)^4\right]^{-1/2} \text{ and } 1730\ \text{Å} = \lambda(0)\left[1 - \left(\frac{7.1\ \text{K}}{T_c}\right)^4\right]^{-1/2}.$$

Taking ratio,

$$\frac{1730}{396} = \left[\frac{1-(7.1/T_c)^4}{1-(3/T_c)^4}\right]^{-1/2} \text{ or, } \left(\frac{1730}{396}\right)^2 = \frac{1-(3/T_c)^4}{1-(7.1/T_c)^4} = 19$$

$$\text{or, } 19\left[1 - \left(\frac{7.1}{T_c}\right)^4\right] = 1 - \left(\frac{3}{T_c}\right)^4 \text{ or, } 19 - 1 = \frac{19(7.1)^4 - 3^4}{T_c^4} \text{ or, } T_c = \mathbf{7.19\ K}$$

and penetration depth at 0 K is

$$\lambda(0) = \lambda(T)\left[1 - \left(\frac{T}{T_c}\right)^4\right]^{1/2} = 396\ \text{Å}\left[1 - \left(\frac{3\ \text{K}}{7.19\ \text{K}}\right)^4\right]^{1/2} = \mathbf{390\ Å}.$$

11.10. The critical temperature of lead is 6 K at zero magnetic field and a critical field of 0.064 Am^{-1} at 0 K. Find the critical field at 4 K.

Solution: We have critical field,

$$H_c = H_0\left[1 - \left(\frac{T}{T_c}\right)^2\right] = (0.064\ \text{Am}^{-1})\left[1 - \left(\frac{4\ \text{K}}{6\ \text{K}}\right)^2\right] = \mathbf{0.0355\ Am^{-1}}.$$

11.11. The critical fields at 6 K and 8 K for NbT_i alloy are 7.616 Am^{-1} and 4.284 Am^{-1} respectively. Find the critical temperature and the critical field at 0 K.

Solution: The critical field,

$$H_c = H_0\left[1 - \left(\frac{T}{T_c}\right)^2\right] \text{ or, } 7.616\ \text{Am}^{-1} = H_0\left[1 - \left(\frac{6\ \text{K}}{T_c}\right)^2\right] \text{ and } 4.284\ \text{Am}^{-1} = H_0\left[1 - \left(\frac{8\ \text{K}}{T_c}\right)^2\right].$$

Dividing, we get

$$\frac{7.616}{4.284} = \frac{1-(6/T_c)^2}{1-(8/T_c)^2} \text{ or, } \frac{1}{4.284} - \frac{1}{7.616} = \left(\frac{8^2}{4.284} - \frac{6^2}{7.616}\right)^2 \frac{1}{T_c^2}.$$

$$\therefore 7.616\ \text{Am}^{-1} = H_0\left[1 - \left(\frac{6\ \text{K}}{10\ \text{K}}\right)^2\right] \text{ or, } H_0 = \frac{7.616}{1-(6/10)^2} = \mathbf{11.9\ Am^{-1}}.$$

11.12. Obtain an expression of London penetration depth using the following data: Each atom gives one conduction electron, density = 7.3 gm-cm^{-3}, T_c = 3.7 K, atomic weight 118.7, effective mass = 1.8 m, *m* = mass of a free electron.

Solution: We have the London penetration depth,

$$\lambda = \left(\frac{m^*}{\mu_0 n_s |e|}\right)^{1/2} \quad ; \quad n_s = \frac{PN_A}{M}.$$

$$\therefore n_s = \frac{[7.3 \times 10\,\text{kg}(1-^{-2}\,\text{m})]\,[6.023 \times 10^{23}\,\text{mol}^{-1}]}{118.7 \times 10^{-3}\,\text{kg-mol}^{-1}}.$$

$$\therefore \lambda = \left[\frac{1.8 \times 9.1 \times 10^{-31}}{(4\pi \times 10^{-7}\,\text{Hm}^{-1})\,(3.7 \times 10^{28}\,\text{m}^{-3})\,(1.6 \times 10^{-19}\,\text{C})^2}\right]^{1/2} = 37 \times 10^{-9}\,\text{m} = \mathbf{37\ nm}.$$

EXERCISE

11(A) Objective (MCQ) Type Questions

Select the correct answer.

11.1. The resistivity of a superconductor as $T \to 0$ becomes

(a) ∞ (b) 0 (c) any value between 0 and ∞ (d) none of these.

11.2. If the temperature $T > T_c$, the material behaves as a

(a) superconductor (b) good conductor (c) semiconductor (d) insulator.

11.3. Below transition temperature (T_c) a superconductor exhibits

(a) only zero resistance (b) only diamagnetic property

(c) zero resistance and diamagnetic property (d) zero resistance and paramagnetic property.

11.4. The magnetic susceptibility χ of a superconductor

(a) has a positive value (b) $\chi \to 0$ as $T \to T_c$ (c) $\chi \to \infty$ as $T \to T_c$ (d) has a negative value.

11.5. Application of critical magnetic field (H_c) to a superconductor changes it to

(a) temperature independent superconducting state (b) normal state

(c) zero conductivity (d) none of these.

11.6. The critical field (H_c)

(a) increases with temperature (T) (b) is independent of T

(c) increases with the decrease of T (d) none of these.

11.7. When a material becomes superconductor

(a) it becomes ferromagnetic (b) the property of lattice structure does not change

(c) magnetic property does not change (d) lattice property does change.

11.8. A superconducting state is associated with

(a) only change of entropy

(b) only change of electronic specific heat

(c) both entropy and electronic specific heat change
(d) none of these.

11.9. The transition temperature of most of the superconductors is in the range
(a) $T > 50$ K (b) $T = 20$ to 50 K (c) $T = 10$ to 20 K (d) $T = 0$ to 10 K.

11.10. Type I superconductors are
(a) Al, Zn and Hg (b) Al, Nb and Ta (c) Ta, V and Nb (d) Al, Ta and V.

11.11. Type II superconductors are
(a) Zr and Al (b) Al and Nb (c) Zr and Nb (d) Ta and Nb.

11.12. The width of the energy gap of a superconductor is zero at
(a) 0 K (b) room temperature (c) transition temperature (d) 100 °C.

11.13. The transition temperature (T_c) depends on average isotopic mass as
(a) $T_c = M^{\alpha}$ (b) $T_c = M^{-\alpha}$ (c) $T_c = \alpha M$ (d) $T_c = \alpha M^{\alpha}$,
where α is a constant.

11.14. The range of critical current in Josephson junction is
(a) 0–10 μ A (b) 100 μ A–1 mA (c) 1 mA–1 A (d) none of these.

11.15. London penetration depth is given by
(a) $\lambda = \dfrac{m}{\mu_0 n_s e^2}$ (b) $\lambda = \left(\dfrac{m}{\mu_0 n_s e^2}\right)^2$ (c) $\lambda = \left(\dfrac{m}{\mu_0 n_s e^2}\right)^{1/2}$ (d) $\lambda = \left(\dfrac{m\mu_0}{n_s e^2}\right)^{1/2}$.

11.16. The penetration depth λ depends on temperature T as
(a) $\lambda(T) = \lambda(0)\left[1 - \left(\dfrac{T}{T_c}\right)^4\right]$ (b) $\lambda(T) = \lambda(0)\left[1 - \left(\dfrac{T}{T_c}\right)^4\right]^{1/2}$
(c) $\lambda(T) = \lambda(0)\left[1 - \left(\dfrac{T}{T_c}\right)^2\right]^{1/2}$ (d) $\lambda(T) = \lambda(0)\left[1 - \left(\dfrac{T}{T_c}\right)^4\right]^{-1/2}$.

11.17. The value of a fluxoid (flux quantum) is
(a) 7.20×10^{15} Wb (b) 2.07×10^{15} Wb (c) 2.07×10^{-15} Wb (d) 7.20×10^{10} Wb.

11.18. Cooper pairs are formed at
(a) $T > T_c$ (b) $T < T_c$ (c) $T = T_c$ (d) none of these.

11.19. A super-current flows across an insulating gap between two superconductors if the gap is small enough in
(a) London effect (b) Meissner effect (c) Josephson effect (d) Isotope effect.

11.20. The coherence length of a paired electron is
(a) 2.50 nm (b) 0.25 nm (c) 0.01 nm (d) 0.001 nm.

11.21. If the applied DC voltage (V_0) across a Josephson junction is zero, the current I_c through the junction is
(a) $I_c = 0$ (b) $I_c = \infty$ (c) I_c = constant (d) I_c varies with time.

11.22. When the applied DC voltage (V_0) across the Josephson junction is greater than critical voltage V_c, the average current through the junction is
(a) $\langle I \rangle \neq 0$ (b) $\langle I \rangle = \langle I \rangle_{max}$ (c) $\langle I \rangle = 0$ (d) $\langle I \rangle = \infty$.

11.23. In a Cooper pair the electrons have

(a) parallel spin (b) anti-parallel spin (c) no spin (d) none of these.

11.24. Cooper pair follows

(a) MB-statistics (b) BE-statistics (c) FD-statistics (d) none of these.

11.25. According to BCS theory the zero resistance of superconductors arises due to

(a) presence of phonon scattering (b) absence of phonon scattering

(c) decrease of phonon scattering (d) increase of phonon scattering.

ANSWERS

11.1. (b) **11.2.** (b) **11.3.** (c) **11.4.** (d) **11.5.** (b) **11.6.** (c) **11.7.** (b) **11.8.** (c) **11.9.** (d) **11.10.** (a) **11.11.** (c) **11.12.** (c) **11.13.** (b) **11.14.** (b) **11.15.** (c) **11.16.** (d) **11.17.** (c) **11.18.** (b) **11.19.** (c) **11.20.** (d) **11.21.** (c) **11.22.** (c) **11.23.** (b) **11.24.** (b) **11.25.** (b).

11(B) Short Answer Type Questions

11.1. What do you understand by superconductivity? What are its main characteristic properties?

11.2. Define critical temperature and critical field. Discuss the effect of magnetic field on superconductivity.

11.3. What is Meissner effect? Show that a superconductor behaves as a diamagnetic body.

11.4. What are Types I and II superconductors. Draw a distinction between them.

11.5. Obtain *Rutgers equation* and discuss the variation of specific heat for superconducting state and normal state of a body.

11.6. Discuss isotope effect in superconductor. What is Energy gap? How does it vary with temperature?

11.7. Write down first and second London equations. What do you mean by *London penetration depth*?

11.8. What is *fluxoid*? What are its applications?

11.9. What is a Cooper pair? Discuss the origin of superconductivity in view of BCS theory.

11.10. What is a *Josephson junction*? How does super-current vary with applied voltage in a superconductor?

11.11. Mention some applications of *Josephson effect.*

11.12. What is a SQUID? Discuss its uses.

11.13. Mention some applications of superconductors.

11.14. Discuss in brief the recent trend in superconductivity research.

11.15. List out some of the characteristic features of superconductors comparing with that of normal conductors.

11.16. Define superconductivity. Write a note on types I and II superconductivity. **[Pune U. 2013]**

11.17. Discuss the effect of temperature and magnetic field on superconductivity. **[Bhabnagar U. 2017]**

11.18. Discuss Meissner effect in types I and II superconductor.

11.19. State and explain Meissner effect. **[Pune U. 2013]**

11.20. What are the important predictions of BCS theory?

11.21. In which way the energy gaps seen in super conductor are different from semiconductor? How does the energy gap vary with temperature? Sketch the variation of specific heat of a superconductor with temperature (Show also the critical transition temperature T_c in the graph). **[CU(H) 2016]**

11.22. Explain briefly the Meissner effect with suitable diagram. Show that the magnetic field decays inside the superconductor exponentially with characteristic length scale. **[CU(H) 2014]**

11.23. (a) Explain the meaning of energy gap in superconductor.

(b) What is isotope effect? What is its importance in the theory of superconductivity? **[CU(H) 2013]**

11(C) Numerical Problems

11.1. The critical fields of a superconductor at 6 K and 8 K are 7.616×10^6 A/m and 4.284×10^6 A/m respectively. Calculate the critical temperature for zero magnetic intensity and the critical field at 0 K.

[*Ans.* 10 K; 11.9×10^6 A/m]

11.2. The critical temperature for Hg is 4.185 K and isotopic mass is 1995. If the isotopic mass changes to 203.4, calculate its critical temperature.

[*Ans.* 4.139 K]

11.3. Calculate the London penetration depth at 0 K for Pb, whose density is 11.3×10^3 kg/m^3 and atomic weight 207.19. [Given, T_c = 7.22 K]. Calculate the increase in penetration depth at 3.61 K from its value at 0 K.

[*Ans.* 1 nm]

11.4. The London penetration depths for Pb at 3 K and 7.1 K are respectively 39.6 nm and 173 nm. Calculate its transition temperature as well as depth at 0 K.

[*Ans.* 7.193 K; 39 nm]

11.5. The penetration depth of Hg at 3.5 K is 750 Å. Estimate the penetration depth at 0 K. [*Ans.* 519 Å]

11.6. The average atomic mass of Hg is 200.59 amu and the critical temperature is 4.153 K. What is the critical temperature of the isotope $_{80}Hg^{204}$?

[*Ans.* 4.118 K]

11.7. What is the critical current density for 1 mm diameter wire of Pb at (i) 4.2 K and (ii) 7 K. [Given, T_c = 7.18 K and $H_c = 6.5 \times 10^4$ A/m]

[*Ans.* 1.71×10^8 A/m^2; 1.285×10^7 A/m^2]

11.8. A superconducting tin has a critical temperature of 8.7 K in zero magnetic field and a critical field of 0.0306 T at zero K. Find the critical magnetic field (in CGS unit) at 7 K. **[Bhabnagar U. 2017]**

11.9. Determine critical field required to destroy superconductivity at 5 K in Pb whose T_c is 7.19 K and $H_c(0)$ = 0.0803 T.

11.10. A super conducting sample has a critical temperature of 4.1 K in zero magnetic field and critical field of 0.0505 T at 0 K. Find the critical field at 2 K. **[Mahatma Gandhi U. (MSc) 2012]**

11.11. Lead in superconducting state has critical temperature 6.2 K at zero magnetic field and a critical field of 0.064 T at 0 K. Determine the critical field at 4 K. **[Delhi U. 2014]**

11(D) Broad Answer Type Questions

11.1. (a) What is a superconductor? How does it differ from a normal conductor?

(b) Discuss the effect of magnetic field on a superconductor.

11.2. (a) What is Meissner effect?

(b) Show that a superconductor behaves as a diamagnet.

(c) Explain the origin of superconductivity from wave mechanical point of view.

11.3. (a) Enumerate the properties of Types I and II superconductors.

(b) Derive the London equation and discuss how does it help in explaining the superconducting state.

(c) What is London penetration depth?

11.4. (a) Explain the formation of a Cooper pair in a superconductor.

(b) Give an account of BCS theory of superconductivity and discuss how does it explain the phenomenon of superconductivity.

(c) What is isotope effect?

11.5. (a) Discuss Josephson effect.

(b) Give formulation of DC and AC Josephson effects.

(c) Discuss how AC Josephson effect can estimate $e/\hbar$.

11.6. (a) Using BCS theory, obtain an expression for the magnitude of coherence length in superconductors and explain the physical significance of this length.

(b) Discuss Josephson tunnelling.

(c) What do you mean by fluxoid?

11.7. (a) What do you mean by a SQUID?

(b) Discuss its important physical applications.

(c) What is magnetic levitation?

11.8. (a) Discuss the effect of magnetic field on the superconducting state of a material.

(b) What is flux exclusion?

(c) Discuss Meissner effect.

11.9. (a) Give a short account of high T_c superconductivity.

(b) Bring out the potential applications of superconductivity.

(c) Prove that

$$C_s - C_n = \mu_0 T\left(\frac{dH_c}{dT}\right)^2,$$

where C_s and C_n are the specific heats of the superconductor and normal conductor respectively and H_c is the critical magnetic field.

11.10. (a) Discuss coherence length in superconductivity.

(b) Discuss the energy gap in superconductivity phenomena. How does it differ from normal band gap in superconductors.

(c) Discuss in brief the applications of superconductivity. **[Bhabnagar U. 2017]**

11.11. (a) Give an account of superconductivity. Explain types I and II superconductors.

(b) Write a note on Meissner effect. **[Mahatma Gandhi U. (MSc) 2012]**

11.12. (a) What are dc and ac Josephson effect?

(b) Give the theory of ac Josephson effect. **[Mahatma Gandhi U. (MSc) 2012]**

Chapter 12

Dielectric Properties of Solids

12.1 Introduction

Dielectrics are basically insulators. All the electrons in dielectrics are bound to their parent molecules and hence, could be affected by externally applied electric fields. There is no *free charge* and as such electrical conduction could not take place in dielectrics.

Dielectrics are *non-metals of high specific resistance and negative temperature coefficients.* Glass, mica, ebonite, paraffin wax and dry air are some of the common examples of dielectric.

12.2 Polar and Non-polar Molecules

The dielectric materials can be classified into the following two categories as regards their behaviour in an electric field:

(i) **Non-Polar Molecules:** *If the centre of gravity of positive charges exactly coincides with that of the negative charges and the net dipole moment is zero, then the molecule is called a non-polar molecule.* H_2, O_2, N_2, CO_2 and CH_4 are the examples of *non-polar molecules.*

(ii) **Polar Molecules:** *If the centre of gravity of positive charges does not coincide with that of the negative charges, then the molecule is called a polar molecule.* Such molecules, therefore, constitute a permanent dipole and have a permanent dipole moment.

Examples of polar molecules are HCl, H_2O, NH_3, etc. These molecules consist of *dissimilar atoms* and their dipole moments are of the order of 10^{-29} C-m.

12.2.1 Polar Molecules in Absence of Electric Field

If a polar dielectric be kept outside the electric field the molecular dipoles are oriented at random with their dipole moment ($\vec{p}$) pointing all possible directions [Fig. 12.1]. As a result, the dipole moments cancel the effect of each causing the dipole moment per unit volume of the dielectric zero. Hence, though the individual molecules possess dipole moment, the dipole moment of the dielectric as a whole is zero.

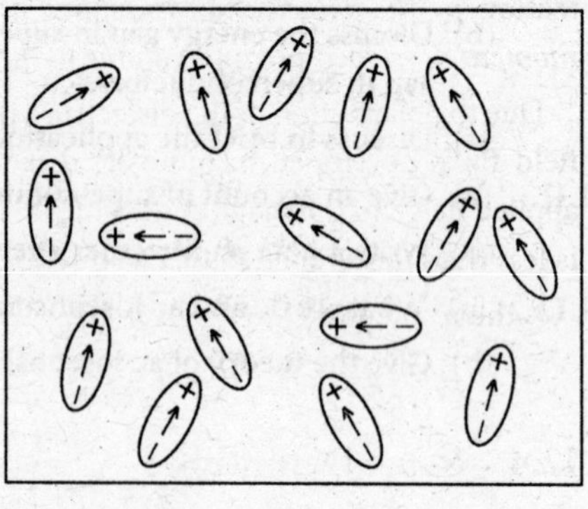

Fig. 12.1: Polar dielectric in absence of $\vec{E}$

12.2.2 Polar Dielectric in Presence of Electric Field

The dipole moments tend to align themselves when an electric field $\vec{E}$ is applied to the dielectric. The alignments take place along the direction of $\vec{E}$ [Fig. 12.2].

The *torque* experienced by a dipole of moment $\vec{p}$ is given by $\vec{\tau} = \vec{p} \times \vec{E}$.

This *torque* tends to align the dipoles along $\vec{E}$ which is opposed by thermal agitation of the molecules.

So the degree of alignment is determined by (i) *the magnitude of* $\vec{E}$ and (ii) *the temperature of the dipole causing thermal agitation.*

If $\vec{E}$ be increased, the separation between the centre of gravity of positive and negative charges will be increased causing an increase in dipole moment. As a result, the magnitude of the torque $\vec{\tau}$ will increase and hence, the *degree of alignment* will be greater.

If the temperature of the dielectric be decreased, then the thermal agitation will be less and the alignment will be greater. When the molecules of a dielectric are completely or partially aligned in the direction of the electric field, the dielectric is said to be polarized [Fig. 12.2].

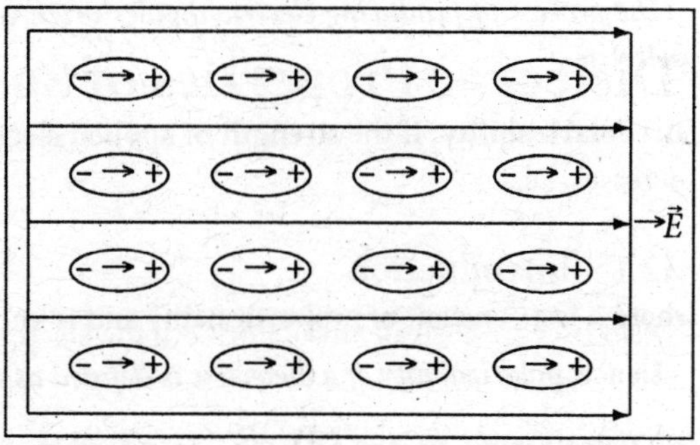

Fig. 12.2: Electric dipole in an electric field $\vec{E}$

12.3 Dielectric Polarization (Concept of Polarization)

When a *non-polar dielectric* is placed in an electric field $\vec{E}_0$, the centres of the positive and negative charges are displaced from their equilibrium positions by a small distance. The separation of the positive and negative charges continues till the force on either of them due to $\vec{E}_0$ is completely balanced by the internal forces due to their relative displacement. The molecules then develope dipole moment (i.e., the induced dipole moment) in the direction of $\vec{E}_0$. The substance is then said to be polarized with a polarization $\vec{p}$. *The polarization is defined as the dipole moment developed per unit volume of the material.*

Due to polarization of the dielectric a polarization field $\vec{E}_p$ is developed in opposite direction of the applied field. So the magnitude of the resultant field is less than the applied field, i.e., $|\vec{E}| < |\vec{E}_0|$.

∴ the resultant field $\vec{E}$ is given by

$$\vec{E} = \vec{E}_0 + \vec{E}_p. \qquad (12.1)$$

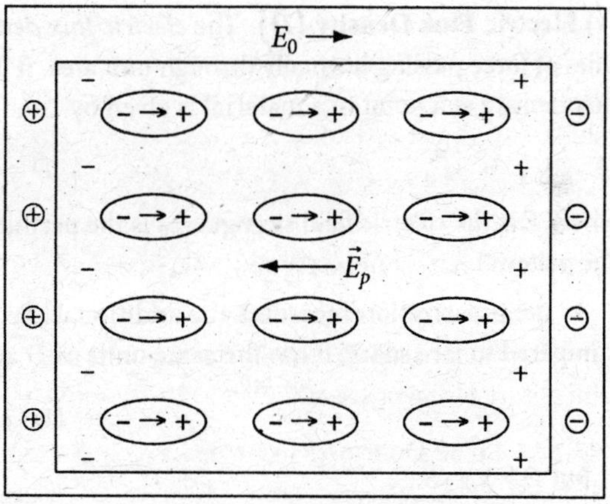

Fig. 12.3: A dielectric slab is placed in $\vec{E}_0$ produced by fixed charges outside the slab. $\vec{E}_p$ is produced by bound charges.

12.4 Some Definitions

(i) Dielectric Constant: *Dielectric constant of a material is defined as the ratio between the permittivity of the medium* (ε) *and the permittivity of the free space* (ε_0), *i.e.,*

$$\varepsilon_r = \varepsilon/\varepsilon_0, \qquad (12.2)$$

ε_r is also called the *relative permittivity* and being *the ratio of two quantities of same nature,* ε_r has no unit. ε_r gives a measure of polarization in the dielectric materials.

(ii) Electric Polarization: If an atom be subjected to an external electric field, the centre of positive charge is displaced along the applied field direction while the centre of the negative charge is displaced in opposite direction. Thus, an electric dipole is produced. When a dielectric material is placed in an electric field such dipoles are created in all the atoms of the material.

The process of producing electric dipoles oriented along the applied electric field is called polarization of dielectric.

(iii) Polarizability: If the strength of applied electric field $\vec{E}$ is increased, then the dipole moment $\vec{p}$ also increases, i.e.,

$$\vec{p} = \alpha \vec{E}, \tag{12.3}$$

where α is a constant of proportionality and is known as the *polarizability.*

Hence, *polarizability of a dielectric is defined as the dipole moment induced per unit electric field.*

(iv) Polarization Vector $(\vec{P})$: *The polarization vector $(\vec{P})$ is defined as dipole moment induced per unit volume of the dielectric material.*

If N be the number molecules per unit volume in a dielectric and $\vec{p}$, the average dipole moment per molecule, then the polarization vector is

$$\vec{P} = N\vec{p}. \tag{12.4}$$

(v) Electric Flux Density $(\vec{D})$: The *electric flux density* or *electric displacement* is the number of electric lines of force passing normally through unit area. It is also known as the *electric displacement* $(\vec{D})$. Electric flux density at a point in a material is given by

$$\vec{D} = \varepsilon_r \varepsilon_0 \vec{E}, \tag{12.5}$$

where $\vec{E}$ is the electric field strength, ε_0 is the permittivity of free space and ε_r is the relative permittivity of the material.

As the polarization measures the additional flux density arising from the presence of the material as compared to free space, it has the same units as $\vec{D}$ and is related to $\vec{D}$ as

$$\vec{D} = \varepsilon_0 \vec{E} + \vec{P}. \tag{12.6}$$

But $\vec{D} = \varepsilon_r \varepsilon_0 \vec{E}$.

$$\therefore \varepsilon_r \varepsilon_0 \vec{E} - \varepsilon_0 \vec{E} = \vec{P}$$

$$\text{or, } \vec{P} = \varepsilon_0 (\varepsilon_r - 1) \vec{E}, \tag{12.7}$$

$\vec{D}$ is similar to magnetic induction $\vec{B}$ in magnetostatic.

(vi) Electric Susceptibility: *The ratio of polarization per unit volume to the net electric field $\vec{E}$ as modified by the induced charges on the surface of the dielectric is called electric susceptibility and is denoted by χ_e*, i.e.,

$$\chi_e = \frac{P}{\varepsilon_0 E} \text{ or, } \chi_e = \frac{\varepsilon_0 (\varepsilon_r - 1) E}{\varepsilon_0 E} \text{ [by eqn (12.7)].} \tag{12.8}$$

$$\therefore \chi_e = (\varepsilon_r - 1). \tag{12.9}$$

12.5 Derivation of $\vec{D} = \varepsilon_0 \vec{E} + \vec{P}$

We consider a parallel plate condenser in which the space between the plates is filled with an *isotropic dielectric.* The electric field at any point between the plates is

$$E = \frac{\sigma_f}{\varepsilon_0}, \tag{12.10}$$

where $\pm\sigma_f$ are the surface density of *free* charges on the plates of the condenser [Fig. 12.4(a)].

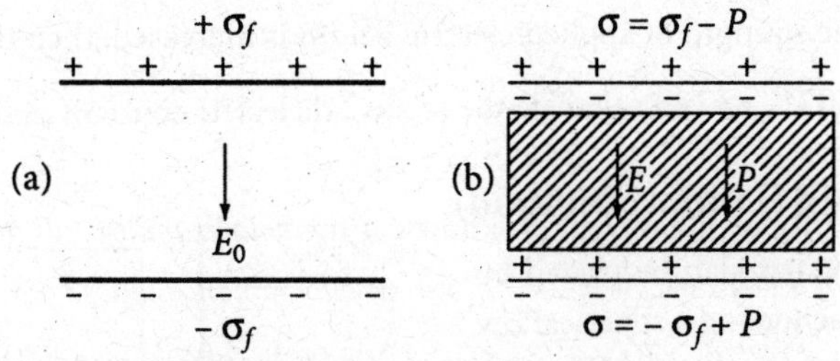

Fig. 12.4: (a) Capacitor without dielectric,
(b) Capacitor with dielectric

When the dielectric is introduced between the plates of the condenser, it gets polarized, and let the density of polarization (i.e., dipole moment per unit volume) be P. A negative charge density $-P$ is induced on the dielectric surface near the positive charged plate [Fig. 12.4(b)] and a similar positive charge $+P$ on the surface near the positive plate. There will be no resultant charge density at any point within the dielectric. This is because all the individual dipoles are aligned parallel to the applied field with negative charge of one dipole being next to the positive charge of the former.

The densities of charges on the condenser plate become $\sigma = \pm\left(\sigma_f - P\right)$.

The new electric field in between the condenser plates is then

$$E = \frac{\sigma}{\varepsilon_0} = \frac{\sigma_f - P}{\varepsilon_0}$$

$$\text{or, } \sigma_f = \varepsilon_0 E + P. \tag{12.11}$$

The quantity on the right-hand side of eqn (12.11) is called the *electric displacement* D.

Thus,

$$D = \varepsilon_0 E + P. \tag{12.12}$$

Since D, E and P are in the same direction, the more general form of eqn (12.12) is expressed as

$$\vec{D} = \varepsilon_0 \vec{E} + \vec{P}. \tag{12.13}$$

12.6 Measurement of Relative Dielectric Constant

The *relative dielectric constant* of a dielectric material can be measured by using a parallel plate capacitor.

Let a voltage V be applied between the plates of the capacitor. When the space between the plates is vacuum, then the capacity of the condenser is given by

$$C = \frac{\varepsilon_0 A}{d}, \tag{12.14}$$

where A is the surface area and d is the separation of the plates.

If now a dielectric material be inserted between the plates, the charge on the plates increases due to polarization in the material. Now the capacitance of the condenser is given by

$$C' = \frac{\varepsilon_0 \varepsilon_r A}{d}. \tag{12.15}$$

From eqns (12.14) and (12.15) the relative dielectric constant is found to be

$$\varepsilon_r = C'/C. \tag{12.16}$$

Thus, measuring C and C', we can calculate the relative dielectric constant of the material.

12.7 The Local Field (Lorentz Method)

We consider that a dielectric slab be placed between the plates of a capacitor and a spherical cavity inside the dielectric around an atom A [Fig. 12.5]. If the diameter of the cavity is larger than the atomic radius, then the *local field* (also called the internal field) at the site of the atom A is given by

$$\vec{E}_{loc} = \vec{E}_1 + \vec{E}_2 + \vec{E}_3 + \vec{E}_4. \tag{12.17}$$

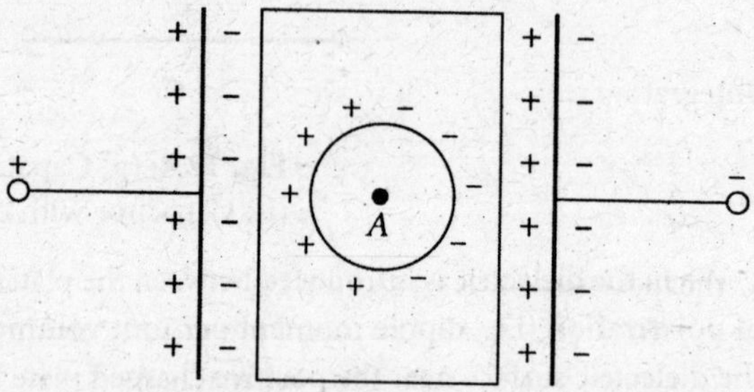

Fig. 12.5: Calculation of $\vec{E}_{loc}$ for a dielectric of cubic structure

Field $\vec{E}_1$: The field $\vec{E}_1$ arises due to the charge density on the capacitor plates. Thus,

$$\vec{E}_1 = \frac{\vec{D}}{\varepsilon_0} = \frac{\vec{P} + \varepsilon_0 \vec{E}}{\varepsilon_0} \quad [\because \vec{D} = \vec{P} + \varepsilon_0 \vec{E}].$$

$$\text{or, } \vec{E}_1 = \vec{E} + \frac{\vec{P}}{\varepsilon_0}. \tag{12.18}$$

Field $\vec{E}_2$: The field $\vec{E}_2$ arises due to the induced charge density on the two sides of the dielectric and is given by

$$\vec{E}_2 = -\vec{P}/\varepsilon_0. \tag{12.19}$$

Field $\vec{E}_3$: The field $\vec{E}_3$ is the field intensity at A due to other atoms in the cavity. Since the structure is cubic, $\vec{E}_3 = 0$ because of symmetry.

Field $\vec{E}_4$: $\vec{E}_4$ is the field intensity at A due to the polarization charges on the surface of the cavity.

Fig. 12.6 shows an enlarged view of the cavity inside the dielectric around A.

If dS be the surface area of the sphere of radius r lying between θ and $\theta + d\theta$, where θ is the angle made by AQ with the direction of the applied field $\vec{E}$, then

$$dS = 2\pi(BQ)(QR).$$

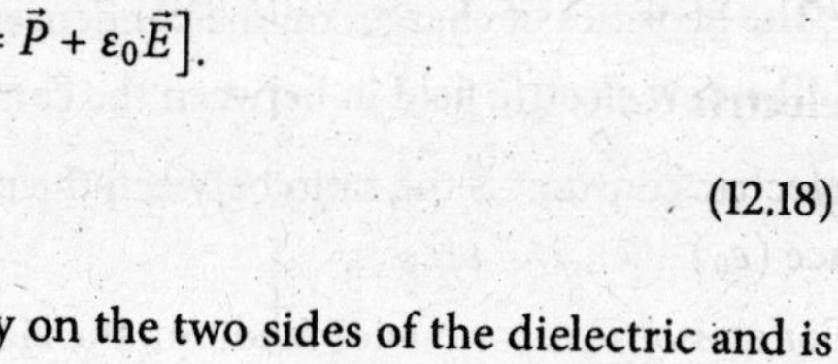

Fig. 12.6: Enlarged cavity around A

Since

$$\sin\theta = BQ/r, \therefore BQ = r\sin\theta$$
$$\text{and } d\theta = QR/r, \therefore QR = rd\theta.$$
$$\therefore dS = 2\pi r\sin\theta \times rd\theta = 2\pi r^2 \sin\theta\, d\theta.$$

Now the charge on the surface dS is

$$dq = \sigma dA = (P\cos\theta)\, dA \quad [\because \sigma = \text{the surface density of charge} = P\cos\theta]$$
$$= P\left(2\pi r^2 \sin\theta\cos\theta\, d\theta\right).$$

$\therefore$ the field at A due to dq is in the direction of $\vec{E}$ (i.e., $\theta = 0$).

$$\therefore dE_4 = \frac{dq\cos\theta}{4\pi\varepsilon_0 r^2} = \frac{P\cos\theta \times 2\pi r^2 \sin\theta\cos\theta\, d\theta}{4\pi\varepsilon_0 r^2} = \frac{P}{2\varepsilon_0}\cos^2\theta\sin\theta\, d\theta.$$

Integrating from $\theta = 0$ to π, we get

$$E_4 = \frac{P}{2\varepsilon_0}\int_0^\pi \cos^2\theta\sin\theta\, d\theta = \frac{P}{2\varepsilon_0}\int_0^\pi \cos^2\theta\, d(-\cos\theta) = \frac{-P}{2\varepsilon_0}\left[\frac{\cos^3\theta}{3}\right]_0^\pi = -\frac{P}{6\varepsilon_0}[-1-1] = \frac{P}{3\varepsilon_0}. \quad (12.20)$$

$\therefore$ the local field (or *Lorentz field*) is

$$\vec{E}_{\text{loc}} = \left(\vec{E} + \frac{\vec{P}}{\varepsilon_0}\right) + \left(-\frac{\vec{P}}{\varepsilon_0}\right) + 0 + \frac{\vec{P}}{3\varepsilon_0} = \vec{E} + \frac{\vec{P}}{3\varepsilon_0}. \quad (12.21)$$

Eqn (12.21) shows that the local field $\vec{E}$ is greater than the applied field $\vec{E}$ by $\vec{P}/e\varepsilon_0$.

12.8 Dielectric Constant and Polarizability

Dielectric Constant

Dielectric constant is the ratio between the permittivity of the medium (ε) to the permittivity of the free space (ε_0), i.e., $\varepsilon_r = \varepsilon/\varepsilon_0$.

For an *isotropic* (or cubic) medium, the electric displacement vector is defined as

$$\vec{D} = \varepsilon_0\varepsilon_r\vec{E} = \varepsilon_0\vec{E} + \vec{P}, \quad (12.22)$$

where ε_r is the relative permeability (or, dielectric constant) of the medium, ε_r is always dimensionless and a constant quantity.

From eqn (12.22),

$$\varepsilon_0\varepsilon_r\vec{E} = \varepsilon_0\vec{E} + \vec{P} \text{ or, } \varepsilon_r = \frac{\varepsilon_0\vec{E} + \vec{P}}{\varepsilon_0\vec{E}} = 1 + \frac{\vec{P}}{\varepsilon_0\vec{E}} \quad (12.23)$$

$$\text{or, } \varepsilon_r = 1 + \chi_e \quad \left[\because \frac{\vec{P}}{\varepsilon_0\vec{E}} = \chi_e\right]. \quad (12.24)$$

Thus, ε_r is also a measure of *degree of polarization* of the medium. If $\vec{P}/\varepsilon_0\vec{E}$ is larger, then ε_r will be larger for the dielectric medium.

From eqn (12.22), we can express the polarization vector as

$$\vec{P} = \varepsilon_0(\varepsilon_r - 1)\vec{E}. \quad (12.25)$$

This shows that for a particular field $\vec{E}$, $\vec{P}$ is greater if ε_r is greater.

Atomic Polarizability

The polarizability (α) of an atom is defined as the dipole moment developed per unit local field at the atom.

$$\therefore \vec{p} = \alpha \vec{E}_{\text{loc}}. \tag{12.26}$$

It is to be noted that ε_r depends on the arrangements of the atoms within the crystal and hence, is a macroscopic property, whereas α is an atomic property.

If there are N atoms per unit volume and all atoms have the same value of α, then the polarization (i.e., dipole moment per unit volume) is given by

$$\vec{P} = N\vec{p} = N\alpha\vec{E}_{\text{loc}}. \tag{12.27}$$

In general,

$$\vec{P} = \sum_j N_j \alpha_j \vec{E}_{\text{loc}}(j), \tag{12.28}$$

where $\sum_j$ represents summation over all the atoms.

From eqns (12.21) and (12.27), we get

$$\vec{P} = N\alpha\left(\vec{E} + \frac{\vec{P}}{3\varepsilon_0}\right). \tag{12.29}$$

12.9 Polarization Processes

When a dielectric is placed in a DC electric field $(\vec{E})$, it gets polarized by the following *four* processes: (i) *Electronic polarization,* (ii) *Ionic polarization,* (iii) *Dipolar* or *oriental polarization* and (iv) *Space charge polarization.*

The contribution of a particular process depends on the nature of the dielectric as well as on the frequency of the applied electric field provided the applied field be an AC field.

12.9.1 Electronic Polarization

When an electric field is applied to the dielectric, the positively charged nucleus and the electrons of an atom are displaced in opposite directions and electronic polarization occurs. The electron cloud shifts towards the positive end of the field [Fig. 12.7(b)]. Dipole moment is created in each atom due to the relative shift of the centre of the electron cloud and the nucleus. The shift is proportional to the applied electric field $\vec{E}$ and the induced dipole moment of the atom:

$$\vec{p}_e = \alpha_e \vec{E}, \tag{12.30}$$

where α_e is the *electronic polarizability* which is independent of temperature of the dielectric.

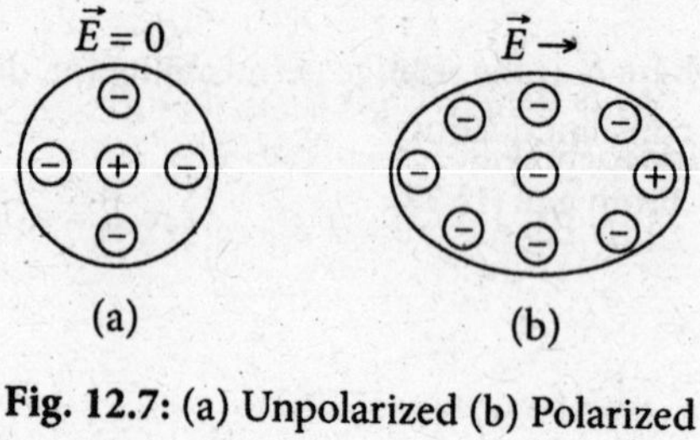

Fig. 12.7: (a) Unpolarized (b) Polarized

Classical Theory of Electronic Polarizability

If ω_0 be the natural frequency of vibration of an electron bound harmonically to the atom, then it exhibits resonant absorption at frequency ω_0 and is given by

$$\omega_0 = \sqrt{\beta/m}, \tag{12.31}$$

where β is the force constant and m, the mass of the electron.

Under equilibrium condition, the force on the electron due to local field $\left(-e\vec{E}_{loc}\right)$ is equal to the force of restitution (βx), where x is the displacement of the electron from the mean position, i.e.,

$$-e\vec{E}_{loc} = \beta x = m\omega_0^2 x. \tag{12.32}$$

The *static electronic polarizability* is given by

$$\alpha_e = \frac{p}{E_{loc}} = \frac{-ex}{\left(\frac{m\omega_0^2 x}{-e}\right)} = \frac{e^2}{m\omega_0^2}. \tag{12.33}$$

To find the *frequency dependance of* α_e we assume the electron motion to be *simple harmonic* and the local field $\vec{E} = \vec{E}_{loc} \sin \omega t$, where ω is the frequency of the field $\vec{E}$. Then the equation of motion of the electron is

$$m\frac{d^2x}{dt^2} + m\omega_0^2 x = -eE_{loc} \sin \omega t. \tag{12.34}$$

Substituting $x = x_0 \sin \omega t$, we get from eqn (12.34)

$$m\frac{d^2}{dt^2}(x_0 \sin \omega t) + m\omega_0^2 x_0 \sin \omega t = -eE_{loc} \sin \omega t$$

$$\text{or, } m\left(\omega_0^2 - \omega^2\right) x_0 = -eE_{loc}. \tag{12.35}$$

The amplitude of the dipole oscillation is

$$p_0 = -ex_0 = \frac{e^2 E_{loc}}{m\left(\omega_0^2 - \omega^2\right)}. \tag{12.36}$$

∴ the frequency dependent electronic polarizability is given by

$$\alpha_e = \frac{p_0}{E_{loc}} = \frac{e^2/m}{\left(\omega_0^2 - \omega^2\right)}. \tag{12.37}$$

The corresponding *quantum mechanical* expression is

$$\alpha_e = \frac{e^2}{m}\sum_j \frac{f_{ij}}{\left(\omega_{ij}^2 - \omega^2\right)}. \tag{12.38}$$

f_{ij} is the *oscillator strength* of the electronic dipole transition between the atomic states i and j. α_e is independent of temperature.

If N be the number of atoms per unit volume of the dielectric, then *static electronic polarization* is

$$P_e = Np_e = N\alpha_e E_{loc}. \tag{12.39}$$

Also from eqn (12.12),

$$P_e = D - \varepsilon_0 E_{loc} = \varepsilon_r \varepsilon_0 E_{loc} - \varepsilon_0 E_{loc} \text{ or, } P_e = \varepsilon_0(\varepsilon_r - 1) E_{loc}.$$

$$\therefore \varepsilon_0(\varepsilon_r - 1) E_{loc} = N\alpha_e E_{loc}$$

$$\text{or, } \alpha_e = \frac{\varepsilon_0(\varepsilon_r - 1)}{N}. \tag{12.40}$$

An alternative approach

Eqn (12.40) can also be derived by assuming that the nucleus of charge $+Ze$ is surround by an electron cloud of charge $-Ze$ distributed in a sphere of radius r. Then the charge density ρ is given by

$$\rho = \frac{-Ze}{\frac{4}{3}\pi r^3} = -\frac{3}{4}\left(\frac{Ze}{\pi r^3}\right). \quad \text{(i)}$$

If an external field $\vec{E}$ be applied, the nucleus and the electrons experience *Lorentz force* of magnitude $F_1 = ZeE$ in opposite directions and as such the electrons and the nucleus are pulled apart. The moment they are separated a *Coulomb force* F_2 develops between them which tends to oppose the displacement.

When the Lorentz force and Coulomb force are equal and opposite equilibrium is reached. Let x be the displacement under this condition.

The nucleus being heavier than the electron cloud, the electron cloud only will be displaced in presence of the applied field $\vec{E}$.

Now the Lorentz force, $F_1 = -ZeE$.

Coulomb force, $F_2 = Ze \times \dfrac{\text{charge enclosed in the sphere of radius } x}{4\pi\epsilon_0 x^2}$.

Now the charge enclosed by the sphere of radius x is

$$= \frac{4}{3}\pi x^3 \rho = \frac{4}{3}\pi x^3\left[-\frac{3}{4}\left(\frac{Ze}{\pi r^3}\right)\right] = -\frac{Zex^3}{r^3}.$$

$$\therefore F_2 = \frac{Ze}{4\pi\epsilon_0 x^2} \times \frac{-Zex^3}{r^3} = -\frac{Z^2e^2x}{4\pi\epsilon_0 r^3}.$$

In equilibrium,

$$-ZeE = -\frac{Z^2e^2x}{4\pi\epsilon_0 r^3} \text{ or, } x = \frac{4\pi\epsilon_0 r^3 E}{Ze}. \quad \text{(ii)}$$

The dipole moment induced in the atom is

$$p_e = Zex = \frac{Ze \times 4\pi\epsilon_0 r^3 E}{Ze} = 4\pi\epsilon_0 r^3 E \quad \text{(iii)}$$

$$\text{or, } p_e = \alpha_e E, \quad \text{(iv)}$$

$\alpha_e = 4\pi\epsilon_0 r^3$ is the *electronic polarizability*.

This gives the dependence of electronic polarizability (α_e) of an atom on its radius (r).

The electronic polarization,

$$P_e = Np_e = N\alpha_e E, \quad \text{(v)}$$

where N is the number of atoms/m^3.

Again, since, $P_e = \epsilon_0 E(\epsilon_r - 1)$ [from eqn (12.7)].

$$\therefore \epsilon_0 E(\epsilon_r - 1) = N\alpha_e E \text{ or, } \epsilon_r - 1 = \frac{N\alpha_e}{\epsilon_0}.$$

$$\therefore \alpha_e = \frac{\epsilon_0(\epsilon_r - 1)}{N}. \quad \text{(vi)}$$

Thus, P_e and α_e are independent of temperature.

12.9.2 Ionic Polarizability

The ionic polarizability arises due to the displacement of cations and anions in opposite directions in an ionic solid.

Let an electric field be applied along the positive directions when the positive ion moves towards right by x_1 and the negative ion by $-x_2$. If each unit cell of the dielectric contains one positive and one negative ions, then the dipole moment due to ionic displacement is $\vec{p} = e(\vec{x}_1 + \vec{x}_2)$.

If $\vec{F}$ be the force due to applied field $\vec{E}$, and β_1 and β_2 the force constants of cation and anion, then $\vec{F} = \beta_1\vec{x}_1 = \beta_2\vec{x}_2$.

$$\therefore \vec{x}_1 = \vec{F}/\beta_1 = \frac{e\vec{E}}{m\omega_0^2},$$

where $\vec{F} = e\vec{E}$ and the restoring force constant $\beta_1 = m\omega_0^2$.

Similarly, $\vec{x}_2 = \frac{e\vec{E}}{M\omega_0^2}$; m and M are the masses of the anion and cation respectively.

$$\therefore \vec{x}_1 + \vec{x}_2 = \frac{e\vec{E}}{\omega_0^2}\left(\frac{1}{M} + \frac{1}{m}\right) \text{ and } \vec{p} = e(x_1 + x_2) = \frac{e^2\vec{E}}{\omega_0^2}\left(\frac{1}{M} + \frac{1}{m}\right).$$

$\therefore$ *ionic polarizability,*

$$\alpha_i = \frac{\vec{p}}{\vec{E}} = \frac{e^2}{\omega_0^2}\left(\frac{1}{M} + \frac{1}{m}\right) \tag{12.41}$$

$$\text{or, } \alpha_i = \frac{e^2}{\omega_0^2\mu} = \frac{e^2}{\beta}, \tag{12.42}$$

where $\beta = \omega_0^2\mu$, and $\mu = \left(\frac{1}{M} + \frac{1}{m}\right)^{-1}$ is the *reduced mass* of the anion and cation.

Thus, α_i is inversely proportional to the square of the natural frequency (ω_0) of the molecule, and its reduced mass is μ.

12.9.3 Dipolar or Oriental Polarizability

The molecules like H_2O, CH_3Cl, etc. having a permanent dipole moment are called *dipolar* or *polar molecules.* The substance comprising of such molecules is called *dipolar substance.*

In absence of external electric field $\vec{E}$, the dipoles are randomly oriented and the net dipole moment of the substance is zero. But in presence of an external electric field $\vec{E}$, the dipoles orient along $\vec{E}$ producing oriental polarizability. The thermal agitation of the molecules tends to disorient the dipoles and in equilibrium condition the different dipoles are oriented at various angles $\theta = 0$ to π radian with the direction of the external field $(\vec{E})$.

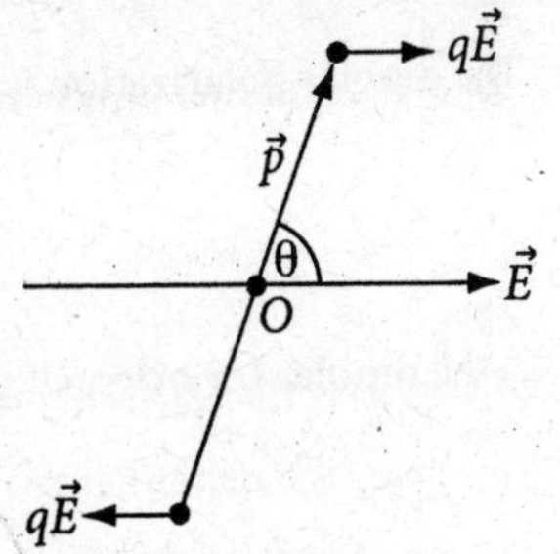

Fig. 12.8: A dipole moment $\vec{p}$ in an electric field $\vec{E}$

The PE of such a dipole oriented at an angle θ with $\vec{E}$ as shown in Fig. 12.8 is given by

$$U = -\vec{p}\cdot\vec{E} = -pE\cos\theta. \tag{12.43}$$

Now the number of dipoles oriented between θ and $d\theta$, i.e., within solid angle $d\Omega = 2\pi\sin\theta\, d\theta$ is proportional to $e^{-\frac{U}{kT}}\, d\Omega = e^{(pE\cos\theta/kT)}\, 2\pi\sin\theta\, d\theta$, where k is the *Boltzmann constant* and T is the absolute temperature of the material. The contribution of the above number of dipoles to the total polarization is given by $2\pi p e^{(pE\cos\theta/kT)}\cos\theta\sin\theta\, d\theta$, where $p\cos\theta$ is the component of the dipole moment along $\vec{E}$.

If there are N number of such dipoles in the material, then the average contribution to polarization is

$$\langle p\rangle = \frac{\text{Total polarization due to } N \text{ dipoles}}{\text{Total number of dipoles } (N)} = p\frac{\int_0^\pi e^{(pE\cos\theta/kT)}\cos\theta\sin\theta\,d\theta}{\int_0^\pi e^{(pE\cos\theta/kT)}\sin\theta\,d\theta}. \tag{12.44}$$

Substituting $\frac{pE}{kT} = x$ and $\cos\theta = y$, $dy = -\sin\theta\,d\theta$.

$$\therefore \langle p\rangle = \frac{p\int_{+1}^{-1} y\,e^{xy}\,dy}{\int_{+1}^{-1} e^{xy}\,dy} \quad \text{or,} \quad \frac{\langle p\rangle}{p} = \frac{e^x + e^{-x}}{e^x - e^{-x}} - \frac{1}{x} = \left(\coth x - \frac{1}{x}\right) \tag{12.45}$$

$$\text{or,} \quad \frac{\langle p\rangle}{p} = L(x), \tag{12.46}$$

where $L(x) = \coth x - \frac{1}{x}$ is the *Langevin function.*

The variation of $L(x)$ with x is shown in Fig. 12.9.

If E is not too large and T is not too small, i.e., $(pE/kT) = x \ll 1$, then

$$L(x) = \frac{x}{3} = \frac{pE}{3kT}. \tag{12.47}$$

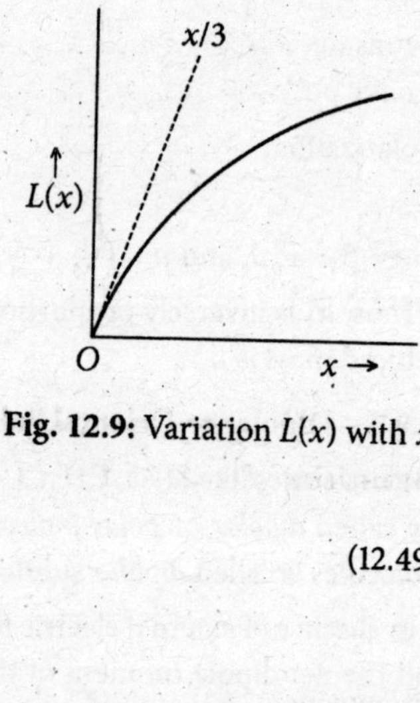

Fig. 12.9: Variation $L(x)$ with x

Then from eqn (12.46),

$$\langle p\rangle = p\left(\frac{pE}{3kT}\right) = \frac{p^2E}{3kT}. \tag{12.48}$$

The dipolar polarization is

$$P = N\langle p\rangle = \frac{Np^2E}{3kT}. \tag{12.49}$$

$\therefore$ the dipolar (or oriental) polarizability is

$$\alpha_d = \frac{\langle p\rangle}{E} = \frac{p^2}{3kT}. \tag{12.50}$$

Thus, the *dipolar polarizability is inversely proportional to the absolute temperature of the dielectric.*

12.9.4 Space Charge Polarizability

The polarizability arising due to the accumulation of charges at the electrodes or at the interfaces in a multiphase material is called *space charge polarizability* (α_s). In presence of an applied electric field the ions diffuse over appreciable distances giving rise to redistribution of charges in the dielectric medium.

The space charge polarization is the slowest process as it involves the diffusion of ions over several interatomic distances.

12.10 Langevin-Debye Equation

The total polarizability of a dielectric is given by

$$\alpha = \alpha_e + \alpha_i + \alpha_d + \alpha_s = \alpha_e + \alpha_i + \alpha_d, \tag{12.51}$$

[since α_s is very small as compared to other polarizabilities]

α decreases with increase in frequency.

$$\text{or, } \alpha = \frac{e^2/m}{\omega_0^2 - \omega^2} + \frac{e^2}{\omega_0^2}\left(\frac{1}{M} + \frac{1}{m}\right) + \frac{p^2}{3kT} \text{ [using eqns (12.37), (12.41) and (12.50)].}$$

Hence, *total polarization* is given by

$$P = N\alpha E = NE\left[\frac{e^2/m}{\omega_0^2 - \omega^2} + \frac{e^2}{\omega_0^2}\left(\frac{1}{M} + \frac{1}{m}\right) + \frac{p^2}{3kT}\right]. \tag{12.52}$$

Eqn (12.52) is called *Langevin-Debye equation.*

12.11 Clausius-Mossotti Relation

We consider a piece of dielectric of cubic structure. As in a dielectric there is no ion and permanent dipole, the ionic polarizability (α_i) and oriental polarizability (α_d) are zero, i.e., $\alpha_i = \alpha_d = 0$. $\therefore \alpha = \alpha_e$.

Polarization,

$$P = N\alpha_e E_{\text{loc}} = N\alpha_e\left(E + \frac{P}{3\varepsilon_0}\right),$$

[where N is the number of molecules per unit volume]

$$\text{or, } P\left[1 - \frac{N\alpha_e}{3\varepsilon_0}\right] = N\alpha_e E \text{ or, } P = N\alpha_e E \Big/ \left[1 - \frac{N\alpha_e}{3\varepsilon_0}\right]. \tag{12.53}$$

Again, since $D = P + \varepsilon_0 E$,

$$\therefore P = D - \varepsilon_0 E \text{ or, } \frac{P}{E} = \frac{D}{E} - \varepsilon_0 = \varepsilon - \varepsilon_0 = \varepsilon_0\varepsilon_r - \varepsilon_0$$

$$\text{or, } P = E\varepsilon_0(\varepsilon_r - 1). \tag{12.54}$$

From eqns (12.53) and (12.54), we get

$$E\varepsilon_0(\varepsilon_r - 1) = N\alpha_e E \Big/ \left(1 - \frac{N\alpha_e}{3\varepsilon_0}\right) \text{ or, } 1 - \frac{N\alpha_e}{3\varepsilon_0} = \frac{N\alpha_e}{\varepsilon_0(\varepsilon_r - 1)}$$

$$\text{or, } 1 = \frac{N\alpha_e}{3\varepsilon_0}\left[1 + \frac{3}{\varepsilon_r - 1}\right] \text{ or, } \frac{N\alpha_e}{3\varepsilon_0} = \frac{1}{1 + \frac{3}{\varepsilon_r - 1}} = \frac{\varepsilon_r - 1}{\varepsilon_r + 2} \text{ or, } \frac{\varepsilon_r - 1}{\varepsilon_r + 2} = \frac{N\alpha_e}{3\varepsilon_0}. \tag{12.55}$$

Eqn (12.55) is known as the *Clausius-Mossotti relation.* It relates dielectric constant to atomic polarizability provided the condition of cubic symmetry holds good. Measuring ε_r we can find out α_e.

In a more general form, it is given by

$$\frac{\varepsilon_r - 1}{\varepsilon_r + 2} = \frac{1}{3\varepsilon_0}\sum_j N_j\alpha_j. \tag{12.56}$$

12.11.1 Debye Equation

If M_A and ρ be the molecular weight and density of the dielectric respectively, then molar volume is M_A/ρ.

$\therefore$ the molar polarization is

$$\left(\frac{\varepsilon_r-1}{\varepsilon_r+2}\right)\frac{M_A}{\rho}=\frac{N\alpha}{3\varepsilon_0}\left(\frac{M_A}{\rho}\right)$$

$$\text{or, }\left(\frac{\varepsilon_r-1}{\varepsilon_r+2}\right)\frac{M_A}{\rho}=\frac{N_A}{3\varepsilon_0}(\alpha_e+\alpha_i+\alpha_d)\quad\left[\because\frac{NM_A}{\rho}=N_A\right]$$

$$\text{or, }\left(\frac{\varepsilon_r-1}{\varepsilon_r+2}\right)\frac{M_A}{\rho}=\frac{N_A}{3\varepsilon_0}(\alpha'+\alpha_d)\ [\alpha'=\alpha_e+\alpha_i].\tag{12.57}$$

Eqn (12.57) is known as the *Debye Equation*.

Eqn (12.57) can be written as

$$\left(\frac{\varepsilon_r-1}{\varepsilon_r+2}\right)\frac{M_A}{\rho}=\frac{N_A}{3\varepsilon_0}\alpha'+\frac{N_Ap^2}{9\varepsilon_0kT}=a+\frac{m}{T},\tag{12.58}$$

where $a=\frac{N_A}{3\varepsilon_0}\alpha'$ and $m=\frac{N_Ap^2}{9\varepsilon_0k}$.

Now the plot of $\left(\frac{\varepsilon_r-1}{\varepsilon_r+2}\right)\frac{M_A}{\rho}$ against $\frac{1}{T}$ will be a straight line of slope m as shown in Fig. 12.10.

Knowing the slope m from the graph we can determine the dipole moment using the equation

$$p=3\left[\sqrt{\frac{mk\varepsilon_0}{N_A}}\right].\left[\because m=\frac{N_Ap^2}{9\varepsilon_0k}\right]\tag{12.59}$$

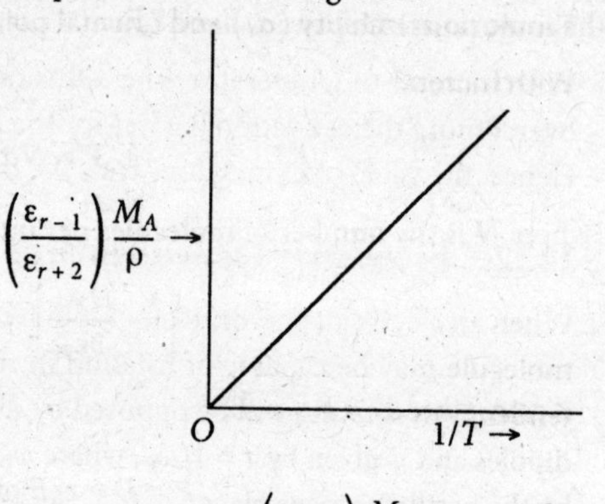

Fig. 12.10: Plot of $\left(\frac{\varepsilon_{r-1}}{\varepsilon_{r+2}}\right)\frac{M_A}{\rho}$ against $\frac{1}{T}$

12.11.2 Refractive Index

We have the velocity of light in vacuum, $c=\frac{1}{\sqrt{\mu_0\varepsilon_0}}$ and that in a dielectric medium, $v=\frac{1}{\sqrt{\mu\varepsilon}}$, where μ_0, ε_0 and μ, ε are respectively the permeability and permittivity of vacuum and the medium.

$$\therefore v=\frac{1}{\sqrt{\mu_r\mu_0\varepsilon_r\varepsilon_0}}.$$

But refractive index,

$$n=\frac{\text{Velocity of light in vacuum }(c)}{\text{Velocity of light in the medium }(v)}=\frac{1}{\sqrt{\mu_0\varepsilon_0}}\Big/\frac{1}{\sqrt{\mu_r\mu_0\varepsilon_r\varepsilon_0}}=\sqrt{\mu_r\varepsilon_r}=\sqrt{\varepsilon_r}\ [\text{as }\mu_r=1].$$

Now from eqn (12.55),

$$\frac{n^2-1}{n^2+2}=\frac{N\alpha_e}{3\varepsilon_0}.\tag{12.60}$$

Eqn (12.60) is known as the *Lorentz-Lorentz relation*

12.11.3 Effect of Temperature on Dielectric Constant

Eqns (12.40) and (12.42) show that the electronic polarizability (α_e) and ionic polarizability (α_i) are independent of temperature for normal temperature range. But oriental polarization and space charge polarization are affected by temperature (T).

Temperature Dependence of Oriental Polarizability

With increase of temperature (i.e., thermal energy) the permanent dipoles are randomly oriented and their tendency to orient along the applied field $(\vec{E})$ reduces. This causes a decrease in dielectric constant with the increase in temperature.

Putting $\alpha_d = \frac{p^2}{3kT}$ in equation,

$$\frac{\varepsilon_r - 1}{\varepsilon_r + 2} = \frac{N\alpha}{3\varepsilon_0}$$

in place of α we get

$$\frac{\varepsilon_r - 1}{\varepsilon_r + 2} = \frac{N\alpha_d}{3\varepsilon_0} = \frac{Np^2}{9\varepsilon_0 kT} = \frac{T_c}{T} \left[\text{where } T_c = \frac{Np^2}{9\varepsilon_0 k}\right]$$

$$\text{or, } \varepsilon_r (T - T_c) = 3T_c + (T - T_c)$$

$$\text{or, } \varepsilon_r = 1 + \frac{3T_c}{T - T_c}. \quad (12.61)$$

Eqn (12.61) shows that *spontaneous polarization* sets in at $T = T_c$ even for $\vec{E} = 0$. T_c is called the *critical temperature for the dielectric.*

Temperature Dependence of Space Charge Polarizability

With increase of temperature the diffusion probability of ions increases. The thermal energy also aids in overcoming the activation barrier for the orientation of the polar molecules along the applied field $(\vec{E})$. Hence, the value of ε_r increases with the increase of temperature.

12.12 Frequency Dependence of Polarization

When an electromagnetic field of radio frequency (ω) region be applied to a dipolar dielectric, the dipolar molecule may be capable of rotation in response to the oscillations of the alternating electric field. The tendency of rotating will be opposed by dissipative forces resulting a characteristic *relaxation time* for the dipoles and is given by $\tau = 1/\omega_0$, where ω_0 is the *resonant frequency* at which maximum power absorption by the oscillator takes place.

If the frequency (ω) of the applied field be very much smaller than resonant frequency, i.e., $\omega \ll \omega_0$, the dipoles being in phase with the field no energy is absorbed. For $\omega \gg \omega_0$, the dipoles cannot respond to the field and again there is no absorption of energy. The contribution of the dipoles to the permittivity is the maximum at low frequencies and minimum at high frequencies.

The time dependence of polarizability may be expressed as

$$P(t) = P_0 e^{-t/\tau}, \quad (12.62)$$

where τ is the *relaxation time.*

Now the frequency dependence of polarization is expressed as

$$P(\omega) = \int_0^\infty P_0 e^{-t/\tau} e^{-i\omega t} \, dt$$

$$\text{or, } P(\omega) = \frac{P_0}{\omega_0 + i\omega} \quad (12.63)$$

with $1/\tau = \omega_0$.

Since P depends directly on the permittivity ε, we write,

$$\varepsilon(\omega) = \frac{A}{\omega_0 + i\omega} + B = \frac{A}{\omega_0\left[1 + i\frac{\omega}{\omega_0}\right]} + B, \tag{12.64}$$

where A and B are constants.

The static permittivity, ε_s, corresponds to $\omega = 0$.

$$\therefore \varepsilon(0) = \frac{A}{\omega_0} + B. \tag{12.65}$$

At $\omega > \omega_0$, $\varepsilon = \varepsilon_\infty$.

$$\therefore \varepsilon_\infty = B. \tag{12.66}$$

From eqns (12.65) and (12.66),

$$\varepsilon_s = \frac{A}{\omega_0} + \varepsilon_\infty$$

$$\text{or, } A = \omega_0\left(\varepsilon_s - \varepsilon_\infty\right). \tag{12.67}$$

Substituting the values of A and B in eqn (12.65), we get

$$\varepsilon(\omega) = \frac{\varepsilon_s - \varepsilon_\infty}{1 + i\frac{\omega}{\omega_0}} + \varepsilon_\infty \text{ or, } \varepsilon(\omega) = \frac{\left(\varepsilon_s - \varepsilon_\infty\right)\left(1 - i\frac{\omega}{\omega_0}\right)}{\left[1 + \left(\frac{\omega}{\omega_0}\right)^2\right]} + \varepsilon_\infty$$

$$\text{or, } \varepsilon(\omega) = \varepsilon'(\omega) - i\varepsilon''(\omega), \tag{12.68}$$

where

$$\varepsilon'(\omega) = \varepsilon_\infty + \frac{\left(\varepsilon_s - \varepsilon_\infty\right)}{\left[1 + \left(\frac{\omega}{\omega_0}\right)^2\right]} \text{ and } \varepsilon''(\omega) = \frac{\left(\frac{\omega}{\omega_0}\right)\left(\varepsilon_s - \varepsilon_\infty\right)}{\left[1 + \left(\frac{\omega}{\omega_0}\right)^2\right]}. \tag{12.69}$$

$\varepsilon''(\omega)$ is the imaginary part of ε which is responsible for the absorption of energy. $\varepsilon''(\omega)$ becomes maximum at $\omega = \omega_0$ and becomes zero for $\omega > \omega_0$ and $\omega < \omega_0$. The real part of $\varepsilon(\omega)$, i.e., ε' falls for $\omega < \omega_0$ to the static dielectric constant ε_s. ε' decreases as ω_0 is approached and reaches half-way between ε_s and ε_∞ at $\omega = \omega_0$. The variations of ε' with $\ln\omega$ and ε'' with $\ln\omega$ are represented in Figs 12.11 and 12.12 respectively.

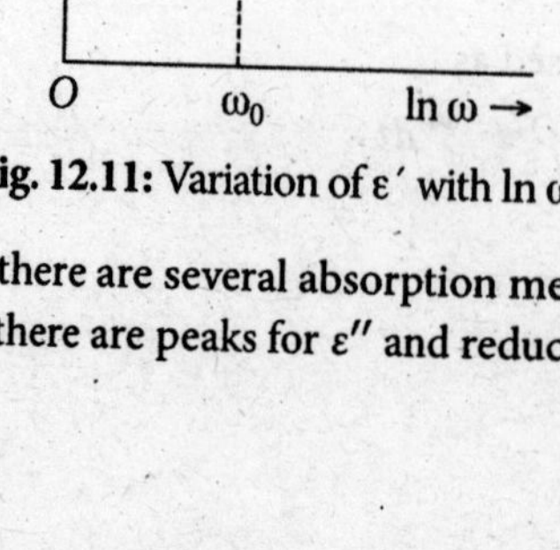

Fig. 12.11: Variation of ε' with $\ln\omega$

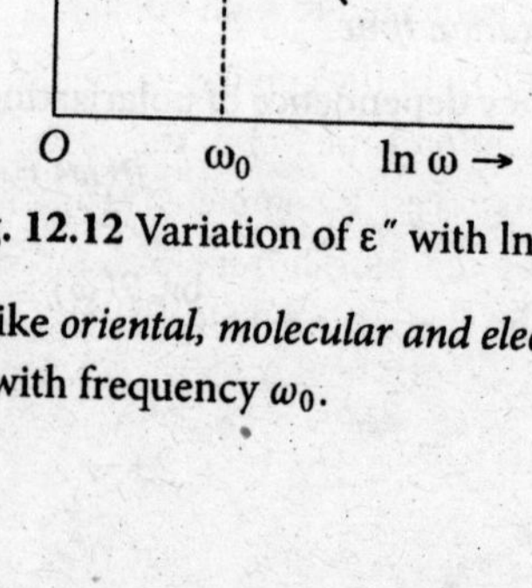

Fig. 12.12 Variation of ε'' with $\ln\omega$

In a dielectric, there are several absorption mechanism like *oriental, molecular and electronic processes.* For each process there are peaks for ε'' and reduction of ε' with frequency ω_0.

Table 12.1: Dielectric strength of some materials

Material	*Dielectric strength* (kV/mm)	*Permittivity*	tan δ
Bakelite	22	6	0.02
Cotton	4	5	0.02
Ebonite	25	2.8	0.01
Glass	8	6.5	0.004
Rubber	17	4	0.016
Dry paper	7	3	0.005
Paraffin wax	8	2.2	0.0003
Porcelain	15	6	0.002

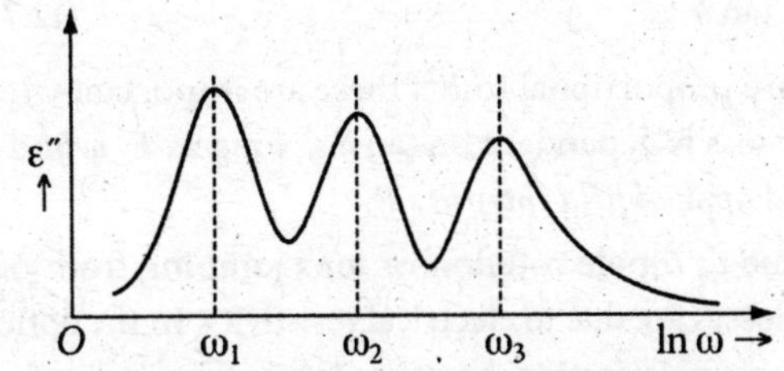

Fig. 12.13: Variation of ε″ with ln ω

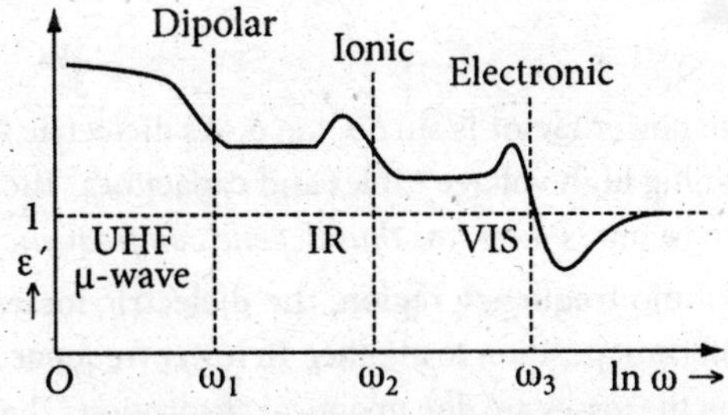

Fig. 12.14: Variation of ε′ with ln ω

The electronic absorption process occurs very quickly in presence of an electrical field and electronic polarization occurs at UV or VIS region of the electromagnetic spectrum. The ionic polarization occurs in IR region as the ionic motion is several thousand times slower than electronic motion. Oriental polarization gives rise to absorption peaks in UHF and microwave frequencies.

The variations of ε'' and ε' with $\ln \omega$ are depicted in Figs 12.13 and 12.14 respectively.

Dielectric Loss

If a dielectric is subjected to AC voltage, electrical energy is absorbed by it which is subsequently dissipated in the form of heat. This dissipation of absorbed energy is known as *dielectric loss* which plays an important role when the material is subjected to high voltage.

If the time period (T) of the applied voltage is much longer than the relaxation time (τ) of the polarization process, polarization process is completed at any instant during each cycle. The charging current leads the voltage by $\pi/2$ [Fig. 12.15] and there is no dielectric loss during charging.

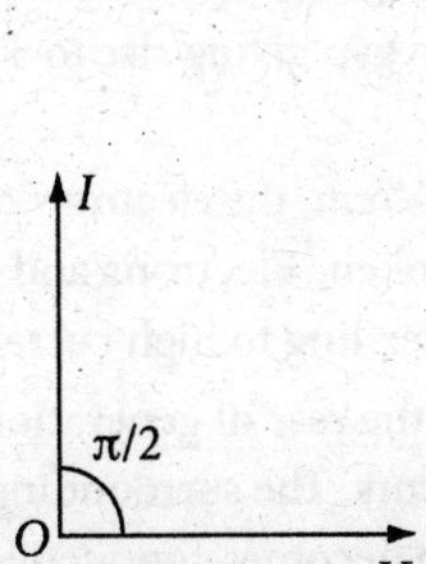

Fig. 12.15: *I* leads *V* by π/2

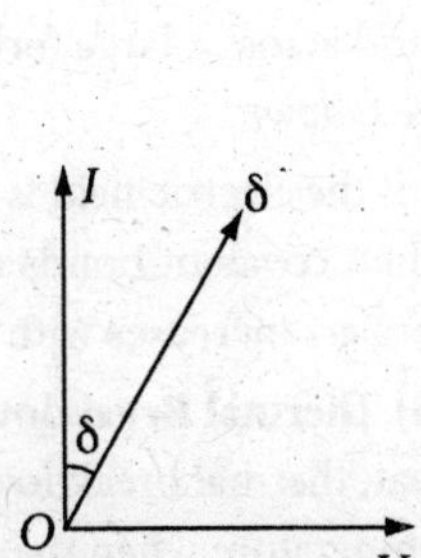

Fig. 12.16: *I* leads *V* by (π/2 – δ)

When the period (T) is equal to the relaxation time, current (I) resonance takes place and I leads V by $\left(\frac{\pi}{2} - \delta\right)$ as shown in Fig. 12.16. Here δ is called *loss angle* and t is called *loss tangent* as it gives a measure of electrical energy loss at resonance.

Considering energy absorption from electromagnetic waves by conducting medium we can show that,

$$k^2 = \mu\varepsilon\omega^2 + i\mu\sigma\omega,$$

where $\varepsilon = \varepsilon_0\varepsilon_r$ and σ is the conductivity. The phase of k^2 being identical,

$$\tan\delta = \frac{\sigma}{\varepsilon\omega} = \frac{\varepsilon''}{\varepsilon}. \tag{12.70}$$

Since ε is real,

$$\varepsilon'' = \frac{\sigma}{\omega}. \tag{12.71}$$

Now power loss per unit volume of the conducting medium is given by

$$W = \sigma E^2 = \varepsilon''\omega E^2 \ [\because \varepsilon'' = \sigma/\omega].$$

Again, $\varepsilon'' = \varepsilon' \tan\delta$.

$$\therefore W = \omega E^2 \varepsilon' \tan\delta. \tag{12.72}$$

The power factor is $\sin\delta \simeq \tan\delta$. As dielectric losses are proportional to E^2, these are important while designing high voltage cables and capacitors. The power loss is dependent on $\tan\delta$ as long as E, ω and ε' are constants. *Power loss thus depends on frequency ω and applied field intensity E.*

In radio frequency region, the dielectric losses are due to dipole rotation or ions jumping from one equilibrium position to another. In lower frequencies the losses are due to electrical resistivity. In the optical region the losses are due to optical absorption. The electronic polarization, however, is free from power loss. In the *infrared region*, the dielectric losses arise due to *ionic vibration.*

12.13 Dielectric Breakdown

Dielectric breakdown is a process in which a dielectric loses its resistivity and permits very large current to flow through it.

The mechanisms responsible for dielectric breakdown are as follows:

(i) Intrinsic Breakdown: Breakdown voltage per unit thickness is called *dielectric strength.* Under the action of high electric field, large number of valence electrons in the valence band jump to the conduction band across a large forbidden gap giving rise to a large current. This is called *intrinsic breakdown* or *Zener breakdown.*

If the electric field is ~ 10^6 V/cm, the electrons collide with high velocity with other atoms and molecules when covalent bonds are broken. Electrons and holes are created in this process and number of charge carriers increases with time leading to high current. This is *Avalanche Breakdown.*

(ii) Thermal Breakdown: If the rate of generation of heat in a dielectric is greater than the rate of loss of heat, thermal breakdown occurs. The surrounding of dissipation of heat suffers a local melting due to high temperature when the region becomes conductive and breakdown occurs.

(iii) Discharge Breakdown: Discharge breakdown is of the following two types, viz.,

(a) *External Breakdown*: It is caused due to *corona discharge. Moisture or dust on the dielectric surface* are also responsible for external breakdown.

(b) *Internal Breakdown*: It occurs when the dielectric contains *occluded gas bubbles.* The ionization energy of the gases being smaller than the solid the gas becomes ionised first. The gaseous atoms then bombard the solid dielectric releasing a large number of electrons and breakdown occurs even at low voltages.

(iv) Defect Breakdown: If the material has defects like porosity, crack, then impurities like dust or moisture are collected leading to breakdown.

(v) Electrochemical Breakdown: The mobility of ions in the material increases with the rise of temperature and electrochemical reaction sets in. With the increase of mobility of the ions, the *leak current* increases leading to breakdown.

12.14 Application of Dielectric Materials

Following are some of the important applications of dielectric materials:

(i) In electric iron: *Mica* or *asbestos* insulators are used to prevent the flow of current to the outer part of the iron.

(ii) In transformer: *Varnished cotton* is used as insulators in transformers, generators and motors.

(iii) In electric distribution lines: *Porcelain structures* are used as insulators in electric distribution lines.

(iv) In electric heater: In order to insulate the outer body and avoid short circuit *ceramic beads* are used in electric heater.

(v) In electric wiring: The copper or aluminium wires used in electric wiring are insulated from one another by *rubber* or *plastic coating.*

12.15 Ferroelectricity

The phenomenon in which spontaneous polarization (i.e., polarization in absence of an electric field) takes place is called *ferroelectricity.* It is analogous to ferromagnetism in which spontaneous magnetization occurs in the materials. The crystals such as barium titanate ($BaTiO_3$), Rochelle salt ($NaKC_4H_4O_6, 4H_2O$) which exhibit ferroelectricity are called *ferroelectric crystals.* The plot of polarization versus electric field exhibits a hysteresis loop.

The essential properties of the ferroelectric crystals are:

(i) The centres of positive and negative charges of the crystal do not coincide with each other even in absence of electric field and as such there is a non-zero value of dipole moment.

(ii) The polarization does not vary linearly with applied electric field, instead forms a *hysteresis loop.*

(iii) Above a certain critical temperature (T_c) the ferroelectricity disappears. T_c is called the *transition temperature* or *Curie point.*

(iv) They have a domain structure visible in polarized light.

(v) They show *piezoelectric* and *pyroelectric properties* below T_c.

(vi) The latent heat phenomenon or the specific heat anomaly is responsible for ferroelectric transition at T_c.

Ferroelectric crystals are divided into *two* main groups, viz., (a) *order-disorder group* (b) *displacive group.*

The order-disorder class crystals are potassium hydrogen phosphate (KH_2PO_4), rubidium hydrogen phosphate (RbH_2PO_4), etc. which contain hydrogen bond. The motion of proton in these crystals gives rise to ferroelectricity. In these crystals ferroelectric transition is related to individual ordering of ions.

In *displacive group,* ferroelectricity arises due to the displacement of whole sub-lattice of ions of one type relative to a lattice of another type. The structure of displacive group is related to *perovskite* and *ilmenite structures.* Examples of this class of crystals are $BaTiO_3$, $KNbO_3$, $LiTaO_3$, etc.

Table 12.2: Transition temperature of some Ferroelectric crystals

Materials	*Transition temperature* T_c *(in K)*
BaT_iO_3	408
KH_2AsO_4	97
GeTe	670
$LiNBO_3$	1480
Triglycine Sulphate	322

12.15.1 Structure of $BaTiO_3$ Crystal

$BaTiO_3$ crystal below the *Curie point* (380 K) shows *perovskite structure* as represented in Fig. 12.17 below.

The unit cell of $BaTiO_3$ crystal is cubic in which Ba^{2+} ions are at the corners, O^{2-} ions at the face-centres and Ti^{4+} ions at the body-centre of the unit cell. So each Ti^{4+} ion in an *octahedral configuration* is surrounded by six O^{2-} ions. At $T > T_c$, the centres of positive and negative charges exactly coincide producing no dipole moment. But at $T < T_c$, there is a net dipole moment. Because for $T < T_c$, the T_i^{4+} ions and Ba^{2+} ions move slightly in the upward direction, whereas O^{2-} ions move slightly in the downward direction. The structure of the crystal becomes tetrahedral with a relative displacement between the centres of the positive and negative charges producing *spontaneous polarization.*

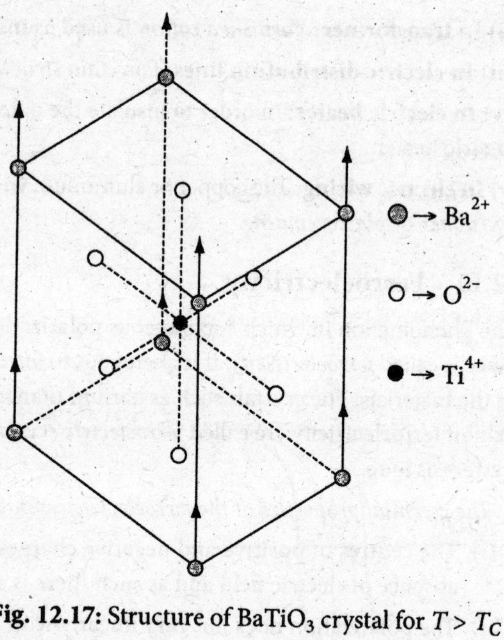

Fig. 12.17: Structure of $BaTiO_3$ crystal for $T > T_C$

12.15.2 Theory of Ferroelectricity

In $BaTiO_3$ ferroelectricity arises due to *induced electronic* and *ionic dipole moments.*

Now the *Clausius-Mossotti relation* in general form is given by

$$\frac{\varepsilon_r - 1}{\varepsilon_r + 2} = \frac{1}{3\varepsilon_0} \sum_j N_j \alpha_j$$

$$\text{or, } \varepsilon_r = \frac{1 + \frac{2}{3\varepsilon_0} \sum_j N_j \alpha_j}{1 - \frac{1}{3\varepsilon_0} \sum_j N_j \alpha_j}, \tag{12.73}$$

where N_j is the number of jth ions per unit volume and α_j electronic plus ionic polarizability of the jth ion.

If

$$\sum_j N_j \alpha_j = 3\varepsilon_0, \tag{12.74}$$

then from eqn (12.73), ε_r becomes infinity and a finite polarization is possible even in absence of applied field $\vec{E}$. This represents the *polarization catastrophe.* In polarization catastrophe, the local electric field due to polarization increases faster than the restoring force on the ion. This causes an asymmetric shift in the position of ions limited to finite displacements.

Putting

$$\frac{1}{3\varepsilon_0}\sum_j N_j \alpha_j = 1 - 3s, \tag{12.75}$$

(where $s \ll 1$) in eqn (12.73), we get

$$\varepsilon_r = \frac{1 + 2(1-3s)}{1 - 1 + 3s} = \frac{1}{s} - 2 \simeq \frac{1}{s}. \tag{12.76}$$

If s varies linearly near T_c, then

$$s \simeq \frac{T - T_c}{\xi}, \tag{12.77}$$

where ξ is a constant.

$\therefore$ from eqn (12.76), we get

$$\varepsilon_r \simeq \frac{\xi}{T - T_c}. \tag{12.78}$$

This type of variation of ε_r with T is experimentally observed in paraelectric state of the crystal.

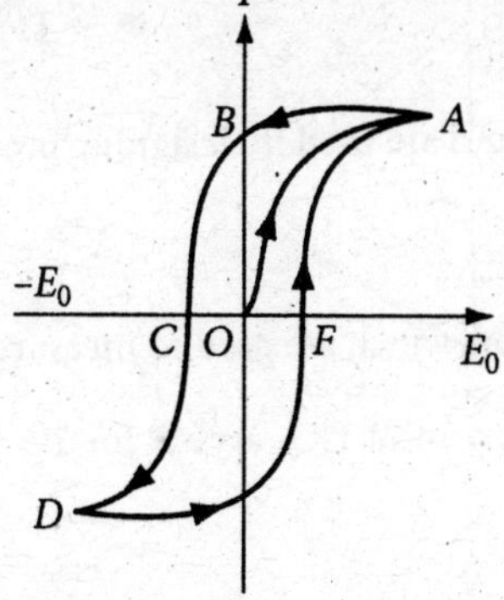

Fig. 12.18 The ferroelectric hysteresis loop

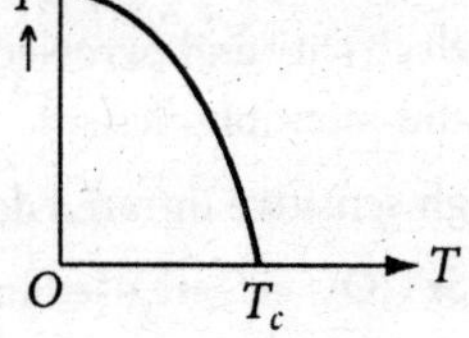

Fig. 12.19 Polarization as a function of temperature

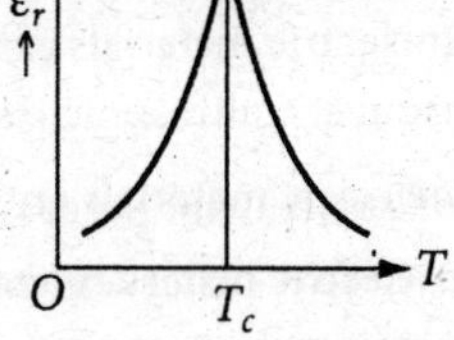

Fig. 12.20 Variation of ε_r with T

Behaviour of Ferroelectric Crystals

Intrinsic ferroelectricity is due to the reversal of orientation of the polarization direction by an electric field. Due to this hysteresis occurs like magnetic hysteresis [Fig. 12.18].

For $T > T_c$, the spontaneous polarization is destroyed by thermal disorder in the system [Fig. 12.19].

For $T = T_c$, the permittivity shows a peak as shows in Fig. 12.20.

12.15.3 Ferroelectric Domains

The term *pseudo-symmetry* is referred to the fact that, the free energy of the *polar* and *non-polar* modification of a dielectric is exactly equal at the *Curie temperature.* Twins are formed due to *pseudo-symmetry* in dielectrics when they involve changes in shape of their unit cell. Each twin has a specific direction of polarization. For this reason a ferroelectric crystal splits up into number of *domains* having specific direction (i.e., the crystallographic directions) of polarization. The dielectric behaviour of a ferroelectric crystal is *hysteric,* since an external electric field E_0 can reverse the permanent polarization in the material [Fig. 12.18].

The plot of polarization P against the applied electric field E_0 (the hysteresis curve) can be explained on the basis of domain theory [Fig. 12.21].

In a particular domain the atomic dipoles are aligned parallel to one another. When an electric field E_0 is applied the favourably oriented domains grow at the expense of the other. At a high field, P attains the maximum value. P, however, follows another path (AB) when E_0 is decreased. At B a certain polarization (OB) is retained even when E_0 is reduced to zero. OB is called the *remanent polarization.* The negative field (OC) necessary to destroy the remanent polarization is called the *coercive force* ($-E_c$). If E_0 is increased further in the reverse direction, again saturation polarization occurs but in the opposite direction. The complete hysteresis curve is obtained on reversing the field E_0.

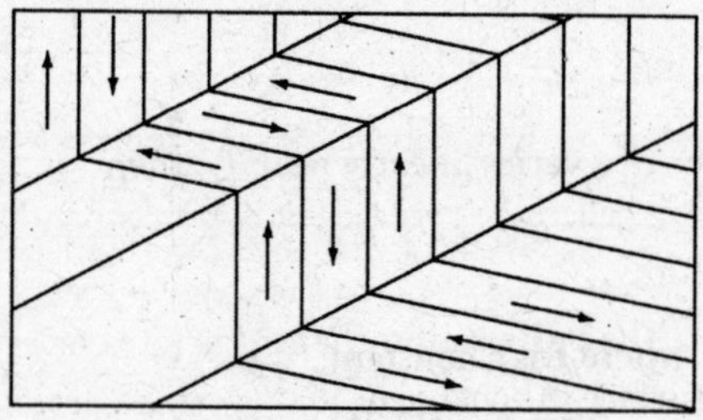

Fig. 12.21: Ferroelectric domains

The ferroelectric domains in the crystal are in the form of very thin elongated platelets with sharp domain walls. Thus, the hysteresis curve establishes the domain structure of ferroelectric crystals.

12.15.4 Applications of Ferroelectric Materials

Ferroelectric materials have important applications as below:

(i) Ferroelectric materials exhibit piezoelectricity and pyroelectricity and are used in designing pressure transducers, ultrasonic transducers and microphones.

(ii) Pyroelectric materials are used in high sensitive infrared detectors.

(iii) Ferroelectric materials like BaT_iO_3, SrT_iO_3, etc. are used in posistors which are used to measure and control temperature.

(iv) Ferroelectric ceramics are used in the manufacture of capacitors.

12.16 Antiferroelectricity

There is a group of solids which have induced ordered dipoles below a characteristic temperature. They do not exhibit spontaneous bulk polarization as each dipole is anti-parallel to its neighbouring dipole. Such materials are called *antiferroelectric.*

Example: Lead zirconite ($PbZrO_3$), tungsten trioxide (WO_3), etc.

The essential difference between the electrical behaviour of an antiferroelectric and a ferroelectric crystals is that the former does not show *hysteresis* but the latter does. The dielectric constant of an antiferroelectric crystal increases with decreasing temperature above the transition temperature.

Due to large dielectric constant the antiferroelectric materials may be used in capacitors.

Materials	*Transition temperature T*(K)
$PbZrO_3$	506
$PbHfO_3$	488
$NH_4H_2PO_4$	148
$NH_4H_2AsO_4$	216
WO_3	1010

12.17 Pyroelectricity

The change in spontaneous polarization with the change in temperature of the materials is known as the pyroelectric effect. If dP be the change in spontaneous polarization for a change of temperature dT, then

$$dP \propto dT$$

$$\text{or, } \lambda = \frac{dP}{dT}. \tag{12.79}$$

λ is called the *pyroelectric coefficient* and is defined as the *change in polarization per unit change of temperature.*

The polarization changes by the change of external electric field and hence, the change of charge on the surface. As charge of 10^{-6} C can be detected by a sensitive electrometer, a change of temperature 10^{-6} K can be measured using the principle of pyroelectric effect.

Example: Rochelle salt ($NaKC_4H_4O_6, 4H_2O$), Potassium hydrogen phosphate (KH_2PO_4), Ammonium hydrate ($NH_4H_2O_4$), α-Quartz (SiO_2), etc.

Applications of Pyroelectric Materials

(i) Pyroelectric materials like BaT_iO_3, $LiNibO_3$, etc. are used in *infrared detectors.*

(ii) Materials like TGS, $NaNO_3$ and PZT ceramics are used in *pyroelectric image tubes.*

12.18 Piezoelectricity

The phenomenon in which certain crystals become electrically polarized under the action of an external stress with the polarization charges appearing on the surfaces of the crystal is called piezoelectricity. The property of piezoelectricity has an inverse. In inverse piezoelectric effect the crystal becomes strained under the action of an electric stress (voltage), the strain being directly proportional to the applied field E.

The crystals exhibiting piezoelectric effect are called *piezoelectric crystals. Rochelle salt* and *tourmaline* are the prominent examples of piezoelectric crystals.

The development of polarization (P) by an applied stress (Z) is given by

$$P = Zd + \varepsilon_0 E\chi, \tag{12.80}$$

where d is the *piezoelectric strain constant,* E the electric field, χ the *electric susceptibility.*

The development of elastic strain (e) by an applied electric field (E) is given by

$$e = Zs + Ed, \tag{12.81}$$

where s is the *elastic compliance constant.* A large electric field ~ 10^4 V/m can produce a strain of about 1 in 10^8 only.

Origin of Piezoelectric Effect

The displacement of ions in certain crystals under the action of external stress is the origin of piezoelectric effect. When such crystals are subjected to stress, the charges are displaced and the original charge distribution is distorted in such a way that it is no longer symmetric. This is shown in Fig. 12.22 for a quartz crystal. Under the compressional stress the angle θ increases but the distance between Si_3 and O_2 remains the same. Comparing the new arrangement with the old one in Fig. 12.23, we note that positive charges have a net displacement w.r.t. the negative charges producing a net dipole moment and hence, piezoelectric effect.

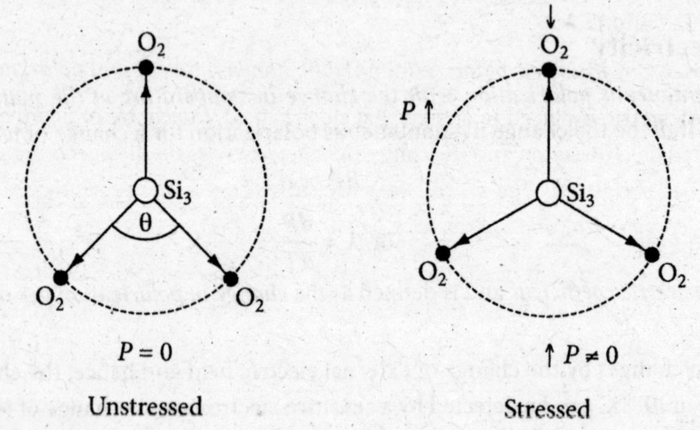

Fig. 12.22: Quartz crystals having no centre of inversion show piezoelectricity.

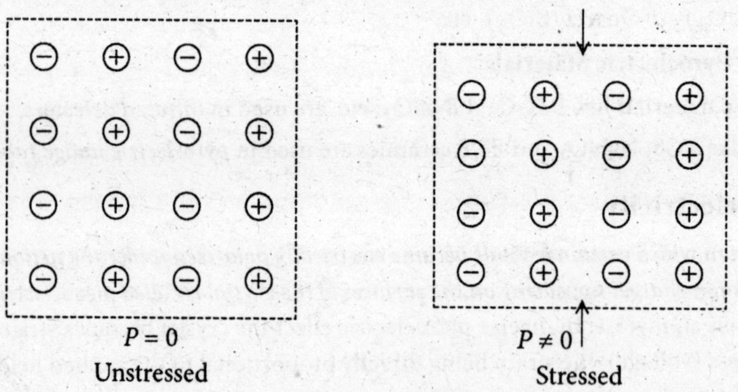

Fig. 12.23: A crystal with centre of inversion does not show piezoelectricity.

But the crystals having centre of inversion do not produce dipole moment and hence, are piezoelectric under stress [Fig. 12.23]. So, the absence of centre of inversion is the essential requirement for piezoelectricity. It is to be remembered that all ferroelectric crystals are piezoelectric but the converse is not true.

Applications of Piezoelectric Crystals

Piezoelectric crystals have the following important applications:

(i) These crystals like *Rochelle salt* are used to convert electrical energy into mechanical energy and vice versa, i.e., these are used as *mechanical transducers*.

(ii) Quartz crystals are used in filter, resonator and delay line applications.

(iii) Barium titanate, lead zirconate, etc. commercial ceramic materials are used in high voltage generation, accelerometers, etc.

(iv) Piezoelectric semiconductors like GaS, ZnO, etc. are applied as amplifiers of ultrasonic waves.

(v) The PVF_2 (polyvinylidene fluoride) are used in monitoring *blood pressure* and *respiration.*

12.19 Answers of Some Important Questions

12.1. *What are dielectrics? What are polar and non-polar dielectrics?*

Ans. See Arts 12.1 and 12.2

12.2. *What do you mean by dielectric polarization? Define dipole moment and polarizability.*

Ans. When a dielectric is placed in an external electric field, the centre of positive charge is displaced along the applied field direction and the centre of negative charge is displaced in the opposite direction. Thus, a dipole is produced. *The process of producing electric dipoles which are oriented along field direction is called polarization in dielectrics.*

The product of one of the charges and their separating distance is called the *dipole moment.* If the centres of $+q$ and $-q$ charges be separated by a distance, then the dipole moment is $\vec{p} = q\vec{d}$.

The dipole moment is a vector quantity and is directed from the negative to positive charge.

If the strength of the applied electric field be increased, then the dipole moment also increases. The induced dipole moment is proportional to the applied field $\vec{E}$, i.e., $\vec{p} \propto \vec{E}$ or, $\vec{p} = \alpha\vec{E}$, where α is a constant of proportionality and is known as the *polarizability* of the material. It is defined as *the induced dipole moment per unit applied electric field.*

12.3. *State the different processes of polarization.*

Ans. In presence of DC electric field $\vec{E}$, a dielectric is polarized by the following four processes:

(i) Electronic polarization
(ii) Ionic polarization
(iii) Dipolar or oriental polarization
(iv) Space charge polarization.

The contribution of a particular process towards polarization depends on the nature of the dielectric and the frequency of the applied field if the applied field be an AC field.

12.4. *What do you mean by electronic polarizability? Write down an expression for the electronic polarizability?*

Ans. When a dielectric is placed in an electric field, the positively charged nucleus and the electrons of an atom are displaced in opposite directions producing electronic polarization. If $\vec{p}_e$ be the dipole moment produced in each atom and $\vec{E}$, the applied electric field, then $\vec{p}_e = \alpha_e \vec{E}$, where α_e *is called the electronic polarizability.* $\therefore \alpha_e = \vec{p}_e / \vec{E}$.

In other words, electronic polarizability is defined as the dipole moment produced in each atom per unit applied field.

12.5. *What is ionic polarizability? On what factors does it depend?*

Ans. If an electric field be applied to an ionic solid the anions and cations get displaced in opposite directions. This phenomenon is called ionic polarization. *The ionic polarizability* (α_i) *is defined as* *the resultant dipole moment* (p) *per unit applied electric field*, i.e.,

$$\alpha_i = \frac{p}{E} = \frac{e^2}{\omega_0^2}\left(\frac{1}{M} + \frac{1}{m}\right),$$

where m, M are the masses of the positive and negative ions respectively.

Thus, ionic polarizability α_i depends:

(i) inversely on the square of the natural frequency (ω_0) of the ionic molecule and

(ii) on the reduced mass $\left(\frac{1}{M} + \frac{1}{m}\right)^{-1}$ of the system.

12.6. *Find a relation between electronic polarizability and relative permittivity.*

Ans. If N be the number of atoms per unit volume of the dielectric, then electronic polarization is given by $P_e = Np_e = N\alpha_e E$.

Again, $P_e = \varepsilon_0 E(\varepsilon_r - 1)$.

$$\therefore N\alpha_e E = \varepsilon_0 E(\varepsilon_r - 1) \text{ or, } \alpha_e = \frac{\varepsilon_0(\varepsilon_r - 1)}{N}.$$

This is the relation between the electronic polarizability and relative permittivity of the material.

12.7. *What is oriental polarizability? How does it depend on the temperature of the dielectric?*

Ans. When an electric field is applied to polar molecules (molecules having permanent dipole moment) they tend to align themselves in the direction of applied field. The polarization due to such alignment of the polar molecules along the applied field is called oriental or dipolar polarization. Oriental polarization is given by

$$P_d = N\vec{p} = Np^2E/3kT = N\alpha_d E \text{ or, } \alpha_d = \frac{P_d}{NE} = \frac{p^2}{3kT}.$$

Thus, *oriental polarizability is defined as the polarization per unit volume per unit applied field.* α_d is inversely proportional to temperature of the material.

12.8. *What is local field? Write down its expression.*

Ans. When a dielectric is placed in an external field $\vec{E}_0$, dielectric polarization takes place and a polarizing field $\vec{E}_p$ is established.

The net electric field at the site of an atom or molecule is significantly different from $\vec{E}_0$ and is called the *local field*. The local field of a dielectric placed in an external electric field ($\vec{E}_0$) is given by

$$\vec{E}_{loc} = \vec{E}_0 + \vec{E}_p + \frac{\vec{P}}{3\varepsilon_0} = \vec{E} + \frac{P}{3\varepsilon_0}.$$

This field is responsible for the polarization of each atom or molecule of solid having cubic symmetry. So $\vec{E}_{loc}$ apart from macroscopic field $\vec{E}$ contains field due to polarization of the remaining molecules. $\vec{E}_{loc}$ depends on the shape of the material.

12.9. *Find a relation between relative permittivity and susceptibility of a dielectric.*

Ans. The electric displacement vector $\vec{D}$ of an isotropic medium is given by $\vec{D} = \varepsilon_0\varepsilon_r\vec{E} = \varepsilon_0\vec{E} + \vec{P}$, where $\varepsilon_r = \varepsilon/\varepsilon_0$ is the *relative permittivity* or *dielectric constant.*

$$\therefore \varepsilon_r = \frac{\varepsilon_0\vec{E} + \vec{P}}{\varepsilon_0\vec{E}} = 1 + \frac{\vec{P}}{\varepsilon_0\vec{E}}.$$

But $\frac{\vec{P}}{\varepsilon_0 \vec{E}}$ is the susceptibility (χ_e) of the dielectric. $\therefore \varepsilon_r = 1 + \chi_e$.

Thus, larger the value of $\vec{P}/\vec{E}$, greater is the value of ε_r, which gives a measure of polarization of the material.

12.10. *What do you understand by static dielectric constant of gases?*

Ans. The molecules of a gas are widely separated from one another and hence, the interaction between them is negligible. The field experienced by a dipole is, therefore, the external applied field $(\vec{E}_0)$. The total polarization $\vec{P}$ may, therefore, be expressed as $\vec{P} = \vec{P}_e + \vec{P}_i + \vec{P}_d = N(\alpha_e + \alpha_i + \alpha_d)\vec{E}_0$, where α_e, α_i and α_d are the electronic, ionic, and dipolar polarizabilities respectively and N is the number of molecules per unit volume.

$$\therefore \alpha = \alpha_e + \alpha_i + \alpha_d = \alpha_e + \alpha_i + \frac{p^2}{3kT}.$$

This is called the *Langevin-Debye equation.*

Again, the polarization $\vec{p}$ is given by

$$\vec{p} = \varepsilon_0(\varepsilon_r - 1)\vec{E}_0.$$

$$\therefore \varepsilon_0(\varepsilon_r - 1) = N\left(\alpha_e + \alpha_i + \frac{p^2}{3kT}\right).$$

This shows that $\varepsilon_0(\varepsilon_r - 1)$ depends on temperature T of the gas. The plot of $\varepsilon_0(\varepsilon_r - 1)$ against $\frac{1}{T}$ is shown in the Fig. 12.24 which is a straight line of slope $Np^2/3k$. Thus, finding slope of the curve, p can be determined and hence, the molecular structure can be known.

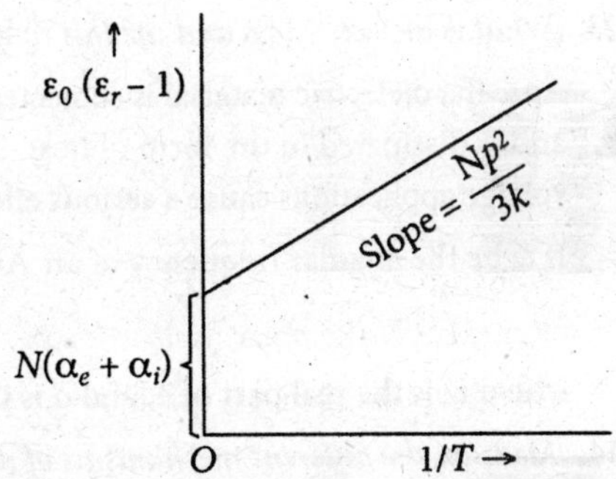

Fig. 12.24: Variation of $\varepsilon_0(\varepsilon_r - 1)$ against $1/T$

12.11. *Starting from Clausius-Mossotti equation explain the origin of spontaneous polarization.*

Ans. If ε_r be the relative permeability and α, the polarizability of a dielectric material, then *Clausius-Mossotti relation* is given by

$$\frac{\varepsilon_r - 1}{\varepsilon_r + 2} = \frac{N\alpha}{3\varepsilon_0},$$

where N is the number of atoms per unit volume and $\alpha = \alpha_e + \alpha_i + \alpha_d + \alpha_s$.

Putting $\alpha_d = \frac{p^2}{3kT}$ and neglecting the contributions of α_e, α_i and α_s, we get

$$\frac{\varepsilon_r - 1}{\varepsilon_r + 2} = \frac{Np^2}{9\varepsilon_0 kT} = \frac{T_c}{T},$$

where $$T_c = \frac{Np^2}{9\varepsilon_0 k} \text{ or, } \varepsilon T - T = \varepsilon T_c + 2T_c \text{ or, } \varepsilon(T - T_c) = 3T_c + (T - T_c)$$

$$\text{or, } \varepsilon = 1 + \frac{3T_c}{T - T_c}.$$

From the above equation we note that $\varepsilon \to \infty$ for $T = T_c$, even for $E = 0$. T_c is called the *critical temperature.* This phenomenon is called *spontaneous polarization.*

Thus, *the phenomenon of dielectric polarization even for zero applied field is called spontaneous polarization.* Spontaneous polarization is the cause of *ferroelectricity.*

12.12. *What is polarization catastrophe?*

Ans. From *Clausius-Mossotti relation,*

$$\frac{\varepsilon_r - 1}{\varepsilon_r + 2} = \frac{1}{3\varepsilon_0} \sum_j N_j \alpha_j,$$

where α_j and N_j are the polarizability and number of ions of j type per unit volume respectively.

$$\therefore \varepsilon_r = \frac{1 + \frac{2}{3\varepsilon_0} \sum_j N_j \alpha_j}{1 - \frac{1}{3\varepsilon_0} \sum_j N_j \alpha_j}.$$

For $\Sigma N_j \alpha_j = 3\varepsilon_0$, $\varepsilon_r \to \infty$, even for $E = 0$.

The condition expressed by $\sum_j N_j \alpha_j = 3\varepsilon_0$ is called *polarization catastrophe* and the phenomenon is called the *spontaneous polarization.*

12.13. *What is dielectric loss and the loss tangent?*

Ans. If a dielectric material is subjected to an AC voltage, electric energy is absorbed by the material and is dissipated in the form of heat. This dissipation of energy is called the *dielectric loss.* The high voltage applications cause a serious effect as it is associated with heat generation and heat dissipation. If ω be the angular frequency of an AC voltage E, then dielectric loss per unit volume is

$$W = \omega E^2 \varepsilon' \tan \delta,$$

where ε' is the real part of ε, and δ is the *loss angle.* The factor $\tan \delta$ is called the *loss tangent.*

12.14. *Mention the different mechanisms of dielectric breakdown.*

Ans. The different mechanisms for the dielectric breakdown are as follows:

(i) Intrinsic breakdown (ii) Thermal breakdown (iii) Discharge breakdown
(iv) Electrochemical breakdown (v) Defect breakdown.

12.15. *What are the essential qualities of a good insulating material?*

Ans. The essential qualities of a good insulator are as follows:

(A) **Electrical**

(i) The dielectric material should have high electrical resistivity and high dielectric strength to withstand high voltage.

(ii) The dielectric loss must be minimum.

(iii) Some liquid and gas dielectrics must have the property of cooling, e.g., hydrogen, helium and transformer oil.

(B) **Mechanical**

(i) Insulators used in electric machines must have sufficient mechanical strength to withstand vibration.

(ii) Certain mechanical properties of insulators are necessary depending on the field of applicability.

(C) **Chemical**

(i) The insulators should have the capacity to resist oil, gas fumes, acids and alkalis,

(ii) The insulators must be water resistant.

12.16. *What are the different classes of insulators?*

Ans. Insulators belong to the following three classes:

(A) *Solid insulators*

Examples: Mica, ceramics, asbestos, rubber and PVC materials.

(B) *Liquid insulators*

Examples: Mineral oil, synthetic oil, miscellaneous oils.

(C) *Gaseous Insulators*

Examples: Air, nitrogen, inert gases, sulphur hexafluoride, etc.

12.17. *What are ferroelectricity, piezoelectricity, pyroelectricity and antiferroelectricity?*

Ans. *Ferroelectricity*: This is a phenomenon in which dielectric polarization occurs even in absence of an applied electric field. The crystals having the property of ferroelectricity are called *ferroelectric crystals*, e.g., BaT_iO_3, $NaKC_4H_4O_6$, $4H_2O$. In ferroelectric crystals,

(i) the centres of positive and negative charges do not coincide with each other even in absence of the field thus creating a non-zero value of dipole moment.

(ii) the variation of polarization with electric field forms a hysteresis loop.

(iii) the ferroelectricity disappears above T_c, T_c is called the *transition temperature* or *Curie point*.

Piezoelectricity: Certain crystals when subjected to external stress induce a net dipole moment producing the electric polarization with the polarization charges appearing on the surface of the crystal. This phenomenon is called the *piezoelectricity*. Quartz, Rochelle salt and tourmaline are some of the examples of piezoelectric crystals.

The inverse effects, i.e., when an electric field is applied, strain is produced in the crystal and the strain is directly proportional to the applied field. Piezoelectric crystals like quartz have no centre of symmetry.

Pyroelectricity: The *pyroelectric effect* is defined as the change in polarization with temperature of the dielectric. The pyroelectric coefficient λ is defined as $\lambda = \frac{dP}{dT}$, where dP is the change in polarization due to a change in temperature dT. A change of temperature of 10^{-6} °C can be measured by suitable electrometer designed on the principle of pyroelectricity.

Antiferroelectricity: In certain substances the neighbouring lines of ions are displaced in opposite directions causing the spontaneous polarization to occur in opposite directions. The crystal does not show spontaneous polarization. This phenomenon is known as *antiferroelectricity*.

Ammonium salt at low temperatures and hydrogen halides are some of the examples of antiferroelectric materials.

12.18. *Compare Ferroelectricity with Piezoelectricity.*

Ans.

(i) In piezoelectricity the dielectric crystal gets polarized by external stress, whereas ferroelectricity arises due to spontaneous polarization in absence of electric field.

(ii) Both piezoelectricity and ferroelectricity occur in non-centrosymmetrical crystals.

(iii) All ferroelectrics are piezoelectrics but all piezoelectrics are not ferroelectrics. For example, tourmaline is piezoelectric but not ferroelectric.

(iv) The piezoelectric coefficient is defined as the ratio of charge developed to stress at crystallographic axis.

12.20 Solved Problems

12.1. *The relative permittivity and square of refractive index of a dielectric material are 4.94 and 2.69 respectively. Find the ratio between electronic and ionic polarizabilities of the material.*

Solution: We have the formula

$$\frac{\varepsilon_r - 1}{\varepsilon_r + 2} = \frac{N\alpha}{3\varepsilon_0},$$

where $\alpha = \alpha_e + \alpha_i + \alpha_d + \alpha_s \simeq \alpha_e + \alpha_i$ [$\because \alpha_d$ and α_s are negligible].

$$\therefore \frac{\varepsilon_r - 1}{\varepsilon_r + 2} = \frac{N(\alpha_e + \alpha_i)}{3\varepsilon_0} \text{ or, } \frac{4.94 - 1}{4.94 + 2} = \frac{N(\alpha_e + \alpha_i)}{3\varepsilon_0} \ [\because \varepsilon_r = 4.94]$$

$$\text{or, } \frac{N(\alpha_e + \alpha_i)}{3\varepsilon_0} = 0.5677.$$

Again, $\varepsilon_r = n^2$.

$$\therefore \frac{n^2 - 1}{n^2 + 2} = \frac{N\alpha_e}{3\varepsilon_0} \text{ or, } \frac{N\alpha_e}{3\varepsilon_0} = \frac{2.69 - 1}{2.69 + 2} = 0.36034 \ [\therefore n^2 = 2.69]$$

Hence,

$$\frac{N(\alpha_e + \alpha_i)}{N\alpha_e} = \frac{0.5677}{0.36034} \text{ or, } 1 + \frac{\alpha_i}{\alpha_e} = 1.575 \text{ or, } \frac{\alpha_i}{\alpha_e} = 0.575.$$

$$\therefore \alpha_e/\alpha_i = \mathbf{1.739}.$$

12.2. *The crystal of sodium chloride has static dielectric constant of 5.6 and optical index of refraction 1.5. Calculate the percentage contribution of ionic polarizability.* **[HPU 1995]**

Solution: We have from Clausius-Mossotti relation

$$\frac{\varepsilon_r - 1}{\varepsilon_r + 2} = \frac{N(\alpha_e + \alpha_i)}{3\varepsilon_0},$$

where α_e and α_i are the electronic and ionic polarizabilities respectively.

At optical frequencies, the above equation takes the form:

$$\frac{n^2 - 1}{n^2 + 1} = \frac{N\alpha_e}{3\varepsilon_0},$$

n is the refractive index of the crystal.

$$\therefore \left(\frac{n^2 - 1}{n^2 + 2}\right) \Big/ \left(\frac{\varepsilon_r - 1}{\varepsilon_r + 2}\right) = \frac{\alpha_e}{\alpha_e + \alpha_i}.$$

Hence, the percentage contribution of ionic polarizability is

$$\left(\frac{\alpha_i}{\alpha_e + \alpha_i}\right) \times 100 = \left(1 - \frac{\alpha_e}{\alpha_e + \alpha_i}\right) \times 100 = \left[1 - \left(\frac{n^2 - 1}{n^2 + 2}\right)\left(\frac{\varepsilon_r + 2}{\varepsilon_r - 1}\right)\right] \times 100$$

$$= \left[1 - \frac{(1.5)^2 - 1}{(1.5)^2 + 2} \times \frac{5.6 + 2}{5.6 - 1}\right] \times 100 = \mathbf{51.4\,\%}.$$

12.3. *Helium gas contains* 2.7×10^{25} atoms/m^3 *and dielectric constant of* He *atom at* NTD *is* 1.0000684. *Calculate the electronic polarizability of* He-*atom.*

Solution: We have electronic polarizability,

$$\alpha_e = \frac{\varepsilon_0 (\varepsilon_r - 1)}{N}.$$

Here $\varepsilon_0 = 8.854 \times 10^{-12}$ F/m, $\varepsilon_r = 1.0000684$, $N = 2.7 \times 10^{25}$ atoms/m^3.

$$\therefore \alpha_e = \frac{8.854 \times 10^{-12} \times (1.0000684 - 1)}{2.7 \times 10^{25}} = \frac{8.854 \times 10^{-12} \times 0.684 \times 10^{-4}}{2.7 \times 10^{25}}$$

$$= \mathbf{2.243 \times 10^{-41}\ Fm^2}.$$

12.4. *The polarizability of ammonium molecule is* 2.42×10^{-39} C^{2}m/N *and* 1.74×10^{-39} C^{2}m/N *at* 309 K *and* 448 K *respectively. Calculate the oriental polarizability for each of the mentioned temperatures.*

Solution: The total molecular polarizability of a polar molecule is given by

$$\alpha = \alpha_e + \alpha_i + \alpha_d = \alpha_e + \alpha_i + \frac{p^2}{3kT},$$

where α_e, α_i and α_d are the electronic, ionic and oriental polarizability respectively.

For $T = 309$ K,

$$2.42 \times 10^{-39} = \alpha_e + \alpha_i + \alpha_d = \alpha_e + \alpha_i + \frac{p_1^2}{3 \times 1.38 \times 10^{-38} \times 309}$$

and for $T = 448$ K,

$$1.74 \times 10^{-39} = \alpha_e + \alpha_i + \frac{p_2^2}{3 \times 1.38 \times 10^{-38} \times 448}.$$

$$\therefore \frac{2.42 \times 10^{-39} - (\alpha_e + \alpha_i)}{1.74 \times 10^{-39} - (\alpha_e + \alpha_i)} = \frac{448}{309}.$$

$$\therefore \alpha_e + \alpha_i = 0.228 \times 10^{-39}\ \text{C}^2\text{m/N}.$$

Now α_e at 309 K is = $2.42 \times 10^{-39} - 0.228 \times 10^{-39}$ = $\mathbf{2.192 \times 10^{-39}}$ **C^{2}m/N** and α_e at 448 K is = $1.74 \times 10^{-39} - 0.228 \times 10^{-39}$ = $\mathbf{1.512 \times 10^{-39}}$ **C^{2}m/N.**

12.5. *For sulphur, the relative dielectric constant at 27 K is 3.75. If the internal field constant* $\gamma = \frac{1}{3}$, *calculate the electronic polarizability of sulphur.* [*Density at 27 °C is 2050 kg/m^3 and At. wt. = 32*]

Solution: We have the number of atoms of sulphur per m^3 is

$$\frac{\text{Avogadro number} \times \text{density}}{\text{Atomic weight}} = \frac{6.02 \times 10^{26} \times 2050}{32} = 385.66 \times 10^{26}\ \text{atoms/m}^3.$$

Now internal field

$$E_i = E + \frac{\gamma P}{\varepsilon_0} = E + \frac{P}{3\varepsilon_0}\left[\because \gamma = \frac{1}{3}\right].$$

$$\therefore \frac{\varepsilon_r - 1}{\varepsilon_r + 2} = \frac{N\alpha_e}{3\varepsilon_0}$$

$$\text{or, } \alpha_e = \left(\frac{\varepsilon_r - 1}{\varepsilon_r + 2}\right) \times \frac{3\varepsilon_0}{N} = \left(\frac{3.75-1}{3.75+2}\right) \times \left(\frac{3 \times 8.55 \times 10^{-12}}{385.66 \times 10^{26}}\right)$$

$$= \left(\frac{2.75}{5.75}\right) \times \left(\frac{25.65}{385.66}\right) \times 10^{-38} = \mathbf{3.181 \times 10^{-40}\ F/m^2}.$$

12.6. *The dielectric constant of* Si *is* 12. *The length of the edge of its cubic unit cell is* 5.43 Å. *Find the electronic polarizability of* Si *atoms.*

Solution: From Clausius-Mossotti relation,

$$\frac{\varepsilon_r - 1}{\varepsilon_r + 2} = \frac{N\alpha_e}{3\varepsilon_0}.$$

For silicon, there are 8 atoms per unit cell of volume $(5.43 \times 10^{-10})^3$ cubic metre.

The number of atoms per unit volume is

$$N = \frac{8}{(5.43 \times 10^{-10})^3} = \mathbf{4.997 \times 10^{28}/m^3}.$$

Putting $\varepsilon_r = 12$ in the Clausius-Mossotti relation, we get

$$\frac{11}{13} = \frac{4.997 \times 10^{28} \times \alpha_e}{3 \times 8.854 \times 10^{-12}} \quad [\because \varepsilon_0 = 8.854 \times 10^{-12}\ \text{F/m}]$$

$$\text{or, } \alpha_e = \frac{11 \times 3 \times 8.854}{13 \times 4.997 \times 10^{40}} = \mathbf{4.5 \times 10^{-40}\ Fm^2}.$$

12.7. *A solid dielectric has electronic polarizability of* 10^{-40} Fm². *If the internal electric field be a Lorentz field, what is the dielectric constant of the material.* [*Given: density* = 3×10^{28} atoms/m³].

Solution: The relative dielectric constant of the material is given by

$$\varepsilon_r - 1 = \frac{N\alpha_e}{\varepsilon_0} \text{ or, } \varepsilon_r = 1 + \frac{N\varepsilon_e}{\varepsilon_0} = 1 + \frac{(3 \times 10^{28}) \times 10^{-40}}{8.854 \times 10^{-12}} = \mathbf{1.339},$$

where the density of the material $N = 3 \times 10^{28}$ atoms/m³ and the dielectric constant of free space $\varepsilon_0 = 8.854 \times 10^{-12}$ F/m³.

12.8. *The relative permittivity of Argon at* 0 °C *at* 1 atm *is* 1.000435. *Find the electronic permittivity of Argon atom.*

Solution: We have electronic susceptibility

$$\chi_e = k_e - 1 = \frac{n\alpha}{\varepsilon_0} \text{ or, } \alpha = \frac{\varepsilon_0}{n}(k_e - 1).$$

22.4 lit = 22.4×10^3 m³ contains 1 mole = 6.023×10^{23} molecules of Argon.

$\therefore$ 1 m³ contains

$$n = \frac{6.023 \times 10^{23}}{22.4 \times 10^{-3}\ \text{m}^3} \text{ molecules} = 2.689 \times 10^{25}\ \text{m}^3.$$

$$\alpha = \left(\frac{10^9 F}{36\pi m}\right) \times \frac{1}{(2.689 \times 10^{25}\ \text{m}^{-3})}(1.000435 - 1) = \mathbf{1.42 \times 10^{-40}\ Fm^2}.$$

12.9. *Suppose, all molecular dipoles in a drop of water of radius* 10^{-3} m *point in the same direction and each molecule has dipole moment* 6×10^{-30} cm. *Find the polarisation, dipole moment per molecule and dipole moment of the drop.*

Solution: Molecular weight of water = 2 × 1 + 16 = 18.

Drop radius = 1×10^{-3} m, Drop volume = $\frac{4\pi}{3} \times (1 \times 10^{-3}\ \text{m})^3$.

Density of water = 1 gm/cc = $\frac{10^{-3}\ \text{kg}}{(10^{-2}\ \text{m})^3} - 10^3\ \text{kg/m}^3$.

No. of molecules per unit volume is

$$n = \frac{\rho N_A}{M} = \left(10^3 \frac{\text{kg}}{\text{m}^3}\right)\left(\frac{6.023 \times 10^{23}}{18 \times 10^{-3}\ \text{kg}}\right) = 3.346 \times 10^{28}\ \text{m}^{-3}.$$

Dipole moment/molecule, $P_{\text{molecule}} = 6 \times 10^{-3}$ cm.

Dipole moment/volume, $P = np = (3.346 \times 10^{28}\ \text{m}^{-3})(6 \times 10^{-30}\ \text{cm})$.

∴ polarisation, $P = 0.2\ \text{cm}^{-2}$.

Hence, dipole moment of the drop is

$$P_{\text{drop}} = \text{Dipole moment/unit volume} \times (\text{volume of a drop})$$
$$= 0.2\ \text{cm}^{-2} \times \frac{4\pi}{3}(1 \times 10^{-3}\ \text{m})^3 = \mathbf{8.38 \times 10^{-10}\ cm}.$$

12.10. *There are* 1.6×10^{26} molecules/m^3 *in* NaCl *vapour and* Na^+ *and* Cl^- *ions are separated by distance* 2.5 Å. *Calculate the oriental polarization at room temperature if the vapour is subjected to a field of* 5×10^6 V/m.

Solution: It is known that dipole moment of NaCl is

$$p = qd = (1.6 \times 10^{-19}\ \text{C})(2.5 \times 10^{-10}\ \text{m}) = 4 \times 10^{-29}\ \text{cm}.$$

∴ oriental polarisability is

$$\alpha_0 = \frac{p^2}{3kT} = \frac{4 \times 10^{-29}\ \text{cm}}{3(1.38 \times 10^{-23}\ \text{JK}^{-1})(300\ \text{K})} = 1.29 \times 10^{-37}\ \text{Fm}^2.$$

∴ Oriental polarisation is

$$P_0 = n\alpha_0 E = (1.6 \times 10^{26}\ \text{m}^{-3})(1.29 \times 10^{-37}\ \text{Fm}^2)\left(5 \times 10^6 \frac{V}{M}\right) = \mathbf{10.32 \times 10^{-5}\ cm^{-2}}.$$

12.11. *One mole of a polar substance is dissolved in* 1000 cm^3 *of a non-polar liquid having dielectric constant* 3 *at* 27 °C. *The solution has dielectric constant* 3.2 *at the same temperature. Find the dipole moment of the polar molecules. (Assume that the volume does not change upon dissociation).*

Solution: The Clausius-Mossotti relation for non-polar liquid is given by

$$\frac{k_e - 1}{k_e + 2} = \frac{n\alpha}{3\varepsilon_0} \text{ or, } n\alpha = 3\varepsilon_0\frac{k_e - 1}{k_e + 2} = 3\varepsilon_0\frac{3-1}{3+2} = 3\varepsilon_0\frac{2}{5},$$

k_e, n and α are for the liquid.

The Clausius-Mossotti relation for the solution is given by

$$\frac{k'_e - 1}{k'_e + 2} = \frac{n\alpha}{3\varepsilon_0} + \frac{n'\alpha'}{3\varepsilon_0} = \frac{1}{3\varepsilon_0}(n\alpha + n'\alpha').$$

k'_e, α' and n' are for the solution.

$$\therefore n\alpha + n'\alpha' = 3\varepsilon_0 \frac{k'_e - 1}{k'_e + 2}.$$

$$\therefore 3\varepsilon_0 \frac{2}{5} + n'\alpha' = 3\varepsilon_0 \frac{3.2 - 1}{3.2 + 2} \text{ or, } n'\alpha' = 3\varepsilon_0 \frac{2.2}{5.2} - 3\varepsilon_0 \frac{2}{5} = 3\varepsilon_0 \left(\frac{2.2}{5.2} - \frac{2}{5}\right) = 3\varepsilon_0 (0.023).$$

$$\therefore \alpha' = \frac{3\varepsilon_0}{n'}(0.023).$$

Now 22.4 lit = 22.4× 1000 cc = 22.4 × 1000 × $(10^{-2}\text{ m})^3$ = 22.4 × 10^{-3} m^3.

22.4 lit = 22.4 × 10^{-3} m^3 contains 1 mole = 6.023 × 10^{23} molecules.

$$\therefore 1\text{ m}^3 \text{ contains } \frac{6.023 \times 10^{23}}{22.4 \times 10^{-3}\text{ m}^3} \text{ molecules} = 2.689 \times 10^{25} \text{ molecules or, } n' = 2.689 \times 10^{25}\text{ m}^{-3}.$$

$$\therefore \alpha' = \frac{p^2}{3kT} = \frac{3\varepsilon_0}{n'}(0.023)$$

$$\text{or, } p^2 = \frac{3kT}{n'} 3\varepsilon_0 (0.023) = \frac{9(1.38 \times 10^{-23}\text{ JK}^{-1})(300\text{ K})\left(\frac{10^{-9}}{36\pi}\frac{F}{m}\right)(0.023)}{2.689 \times 10^{25}\text{ m}^{-3}} = \mathbf{2.8 \times 10^{-58}\ C^2m^2}.$$

∴ **the required dipole moment of the polar molecule is $p = 1.67 \times 10^{-29}$ Cm.**

12.12. *The relative permittivity of Germanium is 16. The edge depth of the conventional cubic cell for Germanium lattice is* 5.65×10^{-10} m. *Calculate the electronic polarisability of Germanium atoms.*

Solution: We have the Clausius-Mossotti relation:

$$\frac{k_e - 1}{k_e + 2} = \frac{n\alpha}{3\varepsilon_0},$$

the different letters have their usual meaning.

$k_e = 16$. Cell volume = $(5.65 \times 10^{-10})^3$.

There are 8 Germanium atoms per unit cell.

$$\therefore n = \frac{8}{(5.65 \times 10^{-10}\text{ m})^3}.$$

∴ Clausius-Mossotti relation becomes

$$\frac{16 - 1}{16 + 2} = \frac{8}{(5.65 \times 10^{-10}\text{ m})^3} \times \frac{\alpha}{3\left(\frac{10^{-9}}{36\pi}\frac{F}{m}\right)}$$

$$\text{or, } \alpha = \frac{16 - 1}{16 + 2} \cdot \frac{(5.65 \times 10^{-10}\text{ m})^3}{8} \cdot 3\left(\frac{10^{-19}}{36\pi}\frac{F}{m}\right) = 4.99 \times 10^{-40}\text{ Fm}^2$$

is the required electronic polarisability of Germanium.

EXERCISE

12(A): Objective (MCQ) Type Questions

Select the correct answer.

12.1. If ε and ε_0 be the permittivity of material and of free space respectively, then the relative permittivity (or dielectric constant) of the material is

(a) $\varepsilon_r = \varepsilon_0/\varepsilon$ (b) $\varepsilon_r = \varepsilon_0\varepsilon$ (c) $\varepsilon_r = \frac{\varepsilon}{\varepsilon_0}$ (d) $\varepsilon_r = \frac{1}{\varepsilon\varepsilon_0}$.

12.2. In polar molecules in absence of an electric field the centre of gravity of the positive and negative charges

(a) coincides (b) does not coincide (c) almost coincides (d) none of these.

12.3. In absence of an electric field the non-polar molecules have a

(a) dipole moment (b) zero dipole moment (c) transient dipole moment (d) none of these.

12.4. Temperature dependence in polarizability is observed in case of

(a) α_e (b) α_i (c) α_d (d) α_s.

12.5. At moderate temperature (T) the electronic polarizability (α_e) depends

(a) linearly on T (b) independent of T (c) inversely dependent on T (d) none of these.

12.6. The unit of dipole moment per unit volume is

(a) coulomb-metre (b) metre/coulomb (c) coulomb (d) coulomb/metre^2.

12.7. Select the correct relation

(a) $D = \varepsilon_0 (\varepsilon_r - 1) E$ (b) $E = \varepsilon_0 (\varepsilon_r - 1) P$ (c) $P = \varepsilon_0 (\varepsilon_r - 1) E$ (d) $\varepsilon_r = (\chi - 1)$.

12.8. Ionic polarizability (α_i)

(a) increases with T (b) decreases with T (c) is independent of T (d) none of these.

12.9. Lorentz field is expressed as

(a) $E_i = E + \frac{P}{3\varepsilon_0}$ (b) $E_i = E + \frac{3P}{\varepsilon_0}$ (c) $E_i = E - \frac{P}{3\varepsilon_0}$ (d) $E_i = \frac{3P}{\varepsilon_0 E}$.

12.10. If α_s and T be respectively the space charge polarizability and temperature of a dielectric, then

(a) α_s increases with T (b) α_s decreases with T (c) α_s is independent of T (d) none of these.

12.11. The dipolar polarizability α_d is equal to

(a) p^2/kT (b) $p^2/3kT$ (c) $3kT/p^2$ (d) p^2kT.

12.12. At optical frequencies (~ 10^{15} Hz) the dielectric constant (ε_r) depends on the refractive index (n) as

(a) $\varepsilon_r \propto n$ (b) $\varepsilon_r \propto n^2$ (c) $\varepsilon_r \propto \frac{1}{n}$ (d) $\varepsilon_r \propto \frac{1}{n^2}$.

12.13. The relation $\frac{N\alpha}{3\varepsilon_0} = \frac{\varepsilon_r - 1}{\varepsilon_r + 2}$ is called

(a) Einstein relation (b) Lorentz relation (c) Clausius-Mossotti relation (d) none of these.

12.14. Dielectrics are basically

(a) conductors (b) semiconductors (c) insulators (d) superconductors.

12.15. Which is the correct relation?

(a) $\varepsilon_r = 1 - \frac{3T_c}{T - T_c}$ (b) $\varepsilon_r = 1 - \frac{3T_c}{T + T_c}$ (c) $\varepsilon_r = 1 + \frac{3T_c}{T - T_c}$ (d) $\varepsilon_r = 1 + \frac{3T}{T - T_c}$.

12.16. The time-dependence of polarizability is expressed as

(a) $P(t) = P_0 e^{-t/\tau}$ (b) $P(t) = P_0 e^{t/\tau}$ (c) $P(t) = P_0 e^{\tau/t}$ (d) $P(t) = P_0 e^{-\tau/t}$.

12.17. The relaxation time (τ) and resonant frequency (ω_0) are given related by

(a) $\tau\omega_0 = 1$ (b) $\tau\omega_0 = 0$ (c) $\tau\omega_0 = \infty$ (d) $\tau + \omega_0 = 1$.

12.18. In a dipolar dielectric, in absence of electric field, the dipoles are

(a) parallel (b) anti-parallel (c) randomly oriented (d) none of these.

12.19. Which is the correct relation of ionic polarizability (α_i) and the angular frequency of a molecule (ω_0)?

(a) $\alpha_i = \frac{e^2}{\omega_0^2}\left(\frac{1}{M} + \frac{1}{m}\right)$ (b) $\alpha_i = \frac{\omega_0^2}{e^2}\left(\frac{1}{M} + \frac{1}{m}\right)$ (c) $\alpha_i = e\omega_0^2\left(\frac{1}{M} + \frac{1}{m}\right)$ (d) $\alpha_i = \omega_0^2\left(\frac{1}{M} + \frac{1}{m}\right)$.

12.20. Space charge polarization occurs at frequencies

(a) ~ 10^{15} Hz (b) ~ 10^3 Hz (c) ~ 10^{13} Hz (d) 10^{10} Hz.

12.21. In piezoelectricity,

(a) electrical energy is converted into mechanical energy

(b) mechanical energy is converted into electrical energy

(c) mechanical energy is converted into thermal energy

(d) electrical energy is converted into thermal energy.

12.22. The remanent polarization in ferroelectrical material is due to

(a) coercive force (b) positive polarization (c) hysteresis (d) none of these.

12.23. The ferroelectric materials show polarization at $T < T_c$

(a) in presence of electric field (b) in absence of electric field

(c) both in presence or absence of electric field (d) none of these.

12.24. The condition of spontaneous polarization is

(a) $\sum_j N_j\alpha_j = 3\varepsilon_0$ (b) $\sum_j N_j\alpha_j = 2\varepsilon_0$ (c) $\sum_j n_j\alpha_j = \varepsilon_0$ (d) $\sum_j N_j\alpha_j = 6\varepsilon_0$.

12.25. A dielectric can be piezoelectric, pyroelectric or ferroelectric, only if it is

(a) centrosymmetry (b) non-centrosysmmery (c) hysteresis (d) none of these

12.26. Posistors are the instruments which measure

(a) resistance (b) humidity (c) voltage (d) control temperature.

12.27. Hysteresis is exhibited by materials which are

(a) piezoelectric (b) ferroelectric (c) pyroelectric (d) magneto-optic.

12.28. Spontaneous polarization takes place due to

(a) atoms (b) free electrons (c) permanent dipoles (d) ions.

12.29. Dielectric loss in a dielectric is proportional to

(a) ω^2 (b) $\frac{1}{\omega}$ (c) ω (d) $\frac{1}{\omega^2}$.

12.30. Which is ferroelectric material?

(a) Cu (b) $BaTiO_3$ (c) H_2O (d) HCl.

12.31. The temperature coefficient of permittivity is given by

(a) $T_{c\varepsilon} = \frac{1}{\varepsilon}\frac{d\varepsilon}{dT}$ (b) $T_{c\varepsilon} = \frac{d\varepsilon}{dT}$ (c) $T_{c\varepsilon} = \frac{dT}{d\varepsilon}\varepsilon$ (d) $T_{c\varepsilon} = \varepsilon d\varepsilon dT$.

ANSWERS

12.1. (c) **12.2.** (b) **12.3.** (b) **12.4.** (c) **12.5.** (b) **12.6.** (d) **12.7.** (c) **12.8.** (c) **12.9.** (b) **12.10.** (a) **12.11.** (b) **12.12.** (b) **12.13.** (c) **12.14.** (c) **12.15.** (c) **12.16.** (a) **12.17.** (a) **12.18.** (c) **12.19.** (a) **12.20.** (b) **12.21.** (b) **12.22.** (a) **12.23.** (b) **12.24.** (a) **12.25.** (b) **12.26.** (d) **12.27.** (b) **12.28.** (c) **12.29.** (c) **12.30.** (b) **12.31.** (a).

12(B): Short Answer Type Questions

12.1. What is a dielectric? State what is the local field in a dielectric and give its decomposition explaining the significance of each part. [BU(H) 1995]

12.2. What are polar and non-polar dielectrics? How would you distinguish them?

12.3. What do you mean by dielectric susceptibility? Explain its physical significance.

12.4. Define polarization and polarizability of a dielectric material. What are the different sources of polarizability? [BU(H) 2000]

12.5. What is oriental polarizability? Derive an expression for the oriental polarizability of a gas of polar molecule. [BU(H) 1997]

12.6. What is electronic polarizabilty? What is the dependence of electronic polarizability of an atom on the radius of the atom? [BU(H) 2004]

12.7. What is dipolar polarizability? Obtain an expression for dipolar polarizability of a dielectric at moderate temperature. [BU(H) 2001]

12.8. Explain how the polarizability of an atom may be determined? Deduce the Clausius-Mossotti relation. [BU(H) 1997]

12.9. Derive Clausius-Mossotti relation between polarizability and the dielectric constant of a solid. [BU(H) 2000, '03]

12.10. Sate the different contributions to the local polarizability of a dielectric. Deduce the Clausius-Mossotti formula for an isotropic dielectric. [BU(H) 1995]

12.11. What is oriental polarizability? How does it differ from atomic polarizability? [BU(H) 1994]

12.12. Explain what is meant by space charge polarization. Does it depend on temperature?

12.13. Discuss frequency dependence of polarization.

12.14. What are dielectric loss and loss tangent? Find an expression for the dielectric loss.

12.15. What is dielectric breakdown? What factors are responsible for it?

12.16. What is piezoelectricity? How does it arise?

12.17. Discuss the origin of ferroelectricity? What is polarization catastrophe?

12.18. Discuss the applications of piezo- and ferroelectric materials.

12.19. Define atomic polarisability and find its dimension. [CU(H) 2011]

12.20. Write down Clausius-Mossotti relation and explain its physical significance. [CU(H) 2010]

12.21. (a) Explain electronic and dipolar polarisability.

(b) Derive an expression for the dipolar polarisability for low applied electric field and high temperature.
[CU(H) 2009]

12.22. What do you understand by oriental polarisation in a dielectric? Using Langevin's theory, show that at high temperature and small electric field, oriental polarisation is inversely proportional to the absolute temperature. **[BU(H) 2015]**

12.23. Derive the Clausius-Mossotti relation expressing the relationship between dielectric constant and atomic polarisability. **[BU(H) 2014]**

12(C) Numerical Problems

12.1. If the susceptibility of CO_2 is 0.985×10^{-3}, calculate the total polarizability of CO_2, if its density is $1.977\ kg/m^3$. [*Ans.* $3.24 \times 10^{-40}\ Fm^2$]

12.2. If the refractive index and dielectric constant of water are 1.33 and 8.1 respectively, then calculate the percentage of ionic polarizability. [*Ans.* 71 %]

12.3. NaCl crystal has a static dielectric constant $\varepsilon_r = 5.6$ and refractive index $n = 1.5$. Calculate the percentage contribution of the ionic polarizability.

12.4. A paramagnetic material having atoms each of magnetic movement $1.2 \times 10^{-23}\ A/m^2$ is placed in magnetic field of induction $1.5\ Wb/m^2$. Calculate the difference in magnetic potential energy for extreme orientations of the atomic dipoles and compare this energy with the thermal energy of translation of the atoms at the temperature of 27 °C. [*Ans.* $R_p = 10.03\ \Omega$; $C_p = 226.56 \times 10^{-12}$ F]

12.5. The structure of $BaTiO_3$ transforms from cubic to tetragonal below the Curie point with lattice parameters $c = 4.03$ Å and $a = 3.98$ Å. The displacement of titanium ion is opposite to that of oxygen ion. The magnitude of displacement is 0.06 Å for titanium, 0.06 Å for oxygen ions on the side faces and 0.08 Å for oxygen ions on the top and bottom faces of the cube. Neglecting the displacement of barium ions calculate the polarization of a $BaTiO_3$ crystal. [*Ans.* $0.16/cm^2$]

12.6. The bakelite is found to have the real part of its complex relative dielectric constant as 4.36 with a loss tangent of 2.8×10^{-2} at frequency 1 MHz. Calculate the complex polarizability of the material assuming Lorentz field. [Given, $N = 4 \times 10^{28}/m^3$]. [*Ans.* $(3.5 - i0.063) \times 10^{-40}$]

12.7. Calculate the value of the radius of electron orbit of helium and hence, the permittivity for 2.7×10^{25} atoms/m³. [Given, electronic polarizability of helium = $0.18 \times 10^{-40}\ Fm^3$.]
[*Ans.* 5.46×10^{-11}; 1.000057776]

12.8. The static dielectric constant of water is 8.125 and its refractive index is 1.33. Calculate the percentage contribution of ionic polarizability. **[BU(H) 2004]**

[*Hints.* cf. Solved problem 12.2]

12.9. Calculate the electronic polarizability of Ar atom having radius 10^{-10} m. [*Ans.* $1.11 \times 10^{-40}\ Fm^2$]

12.10. Consider a system of non-interacting dipoles which are confined to two possible orientations relative to an applied electric field $\vec{E}$, either parallel or antiparallel. Show that a temperature T the average polarisation along the field direction is $\mu^2|\vec{E}|/k_B T$, where μ is the magnitude of the dipole moment of each dipole. **[CU(H) 2012]**

12.11. Argon gas contains 2.7×10^{25} atoms/m³ at 0 °C and one atmosphere. If the diameter of an argon atom is 0.384 mm, find the dielectric constant of argon at 0 °C. **[BU(H) 2014]**

12.12. What is electronic polarisability? Derive an expression for electronic polarisability using the classical theory.

Given, Planck's constant = 6.63×10^{-34} Js, 1 eV = 1.6×10^{-19} J.

Boltzmann constant = 1.38×10^{-23} J/K.

Avogadro number = 6.02×10^{23}/mole.

Atomic mass of carbon = 12 amu.

Bohr magneton = 9.27×10^{-24} Am^2.

$$\int_0^\infty \frac{x^3\,dx}{e^x - 1} = \frac{\pi^4}{15}.$$

[BU(H) 2013]

12.13. One gram molecule of a certain polar substance is dissolved into 1000 cm^3 of a non-polar constant of 3.0 at 27 °C, whereas the solution has a dielectric constant of 3.2 at that temperature. Calculate the dipole moment of the polar molecules. **[BU(H) 2012]**

12.14. A parallel plate condenser has a capacitance of 2 μF. The dielectric has permittivity $\varepsilon_r = 100$.

(a) For an applied voltage of 1 kV, find the energy stored in the condenser as well as the energy stored in polarising dielectric.

(b) Show that the stored energy in the polarised atom is equal to $\frac{1}{2}\alpha E^2$, where α is the polarisability.

[BU(H) 2011]

12(D) Broad Answer Type Questions

12.1. (a) What is a dielectric? What do you mean by dielectric polarization? Distinguish metal and semi-conductor from a dielectric.

(b) What are the various components of the local electric field at an atom in a crystal? Obtain the Lorentz relation for the local electric field. When is this relation valid?

12.2. (a) Distinguish between polar and non-polar dielectric. Give some examples of the above two types of dielectric solids.

(b) What are the different types of polarizability?

(c) Discuss the dependence of the electric polarizability of an atom on the radius of the atom.

[BU(H) 2004]

12.3. (a) Define electric polarization and the local field. Explain how the local field may be expressed as a sum of different components. Find the Lorentz field.

(b) What are the different factors which contribute to polarization in solid? Explain how you would determine the polarizability of an atom. Derive the formula you would use. **[BU(H) 1997]**

(c) What is Lorentz-Lorentz relation?

12.4. (a) Define polarization and polarizability of a dielectric material. Derive Clausius-Mossotti relation between polarizability and dielectric constant of a solid. What are the different sources of polarizability?

[BU(H) 2000, '03]

(b) Distinguish carefully between the following terms: (i) the applied electric field (ii) the macroscopic electric field (iii) the local field in a dielectric.

12.5. (a) What is oriental polarizability? How does it differ from atomic polarizability?

(b) Deduce an expression for the oriental polarizability per molecule of a gas of polar substance at temperature T. [Assume that $pE/kT \ll 1$, where p is the dipole moment of the polar molecule and E is the magnitude of the applied electric field]. **[BU(H) 1994]**

(c) Discuss the temperature dependence of different types of polarizabilities.

12.6. (a) What is a dielectric? State what is local field in a dielectric and give its decomposition explaining the significance of each part.

(b) State the different contributions to the total polarizability of a dielectric material.

(c) Discuss the Clausius-Mossotti formula for an isotropic dielectric. **[BU(H) 1995]**

12.7. (a) Deduce an expression for the electronic polarizability of an atom on the basis of classical theory. **[BU(H) 1996]**

(b) Discuss the temperature dependence of oriental polarizability. What is spontaneous polarization?

(c) What do you mean by dielectric loss and dielectric breakdown?

12.8. (a) Give a schematic sketch of the variation of the total polarizability of a dielectric as a function of frequency. Explaining the physical origin of the various contributions and the relevant frequency ranges. **[JNTU 2004]**

(b) (i) Explain the ionic polarizability of a molecule and its frequency dependence.

(ii) Define piezoelectric effect. Discuss some of the important applications of the piezoelectricity. **[JNTU 2003]**

12.9. (a) What is ferroelectricity? What are the different groups of ferroelectric crystals?

(b) What is polarization catastrophe? Discuss with theoretical reasoning.

(c) Mention some of the applications of ferroelectric crystals.

12.10. (a) Explain the origin of piezoelectricity.

(b) '*All ferroelectric crystals are piezoelectric but the converse is not true*'. Explain with reason.

(c) Mention some important applications of piezoelectricity?

(d) What is pyroelectricity? Define pyroelectric coefficient.

12.11. Write notes on:

(a) Relaxation time of polarization.

(b) Internal field of solids **[JNTU 2003]**

(c) Spontaneous polarization

(d) Behaviour of ferroelectric crystals

(e) Debye equation.

12.12. Consider a gas containing large number of identical molecules each with a permanent dipole moment μ under a static external field $\vec{E}$. The average component of the dipole moment per molecule at a temperature T K is given by

$$\langle \mu \rangle = \mu L(a),$$

where $L(a) = \coth a - \frac{1}{a}$; $a = \frac{\mu E}{k_B T}$.

(a) Sketch $L(a)$ as a function of a.

(b) Evaluate the dipolar polarisability at low field.

(c) How is Clausius-Mossotti relation modified taking into the above contribution? **[CU(H) 2015]**

12.13. Compute the local Lorentz field due to the polarisation charges on the surface of a spherical cavity:

(a) Using this result derive the Clausius-Mossotti relation.

(b) What is the implication of such a relation? **[CU(H) 2013]**

12.14. (a) Obtain expression for oriental polarisability of a dielectric material.

(b) Establish the Clausius-Mossotti material and hence, discuss its physical significance. **[BU(H) 2007]**

Chapter 13

Magnetic Properties of Solids

13.1 Introduction

The magnetic properties of solids were discovered by ancient people of 800 BC. They found that a mineral called *magnetite* which is an oxide of iron (Fe_3O_4) possessed this property and the ancient mariners used the material for finding out direction of a ship on sea. In view of modern concept all materials, viz., metal, semiconductors and insulators exhibit magnetic properties to different extent.

The origin of magnetism in material is the magnetic moment of the magnetic dipole in the same. Materials which can be magnetized by the application of an external magnetic field are called *magnetic materials.* The important magnetic materials in view of practical applications are *diamagnetic, paramagnetic, ferromagnetic, antiferromagnetic and ferrimagnetic.*

13.2 Terms and Definitions in Magnetism

(a) Magnetic Permeability: Magnetic permeability of a material is defined as the ratio of magnetic induction to the magnetic field strength, i.e.,

$$\mu = \frac{B}{H}, \tag{13.1}$$

where B and H are respectively the *magnetic induction* and *magnetic field* at any point in the material.

The permeability of free space is denoted by μ_0 and has value $4\pi \times 10^{-7}$ H/m in SI units.

The ratio of μ/μ_0 is called the *relative permeability* (μ_r) of the solid, i.e.,

$$\mu_r = \frac{\mu}{\mu_0}. \tag{13.2}$$

(b) Magnetization: A magnetic material when placed in a magnetic field gets magnetized.

The magnetic moment developed per unit volume in the material is called the intensity of magnetization or magnetization denoted by M. The unit of M is A/m.

If H be the applied magnetic field, then

$$\chi = \frac{M}{H} \tag{13.3}$$

is called the *magnetic susceptibility* and is defined as below:

The magnetization developed in the material per unit applied magnetic field. Since M and H are measured in A/m, χ is a *dimensionless quantity.*

From eqn (13.1), $$B = \mu H = \mu_0 \mu_r H \left[\because \mu_r = \frac{\mu}{\mu_0} \right]$$

$$= \mu_0 \mu_r H + \mu_0 H - \mu_0 H = \mu_0 H + \mu_0 H (\mu_r - 1) = \mu_0 H + \mu_0 M,$$

where $M = H(\mu_r - 1)$,

$$\text{or, } B = \mu_0 (H + M). \tag{13.4}$$

$\mu_0 H$ and $\mu_0 M$ represent magnetic induction due to external field and due to the magnetization respectively.

$$\text{or, } \mu_0 = \frac{B}{H + M}.$$

$\therefore$ the relative permeability,

$$\mu_r = \frac{\mu}{\mu_0} = \frac{B/H}{B/(H+M)} = \frac{H+M}{H} = 1 + \frac{M}{H} \text{ or, } \mu_r = 1 + \chi \left[\because \chi = \frac{M}{H} \right]. \tag{13.5}$$

In free space, i.e., in absence of material medium, $M = 0$, $\chi = 0$, $\mu = \mu_0$ and hence, $\mu_r = 1$.

$\therefore$ from $B = \mu H = \mu_0 \mu_r H$, we get

$$B = \mu_0 H. \tag{13.6}$$

The unit of B in e.m.u. is *gauss* and in SI unit *tesla*, where 1 tesla = 10^4 gauss.

13.3 Classification of Magnetic Materials

In the atom, the presence of permanent magnetic dipole and the interaction between them is responsible for the origin of different types of magnetic materials. The permanent dipole arises due to the *orbital and spin motion of the electron as well as the spins of the nucleus.* The electron orbit behaves as a current carrying coil and produces magnetic effect. For the production of permanent dipole moment the spins of unpaired valence electrons are mainly responsible. The orientation of the magnetic moments in the material may lead to a zero or non-zero resultant magnetic moments giving rise to various types of magnetism as follows: (i) *Diamagnetism* (ii) *Paramagnetism* (iii) *Ferromagnetism* (iv) *Antiferromagnetism* (v) *Ferrimagnetism.*

Diamagnetism: Diamagnetic solids do not contain permanent magnetic moments, the presence of small non-zero magnetic moment is due to the orbital motion of the electrons and directed oppositely to the applied field. The diamagnetic effect is quite small.

Example: *Copper, Gold, Silver,* He, Ne, *Benzene, Naphthalene, etc.*

Paramagnetism: Paramagnetism arises due to the presence of permanent atomic or electronic magnetic moments. Paramagnetic effect is quite weak but unlike diamagnetism, the magnetic moment is aligned along the direction of the applied field.

Example: *All atoms and molecules with unfilled outer shells like* Ti, *etc., salts of transition elements, rare earths and actinide series containing elements like* Cr^{3+}, Dy^{3+}, *etc. and metals like* Al.

Ferromagnetism: Ferromagnetic effect is very strong and arises due to the alignments of the neighbouring magnetic moments along the same direction in absence of magnetic field below *Curie temperature.* Above Curie temperature the Ferromagnetic materials behave as paramagnetic.

Example: Fe, Co, Ni.

Antiferromagnetism: In antiferromagnetism the adjacent moments being equal and opposite cancel each other completely and thus the total value over a finite volume is zero. Above *Neel temperature* antiferromagnetism vanishes and the material behaves as paramagnetic.

Example: MnO, NiO, MnTe, *etc.*

Ferrimagnetism: In Ferrimagnetic materials, unlike antiferromagnetic, the number of neighbouring moments are unequal and hence, though these are opposite, complete cancellation of moments do not take place. Hence, there is a net magnetic moment.

The magnetic ordering of different kinds of magnetic materials is depicted in Figs 13.1. The ferro- and ferrimagnetic materials are consisting of *domains* of completely magnetized regions. It is interesting to note that *above certain critical temperature the distinction of the various magnetic materials disappears and all magnetic materials become paramagnetic.*

Example: Fe_3O_4, *Nickel ferrite and Yttrium iron garnet.*

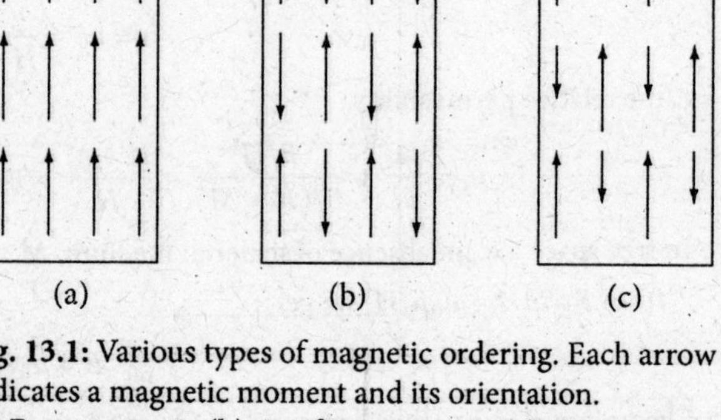

Fig. 13.1: Various types of magnetic ordering. Each arrow indicates a magnetic moment and its orientation. (a) Ferromagnetic (b) Antiferromagnetic (c) Ferrimagnetic

13.4 The Quantum Numbers

The magnetism in materials arises due to the orbital motion of the electron, the spin motion of the electron and the nuclear spin. To understand the origin of the magnetism, specification of the electron in the atom is essential which could be done by the following four quantum numbers.

(i) The Principal Quantum Number (n) : Corresponding to electronic energy levels K, L, M, N the principal quantum number of the electron $n = 1, 2, 3, 4, \ldots$. It determines the size of the electron orbit and the number of electron in a given shell with principal quantum number n, i.e., $N = n^2$.

(ii) The Orbital Quantum Number (l) : For each value of n the orbital quantum number is $l = 0$ to $(n-1)$. Each value of l represents a sub-shell in the orbit. For $l = 0$, the sub-shell is called s sub-shell. Similarly, $p, d, f, \ldots$ sub-shells are for $l = 1$, $l = 2$, $l = 3$ and so on. The number of electrons in each sub-shell is $(2l + 1)$ and the total angular momentum for a given l is

$$L_l = \hbar[l(l+1)]^2 \text{ with } \hbar = \frac{h}{2\pi}. \tag{13.7}$$

(iii) The Orbital Magnetic Quantum Number (m_l) : An atom placed in an external magnetic field precesses about the field direction as axis and can have only specific orientations in space. The projection of L_l along the field direction is given by

$$L_{l,B} = L_l \cos\alpha, \tag{13.8}$$

where B is the magnetic field and α is the angle as shown in Fig. 13.2. Now $L_{l,B}$ can take only integral values in the multiples of $\hbar$, i.e.,

$$L_{l,B} = m_l\hbar = m_l\frac{h}{2\pi}. \tag{13.9}$$

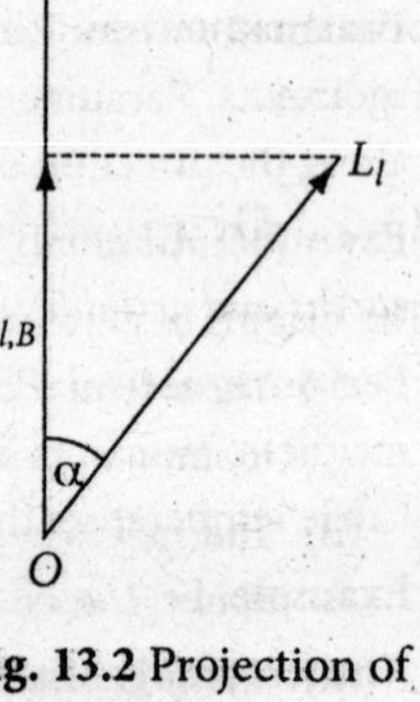

Fig. 13.2 Projection of L_l and the field direction (B)

m_l is called the *orbital magnetic quantum number* and can have values

$$m_l = -l, -(l-1), \ldots, 0, \ldots, (l-1), l$$

$$\text{or, } m_l = 0 \pm 1, \pm 2, \ldots, \pm l. \tag{13.10}$$

Figs 13.3(a) and (b) show the possible orientations of L_l for an electron in p and d states respectively.

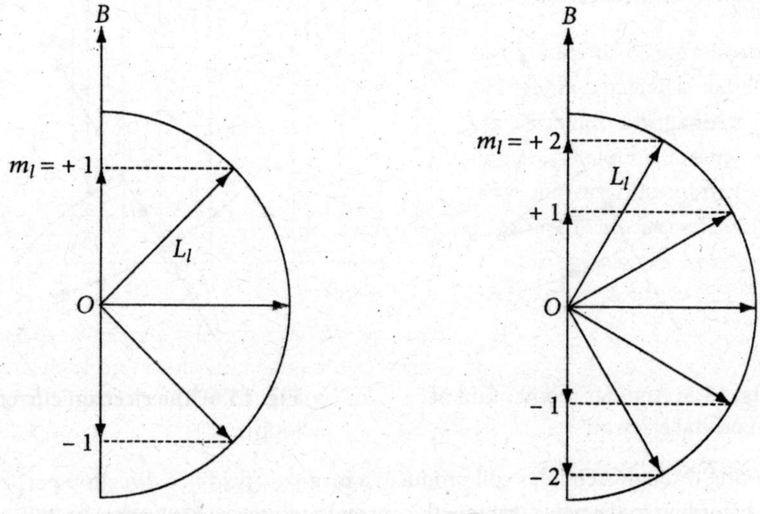

Fig. 13.3: Possible orientations of L_l defined by l in magnetic field B

(iv) The spin magnetic quantum number (m_s) : The spin angular momentum arises due to the spin motion of the electron and has two values corresponding to spin magnetic quantum number, $m_s = \pm\frac{1}{2}$, i.e.,

$$S = m_s\hbar = m_s\frac{h}{2\pi}. \tag{13.11}$$

For clockwise or anticlockwise spin motion, the spin angular momentum points either upward or downward direction respectively as shown in Fig. 13.4.

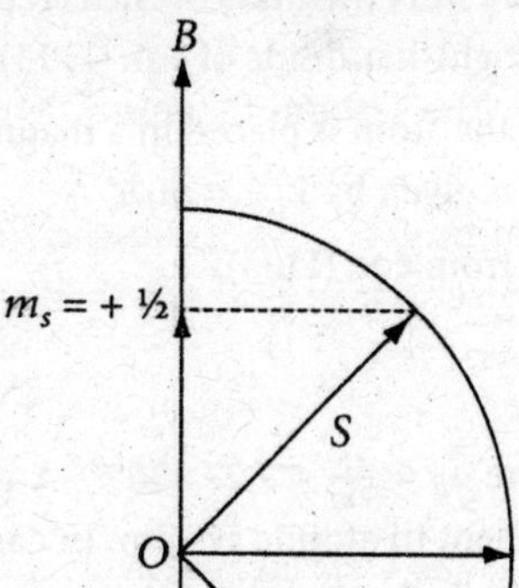

Fig. 13.4: Spin quantisation of the electron

13.5 Origin of Magnetic Moment

The origins of permanent magnetic moment in an atom are as follows:

(i) *The orbital magnetic moment of the electron arising due to its orbital motion around the nucleus.*

(ii) *The spin magnetic moment of the electron due to its spin motion about its own axis.*

(iii) *The spin magnetic moment of the nucleus due to spin motion about its own axis.*

13.5.1 The Orbital Magnetic Moment of the Electrons

If an electron of charge $-e$ revolves round a nucleus in a circular orbit of radius r, then the current in the orbit is

$$I = \frac{-e}{T} = \frac{-e}{2\pi/\omega} = -\frac{e\omega}{2\pi}, \tag{13.12}$$

ω is the *angular velocity* of the electron.

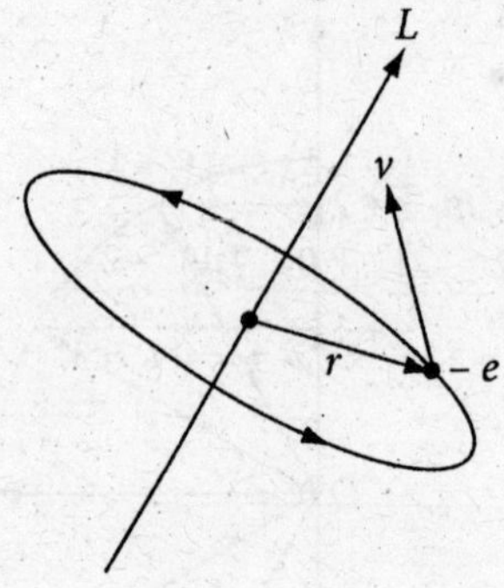

Fig. 13.5: Angular momentum of an orbital electron

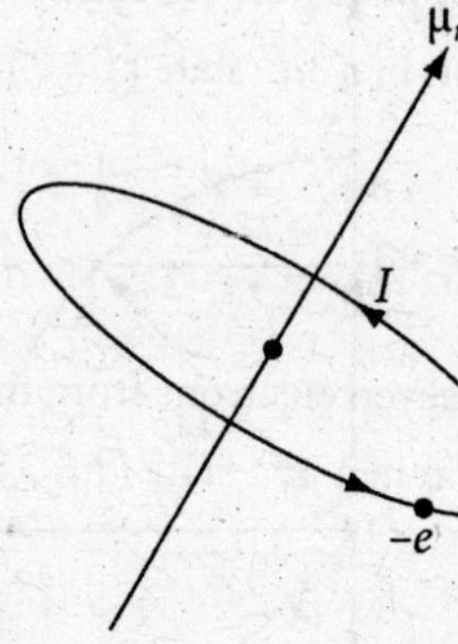

Fig. 13.6: The electron current loop

The current flowing through a circular coil produces a magnetic field in a direction perpendicular to the plane of the coil. According to *Ampere's theorem* the current carrying coil is equivalent to a magnetic dipole of moment [Fig. 13.6],

$$\mu_m = IA = \frac{-e\omega}{2\pi}\left(\pi r^2\right) = -\frac{e\omega r^2}{2} = -\frac{e}{2m}\left(m\omega r^2\right) = -\left(\frac{e}{2m}\right)L, \tag{13.13}$$

where m is the mass of the electron and $L = m\omega r^2$ is its *orbital angular momentum*. The negative sign in the right-hand side of eqn (13.13) indicates that μ_m is *anti-parallel to L.*

If the atom is placed in a magnetic field, then the possible orientation of the angular momentum vector (L) is given by $L_{l,B} = m_l\hbar$.

∴ from eqn (13.13),

$$\mu_m = -\left(\frac{e}{2m}\right)m_l\hbar = -\left(\frac{e\hbar}{2m}\right)m_l = -\mu_B m_l, \tag{13.14}$$

where $\mu_B = \frac{e\hbar}{2m} = 9.27\times10^{-24}$ A-m^2 and is called the *Bohr magneton*. It is also the accepted unit of magnetic moment in atomic system. In case of an atom having many electrons, the total orbital magnetic moment is the algebraic sum of the magnetic moments of individual electrons. The magnetic moment (μ_m) of a completely filled shell is zero but for an atom with partially filled shells will have non-zero orbital magnetic moment (μ_l).

Hund's Rule

Since the filled electron shells do not contribute to the *magnetic moment* of the atom, it arises from the partially filled shells. This is in accordance with the *Pauli's principle,* that no two electrons can have all the quantum numbers n, l, m_l and m_s identical. For the *ground state* of such an atom, *Hund's rule* for the *incomplete shells* may be stated as follows:

1. *The spins of the electron add to give the maximum possible values of S consistent with the Pauli's principle.*
2. *The orbital moments combine to give a maximum value of L that is consistent with* (1) above.
3. *For an incomplete shells,*

$J = L - S$, when the shell is *less than the half* occupied and

$J = L + S$, when the shell is *more than the half* occupied.

Illustration of the application of Hund's Rule

Example 1. CO^{2+} ion in a $3d^7$ state ($l = 2$)

$$m_s \rightarrow \frac{1}{2}\ \frac{1}{2}\ \frac{1}{2}\ \frac{1}{2}\ \frac{1}{2}\quad -\frac{1}{2}\ -\frac{1}{2}\ -\frac{1}{2}\ -\frac{1}{2}\ -\frac{1}{2}$$

$$m_l \rightarrow 2\ \ 1\ \ 0\ -1\ -2\quad 2\ \ 1\ \ 0\ -1\ -2$$

Similarly, for the seven electrons from the left, $S = 3/2; 2S + 1 = 2 \times \frac{3}{2} + 1 = 4; L = 3$.

Example 2. Dy^{3+} ion in a $4f^9$ state ($l = 3$)

$$m_s = \frac{1}{2}\ \frac{1}{2}\ \frac{1}{2}\ \frac{1}{2}\ \frac{1}{2}\ \frac{1}{2}\ \frac{1}{2}\quad -\frac{1}{2}\ -\frac{1}{2}\ -\frac{1}{2}\ -\frac{1}{2}\ -\frac{1}{2}\ -\frac{1}{2}\ -\frac{1}{2}$$

$$m_l = 3\ \ 2\ \ 1\ \ 0\ -1\ -2\ -3\quad 3\ \ 2\ \ 1\ \ 0\ -1\ -2\ -3$$

Similarly, if the nine electrons from the left, $S = 5/2; 2S + 1 = 2 \times \frac{5}{2} + 1 = 6; L = 5$.

Combination of $\vec{L}$ and $\vec{S}$

As the electron is negatively charged $\vec{\mu}_l$ and $\vec{\mu}_s$ are opposite in direction to $\vec{L}$ and $\vec{S}$ respectively. The resultant vector $\vec{J}$ of $\vec{L}$ and $\vec{S}$ is given by $\vec{J} = \vec{L} + \vec{S}$.

Under the action of an external magnetic field $\vec{B}$, $\vec{L}$ tends to precess round $\vec{J}$ with *Larmour frequency* $\vec{\omega}_L$. $\vec{S}$ will precess round $\vec{J}$ with frequency $2\vec{\omega}_L$. If $\vec{B}$ is small, then the coupling between $\vec{L}$ and $\vec{S}$ will be strong *R-S* coupling and $\vec{J}$ will be maintained and will precess around $\vec{B}$ with a compromise frequency. As $\vec{L}, \vec{S}, \vec{\mu}_l$ and $\vec{\mu}_s$ all precess around $\vec{J}$, their normal components to $\vec{J}$ will average to zero. Under such conditions, the resultant moment $\vec{\mu}_j$ will be the sum of their components parallel to $\vec{J}$.

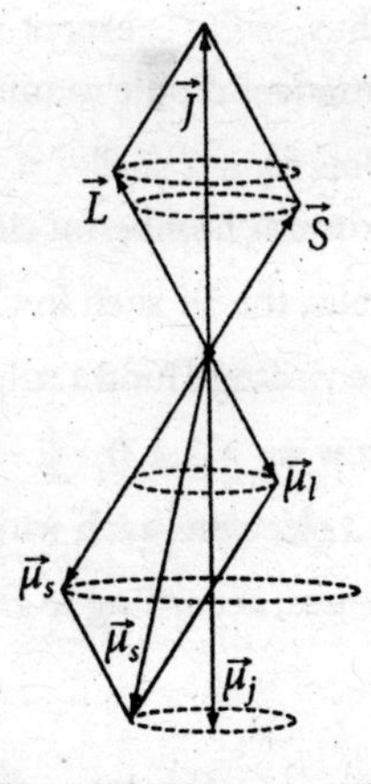

Fig. 13.7: Vector model of the atom

Now the component of μ_l parallel to $\vec{J}$ is

$$\mu_l \cos\left(\vec{L}, \vec{J}\right) = L\frac{e}{2m}\cos\left(\vec{L}, \vec{J}\right). \tag{13.15}$$

Component of $\vec{\mu}_s$ parallel to $\vec{J}$ is

$$S \times \frac{2e}{m}\cos\left(\vec{S}, \vec{J}\right). \tag{13.16}$$

$$\therefore \mu_j = \left[L\cos\left(\vec{L}, \vec{J}\right) + 2S\cos\left(\vec{S}, \vec{J}\right)\right]\frac{e}{2m}. \tag{13.17}$$

From Fig. 13.8,

$$S^2 = L^2 + J^2 - 2LJ\cos\left(\vec{L}, \vec{J}\right) \tag{13.18}$$

$$\text{or, } L\cos\left(\vec{L}, \vec{J}\right) = \frac{J^2 + L^2 - S^2}{2J}. \tag{13.19}$$

Similarly,

$$S\cos\left(\vec{S},\vec{J}\right) = \frac{J^2 - L^2 + S^2}{2J}. \tag{13.20}$$

Substituting eqns (13.19) and (13.20) in eqn (13.17), we get

$$\mu_j = \left[1 + \frac{J^2 - L^2 + S^2}{2J^2}\right] J\frac{e}{2m} = \left[1 + \frac{J(J+1) + S(S+1) - L(L+1)}{2J(J+1)}\right]\sqrt{J(J+1)}\frac{3\hbar}{2m}$$

$$= g\mu_B\sqrt{J(J+1)}, \tag{13.21}$$

where $\mu_B = \frac{e\hbar}{2m}$ is the Bohr magneton and

$$g = \frac{1 + J(J+1) + S(S+1) - L(L+S)}{2J(J+1)}, \tag{13.22}$$

where g is the *Landé g-factor.*

When $\vec{B} = 0$, all the atoms have the same value of μ_j. But when $\vec{B}$ is applied, the energy depends on the statistical occupation of the $(2J+1)$ sub-shells and all the atoms will have M_J value different from μ_j. The magnetic energy of a given atom is then

$$E_M = \vec{\mu}_j \cdot \vec{B} = -g\mu_B M_J B. \tag{13.23}$$

Example: The electron configuration of Cr atom $(Z = 24)$ is

$$1s^2\, 2s^2\, 2p^6\, 3s^2\, 3p^6\, 3d^5\, 4s^1.$$

If 3 electrons from the Cr atom are removed, it becomes Cr^{3+} whose electronic configuration is

$$1s^2\, 2s^2\, 2p^6\, 3s^2\, 3p^6\, 3d^3.$$

Thus, in Cr^{3+}, except $3d$ shell all the shells are filled and hence, the $3d$ shell is responsible for the origin of permanent dipole moment. The $3d$ shell has 3 electrons.

Now for a d shell, $l = 2$ and $m_l = 2, 1, 0, -1, -2$, each can accommodate 2 electrons with $m_s = \pm\frac{1}{2}$. So the maximum number of electrons in the $3d$ shell is $5 \times 2 = 10$.

Thus, the $3d$ shell in Cr^{3+} is less than half occupied (as the atomic number $Z = 24$).

According Hund's rule (1), $S = 3 \times \frac{1}{2} = \frac{3}{2}$ as there are three electrons in the $3d$ shell.

Now $m_l = 2, 1, 0, -1, -2$.

If 3 electrons each with spin $+\frac{1}{2}$ be placed in the first three of these, then $L = 2 + 1 + 0 = 3$.

Hence, according to *Hund's rule* (3),

$$J = L - S = 3 - \tfrac{3}{2} = \tfrac{3}{2} \quad \left[\because L = 3 \text{ and } S = \tfrac{3}{2}\right]$$

∴ the *Landé splitting factor* is

$$g = 1 + \frac{J(J+1) + S(S+1) - L(L+1)}{2J(J+1)} = 1 + \frac{\frac{3}{2}\left(\frac{3}{2}+1\right) + \frac{3}{2}\left(\frac{3}{2}+1\right) - 3(3+1)}{2\cdot\frac{3}{2}\left(\frac{3}{2}+1\right)}$$

$$= 1 + \frac{\frac{3}{2}\times\frac{5}{2} + \frac{3}{2}\times\frac{5}{2} - 3\times 4}{3\times\frac{5}{2}} = 1 + \frac{\frac{15}{4} + \frac{15}{4} - 12}{\frac{15}{2}} = 0.4.$$

13.5.2 Spin Angular Momentum of the Electron

The spin magnetic moment of the electron is

$$\mu_s = -\left(\frac{e}{2m}\right)S = -\left(\frac{e}{2m}\right) \times m_s\hbar \; [\because S = m_s\hbar]$$

$$= \pm\frac{1}{2}\left(\frac{e}{2m}\right)\hbar \left[\because m_s = \pm\frac{1}{2}\right]$$

$$= \pm\frac{1}{2}\mu_B. \tag{13.24}$$

Thus, the magnitude of spin magnetic moment is half of Bohr magneton μ_B.

The magnetic moment component due to electron spin along the field direction is

$$\mu_{sz} = g\left(\frac{e}{2m}\right)m_s\hbar = g\left(\frac{e\hbar}{2m}\right)m_s, \tag{13.25}$$

where g is called the *Landé splitting factor* or *spectroscopic splitting factor* and is given by

$$g = 1 + \frac{J(J+1) + S(S+1) - L(L+1)}{2J(J+1)}, \tag{13.26}$$

where S is the spin angular momentum, L is the orbital angular momentum and J is the total angular momentum of the electron.

For $m_s = \frac{1}{2}$,

$$\mu_{sz} = g\frac{e\hbar}{4m} = g\left(\frac{e}{2m}\right)\frac{\hbar}{2}$$

For the electron spin, $g = 2.0023$.

$\therefore M_{sz} = \mu_B$, i.e., the dipole moment due to electron spin is equal to *Bohr magneton* in the direction of the applied field B.

$\therefore$ the total angular momentum of the atom is

$$\mu = gJ\mu_B, \tag{13.27}$$

where $J\mu_B = \left(\frac{e}{2m}\right)\frac{\hbar}{2}$.

From the knowledge of g and J_{μ_B} we can find the total magnetic moment of the atom.

13.5.3 Why g is called *Splitting Factor*?

Let an electron with spin $\frac{1}{2}$ but no angular momentum be subjected to an external magnetic field $\vec{B}$. Then for $m_s = \frac{1}{2}$, $|\mu_{sz}| = \mu_B$, the direction being anti-parallel to $\vec{B}$. But for $m_s = -\frac{1}{2}$, $|\mu_{sz}| = \mu_B$, the direction being parallel to $\vec{B}$. The PE of magnetic dipole of moment $\vec{\mu}$ when placed in a magnetic field $\vec{B}$ is $-\vec{\mu}\cdot\vec{B}$ and the energy level of the spin $\frac{1}{2}$ electron is split up as shown in Fig. 13.8.

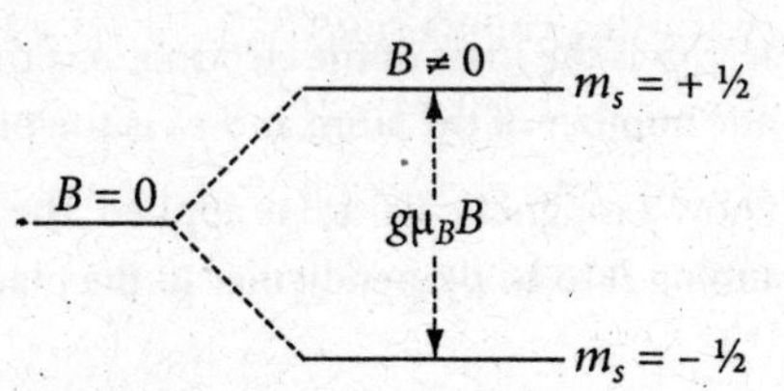

Fig. 13.8: Splitting of energy level of the spin ½ electron for $B = 0$

The separation of the split-up energy levels is

$$\Delta E = 2|\mu_{sz}|B = g\left(\frac{e}{2m}\right)\hbar B = g\mu_B B \left[\because \frac{e\hbar}{2m} = \mu_B\right].$$

So the split-up of the energy levels is determined by g and hence, the name *splitting factor.*

13.5.4 Nuclear Spin Magnetic Moment

Due to its spin motion, the nucleus is associated with a magnetic moment and is called the *nuclear magnetic moment.* The nuclear magnetic moment is expressed in terms of nuclear magneton as

$$\mu_n = \frac{e\hbar}{2M_p} = 5.05 \times 10^{-29}\ \text{A-m}^2, \tag{13.28}$$

where M_p is the mass of a proton.

13.6 Diamagnetism

In an atom, the electrons revolve in closed orbits around the nucleus. The electron orbit may be considered as a current carrying closed coil producing magnetic field perpendicular to its plane. The electrons in an atom revolve in discrete orbits whose planes are not parallel to each other and direction of rotation of electrons is not the same. That is why *the vector sum of the magnetic moments due to all the revolving electrons becomes zero.*

If the electric circuit is subjected to a changing magnetic flux, an induced current is set up in it which, according to *Lenz's law,* opposes the change in flux. The induced magnetic field in the coil opposes the applied field and produces *diamagnetism.* Diamagnetism is produced in all materials but it is suppressed by the stronger *paramagnetic, ferromagnetic effects.*

13.6.1 Langevin's Classical Theory of Diamagnetism

We consider an electron revolving around the nucleus in a circular orbit with frequency ω_0 in absence of any external magnetic field. Then the centrifugal force $\frac{mv^2}{\rho}$ is balanced by the *Coulomb force* of attraction between the nucleus and the electron.

$$\therefore \frac{mv^2}{\rho} = \frac{Ze^2}{4\pi\varepsilon_0\rho^2} \text{ or, } m\rho\omega_0^2 = \frac{Ze^2}{4\pi\varepsilon_0\rho^2}\ [\because v = \rho\omega_0]. \tag{13.29}$$

$$\therefore \omega = \sqrt{\frac{Ze}{4\pi m\varepsilon_0\rho^3}}. \tag{13.30}$$

Here m is the mass of the electron, ρ is the radius of the electron orbit, e is the electronic charge, Z is the atomic number of the atom and ε_0 is the permittivity of the free space.

If now a magnetic field $\vec{B}$ is applied, the additional *Lorentz force,* $\vec{F} = e\left(\vec{v} \times \vec{B}\right)$ will act on the electron. Assuming $\vec{B}$ to be perpendicular to the plane of the orbit, the magnitude of this force is

$$F = e\rho\omega B, \tag{13.31}$$

where ω is the frequency of the electron in presence of the field.

Then

$$m\omega^2\rho = \frac{Ze^2}{4\pi\varepsilon_0\rho^2} - e\rho\omega B = m\omega_0^2\rho - e\rho\omega B. \left[\because \frac{Ze^2}{4\pi\varepsilon_0\rho^2} = m\rho\omega_0^2\right] \tag{13.32}$$

The minus sign with the Lorentz force is due to the negative charge of the electron.

Thus, from eqn (13.32),

$$\omega^2 + \frac{eB}{m}\omega - \omega_0^2 = 0$$

$$\text{or, } \omega = \frac{1}{2}\left[-\frac{eB}{m} \pm \sqrt{\left(\frac{eB}{m}\right)^2 + 4\omega_0^2}\right] = \pm\sqrt{\omega_0^2 + \left(\frac{eB}{2m}\right)^2} - \frac{eB}{2m}.$$

For $\frac{eB}{2m} \ll \omega_0$, because $\omega_0 \simeq 10^{14}$ to 10^{15} rad/s and $B \simeq 1\,\text{Wb/m}^2$,

$$\omega \simeq \pm\omega_0 - \frac{eB}{2m}. \tag{13.33}$$

Eqn (13.33) shows that the frequency of the electron changes by $\left(\frac{eB}{2m}\right)$ in presence of the magnetic field B. The ± sign in eqn (13.33) signifies that *the electrons with orbital moments parallel to the field are slowed down, whereas with moments anti-parallel to the field are speeded up*. This is called *Larmor theorem*. The quantity $\frac{eB}{2m} = \omega_L$ is called the *Larmor frequency.*

This change in frequency gives rise to an additional current,

$$I = (-e)\left(\frac{1}{2\pi}\frac{eB}{2m}\right) = -\frac{e^2B}{4\pi m}. \tag{13.34}$$

The magnetic moment associated with the current loop having one electron is

$$\vec{\mu}_e = I \times \vec{\alpha} = \frac{e^2\vec{B}}{4\pi m} \times \pi\rho^2 = -\frac{e^2\vec{B}}{4m}\rho^2, \tag{13.35}$$

where $\alpha = \pi\rho^2$ is the area of the current loop of radius ρ.

The significance of the negative sign on the right side of eqn (13.35) is that, the induced magnetic moment μ_e is directed opposite to the magnetic field as shown in Fig. 13.9.

If the electron orbit is not perpendicular to the magnetic field $\vec{B}$, then it precesses about $\vec{B}$ with frequency ω_L. In this case, the induced magnetic moment μ_e will have a component opposite to the direction of $\vec{B}$. If the orbits of the electrons of an atom are randomly oriented, then the total induced magnetic moment is given by

$$\mu_a = -\frac{e^2\vec{B}}{4m}\sum_{i=1}^{n'}\langle\rho_i^2\rangle, \tag{13.36}$$

where n' = the total number of electrons in the atom.

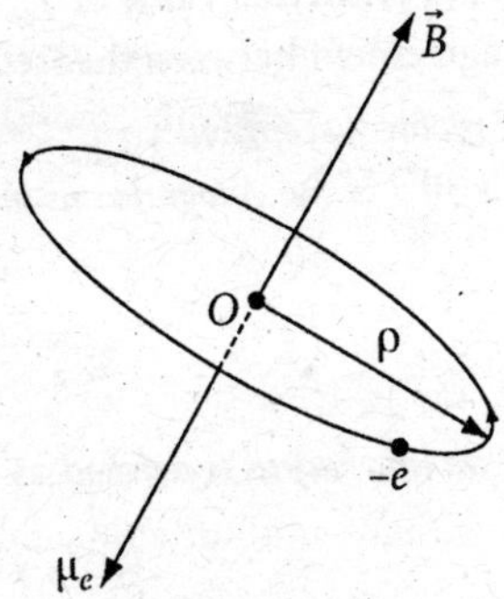

Fig. 13.9: μ_e is directed opposite to the magnetic field $\vec{B}$

If O be the origin of a Cartesian coordinate system OX, OY and OZ coinciding with the centre of the circular orbits and $\vec{B}$ be along the Z-axis, then

$$\langle\rho^2\rangle = \langle x^2\rangle + \langle y^2\rangle + \langle z^2\rangle, \tag{13.37}$$

where x, y, z represent the coordinates of a point on the orbit of radius ρ.

The *mean squared distance* of the electrons from the nucleus is

$$\langle r^2\rangle = \langle x^2\rangle + \langle y^2\rangle + \langle z^2\rangle. \tag{13.38}$$

If the charge distribution is assumed to be *spherically symmetric,* then

$$\langle x^2\rangle = \langle y^2\rangle = \langle z^2\rangle$$

$$\therefore \langle r^2\rangle = \tfrac{3}{2}\langle \rho^2\rangle. \tag{13.39}$$

$\therefore$ eqn (13.36) becomes

$$\vec{\mu}_a = -\frac{e^2\vec{B}}{6m}\sum_{i=1}^{n'}\langle r^2\rangle = -\frac{\mu_0 e^2\vec{H}}{6m}\sum_{i=1}^{n'}\langle r^2\rangle \; [\because \vec{B} = \mu_0\vec{H}]. \tag{13.40}$$

If the magnetic solid contains N atoms/volume and the number of electrons in each atom is Z, then the magnetic moment per unit volume is

$$\vec{M} = -N\vec{\mu}_a = -\frac{N\mu_0 e^2\vec{H}}{6m}\sum_{i=1}^{n'}\langle r^2\rangle. \tag{13.41}$$

The *diamagnetic susceptibility of the material* is

$$\chi_{\text{dia}} = \frac{M}{H} = -\frac{N\mu_0 e^2}{6m}\sum_{i=1}^{n'}\langle r^2\rangle. \tag{13.42}$$

Eqn (13.42) shows that *diamagnetic susceptibility is negative and is independent on temperature of the material.* Eqn (13.42) is known as the *Langevin equation* for diamagnetism.

Assuming one electron of mass $m = 9.1\times10^{-31}$ kg, $\langle r^2\rangle = 0.1\,\text{mm}^2$, $N = 5\times10^{28}/\text{m}^3$ and $e = 1.6\times10^{-19}$ C, we get

$$\chi_{\text{dia}} = \frac{-\left(5\times10^{28}\right)\times\left(4\pi\times10^{-7}\right)\times\left(1.6\times10^{-19}\right)^2\times\left(0.1\times10^{-9}\right)^2}{6\times\left(9.1\times10^{-31}\right)} = \mathbf{-3\times10^{-6}}. \tag{13.43}$$

The experimental value of χ_{dia} for copper at room temperature is -9×10^{-6}. This shows that there is a close agreement between theoretical and experimental values of χ_{dia}.

The *molar susceptibility* $\chi_{\text{dia}}^{(\text{molar})}$ may be obtained by multiplying χ_{dia} by the volume of a mole (N_A/N), $N_A = 6.022\times10^{23}$ is the *Avogadro number.*

So

$$\chi_{\text{dia}}^{(\text{molar})} = -\frac{N_A}{N}\times\frac{N\mu_0 e^2}{6m}\sum_{i=1}^{n'}\langle r^2\rangle = -\frac{N_A\mu_0 e^2}{6m}\sum_{i=1}^{n'}\langle r^2\rangle \tag{13.44}$$

The *atomic radius is defined* as

$$\langle\rho^2\rangle_a = \frac{1}{n'}\sum_{i=1}^{n'}\langle r^2\rangle, \qquad \therefore \chi_{\text{dia}}^{(\text{molar})} = -\frac{n' e^2 N_A\mu_0}{6m}\langle\rho^2\rangle_a. \tag{13.45}$$

In case of He-*atom,* $n' = 2$,

$$\chi_{\text{dia}}^{(\text{molar})} = -1.9\times10^{-6}\ \text{cm}^3/\text{mol}.$$

$$\therefore \sqrt{\langle\rho^2\rangle_a} = 1.64\times10^{-9}\ \text{cm}$$

is the *radius of the* He-atom.

Conclusions: Following conclusions can be drawn from the *Langevin's theory of diamagnetism*:

1. Diamagnetic susceptibility χ_{dia} is independent on temperature of the solid as the orbital areas of the electrons are independent on temperature.
2. Since $\chi_{\text{dia}} \propto \langle\rho^2\rangle_a$, the outer electrons of the atoms contribute in a large scale.

3. The diamagnetic susceptibility increases with the increase in number of atoms.

It is to be noted that, an atom having permanent dipole moment will also possess the diamagnetic property. The permanent magnetic moment in this case tends to rotate into the direction of $\vec{B}$ and the paramagnetic effect results and the diamagnetic effect disappears.

Apart from Cu, Au, Ge, Si, NaCl, etc. *inert gases* are also diamagnetic as they have completely filled electron shells in their atoms.

13.6.2 Quantum Theory of Diamagnetism

The Hamiltonian $\mathcal{H}$ of an electron changes to $\mathcal{H} + \mathcal{H}'$ in presence of a magnetic induction $\vec{B}$, where $\mathcal{H}'$ represents a small perturbation.

Using *quantum mechanics* we can show that

$$\mathcal{H}' = -\frac{ie\hbar}{2m}\left(\nabla\cdot\vec{A} + 2\vec{A}\cdot\nabla\right) + \frac{e^2}{2m}A^2, \tag{13.46}$$

where $\vec{A}$ is the *magnetic vector potential* given by $\vec{B} = \nabla\times\vec{A}$.

If $\vec{B}$ is uniform, then it can be expressed as

$$\vec{A} = \tfrac{1}{2}\vec{B}\times\vec{r} = \tfrac{1}{2}\left[\hat{i}\left(B_y z - B_z y\right) + \hat{j}\left(B_z x - B_x z\right) + \hat{k}\left(B_x y - B_y x\right)\right]. \tag{13.47}$$

If $\vec{B}$ is along the Z-direction, then $B_x = B_y = 0$; $B_z = B$.

$$\therefore A_x = -\tfrac{1}{2}yB,\ A_y = \tfrac{1}{2}xB \text{ and } A_z = 0.$$

Also

$$\nabla\cdot\vec{A} = \frac{\partial A_x}{\partial x} + \frac{\partial A_y}{\partial y} + \frac{\partial A_z}{\partial z} = 0 \tag{13.48}$$

$$\text{and } \vec{A}\cdot\nabla = A_x\frac{\partial}{\partial x} + A_y\frac{\partial}{\partial y} + A_z\frac{\partial}{\partial z} = \tfrac{1}{2}B\left(x\frac{\partial}{\partial x} - y\frac{\partial}{\partial y}\right). \tag{13.49}$$

Now eqn (13.46) becomes

$$\mathcal{H}' = -\frac{ie\hbar B}{2m}\left(x\frac{\partial}{\partial x} - y\frac{\partial}{\partial y}\right) + \frac{e^2B^2}{8m}\left(x^2 + y^2\right). \tag{13.50}$$

The *quantum mechanical angular momentum operator* is

$$\vec{L} = -i\hbar\vec{r}\times\nabla$$

$$\text{or, } \hat{i}L_x + \hat{j}L_y + \hat{k}L_z = -i\hbar\left[\hat{i}\left(y\frac{\partial}{\partial z} - z\frac{\partial}{\partial y}\right) + \hat{j}\left(z\frac{\partial}{\partial x} - x\frac{\partial}{\partial z}\right) + \hat{k}\left(x\frac{\partial}{\partial y} - y\frac{\partial}{\partial x}\right)\right]. \tag{13.51}$$

$$\therefore\quad L_z = \hat{k}\left(x\frac{\partial}{\partial y} - y\frac{\partial}{\partial x}\right). \tag{13.52}$$

$\therefore$ from eqns (13.50) and (13.52) it is clear that the first term on the right-hand side of eqn (13.50) is proportional to L_z.

If now N be the number of atoms per unit volume and n', the number of electrons in each atom, then the second term of eqn (13.50) becomes

$$\frac{Ne^2B^2}{8m}\sum_{i=1}^{n'}\left(x_i^2 + y_i^2\right) = \frac{Nn'e^2B^2}{8m}\langle\rho^2\rangle, \tag{13.53}$$

where $\langle \rho^2 \rangle$ is the mean squared radius of the projection of the orbit on a plane perpendicular to the magnetic field $\vec{B}$. If $\vec{B}$ induces a dipole moment in the specimen, the corresponding energy term should contain B^2. Hence, $\frac{Nn'e^2B^2}{8m}\langle \rho^2 \rangle$ is the energy term associated with the diamagnetism of the material. But the diamagnetic energy is given by $\frac{1}{2}\vec{M}\cdot\vec{B} = -\frac{1}{2\mu_0}\chi_{dia}B^2$.

$$\therefore \frac{Nn'e^2B^2}{8m}\langle \rho^2 \rangle = -\frac{1}{2\mu_0}\chi_{dia}B^2 \text{ or, } \chi_{dia} = -\frac{Nn'\mu_0 e^2}{4m}\langle \rho^2 \rangle \quad (13.54)$$

$$\text{or, } \chi_{dia} = -\frac{Nn'\mu_0 e^2}{6m}\langle r^2 \rangle. \quad (13.55)$$

$$\left[\because \langle r^2 \rangle = \tfrac{3}{2}\langle \rho^2 \rangle\right]$$

Here $\langle r^2 \rangle$ is the mean squared distance of the electron from the nucleus. Determining $\langle r^2 \rangle$ we can calculate the value of χ_{dia} of the material. In quantum mechanical treatment $\langle r^2 \rangle$ has different significance from that in classical treatment.

The experimental value of χ_{dia} agrees fairly with theoretical result specially for light elements and rare gases. For He-atom, $\chi_{dia} = -1.9 \times 10^{-6}$ indicating that *diamagnetism is a very weak effect.*

13.7 Paramagnetism

The paramagnetic property of the materials arise due to the *permanent magnetic moments* of the constituent atoms or molecules. These magnetic moments orient randomly and the net magnetic moment becomes zero. But when a magnetic field is applied these moments orient themselves along the field direction and a net magnetization along the field direction is produced.

The susceptibility of the paramagnetic materials is small but positive and depends on the temperature of the materials. In paramagnetic materials, the permanent magnetic moments arise from (i) *the orbital motion of the electrons* (ii) *the spin and intrinsic motion of the electrons* and (iii) *the spin magnetic moment of the nucleus.*

Some examples of paramagnetic materials and their susceptibilities at room temperature are shown in Table 13.1.

Table 13.1

Material	$\chi = \mu_r - 1$	*Material*	$\chi = \mu_r - 1$
Oxygen	0.19×10^{-5}	$CuSO_4$, H_2O	2.0×10^{-3}
Platinum	35.5×10^{-5}	$MnSO_4$	3.6×10^{-3}
Air	0.035×10^{5}	Fe_2O_3	1.4×10^{-3}
Nitrogen	0.0012×10^{-5}	$FeCl_2$	3.7×10^{-3}
Tungsten	7.5×10^{-5}	$FeSO_4$	2.8×10^{-3}
CoO	585×10^{-5}	$NiSO_4$	1.2×10^{-3}
$CrCl_2$	1.5×10^{-3}		
Cr_2O_3	1.7×10^{-3}		

13.7.1 Langevin's Classical Theory of Paramagnetism

Langevin's *classical theory* of paramagnetism is based on the following assumptions:

(i) *The permanent dipoles of the paramagnetic material may have all possible orientations.*

(ii) *The dipoles are so separated that their mutual interaction force may be neglected.*

If a magnetic dipole of material $\vec{\mu}$ be placed in a magnetic field $\vec{B}$ [Fig. 13.10], then *torque* experienced by the dipole is given by

$$\vec{\tau} = \vec{\mu} \times \vec{B}. \tag{13.56}$$

This torque tends to establish an order by rotating the dipole in the direction of $\vec{B}$. This action is opposed by thermal agitation energy and tends to establish disorder by randomizing the orientations of the dipoles.

The PE of the dipole is

$$E = -\vec{\mu} \cdot \vec{B} = -\mu B \cos\theta, \tag{13.57}$$

where θ is the angle between $\vec{\mu}$ and $\vec{B}$.

According to *Maxwell-Boltzmann statistics,* the probability of finding the dipole at the angle θ is $p(\theta)$ given by

$$p(\theta) \propto e^{-(\vec{\mu}\cdot\vec{B})/kT} \text{ or, } p(\theta) \propto e^{-\mu B\cos\theta/kT},$$

where k is the *Boltzmann constant* and T is the absolute temperature of the material.

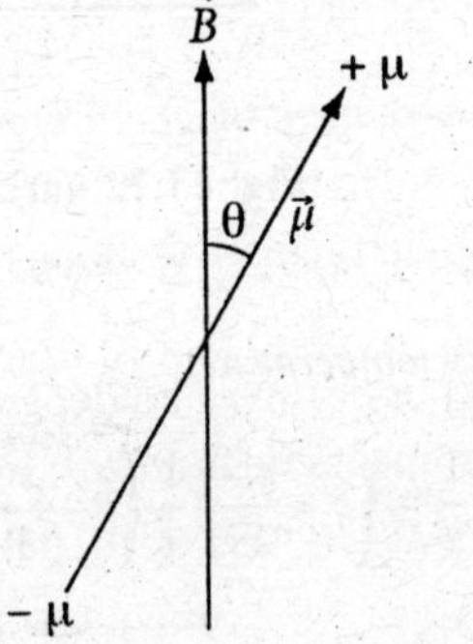

Fig. 13.10: Magnetic dipole in magnetic field $\vec{B}$

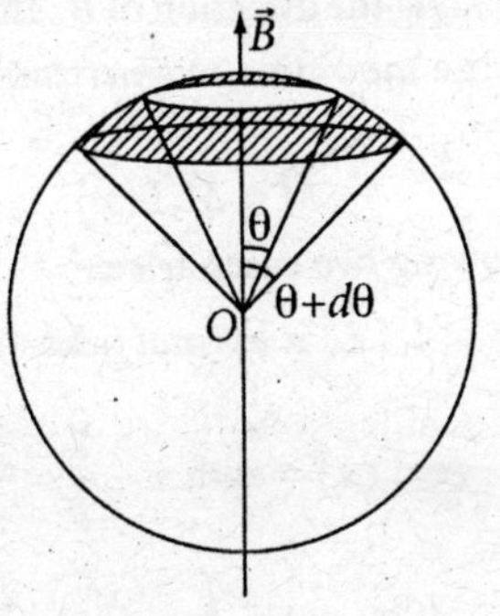

Fig. 13.11: Dipoles lying between θ and (θ + dθ)

If $n(\theta)\,d\theta$ be the number of dipole moments in a direction between θ and $\theta + d\theta$, then

$$n(\theta)\,d\theta \propto p(\theta)\,d\Omega,$$

where $d\Omega$ is the *solid angle* between two hollow cones with semi-vertical angles θ and $\theta + d\theta$ [Fig. 13.11].

$$\therefore n(\theta)\,d\theta = 2\pi A \sin\theta\, e^{-\frac{\mu B\cos\theta}{kT}}\, d\theta, \tag{13.58}$$

where $d\Omega = 2\pi \sin\theta\, d\theta$ and A is a constant of proportionality

$$\text{or, } n(\theta)\,d\theta = C_1 e^{-x\cos\theta} \sin\theta\, d\theta, \tag{13.59}$$

where $C_1 = 2\pi A$ and $x = \frac{\mu B}{kT}$.

If n be the number of dipole moments per unit volume, then

$$n = \int_0^{\pi} n(\theta)\,d\theta = C_1 \int_0^{\pi} e^{x\cos\theta} \sin\theta\,d\theta = C_1 \int_{+1}^{-1} e^{-xy}\,dy = \frac{C_1}{x}(e^x - e^{-x})$$

$$\text{or, } C_1 = \frac{xn}{e^x - e^{-x}}. \tag{13.60}$$

Each of the $n(\theta)\,d\theta$ dipoles has a component of magnetic moment $\mu\cos\theta$ in the direction of the applied field $\vec{B}$, whereas the component perpendicular to $\vec{B}$ cancels each other. Hence, the total magnetic moment of the n dipoles, i.e., the *magnetization* M is

$$M = \int_0^{\pi} (\mu\cos\theta)n(\theta)\,d\theta = \int_0^{\pi} \mu\cos\theta C_1 e^{x\cos\theta}\sin\theta\,d\theta$$

$$= -\mu C_1 \int_1^{-1} y e^{xy}\,dy = -\mu C_1\left[e^x\left(\frac{1}{x}-\frac{1}{x^2}\right)+e^{-x}\left(\frac{1}{x}+\frac{1}{x^2}\right)\right]. \tag{13.61}$$

Substituting the value of C_1 from eqn (13.60) in eqn (13.61), we get

$$M = -\mu\times\frac{xn}{(e^x-e^{-x})}\left[e^x\left(\frac{1}{x}-\frac{1}{x^2}\right)+e^{-x}\left(\frac{1}{x}+\frac{1}{x^2}\right)\right]$$

$$= \mu n\left[\frac{e^x+e^{-x}}{e^x-e^{-x}}-\frac{1}{x}\right] = n\mu\left[\coth x-\frac{1}{x}\right]$$

$$= n\mu L(x), \tag{13.62}$$

where $L(x) = \left[\coth x - \frac{1}{x}\right]$ is called the *Langevin function*. Fig. 13.12 shows the variation of $L(x)$ with x. The maximum value of M is $n\mu$ and is obtained when all the dipoles align along the direction of $\vec{B}$. This is *magnetic saturation*, and the *saturation magnetization* is represented by M_s.

$$\therefore M = M_s L(x). \tag{13.63}$$

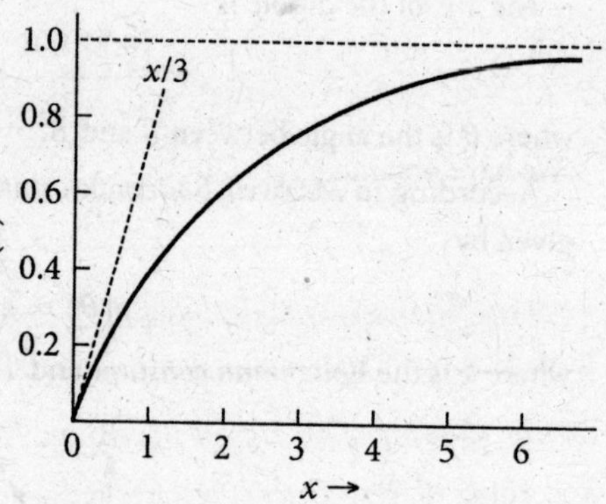

Fig. 13.12 Variation of $L(x)$ with x

Now the following *two* cases arise:

Case I. For $x \ll 1$, i.e., at *normal field strength* and *ordinary temperature*,

$$L(x) = \coth x - \frac{1}{x} = \frac{1+\frac{x^2}{2!}+\frac{x^4}{4!}+\cdots}{x+\frac{x^3}{3!}+\frac{x^5}{5!}+\cdots} - \frac{1}{x} = \left(1+\frac{x^2}{2!}\right)\frac{1}{x}\left(1-\frac{x^2}{3!}\right)-\frac{1}{x}$$

$$\simeq \frac{1}{x}\left(1+\frac{x^2}{2}-\frac{x^2}{6}\right)-\frac{1}{x} = \frac{x}{3}. \tag{13.64}$$

Thus, the variation of $L(x)$ with x near the origin is linear having slope 1/3. Thus,

$$L(x)\Big|_{x=0} = \frac{x}{3} = \frac{\mu B}{3kT}. \tag{13.65}$$

From eqn (13.62),

$$M \simeq \frac{n\mu^2 B}{3kT}. \tag{13.66}$$

The *paramagnetic susceptibility* is given by

$$\chi_{\text{para}} = \frac{\mu_0 M}{B} = \frac{\mu_0 n\mu^2}{3kT}. \tag{13.67}$$

This shows that $\chi_{\text{para}} \propto \frac{1}{T}$.

Putting $\frac{\mu_0 n\mu^2}{3k} = C$, we get

$$\chi_{\text{para}} = \frac{C}{T}, \tag{13.68}$$

where C is a constant, called the *Curie constant*. Expression (13.68) is called the *Curie law* and holds good for $\mu B \ll kT$.

The *molar susceptibility* can be obtained by replacing n by *Avogadro number* N_A in eqn (13.67), i.e.,

$$\chi_{\text{para}}^{(\text{molar})} = \frac{\mu_0 N_A \mu^2}{3kT}. \tag{13.69}$$

The order of magnitude of χ_{para} at room temperature can be estimated by taking

$$\mu = \mu_B = 9.27 \times 10^{-24} \text{ A-m}^2$$
$$n = 5 \times 10^{23}/\text{m}^3$$
$$T = 300 \text{ K}.$$

$$\therefore \chi_{\text{para}} = \frac{(9.27 \times 10^{-24}) \times (1.257 \times 10^{-6}) \times (5 \times 10^{28})}{3 \times (1.38 \times 10^{-23}) \times 300} \simeq \mathbf{10^{-3}}.$$

We have obtained $\chi_{\text{dia}} \sim 10^{-5}$.

$$\therefore \frac{\chi_{\text{para}}}{\chi_{\text{dia}}} = \frac{10^{-3}}{10^{-5}} = \mathbf{10^2}. \tag{13.70}$$

Thus, *paramagnetic susceptibility is roughly hundred times greater than diamagnetic susceptibility.* Table 13.1 shows the paramagnetic susceptibilities of different materials at room temperature.

13.7.2 Quantum Theory of Paramagnetism

According to the *classical theory* of paramagnetism proposed by *Langevin*, the permanent magnetic dipole moment of a given atom or ion can rotate freely and *orient in any direction* to the applied magnetic field. But in view of *quantum theory*, the magnetic moments, on the contrary, can orient along *certain discrete orientations* w.r.t. the magnetic field. Since the magnetic moments $\vec{\mu}$ and its component μ_Z are quantized in the direction of the applied field, they cannot have arbitrary values.

If $\vec{\mu}$ and $\vec{J}$ be the magnetic moment and total angular momentum of an atom (or ion) respectively, then

$$\vec{\mu} = -g\mu_B \vec{J}, \tag{13.71}$$

where $\mu_B = \frac{e\hbar}{2m}$ is the *Bohr magneton* in SI *system*, g is the *Landé splitting factor* having value 2, if $\vec{J}$ is due to orbital motion and electron spin only, i.e.,

$$g = 1 + \frac{J(J+1) + S(S+1) - L(L+1)}{2J(J+1)}, \tag{13.72}$$

where S and L represent respectively spin and angular momentum.

Now the possible orientations of $\vec{\mu}$ along the field direction is given by

$$\mu_Z = -g\mu_B m_J, \tag{13.73}$$

where $m_J = -J, -J+1, \ldots, J-1, J$ is the magnetic quantum number associated with $\vec{J}$. Thus, for each value of J, m_J can have $(2J+1)$ values. This means $\vec{\mu}$ *can have* $(2J+1)$ *orientations relative to the applied field* $\vec{B}$.

The PE of the dipole in presence of $\vec{B}$ is

$$E = -\vec{\mu} \cdot \vec{B} = g\mu_B \vec{J} \cdot \vec{B} = g\mu_B BJ \cos\theta = g\mu_B B m_J = m_J g \mu_B B. \tag{13.74}$$

According to *Maxwell-Boltzmann distribution law*, the number of dipoles having energy E at temperature T is proportional to

$$e^{-E/kT}, \text{ i.e., } e^{\frac{-m_J g \mu_B B}{kT}}.$$

If N be the number of atoms, the magnetization in the direction of $\vec{B}$ is given by

$$M = N \times (\text{statistical average of the magnetic moment component along } \vec{B})$$

$$= N \sum_{m_J=-J}^{+J} \frac{\mu_Z e^{\frac{-m_J g \mu_B B}{kT}}}{e^{-\frac{m_J g \mu_B B}{KT}}} = N \frac{\sum_{-J}^{+J} -m_J g \mu_B e^{-\frac{m_J g \mu_B B}{kT}}}{\sum_{-J}^{+J} e^{-\frac{m_J g \mu_B B}{kT}}}. \tag{13.75}$$

Case I. For *normal values of* $\vec{B}$ and at *ordinary temperature* $\frac{m_J g \mu_B B}{kT} \ll 1$.

$\therefore$ eqn (13.75) reduces to

$$M = N \frac{g\mu_B \sum_{-J}^{+J} -m_J \left(1 - \dfrac{m_J g \mu_B B}{kT}\right)}{\sum_{-J}^{+J} \left(1 - \dfrac{m_J g \mu_B B}{kT}\right)}.$$

But

$$\sum_{-J}^{+J} m_j = 0 \text{ and } \sum_{-J}^{+J} m_J^2 = 2 \sum_{m_J=0}^{J} m_J^2 = \frac{J(J+1)(2J+1)}{3}.$$

$$\therefore M = N \frac{\frac{g^2 \mu_B^2 B}{kT} \frac{J(J+1)(2J+1)}{3}}{(2J+1)} = N \frac{g^2 \mu_B^2 B}{3kT} J(J+1). \tag{13.76}$$

Hence, the paramagnetic susceptibility is

$$\chi_{\text{para}} = \frac{M}{H} = \frac{\mu_0 M}{B} = \frac{\mu_0 N \mu_B^2}{3kT} g^2 J(J+1) \tag{13.77}$$

$$\text{or, } \chi_{\text{para}} = \frac{\mu_0 N p_{\text{eff}}^2 \mu_B^2}{3kT}, \tag{13.78}$$

$$\text{or, } \chi_{\text{para}} = \frac{C}{T} \left[\text{where } C = \mu_0 N p_{\text{eff}}^2 \mu_B^2 / 3k\right] \tag{13.79}$$

where

$$p_{\text{eff}} = \text{the effective number of Bohr magneton} = g\sqrt{J(J+1)}. \tag{13.80}$$

Effective magneton number for trivalent rare earth ions (near room temperature) are shown in Table 13.2.

Case II. If the magnetic field $\vec{B}$ is strong and the temperature T of the material is low, then after some algebraic calculation it can be shown from eqn (13.75) that

$$M = NgJ\mu_B B_J(x), \tag{13.81}$$

where $x = \frac{gJ\mu_B B}{kt}$ and $B_J(x)$ is the *Brillouin function* given by

$$B_J(x) = \frac{2J+1}{2J} \coth\left(\frac{2J+1}{2J}\right) x - \frac{1}{2J} \coth\left(\frac{x}{2J}\right). \tag{13.82}$$

(a) If $x \ll 1$, $\coth \simeq \frac{1}{x} + \frac{x}{3}$

$$\therefore B_J(x) = \frac{x(J+1)}{3J}.$$

Table 13.2

Ion	*Electronic Configuration*	*Spectroscopic ground state*	p_{eff} *(calculated)* $= g\sqrt{J(J+1)}$	$p_{effective}$ *(experimental)*
Ce^{3+}	$4f^1 5s^2 p^6$	$^3F_{5/2}$	2.54	2.4
Pr^{3+}	$4f^2 5s^2 p^6$	$^3H_4$	3.58	3.5
Nd^{3+}	$4f^3 5s^2 p^6$	$^4I_{9/2}$	3.62	3.5
Pm^{3+}	$4f^4 5s^2 p^6$	$^5I_4$	2.68	—
Sm^{3+}	$4f^5 5s^2 p^6$	$^6H_{5/2}$	0.84	1.5
Eu^{3+}	$4f^6 5s^2 p^6$	$^7F_0$	0	3.4
Gd^{3+}	$4f^7 5s^2 p^6$	$^8S_{7/2}$	7.9	8.0
Tb^{3+}	$4f^8 5s^2 p^6$	$^7F_6$	9.72	9.5
Dy^{3+}	$4f^9 5s^2 p^6$	$^6H_{15/2}$	10.63	16.6
Ho^{3+}	$4f^{10} 5s^2 p^6$	$^5I_8$	10.6	10.4
Er^{3+}	$4^{11} 5s^2 p^6$	$^4I_{15/2}$	9.59	9.5
Tm^{3+}	$4f^{12} 5s^2 p^6$	$^3H_6$	7.57	7.3
Y^{3+}	$4f^{13} 5s^2 p^6$	$^2F_{7/2}$	4.54	4.5

Then

$$\chi_{para} = \frac{\mu_0 M}{B} = \frac{\mu_0 N J(J+1) g^2 \mu_B^2}{3kT} \text{ or, } \chi_{para} = \frac{\mu_0 N p_{eff} \mu_B^2}{3kT}. \tag{13.83}$$

This is the same expression for χ_{para} as eqn (13.78).

(b) If $x \gg 1$, i.e., when temperature is low and field $\vec{B}$ is strong, $\coth x \simeq 1$.

$\therefore B_J(x) \simeq 1$.

Eqn (13.77) then takes the form:

$$M = NgJ\mu_B. \tag{13.84}$$

Eqn (13.84) represents the state, where all the dipoles are aligned along the direction of $\vec{B}$. This is the *state of magnetic saturation and takes place at low temperature and strong magnetic field.* Under this limiting condition eqn (13.84) becomes analogous to *Langevin* eqn (13.62).

For large number of allowed orientations of a magnetic dipoles, i.e., $J \to \infty$, we have

$$\coth \frac{x}{2J} \to \frac{2J}{x} \text{ and } \coth\left(1 + \frac{1}{2J}\right) x \to \coth x.$$

$$\therefore B_J(x) \to \left(\coth x - \frac{1}{x}\right) = L(x). \tag{13.85}$$

Eqn (13.85) indicates that for $J \to \infty$, the quantum mechanical value of M approaches that due to classical value. For other values of J the two results differ widely.

Discussion

The paramagnetic susceptibility given by eqn (13.83) is found to depend on the following factors

(i) χ_{para} *is inversely proportional to temperature T of the material. For example, the value of* χ_{para} *calculated from* eqn (13.83) is ~ 10^{-7} at room temperature. But at 1 K, χ_{para} becomes ~ 10^{-5}, i.e., hundred times greater.

(ii) χ_{para} is proportional to the number of atoms N and the strength of the magnetic field $\vec{B}$.

(iii) χ_{para} is proportional to the effective number of Bohr magneton (p_{eff}).

(iv) Eqn (13.83) is unable to account the experimental results on susceptibility for the ions of iron group. This is due to the presence of crystal field which cannot be ignored compared to the applied field.

13.7.3 Pauli's Theory of Paramagnetism

Substances like metals, free atoms or ions with a partially filled inner shell, e.g., *transition elements* and atoms, molecules and lattice defects possessing an odd number of electrons, e.g., sodium atoms, gaseous nitric oxide, F-centres in alkali halides exhibit paramagnetism. *Classical theory* fails to account the paramagnetic susceptibility of the conduction electrons, since it is expected that the magnetization of most normal non-ferromagnetic metals is independent of the temperature of the specimen.

Pauli applied *Fermi-Dirac distribution* and obtained satisfactory result which agrees with experimental observations.

According to *Pauli's exclusion principle* two electrons with opposite spin can occupy the same region in $\vec{k}$ space. So a graph of total density of states versus the KE of two components with

$$g(E) = \frac{V}{2\pi^2}\left(\frac{2m}{\hbar^2}\right)^{3/2} E^{1/2} \tag{13.86}$$

will be of the form as shown in Fig. 13.13(a).

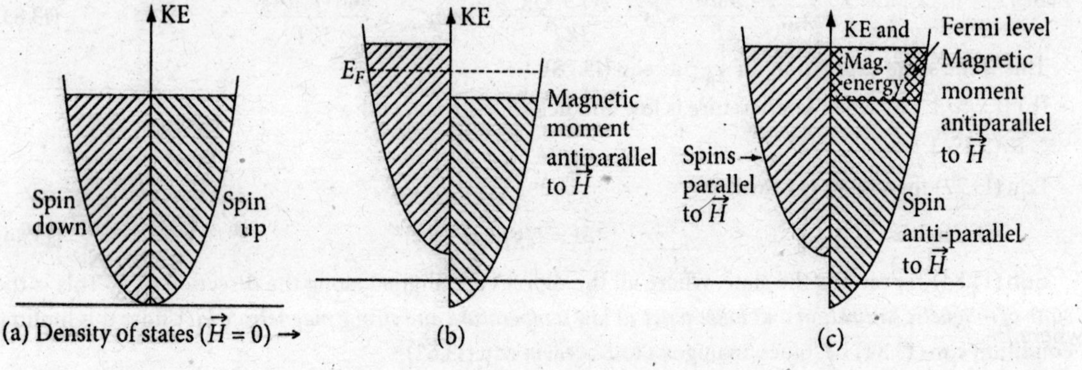

Fig. 13.13: Free electron model to describe paramagnetism in presence of $\vec{H}$

In Fig. 13.13(a), states with spin up and spin down are similarly occupied when the magnetic field $\vec{H} = 0$. Figs 13.13(b) and (c) show the occupancy when $\vec{H}$ is applied. The cross-hatched area represents the excess up-spin electrons.

The energy of a Fermi particle subjected to $\vec{H}$ is

$$E = \frac{p^2}{2m} - \mu H, \tag{13.87}$$

when spin is parallel to $\vec{H}$. For anti-parallel spin

$$E = \frac{p^2}{2m} + \mu H, \tag{13.88}$$

where μ is the *intrinsic moment of the particle*.

All the energy levels up to *Fermi level* are filled at $T = 0$. Then the KE of the first group will be from 0 to $(E_p + \mu H)$ and those of second group will be from 0 to $(E_p - \mu H)$. Some of the electrons, with anti-parallel spin having magnetic moments, parallel to $\vec{H}$, will flip over and orient parallel to $\vec{H}$ to maintain common electrochemical potential of the system. This will give rise to paramagnetism.

Paramagnetic Susceptibility

If N^+ and N^- be the particles in the two groups respectively, then

$$N^+ = \frac{V}{4\pi^2\hbar^3}(2m)^{3/2}\int_0^{E_F+\mu H} E^{1/2}dE \tag{13.89}$$

$$= \frac{V}{4\pi^2\hbar^3}(2m)^{3/2}(E_F+\mu H)^{3/2}$$

$$= \frac{V}{6\pi^2\hbar^3}(2m)^{3/2}(E_F+\mu H)^{3/2}. \tag{13.90}$$

Likewise

$$N^- = \frac{V}{6\pi^2\hbar^3}(2m)^{3/2}(E_F-\mu H)^{3/2}. \tag{13.91}$$

∴ the net magnetic moment is

$$M = \mu(N^+ - N^-) = \frac{\mu V(2m)^{3/2}}{6\pi^2\hbar^3}\left[(E_F+\mu H)^{3/2} - (E_F-\mu H)^{3/2}\right]. \tag{13.92}$$

For $H \to 0$, i.e., low field, $\mu_H < E_F$

$$\therefore M = \frac{\mu V(2m)^{3/2}}{6\pi^2\hbar^2}\left[E_F^{3/2} + \frac{3}{2}E_F^{1/2}\mu H - E_F^{3/2} + \frac{3}{2}E_F^{1/2}\mu H\right] = \frac{\mu V}{6\pi^2\hbar^3}3E_F^{1/2}\mu H. \tag{13.93}$$

Hence, *paramagnetic susceptibility,*

$$\chi_{\text{para}} = \frac{M}{VH} = \frac{\mu^2(2m)^{3/2}}{2\pi^2\hbar^3}\sqrt{E_F} = \frac{\mu^2(2m)^{3/2}}{2\pi^2 E_F\hbar^3}E_F^{3/2} \tag{13.94}$$

$$\text{or, } \chi_{\text{para}} = \frac{3\mu^2 N}{2E_F V} = \frac{3}{2}\frac{\mu^2 N}{kT_F V}, \quad [\because E_F = kT_F] \tag{13.95}$$

where $E_F = \frac{\hbar^2}{2m}\left(\frac{3\pi^2 N}{V}\right)^{2/3}$.

T_F is the *Fermi temperature,* which is constant for a given metal and hence, χ_{para} is independent of temperature (T). Except some deviation, where diamagnetic susceptibility contributes, the *Pauli's theory of paramagnetism agrees closely with experimental results.*

13.7.4 Paragmagnetism of Ionic Solids

The transition group elements (rare earth group) having *incomplete* $4f$ *shell* and the iron group elements having *incomplete* 3d *shell* show paramagnetic properties. The paramagnetic susceptibility (χ_{para}) [Eqn (13.78)] with J values obtained by *Hund's rule,* i.e.,

$J = L - S$, for less than half-filled shells
$J = 0$, for just half-filled shells
and $J = L + S$, for more than half-filled shells

is in excellent agreement with experimental results for many rare earth group of ions.

But for iron group ions, the χ_{para} values obtained from eqn (13.78) differ widely from the experimental values. Why?

The $3d$ shell in iron group ions is the outermost shell and is exposed to *Crystal Field* due to the neighbouring ions. The crystal field destroys the rotational symmetry of such crystals; as a result the total angular momentum differs from the proper quantum number. Also the average values of L_Z is reduced to zero. This effect is known as the *quenching of the orbital angular momentum L.*

Hence, putting $J = 0 + S = S$, we get from eqn (13.77),

$$\chi_{para} = \frac{\mu_0 N \mu_B^2}{3kT} g^2 S(S+1). \tag{13.96}$$

This values of χ_{para} is in good agreement with experimental values for iron group.

Example: χ_{para} for Mn^{3+} ion is zero according to eqn (13.77), but according to eqn (13.96), χ_{para} with $S = 2$ is in good agreement with experimental results.

Table 13.3 shows the calculated values of *effective magneton number* for Iron Group Ions.

Table 13.3: Calculated values of effective magneton number for iron group ions.

Iron Group Ions	*Electronic Configuration*	*Basic level*	p_{eff} *(exptl)*	p_{eff} (calculated) $p_{eff} = g\sqrt{J(J+1)}$	p_{eff} (calculated) $p_{eff} = 2\sqrt{S(S+1)}$	p_{eff} (calculated) $p_{eff} = [L(L+1)+4S(S+1)]^{1/2}$
Ti^{3+}, V^{4+}	$3d^1$	$^2D_{3/2}$	1.8	1.55	1.73	3.01
V^{2+}	$3d^2$	$^3F_2$	2.8	1.63	2.83	4.49
$Cr^{3+}V^{2+}$	$3d^3$	$^4F_{3/2}$	3.8	0.77	3.87	5.20
$Mn^{3+}Cr^{3+}$	$3d^4$	$^5D_0$	4.9	0	4.90	5.50
$Fe^{3+}Mn^{2+}$	$3d^5$	$^6S_{1/2}$	5.9	5.92	5.92	5.92
Fe^{2+}	$3d^6$	$^5D_4$	5.4	6.70	4.90	5.50
CO^{2+}	$3d^7$	$^4F_{9/2}$	4.8	6.63	3.87	5.21
Ni^{2+}	$3d^8$	$^3F_4$	3.2	5.59	2.87	4.49
Cu^{2+}	$3d^9$	$^2D_{5/2}$	1.9	3.55	1.73	3.01

13.7.5 Cooling by Adiabatic Demagnetization

Temperature as low as 0.3 K can be attained by *evaporating liquid helium under low pressure.* But P. Debye and *W.F. Giaugue* suggested that, *adiabatic demagnetization of paramagnetic salt* can produce temperature ~ 0.001 K and lower than this value. The principle of the method is that *at fixed temperature the entropy of a system of magnetic moments is lowered by the application of an external magnetic field.*

The variation of entropy of a system of magnetic moments is shown in Fig. 13.14. The curve OCA represents the variation of entropy (S) with temperature (T) in absence of the magnetic field $\vec{B}$. If at temperature T_1, $\vec{B}$ be applied on the paramagnetic material, the magnetic dipoles will tend to align along $\vec{B}$. The spin system will become more orderly and hence, the entropy of the system will decrease from A to B_1 along the path AB. If now the paramagnetic specimen is isolated from its surroundings and the magnetic field $\vec{B}$ is withdrawn, then the system will suffer an adiabatic transformation and goes from B_1 to C along B_1C with no change in entropy (i.e., $\Delta S = 0$). The temperature of the system at C is T_1, where $T_1 < T_2$. By repeating the process the temperature of the specimen can be reduced to ~ 0.001 K.

From *Maxwell's thermodynamical relations* we can show that the entropy of the system in presence of external magnetic field $\vec{B}$ is

$$S = S_0 + \int_0^B \left(\frac{\partial M}{\partial T}\right)_B dB, \qquad (13.97)$$

where M is the magnetization and S_0 is the entropy at 0 K.

Now from eqn (13.76),

$$M = \frac{Ng^2\mu_B^2 B}{3kT} J(J+1).$$

$$\therefore \frac{\partial M}{\partial T} = -\frac{Ng^2\mu_B^2 B}{3kT^2} J(J+1).$$

$$\therefore \int_0^B \left(\frac{\partial M}{\partial T}\right)_B dB = -\frac{Ng^2\mu_B^2 J(J+1)}{3kT^2} \int_0^B B\,dB$$

$$= -\frac{Ng^2\mu_B^2 B^2}{6kT^2} J(J+1). \quad (13.98)$$

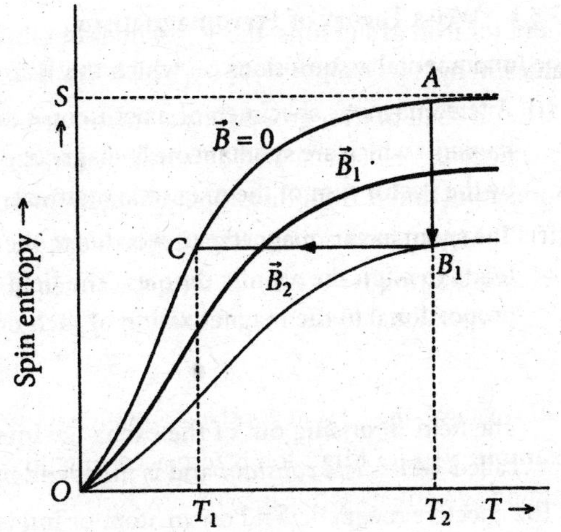

Fig. 13.14: Entropy variation with temperature without and with magnetic field

Substituting eqn (13.98) in eqn (13.97), we get

$$S = S_0 - \frac{Ng^2\mu_B^2 B^2}{6kT} J(J+1). \qquad (13.99)$$

This shows that the entropy S of the system at T_1 with magnetic field B will be lower than the entropy S_0 at T_1 in absence of magnetic field (i.e., $S < S_0$) and will correspond to S_0 at lower temperature T_2. This is obtained by *adiabatic change* along B_1C (i.e., $\Delta S = 0$) in absence of the magnetic field. Hence, by *adiabatic demagnetization* temperature below 1 K can be attained.

13.8 Ferromagnetism

The ferromagnetism also arises due to the presence of permanent magnetic dipole moment like paramagnetism. But in ferromagnetism *the adjacent magnetic dipoles are oriented in a particular direction even if the external magnetic field be zero.* Hence, ferromagnetic materials exhibit magnetic moment in absence of magnetic field — a phenomenon called *spontaneous magnetization.*

The important features of ferromagnetic materials are as follow:

(i) *The magnetization $\vec{M}$ produced in ferromagnetic materials is high even if the applied magnetic field $\vec{B}$ is negligible.*

(ii) *The ferromagnetic susceptibility χ is not constant but varies with the applied field. The variation of χ with $\vec{B}$ exhibits hysteresis curve.*

(iii) *The ferromagnetic susceptibility is given by*

$$\chi = \frac{C}{T - T_c}, \text{ [Curie-Weiss law]}$$

where T_c is the Curie temperature above which the ferromagnetic substance behaves as a paramagnetic one. At $T \le T_c$, spontaneous magnetization occurs.

The important ferromagnetic elements are Fe, Ni, Co, Gd and Dy. A number of alloys and oxides like MnBi, MnAs, CrO_2, etc. also exhibit ferromagnetic property.

13.8.1 Weiss Theory of Ferromagnetism

The fundamental assumptions on which the *Weiss' theory* is based are the following:

(i) A ferromagnetic specimen of macroscopic dimension contains a large number of small regions called *domains* which are spontaneously magnetized. The magnetization of the whole material is determined by the vector sum of the magnetic moments of individual domains.

(ii) The spontaneous magnetization is due to the existence of *molecular field* $\vec{B}_l$ within the material which tends to align the atomic dipoles. The field B_l is quite strong being ~ 1000 T and is assumed to be proportional to the magnetization of each domain, i.e.,

$$\vec{B}_l = \lambda \vec{M}. \tag{13.100}$$

The field $\vec{B}_l$ arising out of the exchange interaction is known as the *Weiss field* and λ is a constant called *Weiss-field constant* and is independent of temperature.

The effective magnetic field on an atom or ion is given by

$$\vec{B}_{\text{eff}} = \vec{B} + \vec{B}_l = \vec{B} + \lambda \vec{M}. \tag{13.101}$$

As $\vec{B}$, $\vec{B}_l$, $\vec{M}$ and $\vec{B}_{\text{eff}}$ are all in the same direction,

$$B_{\text{eff}} = B + \lambda M. \tag{13.102}$$

If N be the number of atoms per unit volume of the ferromagnetic material, then as in case of paramagnetism eqn (13.81), the magnetization

$$M = NgJ\mu_B B_J(x), \tag{13.103}$$

where

$$B_J(x) = \frac{2J+1}{2J}\coth\left(\frac{2J+1}{2J}\right)x - \frac{1}{2J}\coth\left(\frac{x}{2J}\right) \tag{13.104}$$

and

$$x = \frac{gJ\mu_B B_{\text{eff}}}{kT} = \frac{gJ\mu_B}{kT}(B + \lambda M). \tag{13.105}$$

For *spontaneous magnetization*, $B = 0$.

$$\therefore x = \frac{gJ\mu_B \lambda M}{kT} \tag{13.106}$$

$$\text{or, } M(T) = \frac{xkT}{\lambda g J \mu_B} \tag{13.107}$$

For $T \to 0$, $x \to \infty$ and $B_J(x) \to 1$. The magnetic moments of each domain align along the field and magnetization M under this condition becomes *saturated magnetization* $M_s(0)$, given by

$$M_s(0) = NgJ\mu_B. \tag{13.108}$$

From eqns (13.107) and (13.108) we get

$$\frac{M(T)}{M_s(0)} = \frac{xkT}{\lambda N g^2 J^2 \mu_B^2}. \tag{13.109}$$

Also from eqns (13.81) and (13.108), we get

$$\frac{M(T)}{M_s(0)} = B_J(x). \tag{13.110}$$

The simultaneous solutions of eqns (13.109) and (13.110) can be obtained by plotting $\frac{M}{M_s(0)}$ against x for $T > T_c$, $T = T_c$ and $T < T_c$ which is a straight line for different values of T. But the plot of $\frac{M}{M_s(0)}$ against

x according to eqn (13.109) for different T values gives a curve as shown in Fig. 13.15. The point of intersection of the straight line and the curve of Fig. 13.15 gives the solution for $M(T)$.

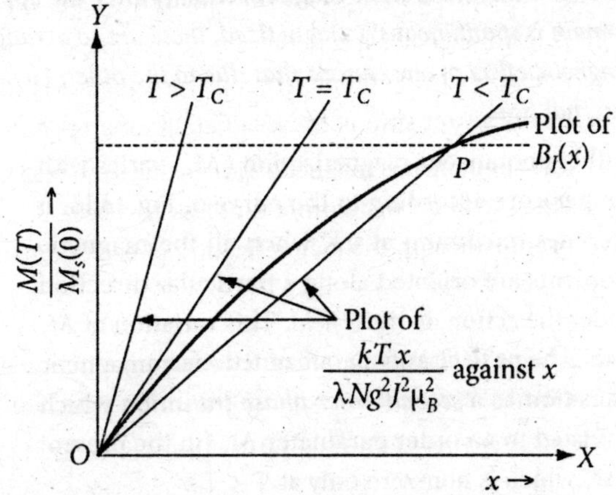

Fig. 13.15: Graphical solution of the simultaneous equations (10.109) and (10.110). The point P determines the spontaneous magnetization $M_s(T)$ at a given temperature. For $T > T_C$, no solution exists.

Discussion

(i) At critical temperature or Curie Temperature (T_c): From Fig. 13.15 it is evident that at T_c the straight line given by eqn (13.109) is tangent to the curve given by eqn (13.110) at the origin O.

The straight line has slope given by

$$\frac{M(T)}{M_s(0)} = \frac{kT}{\lambda N g^2 J^2 \mu_B^2} x. \tag{13.111}$$

At $T = T_c$,

$$\frac{\left(\frac{M(T)}{M_s(0)}\right)}{x} = \frac{kT_c}{\lambda N g^2 J^2 \mu_B^2}. \tag{13.112}$$

Again, for $x \ll 1$,

$$B_J(x) = \frac{J(J+1)}{3J} x \text{ [from eqn (13.104)].}$$

So the tangent to the curve at the origin has a slope

$$\frac{B_J(x)}{x} = \frac{J(J+1)}{3J}. \tag{13.113}$$

Equating the two slopes from eqns (13.112) and (13.113), we get

$$\frac{kT_c}{\lambda N g^2 J^2 \mu_B^2} = \frac{(J+1)}{3J} \text{ or, } T_c = \frac{\lambda N g^2 J^2 \mu_B^2}{3k} J(J+1) = \frac{N\mu_j^2 \lambda}{3k}, \tag{13.114}$$

where

$$\mu_j = \text{the total magnetic moment per atom} = g_j \mu_B \sqrt{J(J+1)}. \tag{13.115}$$

Eqn (13.114) shows that, *T_c is proportional to the Weiss field constant λ.*

(ii) Below Curie Temperature (T_c): At $T < T_c$, the straight line given by eqn (13.109) and the curve given by eqn (13.110) intersects at $x = 0$ and at P. The intersection at $x = 0$ represents *unstable magnetization*, whereas the intersection at P represents non-zero value of M for $\vec{B} = 0$ and corresponds to *spontaneous magnetization.*

Substituting

$$\frac{k}{\lambda N g^2 J^2 \mu_B^2} = \frac{1}{T_c} \frac{(J+1)}{3J}$$

in eqn (13.111), we get

$$\frac{M(T)}{M_s(0)} = \frac{T}{T_c}\left(\frac{J+1}{3J}\right) x. \tag{13.116}$$

Eqn (13.116) gives the maximum possible values of $M_s(0)$ and shows that $M_s(0)$ is inversely proportional to temperature (T).

Practically, a specimen of ferromagnetic at $T < T_c$ is not magnetized in absence of external magnetic field ($\vec{B}$). But it becomes magnetized only after the application of $\vec{B}$. *This is due to the fact that though each domain is spontaneously magnetized, these are so arranged to form closed chains within the material and the magnetic effect of one cancels that due to the other. Hence, the material becomes unmagnetized in absence of external field.*

The spontaneous magnetization (M_s) varies with temperature according to the curve of Fig. 13.16. It becomes maximum at 0 K when all the magnetic moments are oriented along a particular direction under the action of *Weiss field*. This variation of M_s with T helps to classify paramagnetic-ferromagnetic transition as a *second order phase transition* which is related to an order parameter M_s (in the present case) which is non-zero only at $T < T_c$.

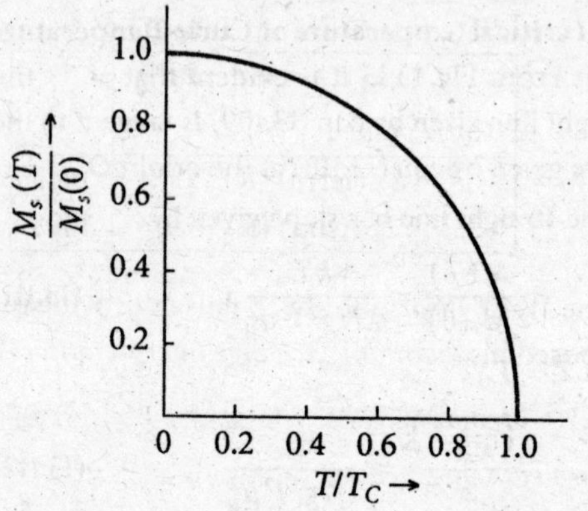

Fig. 13.16: Spontaneous magnetization versus temperature at $T < T_C$

(iii) Above Curie Temperature ($T > T_c$): Above Curie temperature, spontaneous magnetization is not possible. An applied magnetic field can produce only magnetization in this case. In this region, ($T > T_c$), we find from eqn (13.105) that $x \ll 1$.

$$\therefore B_J(x) \simeq \left(\frac{J+1}{3J}\right) x \text{ [from eqn (13.104)]}$$

Eqns (13.103) and (13.105) reduce to

$$M = \frac{Ng\mu_B(J+1)x}{3}, \left[x = \frac{gJ\mu_B(B+\lambda M)}{kT}\right]. \tag{13.117}$$

$$\therefore M = \frac{Ng^2\mu_B^2 J(J+1)}{3kT}(B+\lambda M).$$

Then susceptibility,

$$\chi = \frac{\mu_0 M}{B} = \frac{\mu_0 T_c/\lambda}{T-T_c} \text{ or, } \chi = \frac{C}{T-T_c}, \tag{13.118}$$

where

$$C = \frac{\mu_0 T_c}{\lambda} \tag{13.119}$$

$$\text{and } T_c = \frac{\lambda N g^2 \mu_B^2 J(J+1)}{3k}. \tag{13.120}$$

Eqn (13.118) is known as the *Curie-Weiss law*. Curie-Weiss law describes the temperature dependence of susceptibility in the paramagnetic region with satisfaction, if temperature is well above the Curie temperature.

From eqn (13.118)

$$\frac{1}{\chi} = \frac{T}{C} - \frac{T_c}{C}.$$

Thus, the graph between $\frac{1}{\chi}$ versus T is a straight line as shown in Fig. 13.17.

For $\chi \to \infty$, $\frac{1}{\chi} = 0$, then $T = T_c$. *Curie-Weiss law* agrees quite accurately with experimental results. The slight curvature near T_c represents the distinction between ferromagnetic and paramagnetic materials. The difference between ferromagnetic Curie point T_c and paramagnetic Curie point T_c' is, however, very small as shown in Table 13.4.

Table 13.4

Material → / *Curie point ↓*	Fe	CO	Ni
T_c (ferro) (K)	1043	1393	631
T_c' (para) (K)	1093	14.28	650

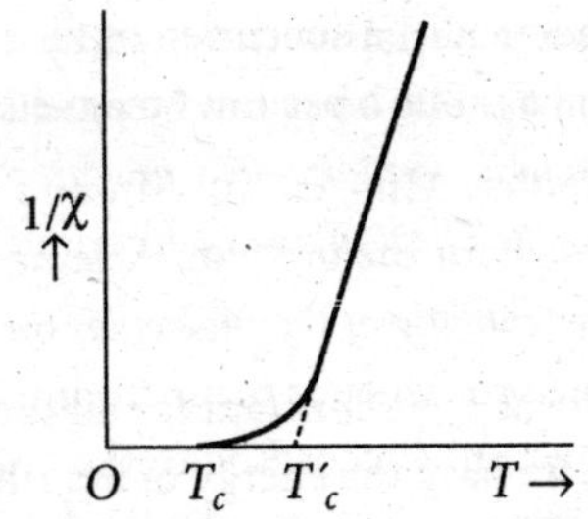

Fig. 13.17: Distinction between ferromagnetic and paramagnetic Curie points

The *Weiss' theory*, however, fails to provide the explanation of the origin of *Weiss field* until *Heisenberg* proposed his *quantum exchange interpretation* (1928) for the origin of Weiss field in ferromagnetism.

13.8.2 Origin of Weiss Molecular Field (Heisenberg's Exchange Interaction)

Weiss field calculated on the basis of magnetic dipole-dipole interaction is only of the order of 10^3 G, whereas the actual field strength is observed to be to the order of 10^7 G. Hence, Weiss field cannot be simply due to magnetic dipole-dipole interaction.

Heisenberg in 1928 first proposed that the large Weiss field is due to *quantum mechanical exchange interaction* between the electron spin. The exchange interaction is due to the *Pauli's exclusion principle.* According to this principle, *the spatial charge distribution is disturbed due to any change in the relative orientation of the two spins, and interaction between two atoms. The extent of overlap of the wave functions of two neighbouring atoms and the relative orientation of the electron spin determine the strength of the exchange interaction.* Hence, exchange interaction is *electrostatic and non-magnetic in nature.*

Heisenberg formulated his theory on the analogy of *Heitler-London* model of chemical bonding. In view of this theory, it can be shown that the exchange energy between two atoms is

$$U_{ij} = -kJ_e\vec{S}_i \cdot \vec{S}_j, \tag{13.121}$$

where k is a constant, J_e is the exchange integral and $\vec{S}_i$ and $\vec{S}_j$ are the spin angular momenta of the ith and jth atoms respectively. The value of U_{ij} depends on the overlap of charge distribution of two atoms and hence, their interatomic distance is r_{ij}. The plot of J_e against r_{ij}/r_{3d}, where r_{3d} is the radius of the $3d$ orbital, is shown in Fig. 13.18. It is evident from this figure that J_e is positive for large interatomic distances and negative for $\left(r_{ij}/r_{3d}\right) < 3$.

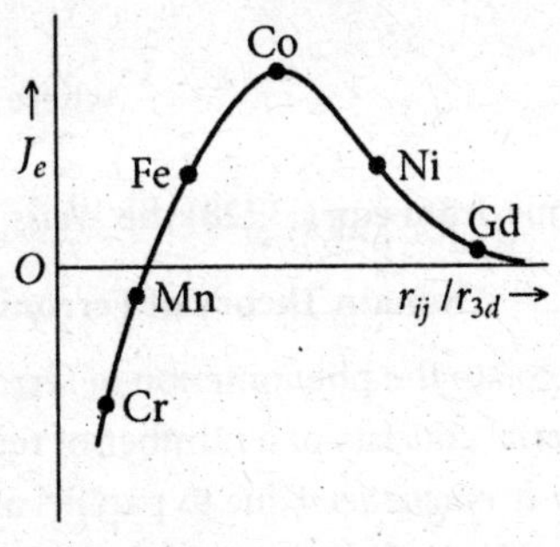

Fig. 13.18: Variation of exchange interaction J_e with r_{ij}/r_{3d}

N.B. *It was conclusively proved through the gyromagnetic experiment conducted by Einstein, de Haas and Barnett that ferromagnetism is due to spin moments and not orbited moments.*

For J_e value positive, the interaction energy U_{ij} is negative and hence, minimum for parallel arrangement of spin and is more stable than anti-parallel spins. The atom under this condition exhibits ferromagnetism. In view of this, Fe, CO, Ni and Gd are ferromagnetic. But $\left(r_{ij}/r_{3d}\right)$ is less than 3 for Cr and Mn and hence, these are not ferromagnetic.

Two materials with $(r_{ij}/r_{3d}) < 3$ can be combined to a compound for which $(r_{ij}/r_{3d}) > 3$ and thus a ferromagnetic body can be produced.

Examples: Mn–As, Cu–Mn and Mn–Sb alloys are ferromagnetic as r_{ij}/r_{3d} for them is greater than 3, although for the individual elements $(r_{ij}/r_{3d}) < 3$. *This is the greatest triumph for solid state theory that by proper cooking of the molecules we can fabricate materials of any desired characteristics.*

Now if J_e be constant for all the neighbouring pairs, then considering the effect of neighbouring pairs only, we have the energy of the ith atom as

$$U_i = -kJ_e \sum_j \vec{S}_i \cdot \vec{S}_j. \tag{13.122}$$

$\sum_j$ represents the summation over the nearest neighbours of the ith atom. If the number of nearest neighbours be Z, then

$$U_i = -kZJ_e\left(S_{xi}\langle S_{xj}\rangle + S_{yi}\langle S_{yj}\rangle + S_{zi}\langle S_{zj}\rangle\right), \tag{13.123}$$

where the instantaneous values of neighbouring spins have been replaced by their time averages. Considering magnetization M along Z-axis only,

$$\langle S_{xj}\rangle = \langle S_{yj}\rangle = 0 \text{ and } \langle S_{zj}\rangle = \frac{M}{g\mu_B N}. \tag{13.124}$$

$$\therefore U_i = -\frac{kZJ_e S_{zi} M}{gN\mu_B} \tag{13.125}$$

$$\text{or, } U_i = -gS_{zi}\mu_B B_1, \tag{13.126}$$

$$\text{where } B_1 = \frac{kZJ_e}{Ng^2\mu_B^2} M \tag{13.127}$$

$$\text{or, } B_1 = \lambda M, \tag{13.128}$$

$$\text{where } \frac{kZJ_e}{Ng^2\mu_B^2} = \lambda \text{ is a constant.} \tag{13.129}$$

Thus, from eqn (13.128) the *Weiss field B_1 is proportional to magnetization M.*

13.8.3 Domain Theory of Ferromagnetism

To explain the phenomenon of ferromagnetism *Weiss* proposed that a virgin specimen of ferromagnetic material consists of a number of region (~ 10^{-6} m or larger) called *domains*. These domains are *spontaneously magnetized* due to parallel alignment of all spin magnetic moments in a particular direction. The direction of alignments of the spin magnetic moments varies from domain to domain and as such the resultant magnetic moment of the specimen may be zero or nearly zero. In presence of an external magnetic field, the random domains align in the following ways:

(i) By the motion of domain walls: The size of the domains in which the magnetic moments subtend the minimum angle w.r.t. the external field $\vec{B}$ increases with the increase of B [Fig. 13.19(b)]. The domain in which the magnetic moments are oriented at larger angles w.r.t. $\vec{B}$ suffers a decrease in size with the increase of $\vec{B}$ [Fig. 13.19(c)]. This type of change in domain size leads to a diminution in the magnetic energy, producing more stable configuration. The process is known as *displacement process of magnetization* as the changes in domain size are produced due to the displacement of *domain boundaries.*

(ii) By the rotation of domains: If the applied magnetic field $\vec{B}$ be strong, the rotation of the direction of the magnetic moment towards $\vec{B}$ occurs as shown in Fig. 13.19(c). When the magnetic moment of the domain turns fully in the direction of $\vec{B}$, *technical saturation* occurs.

(iii) By the Para-Process: If $\vec{B}$ be further increased, then the magnetization M increases by a small amount. We assume that in the spontaneously magnetized domains the spin moments of all the atoms are aligned along $\vec{B}$. But this occurs strictly at $T = 0$ K. At higher temperature complete magnetic saturation in a domain cannot be achieved due to thermal agitation and some of the spin moments may align anti-parallel to $\vec{B}$. At higher magnetic field these magnetic moments may be made to align parallel to $\vec{B}$, whereby magnetization increases slightly beyond technical saturation. *This is the para-process.*

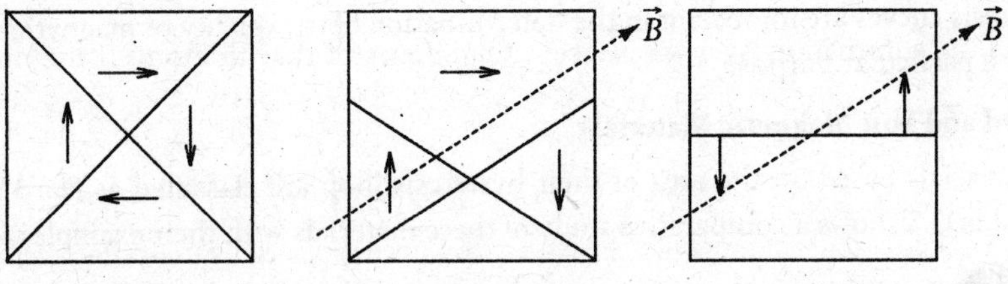

Fig. 13.19: (a) Ferromagnetic domain, (b) Domain magnetized by wall moment, (c) Domain magnetized by rotation

13.8.4 Hysteresis

Hysteresis in ferromagnetic materials is referred to the lag of magnetization behind the magnetizing field. Ferromagnetic materials show hysteresis below the ferromagnetic Curie temperature (T_c). The phenomenon of hysteresis can be well understood by the concept of *domain*. The increase of resultant magnetic moment of a ferromagnetic specimen under the action of an applied magnetic field (B) is due to (i) *the motion of domain walls* and (ii) *the rotation of the domains.*

If a *weak magnetic field B* be applied, then the domains whose spin magnetic moments are aligned along B and in the easy direction of magnetization, grow in size at the expense of less favourably oriented domains due to the motion of the domain walls. After the removal of B, the domains reverse back to the original size. This reversible domain wall displacement is indicated by the portion OA of the magnetization curve [Fig. 13.20].

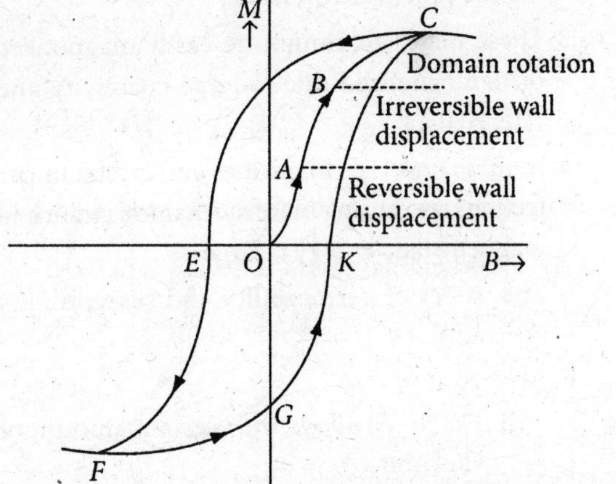

Fig. 13.20: Hysteresis loop of a ferromagnetic material

If the *applied field B be stronger,* then the movement of domain walls becomes irreversible as indicated by the portion AB of the magnetization curve. This irreversible domain wall movement is responsible for the phenomenon of hysteresis.

The point B represents the situation, where all the domains are aligned along their easy directions. If the value of B be still further increased, then the rotation of the domains takes place in a direction away from the easy direction and thereby an *isotropic energy* is stored in the specimen. At the completion of rotation of the domains, *magnetic saturation* occurs as represented by the point *C* of the curve. If now the applied field *B* be withdrawn, the specimen tends to attain the original configuration by the motion of domain walls. But this tendency is opposed by the *impurities, lattice imperfection,* etc. present in it. So more energy is required to overcome the opposite action. This is equivalent to say that, a negative field, called the *coercive force,* is required to reduce the magnetization to zero. This energy is actually a loss and increases the temperature of the specimen.

Hence, ***hysteresis is the loss of energy in taking a ferromagnetic body through a complete cycle of magnetization and is proportional to the area enclosed by the hysteresis loop.***

The hysteresis curves are important in the determination of the quality of magnetic materials and their selection for a particular purpose.

13.8.5 Hard and Soft Magnetic Materials

Magnetic materials based on the area of their hysteresis loop are classified as Hard and Soft magnetic materials. Table 13.5 shows a comparative study of these materials with their examples and important field of applications.

Table 13.5: Distinction between Hard and Soft magnetic materials

Hard Magnetic Materials	*Soft Magnetic Materials*
1. These materials have large area of the hysteresis loop and hence large hysteresis loss.	1. The hysteresis loss is small as the area of hysteresis loop is less.
2. Here the impurities, imperfections in the lattice oppose the domain wall movement making the process irreversible.	2. Here the domain wall movement is easier. The change of magnetization even for low field is large.
3. These materials cannot be easily magnetized or demagnetized due to large coercivity and retentivity.	3. These materials can be easily magnetized or demagnetized due to small coercivity and retentivity.
4. In these materials impurities and crystal imperfections are causing more mechanical strain and magnetostatic energy is large.	4. The magnetostatic energy in these materials is small due to the absence of irregularities.
5. The values of permeability and susceptibility are small.	5. Permeability and susceptibility of these materials are large.
6. *Examples*: (i) Fe–Ni–Al alloys with certain amount of Co (Alnico). (ii) Cu–Ni–Fe alloys (cunif) (iii) Cu–Ni–Co alloys (cunico)	6. *Examples*: (i) Fe–Si alloys (ii) Ferrites (iii) Garnets
7. Used for production of permanent magnets which are used in magnetic detectors, flux meters, magnetic separators, etc.	7. Used in electromagnets, which are used in electromagnetic machinery and transformer's cores, matrix storage of computers, etc.

13.9 Antiferromagnetism

Heisenberg showed that, *quantum mechanical exchange interactions* produce large internal fields and the neighbouring magnetic dipoles tend to align in the same direction [Fig. 13.21(a)]. It can also be proved quantum mechanically that, for small separation of the interacting dipoles, the dipole moments of the neighbouring atoms have a tendency of anti-parallel alignments [Fig. 13.21(b)] by the action of the exchange forces. This occurs below a certain temperature T_N, called the *Neel temperature,* and the corresponding material is known *as antiferromagnetic material.*

Fig.13.21: (a) Ferromagnetism produced by parallel spin moments of neighbouring dipoles

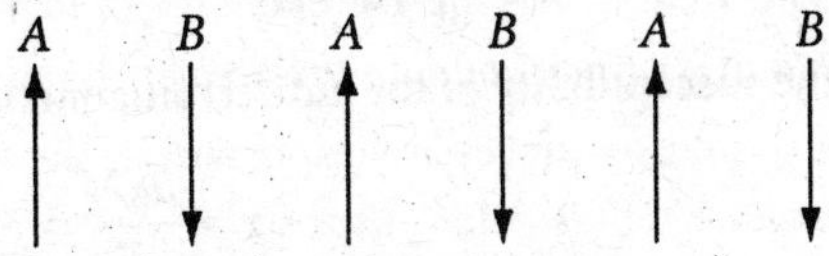

Fig.13.21: (b) Neighbouring antiparallel spin moments of *A* and *B* sublattices in antiferromagnetism

Theory of Antiferromagnetism

We consider that, an antiferromagnetic material consists of two sub-lattices *A* and *B* so that an atom of one type has all the nearest neighbours of the opposite type. In antiferromagnetic materials besides nearest neighbour negative *AB* interactions, nearest neighbour *AA* (or *BB*) weak negative interaction is also present.

The internal fields at *A*- and *B*-sites can be expressed as

$$\left.\begin{aligned} B_{mA} &= -\alpha M_A - \beta M_B \\ B_{mB} &= -\beta M_A - \alpha M_B \end{aligned}\right\}, \tag{13.130}$$

where α and β are *Weiss field constants for AA* (or *BB*) and *AB* interaction respectively and M_A and M_B are the corresponding magnetization.

If *B* be the *external field,* then

$$\left.\begin{aligned} B_A &= B - \alpha M_A - \beta M_B \\ \text{and } B_B &= B - \beta M_A - \alpha M_B \end{aligned}\right\}. \tag{13.131}$$

Now we consider the situation in two different cases:

Case I. $T > T_N$, *i.e.,* above Neel temperature (T_N).

If the temperature of the material be above Neel temperature, it behaves as a paramagnetic. The magnetization of sub-lattices *A* and *B* can be written as

$$\left.\begin{aligned} M_A &= \frac{N\mu^2}{3kT} B_A \\ M_B &= \frac{N\mu^2}{3kT} B_B \end{aligned}\right\}, \tag{13.132}$$

by eqn (13.66), where $\mu^2 = \mu_B^2 g^2 J(J+1)$.

Here we have assumed $N_A = N_B = N$.

$\therefore$ the total magnetization is given by

$$M = M_A + M_B = \frac{N\mu^2}{3kT} B_A + \frac{N\mu^2}{3kT} B_B = \frac{N\mu^2}{3kT}(B_A + B_B) = \frac{N\mu^2}{3kT}[2B - (\alpha + \beta)(M_A + M_B)].$$

$\because$ from eqn (13.131),

$$(B_A + B_B) = [2B - (\alpha+\beta)(M_A + M_B)] = \frac{N\mu^2}{3kT}[2B - (\alpha+\beta)M]\ [\because M_A + M_B = M]$$

$$\text{or, } M\left[1 + \frac{N\mu^2}{3kT}(\alpha+\beta)\right] = \frac{N\mu^2}{3kT}2B \tag{13.133}$$

$$\text{or, } M = \frac{\frac{2}{3}\frac{N\mu^2 B}{k}}{T + \frac{N\mu^2}{3k}(\alpha+\beta)}, \tag{13.134}$$

$\therefore$ the susceptibility of the antiferromagnetic material can be expressed as

$$\chi = \frac{\mu_0 M}{B} = \frac{2\mu_0 N\mu^2/3k}{T + \frac{N\mu^2(\alpha+\beta)}{3k}} \text{ or, } \chi = \frac{C}{T+\theta}, \tag{13.135}$$

where

$$C = \frac{2\mu_0 N\mu^2}{3k} \tag{13.136}$$

$$\text{and } \theta = \frac{N\mu^2(\alpha+\beta)}{3k}, \tag{13.137}$$

where θ is called the *asymptotic Curie temperature.*

Eqn (13.135) is comparable with $\chi = \frac{C}{T-\theta}$ for ferromagnetic materials above *Curie temperature* except eqn (13.135) contains $(T+\theta)$ instead of $(T-\theta)$ [or $(T-T_c)$]. The *Curie constant* C in eqn (13.135) is twice the value of Curie constant for individual A and B lattices. To illustrate the difference of paramagnetic, ferromagnetic and antiferromagnetic we draw curves of χ against T in the above three cases [Fig. 13.22].

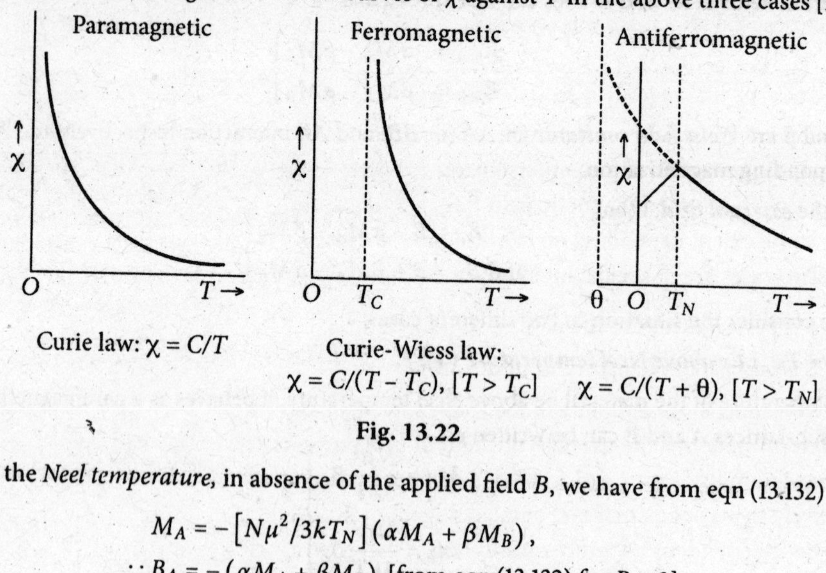

Fig. 13.22

At $T = T_N$, the *Neel temperature*, in absence of the applied field B, we have from eqn (13.132)

$$M_A = -\left[N\mu^2/3kT_N\right](\alpha M_A + \beta M_B),$$

$$\because B_A = -(\alpha M_A + \beta M_B)\ [\text{from eqn (13.132) for } B = 0]$$

$$\text{or, } M_A\left[1 + \frac{N\mu^2\alpha}{3kT_N}\right] + \left[\frac{N\mu^2\beta}{3kT_N}\right]M_B = 0. \tag{13.138}$$

Similarly,

$$M_B\left[1 + \frac{N\mu^2\alpha}{3kT_N}\right] + \left[\frac{N\mu^2\beta}{3kT_N}\right]M_A = 0. \tag{13.139}$$

The above equations have the trivial solution for $M_A = M_B = 0$. This means, there is no spontaneous magnetization for above *Neel temperature*. For spontaneous magnetization to occur there should exist a solution of the simultaneous equations for a finite value of M_A and M_B. This is possible, if

$$\begin{vmatrix} \left(1+\frac{C\alpha}{2T_N}\right) & \frac{C\beta}{2T_N} \\ \frac{C\beta}{2T_N} & \left(1+\frac{C\alpha}{2T_N}\right) \end{vmatrix} = 0 \text{ or, } \left(1+\frac{C\alpha}{2T_N}\right)^2 - \left[\frac{C^2\beta^2}{4T_N^2}\right] = 0$$

$$\text{or, } 1+\frac{C^2\alpha^2}{4T_N^2}+\frac{C\alpha}{T_N}-\frac{C^2\beta^2}{4T_N} = 0 \text{ or, } 4T_N^2 + C^2\alpha^2 = 4C\alpha T_N - C^2\beta^2 = 0$$

$$\text{or, } T_N^2 + C\alpha T_N + \frac{C^2}{4}(\alpha^2-\beta^2) = 0$$

$$\text{or, } T_N = \frac{-C\alpha \pm \sqrt{C^2\alpha^2 - C^2(\alpha^2-\beta^2)}}{2} = \frac{-C\alpha \pm C\beta}{2}$$

$$\text{or, } T_N = \frac{C(\beta-\alpha)}{2}. \tag{13.140}$$

Eqn (13.140) shows that the *Neel temperature* T_N increases as *AB* interaction (β) is stronger, but T_N decreases as *AA* or *BB* interaction (α) is stronger.

The difference between para-, ferro- and antiferromagnetic materials above critical temperature is represented in Fig. 13.23.

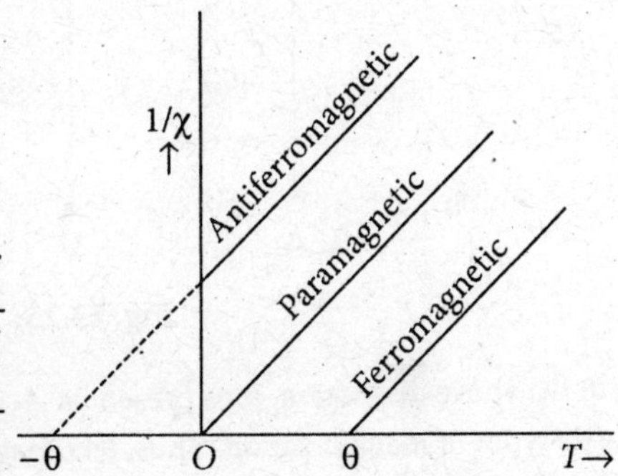

Fig. 13.23: $1/\chi$ versus T plot for para-, ferro- and antiferromagnetic materials below critical temperature

Case II. $T < T_N$, *i.e., below Neel temperature*: In the temperature region below T_N, the two sub-lattices *A* and *B* are *spontaneously magnetized* along their preferred direction of magnetization. The net magnetization of the antiferromagnetic material is, however, zero. The applied magnetic field $\vec{B}$ magnetize the specimen to some extent depending on the direction of application of $\vec{B}$. Two cases are of special interest:

(i) When $\vec{B}$ is *perpendicular* to the natural spin direction.

(ii) When $\vec{B}$ is *parallel* to the natural spin direction.

The antiferromagnetic susceptibility in the above two cases were separately determined and the average of the two is found out for the net susceptibility, i.e., $\chi = \frac{1}{2}(\chi_\perp + \chi_\parallel)$, where $\chi_\perp$ and $\chi_\parallel$ refer respectively to perpendicular and parallel directions of $\vec{B}$. $\chi_\perp$ is found to be independent of temperature. But $\chi_\parallel$ is found to decrease continuously with temperature and $\chi_\parallel \to 0$ as $T \to 0$. This implies that for $T > T_N$, the internal field is very strong which restricts the change of orientation of spins of one sub-lattice to the orientation of spins of the other sub-lattice even in presence of $\vec{B}$. Fig. 13.24 shows the variation of $\chi_\perp$, $\chi_\parallel$ and χ with temperature, where $\chi_\perp = \chi_\parallel = \chi$ at $T = T_N$.

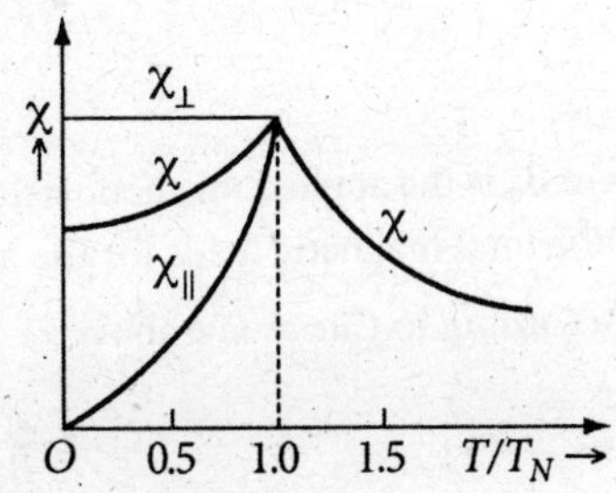

Fig. 13.24: Variation of χ with T/T_N for an antiferromagnetic material

13.10 Ferrimagnetism

Ferrimagnetism is identical with antiferromagnetism, where the opposing magnetic moments of two sets of atoms in the sub-lattices are of different magnitude resulting a large net magnetization. Ferrites which are basically the oxides of various metal elements are the examples of ferromagnetic. The chemical formula for the ferrites may be written as XY_2Z_4, where X is a divalent negative ion, Z is mostly divalent oxygen ion O^{2-} and Y is Fe^{3+}ion. Chemical composition of ferrites may be written as $M_e^{2+}Fe_2^{3+}O_4^{2-}$, where M_e^{2+} represents variety of divalent ions like F_e^{2+}, CO^{2+}, Mn^{2+}, Cd^{2+}, Mg^{2+}, etc. The most common example of ferrite is magnetite or ferrous ferrite, Fe_3O_4 or FeO, Fe_2O_3.

The state of ferric ions Fe^{3+} with spin is $S = 5/2$ and zero orbital moment. The contribution to saturation magnetic moment is $5\mu_B$. Similarly, for Fe^{2+} ferrous ions the spin state is $S = 2$ and the contribution to magnetic moment is $4\mu_B$ in addition to any residual orbital magnetic moment, where μ_B is the *Bohr magneton.* If all the ions of $Fe^{2+}Fe^{3+}O_4^{2-}$ are aligned in parallel, then the magnetic moment per mole $= (5 \times 2 + 2 \times 2)\,\mu_B = 14\mu_B$. But the observed value is $4.1\mu_B$. This discrepancy can be explained if the two Fe^{3+} ions are anti-parallel so that their net moment is zero. The observed value of magnetic moment will then be due Fe^{2+} ions having value $4\mu_B$ [Fig. 13.25].

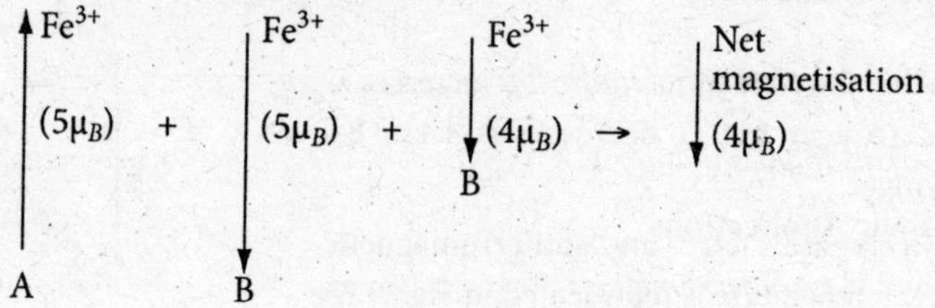

Fig. 13.25: The net magnetization of Fe_3O_4

In the above discussion, ions present at *A*- and *B*-sites constitute *A* and *B* sub-lattices respectively with opposite types of magnetization. Thus, ferrimagnetics like ferromagnetics possess spontaneous magnetization below $T < T_N$ and exhibit hysteresis. *Neel* described the ferrimagnetic property by attributing anti-parallel arrangement of spins of *A*- and *B*-sites to the negative *AB* interaction. Apart from that negative *AA* and *BB* interactions which are much weaker than *AB* interaction are also present.

An expression for the ferrimagnetic susceptibility can be obtained as follows:

Let the atoms *A* and *B* are magnetically not equivalent. Then the magnetic moments of *A* and *B* for anti-parallel alignment may be written as

$$\vec{B}_A = \vec{B}_0 - \lambda_A \vec{M}_B$$

$$\text{and } \vec{B}_B = \vec{B}_0 - \lambda_B \vec{M}_A,$$

where $\vec{B}_0$ is the applied magnetic field and λ_A and λ_B are constants independent of temperature. $\vec{B}_A$ and $\vec{B}_B$ are the total magnetic fields for the atoms *A* and *B* respectively.

According to *Curie's law* above, $T > T_N$.

$$\frac{C_A}{T} = \frac{\vec{M}_A}{\vec{B}_0 - \lambda_A \vec{M}_B}, \quad \therefore \vec{M}_A = \frac{C_A}{T}\left(\vec{B}_0 - \lambda_A \vec{M}_B\right). \tag{13.141}$$

Similarly,

$$\vec{M}_B = \frac{C_B}{T}\left(\vec{B}_0 - \lambda_B \vec{M}_A\right), \tag{13.142}$$

where

$$C_A = N_A\left(\frac{e\hbar g_A}{2m}\right)^2 \frac{J_A(J_A+1)}{3k} \text{ and } C_B = N_B\left(\frac{e\hbar g_B}{2m}\right)\frac{J_B(J_B+1)}{3k}.$$

Using the value of M_B from eqn (13.142) in eqn (13.141), we get

$$\vec{M}_A = \frac{C_A}{T}B_0 - \frac{C_A\lambda_A}{T}\left[\frac{C_B}{T}(\vec{B}_0 - \lambda_B\vec{M}_A)\right] = \left(\frac{TC_A - C_AC_B\lambda_A}{T^2}\right)\vec{B}_0 + \frac{C_AC_B\lambda_A\lambda_B}{T^2}\vec{M}_A.$$

Putting $C_AC_B\lambda_A\lambda_B = T_C^2$, we get after rearranging

$$\vec{M}_A\left(1-\frac{T_C^2}{T^2}\right) = \frac{TC_A - C_AC_B\lambda_A}{T^2}\vec{B}_0. \tag{13.143}$$

Similarly,

$$\vec{M}_B\left(1-\frac{T_C^2}{T^2}\right) = \frac{TC_B - C_AC_B\lambda_B}{T^2}\vec{B}_0. \tag{13.144}$$

Adding eqns (13.143) and (13.144), we get

$$(M_A + M_B)(T^2 - T_C^2) = [T(C_A + C_B) - C_AC_B(\lambda_A + \lambda_B)]\vec{B}_0.$$

Now the ferrimagnetic susceptibility,

$$\chi = \frac{\mu_0(M_A+M_B)}{B_0} = \frac{\mu_0[T(C_A+C_B) - C_AC_B(\lambda_A+\lambda_B)]}{(T+T_C)(T-T_C)}. \tag{13.145}$$

$\therefore \chi \to \infty$ for $T \to T_C$. *This implies that for $T < T_N$, there is a spontaneous magnetization.*

13.10.1 Ferrites and Their Applications

Ferrite forms an important class of *ferrimagnets* having large magnetization at room temperature and high resistivity. The net magnetization due to the *antiferromagnetic interaction* is zero as the sites *A* and *B* are equal and anti-parallel. In special cases the spins in these two sites do not cancel each other and we get a class of substances called *ferrimagnets* or ferrites exhibiting spontaneous magnetization below *Neel temperature.*

Ferrites have spinel structure same as the structure of $MgAl_2O_4$. In this structure the oxygen atoms are closely packed in face-centred lattices into the interstices of which the metal ions are distributed. In a unit cell of *magnetite*, there are 8 molecules with 32 divalent oxygen ions, 16 trivalent iron ions and 8 divalent iron ions per unit cell. Due to the arrangement of oxygen atoms in face-centred structure, there are 8 tetrahedral voids (*A*-sites) and 16 octahedral voids (*B*-sites). In the *spinel structure* of $Mg\,Al_2O_4$, the *divalent ions* occupy the sites *A* and the *trivalent ions* sites *B*. But magnetite (the natural ferrite) has an inverse spinel structure. Here the $8Fe^{3+}$ ions occupy all the *A*-sites and the other $8Fe^{3+}$ ions half the number of *B*-sites. The remaining *B*-sites are filled by Fe^{++} ions. In *nickel ferrite*, the divalent nickel ions occupy half of the *B*-sites and the remaining *B*-sites as well as the *A*-sites are filled by trivalent iron ions. As several ion-combinations in *A*- and *B*-sites are possible, a large variety of ferrite may exist.

As it has been discussed earlier, in magnetite, Fe_3O_4 (or, FeO, Fe_2O_3), the Fe^{3+} ions have spin state with spin $S = \frac{5}{2}$ and zero orbital moment. Thus, each ion contributes $5\mu_B$ to the saturation magnetic moment, whereas the Fe^{2+} ions have spin 2 and the contribution to the magnetic moment is $4\mu_B$. Hence, the effective number of Bohr magnetons per Fe_3O_4 unit is $(2 \times 5 + 4)\mu_B = 14\mu_B$, if all the spins are parallel. The observed magnetic moment is $4.1\mu_B$, μ_B being *Bohr magneton*. This discrepancy between theory and experiment, can, however, be resolved if we assume that, the moments of two Fe^{3+} ions are anti-parallel so that their net magnetic moment is zero. Then the magnetic moment of Fe_3O_4 will be due to the Fe^{2+} ions which is $4\mu_B$. Thus, the discrepancy is explained.

Mixed ferrites like Mn Zn ferrites (Ferroxcube IV) have great technical importance under the trade name *Ferroxcube.* Mixferrites have higher saturation magnetization than simple ferrites.

Application of Ferrites

Ferrites have great technical importance as follows:

(i) Ferrites are insulators having resistivity 10^{10} Ω-m and do not have eddy current loss problem as in iron specially at high frequencies. For this reason ferrites are used in *high frequency transformers* and *high frequency generators.*

(ii) The ferrites are used as core computer memories because of their low coercive force and short switching time.

(iii) Gyromagnetic properties are used in developing *isolators* and *circulators* having application in radar communication.

(iv) Ferrites are used in microwave devices based on *Faraday effect* in a ferromagnetic medium.

13.10.2 Ferrites and Their Properties

Ferrites are ferrimagnetic materials which are prepared from electrically non-conducting materials. Magnetite $\left[Fe^{2+}O^{2-}\cdot Fe^{3+}, O_3^{2}\right]$ is a common ferrite which forms a class of mixed oxides [MO, Fe_2, O_3, where M stands for a divalent material like Mn, Fe, Ni, Cu, Zn].

Ferrites have the following general electrical and magnetic properties:

(i) They are mechanically *hard, brittle* and difficult to work with machine.

(ii) They are prepared using *ceramic method.*

(iii) Although they show similar hysteresis curve, domain structure and motion similar to ferromagnets, the coercive force and saturation magnetization are smaller.

(iv) Curie temperature for ferrites varies from 100 °C to several hundred degrees centigrade.

(v) Ferrites have resistivity $(\sim 10^{12}\ \Omega\text{-cm})$ and hence, are insulators without eddy current loss.

(vi) The permeability is high (500–1000) for mixed Ni–Zn ferrites $(Ni_{0.36}Zn_{0.64}Fe_2O_3)$ at low frequencies. At high frequencies ≥ 300 MHz, permeability is ~ 10.

(vii) They have *low dielectric loss* as microwave dielectric constant is ~ 10 to 12.

(viii) Saturation magnetization and Curie temperature of some ferrimagnetics (ferrites) are shown in Table 13.6.

Table 13.6

Ferrites	*Saturation magnetization* (Wb/m^2)	*Curie temperature* (°C)
Fe_2O_3	0.60	585
$MnFe_2O_4$	0.49	300
$NiFe_2O_4$	0.34	585
$MgFe_2O_4$	0.15	440

Table 13.7: Distinction Between Magnetic Materials

Type	*Magnitude and sign of* χ	*Temperature dependence*	*Remarks*	*Examples*
Diamagnetism	Small, negative ($\sim 10^{-6}$)	Independent	No permanent dipoles, small magnetic effects.	Organic materials, light elements, Alkali earth, Bismuth.
Paramagnetism	Small, positive	Independent	Permanent dipoles randomly oriented, in absence of B net magnetization in any given direction is zero.	Alkali and transition metals.
	large, positive	$\chi = \frac{C}{T-\theta}$		Rare earths.
Ferromagnetism	Very large, positive	For $T > \theta_c$, $\chi = \frac{C}{T-\theta}$. For $T < \theta_c$, $\chi \rightarrow$ complex, $\theta_c \rightarrow$ Curie temperature	Due to large internal field, permanent dipoles align in the same direction. A large spontaneous magnetization results in absence of $\vec{B}$.	Some transition and rare earth metals
Antiferro-magnetism	Small, positive	When $T > T_N$, $\chi = \frac{C}{T+\theta}$; when $T < T_N$, $\chi \propto T$; $T_N \rightarrow$ Neel temperature	Due to negative molecular field, magnetic ions on lattice site–A align anti-parallel to those on lattice site–B. Also $M_A = M_B$.	Salts of transition elements
Ferrimagnetism	Very large, positive	For $T > T_N$, $\chi = \frac{C}{T \pm \theta}$. For $T < T_N$; $\chi \rightarrow$ complex	Special case of antiferromagnetism. Opposed moments are of different magnitudes and a large magnetization occurs.	Ferrites

13.11 Answers of Some Important Questions

13.1. *What is a magnetic material? What are its different classes?*

Ans. Any material that can be magnetized by the application of an external magnetic field is called a *magnetic material.*

The different types of magnetic materials are:

(i) Diamagnetic materials, viz., Cu, Si, He, Au, etc.

(ii) Paramagnetic materials, viz., O, N, $FeCl_2$, etc.

(iii) Ferromagnetic materials, viz., Fe, Co, Ni, etc.

(iv) Antiferromagnetic materials, viz., MnO, FeO, NiO, etc.

(v) Ferimagnetic materials, viz., Fe_3O_4, Mg Al_2O_4, etc.

13.2. *Where are the origins of magnetic moment in an atom?*

Ans. The origins of permanent magnetic moments in an atom are as follows:

(i) The orbital magnetic moment of the electron arising out of the orbital motion of the electron around the nucleus.

(ii) The spin magnetic moment of the electron due to its orbital motion around the nucleus.

(iii) The spin magnetic moment of the nucleus about its own axis.

13.3. *What is a Bohr magneton?*

Ans. Bohr magneton is the unit of magnetic movements of atomic system.

If *e*, *m* be the charge and mass of an orbiting electron in the first Bohr orbit, then its magnetic moment is *Bohr magneton* given by

$$\mu_B = \frac{e\hbar}{2m} = 9.27 \times 10^{-24} \text{ A-m}^2,$$

where $\hbar = \frac{h}{2\pi}$, h is the *Planck's constant.*

13.4. *What is Landé splitting factor? Why is it so called?*

Ans. The energy level of an electron splits up by the action of an external magnetic field B by an amount $\Delta E = g\mu_B B$, since the split up energy ΔE of the energy level of the electron is determined by the factor g and hence, the name *splitting factor*. The concept was introduced for the first time by spectroscopist *Landé* and hence, the name *Landé splitting factor.*

13.5. *What is Hund's rule? Explain with example.*

Ans. See Art. 13.5.2.

13.6. *Discuss the origin of Diamagnetism. Does diamagnetic susceptibility depend on temperature?*

Ans. In an atom the electrons revolve in closed orbits around the nucleus. The electron orbit may be considered as current carrying closed coil producing magnetic field perpendicular to its plane. The electrons in atom revolve in discrete orbits whose planes are not parallel to each other and the direction of rotation of the electrons is not the same. This is why the vector sum of the magnetic moments due to all the revolving electrons on an average becomes zero.

If the electron circuit be subjected to a changing magnetic flux, an induced current is set up in it, which according to *Lenz's law,* opposes the change in flux. The induced magnetic field in the coil opposes the applied field and produces diamagnetism. Diamagnetism is produced in all materials but it is suppressed by the stronger paramagnetic, ferromagnetic effects.

Diamagnetic susceptibility is of the order of -10^{-16} (very small) and independent of temperature.

13.7. *What is the origin of paramagnetism? Discuss the temperature dependence of paramagnetic susceptibility.*

Ans. The paramagnetism originates from the permanent magnetic moment of its constituent atoms or molecules. The completely filled electronic shells contribute to diamagnetism. But as the magnetic moments due to orbital and spin motions are self-neutralizing, they cannot give rise to paramagnetism. Paramagnetism originates due to some of the incomplete electronic shells in atoms or molecules in which permanent dipole moment exists.

The paramagnetic susceptibility χ_{para} is small but positive and depends on temperature of material. The temperature dependence of χ_{para} is expressed by Curie law which holds good for $\mu_B \ll kT$:

$$\chi_{\text{para}} = \frac{C}{T}, \text{ i.e., } \chi_{\text{para}} \propto \frac{1}{T},$$

where C is the Curie constant and T is the temperature. The order of χ_{para} at room temperature is $\sim 10^{-5}$.

13.8. *What is magneto-caloric effect? Discuss how temperature as low as* 10^{-3} K *can be produced.*

Ans. The process of production of low temperature of the order of 10 K by the adiabatic demagnetization is called the *magneto-caloric effect.*

For the answer of the second part of the question, see Art. 13.7.5.

13.9. *Discuss the origin of ferromagnetism and point out its important features.*

Ans. See Art. 13.8.

13.10. *How does internal field (Weiss-field) arise in ferromagnetic crystals?*

Ans. According to Weiss a ferromagnetic material of macroscopic dimension contains a number of small regions called *domain* which are spontaneously magnetized.

In a domain, the spontaneous magnetization arises due to the alignment of the magnetic moments of the adjacent atoms in the same direction. The alignments of all the magnetic moments in a domain is caused due to the *quantum mechanical exchange interaction* of the electrons which is electrostatic in nature. This exchange interaction is called *Weiss field* or *internal field* which arises due to the overlap of the wave functions of two neighbouring atoms. As the electrons are indistinguishable they belong to both the neighbouring atoms and as such the symmetry or antisymmetry of the wave function strongly influences the energy of the system. The nature and extent of exchange interaction depends on the exchange symmetry between the spin and extent of overlap of the wave functions.

The internal field $\vec{B}_I$ is given by $\vec{B}_I = \lambda \vec{M}$, where λ is Weiss constant and $\vec{M}$ is the magnetization. So $\vec{B}_I \propto \vec{M}$. $\vec{B}_I$ is generally very strong being $\sim$ 1000 T.

13.11. *What is Curie-Weiss law? Discuss the significance of Curie temperature.*

Ans. The Curie-Weiss law for ferromagnetic material is given by $\chi = \frac{C}{T-T_C}$, where C is Curie constant and T_C is the Curie Temperature. The above law shows that for

(i) $T > T_C$, χ is positive and the material behaves as paramagnetic.

(ii) $T = T_C$, the transition from ferromagnetism to paramagnetism occurs as χ is ∞.

(iii) $T < T_C$, χ is negative, *spontaneous magnetization* takes place.

Thus, Curie temperature T_C is that temperature above which the spontaneous magnetization vanishes. It separates the disorder paramagnetic state at $T < T_C$ from the order ferromagnetic phase at $T > T_C$ of the material.

13.12. *Discuss the variation of spontaneous magnetization with temperature for ferromagnetic materials.*

Ans. We know that spontaneous magnetization arises due to the interaction between the neighbouring atoms which tend to align their spins in the same direction. Magnetization of the material becomes maximum when all spins are aligned parallel to a particular direction. This long range alignment takes place due to the strong nearest neighbour interactions and the continuity of the crystal. This alignment of spins is, however, obstructed by thermal vibration of the atoms of crystal. Hence, maximum magnetization occurs only when thermal vibration is stopped, i.e., at 0 K. If the temperature T is increased, magnetization is decreased due to thermal vibration and becomes zero at Curie temperature (θ). The variation of spontaneous magnetization $M_S(T)$ with temperature is shown in Fig. 13.26.

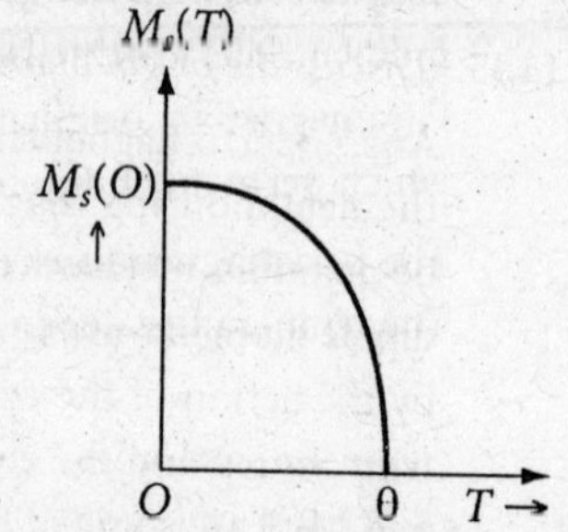

Fig. 13.26: Temperature dependence of spontaneous magnetization

At $T > \theta$, the ferromagnetic material behaves like paramagnetic having susceptibility

$$\chi = \frac{C}{T-\theta},$$

which is the Curie-Weiss law. Here C is the Curie constant and θ is the paramagnetic Curie temperature.

Now the ferromagnetic Curie temperature θ_f is less than paramagnetic Curie temperature $(\theta_f < \theta)$ and $(\theta - \theta_f)$ is small. These elements have large value of χ above θ_f and they do not obey *Curie-Weiss law.* Fe, CO, etc. obey *Curie-Weiss law* but above θ_f, χ deviates from exact agreement of Curie-Weiss' law. The behaviour of the aforesaid metals above θ_f is interesting. Because though the thermal motion of the atoms has destroyed the long range spin order, some short range spin order still exists giving rise to magnetization.

13.13. *How the domain structure can be experimentally demonstrated?*

Ans. The domain structure proposed by Weiss in ferromagnetics can be demonstrated by microphotographs of domain boundaries by *magnetic powder pattern.* In this method a drop of colloidal suspension of finely divided ferromagnetic material (e.g., magnetite) is placed on the surface of a ferromagnetic crystal. It is observed through microscope that the colloidal particles become strongly concentrated about certain well defined lines, which are the boundaries between the domains that are magnetized in different directions. This is due to the fact that the strong magnetic field in the vicinity of the colloidal particles attracts the magnetic particles of the solution.

13.14. *What is magnetostriction? How does the magnetostriction energy originate?*

Ans. A ferromagnetic material subjected to external magnetic field undergoes a change in dimension. This phenomenon is known as *magnetostriction.* The magnitude of deformation is different for different direction of the crystal but independent of the direction of the applied magnetic field. The dimension may increase or decrease depending on the nature of the material. As for example, the length of a nickel rod undergoes a decrease, whereas that for a permalloy undergoes a decrease due to the application of a magnetic field. If an iron rod be placed in an alternating magnetic field, the rod vibrates with twice the frequency of the applied field. The amplitude of vibration assumes the maximum value when the frequency of the field be varied to the natural frequency of the rod. Magnetostriction is successfully applied in ultrasonic generators.

The energy due to the mechanical stresses generated by the rotation of the domains is called the *magnetostriction energy.*

13.15. *Discuss the origin of antiferromagnetism.*

Ans. In ferromagnetism, the quantum mechanical exchange forces produce large internal field and the neighbouring magnetic dipoles tend to align in the same direction [Fig. 13.19(a)]. It can also be quantum mechanically proved that for small separation of the interacting magnetic dipoles, the dipole moments of the neighbouring atoms have a tendency of anti-parallel alignments [Fig. 13.19(b)] by the action of the exchange forces. This occurs below a certain temperature T_N called the *Neel temperature* and the corresponding material is called *antiferromagnetic material.* MnO, FeO, CoO and MnS are some of the examples of antiferromagnetic materials. Above Neel temperature these materials exhibit paramagnetism. In absence of external magnetic field and below Neel temperature the net magnetic moment of the antiferromagnetic is zero. Magnetization develops by the application of external magnetic field.

13.16. *Explain Neel's model for ferrimagnetism?*

Ans. The ferrimagnetic materials are defined as antiferromagnetic in which the sub-lattice moments are unequal. This leads to a net spontaneous magnetization in the direction of the greater sub-lattice moments. The non-vanishing net magnetization is attributed to the following two main cases:

(i) *the two sub-lattices, A and B, can be occupied by different types and number of magnetic ions and*

(ii) *the two sub-lattices may be related to two different crystallographic sites.*

The ferrimagnetics turn to be paramagnetics if $T > T_N$, T_N being the Neel temperature.

Fe_3O_4, the natural magnetite, is a common example of ferrite. These materials show spontaneous magnetization below Neel temperature.

13.17. *What are ferrites?*

Ans. The non-metallic ferrimagnetic materials are called *ferrites.* They form class of solid oxides of iron. The general chemical formula is $Me^{2+}Fe^{3+}O_4$, where Me^{2+} is a divalent metal ion like Mn, Ni, Co, Mg, Cu, etc. Mixed ferrites are called *Feroxcube,* e.g., Mn Zn.

Ferrites have important electrical and electronic applications.

13.18. *Give the units of the magnetic quantities.*

Ans. The following table shows the units of various magnetic quantities:

Table 13.8

Name of the magnetic quantities	emu-cgs *units*	*Rationalized* MKS (SI) *units*	*Conversion*
Magnetic Induction (B)	gausses = $g^{1/2}/cm^{1/2}$-s	Wb/m^2 = kg/(s-C) = tesla (T) = kg/(s-C)	1 $Wb/m^2 = 10^4$ gauss
Magnetic field strength (H)	oersted = $g^{1/2}/(cm^{1/2}$-s)	ampere/metre	1 A/m = $\frac{4\pi}{10^3}$ Oe
Magnetization (M)	maxwell/cm^2 = $g^{1/2}/(cm^{1/2}$-s)	ampere/metre	1 A/m = $\frac{4\pi}{10^3}$ maxwell/cm^2
Magnetic flex (ϕ)	maxwell ($cm^{3/2} - g^{1/2}$)/s	Wb = (kg-m^2)/s-C = V-s	1 Wb = 10^8 maxwell
Susceptibility (χ)	Unitless	Unitless	$\chi_{SI} = 4\pi\chi_{cgs}$
Permeability (μ_r)	Unitless	Unitless	Same value

1 gauss = 10^{-4} T (telsla)

13.12 Solved Problems

13.1. *A paramagnetic material has a magnetic field intensity of 10^4 A/m. If the susceptibility of the material at room temperature is 3.7×10^{-3}, calculate the magnetization and flux density of the material.*

Solution: We have the susceptibility, $\chi = \frac{M}{H}$ and flux density, $B = \mu_0(M + H)$.

$\therefore M = \chi H = (3.7 \times 10^{-3}) \times 10^4 = \mathbf{3.7\ A\text{-}m^{-1}}$.

Flux density, $B = \mu_0(M + H) = (4\pi \times 10^{-7}) \times 10037 = \mathbf{0.0126\ Wb/m^2}$.

13.2. *The magnetic field strength of a piece of metal is 10^6 A/m. Calculate the magnetization and flux density of the material. [Given: $\chi = 1.5 \times 10^{-3}$]*

Solution: We have the magnetization of the metal.

$M = \chi H = (1.5 \times 10^{-3}) \times 10^6 = \mathbf{1.5 \times 10^3\ A/m}$.

Flux density $B = \mu_0(M + H) = 4\pi \times 10^{-7}(1.5 \times 10^3 + 10^6) = \mathbf{1.257\ T}$.

13.3. *Calculate the diamagnetic susceptibility of atomic hydrogen in the ground state at STP. [Assume the mean square distance of electronic charge distribution of atomic hydrogen from the nucleus $\langle r^2 \rangle = 3a_0^2$, a_0 being the radius of the first Bohr orbit of hydrogen.]*

Solution: We have diamagnetic susceptibility of atomic hydrogen,

$$\chi_{\text{dia}} = -\frac{N\mu_0 Ze^2}{6m}\langle r^2 \rangle = -\frac{N\mu_0 Ze^2}{2m}a_0^2.$$

Here $N = \frac{6.023 \times 10^{26}}{2.24 \times 10^{-2}} = 2.69 \times 10^{28}\ \text{m}^{-3}$, $Z = 1$, $a_0 = 0.46 \times 10^{-10}$.

$$\therefore \chi_{\text{dia}} = -\frac{(2.69 \times 10^{28}) \times (4\pi \times 10^{-7}) \times (1.6 \times 10^{-19})^2 \times (0.46 \times 10^{-10})^2}{2 \times 9.1 \times 10^{-31}} = \mathbf{-1.01 \times 10^{-6}}.$$

13.4. *An atom with L = 2 and 0 spin angular momentum is placed in a uniform magnatic field of induction 2 Wb/m^2. Calculate the rate of precession of the resultant magnetic moment. Also calculate the corresponding linear frequency and the radius of an electron in the ground state.*

Solution: The rate of precession is the Larmor frequency given by $\omega_L = g_L\left(\frac{eB}{2m}\right)$, where

$$g_J = \frac{J(J+1) + S(S+1) - L)L+1)}{2J(J+1)} + 1.$$

Here $L = 2$, $S = 0$, $J = L + S = 2$.

$$\therefore g_J = g_L = \frac{2 \times 3 + 0 - 2 \times 3}{2 \times 2 \times 3} + 1 = 1.$$

So

$$\omega_L = \frac{1 \times (1.6 \times 10^{-19}) \times 2}{2 \times (9.1 \times 10^{-31})} = \mathbf{1.76 \times 10^{11}\ rad/s}.$$

The corresponding linear frequency,

$$\nu_L = \frac{\omega_L}{2\pi} = \frac{1.76 \times 10^{11}}{2 \times 3.14} = \mathbf{2.8 \times 10^{10}\ Hz}.$$

The mean radius of an electron in the atom in the ground state (1s) is

$$\langle r \rangle = \frac{\hbar^2}{me^2} = \mathbf{0.53 \times 10^{-8}\ cm}.$$

13.5. *If $N = 27 \times 10^{23}/\text{cm}^3$ be the atomic density of helium gas and $e^2/mc^2 = 2.8 \times 10^{-13}$ cm, calculate the susceptibility of the He-atom. [$\langle r \rangle = 0.53 \times 10^{-8}$ cm.]*

Solution: The susceptibility of the He-atom is

$$\chi = -2\left(\frac{Ne^2}{6mc^2}\right)\langle r\rangle^2 \text{ [}\therefore \text{ He is diamagnetic and there are two electrons per atom]}$$

$$\text{or, } \chi = \frac{-(27\times 10^{23})\times(2.8\times 10^{-13})\times\left(0.053\times 10^{-8}\right)^2}{3}$$

$$= \mathbf{-0.71\times 10^{-7}\ (emu).}$$

13.6. *If the magnetization and flux density of a magnetic material be* 3200 A/m *and* 0.005 Wb/m^2, *calculate the relative permeability of the material.*

Solution: We have magnetic induction $B = \mu_0(M + H)$, where M and H are the magnetization and magnetizing field respectively.

Here $B = 0.005$ Wb/m^2, $\mu_0 = 4\pi\times 10^7$ and $M = 3200$ A/m.

$$\therefore 0.005 = \left(4\pi\times 10^7\right)(3200 + H) \text{ or, } H = \frac{0.005}{4\pi\times 10^7} - 3200$$

$$\therefore H = \mathbf{780.9\ A/m.}$$

Again, the relative permeability μ_r is given by

$$\mu_r - 1 = \frac{M}{H} = \frac{3200}{780.9} = 4.1 \text{ or, } \boldsymbol{\mu_r = 5.1}$$

13.7. *Calculate the change in magnetic moment of a circulating electron in an applied field of* 2 T *acting perpendicular to the plane of the orbit.* $\left[r = 5.29\times 10^{-11}\text{ m } \textit{for the radius of the orbit}\right]$ **[JNTU May 2004]**

Solution: The induced magnetic moment in the orbit of the electron is

$$\mu_i = \left(\frac{e^2}{4m}\right)r^2 B.$$

Here $r = 5.29\times 10^{-11}$ m, $B = 2$ T, electronic charge $e = 1.602\times 10^{-19}$ C and $m = 9.108\times 10^{-31}$ kg.

$$\therefore \mu_i = \frac{\left(1.602\times 10^{-19}\right)^2\times\left(5.29\times 10^{-11}\right)^2\times 2}{4\times 9.108\times 10^{-31}} = \mathbf{3.943\times 10^{-29}\ Am^2.}$$

13.8. *A paramagnetic system of electronic spin dipole moment is placed in an external field of* 10^5 A/m. (a) *What is the average magnetic moment per dipole at 300 K?* (b) *What is the fractional number of spins parallel to the external field?*

Solution: (a) We have the average magnetic moment per dipole at the temperature T is

$$\langle\mu_m\rangle = \frac{M}{N} = \frac{\mu_0 H\mu_B^2}{kT}.$$

Here $\mu_0 = 4\pi\times 10^{-7}$ henry/m, $H = 10^5$ A/m, $\mu_B = 9.27\times 10^{-24}$ Am^2, $k = 1.38\times 10^{-23}$ and $T = 300$ K.

$$\therefore \langle\mu_m\rangle = \frac{(4\pi\times 10^{-7})\times 10^5\times\left(9.27\times 10^{-24}\right)^2}{(1.38\times 10^{-23})\times 300} = \mathbf{2.6\times 10^{-27}\ Am^2.}$$

(b) If N be the total number and N_p the number of parallel spins, the fractional number of spins which are parallel to the field is

$$\frac{N_p}{N} = \frac{e^{(\mu_0\mu_B H/kT)}}{\left[e^{(\mu_0\mu_B H/kT)} + e^{-(\mu_0\mu_B H/kT)}\right]} = \frac{e^{(2.8\times 10^{-4})}}{\left[e^{(2.8\times 10^{-4})} + e^{-(2.8\times 10^{-4})}\right]} = 0.5.$$

$$\therefore \frac{N_P}{N} = \mathbf{0.5}.$$

13.9. *A paramagnetic material has 10^{28} atoms/m^3. Its susceptibility at 350 K is 2.8×10^{-4}. Calculate the susceptibility at 300 K.*

[JNTU Dec. 2004]

Solution: If C be the Curie constant and T, the temperature of the paramagnetic material, then its susceptibility is given by

$$\chi = \frac{C}{T}.$$

$\therefore$ χ_1 and χ_2 be the susceptibilities at temperatures T_1 and T_2 respectively, then

$$\frac{\chi_1}{\chi_2} = \frac{T_2}{T_1}.$$

Here $\chi_1 = 2.8 \times 10^{-4}$, $T_1 = 350$ K, $T_2 = 300$ K, χ_2 =?

$$\therefore \chi_2 = \frac{\chi_1 T_1}{T_2} = \frac{\left(2.8 \times 10^{-4}\right) \times 350}{300} = \mathbf{3.267 \times 10^{-4}}.$$

13.10. *Assuming that iron has a body-centred cubic structure with an edge length of 2.87 Å, calculate the average number of Bohr magnetons which contributes to the saturation magnetization per atom. [Saturation magnetization of iron = 1.75×10^6 A/m].*

Solution: If n be the total number of atoms per unit volume and μ, the dipole moment per atom, then saturation magnetization is $M_s = n\mu$.

Now

$$n = \frac{2}{\left(2.87 \times 10^{-10}\right)^3}.$$

Since there are two atoms per unit cell in a body-centred cubic crystal having volume of each cubic cell, $V = \left(2.87 \times 10^{-10}\right)^3$ m^3.

$$\therefore n = \frac{2}{\left(2.87 \times 10^{-10}\right)^3} = \frac{2 \times 10^{30}}{(2.87)^3} \text{ m}^{-3}$$

$$\text{and } \mu = \frac{M_s}{n} = \frac{\left(1.75 \times 10^6\right) \times (2.87)^3}{\left(2 \times 10^{30}\right) \times \left(9.27 \times 10^{-24}\right)} = \mathbf{2.23}.$$

[$\because$ 1 Bohr magneton $\mu_B = 9.27 \times 10^{-24}$ JT^{-1}]

$\therefore$ the average number of Bohr magneton which contributes to the saturation magnetization per atom is μ = **2.23.**

13.11. *Gadolinium is ferromagnetic below 16 °C with average magnetic moment per atom 7.1 Bohr magneton.* (a) *What is the magnetic moment per gram?* (b) *What is the value of saturation magnetization? Given that the atomic weight of gadolinium is 157.26 and its density is 7.8×10^3 kg/m^3?*

Solution: (a) We know that 157.26 kg of gadolinium contains = 6.025×10^{26} atoms.

$$\therefore 7.8 \times 10^3 \text{ kg of gadolinium contains} = \frac{7800 \times 6.025 \times 10^{26}}{157.26} \text{ atoms.}$$

So 1 kg of gadolinium contains = 298.836×10^{26} atoms.

1 gm of the element contains = 298.836×10^{23} atoms.

∴ magnetic moment per gram

$$= (298.836 \times 10^{23}) \times 7.1 \times \text{Bohr magneton}$$
$$= (298.836 \times 10^{23}) \times 7.1 \times 9.27 \times 10^{-24}\ \text{Am}^2$$
$$= 1966.849\ \text{Am}^2.$$

Hence, saturation magnetization

$$M_s = N\mu_0\mu_m = (298.836 \times 10^{26}) \times (4\pi \times 10^{-7}) \times (9.27 \times 10^{-24})\ \text{Wb/m}^2$$
$$= \mathbf{2.4726}\ \text{Wb/m}^2.$$

13.12. *Find the order of magnitude of the exchange integral of a ferromagnet having Curie temperature* 727 °C. *What is the internal field?* [*Given:* $\mu_B = 9.3 \times 10^{-21}$ ergs/gauss]

Solution: The exchange integral is given by

$$J = \frac{kT_c}{3},$$

where the Boltzmann constant $k = 1.4 \times 10^{-16}$ ergs/K

$$T_c = 273 + 727 = 1000\ \text{K}$$

$$\therefore J = \frac{(1.4 \times 10^{-16}) \times 1000}{3} = \mathbf{4.66 \times 10^{-14}\ ergs.}$$

Again, internal field in the ferromagnet is

$$H = \frac{kT_c}{\mu_B} = \frac{(1.4 \times 10^{-16}) \times 1000}{9.3 \times 10^{-21}} = \mathbf{1.5 \times 10^7\ gauss.}$$

13.13. *The outer electronic configuration of* Dy^{3+} *ion is* $4f^9\,6s^0$. *Calculate the magnetic susceptibility for a salt containing one* kg-mole *of* Dy^{3+} *ions at* 300 K.

Solution: Unpaired electrons in Dy^{3+} is 5.

Applying Hund's rules,

$$S = 5 \times \tfrac{1}{2} = \tfrac{5}{2},\ L = 3 + 2 + 1 + 0 - 1 = 5,\ \text{and } J = L + S = 5 + \tfrac{5}{2} = \tfrac{15}{2}.$$

$$\therefore g = 1 + \frac{J(J+1) + S(S+1) - L(L+1)}{2J(J+1)} = 1 + \frac{(\frac{15}{2} \times \frac{17}{2}) + (5 \times 6) - 5 \times 6}{2 \times \frac{15}{2} \times \frac{17}{2}}$$
$$= 1 + 0.33 = 1.33.$$

The susceptibility of the material is

$$\chi = \frac{\mu_0 N \mu_B^2}{3kT} g^2 J(J+1)$$

$$\text{or, } \chi = \frac{(4\pi \times 10^{-7}) \times (6.023 \times 10^{26}) \times (9.27 \times 10^{-24})^2 \times (1.33)^2}{3 \times (1.38 \times 10^{-23}) \times 300} \times \frac{15}{2} \times \frac{17}{2} = \mathbf{5.9 \times 10^{-4}},$$

where $N = 6.023 \times 10^{26}$ for 1 kg mole of the salt, $\mu_B = (9.27 \times 10^{-24})\ \text{Am}^2$ and $T = 300$ K.

13.14. *Calculate the internal field for iron having Curie temperatures* 1043 K *and effective magnetic moment of* 2.2 *Bohr magneton.*

Solution: If B_1 be the internal field and M the magnetization, then $B_1 = \lambda M$, λ is a constant of proportionality given by $\lambda = \frac{\mu_0 T_c}{C}$,

$$\text{where } C = \frac{\mu_0 g^2 N\mu_B^2 S(S+1)}{3k} \text{ with } g = 2 \text{ and } M = N\mu_m.$$

Here T_c is the Curie temperature and C is the Curie constant. The other symbols have their usual significances.

The magnetic moment of an iron atom is $= 2.2\mu_B$. So the magnetic moment $\mu_m = g\mu_B[S(S+1)]^{1/2}$ or, $2.2\mu_B = g\mu_B[S(S+1)]^{1/2}$ or, $1.1 = [S(S+1)]^{1/2}$ $[\because g = 2]$.

$\therefore S(S+1) = (1.1)^2$.

Hence,
$$B_1 = \frac{\mu_0 T_c M}{C} = \frac{\mu_0 T_c M \times 3k}{\mu_0 g^2 N\mu_B^2 S(S+1)} = \frac{\mu_0 T_c (N\mu_m) \times 3k}{\mu_0 g^2 N\mu_B^2 S(S+1)}$$

$$= \frac{3kT_c (2.2\mu_B)}{g^2 S(S+1)\mu_B^2} \quad [\because \mu_m = 2.2\mu_B]$$

$$= \frac{3 \times (1.38 \times 10^{-23}) \times 2.2 \times (9.27 \times 10^{-24})}{4 \times (9.27 \times 10^{-24})^2 \times (1.1)^2} \quad [\because \mu_B = 9.27 \times 10^{-24}\ \text{Am}^2]$$

$= \mathbf{2117\ T.}$

13.15. *The adiabatic demagnetization of gadolinium sulphate from an initial magnetic field of* 6200 gauss *to zero yields a final temperature of* 0.346 K. *What is the initial temperature of the specimen.* [*Given: Curie constant* $C = 7.85\ \text{cm}^2$ deg/(gm-ion); *specific heat constant* = 0.32 R].

Solution: If T_1 and T_2 be the initial and final temperatures of the specimen, then for adiabatic demagnetization

$$\left(\frac{T_1}{T_2}\right)^2 = 1 + \frac{H_1^2 \times \text{Curie const}}{\text{specific constant}},$$

where H_1 is the initial field = 6200 gauss.

$$\therefore \left(\frac{T_1}{T_2}\right)^2 = 1 + \frac{(6200)^2 \times (7.85)}{(1.38 \times 10^{-16})^2 \times (6.02 \times 10^{23}) \times 0.32}$$

$$= 1 + \frac{6200^2 \times 7.85}{1.38 \times 6.02 \times 0.32 \times 10^7} = 11.3 + 1 = 12.3$$

$$\text{or, } T_1^2 = 12.3 \times T^2 = 12.3 \times (0.346)^2 \text{ or, } T_1 = \mathbf{1.213\ K.}$$

13.16. *Calculate* (a) *the intrinsic flux density near* 0 K *and* (b) *the ratio of magnetization at* 300 K *in presence of an external field of* 1 mT *to the spontaneous magnetization at* 0 K *of a ferromagnetic material.* [*Given:* $J = \frac{3}{2}$ *and* $g = 2$; *Curie temperature,* $T_c = 125$ K]

Solution: (a) The intrinsic flux density B_1 is given by

$$B_1 = \lambda M = \lambda M_s(0) = \lambda N g J \mu_B \quad [\because M_s(0) = NgJ\mu_B]$$

Again, from

$$T_c = \frac{\lambda N g^2 \mu_B^2 J(J+1)}{3k} = \frac{B_1 g\mu_B (J+1)}{3k},$$

$$B_1 = \frac{3kT_c}{g\mu_B(J+1)} = \frac{3 \times (1.38 \times 10^{-23}) \times 125}{2 \times (9.27 \times 10^{-24}) \times \frac{5}{2}} = \mathbf{111.6\ T.}$$

From equation

$$\chi = \frac{\mu_0 M}{B} = \frac{C}{T - T_c},$$

we have the magnetization

$$M = \frac{CB}{\mu_0 (T - T_c)} = \frac{Ng^2 \mu_B^2 J(J+1)B}{3k(T - T_c)} \left[\because C = \frac{\mu_0 T_c}{\lambda} \text{ and } T_c = \frac{\lambda N g^2 \mu_B^2 J(J+1)}{3k}\right].$$

$$\therefore \frac{M}{M_s(0)} = \frac{g\mu_B B(J+1)}{3k(T - T_c)} \quad [\because M_s(0) = NgJ\mu_B]$$

$$\text{or, } \frac{M}{M_s(0)} = \frac{2 \times (9.27 \times 10^{-24}) \times 10^{-3} \times (5/2)}{3 \times (1.38 \times 10^{-23}) \times (300 - 125)} = \mathbf{6.4 \times 10^{-6}}.$$

13.17. *Show that for europium oxide the paramagnetism at 300 K is much smaller than ferromagnetism at 0 K for an applied field 0.01 T. [Given: The europium ion has $g = 2$, the Curie temperature = 70 K and $J = 7/2$]*

Solution: If M and B be magnetization and magnetic field respectively, then the susceptibility of the specimen is

$$\chi = \frac{\mu_0 M}{B} = \frac{C}{T - T_c} = \frac{\mu_0 N g^2 \mu_B^2 J(J+1)}{3k(T - T_c)}, \therefore M = \frac{Ng^2 \mu_B^2 J(J+1)B}{3k(T - T_c)}.$$

At $T \to 0$, $M(0) = Ng\mu_B J$.

$$\therefore \frac{M(300)}{M(0)} = \frac{g\mu_B (J+1)B}{3k(T - T_c)} = \frac{2 \times (9.27 \times 10^{-24}) \times (4.5) \times (0.01)}{3 \times (1.38 \times 10^{-23}) \times (300 - 70)} = \frac{2 \times 9.27 \times 4.5 \times 0.01}{3 \times 1.38 \times 2300}$$

$$\text{or, } \frac{M(300)}{M(0)} = \mathbf{8.76 \times 10^{-5}}.$$

The specimen behaves as paramagnetic and ferromagnetic at 300 K and 0 K respectively. Since $M(300) = (8.76 \times 10^{-5})M(0)$, therefore, the paramagnetic magnetization is $\sim 10^{-5}$ smaller than the ferromagnetic magnetization. Hence, paramagnetism is much smaller than ferromagnetism.

13.18. *For Cu^{2+} it is given that spectroscopic term is $^2D_{5/2}$ and $(P_{\text{eff}})_{\text{observed}} = 19$.*

(a) *Find the theoretical value of P_{eff}.*

(b) *Comment on the agreement with experimental value.*

Solution: (i) The spectroscopic term of Cu^{+2} is $^3D_{5/2} \Rightarrow D$ state or, $L = 2$ state, $2S + 1 = 2$ or, $S = \frac{1}{2}$, $J = \frac{5}{2}$.

For L-S coupling, $P_{\text{eff}} = g\sqrt{J(J+1)}$

$$g = 1 + \frac{J(J+1) + S(S+1) - L(L+1)}{2J(J+1)} = 1 + \frac{\frac{5}{2}(\frac{5}{2}+1) + \frac{1}{2}(\frac{1}{2}+1) - 2(2+1)}{2 \cdot \frac{5}{2}(\frac{5}{2}+1)} = 1.2.$$

$$P_{\text{eff}} = (1.2)\sqrt{\tfrac{5}{2}(\tfrac{5}{2}+1)} = 3.55 = (P_{\text{eff}})_{\text{observed}} = 1.9.$$

Supposing L-S coupling to be replaced by coupling L and B; S and B,

$$p_{\text{eff}} = \sqrt{g_L^2 L(L+1) + g_S^2 S(S+1)} = \sqrt{L(L+1) + 4S(S+1)}$$
$$= \sqrt{2(2+1) + 4\tfrac{1}{2}(\tfrac{1}{2}+1)} = 3 = (p_{\text{eff}})_{\text{observed}} = 1.9.$$

(ii) The theoretically calculated value of p_{eff} does not tally with observed value of p_{eff} = 1.9. If quenching is considered to take place, then

$$(p_{eff})_{theoretical} = g_s\sqrt{S(S+1)} = 2\sqrt{\tfrac{1}{2}\left(\tfrac{1}{2}+1\right)} = \sqrt{3} = 1.732.$$

This value is close to the experimental value of $(p_{eff})_{observed}$ = 1.9, which indicates parallel quenching of orbital angular momenta.

13.19. *Show that* $\chi_m = \dfrac{z|e|^2\mu_0 n(r^2)}{6m}$ *is dimensionless. Find units of* $\chi_m^{mass}, \chi_m^{molar}$.

Solution: We have

$$\chi_m \sim \frac{C^2(Hm^{-1})(m^{-3})(m^2)}{kg}.$$

H is henry = unit of inductance (L).

By Faraday law,

$$e = -L\frac{di}{dt} \text{ or, } L = e\frac{dt}{di} = \frac{V_s}{A} = \frac{(JC^{-1})s}{C(s^{-1})} = JC^{-2}s^2 = H.$$

$$\therefore \chi_m \sim \frac{C^2m^{-2}}{kg}(JC^{-2}s^2) = Jm^{-2}s^2\,kg^{-1} = (kg\,ms^{-1}\,m)\,m^{-2}\,s^2\,kg^{-1} = \text{no unit.}$$

$$\chi_m^{mass} = \frac{\chi_m}{\rho} = \frac{\chi_m}{kg\,m^{-3}} \to m^3\,kg^{-1}.$$

$$\chi_m^{molar} = \frac{\chi_m}{\rho}M = \frac{\chi_m}{kg\,m^{-3}}\,kg\text{-}mol^{-1}\,m^3\,mol^{-3}.$$

EXERCISE

13(A) Objective (MCQ) Type Questions

Select the correct answer.

13.1. The chemical formula for magnetite is

(a) Fe_2O_3 (b) FeO (c) Fe_3O_4 (d) $Fe(OH)_2$.

13.2. The origin of magnetism in material is the presence of

(a) electric dipole (b) magnetic dipole (c) magnetic monopole (d) electric monopole.

13.3. All materials exhibit

(a) ferromagnetism (b) ferrimagnetism (c) paramagnetism (d) diamagnetism.

13.4. Diamagnetic susceptibility is

(a) small and positive (b) small and negative (c) large and positive (d) large and negative.

13.5. If a diamagnetic material is subjected to an external magnetic field, the induced dipole moment acts along the

(a) applied field (b) opposite to the applied field
(c) perpendicular to the applied field (d) none of these.

13.6. Bohr magneton is the magnetic moment of

(a) an electron spin (b) a nuclear spin (c) an electron orbital motion (d) none of these.

13.7. If the susceptibility of a material is independent of temperature, then it is

(a) ferromagnetic (b) paramagnetic (c) ferrimagnetic (d) diamagnetic.

13.8. The materials in which the permanent dipoles are aligned due to bonding forces are called

(a) diamagnetic (b) paramagnetic (c) ferromagnetic (d) ferrimagnetic.

13.9. The paramagnetic susceptibility varies as

(a) T (b) T^2 (c) $\frac{1}{T}$ (d) $\frac{1}{T^2}$.

13.10. The induced dipole moment (μ_i) of an orbiting electron in presence of an external magnetic field (B) is

(a) $\frac{e^2r^2B}{4m}$ (b) $\frac{e^2B}{4mr^2}$ (c) $\frac{e^2Bm}{4\pi r^2}$ (d) $\frac{e^2B}{4\pi r^2}$.

13.11. In a ferrimagnetic material, the number of atoms with opposite spins is

(a) equal (b) unequal (c) zero (d) none of these.

13.12. Interaction between the neighbouring dipoles is negligible in

(a) antiferromagnetics (b) ferrimagnetics (c) diamagnetics (d) paramagnetics.

13.13. The relative permeability is related to magnetic susceptibility by

(a) $\mu_r = 1 - x$ (b) $\mu_r = \frac{1}{x}$ (c) $\mu_r = x - 1$ (d) $\mu_r = x + 1$.

13.14. The ferromagnetic susceptibility is given by

(a) $\chi = \frac{C}{T+T_c}$ (b) $\chi = C(T + T_c)$ (c) $\chi = \frac{C}{T-T_c}$ (d) $\chi = \frac{CT}{T+T_c}$.

13.15. The paramagnetic-ferromagnetic transition is an example of

(a) first order phase transition (b) second order phase transition

(c) no phase transition (d) neither of these.

13.16. If T_c and T_c' represent respectively ferromagnetic and paramagnetic Curie temperatures, then

(a) $T_c = T_c'$ (b) $T_c > T_c'$ (c) $T_c < T_c'$ (d) $T_c = \frac{1}{T_c'}$.

13.17. The experiment which conclusively proves that ferromagnetism is due to spin magnetic moments and not due to orbital magnetic moment is

(a) Davisson-Germer experiment

(b) Stern-Gerlach experiment

(c) Gyromagnetic experiment by Einstein, de Haas and Barnett.

13.18. Magnetically saturated ferrites can

(a) produce low eddy current (b) interact with electromagnetic waves

(c) cause large hysteresis loss (d) interact with ultrasonic waves.

13.19. The susceptibility of antiferromagnetics is given by

(a) $\chi = \frac{C}{T+\theta}$ (b) $\chi = C(T + \theta)$ (c) $\chi = \frac{C}{T-\theta}$ (d) $\chi = \frac{C}{\theta-T_c}$.

13.20. For simple ferrite, the chemical formula is

(a) $Me^{2+}Fe_2^{3+}O_4^{3+}$ (b) $Me^{3+}Fe^{4+}O_4^{2+}$ (c) $Me^{4+}Fe^{2+}O_4^{3+}$ (d) none of these.

13.21. Ferrimagnetic materials show spontaneous magnetization for

(a) $T > T_N$ (b) $T = T_N$ (c) $T < T_N$ (d) $T = 1/T_N$.

13.22. If χ and T be the antiferromagnetic susceptibility and temperature of the specimen respectively, then

(a) χ increases with T (b) χ decreases with T (c) χ is independent of T (d) all are true.

13.23. The temperature at which antiferro- to paramagnetism takes place is called

(a) Debye temperature (b) Curie-Weiss temperature

(c) Neel temperature (d) Curie temperature.

13.24. The striking feature of antiferromagnetic is

(a) magnetic domains

(b) hysteresis loss

(c) decrease of susceptibility with increase of temperature

(d) sharp maximum in the χ versus T curve.

13.25. Materials which can be readily magnetized in either direction are called

(a) soft magnetic materials (b) hard magnetic materials

(c) low hysteresis loss materials (d) high hysteresis loss materials.

13.26. The dimension of ferromagnetic domain is of the order of

(a) 10^6 m (b) 10^{-9} m (c) 10^{-6} m (d) 10^{-3} cm.

13.27. The susceptibility of diamagnetic material is about

(a) 10^5 (b) -10^{-5} (c) 10^7 (d) 10^{-5}.

13.28. The material used to make permanent magnet is

(a) carbon steel (b) Alnico V (c) Platinum cobalt (d) all of the three.

13.29. The ferromagnetic Curie temperature of iron is

(a) 631 K (b) 1043 K (c) 900 K (d) 1428 K.

13.30. The spontaneous magnetization of ferromagnetic at Curie temperature is

(a) 1 (b) 0 (c) ∞ (d) none of these.

13.31. The permanent magnetic moment in ferromagnetics arises due to

(a) partially filled electronic shells (b) completely filled electronic shells

(c) innermost electronic shells (d) none of these.

ANSWERS

13.1. (c) **13.2.** (b) **13.3.** (d) **13.4.** (b) **13.5.** (b) **13.6.** (a) **13.7.** (d) **13.8.** (c) **13.9.** (c) **13.10.** (a) **13.11.** (b) **13.12.** (d) **13.13.** (d) **13.14.** (c) **13.15.** (b) **13.16.** (c) **13.17.** (c) **13.18.** (b) **13.19.** (a) **13.20.** (a) **13.21.** (c) **13.22.** (a) **13.23.** (c) **13.24.** (d) **13.25.** (a) **13.26.** (c) **13.27.** (b) **13.28.** (d) **13.29.** (b) **13.30.** (b) **13.31.** (a).

13(B) Short Answer Type Questions

13.1. What is the origin of magnetism in materials? How does it depend on temperature? **[JNTU May 2003]**

13.2. Discuss how diamagnetism, paramagnetism, ferromagnetism, antiferromagnetism and ferrimagnetism appear in materials?

13.3. Write down the names of three kinds of magnetic materials and discuss their differences in respect of magnetic susceptibilities. **[BU(H) 2006]**

13.4. Give a comparative study of dia-, para- and ferromagnetic materials. Give some examples of each kind. **[BU(H) 2002]**

13.5. Show that diamagnetic susceptibility increases with the number atoms per unit volume but is independent of temperature. **[BU(H) 2002]**

13.6. Why does a paramagnetic sample display greater magnetization when cooled? Why is it not so in diamagnetism? **[BU(H) 2001]**

13.7. Define magnetization and show that $B = \mu_0(H + M)$. **[JNTU 2004]**

13.8. In hydrogen atom an electron e revolves round the nucleus at a distance of r metre with an angular velocity ω radians. Obtain expression for magnetic moment associated with it due to its orbital motion. **[JNTU May 2003, Dec. '04]**

13.9. Explain the detailed classification of magnetic materials on the basis of electron spin. **[JNTU Jan. 2003]**

13.10. What is Bohr magneton? How is it related to magnetic moment of electron?

13.11. What is a ferromagnet? How do you account for the alignment of spins in such systems? **[BU(H) 1998]**

13.12. Point out the essential difference between the classical theory and quantum theory of paramagnetism.

13.13. What is spontaneous magnetization? How does it vary with temperature of the material?

13.14. State Curie-Weiss law and discuss its application to ferromagnetic materials.

13.15. Explain how the exchange interaction helps to explain the phenomenon of iron group of elements.

13.16. What is hysteresis? What is the significance of hysteresis loop?

13.17. How does susceptibility vary with temperature in case of antiferromagnetic materials?

13.18. Define Curie temperature and Neel temperature. Point out the essential difference between the two.

13.19. What is internal field (or Weiss field) in ferromagnetics? Find an expression for it.

13.20. How can the domain structure be experimentally demonstrated?

13.21. What is ferromagnetism? What are ferrites? Give some applications of ferrites.

13.22. Explain the B-H loop with the help of domain theory.

13.23. How ferrites are superior to ferromagnetic materials? **[JNTU June 2005]**

13.24. Explain the difference of ferro-, antiferro- and ferrimagnetic materials. **[JNTU May 2003]**

13.25. Explain clearly the difference between hard and soft magnetic materials. What are mixed ferrites? Mention their uses. **[JNTU June 2005]**

13.26. Write the importance of hard magnetic materials in engineering applications. **[JNTU Jan. 2003]**

13(C) Numerical Problems

13.1. A paramagnetic system of electronic spin magnetic dipole moment is placed in an external field of 10^5 A/m. Calculate the average magnetic moment per dipole at 300 K. [*Ans.* 2.6×10^{-27} A/m^2]

13.2. Given the diamagnetic susceptibility of He to be -1.9×10^{-6} cm/mole. Estimate the radius of the He atom. **[BU(H) 1999]**

[*Ans.* 0.0164 nm]

13.3. The earth has a magnetic dipole moment of 6.4×10^{21} Am^2. What current would have to be set up in a single turn of wire going around the earth at its magnetic equator, if we wish to set up such a dipole? [The mean radius of the earth = 6.4×10^6 m] [*Ans.* 5×10^7 A]

13.4. Determine the flux density and magnetization in Fe_2O_3 when placed in a magnetic field of strength 1.2×10^6 A/m. [Given, the susceptibility of Fe_2O_3 at 300 K is 1.4×10^{-3}] [*Ans.* 1.51 T; 1.68×10^3 A/m]

13.5. Calculate p_{eff} of Nd^{3+} with electronic configuration $[\text{Xe}]4f^3$. **[BU(H) 2004]**

13.6. Calculate p_{eff} for the following ions in the Hund ground state:

$$Pr^{3+}\left[4f^3 5s^2 p^6; {}^4I_{9/2}\right] \text{ and } Ni^{2+}\left[3d^8; 3F_4\right].$$ **[BU(H) 2000]**

13.7. Calculate the Landé g-factor and the total magnetic moment of an atom in the ${}^8S_{7/2}$ state. **[BU(H) 1999]**

13.8. State Hund's rule. Find the ground states of the following ions:

(a) $Gd^{3+}([\text{Xe}]4f^7)$

(b) $Ti_i^{3+}([\text{Ar}]3d^1)$

(c) $Co^{2+}([\text{Ar}]3d^7)$. **[BU(H) 1997]**

13.9. Compute the internal field for iron having Curie temperature 1043 K and an effective moment of $2.2\mu_B$ per atom. [*Ans.* 2100 T]

13.10. What magnetic field must be applied to a salt containing Cu^{2+} ions at 1 K so that 90 % of the ions are in the ground state. [Cu^{2+} has 9 electrons in the $3d$ shell] [*Ans.* 3 T]

13.11. Compute the magnetic susceptibility for a salt containing one kg-mole of Cr^{2+} ions at 300 K. [The electronic configuration of Cr^{2+} is $3d^4 4s^\circ$] [*Ans.* 1.25×10^{-4}]

13.12. The magnetic intensity of a ferric oxide piece is 10^6 A/m. If the susceptibility of the material at room temperature is 1.5×10^{-3}, calculate the flux density and magnetization of the material.

[*Ans.* 1.259 T and 1500 A/m]

13.13. Anhydrous nickel chloride ($NiCl_2$) has a paramagnetic susceptibility of 78.5×10^{-8} kg^{-1} at 20 °C. Using simplified Langevin formula, calculate the magnetic moment of the Ni^{2+} ion in Bohr magneton. [Atomic weight of Ni is 58.7 and of chlorine 35.5] [*Ans.* 4.36 Bohr magneton]

13.14. A system of electron spins is placed in a magnetic field of 2 Wb/m^2 at temperature T. If the number of spins parallel to the field be twice as large as the number of anti-parallel spins, calculate T.

[*Ans.* $T = 3.8$ K]

13.15. A paramagnetic material is subjected to a homogeneous field of 10^6 A/m at 37 °C. Calculate the average magnetic moment along the field direction per spin in Bohr magneton.

[*Ans.* 2.72×10^{-3} Bohr magneton/spin]

13.16. Calculate the theoretical saturation magnetization in pure iron if iron has BCC structure with lattice parameter $a = 0.286$ nm and $\mu_B = 9.27 \times 10^{-24}$ Am^2. [*Ans.* 1.7×10^5 A/m^2]

13.17. A typical magnetic field achievable with an electromagnet of iron core is 1 T. Compare the magnetic interaction energy of an electron spin magnetic dipole moment with kT at room temperature.

[*Ans.* 2.3×10^{-3}]

13.18. A ferromagnetic material with $J = \frac{3}{2}$ and $g = 2$ has a transition temperature $T_c = 200$ K. Estimate the internal field near 0 K. **[BU(H) 1995]**

13.19. A ferromagnetic material with $J = \frac{2}{3}$ and $g = 2$ has a Curie temperature of 125 K. Calculate the intrinsic flux density near 0 K. Also calculate the ratio of the ferromagnetisation at 300 K in the presence of an external field of 1 mT to the spontaneous magnetisation at 0 K. **[BU(H) 2013]**

13.20. Consider a helium atom in its ground state $(1s)$. The mean radius in the Langevin formula may be approximated by Bohr radius. Calculate the diamagnetic susceptibility of a helium atom. (Atomic density of helium = 27×10^{23}). **[BU(H) 2012]**

13.21. Atomic weight and density of iron are 55.847 and 8.87×10^3 kg/m^3 respectively. If iron has a magnetic moment of 2.2 Bohr magneton, determine its spontaneous magnetisation. **[CU(H) 2017]**

13.22. Consider a single spin $\frac{1}{2}$ particle with no orbital motion in uniform magnetic field B. How many energy levels are possible in such a case? If total number of spins is N, then how the populations in the levels are varied with magnetic field B at a given temperature T? Hence, compute the magnetisation and susceptibility of the system. **[CU(H) 2016]**

13.23. (a) Derive the quantum expression for the paramagnetic susceptibility of a sample.

(b) The Curie temperature of the iron is 1043 K. Assume that iron atoms, when in metallic form, have moments of two Bohr magneton per atom. Iron is BCC with lattice parameter $a = 0.286$ nm. Given, $\mu_B = 9.2741 \times 10^{24}$ J/T. Calculate:

(i) the saturation magnetisation

(ii) the Curie constant

(iii) the Weiss field constant. **[CU(H) 2012]**

13(D) Broad Answer Type Questions

13.1. Deduce an expression for the molar diamagnetic susceptibility of atoms. Does the diamagnetic susceptibility vary with temperature? Give reasons for your answer. **[BU(H) 1994]**

13.2. (a) Calculate the magnetic moments of the following: (a) Na^+ (b) $Fe^{3+}([Ar]3d^5)$ (b) $Ce^{3+}([Xe]4f^1)$.

(b) Find an expression for the susceptibility of a ferromagnet in the paramagnetic phase. **[BU(H) 2005]**

13.3. (a) Deduce Langevin's formula for the molar diamagnetic susceptibility arising from the core electrons in an atom.

(b) How would you determine whether a material is dia-, para- or ferromagnetic? Derive an expression or the molar magnetic susceptibility of a paramagnetic compound, where each molecule contains one ion with total angular momentum $\hbar\sqrt{(J(J+1)}$. **[BU(H) 1996]**

13.4. (a) Define effective Bohr magneton number. Write the expressions for the molar susceptibility of the compounds of Gd^{3+}, Ti^{3+} and Co^{2+}, where a molecule contains only one paramagnetic ion.

(b) Derive the expression for the susceptibility of a ferromagnetic material much above the transition temperature (T_c). **[BU(H) 1997]**

13.5. What do you mean by ferromagnetism and the ferromagnetic domains? Discuss the Weiss' theory of ferromagnetism and explain how magnetic susceptibility varies with temperature. **[BU(H) 2008]**

13.6. Find an expression for paramagnetic susceptibility from quantum theory. Under what condition does it reduce to classical expression of Langevin? **[CU(H) 2005]**

13.7. (a) Discuss the characteristics of dia-, para- and ferromagnetic substances. How would you distinguish between them experimentally? **[CU(H) 1990]**

(b) Name an element which exhibits paramagnetism. Explain why inert gases do not show paramagnetism. **[CU(H) 1997]**

13.8. (a) Discuss Langevin's classical theory on paramagnetism and its physical significance. **[BU(H) 2006]**

(b) Derive Curie's law of paramagnetism from Langevin's theory.

13.9. (a) Give a comparative study of dia-, para- and ferromagnetic materials. Give some examples of each kind of material.

(b) Show that diamagnetic susceptibility increases with number of atoms per unit volume but is independent of temperature. **[BU(II) 2002]**

13.10. (a) Deduce Langevin's formula for the molar diamagnetic susceptibility. What is the diamagnetic susceptibility of a perfect diamagnet?

(b) Discuss briefly the Weiss' theory of ferromagnetism. **[BU(II) 1999]**

13.11. What is meant by ferromagnetic materials? Give some examples. What are their characteristics? Give an account of Weiss' theory of ferromagnetism. Discuss the temperature variation of saturation magnetization. **[BU(II) 2001]**

13.12. (a) Give an account of the quantum theory of paramagnetism and explain how it removes the shortcomings of Langevin's theory.

(b) Give an account of domain theory of hysteresis.

(c) Is the permeability of a ferromagnetic material independent of magnetic field? If not, is it more for lower or higher fields? **[BU(II) 2000]**

13.13. (a) Describe the Heisenberg's exchange interaction. How does it explain ferromagnetism?

(b) Relate exchange integral to the Weiss constant and the ferromagnetic Curie temperature.

13.14. What is ferromagnetism? Discuss the Weiss-Field theory of ferromagnetism and explain how magnetic susceptibility varies with temperature. **[BU(II) 2003]**

13.15. (a) Draw the *B-H* curve for a ferromagnetic material and identify the retentivity and the coercive field on the curve. **[JNTU 2005]**

(b) Explain ferromagnetism and antiferromagnetism. **[JNTU 2003, '04]**

13.16. (a) Deduce Langevin's formula for the molar diamagnetic susceptibility. What is the diamagnetic susceptibility of a perfect diamagnet?

(b) Calculate the diamagnetic susceptibility of atomic hydrogen in the ground state at STP using the wave function

$$\psi(r) = \frac{1}{(\pi a_0^3)^{1/2}} e^{-r/a_0},$$

where $a_0(= 0.053 \text{ nm})$ is the atomic radius. **[BU(II) 2004]**

[*Hints.* We have the wave function for the ground state of hydrogen atom

$$\psi(r) = \frac{1}{(\pi a_0^3)^{1/2}} e^{-r/a_0}.$$

Now the mean square distance of the electronic charge distribution from the nucleus is

$$\langle r^2 \rangle = \int_0^\infty \psi^* r^2 \psi d\tau = 4\pi \int_0^\infty \psi^* r^2 \psi r^2 dr = \frac{4\pi}{\pi a_0^3} \int_0^\infty r^4 e^{-2r/a_0} dr.$$

Let $-\frac{2r}{a_0} = x$.

$$\therefore r = -\frac{a_0}{2} x \text{ and } dr = -\frac{a_0}{2} dx, \therefore \langle r^2 \rangle = \frac{4}{a_0^3} \left(\frac{a_0^4}{16}\right)\left(-\frac{a_0}{2}\right) \int_0^\infty x^4 e^{-x} dx.$$

But

$$\int_0^\infty x^4 e^{-x}\,dx = 24, \therefore \langle r^2\rangle = 3a_0.$$

Now the susceptibility of the diamagnet is

$$\chi_{\text{dia}} = -\frac{N\mu_0 Ze^2}{6m}\langle r^2\rangle = -\frac{N\mu_0 Ze^2}{2m}a_0^2.$$

But here

$$N = \frac{6.023\times 10^{26}}{2.24\times 10^{-2}} = 2.69\times 10^{28}\ \text{m}^{-3}.$$

Here $Z = 1$

and $a_0 = 0.053\,\text{nm} = 0.53\times 10^{-10}$ m.

$$\therefore \chi_{\text{dia}} = -\frac{(2.69\times 10^{28})\times(4\pi\times 10^{-7})\times(1.6\times 10^{-19})^2\times(0.53\times 10^{-10})^2}{2\times 9.1\times 10^{-31}}$$

$$= -\frac{2.69\times(4\times 3.14)\times(1.6\times 1.6)\times(0.53\times 0.53)}{2\times 9.1}\times 10^{-6}$$

$$= -\frac{2.69\times 12.56\times 2.56\times 0.28}{18.2}\times 10^{-6} = \mathbf{-1.33\times 10^{-6}}\ \textbf{(SI units)}\,]$$

13.17. (a) What is the difference between diamagnetic and paramagnetic atoms? Describe a method for measuring magnetic susceptibility of a powder sample. Derive the expression you use.

(b) What is Curie law of magnetic susceptibility?

Derive the susceptibility expression for a paramagnetic compound having one Mn^{2+} ([Ar]$3d^5$) molecule per molecule.

(c) What is a ferromagnet? How do you account for the alignment of spins in such a system?

[BU(H) 1998]

13.18. (a) Describe the two-sub-lattice model to explain antiferromagnetism.

(b) How does this model account for the difference between the Neel temperature (T_N) and Curie-Weiss temperature (θ)?

13.19. (a) Discuss the phenomenon of antiferromagnetism.

(b) Show that the susceptibility of antiferromagnetic materials is inversely proportional to ($T + T_N$), where T_N is the Neel temperature.

13.20. (a) Explain Ferrimagnetism.

(b) What are ferrites? Discuss their applications.

13.21. (a) Give an account of Neel's theory of antiferromagnetism and show how the ferromagnetic behaviour of ferrite can be explained from Neel's theory.

(b) Describe the structure of ferrite. How is the magnetic moment of ferrite molecule calculated?

13.22. (a) What are the distinguishing characteristics of ferromagnetic materials?

(b) What do you mean by magnetic susceptibility? At a particular temperature T, below the Curie temperature, does the susceptibility of a ferromagnetic material depend on the measuring field?

(c) According to Curie-Weiss law, for a ferromagnetic material

$$\chi = \frac{C}{T - \theta}$$

In an actual measurement, what value of χ do you except at $T = 0$ and why?

(d) Explain hysteresis phenomenon for a ferromagnetic material. **[BU(H) 2015]**

13.23. (a) What do you mean by ferromagnetism and the ferromagnetic domains?

(b) Give an account of Weiss theory of ferromagnetism.

(c) Discuss the temperature variation of spontaneous magnetization. Explain Curie point on the basis of this theory. **[BU(H) 2007, '13]**

13.24. (a) Explain the origin of diamagnetism in materials.

(b) Derive an expression for diamagnetic susceptibility using Langevin's theory.

(c) Why do diamagnetic materials have negative susceptibility? **[BU(H) 2012]**

13.25. Give an account of Weiss field theory of ferromagnetism. Hence, discuss spontaneous magnetisation. **[BU(H) 2010]**

13.26. (a) What is paramagnetism? Derive an expression for the paramagnetic susceptibility using Langevin classical theory of paramagnetism.

(b) In this context discuss the limitations of the theory.

(c) What is meant by anti-ferromagnetism? **[BU(H) 2008, '14]**

13.27. (a) What materials are called *diamagnetism* in respect of magnetic susceptibility?

(b) Obtain classical Langevin result for expressing the diamagnetic susceptibility of material.

(c) Give the reason of exchange field and obtain Curie-Weiss law of simple ferromagnetism. Why is the spontaneous magnetisation observed at Curie temperature? **[BU(H) 2007]**

13.28. (a) Discuss Langevin's classical theory of paramagnetism and its physical significance.

(b) Give the quantum theory of paramagnetism.

(c) Describe in detail the Gouy method of measuring susceptibility.

(d) Why is energy loss observed in case of magnetic hysteresis? **[BU(H) 2005]**

13.29. (a) Suppose, a paramagnetic atom having permanent moment $\vec{\mu}$ with given resultant quantum number $\vec{J}$, is placed in a uniform magnetic field $\vec{B}$. Obtain an expression of magnetisation as a function of B and temperature T. Hence, obtain Curie-Weiss law in the appropriate limit.

(b) Obtain the spontaneous magnetisation of a ferromagnet as a function of temperature. Indicate the universal feature associated with the graph. **[CU(H) 2017]**

13.30. (a) What is Weiss molecular field?

(b) Derive a relation between Curie temperature and the molecular field constant of a ferromagnet. (Assume the relation for paramagnet $M = Ng\mu_B B_J(x)$, where the symbols have their usual meanings). **[CU(H) 2016]**

13.31. (a) Suppose, a paramagnetic atom with a given resultant quantum number $\vec{j}$ is placed in a uniform magnetic field $\vec{B}$.

(i) How many energy levels are possible in such a situation?

(ii) Find an expression of the magnetisation as a function of magnetic field $\bar{B}$ and expression for $J = \frac{1}{2}$? Obtain Curie's law for $J = \frac{1}{2}$ in the appropriate limit.

(b) Arrange in increasing order the absolute magnitude of diamagnetic susceptibility of He, Ne and Ar. Indicate the reason behind your arrangement. **[CU(H) 2015]**

13.32. (a) What is Weiss molecular field? Derive a relation between Curie temperature and the molecular field constant of a ferromagnet. (Assume the relation for paramagnet $M = Ng\mu_B J B_J(x)$, where the symbols have their usual meanings).

(b) Find the magnetic field B_0, which has to be applied to a paramagnetic salt containing ions with $\mu_{mj} = \pm\mu_B$, so that x per cent of these ions are in the lowest energy state.

(c) What is hysteresis? What type of hysteresis curve is preferable for permanent magnet? **[CU(H) 2014]**

13.33. (a) When a paramagnetic atom with resultant quantum number $\vec{J}$ is placed in a magnetic field $\vec{B}$, the energy levels are given by $E_J = -m_J g\beta|\vec{B}'$, where m_J = magnetic total angular momentum, g = Landé g-factor and β = Bohr magneton.

(i) Find the magnetisation of N such atoms of finite temperature T as a function of magnetic field $\vec{B}$. What happens to the expression in the limit of J values? Explain.

(ii) Obtain Curie's law from the above expression.

(b) Explain why diamagnetism is an inherent property of an atom. **[CU(H) 2013]**

13.34. (a) Distinguish between paramagnetic and ferromagnetic substances with examples.

(b) Consider a system of N non-interacting atoms having spin $\frac{1}{2}$. Show that the magnetisation of the system in a magnetic field B at any temperature T is given by

$$M = n\mu \tanh\left(\frac{\mu B}{k\beta T}\right),$$

where $\mu = \frac{1}{2}g\mu_B$.

(c) Starting from the basic assumptions of the molecular field theory of magnetisation derive Curie-Weiss law of ferromagnetism. **[CU(H) 2011]**

13.35. (a) Outline the Langevin's theory for a paramagnetic gas and obtain an expression for paramagnetic susceptibility of the gas.

(b) What is Curie point?

(c) What is hysteresis? Explain its origin from domain theory. **[CU(H) 2010]**

13.36. Write down the basic assumptions of molecular field theory of magnetism. Use this method to derive the Curie-Weiss law of ferromagnetism. **[CU(H) 2009]**

Chapter 14

Elastic Properties of Solids

14.1 Introduction

The elastic properties of solid is related to elasticity which is defined as *the property by virtue of which the bodies regain their original shape and size when the deforming forces are removed.* This is a common property of all classes of solids. In technology it would tell us about the strength of the materials. In fundamental research, it provides into the nature of the binding forces in solids. The relevant elastic constants are also interrelated themselves to thermal properties like Debye temperature.

In order to study the elastic properties of solids, the crystal is supposed to be a homogeneous and continuous medium instead of a periodic array of atoms. The continuum approximation is valid for elastic waves of frequencies below 10^{11} or 10^{12} cycles per second which are actually ultrasonic waves.

The Hooke's law for an anisotropic crystal introduces elastic constants as the components of a tensor of fourth order because both the stress and strain are components of the relevant second order tensor. In general, the number of elastic constants appropriate for a body is large. But this is considerably reduced by assuming the existence of elastic potential, and the isotropic elastic constants are sufficient to characterise the elastic behaviour. The three elastic constants may be experimentally found by propagating acoustic waves through the cubic crystal along the symmetry directions [100], [110] and [111].

14.2 Stress components

When an elastic body is subjected to deforming force, forces of reaction are developed internally inside the body trying to resist further deformation of body, the tending to restore the original condition of the body. This restoring force acting per unit area inside the body is termed as stress. In equilibrium condition of the body, the two sets of forces (viz., externally applied forces and internal force developed in the body) being equal and opposite cancel each other. Hence, stress is equal to the magnitude of the force divided by the area normal to the force over which it is applied.

There are nine stress components: X_x, X_y, X_z; Y_x, Y_y, Y_z; Z_x, Z_y, Z_z. The capital letter indicates the direction of the force and the subscript indicates the normal to the plane to which the force is applied. In Fig 14.1, X_x represents the force applied in the $\hat{i}$ direction to a unit area of a plane whose normal lise in the $\hat{i}$ direction; the stress component X_y represents a force applied in the X-direction to a unit area of a plane whose normal lies in the $\hat{j}$ direction.

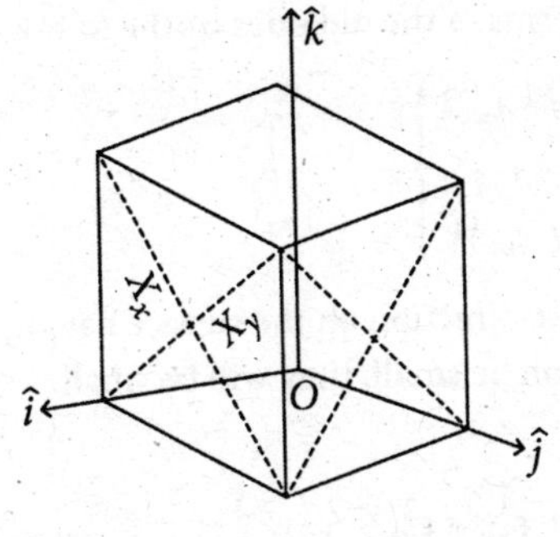

Fig. 14.1: The stress component X_x is a force applied in the $\hat{i}$-direction to a unit area of a plane whose normal lies in the $\hat{i}$-direction. X_y is applied in the $\hat{i}$-direction to a unit area of a plane whose normal lies in the $\hat{j}$-direction.

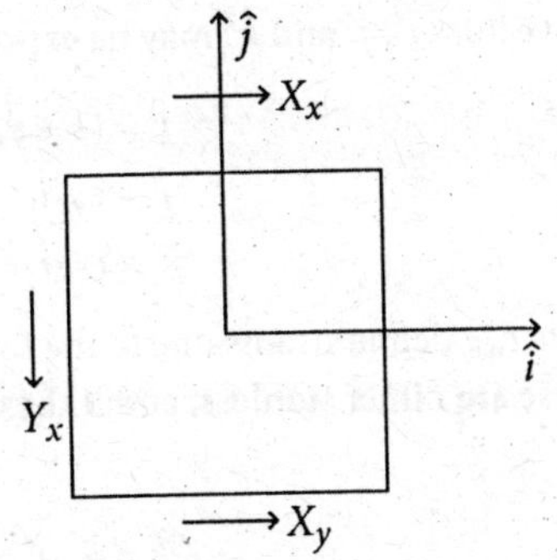

Fig. 14.2: Demonstration of a body in static equilibrium $Y_x = X_y$. The sum of the forces along $\hat{i}$ and $\hat{j}$ directions are zero. The total force vanishes. The total torque about the origin is also zero if $Y_x = X_y$.

The number of independent stress components is reduced from nine to six by applying to an elementary cube (Fig. 14.2), the condition that the angular acceleration vanishes and hence, the total torque must be zero. It follows that

$$Y_z = Z_y;\ Z_x = X_z;\ X_y = Y_x. \tag{14.1}$$

The six independent stress components are taken as X_x, Y_y; $Z_z Y_z$; Z_x, X_y. The stress components have the dimensions of force per unit area or energy per unit volume.

14.3 Strain components

The local strain in a deformed body may be specified by six components: $e_{xx}, e_{yy}, e_{xy}, e_{yz}$ and e_{zx} defined as below:

Let $\hat{i}, \hat{j}, \hat{k}$ be the orthogonal unit vectors in the unstrained solid as shown in Fig. 14.3(a). After a small deformation (the deformation in which each primitive cell of the crystal is deformed in the same way), the axes are distorted in the orientation as well as in length [Fig. 14.3(b)].

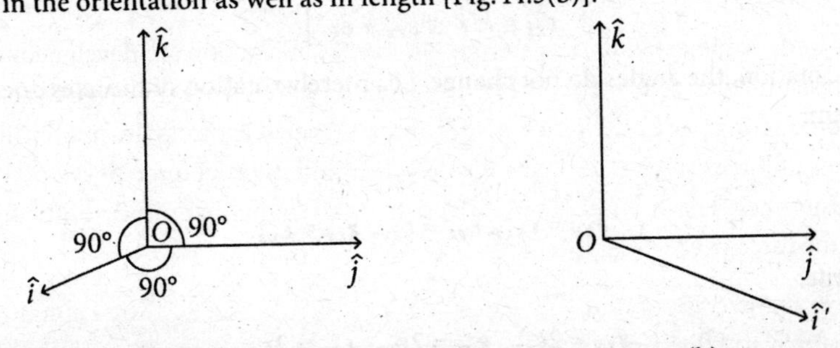

Fig. 14.3: The coordinate axes for the description of the state of strain, the orthogonal unit axes in the unstrained state (a) are deformed in the strained state (b).

The new unit vectors $\hat{i}'$, $\hat{j}'$ and $\hat{k}'$ may be expressed in terms of the old ones by the following relations:

$$\left.\begin{aligned}\hat{i}' &= (1+\varepsilon_{xx})\hat{i} + \varepsilon_{xy}\hat{j} + \varepsilon_{xz}\hat{k}\\ \hat{j}' &= \varepsilon_{yz}\hat{i} + (1+\varepsilon_{tt})\hat{j} + \varepsilon_{yz}\\ \hat{k}' &= \varepsilon_{zx}\hat{i} + \varepsilon_{zy}\hat{j} + (1+\varepsilon_{zz})\hat{k}\end{aligned}\right\}. \tag{14.2}$$

The coefficients $\varepsilon_{\alpha\beta}$ define strains due to the force in the α-direction on the surface having normal in the β-direction. These are dimensionless, and if the deformation be small, they will be small.

Now

$$|i'|^2 = \hat{i}' \cdot \hat{i}' = 1 + 2\varepsilon_{zx} + \varepsilon_{xx}^2 + \varepsilon_{zz}^2. \tag{14.3}$$

This shows that the new axes are not necessarily of unit length. The new axes are not necessarily orthogonal as well, since $\hat{i}' \cdot \hat{j}' \neq 0$.

$$(\hat{i}' \cdot \hat{j}') = (1+\varepsilon_{xx})\varepsilon_{yz} + \varepsilon_{xy}(1+\varepsilon_{yy}) + \varepsilon_{yz}\varepsilon_{zz} = \varepsilon_{yz} + \varepsilon_{xy}, \tag{14.4}$$

neglecting the product terms as they are very small.

Neglecting the square terms in eqn (14.2), we get

$$|i'|^2 = (1+2\varepsilon_{xx})^{1/2} = 1 + \varepsilon_{xx} \text{ (to first order)}. \tag{14.5}$$

So, ε_{xx} is the fractional change of length of $\hat{i}$-axes and thus represents the linear strain components. Similarly, ε_{yy} and ε_{zz} represents the linear strain components.

Now, we have

$$e_{xx} \equiv \varepsilon_{xx},\ e_{yy} \equiv \varepsilon_{yy},\ e_{zz} \equiv \varepsilon_{zz}. \tag{14.6}$$

The strain components e_{xx}, e_{yy} and e_{zz} are defined in terms of the changes in angle between the axes and are known as *shearing strains.* Since $\hat{i} \cdot \hat{j} = 0$, $\hat{i}' \cdot \hat{j}$ gives a measure of the change in orientation between $\hat{i}$ and $\hat{j}$ axes as a result of deformation.

We have

$$\left.\begin{aligned}\varepsilon_{xy} &\cong \hat{i}'\hat{j}' \cong \varepsilon_{yx} + \varepsilon_{xy}\\ \varepsilon_{yz} &\cong \hat{j}'\hat{k}' \cong \varepsilon_{yz} + \varepsilon_{zy}\\ \varepsilon_{zx} &\cong \hat{k}'\hat{i}' \cong \varepsilon_{zx} + \varepsilon_{xz}\end{aligned}\right\}. \tag{14.7}$$

In case of rigid rotation, the angles do not change, i.e., merely rotation of the axes does not change the angle between them.

Hence,

$$\varepsilon_{xy} = \varepsilon_{yx},\ \varepsilon_{yz} = \varepsilon_{zy},\ \varepsilon_{zx} = \varepsilon_{xz}. \tag{14.8}$$

Then we can write,

$$\varepsilon_{xy} = 2\varepsilon_{xy},\ \varepsilon_{yz} = 2\varepsilon_{yz},\ \varepsilon_{zx} = 2\varepsilon_{zx}. \tag{14.9}$$

We now consider the effect of deformation on a point or atom originally at (x, y, z). The position vector $\vec{r}$ is given by

$$\vec{r} = \hat{i}x + \hat{j}y + \hat{k}z. \tag{14.10}$$

If the deformation is uniform, then in strained conditions its position is given by

$$\vec{r}' = \hat{i}'x + \hat{j}'y + \hat{k}'z. \tag{14.11}$$

The displacement $\vec{R}$ of the deformation is given by

$$\vec{R} = \vec{r}' - \vec{r} = (\hat{i}' - \hat{i})x + (\hat{j}' - \hat{j})y + (\hat{k}' \cdot \hat{k}). \tag{14.12}$$

Now substituting the value of $(\hat{i}' - \hat{i})$, etc. from equation (14.1), we get

$$\vec{R} = (x\varepsilon_{xx} + y\varepsilon_{yx} + z\varepsilon_{xz})\hat{i} + (x\varepsilon_{xy} + y\varepsilon_{yy} + z\varepsilon_{zy})\hat{j} + (x\varepsilon_{xz} + y\varepsilon_{yz} + z\varepsilon_{zz})\hat{k}. \tag{14.13}$$

Substituting the values of $\varepsilon_{\alpha\beta}$ in terms of $e_{\alpha\beta}$ by making use of eqns (14.5) and (14.6).

$$\therefore \vec{R} = \left(xe_{xx} + \tfrac{1}{2}ye_{xy} + \tfrac{1}{2}ze_{xz}\right)\hat{i} + \left(\tfrac{1}{2}xe_{xy} + ye_{yy} + \tfrac{1}{2}ze_{yz}\right)\hat{j} + \left(\tfrac{1}{2}xe_{xz} + \tfrac{1}{2}ye_{yz} + ze_{zz}\right)\hat{k}. \tag{14.14}$$

Now, we put

$$\vec{R} = \hat{i}u + \hat{j}v + \hat{k}w, \tag{14.15}$$

where u, v and w are the components of displacement $\vec{R}$ along the three ordinate axes.

Comparing eqns (14.3) and (14.4), we get

$$\left.\begin{aligned} u &= xe_{xx} + \tfrac{1}{2}ye_{xy} + \tfrac{1}{2}ze_{zz} \\ v &= \tfrac{1}{2}xe_{xy} + ye_{yy} + \tfrac{1}{2}ze_{yz} \\ w &= \tfrac{1}{2}xe_{xx} + \tfrac{1}{2}ye_{yz} + ze_{zz} \end{aligned}\right\}. \tag{14.16}$$

Taking derivatives,

$$\left.\begin{aligned} e_{xx} &= \frac{\partial u}{\partial x}, & e_{yy} &= \frac{\partial v}{\partial y}, & e_{zz} &= \frac{\partial w}{\partial z} \\ e_{xy} &= \frac{\partial u}{\partial y} + \frac{\partial v}{\partial x}, & e_{yz} &= \frac{\partial v}{\partial z} + \frac{\partial w}{\partial y}, & e_{zz} &= \frac{\partial u}{\partial z} + \frac{\partial w}{\partial x} \end{aligned}\right\}. \tag{14.17}$$

These provide alternate definition of strain components. Here, e_{xx}, e_{yy} and e_{zz} are linear components, i.e., changes in length per unit length along the three axes respectively. We may interpret the strain components of the type

$$e_{xy} = \frac{\partial u}{\partial t} + \frac{\partial v}{\partial x} \tag{14.18}$$

as consisting of simple shears. In one of them the planes normal to the X-axis slide in the Y-direction, while the other planes normal to Y-axis slide in the X-direction.

14.4 Delation

The fractional increase in volume due to deformation of a body is called *delation*. If original volume of a body be V and after deformation it becomes V', then delation δ is defined by

$$\delta = \frac{V' - V}{V}, \tag{14.19}$$

Let us consider a unit cube with edges $\hat{i}, \hat{j}, \hat{k}$ and after deformation its edges are $\hat{i}', \hat{j}, \hat{k}$. Then after deformation the volume V' is given by $V' = \hat{i}' \cdot \hat{j}' \times \hat{k}'$.

14.8 Elastic constants for cubic isotropic crystals

For a cubic crystal, we choose the coordinate axes parallel to the cubic edges. A cube has got a four-fold axis normal to each face, three axes will be equivalent and can be replaced by one another by a rotation of 90°.

If we interchange Z- and Y-axis, this will not have any effect on left-hand side of equation for X_x in expression (14.32) but on right-hand side elongation e_{yy} and e_{zz} will be interchanged.

As a result, the equation for X_x will be $X_x = C_{11}e_{xx} + C_{12}e_{zz} + C_{13}e_{yy}$ giving $C_{12} = C_{13}$.

Interchanging X- and Z-axes, $C_{31} = C_{32}$.

Now the expression for stresses can be written as

$$\left.\begin{aligned} X_x &= C_{11}e_{xx} + C_{12}e_{yy} + C_{12}e_{zz} \\ Y_y &= C_{12}e_{xz} + C_{22}e_{yy} + C_{12}e_{zz} \\ Z_z &= C_{12}e_{xx} + C_{12}e_{yy} + C_{33}e_{zz} \\ X_y &= C_{44}e_{xy} \\ Y_z &= C_{55}e_{yz} \\ Z_x &= C_{66}e_{zx} \end{aligned}\right\}. \tag{14.33}$$

If we now change X-axis by Y-axis, Y-axis by Z-axis and Z-axis by X-axis, the form of eqn (14.33) will not change. This gives $Y_y = C_{11}e_{yy} + C_{12}e_{zx} + C_{12}e_{xx}$. $\therefore C_{11} = C_{22}$.

Similarly, $C_{11} = C_{32}$.

$$\therefore C_{11} = C_{22} = C_{33} \tag{14.34}$$

and $Y_z = C_{44}e_{yz}$, $Z_x = C_{55}e_{zx}$, $X_y = C_{66}e_{xy}$.

$$\therefore C_{44} = C_{55} = C_{66}. \tag{14.35}$$

Using the eqns (14.34) and (14.35) in eqn (14.33), we get the values of the stress components as

$$\left.\begin{aligned} X_x &= C_{11}e_{xx} + C_{12}e_{yy} + C_{12}e_{zz} \\ Y_y &= C_{11}e_{yy} + C_{12}e_{xx} + C_{12}e_{zz} \\ Z_z &= C_{11}e_{zz} + C_{12}e_{xx} + C_{12}e_{yy} \\ X_x &= C_{44}e_{xy} \\ Y_z &= C_{55}e_{yz} \\ Z_x &= C_{66}e_{zx} \end{aligned}\right\}. \tag{14.36}$$

Thus, the values of the elastic stiffness constants for a cubic crystal are reduced to three only, i.e., C_{11}, C_{12} *and* C_{44}.

The elastic constants for an elastic cubic isotropic body can be expressed in the form of a matrix as below:

$$|e_{ij}| = \begin{vmatrix} C_{11} & C_{12} & C_{12} & 0 & 0 & 0 \\ C_{12} & C_{11} & C_{12} & 0 & 0 & 0 \\ C_{12} & C_{12} & C_{11} & 0 & 0 & 0 \\ 0 & 0 & 0 & C_{44} & 0 & 0 \\ 0 & 0 & 0 & 0 & C_{44} & 0 \\ 0 & 0 & 0 & 0 & 0 & C_{44} \end{vmatrix} \tag{14.37}$$

Elastic compliances constants for cubic isotropic crystal are also reduced to three only, i.e., S_{11}, S_{12} and S_{44}. Hence, S_{ij} can be written as below:

$$|S_{ij}| = \begin{vmatrix} S_{11} & S_{12} & S_{12} & 0 & 0 & 0 \\ S_{12} & S_{11} & S_{12} & 0 & 0 & 0 \\ S_{12} & S_{12} & S_{11} & 0 & 0 & 0 \\ 0 & 0 & 0 & S_{44} & 0 & 0 \\ 0 & 0 & 0 & 0 & S_{44} & 0 \\ 0 & 0 & 0 & 0 & 0 & S_{44} \end{vmatrix}. \tag{14.38}$$

14.9 Relation between elastic stiffness constants and elastic compliance constants

The stress components for a cubic crystal are given by eqn (14.36), the strain components can be written as

$$\left.\begin{aligned} e_{xx} &= S_{11}X_x + S_{12}Y_y + S_{12}Z_z \\ e_{yy} &= S_{12}X_x + S_{11}Y_y + S_{12}Z_z \\ e_{zz} &= S_{12}X_x + S_{12}Y_y + S_{11}Z_z \\ e_{xy} &= S_{44}X_y \\ e_{yz} &= S_{44}Y_z \\ e_{zx} &= S_{44}Z_x \end{aligned}\right\} \tag{14.39}$$

Now substituting the values of e_{xx}, e_{yy} and e_{zz} from eqn (14.37) in equation x_x of eqn (14.36), we get

$$\begin{aligned} O = {} & X_x(C_{11}S_{12} + 2C_{12}S_{12} - 1) \\ & + Y_y(C_{11}S_{12} + C_{12}S_{11} + C_{12}S_{12}) \\ & + Z_z(C_{11}S_{12} + C_{12}S_{12} + C_{12}S_{11}). \end{aligned} \tag{14.40}$$

Equating the coefficients of X_x and Y_y equal to zero, we get

$$C_{11}S_{11} + 2C_{12}S_{12} - 1 = 0 \tag{14.41}$$

$$\text{and } C_{11}S_{12} + C_{12}S_{11} + C_{12}S_{12} = 0. \tag{14.42}$$

Solving eqn (14.39) and (14.40) for S_{11} and S_{12}, we get

$$\left.\begin{aligned} S_{11} &= \frac{C_{11} + C_{12}}{(C_{11} - C_{12})\,(C_{11} + 2C_{12})} \\ S_{12} &= \frac{C_{12}}{(C_{11} - C_{12})\,(C_{11} + 2C_{12})} \\ S_{44} &= \frac{1}{C_{44}} \end{aligned}\right\}. \tag{14.43}$$

We conclude from expressions (14.36), (14.37) and (14.41) that the term S_{11} measures the amount of tensile strain in X-direction produced by an X-axis tensile stress of unit magnitude and term S_{12} measures the amount of Y-axis (or, Z-axis) strain which accompanies the X-axis tensile stress. The opposite sign of S_{12} indicates that a tensile strain is always accompanied by a compressive strain of opposite kind in all directions at right angles to the direction of the applied tensile stress. Since, there is no distinction between X-, Y-

and Z-axis in actual materials, we may also interpret S_{11} as measuring the Y-axis strain produced by a unit Y-axis tensile stress and S_{12} as measuring the Z-axis (or X-axis) strain, etc.

In view of the above understanding, it is now clear that S_{11}^{-1} is equivalent to E, the Yong's modulus, $\left(-\frac{S_{12}}{S_{11}}\right)$ equivalent to Poisson's ratio σ and S_{44}^{-1} equivalent to shear modulus G.

In terms of CS, we have

$$\left.\begin{aligned} E &= \frac{1}{S_{11}} = \frac{(C_{11}-C_{12})(C_{11}+2C_{11})}{(C_{11}+C_{12})} \\ \sigma &= -\frac{S_{12}}{S_{11}} = \frac{C_{12}}{(C_{11}+C_{12})} \\ G &= \frac{1}{S_{44}} = C_{44} \end{aligned}\right\} . \tag{14.44}$$

14.10 Elastic waves

When a system of stresses to which the crystal is subjected, is suddenly altered, the elastic waves or shear waves of small amplitude can be generated. This sudden changes of stress will cause the stress to propagate through it. These are identical to acoustical waves in solids which may be studied using newtonian mechanics. In case of small amplitude theories, the total deformation must all times be small compared to the overall dimensions of the crystal.

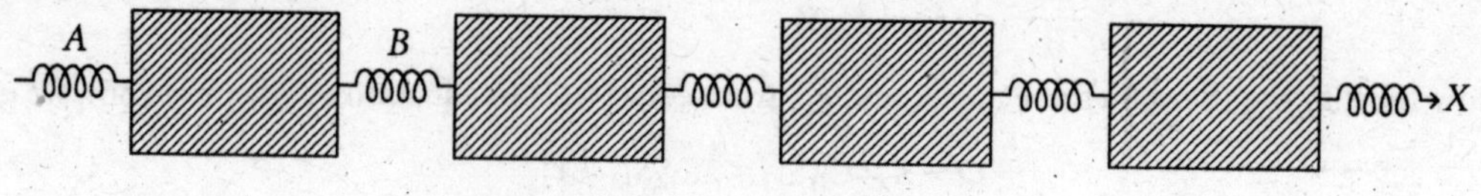

Fig. 14.6: Blocks are connected by springs.

In order to treat these waves we consider a cube of volume $\Delta x\,\Delta y\,\Delta z$ [Fig. 14.6] acted on by a stress $X_x(x)$ on the face at x and

$$X_x(x+\Delta x) \simeq X_x + \frac{\partial X_x}{\partial x}\Delta x$$

on the parallel face at $(x+\Delta x)$.

If now in Fig. 14.6 springs A and B are stretched equally, the block between them experiences no net force. This illustrates that a uniform stress X_x in a solid does not give any net force on volume element.

In the spring at B is stretched more than the spring at A, the block will be accelerated by a force

$$X_x(B) - X_x(A) = \left(X_x + \frac{\partial X_x}{\partial x}\right)\Delta x\,\Delta y - (X_x)\Delta x\,\Delta y = \frac{\partial X_x}{\partial x}\Delta x\,\Delta y.$$

Along the X-direction, the other forces arise from the variation across the cube of stresses X_y and X_z (not shown in the figure). The net X-component of the force on the cube is

$$F_x = \left(\frac{\partial X_x}{\partial x} + \frac{\partial X_y}{\partial y} + \frac{\partial X_z}{\partial z}\right)\Delta x\,\Delta y\,\Delta z.$$

If ρ be the density of the cube, then its mass is $\rho\,\Delta x\,\Delta y\,\Delta z$.

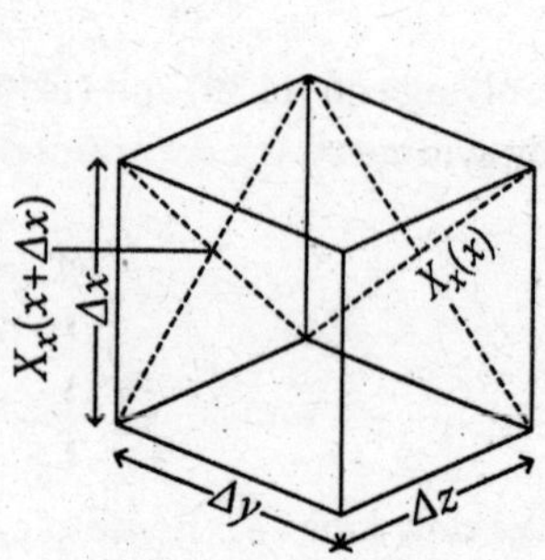

Fig. 14.7: Cube of volume $\Delta x\,\Delta y\,\Delta z$ acted on by a stress—$X_x(x)$ on the face at x and $X_x(x+\Delta x)$ on the parallel face at $x+\Delta x$.

If $\frac{\partial^2 u}{\partial t^2}$ be the acceleration of the cube, then the equation of motion along the X-direction is

$$\rho \frac{\partial^2 u}{\partial t^2} = \frac{\partial X_x}{\partial x} + \frac{\partial X_y}{\partial y} + \frac{\partial X_z}{\partial z}. \tag{14.45}$$

Now differentiating eqn (14.32), we get

$$\left.\begin{aligned} \frac{\partial X_x}{\partial x} &= C_{11}\frac{\partial e_{xx}}{\partial x} + C_{12}\left(\frac{\partial e_{yy}}{\partial x} + \frac{\partial e_{zz}}{\partial x}\right) \\ \frac{\partial X_y}{\partial y} &= C_{44}\frac{\partial e_{xy}}{\partial y} \\ \frac{\partial X_z}{\partial z} &= C_{44}\frac{\partial e_{xz}}{\partial z} \end{aligned}\right\}. \tag{14.46}$$

Substituting these in eqn (14.43), we get

$$\rho \frac{\partial^2 u}{\partial t^2} = C_{11}\frac{\partial e_{xx}}{\partial x} + C_{12}\left(\frac{\partial e_{yy}}{\partial x} + \frac{\partial e_{zz}}{\partial x}\right) + C_{44}\left(\frac{\partial e_{xy}}{\partial y} + \frac{\partial e_{xz}}{\partial z}\right). \tag{14.47}$$

Differentiating eqn (14.17), we get

$$\frac{\partial e_{xx}}{\partial x} = \frac{\partial^2 u}{\partial x^2}, \quad \frac{\partial e_{yy}}{\partial y} = \frac{\partial^2 v}{\partial y\,\partial x}, \quad \frac{\partial e_{zz}}{\partial z} = \frac{\partial^2 w}{\partial x\,\partial z},$$

$$\frac{\partial e_{xy}}{\partial y} = \frac{\partial^2 u}{\partial y^2} + \frac{\partial^2 v}{\partial x\,\partial y}.$$

$$\frac{\partial e_{xz}}{\partial z} = \frac{\partial^2 u}{\partial z^2} + \frac{\partial^2 w}{\partial x\,\partial z},$$

where u, v, w are the components of displacement vector $\vec{R}$ defined by $\vec{R} = \hat{i}x + \hat{j}v + \hat{k}z$. Substituting the values of $\frac{\partial e_{xx}}{\partial x}$, etc. in eqn (14.47), we get

$$\rho \frac{\partial^2 u}{\partial t^2} = C_{11}\frac{\partial^2 u}{\partial x^2} + C_{44}\left(\frac{\partial^2 u}{\partial y^2} + \frac{\partial^2 u}{\partial z^2}\right) + (C_{12} + C_{44})\left(\frac{\partial^2 v}{\partial x\,\partial y} + \frac{\partial^2 w}{\partial x\,\partial z}\right). \tag{14.48}$$

By symmetry, the corresponding equations of motion along Y- and Z-direction are

$$\rho \frac{\partial^2 v}{\partial t^2} = C_{11}\frac{\partial^2 v}{\partial y^2} + C_{44}\left(\frac{\partial^2 v}{\partial x^2} + \frac{\partial^2 v}{\partial z^2}\right) + (C_{12} + C_{44})\left(\frac{\partial^2 w}{\partial y\,\partial z} + \frac{\partial^2 u}{\partial x\,\partial y}\right) \tag{14.49}$$

$$\text{and } \rho \frac{\partial^2 w}{\partial t^2} = C_{11}\frac{\partial^2 w}{\partial z^2} + C_{44}\left(\frac{\partial^2 w}{\partial x^2} + \frac{\partial^2 w}{\partial y^2}\right) + (C_{12} + C_{44})\left(\frac{\partial^2 u}{\partial x\,\partial z} + \frac{\partial^2 v}{\partial y\,\partial z}\right). \tag{14.50}$$

We now look for the simple special solutions of the above waves. Once the direction of propagation is fixed there will be three basic directions of the displacement of wave. If the displacement is along the direction of propagation, the wave is termed as longitudinal or compressive wave. If the motion is transverse, the waves are termed as transverse or shear waves. In isotropic materials, there may be two shear waves with orthogonal displacement. In more complex crystalline materials several shear and tensile strains accompany one another in waves travelling in certain directions through the material. By complicated analysis it is possible to isolate the non-interacting waves having different wave velocities.

14.11 Waves in [100] direction

The solution of eqn (14.46) is given by a longitudinal wave of the form:

$$u = u_0 e^{i(kx-\omega t)}, \tag{14.51}$$

where u is the X-component of the particle displacement. The wave vector k and the particle motion are along the edge of the cube in X-direction. Here, $k = 2\pi/\lambda$ and $\omega = 2\pi\nu$ is the angular frequency.

Substituting eqn (14.46) in eqn (14.48), we get $\omega^2\rho = C_{11}k^2$. Then the velocity of ω/k of longitudinal wave along [100] direction is

$$v_l = \nu\lambda = \frac{\omega}{k} = \left(\frac{C_{11}}{\rho}\right)^{1/2}. \tag{14.52}$$

A second solution of eqn (14.49) is given by transverse or shear wave with wave vector along the X-direction and the particle displacement in the Y-direction with velocity

$$v = v_0 e^{i(kx-\omega t)}. \tag{14.53}$$

Substituting the above equation in eqn (14.49), we get the dispersion relation, $\omega^2\rho = C_{44}k^2$.

Then the velocity of the transverse waves, $v_s = \omega/k$ in [100] direction is

$$v_s = \left(\frac{C_{44}}{\rho}\right)^{1/2}. \tag{14.54}$$

We obtain identical velocity if the particle displacement is in the Z-direction.

Hence, for $\vec{k}$ parallel to [110] the two independent shear waves have equal velocities as shown in Fig. 14.8. This is, however, not true for $\vec{k}$ in general direction in the crystal.

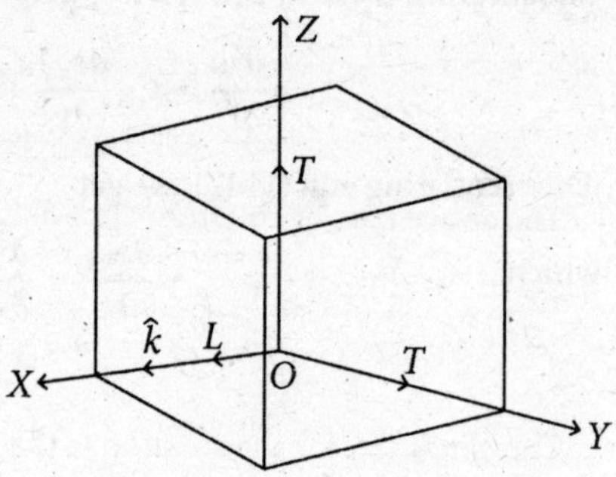

Fig. 14.8: Waves propagating with $\vec{k}$ in X-direction [100] with one longitudinal mode and two degenerate transverse modes.

14.12 Waves in [110] direction

There is a special interest in waves that propagate in a face diagonal direction of a cubic crystal [Fig. 14.8]. Because the three elastic constants can be found simply from the three propagation velocities in this direction.

We consider a shear wave that propagates in the XY-plane with particle displacement ω in Z-direction. The solution of eqn (14.50) can be written as

$$\omega = \omega_0 \exp[i(k_x x + k_y y - \omega t)]. \tag{14.55}$$

Substituting in eqn (14.50), we get

$$\omega^2\rho = C_{44}\left(k_x^2 + k_y^2\right) = C_{44}k^2 \text{ or, } v_s = \frac{\omega}{k} = \left(\frac{C_{44}}{\rho}\right)^{1/2}. \tag{14.56}$$

Equation (14.56) shows that there is no propagation in the direction of the XY-plane and that the velocity is the same as given by eqn (14.54).

Let us now consider other waves that propagate in the XY-plane with particle motion in XY-plane.

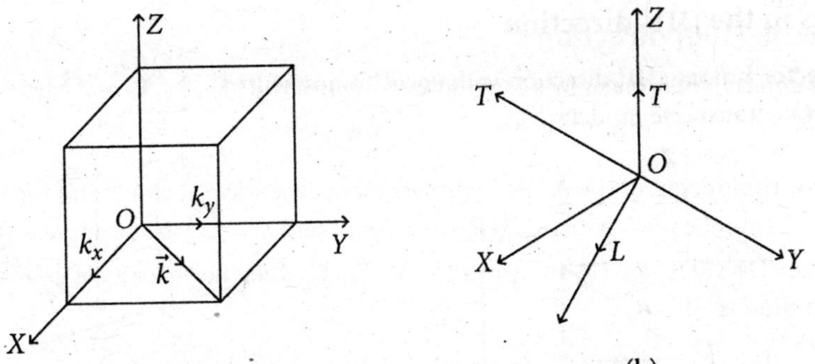

Fig. 14.9: Wave propagating with vector $\vec{k}$ in XY-plane [110] direction (a) with one longitudinal mode and two degenerate transverse mode (b).

Let

$$\left.\begin{aligned} u &= u_0 \exp\left[i(k_x x + k_y y - \omega t)\right] \\ v &= v_0 \exp\left[i(k_x x + k_y y - \omega t)\right] \end{aligned}\right\} \tag{14.57}$$

The above pair of equation (14.57) has a particular simple solution for a wave in the [110] direction for which

$$k_x = k_y = \frac{k}{\sqrt{2}}.$$

The condition for the solution is that the determinant of the coefficients of u and v in eqn (14.57) should vanish. So,

$$\begin{vmatrix} -\omega^2\rho + (C_{11} + C_{44})\dfrac{k^2}{2} & (C_{12} + C_{44})\dfrac{k^2}{2} \\ (C_{12} + C_{44})\dfrac{k^2}{2} & -\omega^2\rho + (C_{11} + C_{44})\dfrac{k^2}{2} \end{vmatrix}. \tag{14.58}$$

The roots of the above equation are

$$\left.\begin{aligned} \omega^2\rho &= \tfrac{1}{2}(C_{11} + C_{12} + 2C_{44})k^2 \\ \omega^2\rho &= \tfrac{1}{2}(C_{11} + C_{12})k^2 \end{aligned}\right\}. \tag{14.59}$$

The first root when substituted in eqn (14.53) gives $u = v$ describing a longitudinal wave. Thus, the particle displacement is along [110] and parallel to $\vec{k}$ vector. The velocity v_L of the longitudinal wave is given by

$$v_L = \frac{\omega}{k}\left[\frac{C_{11} + C_{12} + 2C_{44}}{2\rho}\right]^{1/2}. \tag{14.60}$$

The second root of eqn (14.53) describes a transverse wave, as its substitution in eqn (14.53) gives $u = -v$. Thus, the particle displacement is along $[1\bar{1}0]$ direction and perpendicular $\vec{k}$ vector. The velocity of the transverse wave v_s in X-Y plane is given by

$$v_s = \left[\frac{C_{11} - C_{12}}{2\rho}\right]^{1/2}. \tag{14.61}$$

14.13 Waves in the [111] direction

Fig. 14.9 wave vector $\vec{k}$ along [111] direction with equal components k_x, k_y, k_z and each equal to $k^2/\sqrt{3}$ (a) and two degenerate transverse models (b).

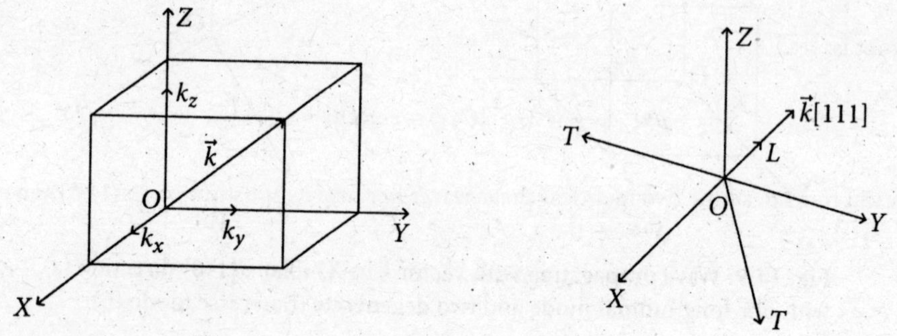

(a) (b)

Fig. 14.10: Wave vector $\vec{k}$ along [111]-direction with equal components k_x, k_y, k_z and each equal to $k^2/\sqrt{3}$ (a) and two degenerate transverse modes (b).

The solution of the wave eqns (14.48), (14.49) and (14.50) can be written as

$$\left.\begin{aligned} u &= u_0 e^{i}\left\{\frac{k}{\sqrt{3}}(x+y+z)-\omega t\right\} \\ v &= v_0 e^{i}\left\{\frac{k}{\sqrt{3}}(x+y+z)-\omega t\right\} \\ w &= w_0 e^{i}\left\{\frac{k}{\sqrt{3}}(x+y+z)-\omega t\right\} \end{aligned}\right\}. \tag{14.62}$$

Now substituting eqn (14.62) in eqns (14.48), (14.49) and (14.50), we get

$$\left\{\frac{k^2}{3}(C_{11}+2C_{44})-\rho^2 w\right\}u+(C_{12}+C_{44})\frac{k^2}{3}v+(C_{12}+C_{44})\frac{k^2}{3}w \tag{14.63}$$

$$\left\{\frac{k^2}{3}(C_{11}+2C_{44})-\rho^2 w\right\}v+(C_{12}+C_{44})\frac{k^2}{3}w+(C_{12}+C_{44})\frac{k^2}{3}u \tag{14.64}$$

$$\left\{\frac{k^2}{3}(C_{11}+2C_{44})-\rho^2 w\right\}w+(C_{12}+C_{44})\frac{k^2}{3}u+(C_{12}+C_{44})\frac{k^2}{3}v \tag{14.65}$$

The condition for non-vanishing solution of eqns (14.48), (14.49) and (14.50) is that the determinant formed by the coefficients of u, v, w must vanish.

Hence,

$$\begin{vmatrix} \left\{\frac{k^2}{3}(C_{11}+2C_{44}-\rho w)\right\} & (C_{12}+C_{44})\frac{k^2}{3} & (C_{12}+C_{44})\frac{k^2}{3} \\ \left\{\frac{k^2}{3}(C_{12}+C_{44})\right\} & \left\{\frac{k^2}{3}(C_{11}+2C_{44})-w^2\right\} & (C_{12}+C_{44})\frac{k^2}{3} \\ \left\{\frac{k^2}{3}(C_{12}+C_{44})\right\} & \left\{\frac{k^2}{3}(C_{11}+C_{44})\right\} & \left\{\frac{k^2}{3}(C_{11}+2C_{44})-\rho w^2\right\} \end{vmatrix} = 0. \tag{14.66}$$

Eqn (14.66) has three roots out of which two are identical. The identical roots are

$$\rho w^2 + \frac{k^2}{3}(C_{11} + 2C_{44}) = (C_{12} + C_{44})\frac{k^2}{3} \tag{14.67}$$

and other root is

$$-\rho w^2 + \frac{k^2}{3}(C_{11} + 2C_{44}) = -2(C_{12} + C_{44})\frac{k^2}{3}. \tag{14.68}$$

The identical roots describe two identical shear waves because its substitution in (14.63) gives

$$\frac{k^2}{3}(C_{12} + C_{44})(u + v + w) = 0 \text{ or, } u + v + w = 0.$$

This shows that the particle displacement is in a direction perpendicular to [111] and thus perpendicular to $\vec{k}$. The velocity of the shear waves v_s is given by

$$v_s = \frac{w}{k} = \left\{\frac{C_{11} - C_{12} + C_{44}}{3\rho}\right\}^{1/2}. \tag{14.69}$$

The other root is

$$-\rho w^2 + (C_{11} + C_{12} + 4C_{44})\frac{k^2}{33} = 0 \tag{14.70}$$

and describes a longitudinal wave because its substitution in eqn (14.63) gives

$$(C_{12} + 2C_{44})\frac{k^2}{3}(-2u + v + w) = 0. \tag{14.71}$$

The possible solution of the eqn (14.70) is $u = v = w$ indicating that particle displacement in this case is along [111] and parallel to $\vec{k}$ vector. The velocity of propagation of this wave is

$$v_L = \frac{w}{k} = \left\{\frac{C_{11} + C_{12} + 4C_{44}}{3\rho}\right\}^{1/2}. \tag{14.72}$$

14.14 Determination of elastic constants by ultrasonic pulse method

The elastic constants of a specimen is determined by using the relation between velocities of propagation of elastic waves and the moduli of elasticity. The elastic waves are excited in a crystal and the velocity is measured. Since the crystal acts like a continuous and homogeneous elastic medium in the ultrasonic frequency range the measurements are often in this frequency range. As the waves are usually pulsed, the method is known as *ultrasonic pulse method.*

The two ends of the crystal specimen are plane. A piezoelectric quartz transducer is glued to one end surface of the specimen. Now an ultrasonic pulse is reflected from rear surface of the crystal back to transducer as shown in Fig. 14.11.

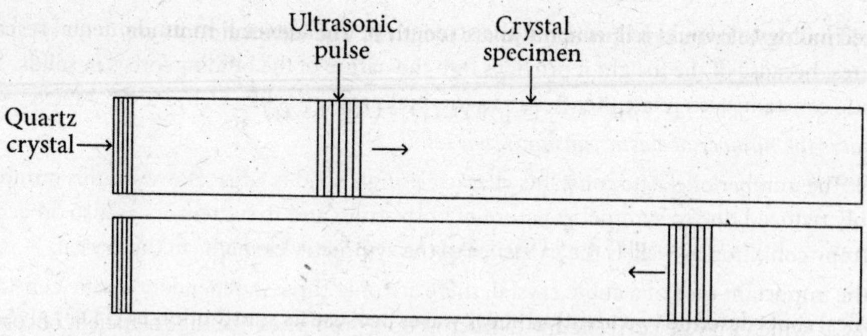

Fig. 14.11: Ultrasonic method for determination of elastic wave velocities.

The pulse is transmitted through the test crystal and reflected from the rear surface of the crystal back to the transducer. The time interval between initiation and receipt of the pulse is measured by standard digital device. The velocity is obtained by dividing the round trip distance by the elapsed time.

The experiment

The experimental arrangement is shown in Fig. 14.12. The trigger generator starts the process by closing the thyratron switch. This discharges a condenser through *LC* network and imposes the indicated voltage waveform on the transducer glued to the specimen to produce the train of echoes shown on the oscilloscope screen. The time mark generator and sweep delay circuit allow the time of arrival of some latter echo to be measured accurately.

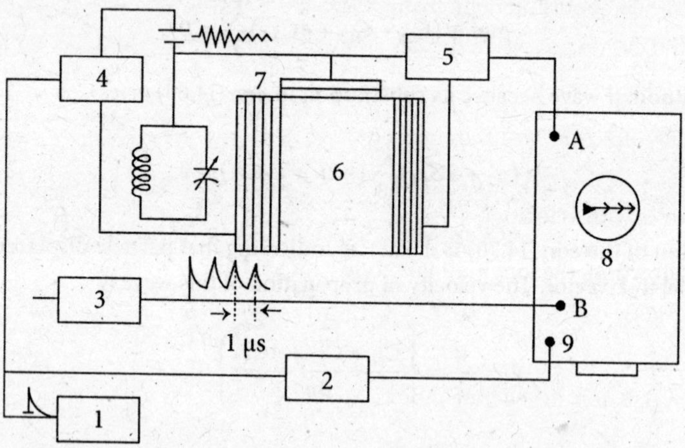

Fig. 14.12: Block diagram of an electronic apparatus to measure velocity of sound by ultrasonic pulse echo technique.
1. Trigger generator (100 μs period. 2. Delay circuit. 3. Time mark generator. 4. Thyratron switch. 5. Wideband *r-f* amplifier. 6. Specimen. 7. Quartz. 8. Oscilloscope. 9. Sweep trigger.

14.15 Answers of Some Important Questions

14.1. *What is the importance of the study of elastic behaviour of solids?*

Ans. The study of elastic behaviour of a solid is very important in fundamental and technical researches.

In technology, it would tell us about the strength of the material. In fundamental research, it is of interest because of the insight it provides into the nature of the binding forces in solids. The relevant elastic constants also relate themselves to thermal properties like Debye temperature.

14.2. *What is the number of elastic constants in a solid?*

Ans. The number of elastic constants characterising a solid is large. However, this number is considerably reduced due to symmetry nature of both strain and stress tensors as also on account of the isotropy condition as well as the existence of the symmetry elements in the crystal.

In the important case of a cubic crystal, there are only three independent elastic constants C_{11}, C_{12} and C_{44} which may be correlated with microscopic properties like the coefficient of compressibility. The existence of the centre of symmetry of a cubic crystal stable under the central interatomic forces leads to the *Cauchy relation* $C_{12} = C_{44}$. This reduces the number of independent elastic constants of a cubic crystal to two only. Further, if the crystal is elastically isotropic, we have $C_{11} - C_{12} = 2C_{44}$. This is equivalent to the statement that the elastic behaviour of such a crystal may be completely describable by a single independent elastic constant. Such circumstances corresponds to only a hypothetical and ideal crystal.

14.3. *What is Cauchy's relation:*

Cauchy's relation states that

$$C_{12} = C_{44} = \frac{2\beta}{a},$$

where β is next neighbour interaction and a the lattice constants.

14.4. Write down the expressions of velocities of elastic waves propagating along [100], [110] and [111] directions of an elastic solid.

Ans. (i) For [100] direction, we have one longitudinal mode with velocity $\left(\frac{C_{11}}{\rho}\right)^{1/2}$ and two degenerate transverse modes with velocity $\left(\frac{C_{44}}{\rho}\right)^{1/2}$.

(ii) For [110] direction, the longitudinal mode with velocity

$$\left(\frac{C_{11} + C_{12} + 2C_{44}}{2\rho}\right)^{1/2}$$

and two transverse non-degenerate modes having velocity

$$\left(\frac{C_{44}}{\rho}\right)^{1/2} \text{ and } \left(\frac{C_{11} - C_{12}}{2\rho}\right)^{1/2}.$$

(iii) For [111] direction, we have one longitudinal mode with velocity

$$\left\{\frac{C_{11} + 2C_{12} + 4C_{44}}{3\rho}\right\}^{1/2}$$

and two degenerate transverse modes with velocity

$$\left\{\frac{C_{11} - C_{12} + C_{44}}{3\rho}\right\}^{1/2}.$$

14.5. *Write down the expressions of Young's modulus, Poisson's ratio and modulus of rigidity.*

Ans. The Young's modulus, $E = S_{11}^{-1}$.

Poisson's ratio,

$$\sigma = \left(-\frac{S_{12}}{S_{11}}\right)$$

and modulus of rigidity, $G = Su^{-1}$

$$\therefore G = \frac{E}{2(\sigma+1)},$$

which gives $C_{11} - C_{12} = 2C_{44}$.

14.16 Solved Problems

14.1. *Show that for a cubic isotropic substance these are only two moduli of elasticity.*

Solution: From eqn (14.48), we have

$$\rho\frac{\partial^2 u}{\partial t^2} = C_{11}\frac{\partial^2 u}{\partial x^2} + C_{44}\left(\frac{\partial^2 u}{\partial y^2} + \frac{\partial^2 u}{\partial z^2}\right) + (C_{12}+C_{44})\left(\frac{\partial^2 v}{\partial y\partial x} + \frac{\partial^2 w}{\partial x\partial z}\right)$$

$$\text{or, } \rho\frac{\partial^2 u}{\partial t^2} = (C_{11} - C_{12} - 2C_{44})\frac{\partial^2 u}{\partial x^2} + C_{44}\left(\frac{\partial^2 u}{\partial x^2} + \frac{\partial^2 u}{\partial y^2} + \frac{\partial^2 u}{\partial z^2}\right) + (C_{12}+C_{44})\frac{\partial}{\partial x}\left(\frac{\partial u}{\partial x} + \frac{\partial v}{\partial y} + \frac{\partial w}{\partial z}\right)$$

$$\text{or, } \rho\frac{\partial^2 u}{\partial t^2} = (C_{11} - C_{12} - 2C_{44})\frac{\partial^2 u}{\partial x^2} + C_{44}\nabla^2 u + (C_{12}+C_{44})\frac{\partial}{\partial x}(\nabla\cdot\vec{R}), \quad \text{(i)}$$

where $\vec{R} = \hat{i}u + \hat{j}v + \hat{k}w$.

Writing eqns (14.49) and (14.50) in the similar form, we have

$$\rho\frac{\partial^2 v}{\partial t^2} = (C_{11} - C_{12} - 2C_{44})\frac{\partial^2 v}{\partial y^2} + C_{44}\nabla^2 v + (C_{12}+C_{44})\frac{\partial}{\partial y}(\nabla\cdot\vec{R}) \quad \text{(ii)}$$

$$\text{and } \rho\frac{\partial^2 w}{\partial t^2} = (C_{11} - C_{12} - 2C_{44})\frac{\partial^2 w}{\partial z^2} + C_{44}\nabla^2 w + (C_{12}+C_{44})\frac{\partial}{\partial z}(\nabla\cdot\vec{R}). \quad \text{(iii)}$$

Now combining eqns (i), (ii) and (iii) and making use of equation

$$\vec{R} = \hat{i}u + \hat{j}v + \hat{k}w,$$

we get

$$\rho\frac{\partial^2 \vec{k}}{\partial t^2} = (C_{11} - C_{12} - 2C_{44})\left(\hat{i}\frac{\partial^2 u}{\partial x^2} + \hat{j}\frac{\partial^2 v}{\partial y^2} + \hat{k}\frac{\partial^2 w}{\partial z^2}\right)$$
$$+ (C_{12} + 2C_{44})\nabla(\nabla\cdot\vec{R}) - C_{44}\nabla\times(\nabla\times\vec{R}). \quad \text{(iv)}$$

As the only variant term in eqn (iv) is the first member on right-hand side, the equation will be invariant, if and only if this term vanishes.

Thus, the condition for the crystal to be elastically isotropic is $C_{11} - C_{12} - 2C_{44} = 0$ or, $C_{11} - C_{12} = 2C_{44}$.

This shows that for a isotropic cubic crystal there are only two moduli of elasticity.

14.2. *Obtain an expression for the energy of strained cubic crystal. Show that when the crystal is under a hydrostatic pressure, the energy per unit volume can be expressed as*

$$u = \frac{1}{2}\frac{(C_{11} + C_{13})}{3}(\Delta v)^2,$$

where Δv is the change in volume per unit volume. Hence, prove that the Bulk modulus of the solid is given by

$$k = \tfrac{1}{2}(C_{11} + 2C_{12}).$$

Solution: The elastic energy density is given by

$$u = \tfrac{1}{2}(\text{stress} \times \text{strain}) \text{ or, } u = \tfrac{1}{2}(X_x e_{xx} + Y_y e_{yy} + Z_z e_{xz}) + X_y e_{xy} + Y_z e_{yz} + Z_x e_{zx}.$$

Putting the values of X_x, Y_y, Z_z, etc. from eqn (14.36), we get

$$u = \tfrac{1}{2}\{(C_{11}e_{xx} + C_{12}e_{yy} + C_{12}e_{zz})e_{xx} + (C_{11}e_{yy} + C_{12}e_{zz} + C_{12}e_{xx})e_{yy}$$
$$+ (C_{11}e_{zz} + C_{12}e_{yy} + C_{12}e_{xx})e_{zx} + C_{44}e_{xy}e_{xy} + C_{55}e_{yz}e_{yz} + C_{66}e_{zx}e_{zx}\}$$
$$\text{or, } u = \tfrac{1}{2}C_{11}(e_{xx}^2 + e_{yy}^2 + e_{zz}^2) + C_{12}(e_{xx}e_{yy} + e_{xx}e_{zz} + e_{zz}e_{zx}) + \tfrac{1}{4}C_{44}(e_{xy}^2 + e_{yz}^2 + e_{zx}^2).$$

For cubic crystal $e_{xx} = e_{yy} = e_{zz} = \frac{1}{3}(\Delta v)$ and for hydrostatic pressure.

$$X_y = Y_z = Z_x = 0 \text{ or, } e_{xy} = e_{yz} = e_{zx} = 0.$$

$$\therefore u = \frac{1}{2}\frac{(C_{11} + 2C_{12})}{3}(\Delta v)^2.$$

Now Bulk modulus k is defined by the equation

$$u = \frac{1}{2}k(\Delta v)^2 \text{ or, } \frac{1}{2} = \frac{(C_{11} + 2C_{12})}{3}(\Delta v)^2 = \frac{1}{2}k(\Delta v)^2 \text{ or, } \mathbf{k} = \frac{\mathbf{C_{11} + 2C_2}}{\mathbf{3}}.$$

14.3. *Deduce the relation between the compressibility and elastic constants of a cubic crystal.*

Solution: The compressibility is defined as the reciprocal of Bulk modulus. Mathematically, compressibility $\beta = 1/k$.

But

$$k = \frac{C_{11} + 2C_{12}}{3}. \quad \therefore \boldsymbol{\beta} = \frac{\mathbf{3}}{\mathbf{C_{11} + 2C_{12}}}.$$

14.4. *Prove that for a cubic lattice, $C_{11} - C_{12} = 2C_{44}$, where symbols have their usual meaning.*

Solution: We have Young's modulus,

$$E = \frac{(C_{11} - C_{12})(C_{11} + 2C_{11})}{C_{11} + C_{12}} \text{ [from eqn (14.44)].}$$

Poissons's ratio,

$$\sigma = \frac{C_{12}}{C_{11} + C_{12}}$$

and modulus of rigidity $G = C_{44}$.

The relation connecting E, σ and G is

$$G = \frac{E}{2(\sigma + 1)}.$$

Substituting the values of E, σ and G, we get

$$C_{44} = \frac{(C_{11} - C_{12})(C_{11} + 2C_{12}) \times (C_{11} + C_{12})}{(C_{11} + C_{12})2(C_{12} + C_{11} + C_{12})} \text{ or, } \mathbf{2C_{44} = C_{11} - C_{12}}.$$

14.5. *Find the values of Young's modulus, bulk modulus and compressibility for one-dimensional lattice in terms of elastic constants.*

Solution: A tensile stress in X-direction of a one-dimensional lattice will produce no contraction in the Y- and Z-directions, i.e., $C_{12} = 0$.

$$\therefore \text{Young's modulus } E = C_{11}; \text{ Bulk modulus } K = \frac{C_{11}}{3} \text{ and Compressibility } \beta = \frac{3}{C_{11}}.$$

This corresponds to the case of no transverse contraction accompanying longitudinal expansion.

EXERCISE

14(A): Objective (MCQ) Type Questions

Select the correct answer.

14.1. While discussing elastic properties of a crystal it is considered as

(a) a periodic array of atoms (b) a homogeneous continuous medium

(c) an inhomogeneous continuous medium (d) a discrete medium.

14.2. In the important case of cubic crystal, there are only three elastic constants, viz.,

(a) C_{11}, C_{12}, C_{18} (b) C_{11}, C_{13}, C_{14} (c) C_{11}, C_{12}, C_{44} (d) C_{22}, C_{23}, C_{24}.

14.3. The Cauchy relation for a nonatomic cubic crystal is

(a) $C_{12} = C_{44}$ (b) $C_{12} \neq C_{44}$ (c) $C_{12} > C_{44}$ (d) $C_{12} < C_{44}$.

14.4. The poisson's ratio σ of an elastic body is given by

(a) $\sigma = \dfrac{C_{12}}{(C_{11} + C_{12})}$ (b) $\sigma = \dfrac{C_{11}}{(C_{11} + C_{12})}$ (c) $\sigma = \dfrac{C_{11} + C_{12}}{C_{11}}$ (d) $\sigma = \dfrac{C_{11} + C_{12}}{C_{12}}$.

14.5. The compressibility of an elastic body is given by

(a) $\beta = \dfrac{C_{12}}{C_{11} + 3}$ (b) $\beta = \dfrac{C_{11}}{C_{12} + 3}$ (c) $\beta = \dfrac{3}{C_{11} + C_{12}}$ (d) $\dfrac{1}{3}(C_{11} + C_{12})$.

14.6. The velocity of transverse elastic wave in an elastic solid along [100] direction is

(a) $v_s = \left(\dfrac{\rho}{C_{44}}\right)^{1/2}$ (b) $\left(\dfrac{C_{44}}{\rho}\right)^{1/2}$ (c) $\left(\dfrac{C_{44}}{\rho}\right)^{3/2}$ (d) $\left(\dfrac{\rho}{C_{44}}\right)^{3/2}$.

14.7. The velocity of longitudinal wave in an elastic solid is

(a) $v_L = \left(\dfrac{C_{11} + C_{12} + C_{44}}{2\rho}\right)^{1/2}$ (b) $v_L = \left(\dfrac{C_{11} + C_{12} + 2C_{44}}{2\rho}\right)^{1/2}$

(c) $v_L = \left(\dfrac{C_{11} + 2C_{44}}{2\rho}\right)^{1/2}$ (d) $v_L = \left(\dfrac{C_{11} + C_{12} + 2C_{44}}{\rho}\right)^{1/2}$.

14.8. The velocity of propagation of longitudinal wave in an elastic solid is

(a) $v_L = \left(\dfrac{C_{11} + 2C_{12} + 4C_{44}}{3\rho}\right)^{1/2}$ (b) $v_L = \left(\dfrac{C_{11} + C_{12} + 4C_{44}}{3\rho}\right)^{1/2}$

(c) $v_L = \left(\dfrac{C_{11} + 2C_{44} + 4C_{44}}{\rho}\right)^{1/2}$ (d) $v_L = \left(\dfrac{2C_{12} + 4C_{44}}{3\rho}\right)^{1/2}$.

14.9. The velocity of propagation of a shear wave moving along [110] direction with particle motion along [110] direction and perpendicular to wave vector $\vec{k}$ in a cubic crystal is

(a) $v = \left(\dfrac{C_{11} - C_{12}}{2\rho}\right)^{1/2}$ (b) $v = \left(\dfrac{C_{11} - C_{12}}{\rho}\right)^{1/2}$ (c) $v = \dfrac{2\rho}{(C_{11} - C_{12})^{1/2}}$ (d) $v = \left(\dfrac{C_{11} - C_{12}}{2\rho}\right)^{3/2}$.

14.10. For an isotropic cubic crystal, there are only two moduli of elasticity, viz.,

(a) $C_{11} - C_{12} = 2C_{44}$ (b) $C_{11} + C_{12} = 2C_{44}$ (c) $C_{11} - C_{12} = C_{44}$ (d) $C_{11}+C_{12}=C_{44}$.

14.11. Modulus of rigidity for an elastic body is

(a) $G = \dfrac{E}{2(\sigma+1)}$ (b) $\dfrac{E}{2(\sigma-1)}$ (c) $\dfrac{\sigma}{2(E-1)}$ (d) $\dfrac{\sigma}{(E-1)}$.

ANSWERS

14.1. (b) **14.2.** (c) **14.3.** (a) **14.4.** (a) **14.5.** (c) **14.6.** (b) **14.7.** (b) **14.8.** (a) **14.9.** (b) **14.10.** (a) **14.11.** (a).

14(B): Short Answer Type Questions

14.1. Define stress and strain tensors and write down the equation connecting them for a cubic crystal with the elastic stiffness constants.

14.2. Discuss elastic constants for an FCC metal.

14.3. Show that in a cubic crystal there are only three elastic constant.

14.4. Prove that the coefficient of compressibility is given by

$$\beta = \frac{3}{C_{11} + 2C_{12}}.$$

14.5. Prove that in a cubic crystal, the effective elastic constant for a shear across (110) plane in the $[1\bar{1}0]$ direction is $\dfrac{C_{11} - C_{12}}{2}$.

14(C): Numerical Problems

14.1. The longitudinal and transverse velocities of sound wave in diamond — a cubic crystal, along [100]-direction are respectively 1.76×10^4 and 1.28×10^4 cm/s. The longitudinal velocity in the [111]-direction is 1.86×10^4 cm/s. From these data and the fact that the density is 3.52 gm/cm^3, calculate the elastic constants C_{11}, C_{12} and C_{44} for diamond.

14.2. What form does Hook's law take in crystalline medium? Obtain an expression for strain energy per unit volume in a crystal.

14.3. Prove that for a cubic lattice $C_{11} - C_{12} = 2C_{44}$, where the symbols have their usual meaning.

14(D): Broad Answer Type Questions

14.1. Obtain an expression for longitudinal and transverse sound wave velocities through a cubic crystal propagating along the $[k, K, K]$-direction in terms of the elastic constants of the crystal.

14.2. Define moduli of elasticity for a crystal. Show that for a cubic crystal there are only three independent moduli of elasticity and for isotropic substance there are only two.

14.3. Establish relation between elastic compliances and stiffness constants of cubic crystal.

14.4. (a) Derive wave equation for elastic waves in a cubic crystal.

(b) Solve it for longitudinal and transverse waves moving in the [100]-direction and calculate the ratio of velocities of two waves.

14.5. A cubic crystal is subjected to tension in [100]-direction. Find the expression in terms of the elastic stiffness constants for Young's modulus and Poisson's ratio.

14.6. How does one experimentally determine the elastic constants of a crystal? Give a very brief description of the experimental set-up and mention the principle one uses for the determination of elastic constants.

14.7. (a) What is Delation? Find out an expression for it.

(b) Derive an expression for Elastic Energy Density of an elastic body.

APPENDICES

Appendix A

Five-Fold Symmetry is Impossible

Let a two-dimensional lattice be invariant under the rotation operation.

The translation vector is given by

$$\vec{t} = u\vec{a} + v\vec{b}, \tag{A.1}$$

where $\vec{a}$ and $\vec{b}$ are the fundamental translation vectors, and u and v are integers.

If a point $P(x, y)$ with reference to a fixed origin be rotated through an angle θ to $P'(x', y')$, then

$$x' = x\cos\theta - y\sin\theta$$
$$y' = x\sin\theta + y\cos\theta.$$

If $\vec{a}$ be the shortest non-vanishing translation vector in the translation group and $\vec{a}$ is parallel to X-axis, then rotating $\vec{a}$ by an angle θ we get a new vector $\vec{a}'$ having components,

$$\vec{a}_x = a\cos\theta$$
$$\vec{a}_y = a\sin\theta.$$

If the lattice is invariant under rotation θ, then the vector $(\vec{a}' - \vec{a})$ must be of the form $\vec{t}$ in Eqn (A.1).

$$\therefore a'_x - a_x = a(\cos\theta - 1)$$
$$a'_y - a_y = a\sin\theta$$

and $|\vec{a}' - \vec{a}|^2 = (a'_x - a_x)^2 + (a'_y - a_y)^2 = a^2(\cos\theta - 1)^2 + a^2\sin^2\theta = a^2(2 - 2\cos\theta)$.

Since, $\vec{a}$ is the shortest vector in the translation group $|\vec{a}' - \vec{a}|^2 \geq a^2$,

$$\therefore (2 - 2\cos\theta)$$

$$\text{or, } \cos\theta \leq \frac{1}{2}, \text{ so that } \pi/2 \leq \theta \leq \frac{5\pi}{2}. \tag{A.2}$$

This is the first condition of the acceptable angle θ. As $\theta = 0$ makes $\vec{a}' = \vec{a}$ so that $\vec{t} = \vec{0}$, the rotation $\theta = 0$ is allowed.

Again, if $(\vec{a}' + \vec{a})$ is of the form $\vec{t}$ and not smaller in magnitude than $\vec{a}$, then

$$(2 + 2\cos\theta) \geq 1 \text{ or, } \cos\theta \geq -\frac{1}{2}$$

so that

$$-\frac{2\pi}{3} \leq \theta \leq \frac{2\pi}{3}. \tag{A.3}$$

This represents the second condition of the acceptable angle θ. The rotation $\theta = \pi$ is allowed as $\vec{a}' = -\vec{a}$, so that $t = 0$.

The third restriction is when inverse rotation $-\theta$ is applied to $\vec{a}$ so that

$$a'_x + a''_x = 2a\cos\theta$$

$$a'_y + a_y = 0 \text{ and } |a' + a''|^2 = 4a^2\cos^2\theta.$$

Since, $\vec{a}$ is the shortest vector,

$$4a^2\cos^2\theta \geq 1 \text{ or, } 2a\cos\theta \geq 1$$

$$\left.\begin{aligned} \text{or,} \quad -\frac{\pi}{3} \leq \theta \leq \frac{\pi}{3} \\ \frac{2\pi}{3} \leq \theta \leq \frac{4\pi}{3} \end{aligned}\right\} \tag{A.4}$$

$\theta = \frac{\pi}{2}$ and $\frac{3\pi}{2}$ are also allowed as $a' + a'' = 0$. From Eqns (2), (3) and (4) we conclude that,

$$\theta = 0, \frac{\pi}{3}, \frac{\pi}{2}, \frac{2\pi}{3}, \pi, \frac{3\pi}{2}, \frac{5\pi}{2}.$$

$\therefore$ the permissible θ values are multiples of $\frac{2\pi}{n}$, where $n = 1, 2, 3, 4$ or 6. Hence, five-fold symmetry rotation is impossible.

Appendix B

Excitons

An *exciton* is a neutral excited mobile state of a crystal that corresponds to the formation of a bound *electron-hole pair.* It may also be defined as a non-conducting electronic state in a perfect insulator, usually a *non-magnetic insulator.*

If a photon of energy $\hbar\omega_{\text{photon}} \rangle E_g$ is absorbed by an insulating crystal a free hole and a free electron are created. The threshold condition for the above mechanism is $\hbar\omega_{\text{photon}} \rangle E_g$ in a direct process. In the indirect phonon assisted process the threshold is $\hbar\omega_{\text{photon}} \rangle (E_g - \hbar\omega_{\text{phonon}})$. The exciton formation involves lowering of energy w.r.t. the finding energy of the exciton (i.e., 1 meV to 1 eV).

An exciton is electrically neutral and has no role in electrical conductivity. It can travel through the crystal transporting exciton energy but not transporting charge. It can have a polarization vector when generated by light. Frenkel exciton is more tightly bound than *Mott-Wannier exciton.* Exitons can group themselves forming an *electron-hole droplet.*

Energy states: The energy of an exciton in the nth quantum state is

$$E_n = -\frac{\mu e^4}{2\hbar^2 \epsilon^2 n^2},$$

where n is the principal quantum number and $\mu = \frac{m_e m_h}{(m_e + m_h)}$ is the reduced mass of the exciton.

The allowed exciton states in the forbidden energy band is shown in Fig. A.1.

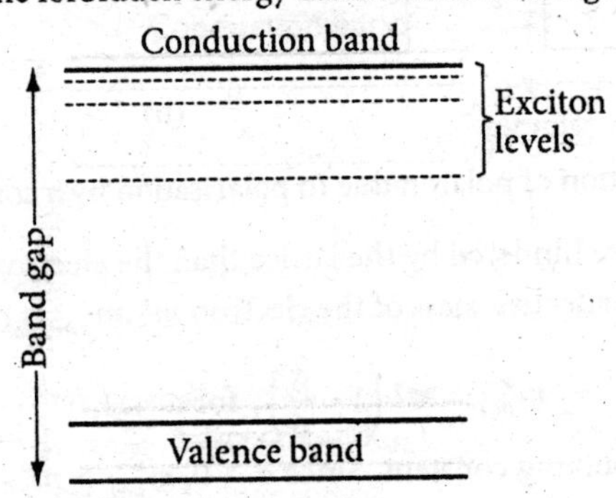

Fig. A-1: Exciton energy levels

32. Levy R.A., *Principles of Solid State Physics,* Academic Press (1968).
33. Liebfried G. and Ludwig W., *Theory of Anharmonic Effects in Crystals: Solid State Physics,* **12**, 276–44 (1961).
34. Long D., *Energy Bands in Semiconductors,* Wiley (1968).
35. Lynton E.A., *Superconductivity,* 3rd edn, Halsted (1971).
36. Madelung O., *Introduction to Solid State Theory,* Springer Verlag, German (1996).
37. Maradudin A.A., Montroll E.W. and Weiss G.H., *Theory of Lattice Dynamics in the Harmonic Approximation: Solid State Physics,* Supplement **3**, Academic Press (1963).
38. Mckelvy John P., *Solid State and Semiconductor Physics,* Harper & Row v., NY (1960).
39. Morrish A.H., *Physical Principles of Magnetism,* Wiley (1965).
40. Nye J.F., *Physical Properties of Crystals,* OUP (1985).
41. Omar M.A., *Elementary Solid State Physics,* Academic Press (1968).
42. Pauling I., *Nature of the Chemical Bond,* 3rd edn, Cornel (1960).
43. Pearson M.D., *Crystal Chemistry and Physics of metals and alloys,* Wiley-Interscience (1972).
44. Phillips C., *Bonds and Bands in Semiconductors,* Academic Press (1973).
45. Phillips F.C., *An Introduction to Crystallography,* 4th edn, Wiley (1971).
46. Rose A.C., Innes and Rhodrick E.H., *Introduction to Superconductivity,* Pergamon Press, NY (1969).
47. Schrieffer J.R., *Theory of Superconductivity,* Benjamin (1964).
48. Schriffer J.R., *Theory of Superconductivity,* Revised edn. Benjamin, Cummings (1983).
49. Seitz F., *Modern Theory of Solids,* McGraw-Hill Book Co. Inc. NY (1940).
50. Slator J.C., *Insulator, Semiconductors and Metals,* McGraw-Hill, NY (1967).
51. Smyth C.P., *Dielectric Behaviour and Structure,* McGraw-Hill (1955).
52. Sze S.M., *Physics of Semiconductor Devices,* Wiley (1969).
53. Taylor P.L., *A Quantum Approach to Solid State,* 2nd edn, Chapman and Hall, London (1970).
54. Tinkham M. and Lobb C.J., *Physical Properties of the New Superconductors: Solid State Physics,* **42**, 90 (1989).
55. Tinkham., *Introduction to Superconductivity,* Krieger (1980).
56. Tosi M.P., *Cohesion of Ionic Solids in the Born Model, Solid State Physics,* 16, 1 (1964).
57. Van Vleck J.H., *Theory of Electric and Magnetic Susceptibilities,* OUP, London (1932).
58. Wagner D., *Introduction to the Theory of Magnetism,* Pergamon Press, OUP (1972).
59. White R.M., *Quantum Theory of Magnetism,* McGraw-Hill (1970).
60. Wilson M.N., *Superconducting Magnets,* OUP (1983).
61. Wyckoff R.W.G., *Crystal structure,* Wiley (1963).
62. Ziel Aldert Vander., *Solid State Physics Electronics,* Prentice Hall of India Pvt. Ltd, New Delhi (1971).
63. Ziman J.M., *Principles of the Theory of Solids,* Cup-Vikas Student edn (1979).

Index